# MOMENT AND POLYNOMIAL OPTIMIZATION

# MOMENT AND POLYNOMIAL OPTIMIZATION

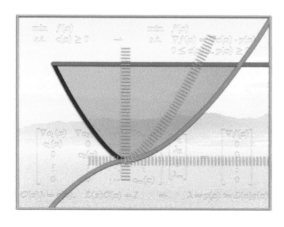

**Jiawang Nie**
University of California, San Diego
San Diego, California

Society for Industrial and Applied Mathematics
Philadelphia

Mathematical Optimization Society

Mathematical Optimization Society
Philadelphia

# Contents

# Preface

Moment and polynomial optimization has received high attention in recent decades. It has beautiful theory and efficient methods, as well as broad applications for various mathematical, scientific, and engineering fields. The research status of optimization has been enhanced extensively due to its recent developments. Nowadays, moment and polynomial optimization is an important technique in many fields.

The core technique in moment and polynomial optimization is the hierarchy of Moment-SOS relaxations. Its magic power is to transform a nonconvex optimization problem, which is given by polynomial or rational functions, into a linear convex optimization problem with the moment cone. The latter one can be solved by a hierarchy of semidefinite programming relaxations. The dual of the moment cone is the nonnegative (or positive semidefinite) polynomial cone, which can be well approximated by sum-of-squares (SOS) type representations. Interestingly, semidefinite programming relaxations for the moment cone are dual to the SOS approximations for the nonnegative polynomial cone. The sequence of these primal-dual pairs is the called Moment-SOS hierarchy. This view makes the book entitled *Moment and Polynomial Optimization*. An amazing power of the Moment-SOS hierarchy is that essentially all polynomial optimization problems can be solved exactly, for both their optimal values and optimizers, by applying Moment-SOS relaxations.

The main contents of this book are outlined as follows.

**Chapter 1** Convex relaxations are the foundation of moment and polynomial optimization. This chapter reviews classical backgrounds such as vector spaces, convex sets and convex cones, positive semidefinite cones, linear matrix inequalities, semidefinite programming, and linear conic optimization. We also give expositions for how to formulate polynomial optimization equivalently as linear optimization with moment cones.

**Chapter 2** The theory and methods for moment and polynomial optimization highly rely on algebra, Positivstellensätze, and truncated moment problems. This chapter reviews classical backgrounds, such as computational algebra, sum-of-squares polynomials, quadratic modules and preordering, and moment problems.

**Chapter 3** Univariate polynomial and moment problems are particularly interesting, because there exist perfect semidefinite programming representations for univariate moment cones and nonnegative polynomial cones. Their methodology and theory are quite similar to the multivariate case, although they have different mathematical properties. This chapter introduces some classical results about the univariate case.

**Chapter 4** A foundationally important question in optimization is to find global optimal values and optimizers for polynomial functions without any constraints. Indeed, the field of polynomial optimization originated from the research on this basic question. This chapter presents classical and current results for unconstrained polynomial optimization.

**Chapter 5** The Moment-SOS hierarchy is a powerful tool for solving constrained polynomial optimization. Interestingly, the convergence theory of the Moment-SOS hierarchy is closely related to the classical optimality condition theory in nonlinear programming. This chapter presents the framework of Moment-SOS relaxations, their asymptotic and finite convergence, extraction of optimizers, optimality certificates, and related topics.

**Chapter 6** The classical Moment-SOS hierarchy may not be tight for solving some polynomial optimization problems. However, optimality conditions and Lagrange multiplier expressions can be used to get tight Moment-SOS hierarchies. This chapter shows how to use optimality conditions and Lagrange multiplier expressions to construct tight Moment-SOS hierarchies.

**Chapter 7** Convexity is a topic of foundational importance in optimization. This chapter introduces basic results about convex polynomials, convex polynomial optimization, and semidefinite representations for convex semialgebraic sets.

**Chapter 8** It is interesting to observe that many hard nonconvex optimization problems can be equivalently formulated as linear convex optimization with moment or nonnegative polynomial cones. This chapter presents the basic theory and methods for solving these linear conic optimization problems. The topics of infeasibility, unboundedness, and truncated moment problems are also discussed in this chapter.

**Chapter 9** An especially interesting class of optimization problems is that of polynomials that are nonnegative over the nonnegative orthant. These polynomials are called copositive polynomials. They are dual to the cone of completely positive (CP) moments, which are admitting measures supported in the nonnegative orthant. This chapter introduces basic theory and methods for copositive and CP optimization.

**Chapter 10** Some optimization problems are given by polynomial matrix inequality constraints. This chapter introduces topics of positive semidefinite matrix polynomials, various matrix Positivstellensätze, matrix-valued moment problems, optimization, and convex sets given by polynomial matrix inequalities.

**Chapter 11** Polynomials and tensors are closely related. Tensor optimization problems are usually formulated as homogeneous polynomial optimization with sphere constraints. Most of the theory and methods for tensor computation can be given in the framework of polynomial optimization. This chapter introduces basic results about tensor theory, spectral and nuclear norms, quantum entanglements, tensor positivity, tensor eigenvalues, and singular values.

**Chapter 12** Moment and polynomial optimization has broad applications. This chapter introduces special topics such as sparse optimization, rational optimization, saddle points, Nash equilibrium, bilevel optimization, distributionally robust optimization, and multi-objective optimization.

The connections between these chapters are depicted in the following diagram.

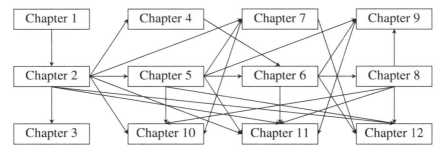

Moment and polynomial optimization is an active vivid research field. It becomes a more and more important tool for pure and applied mathematicians, scientists, and engineers. Optimization about moments and polynomials is highly challenging. It concerns optimality certificates, nonnegativity certificates, characterization of moments, representations for convex sets, and various Positivstellensätze. To do research in this field requires backgrounds in real and complex algebraic geometry, convex optimization, matrix computations, and nonlinear programming. Many traditionally hard problems can be solved successfully now, due to advances in moment and polynomial optimization. For cleanness of the book, numerical results are displayed with four decimal digits, unless specified in the context.

This book gives a quick, efficient introduction to the field of moment and polynomial optimization. It is especially suitable for graduate students, junior scholars, applied mathematicians, engineers and researchers to enter the field. The book can also be used as a textbook for graduate students or as a monograph for scholars working in relevant fields.

I would like to thank Jiyoung Choi, Xiaomeng Hu, Lei Huang, Igor Klep, Xindong Tang, Jiajia Wang, Li Wang, Zi Yang, Yang Zheng, Zequn Zheng, Suhan Zhong, and Jinling Zhou for their careful checking and reading. With love, I also thank Ping, Christopher, and Scarlett for their heartfelt help while this book was written.

Jiawang Nie,
San Diego, USA, 2023.

# Part I

# Mathematical Backgrounds

# Chapter 1

# Basics of Linear Convex Optimization

*Convex relaxations are basic tools for solving moment and polynomial optimization problems. This chapter reviews some backgrounds for convex optimization. We introduce basic topics such as vector spaces, convex sets and convex cones, positive semidefinite cones, linear matrix inequalities, semidefinite programming, linear conic optimization, and linear convex formulation for moment and polynomial optimization.*

## 1.1 ▪ Vector Spaces

Let $\mathbb{F}$ be a field, mostly the real field $\mathbb{R}$ or complex field $\mathbb{C}$ in this book. Suppose $V$ is a nonempty set for which two operations are defined for elements in $V$: the addition $+$ (i.e., $a + b$ is defined for all $a, b \in V$) and the scalar multiplication $\cdot$ (i.e., $\tau \cdot c$ is defined for all $\tau \in \mathbb{F}$ and all $c \in V$). If the meaning is clear in the context, we often simply write $\tau \cdot c$ as $\tau c$. We assume that the operations $+, \cdot$ are closed, i.e., $a + b \in V$ for all $a, b \in V$ and $\tau c \in V$ for all $c \in V$ and $\tau \in \mathbb{F}$. The set $V$ is called a *vector space* if the following conditions are satisfied for all $a, b, c \in V$ and for all scalars $\alpha, \beta \in \mathbb{F}$: (i) additive commutativity $a + b = b + a$; (ii) additive associativity $(a + b) + c = a + (b + c)$; (iii) there exists the zero vector $0 \in V$ such that $0 + a = a + 0 = a$; (iv) for every $x \in V$, there exists $y \in V$ such that $x + y = 0$; such a vector $y$ is called the additive inverse of $x$ and is denoted as $-x$; (v) associativity of scalar multiplications $\alpha(\beta c) = (\alpha\beta)c$; (vi) distributivity of scalar additions $(\alpha + \beta)c = \alpha c + \beta c$; (vii) distributivity of vector additions $\alpha(a + b) = \alpha a + \alpha b$; (viii) scalar multiplication identity $1a = a$. If $V$ is a vector space, each element in $V$ is called a vector. More detailed introductions to vector spaces can be found in [173]. Typical examples of vector spaces are the following:

- $\mathbb{F}^n$: the $n$-dimensional Euclidean space over $\mathbb{F}$.

- $\mathbb{F}^{m \times n}$: the space of $m$-by-$n$ matrices over $\mathbb{F}$.

- $\mathbb{F}[x_1, \ldots, x_n]$: the space of polynomials in $x_1, \ldots, x_n$ and coefficients in $\mathbb{F}$. Moreover, the notation $\mathbb{F}[x_1, \ldots, x_n]_d$ stands for the subset of polynomials in $\mathbb{F}[x_1, \ldots, x_n]$ with degrees less than or equal to $d$, while $\mathbb{F}[x_1, \ldots, x_n]_d^{hom}$ denotes the subset of homogeneous polynomials of degree $d$ in $\mathbb{F}[x_1, \ldots, x_n]$.

Let $V$ be a vector space. A set of vectors $v_1, \ldots, v_k \in V$ is said to be *linearly independent* if the scalars $\lambda_1, \ldots, \lambda_k$ satisfying

$$\lambda_1 v_1 + \cdots + \lambda_k v_k = 0$$

are only zeros, i.e., $\lambda_1 = \cdots = \lambda_k = 0$. The sum like $\lambda_1 v_1 + \cdots + \lambda_k v_k$ is called a linear combination of $v_1, \ldots, v_k$. For a set $T \subseteq V$, the set of all possible linear combinations of vectors in $T$ is called the span of $T$, for which we denote span $\{T\}$. A subset $B$ of $V$ is a *basis* if $B$ is linearly independent and $V = \text{span}\{B\}$. In the Euclidean space $\mathbb{F}^n$ ($\mathbb{F} = \mathbb{R}$ or $\mathbb{C}$), denote by $e_i$ the standard unit vector whose $i$th entry is one and all other entries are zeros, while $e$ denotes the vector of all ones. The set $\{e_1, \ldots, e_n\}$ is called the *canonical basis* of $\mathbb{F}^n$. The cardinality of a basis of $V$ is called the *dimension* of $V$. If the cardinality is finite, $V$ is called *finitely dimensional*; otherwise, $V$ is called *infinitely dimensional*.

### 1.1.1 ▪ Subspaces and linear maps

For a vector space $V$, a subset $U \subseteq V$ is called a *subspace* if $U$ itself is a vector space, under the same operations of addition $+$ and scalar multiplication $\cdot$ for $V$. A subspace $U$ of $V$ is called *proper* if $U \neq V$. Bases and dimensions can be similarly defined for subspaces. The gap between the dimensions of $V$ and $U$ is called the *codimension* of $U$.

Let $V, W$ be two vector spaces over a field $\mathbb{F}$. A function $\varphi : V \to W$ is said to be a *linear map* if it satisfies that

$$\varphi(\alpha u + \beta v) = \alpha \varphi(u) + \beta \varphi(v) \quad \text{for all } u, v \in V, \alpha, \beta \in \mathbb{F}.$$

The vector $\varphi(u)$ is called the image of $u$ under $\varphi$. The image of a subset $S \subseteq V$ is then denoted as

$$\varphi(S) := \{\varphi(u) : u \in S\}.$$

In particular, the entire image set $\varphi(V)$ is called the *range* of $\varphi$, for which we denote Range $(\varphi)$, i.e.,

$$\text{Range}(\varphi) = \{w \in W : w = \varphi(u) \text{ for some } u \in V\}.$$

For a vector $b \in W$, its *preimage* is the set

$$\varphi^{-1}(b) := \{u \in V : \varphi(u) = b\}.$$

For a subset $T \subseteq W$, its preimage is then defined as

$$\varphi^{-1}(T) := \{u \in V : \varphi(u) \in T\}.$$

In particular, the preimage of the zero vector, i.e., $\varphi^{-1}(0)$, is called the *null space* or *kernel* of $\varphi$, for which we denote Null $(\varphi)$ or ker $\varphi$, i.e.,

$$\text{Null}(\varphi) = \ker \varphi = \{v \in V : \varphi(v) = 0\}.$$

It can be verified that Null $(\varphi)$ is a subspace of $V$ and Range $(\varphi)$ is a subspace of $W$. The linear map $\varphi$ is called *one-to-one* (or just 1-1) if Null $(\varphi) = \{0\}$, and $\varphi$ is called *onto* if Range $(\varphi) = W$. If $\varphi$ is simultaneously one-to-one and onto, then $\varphi$ is called an *isomorphism* and $V$ is said to be *isomorphic* to $W$. For such a case, we write $V \cong W$. If two vector spaces are isomorphic, they can be considered to be the same for linear operations on them.

When $W = \mathbb{F}$ is a field, a linear map $f : V \to \mathbb{F}$ is called a *linear functional*. The space of all linear functionals on $V$ is called the *dual space* of $V$, for which we denote $V^*$. When $V$ is finitely dimensional, $V \cong V^*$ and they have the same dimension. For a basis $B = \{v_1, \ldots, v_n\}$ of $V$, a basis $D = \{f_1, \ldots, f_n\}$ of $V^*$ is said to be *dual* to $B$ if $f_i(v_j) = 1$ for $i = j$ and zero otherwise. When $B$ is clear in the context, we just simply call $\{f_1, \ldots, f_n\}$ the *dual basis*.

In some applications, people need to work on quotient spaces. Let $U$ be a subspace of a vector space $V$. Define an equivalence relation $\sim$ such that for $a, b \in V$, $a \sim b$ if and only if

$a - b \in U$. If $a \sim b$, we say that $a$ is equivalent to $b$. The set of all vectors that are equivalent to $a$ is called the *equivalent class* of $a$, for which we denote $[a]$. Equivalently, $[a] = [b]$ if and only if $a - b \in U$. The addition $+$ for equivalent classes is defined such that

$$[a] + [b] := [a + b].$$

The scalar multiplication $\cdot$ can be defined as $\tau \cdot [a] = [\tau a]$. The operations $+$ and $\cdot$ are well-defined for equivalent classes. Under these operations of additions and scalar multiplications, the set of equivalent classes is also a vector space. It is called the *quotient space* of $V$ over $U$, and we denote it as $V/U$. For two vectors $a, b \in V$, we write $a \equiv b \mod U$ to mean that $a - b \in U$. Clearly, if $U = \{0\}$ is the trivial zero subspace, then the quotient space $V/U \cong V$; if $U = V$, then $V/U \cong \{0\}$.

## 1.1.2 • Inner products and norms

A vector space $V$ is often equipped with an inner product

$$\langle \cdot, \cdot \rangle : V \times V \quad \to \quad \mathbb{F},$$

i.e., it is a scalar-valued function satisfying (i) $\langle \cdot, \cdot \rangle$ is linear in the first argument and conjugate linear in the second argument, i.e.,

$$\langle \alpha x + \beta y, z \rangle = \alpha \langle x, z \rangle + \beta \langle y, z \rangle,$$
$$\langle z, \alpha x + \beta y \rangle = \bar{\alpha} \langle z, x \rangle + \bar{\beta} \langle z, y \rangle$$

for all $x, y, z \in V$ and for all $\alpha, \beta \in \mathbb{F}$. (The bar ‾ here means complex conjugate.) (ii) $\langle x, x \rangle \geq 0$ for all $x \in V$ and $\langle x, x \rangle = 0$ only if $x = 0$. (iii) $\langle x, y \rangle = \overline{\langle y, x \rangle}$, the complex conjugate of $\langle x, y \rangle$, for all $x, y \in V$. When an inner product is defined, $V$ is also called an inner product space. The following are some typical inner product spaces.

- For the space $\mathbb{F}^n$ ($\mathbb{F} = \mathbb{R}$ or $\mathbb{C}$), the Euclidean inner product is

$$\langle x, y \rangle := \sum_{i=1}^{n} x_i \overline{y_i} = y^* x$$

  for vectors $x = (x_1, \ldots, x_n)$ and $y = (y_1, \ldots, y_n)$. Here, the notation $y^*$ stands for the *conjugate* (or *Hermitian*) transpose[1] of the vector $y$.

- For the matrix space $\mathbb{F}^{m \times n}$, the Euclidean inner product $\langle \cdot, \cdot \rangle$ is

$$\langle X, Y \rangle := \sum_{i=1}^{m} \sum_{j=1}^{n} X_{ij} \overline{Y_{ij}} = \mathrm{trace}(XY^*)$$

  for matrices $X = (X_{ij})$ and $Y = (Y_{ij})$ in $\mathbb{F}^{m \times n}$. Here, the trace of a matrix means the sum of its diagonal entries. For notational convenience, the Euclidean inner product of matrices is also denoted as

$$X \bullet Y := \mathrm{trace}(XY^*). \tag{1.1.1}$$

- For the space $C[a, b]$ of complex continuous functions on an interval $[a, b]$, a standard inner product $\langle \cdot, \cdot \rangle$ is

$$\langle f, g \rangle := \int_a^b f(t) \overline{g(t)} \mathrm{d}t \quad \text{for } f, g \in C[a, b].$$

---

[1]The superscript $A^T$ denotes the classical transpose of a vector or matrix $A$.

If $\langle \cdot, \cdot \rangle$ is an inner product for $V$, then it gives a norm function on $V$

$$\|x\| := \sqrt{\langle x, x \rangle}, \quad x \in V.$$

Throughout the book, for a vector $x \in \mathbb{F}^n$, the notation $\|x\|$ means the norm given by the Euclidean inner product, unless it is otherwise specified. In particular, under this norm for $\mathbb{F}^n$, the ball with center $c$ and radius $r$ is the set

$$B(c, r) := \{x \in \mathbb{F}^n : \|x - c\| \leq r\}. \tag{1.1.2}$$

The set $B(0, 1)$ is called the unit ball, centered at the origin. For the matrix space $\mathbb{F}^{m \times n}$, the *Frobenius norm* of a matrix $A$ is

$$\|A\|_F := \sqrt{A \bullet A}.$$

The conventional notation $\|A\|$ stands for the spectral norm of $A$ (i.e., the largest singular value of $A$; see Exercise 1.1.9), unless its meaning is specified in the context. Sometimes, we also denote by $\|A\|_\sigma$ the spectral norm of $A$ and denote by $\|A\|_*$ the nuclear norm of $A$ (i.e., the sum of all its singular values).

### 1.1.3 ▪ Exercises

**Exercise 1.1.1.** *For given vectors $a_1, \ldots, a_m \in \mathbb{C}^n$ and constants $b_1, \ldots, b_m \in \mathbb{C}$, show that the linear system*

$$a_1^T x - b_1 = \cdots = a_m^T x - b_m = 0$$

*has no common solutions if and only if there exist scalars $\lambda_1, \ldots, \lambda_m$ such that the following polynomial identity holds (i.e., it holds for all $x$):*

$$\lambda_1(a_1^T x - b_1) + \cdots + \lambda_m(a_m^T x - b_m) = 1.$$

*(The above is equivalent to $\sum_{i=1}^m \lambda_i a_i = 0$, $\sum_{i=1}^m \lambda_i b_i = -1$.)*

**Exercise 1.1.2.** *For $\mathbb{F} = \mathbb{R}$ or $\mathbb{C}$, let $\mathbb{F}[x_1, \ldots, x_n]_d^{hom}$ denote the subspace of homogeneous polynomials of degree d in $\mathbb{F}[x_1, \ldots, x_n]_d$. Show that*

$$\dim \mathbb{F}[x_1, \ldots, x_n]_d = \binom{n+d}{d},$$

$$\dim \mathbb{F}[x_1, \ldots, x_n]_d^{hom} = \binom{n+d-1}{d}.$$

**Exercise 1.1.3.** *Let $\varphi : V \to W$ be a linear map between two finitely dimensional vector spaces $V, W$. Show that*

$$\dim V = \dim \mathrm{Null}(\varphi) + \dim \mathrm{Range}(\varphi).$$

**Exercise 1.1.4.** *Let $U$ be a subspace of a finitely dimensional vector space $V$. Show that the dimension of the quotient space $V/U$ is equal to the codimension of $U$.*

**Exercise 1.1.5.** *Show that every finitely dimensional vector space is isomorphic to its dual space.*

**Exercise 1.1.6.** *For given sets $S_1, \ldots, S_m$, their Cartesian product $S_1 \times \cdots \times S_m$ is defined to be the set*

$$S_1 \times \cdots \times S_m := \{(u_1, \ldots, u_m) : \quad \text{each } u_i \in S_i\}.$$

*If $U_1, \ldots, U_m$ are vector spaces over a common field, show that $U_1 \times \cdots \times U_m$ is also a vector space under the usual addition and scalar multiplication:*

$$
\begin{aligned}
(u_1, \ldots, u_m) + (v_1, \ldots, v_m) &= (u_1 + v_1, \ldots, u_m + v_m), \\
\alpha(u_1, \ldots, u_m) &= (\alpha u_1, \ldots, \alpha u_m).
\end{aligned}
$$

*Also show that $\dim U_1 \times \cdots \times U_m = \dim U_1 + \cdots + \dim U_m$.*

**Exercise 1.1.7.** *For any inner product $\langle \cdot, \cdot \rangle$ on a vector space $V$, show that the induced norm $\|x\| := \sqrt{\langle x, x \rangle}$ satisfies the triangle inequality:*

$$\|x + y\| \leq \|x\| + \|y\| \quad \forall x, y \in V.$$

**Exercise 1.1.8.** *For $p \geq 1$, the p-norm of $x := (x_1, \ldots, x_n)$ is*

$$\|x\|_p := \Big( \sum_{i=1}^{n} |x_i|^p \Big)^{1/p}.$$

*For all $x, y \in \mathbb{R}^n$, show that $\|x + y\|_p \leq \|x\|_p + \|y\|_p$.*

**Exercise 1.1.9.** *For every matrix $A \in \mathbb{R}^{m \times n}$, there exist two orthogonal matrices[2] $U \in \mathbb{R}^{m \times m}$, $U \in \mathbb{R}^{n \times n}$ and a diagonal matrix*

$$\Sigma = \begin{bmatrix} \sigma_1 & & & & \\ & \ddots & & & \\ & & \sigma_r & & \\ & & & 0 & \\ & & & & \ddots \end{bmatrix} \in \mathbb{R}^{m \times n}, \quad \sigma_1 \geq \cdots \geq \sigma_r > 0$$

*such that $A = U \Sigma V^T$. This is called the singular value decomposition (SVD) of A. Show that*

$$\|A\| := \max_{x \neq 0} \frac{\|Ax\|}{\|x\|} = \sigma_1, \quad \|A\|_F = \sqrt{\sigma_1^2 + \cdots + \sigma_r^2}.$$

**Exercise 1.1.10.** *Let A be the matrix having the SVD as in Exercise 1.1.9. Denote the diagonal matrix*

$$\Sigma^\dagger = \begin{bmatrix} \sigma_1^{-1} & & & & \\ & \ddots & & & \\ & & \sigma_r^{-1} & & \\ & & & 0 & \\ & & & & \ddots \end{bmatrix} \in \mathbb{R}^{n \times m}.$$

*The matrix $A^\dagger := V \Sigma^\dagger U^T$ is called the Moore-Penrose Pseudoinverse. Show that*

$$(AA^\dagger)^T = (AA^\dagger), \quad (A^\dagger A)^T = (A^\dagger A), \quad AA^\dagger A = A, \quad A^\dagger A A^\dagger = A^\dagger.$$

---

[2] A matrix $Q$ is orthogonal if the inverse $Q^{-1} = Q^T$. Similarly, the $Q$ is unitary if the inverse $Q^{-1} = Q^*$, the conjugate transpose of $Q$.

## 1.2 ▪ Convex Sets and Convex Cones

Convexity is a basic concept in optimization. More detailed introductions to convexity can be found in books [12, 23, 24, 37, 285]. This section reviews some basic knowledge for convexity. Throughout this section, assume $V$ is a finitely dimensional vector space over $\mathbb{R}$.

A set $C \subseteq V$ is *convex* if $\lambda x + (1 - \lambda)y \in C$ for all points $x, y \in C$ and for all scalars $\lambda \in [0, 1]$. For convenience, denote the intervals

$$
\begin{aligned}
[x, y] &:= \{\lambda x + (1 - \lambda)y : 0 \le \lambda \le 1\}, \\
(x, y) &:= \{\lambda x + (1 - \lambda)y : 0 < \lambda < 1\}.
\end{aligned}
\tag{1.2.1}
$$

A frequently appearing class of convex sets are *ellipsoids*. Let $P$ be a $n$-by-$n$ real symmetric positive definite matrix (see §1.4). For a vector $c \in \mathbb{R}^n$, the set

$$
E := \{x \in \mathbb{R}^n : (x - c)^T P^{-1}(x - c) \le 1\}
$$

is called an ellipsoid. The point $c$ is called the *center* and $P$ is called the shape matrix. In particular, for $P = r^2 I_n$ with $r > 0$ (the symbol $I_n$ denotes the $n$-by-$n$ identity matrix), we get the ball $B(c, r)$. The volume of $E$ equals the product of $\sqrt{\det P}$ and the volume of the unit ball, where $\det P$ is the determinant of $P$.

For two sets $A, B \subseteq V$, their *Minkowski sum* is

$$
A + B := \{u + v : u \in A, v \in B\}.
\tag{1.2.2}
$$

If $A, B$ are both convex, then their Minkowski sum is also convex, as well as their intersection $A \cap B$. Note that if one of $A, B$ is the empty set, then $A + B$ is also empty. If $B = \{b\}$ is a singleton, then $A + \{b\}$ is called a *translation* of $A$. For convenience, we sometimes also write that

$$
b + A := A + \{b\}.
$$

For a scalar $s$, the product $sA := \{sx : x \in A\}$ is called a *scaling* of $A$. If $s > 0$, the set $sA$ is called a *dilation* of $A$.

### 1.2.1 ▪ Supporting hyperplanes

Let $C \subseteq V$ be a set. For a linear functional $0 \ne \ell \in V^*$, the hyperplane

$$
H := \{x \in V : \ell(x) = b\}
$$

is called a *supporting hyperplane* for $C$ if $C$ is contained in the half space

$$
H^+ := \{x \in V : \ell(x) \ge b\}
$$

and the distance between $H$ and $C$ is zero. For instance, $x_1 = 0$ is a supporting hyperplane for the hyperboloid $\{x_1 \cdots x_n \ge 1, x_1 > 0, \ldots, x_n > 0\}$. For two sets $C_1, C_2 \subseteq V$, the hyperplane $H$ is called a *separating hyperplane* for them if $C_1$ and $C_2$ lie in opposite sides of $H$. The following is the classical separating hyperplane theorem for disjoint convex sets.

**Theorem 1.2.1.** *Let $C_1, C_2$ be two nonempty convex sets of a vector space $V$. If $C_1 \cap C_2 = \emptyset$, then there exists a nonzero linear functional $\ell \in V^*$ such that*

$$
\ell(x) \le \ell(y) \quad \text{for all } x \in C_1, \, y \in C_2.
\tag{1.2.3}
$$

*Moreover, if one of $C_1, C_2$ is open, or if $C_1, C_2$ are closed and one of them is compact, then the above $\ell$ can be chosen such that*

$$\ell(x) < \ell(y) \quad \text{for all } x \in C_1,\, y \in C_2. \tag{1.2.4}$$

The inequality (1.2.4) means that $C_1, C_2$ are *strictly separated*. When $V$ has an inner product $\langle \cdot, \cdot \rangle$, the nonzero functional $\ell \in V^*$ can be expressed such that $\ell(x) = \langle \lambda, x \rangle$ for some vector $0 \neq \lambda \in V$. Then, the separating hyperplane inequality (1.2.3) reads as

$$\langle \lambda, x \rangle \leq \langle \lambda, y \rangle \quad \text{for all } x \in C_1,\, y \in C_2$$

and the strict one (1.2.4) reads as

$$\langle \lambda, x \rangle < \langle \lambda, y \rangle \quad \text{for all } x \in C_1,\, y \in C_2.$$

Next, we review the concepts of closures and closedness. The closure of a set $S \subseteq V$ is

$$\operatorname{cl}(S) := \Big\{ a \in V : \lim_{k \to \infty} a_k = a, \quad \text{each } a_k \in S \Big\}.$$

The set $S$ is *closed* if $\operatorname{cl}(S) = S$. The closure of a convex set can be characterized by supporting hyperplanes.

**Theorem 1.2.2.** *If $C \subseteq V$ is a convex set, then*

$$\operatorname{cl}(C) = \bigcap_{\substack{H \text{ is a supporting} \\ \text{hyperplane for } C}} H^+.$$

*In the above, $H^+$ denotes the half space of $H$ containing $C$.*

The proofs of the above theorems can be found in the books [12, 23, 24, 37, 285].

## 1.2.2 ▪ Geometric properties

For a set $S$, its *affine hull* is the set

$$\operatorname{aff}(S) := \Big\{ \sum_{i=1}^{k} \lambda_i u_i \,\Big|\, \lambda_1 + \cdots + \lambda_k = 1, u_i \in S, k \in \mathbb{N} \Big\}. \tag{1.2.5}$$

For instance, for the parabola curve set $S = \{x_2 = x_1^2\}$, its affine hull is the entire space $\mathbb{R}^2$. The translation of a subspace is called an *affine subspace*. Equivalently, the affine hull of a set is the smallest affine subspace containing it. If an affine subspace $A = a + U$, for some vector $a \in V$ and a subspace $U \subseteq V$, the dimension of $A$ is equal to the dimension of $U$.

In the following, assume a norm is defined for the vector space $V$ and the balls are defined under that norm. For a convex set $C \subseteq V$, the dimension of $C$ is defined to be the dimension of the smallest affine subspace containing $C$. The *interior* of $C$ is

$$\operatorname{int}(C) := \Big\{ u \in C : \exists \epsilon > 0,\, B(u, \epsilon) \subseteq C \Big\}. \tag{1.2.6}$$

Each point in $\operatorname{int}(C)$ is called an interior point of $C$. If $\operatorname{int}(C) = C$, then $C$ is said to be an *open* set. A convex set with nonempty interior is called a *convex domain*, while a compact convex domain is called a *convex body*. The boundary of $C$ is the subset of points that are not in the interior of $C$ and we denote it as $\partial C$. It is possible that a convex set has empty interior. For instance, any hyperplane has no interior. For such a case, the *relative interior* can be defined. The relative interior of a convex set $C$ is the set

$$\mathrm{ri}\,(C) := \left\{ u \in C : \exists \epsilon > 0, \, B(u, \epsilon) \cap \mathrm{aff}\,(C) \subseteq C \right\}. \tag{1.2.7}$$

For a nonempty convex set, the relative interior always exists. If $\mathrm{ri}\,(C) = C$, then $C$ is said to be *relatively open*.

For a convex set $C$, a point $u \in C$ is called *extreme* if

$$u \in (x, y), \, x \in C, \, y \in C \quad \Rightarrow \quad u = x = y.$$

For instance, for the convex hyperboloid given by

$$x_1 \cdots x_n \geq 1, \, x_1 > 0, \ldots, x_n > 0,$$

all its points lying on the hypersurface $x_1 \cdots x_n = 1$ are extreme. A subset $F \subseteq C$ is called a *face* if $F$ itself is convex and

$$u \in (x, y), \, u \in F, \, x, y \in C \quad \Rightarrow \quad x, y \in F.$$

Clearly, the empty set $\emptyset$ and the entire set $C$ are faces of $C$. They are called *trivial faces*. The face $F$ is said to be *proper* if $F \neq C$. The set of any single extreme point is also a face. For a subset $T \subseteq C$, we denote by $\mathrm{face}\,(T, C)$ the minimum face of $C$ containing $T$. A face $F$ of $C$ is called *exposed* if there exists a supporting hyperplane $H$ for $C$ such that

$$F = H \cap C.$$

**Theorem 1.2.3.** *([299, Theorem 1.3.3]) Let $K \subseteq \mathbb{R}^n$ be a closed set having nonempty interior. If for every boundary point $u$ of $K$ there exists a supporting hyperplane through $u$, then $K$ must be convex.*

**Proof.** Suppose otherwise $K$ is not convex, then there must exist $a, b \in K$ such that $[a, b] \not\subseteq K$, say, $c \in (a, b)$ but $c \notin K$. Since $K$ has nonempty interior, there is a point $d \in \mathrm{int}\,(K)$. The line segment $[c, d]$ must intersect the boundary of $K$, say, $f \in [c, d] \cap \partial K$. By the assumption, there exists a supporting hyperplane $H$ for $K$ through $f$, say, $H := \{\lambda^T x = \lambda^T f\}$ $(\lambda \neq 0)$ such that

$$\lambda^T x \geq \lambda^T f \quad \forall x \in K.$$

Since $d \in \mathrm{int}\,(K)$, we must have that $\lambda^T d > \lambda^T f$. Since $f$ is an inner point of $[c, d]$ with $f \neq c, d$, we can get

$$\lambda^T c < \lambda^T f.$$

However, since $a, b \in K$, we know $\lambda^T a \geq \lambda^T f$ and $\lambda^T b \geq \lambda^T f$. The $c$ is an inner point of $[a, b]$, so $\lambda^T c \geq \lambda^T f$, which is a contradiction to $\lambda^T c < \lambda^T f$. Therefore, $[a, b] \subseteq K$ for all $a, b \in K$, i.e., $K$ must be convex. $\qquad\square$

For a set $S$ that is not convex, its *convex hull* is the smallest convex set containing $S$. The convex hull of $S$, denoted as $\mathrm{conv}\,(S)$, is the set

$$\mathrm{conv}\,(S) := \left\{ \sum_{i=1}^{k} \lambda_i u_i \,\middle|\, \begin{array}{l} u_i \in S, \, \lambda_i \geq 0, \, k \in \mathbb{N}, \\ \lambda_1 + \cdots + \lambda_k = 1 \end{array} \right\}. \tag{1.2.8}$$

In the above, the sum $\sum_{i=1}^{k} \lambda_i u_i$ is called a convex combination of $u_1, \ldots, u_k$. If we remove $\lambda_1 + \cdots + \lambda_k = 1$ in (1.2.8), then we get the conic hull

$$\text{cone}(S) := \left\{ \sum_{i=1}^{k} \lambda_i u_i \,\middle|\, u_i \in S, \, \lambda_i \geq 0, \, k \in \mathbb{N} \right\}. \tag{1.2.9}$$

For every set $S \subseteq V$, each point in $\text{conv}(S)$ can be expressed as a convex combination of at most $\dim(V) + 1$ points from $S$. This conclusion is referenced as *Carathéodory's Theorem*.

**Theorem 1.2.4.** *(Carathéodory's Theorem) Let $S$ be a nonempty set of a vector space $V$. Then, every point in the convex hull $\text{conv}(S)$ can be expressed as a convex combination of at most $\dim V + 1$ points in $S$, and every point in the conic hull $\text{cone}(S)$ can be expressed as a conic combination of at most $\dim V$ points in $S$.*

Carathéodory's Theorem implies the following proposition.

**Proposition 1.2.5.** *([23]) Let $S$ be a compact set of a vector space $V$. Then, the convex hull $\text{conv}(S)$ is compact.*

The conic hull of a compact set is never bounded, and hence it is never compact. However, it is closed under an additional condition.

**Proposition 1.2.6.** *Let $S$ be a compact set of a vector space $V$. If there exists a linear functional $\ell \in V^*$ such that $\ell(x) > 0$ for all $x \in S$, then the conic hull $\text{cone}(S)$ is closed.*

***Proof.*** Suppose $\{z_k\} \subseteq \text{cone}(S)$ is a convergent sequence and $z^*$ is the limit. Let $D := \dim V$. By Carathéodory's Theorem, one can write that

$$z_k = \lambda_{k,1} u_{k,1} + \cdots + \lambda_{k,D} u_{k,D},$$

with all $u_{k,j} \in S$ and $\lambda_{k,j} \geq 0$. Since $S$ is compact, we can generally assume $u_{k,j} \to u_j \in S$ as $k \to \infty$, for $j = 1, \ldots, D$, up to choosing convergent subsequences. Since $\ell(x) > 0$ on $S$ and $S$ is compact, there exists $\epsilon > 0$ such that $\ell(x) \geq \epsilon$ for all $x \in S$. So, we get

$$\ell(z_k) = \lambda_{k,1} \ell(u_{k,1}) + \cdots + \lambda_{k,D} \ell(u_{k,D}) \geq \epsilon(\lambda_{k,1} + \cdots + \lambda_{k,D}).$$

This implies that each scalar sequence $\{\lambda_{k,j}\}_{k=1}^{\infty}$ is bounded. Hence, we can also generally assume $\lambda_{k,j} \to \lambda_j$ as $k \to \infty$, with the limit $\lambda_j \geq 0$. Applying the limit, we get

$$z^* = \lambda_1 u_1 + \cdots + \lambda_D u_D \in \text{cone}(S).$$

Therefore, the conic hull $\text{cone}(S)$ is closed. $\qquad\square$

When there is no linear functional $\ell \in V^*$ such that $\ell > 0$ on $S$, the conic hull $\text{cone}(S)$ may not be closed, even if $S$ is compact. Such an example can be found in the exercises. The following theorem connects convex sets and their extreme points, which is known as the *Krein-Milman Theorem* or *Minkowski's Theorem*.

**Theorem 1.2.7.** *([12]) If $C \subseteq V$ is a compact convex set, then $C$ is the convex hull of its extreme points.*

## 1.2.3 ▪ Polar and dual cones

In the following, assume $V$ is equipped with an inner product $\langle \cdot, \cdot \rangle$. For a nonempty set $T \subseteq V$, its *polar* is the set

$$T^\circ := \{y \in V : \langle x, y \rangle \leq 1 \text{ for all } x \in T\}. \tag{1.2.10}$$

The polar $T^\circ$ is a closed convex set and contains the origin. Note that $T$ is always contained in the polar of $T^\circ$, i.e., $T \subseteq (T^\circ)^\circ$. Indeed, they are equal under some assumptions. This is the bipolar theorem.

**Theorem 1.2.8.** *(Bipolar Theorem [12]) If $T \subseteq V$ is a closed convex set containing the origin, then $(T^\circ)^\circ = T$.*

A set $K \subseteq V$ is a *cone* if $tx \in K$ for all $x \in K$ and for all $t > 0$. The cone $K$ is *pointed* if $K \cap -K = \{0\}$. It is said to be *solid* if its interior $\text{int}(K) \neq \emptyset$. Any nonzero point of a cone cannot be extreme. If a cone is pointed, the only extreme point is the origin. The analogue of extreme point for convex cones is the *extreme ray*. For a convex cone $K$ and $0 \neq u \in K$, the line segment

$$u \cdot [0, \infty) := \{tu : t \geq 0\} \tag{1.2.11}$$

is called an extreme ray of $K$ if

$$u \in (x, y),\ x, y \in K \quad \Rightarrow \quad u, x, y \text{ are parallel to each other.}$$

If $u \cdot [0, \infty)$ is an extreme ray, we say that $u$ gives the extreme ray.

**Definition 1.2.9.** *A cone $K$ is proper if it is closed, convex, pointed, and solid.*

For a cone $K \subseteq V$, one can show that

$$K^\circ = \{y \in V : \langle x, y \rangle \leq 0 \text{ for all } x \in K\}. \tag{1.2.12}$$

The set $K^\circ$ is called the *polar cone* of $K$. The negative of $K^\circ$ is the *dual cone*

$$K^\star := \{y \in V : \langle x, y \rangle \geq 0 \text{ for all } x \in K\}. \tag{1.2.13}$$

Clearly, the dual cone $K^\star$ is always closed and convex. The following are basic properties about dual cones.

**Theorem 1.2.10.** *([19]) Let $K \subseteq V$ be a nonempty set.*

   (i) *If $\text{int}(K) \neq \emptyset$, then $K^\star$ is pointed.*

  (ii) *If $K$ is a closed convex cone that is pointed, then $K^\star$ is solid.*

 (iii) *If $K$ is a closed convex cone, then $(K^\star)^\star = K$.*

 (iv) *A closed convex cone $K$ is proper if and only if its dual cone $K^\star$ is proper.*

**Definition 1.2.11.** *A cone $K$ is self-dual if $K^\star = K$.*

For instance, the nonnegative orthant $\mathbb{R}^n_+$ is a self-dual proper cone. Moreover, the *Lorentz cone* (also called the second order cone, or the ice cream cone in some literature)

$$\mathcal{Q}_n := \left\{ (x_0, x_1, \ldots, x_n) \in \mathbb{R}^{n+1} : x_0 \geq \sqrt{x_1^2 + \cdots + x_n^2} \right\}$$

is a proper cone of $\mathbb{R}^{n+1}$. It is also self-dual. This is because if

$$x_0 y_0 + x_1 y_1 + \cdots + x_n y_n \geq 0 \quad \forall (x_0, x_1, \ldots, x_n) \in \mathcal{Q}_n,$$

we must have $(y_0, y_1, \ldots, y_n) \in \mathcal{Q}_n$. However, not every cone is self-dual. For instance, the set

$$T = \{(x_1, x_2) : 2x_1 - x_2 \geq 0, 2x_2 - x_1 \geq 0\}$$

is a proper cone of $\mathbb{R}^2$. However, its dual cone is

$$T^\star = \{(y_1, y_2) : 2y_1 + y_2 \geq 0, 2y_2 + y_1 \geq 0\}.$$

This cone $T$ is not self-dual.

## 1.2.4 ▪ Exercises

**Exercise 1.2.1.** *Let $V, W$ be two vector spaces over $\mathbb{R}$ and let $f : V \to W$ be an affine linear function, i.e.,*

$$f(\lambda_1 u_1 + \cdots + \lambda_k u_k) = \lambda_1 f(u_1) + \cdots + \lambda_k f(u_k)$$
$$\text{for all } \lambda_i \in \mathbb{R}, \ u_i \in V, \lambda_1 + \cdots + \lambda_k = 1.$$

*Show that (i) if $T \subseteq V$ is convex, then its image $f(T) := \{f(u) : u \in T\}$ is convex; (ii) if $R \subseteq W$ is convex, then its preimage $f^{-1}(R) := \{v \in V : f(v) \in R\}$ is convex.*

**Exercise 1.2.2.** *Let $S_1, \ldots, S_m$ be nonempty sets of $\mathbb{R}^n$. Show that*

$$(S_1 \cup \cdots \cup S_m)^\circ = S_1^\circ \cap \cdots \cap S_m^\circ.$$

**Exercise 1.2.3.** *Express the polar set of*

$$\{x \in \mathbb{R}^n : -1 \leq x_1 \leq \cdots \leq x_n \leq 1\}$$

*by finitely many linear inequalities.*

**Exercise 1.2.4.** *The conic hull of a compact set may not be closed. For the unit circle $\{x_1^2 + (x_2 - 1)^2 = 1\}$, show that its conic hull is not closed.*

**Exercise 1.2.5.** *For a convex set $C \subseteq \mathbb{R}^n$, show that $C$ and its closure $\mathrm{cl}\,(C)$ have the same relative interior.*

**Exercise 1.2.6.** *Let $K$ be the cone $\{(x_1, x_2) : x_1 \in \mathbb{R}^1, x_2 > 0\}$. Determine whether or not $K$ and $K^\star$ are solid (or pointed).*

**Exercise 1.2.7.** *Let $K_1, K_2$ be two nonempty convex cones in $V$. Show that*

$$(K_1 + K_2)^\star = K_1^\star \cap K_2^\star.$$

*Moreover, (i) if one of $K_1$ and $K_2$ is solid, show that $K_1 + K_2$ is solid; (ii) if $K_1 + K_2$ is pointed, show that both $K_1$ and $K_2$ are pointed.*

**Exercise 1.2.8.** *Prove Proposition 1.2.5 and Theorem 1.2.8.*

**Exercise 1.2.9.** *Let $K$ be a closed convex cone in $V$. If $K$ is pointed, show that $K$ equals the convex hull of its extreme rays. Is this conclusion still true if $K$ is not pointed?*

**Exercise 1.2.10.** *(Helly's Theorem [19]) Let $C_1, \ldots, C_m$ be nonempty convex sets in $\mathbb{R}^d$, with $m \geq d + 1$. If every $d + 1$ of $C_1, \ldots, C_m$ has a common point, show that all $C_1, \ldots, C_m$ has a common point.*

## 1.3 ▪ Polyhedra and Linear Programs

In a vector space $V$, a *polyhedron* is a set that is defined by finitely many linear inequalities. A polyhedron $P \subseteq \mathbb{R}^n$ can be expressed as

$$P = \{x \in \mathbb{R}^n : Ax \geq b\}$$

for a matrix $A \in \mathbb{R}^{m \times n}$ and a vector $b \in \mathbb{R}^m$. For two vectors $a, b \in \mathbb{R}^n$, the inequality $a \geq b$ means that all entries of $a - b$ are nonnegative, i.e., $a - b \in \mathbb{R}^n_+$. In particular, if $m = 1$ and $A \neq 0$, then $P$ is called a *half space*, which is given by a single linear inequality.

The convex hull of a finite set of points is called a *polytope*. In fact, every polytope is a polyhedron and every bounded polyhedron is a polytope. An extreme point of a polytope is called a *vertex*. A 1-dimensional face of a polytope is called an *edge*. A $(d-1)$-dimensional face of a $d$-dimensional polytope is called a *facet*, while a $(d-2)$-dimensional face of it is called a *ridge*. They are also defined in the same way for convex sets that are not polyhedra.

### 1.3.1 ▪ Strong duality for LP

A linear program (LP) can be expressed in the form

$$\begin{cases} \min & c^T x \\ s.t. & Ax = b, x \geq 0, \end{cases} \tag{1.3.1}$$

for given $A \in \mathbb{R}^{m \times n}$, $b \in \mathbb{R}^m$, $c \in \mathbb{R}^n$. It is closely related to the following maximization problem:

$$\begin{cases} \max & b^T y \\ s.t. & c - A^T y \geq 0. \end{cases} \tag{1.3.2}$$

The problem (1.3.1) is called the primal form, while (1.3.2) is called the dual form. They are called the primal-dual pair. Denote by $\vartheta_1, \vartheta_2$ the optimal values of (1.3.1)-(1.3.2) respectively. The dual relationship between them can be seen as follows. For every $x$ that is feasible for (1.3.1) and for every $y$ that is feasible for (1.3.2),

$$c^T x - b^T y = c^T x - x^T A^T y = (c - A^T y)^T x \geq 0,$$

so $\vartheta_1 \geq \vartheta_2$. This is the *weak duality*. If $\vartheta_1 = \vartheta_2$, the *strong duality* is said to hold. The LP (1.3.1) is said to be *feasible* if there exists $x \geq 0$ satisfying the constraint $Ax = b$. If there is no such $x$, then (1.3.1) is said to be *infeasible*. The LP (1.3.1) is said to be unbounded below if there exists a sequence $\{u_k\}_{k=1}^{\infty} \subseteq \mathbb{R}^n_+$ such that $c^T u_k \to -\infty$ and $Au_k = b$. If (1.3.1) is infeasible, we set the optimal value $\vartheta_1 := +\infty$. Similar characteristics are defined for the dual problem (1.3.2). In particular, if (1.3.2) is infeasible, we set $\vartheta_2 := -\infty$. The following is the well-known strong duality theorem for LPs.

**Theorem 1.3.1.** *For the above primal-dual pair, we have the following:*

- *If one of (1.3.1) and (1.3.2) is feasible, then $\vartheta_1 = \vartheta_2$.*

- *If one of $\vartheta_1$ and $\vartheta_2$ is finite, then $\vartheta_1 = \vartheta_2$ and both (1.3.1) and (1.3.2) achieve the same optimal value (i.e., they both have optimizers).*

- *A primal feasible point $x^*$ of (1.3.1) is a minimizer if and only if there exists a dual feasible point $y^*$ of (1.3.2) such that $c^T x^* = b^T y^*$.*

To illustrate the strong duality, consider the following primal-dual pair:

$$\begin{cases} \min\limits_{x \in \mathbb{R}^2} & x_1 + x_2 + 2x_3 \\ s.t. & \begin{bmatrix} -1 & 1 & 1 \\ 1 & 1 & 2 \end{bmatrix} \begin{bmatrix} x_1 \\ x_2 \\ x_3 \end{bmatrix} = \begin{bmatrix} 0 \\ 1 \end{bmatrix}, \\ & x_1 \geq 0, x_2 \geq 0, x_3 \geq 0, \end{cases} \qquad \begin{cases} \max\limits_{y \in \mathbb{R}^2} & y_2 \\ s.t. & 1 + y_1 - y_2 \geq 0, \\ & 1 - y_1 - y_2 \geq 0, \\ & 2 - y_1 - 2y_2 \geq 0. \end{cases}$$

One can observe that $(\frac{1}{2}, \frac{1}{2}, 0)$ is a minimizer for the primal and $y^* = (0, 1)$ is a maximizer of the dual, since they are both feasible and $c^T x^* = b^T y^*$. An important property of a LP is that if the primal problem (1.3.1) is feasible and bounded below, then it must have an optimizer that is a *basic feasible* point, i.e., (1.3.1) has a minimizer $x^*$ that has at most $m$ nonzero entries. The simplex method for solving LPs is searching for optimizers among basic feasible points.

The following are the well-known results for detecting infeasibility and unboundedness for LPs.

**Theorem 1.3.2.** *Infeasibility and unboundedness of a LP can be certified by existence of an improving/decreasing ray for the dual, as follows:*

- *When (1.3.1) is feasible, the primal problem (1.3.1) is unbounded below if and only if it has a decreasing ray, i.e., there exists $d \in \mathbb{R}^n$ such that*

$$c^T d < 0, \quad Ad = 0, \quad d \geq 0.$$

- *When (1.3.2) is feasible, the dual problem (1.3.2) is unbounded above if and only if it has an improving ray, i.e., there exists $p \in \mathbb{R}^m$ such that*

$$b^T p > 0, \quad A^T p \leq 0.$$

- *The primal problem (1.3.1) is infeasible if and only if the dual (1.3.2) has an improving ray, i.e., there exists $\nu \in \mathbb{R}^m$ such that*

$$b^T \nu > 0, \quad A^T \nu \leq 0.$$

- *The dual problem (1.3.2) is infeasible if and only if the primal (1.3.1) has a decreasing ray, i.e., there exists $\lambda \in \mathbb{R}^n$ such that*

$$c^T \lambda < 0, \quad A\lambda = 0, \quad \lambda \geq 0.$$

For instance, consider the following primal-dual pair:

$$\begin{cases} \min\limits_{x \in \mathbb{R}^3} & -x_1 + x_3 \\ s.t. & \begin{bmatrix} -1 & 0 & 2 \\ -1 & -1 & 3 \end{bmatrix} \begin{bmatrix} x_1 \\ x_2 \\ x_3 \end{bmatrix} = \begin{bmatrix} 0 \\ 0 \end{bmatrix}, \\ & x_1 \geq 0, x_2 \geq 0, x_3 \geq 0, \end{cases} \qquad \begin{cases} \max\limits_{y \in \mathbb{R}^2} & 0y_1 + 0y_2 \\ s.t. & -1 + y_1 + y_2 \geq 0, \\ & y_2 \geq 0, \\ & 1 - 2y_1 - 3y_2 \geq 0. \end{cases}$$

The primal is unbounded below while the dual is infeasible. This is certified by the following:

$$\begin{bmatrix} -1 & 0 & 1 \end{bmatrix} \begin{bmatrix} 2 \\ 1 \\ 1 \end{bmatrix} < 0, \qquad \begin{bmatrix} -1 & 0 & 2 \\ -1 & -1 & 3 \end{bmatrix} \begin{bmatrix} 2 \\ 1 \\ 1 \end{bmatrix} = \begin{bmatrix} 0 \\ 0 \end{bmatrix}.$$

Another example is the following LP:

$$\begin{cases} \min_{x \in \mathbb{R}^2} & -x_1 + x_2 \\ s.t. & \begin{bmatrix} 1 & 2 \\ 1 & 1 \end{bmatrix} \begin{bmatrix} x_1 \\ x_2 \end{bmatrix} = \begin{bmatrix} 1 \\ 3 \end{bmatrix}, \\ & x_1 \geq 0,\ x_2 \geq 0, \end{cases} \qquad \begin{cases} \max_{y \in \mathbb{R}^2} & y_1 + 3y_2 \\ s.t. & \begin{bmatrix} -1 \\ 1 \end{bmatrix} - \begin{bmatrix} 1 & 1 \\ 2 & 1 \end{bmatrix} \begin{bmatrix} y_1 \\ y_2 \end{bmatrix} \geq \begin{bmatrix} 0 \\ 0 \end{bmatrix}. \end{cases}$$

The above primal is infeasible while its dual is unbounded above. This is certified by the following:

$$\begin{bmatrix} 1 & 3 \end{bmatrix} \begin{bmatrix} -1 \\ 1 \end{bmatrix} > 0, \qquad \begin{bmatrix} 1 & 1 \\ 2 & 1 \end{bmatrix} \begin{bmatrix} -1 \\ 1 \end{bmatrix} \leq \begin{bmatrix} 0 \\ 0 \end{bmatrix}.$$

We would like to remark that the primal (1.3.1) and the dual (1.3.2) are possibly both infeasible. For instance, the following primal-dual problems are both infeasible:

$$\begin{cases} \min_{x \in \mathbb{R}^2} & -x_1 - x_2 \\ s.t. & \begin{bmatrix} -1 & 1 \\ -1 & 1 \end{bmatrix} \begin{bmatrix} x_1 \\ x_2 \end{bmatrix} = \begin{bmatrix} 2 \\ 3 \end{bmatrix}, \\ & x_1 \geq 0,\ x_2 \geq 0, \end{cases} \qquad \begin{cases} \max_{y \in \mathbb{R}^2} & 2y_1 + 3y_2 \\ s.t. & -1 + y_1 + y_2 \geq 0, \\ & -1 - y_1 - y_2 \geq 0. \end{cases}$$

## 1.3.2 ▪ Farkas Lemma and Theorem of the Alternatives

A foundational result in linear programming is the *Farkas Lemma*.

**Theorem 1.3.3.** *(Farkas Lemma) For given $A \in \mathbb{R}^{m \times n}$ and $c \in \mathbb{R}^n$, if $c^T x \geq 0$ for all $x$ satisfying $Ax \geq 0$, then there exists $\lambda \in \mathbb{R}^m$ such that*

$$c = A^T \lambda, \quad \lambda \geq 0.$$

The Farkas Lemma can be equivalently stated as follows. If the linear function $c^T x \geq 0$ on the set $\{Ax \geq 0\}$, then there exists $\lambda \in \mathbb{R}^m_+$ such that $c^T x = \lambda^T Ax$, which is a polynomial identity. Interestingly, this result can be generalized to inhomogeneous linear functions. The inhomogeneous version of the Farkas Lemma is also referenced as the *linear Positivstellensatz*.

**Theorem 1.3.4.** *(Inhomogeneous Farkas Lemma) Suppose the set $P := \{Ax \geq b\}$ is nonempty. If a linear function $c^T x - d \geq 0$ on $P$, then there exist $\lambda \in \mathbb{R}^m$ and $\nu \in \mathbb{R}$ such that*

$$c^T x - d = \lambda^T (Ax - b) + \nu, \quad \lambda \geq 0, \quad \nu \geq 0.$$

*This is also equivalent to that*

$$c = A^T \lambda, \quad d \leq b^T \lambda, \quad \lambda \geq 0.$$

More general than the above is the *Theorem of the Alternatives*. It shows that a polyhedral set is empty if and only if another polyhedral set is nonempty.

**Theorem 1.3.5.** *(Theorem of the Alternatives) For given $A_1 \in \mathbb{R}^{m_1 \times n}$, $A_2 \in \mathbb{R}^{m_2 \times n}$, $b_1 \in \mathbb{R}^{m_1}$, and $b_2 \in \mathbb{R}^{m_2}$, the set*

$$\{x \in \mathbb{R}^n : A_1 x > b_1, \; A_2 x \geq b_2\}$$

*is empty if and only if the following set*

$$\left\{ (\lambda_1, \lambda_2) \in \mathbb{R}^{m_1} \times \mathbb{R}^{m_2} \;\middle|\; \begin{array}{l} \lambda_1 \geq 0, \; \lambda_2 \geq 0, \\ b_1^T \lambda_1 + b_2^T \lambda_2 \geq 0, \\ A_1^T \lambda_1 + A_2^T \lambda_2 = 0, \\ (e + b_1)^T \lambda_1 + b_2^T \lambda_2 = 1 \end{array} \right\}$$

*is nonempty. (The above $e$ is the vector of all ones.)*

### 1.3.3 ▪ Exercises

**Exercise 1.3.1.** *For a matrix $A \in \mathbb{R}^{m \times n}$, show that the dual cone of the polyhedral cone $P = \{x \in \mathbb{R}^n : Ax \geq 0\}$ is the set*

$$Q := \left\{ y \in \mathbb{R}^n : y = A^T z \text{ for some } z \in \mathbb{R}^m_+ \right\}.$$

**Exercise 1.3.2.** *Show that every polytope is a polyhedron and every bounded polyhedron is a polytope.*

**Exercise 1.3.3.** *Prove that a polyhedron has finitely many faces and all its faces are exposed.*

**Exercise 1.3.4.** *If $P_1, P_2 \subseteq \mathbb{R}^n$ are two polyhedrons, show that $P_1 + P_2$ is also a polyhedron.*

**Exercise 1.3.5.** *Show that the linear inequality system*

$$a_1^T x - b_1 \geq 0, \quad i = 1, \ldots, m$$

*does not have a solution if and only if there exist scalars $\lambda_1, \ldots, \lambda_m$ satisfying*

$$-1 = \sum_{i=1}^{m} \lambda_i (a_i^T x - b_i), \quad \lambda_1 \geq 0, \ldots, \lambda_m \geq 0.$$

**Exercise 1.3.6.** *([19]) For given matrices/vectors $A_1, A_2, b_1, b_2$, show that the set*

$$\{x \in \mathbb{R}^n : A_1 x > b_1, \; A_2 x \geq b_2\}$$

*is empty if and only if the following set*

$$\left\{ (x, t, s) \in \mathbb{R}^n \times \mathbb{R}^1 \times \mathbb{R}^1 \;\middle|\; \begin{array}{l} A_1 x - t b_1 - s e \geq 0, \\ A_2 x - t b_2 \geq 0, \\ t - s \geq 0, \; s > 0 \end{array} \right\}$$

*is empty. Use this fact and Farkas Lemma to prove the inhomogeneous Farkas Lemma and Theorem of the Alternatives.*

**Exercise 1.3.7.** *For given $c, a_1, \ldots, a_m \in \mathbb{R}^n$, show that if*

$$c^T x \leq \max_{1 \leq i \leq m} a_i^T x \quad \text{for all } x \geq 0,$$

*then there exists* $\lambda := (\lambda_1, \ldots, \lambda_m) \in \mathbb{R}^m$ *satisfying*

$$\lambda \geq 0, \quad e^T \lambda = 1, \quad c \leq \sum_{i=1}^{m} \lambda_i a_i.$$

## 1.4 ▪ PSD Matrices and Spectrahedra

A real matrix $A = (A_{ij}) \in \mathbb{R}^{N \times N}$ is symmetric if it equals its transpose $A^T$, i.e., $A_{ij} = A_{ji}$ for all $i, j$. Let $\mathcal{S}^N$ denote the space of all real $N$-by-$N$ symmetric matrices. A matrix $A \in \mathcal{S}^N$ is said to be *positive semidefinite* (psd or PSD) if

$$x^T A x \geq 0 \quad \text{for all } x \in \mathbb{R}^N.$$

It is *positive definite* (pd) if the above is a strict inequality for all $x \neq 0$. When $A$ is psd (resp., pd), we write that $A \succeq 0$ (resp., $A \succ 0$). Recall that the Euclidean inner product for the matrix space $\mathbb{R}^{N \times N}$ is such that

$$\langle A, B \rangle := \text{trace}(AB^T) \tag{1.4.1}$$

for $A, B \in \mathbb{R}^{N \times N}$. For notational convenience, recall that we also use the notation

$$A \bullet B := \text{trace}(AB^T).$$

The set of all psd matrices in $\mathcal{S}^N$ is denoted as $\mathcal{S}_+^N$, and $\mathcal{S}_{++}^N$ stands for the set of all positive definite matrices in $\mathcal{S}^N$. The set $\mathcal{S}_+^N$ is a proper cone. Its interior is $\mathcal{S}_{++}^N$. Under the inner product (1.4.1), the dual cone of $\mathcal{S}_+^N$ equals itself, i.e., the cone $\mathcal{S}_+^N$ is self-dual. This is because if $\langle X, Y \rangle \geq 0$ for all $X \in \mathcal{S}_+^N$, then $Y$ must also be psd. A real symmetric matrix is psd if and only if all its eigenvalues are nonnegative, which is equivalent to all its principal minors being nonnegative. Similarly, a real symmetric matrix is pd if and only if all its eigenvalues are positive, which is equivalent to all its leading principal minors being positive. We refer to the books [74, 130] for basic properties of psd matrices.

In the following, we characterize faces of psd cones. For two symmetric matrices $A, B \in \mathcal{S}^N$, the inequality $A \succeq B$ means that $A - B$ is psd, while $A \succ B$ means that $A - B$ is positive definite, i.e.,

$$A \succeq B \Leftrightarrow A - B \in \mathcal{S}_+^N, \qquad A \succ B \Leftrightarrow A - B \in \mathcal{S}_{++}^N.$$

The inequality $\succeq$ gives a partial ordering in $\mathcal{S}^N$. Recall that Range $(A)$ denotes the range space of $A$. Note that if $A \succeq B \succeq 0$, then Range $(B) \subseteq$ Range $(A)$. For a subset $T \subseteq \mathcal{S}_+^N$, recall that face $(T, \mathcal{S}_+^N)$ denotes the smallest face of $\mathcal{S}_+^N$ that contains $T$. We first characterize the smallest face that contains a psd matrix $A$, which is face $(A, \mathcal{S}_+^N)$. Clearly, if $A$ is positive definite, then face $(A, \mathcal{S}_+^N)$ is the entire cone $\mathcal{S}_+^N$. If $A$ is psd but singular, then it has the spectral decomposition

$$Q^T A Q = \begin{bmatrix} \Lambda_1 & 0 \\ 0 & 0 \end{bmatrix},$$

where $\Lambda_1 \in \mathcal{S}_{++}^r$ is diagonal, $r = \text{rank } A$, and $Q$ is orthogonal. If $A = \lambda B + (1 - \lambda)C$ with $B, C \in \mathcal{S}_+^N$ and $\lambda \in (0, 1)$, then $B, C$ can be expressed in the block form

$$B = Q \begin{bmatrix} B_1 & 0 \\ 0 & 0 \end{bmatrix} Q^T, \quad C = Q \begin{bmatrix} C_1 & 0 \\ 0 & 0 \end{bmatrix} Q^T$$

for some $B_1, C_1 \in \mathcal{S}_+^r$. Therefore, face $(A, \mathcal{S}_+^N)$ must contain the set

$$G := \left\{ Q \begin{bmatrix} X & 0 \\ 0 & 0 \end{bmatrix} Q^T : X \in \mathcal{S}_+^r \right\}.$$

One can check that $G$ is a face, so $G = \text{face}\,(A, \mathcal{S}_+^N)$. More general faces of the psd cone $\mathcal{S}_+^N$ can be characterized as follows.

**Theorem 1.4.1.** *([321, Theorem. 3.7.1]) A set $F \subseteq \mathcal{S}_+^N$ is a face if and only if there exists a subspace $L \subseteq \mathbb{R}^n$ such that*

$$F = \left\{ X \in \mathcal{S}_+^N : \text{Range}\,(X) \subseteq L \right\}.$$

## 1.4.1 ▪ Affine sections of psd cones

For given symmetric matrices $A_1, \ldots, A_m \in \mathcal{S}^N$, define the linear map

$$\mathcal{A} : \mathcal{S}^N \to \mathbb{R}^m, \quad X \mapsto \mathcal{A}(X) := (\langle A_1, X \rangle, \ldots, \langle A_m, X \rangle). \tag{1.4.2}$$

For a given $b = (b_1, \ldots, b_m) \in \mathbb{R}^m$, consider the set

$$\Omega = \{ X \in \mathcal{S}_+^N : \mathcal{A}(X) = b \}. \tag{1.4.3}$$

It is the intersection of $\mathcal{S}_+^N$ with the affine subspace

$$\mathcal{A}^{-1}(b) := \{ X \in \mathcal{S}^N : \mathcal{A}(X) = b \}.$$

Clearly, $\Omega$ is closed and convex. Geometric properties of $\Omega$ are given as follows. For a convex set $K$ and $T \subseteq K$, note that face $(T, K)$ denotes the minimum face of $K$ containing $T$. If $K$ is a convex cone containing the origin, then every nonempty face $F$ of $K$ also contains the origin. For such a case, we use $\text{lin}\,(F)$ to denote the subspace spanned by $F$, which also equals the affine hull of $F$. The following is implied by Theorem 3.3.1 of [321].

**Theorem 1.4.2.** *Let $\mathcal{A}, b, \Omega$ be as above. For a matrix $X \in \Omega$, let*

$$F_1 = \text{face}\,(X, \Omega), \quad F_2 := \text{face}\,(X, \mathcal{S}_+^N).$$

*Let $\text{lin}\,(F_2)$ denote the subspace spanned by $F_2$. Then, we have the following:*

- *$F_2 = \text{face}\,(F_1, \mathcal{S}_+^N)$, $F_1 = F_2 \cap \mathcal{A}^{-1}(b)$, and*

$$\text{aff}\,(F_1) = \text{lin}\,(F_2) \cap \mathcal{A}^{-1}(b) = X + \text{lin}\,(F_2) \cap \ker \mathcal{A}.$$

  *In the above, $\ker \mathcal{A}$ denotes the kernel (or null space) of $\mathcal{A}$.*

- *$\dim F_1 = \dim \big( \text{lin}\,(F_2) \cap \ker \mathcal{A} \big)$, and*

$$\dim F_1 = \dim F_2 - \dim \text{Range}\,(\mathcal{A}) + \dim \big( F_2^{\perp} \cap \text{Range}\,(\mathcal{A}^*) \big).$$

  *In the above, $\mathcal{A}^*$ is the adjoint operator of $\mathcal{A}$ and $\text{Range}\,(\mathcal{A})$ denotes the range space of $\mathcal{A}$.*

- *The matrix $X$ is an extreme point of $\Omega$ if and only if*

$$\text{lin}\,(F_2) \cap \ker \mathcal{A} = \{0\}.$$

## 1.4.2 ▪ Spectrahedra

For the linear map $\mathcal{A}$ in (1.4.2), its adjoint map $\mathcal{A}^* : \mathbb{R}^m \to \mathcal{S}^N$ is given as

$$\mathcal{A}^*(y) := y_1 A_1 + \cdots + y_m A_m$$

in the variable $y := (y_1, \ldots, y_m)$. For a matrix $A_0 \in \mathcal{S}^N$, the set

$$\Psi := \{y \in \mathbb{R}^m : A_0 + \mathcal{A}^*(y) \succeq 0\} \tag{1.4.4}$$

is called a *spectrahedron*. Clearly, $\Psi$ is also closed and convex. The inequality $A_0 + \mathcal{A}^*(y) \succeq 0$ is called a *linear matrix inequality* (LMI). If $A_0 \succ 0$ (e.g., $A_0 = I_N$, the $N$-by-$N$ identity matrix), then the origin $0$ is an interior point of $\Psi$. When $A_0 = 0$, $\Psi$ is called a *spectrahedral cone*. Linear matrix inequalities have broad applications (see [19, 36, 70, 321]). To study geometric properties of the spectrahedron $\Psi$, it is convenient to consider the set

$$\mathcal{D} := \left\{(y, Z) \in \mathbb{R}^m \times \mathcal{S}_+^N : \mathcal{A}^*(y) + A_0 = Z\right\}. \tag{1.4.5}$$

Note that a subset $F \subseteq \Psi$ is a face of $\Psi$ if and only if the set

$$\left\{(y, Z) : \mathcal{A}^*(y) + A_0 = Z, \, y \in F\right\} \tag{1.4.6}$$

is a face of $\mathcal{D}$. Thus, to investigate the facial structure of $\Psi$, it is enough to investigate the face structure of $\mathcal{D}$. For a face $G$ of $\mathcal{S}_+^N$, $\mathrm{lin}\,(G)$ denotes the subspace generated by $G$. The following is implied by Theorem 3.3.2 of [321].

**Theorem 1.4.3.** *Let $A_0, \mathcal{A}, \Psi, \mathcal{D}$ be as above. For a point $\hat{y} \in \Psi$, let $\hat{Z} = A_0 + \mathcal{A}^*(\hat{y})$ and*

$$G_1 = \mathrm{face}\left((\hat{y}, \hat{Z}), \mathcal{D}\right), \quad G_2 = \mathrm{face}\left(\hat{Z}, \mathcal{S}_+^N\right).$$

*Then, we have the following:*

- $\mathbb{R}^m \times G_2 = \mathrm{face}\left(G_1, \mathbb{R}^m \times \mathcal{S}_+^N\right).$

- $G_1 = \left\{(y, Z) : Z = A_0 + \mathcal{A}^*(y) \in G_2\right\}$ *and*

$$\begin{aligned}\mathrm{aff}\,(G_1) &= \left\{(y, Z) : Z = A_0 + \mathcal{A}^*(y) \in \mathrm{lin}\,(G_2)\right\} \\ &= (\hat{y}, \hat{Z}) + \{(\Delta y, \Delta Z) : \mathcal{A}^*(\Delta y) = \Delta Z \in \mathrm{lin}\,(G_2)\}.\end{aligned}$$

- $\dim G_1 = \dim\left(\mathrm{lin}\,(G_2) \cap \mathrm{span}\,\{A_1, \ldots, A_n\}\right)$ *and*

$$\dim G_1 = \dim G_2 - \dim \mathcal{S}^N + \dim \mathrm{Range}\,(\mathcal{A}) + \dim\left(G_2^\perp \cap \ker \mathcal{A}\right).$$

  *The above superscript $\perp$ denotes the orthogonal complement.*

- *The point $\hat{y}$ is an extreme point of $\Psi$ if and only if*

$$\mathrm{lin}\,(G_2) \cap \mathrm{span}\,\{A_1, \ldots, A_n\} = \{0\}.$$

### 1.4.3 ▪ Nondegeneracy and complementarity

In convex optimization about psd matrix variables, the *degeneracy* or *nondegeneracy* often needs to be considered. For a face $F$ of a cone $K$, its complementary face is the set

$$F^{\triangle} := \{u \in K^{\star} : \langle u, v \rangle = 0 \,\forall\, v \in F\}.$$

The degeneracy or nondegeneracy for a point in $\Omega, \Psi$ can be defined as follows (also see [7, 321]).

**Definition 1.4.4.** *([321, Definition 3.3.1]) Let $\mathcal{A}, \Omega, \Psi$ be as above. Consider a point $X \in \Omega$ and $y \in \Psi$. Let $Z := A_0 + \mathcal{A}^*(y)$ and*

$$F = \text{face}\left(X, \mathcal{S}_+^N\right), \quad G = \text{face}\left(Z, \mathcal{S}_+^N\right).$$

- *The point $X$ is called nondegenerate if*

$$\text{lin}\left(F^{\triangle}\right) \cap \text{Range}\left(\mathcal{A}^*\right) = \{0\}.$$

  *Otherwise, $X$ is called degenerate.*

- *The point $y$ is called nondegenerate if*

$$\text{lin}\left(G^{\triangle}\right) \cap \ker \mathcal{A} = \{0\}.$$

  *Otherwise, $y$ is called degenerate.*

- *The pair $(X, Z)$ is said to be strictly complementary if $F^{\triangle} = G$.*

The degeneracy and nondegeneracy can be defined for arbitrary cones. This is done in [321, §3.3]. The following are some conclusions about ranks and dimensions for faces of $\Omega, \Psi$.

**Theorem 1.4.5.** *([321, Corollary 3.3.1]) Let $\mathcal{A}, \Omega, \Psi$ be as above. Consider a point $X \in \Omega$ and $y \in \Psi$. Let $Z := A_0 + \mathcal{A}^*(y)$ and*

$$\mathcal{F} = \text{face}\left(X, \Omega\right), \quad \mathcal{G} = \text{face}\left(Z, \Psi\right).$$

*Then, the following hold:*

- *Let $r = \text{rank } X$, then $\binom{r+1}{2} \leq m + \dim \mathcal{F}$.*

- *If $X$ is nondegenerate, then $\binom{N-r+1}{2} \leq \binom{N+1}{2} - m$.*

- *Let $s = \text{rank } Z$, then $\binom{s+1}{2} \leq \binom{N+1}{2} - m + \dim \mathcal{G}$.*

- *If $y$ is nondegenerate, then $\binom{N-s+1}{2} \leq m$.*

### 1.4.4 ▪ Exercises

**Exercise 1.4.1.** *For given $A \in \mathcal{S}^n$ and $b \in \mathbb{R}^n, c \in \mathbb{R}$, show that $x^T A x + 2b^T x + c \geq 0$ for all $x \in \mathbb{R}^n$ if and only if*

$$\begin{bmatrix} c & b^T \\ b & A \end{bmatrix} \succeq 0.$$

**Exercise 1.4.2.** *(Schur complements) For $A \in \mathcal{S}_+^{n_1}$, $B \in \mathbb{R}^{n_1 \times n_2}$ and $C \in \mathcal{S}^{n_2}$, show that*

$$\begin{bmatrix} A & B \\ B^T & C \end{bmatrix} \succeq 0 \quad \Leftrightarrow \quad \text{Range}(B) \subseteq \text{Range}(A), \, C \succeq B^T A^\dagger B.$$

*The above superscript $\dagger$ denotes the Moore-Penrose Pseudoinverse (see Exercise 1.1.10).*

**Exercise 1.4.3.** *([321]) For all $X, Y \in \mathcal{S}_+^N$, show the following:*

(i) $\text{Range}(X + Y) = \text{Range}(X) + \text{Range}(Y)$.

(ii) *If $\text{Range}(Y) \subseteq \text{Range}(X)$, then there exists $Z \in \mathcal{S}_+^N$ such that $X = \lambda Y + (1 - \lambda)Z$ for some $\lambda \in (0, 1)$.*

(iii) *The inner product $\langle X, Y \rangle = 0$ if and only if $\text{Range}(X) \perp \text{Range}(Y)$.*

**Exercise 1.4.4.** *Let $F \in \mathcal{S}^n$ and $G \in \mathbb{R}^{m \times n}$. Show that $x^T F x > 0$ for all $x \neq 0$ satisfying $Gx = 0$ if and only if there exists a scalar $\lambda$ such that*

$$F + \lambda G^T G \succ 0.$$

**Exercise 1.4.5.** *Consider the spectrahedron $H$ consisting of all $y = (y_1, y_2, y_3)$ satisfying*

$$\begin{bmatrix} 1 & y_1 & y_2 \\ y_1 & y_2 & y_3 \\ y_2 & y_3 & 8 \end{bmatrix} \succeq 0.$$

*Determine all the faces of $H$. Does $H$ have a degenerate point?*

**Exercise 1.4.6.** *The set of psd matrices whose diagonal entries are ones is called an elliptope. The $N$-by-$N$ elliptope is*

$$E_N = \{X \in \mathcal{S}_+^N : X_{11} = \cdots = X_{NN} = 1\}.$$

*For instance, when $N = 4$, the elliptope is defined as*

$$\begin{bmatrix} 1 & X_{12} & X_{13} & X_{14} \\ X_{21} & 1 & X_{23} & X_{24} \\ X_{31} & X_{32} & 1 & X_{34} \\ X_{41} & X_{42} & X_{43} & 1 \end{bmatrix} \succeq 0.$$

*Let $W$ be the matrix of all ones and $F = \text{face}(W, E_N)$. Determine $\text{ri}(F)$, $\text{aff}(F)$, $F^\triangle$, $\dim F$. What are the extremal points of $F$? Is $W$ degenerate or nondegenerate?*

**Exercise 1.4.7.** *Prove Theorem 1.4.1. Use Theorems 1.4.2 and 1.4.3 to prove Theorem 1.4.5.*

**Exercise 1.4.8.** *Show that every face of a spectrahedron is exposed.*

## 1.5 ▪ Semidefinite Optimization

A semidefinite program (SDP) is a major convex relaxation tool for solving moment and polynomial optimization. In applications, there is often more than one psd matrix variable. This section reviews basic knowledge about semidefinite optimization with a tuple of psd matrix variables.

For integers $N_1, \ldots, N_\ell > 0$, we denote the Cartesian product

$$\mathcal{S}_+^{N_1,\ldots,N_\ell} := \mathcal{S}_+^{N_1} \times \cdots \times \mathcal{S}_+^{N_\ell}. \tag{1.5.1}$$

It is a proper cone of the matrix space

$$\mathcal{S}^{N_1,\ldots,N_\ell} := \mathcal{S}^{N_1} \times \cdots \times \mathcal{S}^{N_\ell}. \tag{1.5.2}$$

Every $X \in \mathcal{S}^{N_1,\ldots,N_\ell}$ is a tuple

$$X = (X_1, \ldots, X_\ell)$$

with each $X_i \in \mathcal{S}^{N_i}$. So $X$ can be viewed as a block diagonal symmetric matrix. For $X \in \mathcal{S}_+^{N_1,\ldots,N_\ell}$, we write that $X \succeq 0$. If each $X_i \succ 0$, we write that $X \succ 0$. Clearly, $X$ is in the interior of $\mathcal{S}_+^{N_1,\ldots,N_\ell}$ if and only if each $X_i$ is positive definite.

For two tuples $X = (X_1, \ldots, X_\ell)$, $Y = (Y_1, \ldots, Y_\ell)$ in the space $\mathcal{S}^{N_1,\ldots,N_\ell}$, their inner product is then defined as

$$\langle X, Y \rangle = \text{trace}(X_1 Y_1) + \cdots + \text{trace}(X_\ell Y_\ell).$$

This induces the norm

$$\|X\| = \left(\text{trace}(X_1^2) + \cdots + \text{trace}(X_\ell^2)\right)^{1/2}.$$

Under the above inner product, the cone $\mathcal{S}_+^{N_1,\ldots,N_\ell}$ is self-dual, i.e.,

$$\left(\mathcal{S}_+^{N_1,\ldots,N_\ell}\right)^* = \mathcal{S}_+^{N_1,\ldots,N_\ell}.$$

Let $\mathcal{A} : \mathcal{S}^{N_1,\ldots,N_\ell} \to \mathbb{R}^m$ be a linear map. For $C \in \mathcal{S}^{N_1,\ldots,N_\ell}$ and $b \in \mathbb{R}^m$, consider the linear optimization problem

$$\left\{ \begin{array}{ll} \min\limits_{X} & \langle C, X \rangle \\ s.t. & \mathcal{A}(X) = b, \quad X \in \mathcal{S}_+^{N_1,\ldots,N_\ell}. \end{array} \right. \tag{1.5.3}$$

The $C$ is a given matrix tuple. We can write it as

$$C = (C_1, \ldots, C_\ell).$$

Since the cone $\mathcal{S}_+^{N_1,\ldots,N_\ell}$ is self-dual, the dual problem of (1.5.3) is

$$\left\{ \begin{array}{ll} \max & b^T y \\ s.t. & C - \mathcal{A}^*(y) \in \mathcal{S}_+^{N_1,\ldots,N_\ell}, \quad y \in \mathbb{R}^m. \end{array} \right. \tag{1.5.4}$$

The above $\mathcal{A}^*$ is the adjoint map of $\mathcal{A}$, i.e., $\mathcal{A}^*$ is the linear map such that

$$\mathcal{A}^* : \mathbb{R}^m \to \mathcal{S}^{N_1,\ldots,N_\ell}, \quad \langle \mathcal{A}(X), y \rangle = \langle X, \mathcal{A}^*(y) \rangle$$

for all $X$ and $y$. If $\mathcal{A}$ can be expressed as

$$\mathcal{A}(X) = (\langle A_1, X \rangle, \ldots, \langle A_m, X \rangle)$$

for symmetric matrix tuples $A_1, \ldots, A_m$, then $\mathcal{A}^*(y)$ is given as

$$\mathcal{A}^*(y) = y_1 A_1 + \cdots + y_m A_m \quad \text{for } y = (y_1, \ldots, y_m).$$

If (1.5.3) has a minimizer, then (1.5.3) is called *solvable*. The solvability of (1.5.4) is similarly defined. If there exists $X \succ 0$ such that $\mathcal{A}(X) = b$, then (1.5.3) is called *strictly feasible*. Similarly, if there exists $y$ such that $C - \mathcal{A}^*(y) \succ 0$, then (1.5.4) is called strictly feasible. Let $c^*$ be the optimal value of (1.5.3) and $b^*$ be the optimal value of (1.5.4). If (1.5.3) (resp., (1.5.4)) is infeasible, then we set $c^* = +\infty$ (resp., $b^* = -\infty$) by convention. If $c^* > -\infty$, then (1.5.3) is called *bounded below*; if $b^* < +\infty$, then (1.5.4) is called *bounded above*. The notions of unbounded below and unbounded above are similarly defined.

### 1.5.1 ▪ Duality theory

The duality relation between (1.5.3) and (1.5.4) can be seen as follows. For all $X$ that are feasible for (1.5.3) and for all $y$ that are feasible for (1.5.4), one can see that

$$\langle C, X \rangle - b^T y = \langle C, X \rangle - \langle \mathcal{A}^*(y), X \rangle = \langle C - \mathcal{A}^*(y), X \rangle \geq 0.$$

Therefore, we always have

$$c^* \geq b^*.$$

This is the so-called *weak duality* relation. The difference $c^* - b^*$ is called the *duality gap* between (1.5.3) and (1.5.4). When $c^* = b^*$, the duality gap is said to be zero, or we say that there is no duality gap. When $c^* = b^*$ and one of (1.5.3) and (1.5.4) has an optimizer, then the *strong duality* is said to hold. We refer to [19, 313, 321] about the duality theory for semidefinite optimization. The following is the classical strong duality theorem.

**Theorem 1.5.1.** *For the semidefinite optimization (1.5.3)-(1.5.4), we have the following:*

(i) *If (1.5.3) is bounded below and is strictly feasible, then $c^* = b^*$ and (1.5.4) is solvable.*

(ii) *If (1.5.4) is bounded above and is strictly feasible, then $c^* = b^*$ and (1.5.3) is solvable.*

(iii) *If one of (1.5.3) and (1.5.4) is bounded and strictly feasible, then a primal-dual feasible pair $(X, y)$ is a pair of optimal solutions to the respective problems if and only if*

$$\langle C, X \rangle = b^T y,$$

*which then holds if and only if $\langle C - \mathcal{A}^*(y), X \rangle = 0$.*

Therefore, solving (1.5.3) and (1.5.4) requires finding feasible $X$ and $y$ such that $\langle C - \mathcal{A}^*(y), X \rangle = 0$. It is equivalent to the matrix product

$$(C - \mathcal{A}^*(y))X = 0.$$

When neither of (1.5.3) and (1.5.4) is strictly feasible, the strong duality may not hold, as shown in the following examples from [313].

- If the dual problem (1.5.4) is

$$\left\{ \begin{array}{ll} \max & -y_1 \\ s.t. & \begin{bmatrix} y_1 & 1 \\ 1 & y_2 \end{bmatrix} \succeq 0, \end{array} \right.$$

  the duality gap is still zero. The primal (1.5.3) has a minimizer, but (1.5.4) does not have a maximizer.

- If the dual problem (1.5.4) is

$$\left\{ \begin{array}{ll} \max & 2y_2 \\ s.t. & \begin{bmatrix} -y_1 & -y_2 & 0 \\ -y_2 & 0 & 0 \\ 0 & 0 & 1 - 2y_2 \end{bmatrix} \succeq 0, \end{array} \right.$$

  the duality gap is 1. Both the primal and dual have optimizers.

- If the dual problem (1.5.4) is

$$\begin{cases} \max & 2y_2 \\ s.t. & \begin{bmatrix} -y_1 & -y_2 \\ -y_2 & 0 \end{bmatrix} \succeq 0, \end{cases}$$

the duality gap is infinite, because the primal problem is infeasible and $c^* = +\infty$. The dual has the maximizer $(y_1, 0)$, for all $y_1 \leq 0$, and so $b^* = 0$.

The following is another example of semidefinite optimization.

**Example 1.5.2.** *For given points* $(\alpha_1, \beta_1), \ldots, (\alpha_m, \beta_m) \in \mathbb{R}^2$, *we are looking for the smallest circle* $\sqrt{(x_1 - y_1)^2 + (x_2 - y_2)^2} = y_3$, *with center* $(y_1, y_2)$ *and radius* $y_3$, *containing all of them. The corresponding optimization problem is*

$$\begin{cases} \min & y_3 \\ s.t. & \sqrt{(\alpha_i - y_1)^2 + (\beta_i - y_2)^2} \leq y_3, \ i = 1, \ldots, m. \end{cases}$$

*It can be equivalently written as*

$$\begin{cases} \max & -y_3 \\ s.t. & \begin{bmatrix} y_3 + y_1 - \alpha_i & \beta_i - y_2 \\ \beta_i - y_2 & y_3 - y_1 + \alpha_i \end{bmatrix} \succeq 0, \ i = 1, \ldots, m. \end{cases}$$

*The above is in the dual form (1.5.4) with*

$$b = (0, 0, -1), \ N_1 = \cdots = N_m = 2,$$

$$C = \left( \begin{bmatrix} -\alpha_1 & \beta_1 \\ \beta_1 & \alpha_1 \end{bmatrix}, \ldots, \begin{bmatrix} -\alpha_m & \beta_m \\ \beta_m & \alpha_m \end{bmatrix} \right),$$

$$A_1 = \left( \begin{bmatrix} -1 & 0 \\ 0 & 1 \end{bmatrix}, \ldots, \begin{bmatrix} -1 & 0 \\ 0 & 1 \end{bmatrix} \right),$$

$$A_2 = \left( \begin{bmatrix} 0 & -1 \\ -1 & 0 \end{bmatrix}, \ldots, \begin{bmatrix} 0 & -1 \\ -1 & 0 \end{bmatrix} \right),$$

$$A_3 = \left( \begin{bmatrix} -1 & 0 \\ 0 & -1 \end{bmatrix}, \ldots, \begin{bmatrix} -1 & 0 \\ 0 & -1 \end{bmatrix} \right).$$

A semidefinite program is a powerful relaxation tool in convex optimization. It can be solved efficiently by numerical methods. There exist numerous software packages for solving SDPs. Frequently used ones are MOSEK [197], SeDuMi [308], SDPT3 [314], SDPA [324], SDPLR [41], and SDPNAL+ [325]. A concrete example for coding by SeDuMi is Example 1.6.2. References and software for SDPs can be found in Helmberg's page for semidefinite programs, https://www-user.tu-chemnitz.de/~helmberg/semidef.html. We refer to the books [19, 70, 313, 321] for more comprehensive introductions to the theory, algorithms, and applications of semidefinite programming.

## 1.5.2 ▪ Exercises

**Exercise 1.5.1.** *Consider the question of finding scalars* $a, b, c$ *such that the maximum eigenvalue of the following matrix*

$$\begin{bmatrix} 1 & 2 & a & b & c \\ 2 & 1 & 2 & a & b \\ a & 2 & 1 & 2 & a \\ b & a & 2 & 1 & 2 \\ c & b & a & 2 & 1 \end{bmatrix}$$

*is the smallest. Formulate this question equivalently as a semidefinite program and then solve it by the software* SeDuMi.

**Exercise 1.5.2.** *([321, §3.3]) Consider the semidefinite program*

$$\begin{cases} \min & X_{33} \\ s.t. & X_{11} = 1,\ X_{13} + X_{22} + X_{31} = 0, \\ & X_{12} + X_{21} + X_{33} = 0,\ X \in \mathcal{S}_+^3. \end{cases}$$

*Formulate the dual problem explicitly. Find the primal and dual optimizers. Are they degenerate or nondegenerate (see Definition 1.4.4)? Does the strict complementarity hold for them?*

**Exercise 1.5.3.** *Let* $S$ *be the spectrahedron such that*

$$S = \{x \in \mathbb{R}^n \mid A(x) := x_1 A_1 + \cdots + x_n A_n \succeq 0\}$$

*for given matrices* $A_i \in \mathcal{S}^N$. *Suppose there exists* $u \in \mathbb{R}^n$ *such that* $A(u) \succ 0$. *Show that the dual cone* $S^\star$ *is given as*

$$S^\star = \{(\langle A_1, X \rangle, \ldots, \langle A_n, X \rangle) : X \in \mathcal{S}_+^N\}.$$

**Exercise 1.5.4.** *([183, 313]) Let* $G = (V, E)$ *be a graph with the vertex set* $V = \{1, \ldots, n\}$ *and the edge set* $E$, *i.e.,* $(i, j) \in E$ *if and only if* $(i, j)$ *is an edge. A stable set of* $G$ *is a subset of vertices that are not connected by any edges. The maximum size of a stable set is called the stability number of* $G$, *for which we denote* $\alpha(G)$. *A clique of* $G$ *is a subset of vertices that are all connected to each other by edges. The smallest size of a clique cover, i.e., a collection of cliques that together includes all the vertices of* $G$, *is denoted as* $\bar{\chi}(G)$. *Equivalently,* $\bar{\chi}(G)$ *equals the chromatic number of the complementary graph of* $G$. *The Lovász number of* $G$, *denoted as* $\vartheta(G)$, *is the optimal value of the following optimization problem (e is the vector of all ones):*

$$\begin{cases} \max & e^T X e \\ s.t. & \operatorname{trace}(X) = 1, \\ & X_{ij} = 0\ \forall (i, j) \in E, \\ & X \in \mathcal{S}_+^n. \end{cases}$$

*Prove Lovász's sandwich inequality:*

$$\alpha(G) \le \vartheta(G) \le \bar{\chi}(G).$$

**Exercise 1.5.5.** *([260]) Consider the semidefinite program*

$$\begin{cases} \min & \langle C, X \rangle \\ s.t. & \langle A_i, X \rangle = b_i\ (i = 1, \ldots, m), \\ & X \in \mathcal{S}_+^n. \end{cases}$$

*If it has an optimizer, use Theorem 1.4.5 to show that there must exist an optimizer matrix $X^*$ such that $r = \operatorname{rank} X^*$ satisfies the inequality*

$$\binom{r+1}{2} \leq m.$$

# 1.6 ▪ Linear Optimization with LP, SOCP, and PSD Constraints

In applications, we often need to solve various convex relaxations that have a mixture of linear, quadratic, and semidefinite constraints. Interestingly, they can be cast in a uniform format as in the software SeDuMi [308]. This section introduces the SeDuMi format.

Consider the following general linear convex optimization problem:

$$\begin{cases} \max\limits_{y \in \mathbb{R}^m} & b^T y \\ s.t. & Fy = g, \\ & f \geq Gy, \\ & h_i^T y + \tau_i \geq \|H_i y + p_i\|_2 \, (i = 1, \ldots, r), \\ & B_{j,0} + \sum\limits_{k=1}^m y_k B_{j,k} \succeq 0 \, (j = 1, \ldots, s) \end{cases} \tag{1.6.1}$$

for given matrices/vectors

$$F \in \mathbb{R}^{\ell_1 \times m}, \, G \in \mathbb{R}^{\ell_2 \times m}, \, H_k \in \mathbb{R}^{l_k \times m}, \, B_{j,k} \in \mathcal{S}^{n_k}, \, h_k \in \mathbb{R}^m, p_k \in \mathbb{R}^{l_k}, \, \tau_k \in \mathbb{R}^1.$$

Define the linear function

$$\varphi(y) := \left( Fy, \, Gy, \, \begin{bmatrix} -h_1^T y \\ -H_1 y \end{bmatrix}, \ldots, \begin{bmatrix} -h_r^T y \\ -H_r y \end{bmatrix}, \, -\sum_{k=1}^m y_k B_{1,k}, \ldots, -\sum_{k=1}^m y_k B_{s,k} \right). \tag{1.6.2}$$

It is a linear map from $\mathbb{R}^m$ to the vector space of Cartesian products

$$V := \mathbb{R}^{\ell_1} \times \mathbb{R}^{\ell_2} \times \mathbb{R}^{l_1+1} \times \cdots \times \mathbb{R}^{l_r+1} \times \mathcal{S}^{n_1} \times \cdots \times \mathcal{S}^{n_s}. \tag{1.6.3}$$

A vector $X \in V$ can be written as the tuple

$$X := \left( x_1, x_2, \mathbf{x}_1, \ldots, \mathbf{x}_r, X_1, \ldots, X_s \right)$$

for subvectors

$$x_1 \in \mathbb{R}^{\ell_1}, \, x_2 \in \mathbb{R}^{\ell_2}, \, \mathbf{x}_i \in \mathbb{R}^{l_i+1}, \, X_j \in \mathcal{S}^{n_j}.$$

The inner product for $V$ can be defined as the sum of inner products of each component space. For another vector $Y \in V$, we can similarly write it as

$$Y = \left( y_1, y_2, \mathbf{y}_1, \ldots, \mathbf{y}_r, Y_1, \ldots, Y_s \right),$$

then the inner product of $X$ and $Y$ is

$$\langle X, Y \rangle = \langle x_1, y_1 \rangle + \langle x_2, y_2 \rangle + \sum_{i=1}^r \langle \mathbf{x}_i, \mathbf{y}_i \rangle + \sum_{j=1}^r \langle X_i, Y_i \rangle.$$

The space $\mathbb{R}^{\ell_1}$ itself is a cone, called a *free cone*. Its dual cone is then just the trivial subspace of the zero vector. The nonnegative orthant $\mathbb{R}_+^{\ell_2}$ is a proper cone in the space $\mathbb{R}^{\ell_2}$. For each $k = 1, \ldots, r$, the Lorentz cone

$$\mathcal{Q}_{l_k} := \left\{ (z_0, z_1, \ldots, z_{l_k}) \in \mathbb{R}^{l_k+1} : z_0 \geq \sqrt{z_1^2 + \cdots + z_{l_k}^2} \right\}$$

is a proper cone in $\mathbb{R}^{l_k+1}$. For each $t = 1, \ldots, s$, the psd matrix cone $\mathcal{S}_+^{n_j}$ is a proper cone in the symmetric matrix space $\mathcal{S}^{n_j}$. Except for the free cone, all of them are self dual cones. Let $\mathcal{K}$ be the Cartesian product of these cones:

$$\mathcal{K} := \mathbb{R}^{\ell_1} \times \mathbb{R}_+^{\ell_2} \times \mathcal{Q}_{l_1} \times \cdots \times \mathcal{Q}_{l_r} \times \mathcal{S}_+^{n_1} \times \cdots \times \mathcal{S}_+^{n_s}.$$

Its dual cone is then

$$\mathcal{K}^\star = \{0\}^{\ell_1} \times \mathbb{R}_+^{\ell_2} \times \mathcal{Q}_{l_1} \times \cdots \times \mathcal{Q}_{l_r} \times \mathcal{S}_+^{n_1} \times \cdots \times \mathcal{S}_+^{n_s}.$$

Let $C$ be the vector in the space $V$ such that

$$C := \left( g, f, \begin{bmatrix} \tau_1 \\ p_1 \end{bmatrix}, \ldots, \begin{bmatrix} \tau_r \\ p_r \end{bmatrix}, B_{1,0}, \ldots, B_{s,0} \right).$$

Then, the optimization problem (1.6.1) can be equivalently written as

$$\begin{cases} \max\limits_{y \in \mathbb{R}^m} & b^T y \\ \text{s.t.} & C - \varphi(y) \in \mathcal{K}^\star. \end{cases} \tag{1.6.4}$$

The linear map $\varphi(y)$ can be written as

$$\varphi(y) = y_1 A_1 + \cdots + y_m A_m$$

for some vectors $A_1, \ldots, A_m \in V$. The primal optimization problem for (1.6.4) is

$$\begin{cases} \min\limits_{X \in V} & \langle C, X \rangle \\ \text{s.t.} & \langle A_i, X \rangle = b_i, \ i = 1, \ldots, m, \\ & X \in \mathcal{K}. \end{cases} \tag{1.6.5}$$

Like semidefinite optimization, the primal-dual pair (1.6.4)-(1.6.5) has the following property. As in §1.5, an optimization problem is said to be solvable if it has an optimization and it is called strictly feasible if it has a feasible point lying in the interior of the cone.

**Theorem 1.6.1.** *For the optimization (1.6.4)-(1.6.5), we have the following:*

(i) *If (1.6.5) is bounded below and is strictly feasible, then (1.6.4) is solvable and they have the same optimal value.*

(ii) *If (1.6.4) is bounded above and is strictly feasible, then (1.6.5) is solvable and they have the same optimal value.*

(iii) *If one of (1.6.4)-(1.6.5) is bounded and strictly feasible, then a primal-dual feasible pair $(X, y)$ is a pair of optimizers to the respective optimization if and only if $\langle C, X \rangle = b^T y$, which is equivalent to*

$$\langle C - \varphi(y), X \rangle = 0.$$

The following is an example of linear conic optimization (1.6.1).

**Example 1.6.2.** *Consider a point* $(y_1, y_2, y_3)$ *in the plane* $y_1 + y_2 + y_3 = 3$ *such that it lies in the half spaces* $y_1 + y_2 \geq 1$, $y_2 + y_3 \geq 1$ *and its distance to* $(1, 0, 1)$ *is at most* $y_1 + y_3$, *and it lies inside the elliptope*

$$\begin{bmatrix} 1 & y_1 & y_2 \\ y_1 & 2 & y_3 \\ y_2 & y_3 & 3 \end{bmatrix} \succeq 0.$$

*We are looking for such a vector* $(y_1, y_2, y_3)$ *so that the difference* $y_3 - y_1$ *is maximum. This problem can be written in the dual form:*

$$\begin{cases} \max_{y \in \mathbb{R}^3} & y_3 - y_1 \\ s.t. & y_1 + y_2 + y_3 = 3 \\ & -1 \geq -y_1 - y_2, \quad -1 \geq -y_2 - y_3, \\ & y_1 + y_3 \geq \sqrt{(y_1 - 1)^2 + y_2^2 + (y_3 - 1)^2}, \\ & \begin{bmatrix} 1 & y_1 & y_2 \\ y_1 & 2 & y_3 \\ y_2 & y_3 & 3 \end{bmatrix} \succeq 0. \end{cases} \qquad (1.6.6)$$

*The vector* $b = (-1, 0, 1)$. *The linear map* $\varphi(y)$ *is*

$$\varphi(y) = \left( y_1 + y_2 + y_3, \begin{bmatrix} -y_1 - y_2 \\ -y_2 - y_3 \end{bmatrix}, \begin{bmatrix} -y_1 - y_3 \\ -y_1 \\ -y_2 \\ -y_3 \end{bmatrix}, \begin{bmatrix} 0 & -y_1 & -y_2 \\ -y_1 & 0 & -y_3 \\ -y_2 & -y_3 & 0 \end{bmatrix} \right).$$

*The vector space* $V = \mathbb{R}^1 \times \mathbb{R}^2 \times \mathbb{R}^4 \times \mathcal{S}^3$ *and the cone* $\mathcal{K}$ *is*

$$\mathbb{R}^1 \times \mathbb{R}_2^+ \times \mathcal{Q}_3 \times \mathcal{S}_+^3,$$

*where* $\mathcal{Q}_3$ *is the Lorentz cone in* $\mathbb{R}^4$. *The vectors* $A_1, A_2, A_2$ *are respectively*

$$A_1 = \left( 1, \begin{bmatrix} -1 \\ 0 \end{bmatrix}, \begin{bmatrix} -1 \\ -1 \\ 0 \\ 0 \end{bmatrix}, \begin{bmatrix} 0 & -1 & 0 \\ -1 & 0 & 0 \\ 0 & 0 & 0 \end{bmatrix} \right),$$

$$A_2 = \left( 1, \begin{bmatrix} -1 \\ -1 \end{bmatrix}, \begin{bmatrix} 0 \\ 0 \\ -1 \\ 0 \end{bmatrix}, \begin{bmatrix} 0 & 0 & -1 \\ 0 & 0 & 0 \\ -1 & 0 & 0 \end{bmatrix} \right),$$

$$A_3 = \left( 1, \begin{bmatrix} 0 \\ -1 \end{bmatrix}, \begin{bmatrix} -1 \\ 0 \\ 0 \\ -1 \end{bmatrix}, \begin{bmatrix} 0 & 0 & 0 \\ 0 & 0 & -1 \\ 0 & -1 & 0 \end{bmatrix} \right).$$

*The vector* $C$ *is accordingly*

$$C = \left( 3, \begin{bmatrix} -1 \\ -1 \end{bmatrix}, \begin{bmatrix} 0 \\ -1 \\ 0 \\ -1 \end{bmatrix}, \begin{bmatrix} 1 & 0 & 0 \\ 0 & 2 & 0 \\ 0 & 0 & 3 \end{bmatrix} \right).$$

*The syntax for solving this optimization by the software* SeDuMi *is*

```
K.f = 1; K.l = 2; K.q = 4; K.s = 3;
b =[-1 0 1];
c = [3,  -1,-1,  0,-1,0,-1,  1,0,0, 0, 2,0, 0, 0, 3];
A = [
1,  -1  0,  -1 -1  0  0,   0 -1  0 -1 0  0  0  0 0;
1,  -1 -1,   0  0 -1  0,   0  0 -1  0 0  0 -1  0 0;
1,   0 -1,  -1  0  0 -1,   0  0  0  0 0 -1  0 -1 0];
[xopt, yopt, info] = sedumi(A,b,c,K);
```

*The computed optimizer is yopt = (0.0000, 1.0000, 2.0000).*

### 1.6.1 ▪ Exercises

**Exercise 1.6.1.** *Use the software* SeDuMi *to solve the following optimization:*

$$\begin{cases} \min_{y \in \mathbb{R}^6} & -y_1 + y_2 - y_3 + y_4 - y_5 + y_6 \\ s.t. & y_1 + y_3 + y_5 = 0, \ y_2 + y_4 \le y_6, \\ & 2y_4 \ge \sqrt{y_1^2 + y_2^2 + y_3^2}, \\ & \begin{bmatrix} 1 - y_4 & y_1 - y_5 \\ y_1 - y_5 & y_2 - y_6 \end{bmatrix} \succeq 0, \\ & \begin{bmatrix} 1 & y_1 & y_2 & y_3 \\ y_1 & y_2 & y_3 & y_4 \\ y_2 & y_3 & y_4 & y_5 \\ y_3 & y_4 & y_5 & y_6 \end{bmatrix} \succeq 0. \end{cases} \tag{1.6.7}$$

**Exercise 1.6.2.** *For given $A \in \mathbb{R}^{m \times n}$, $b \in \mathbb{R}^n$, and a scalar $\lambda > 0$, formulate the following optimization*

$$\min_{x \in \mathbb{R}^n} \ \|Ax - b\|_\infty + \lambda \|x\|_2$$

*as a linear conic optimization problem, in the standard dual form, with the Cartesian product of the nonnegative orthant cone and the second order cone. Generate random $A, b, \lambda$ and solve it with the software* SeDuMi.

**Exercise 1.6.3.** *Let $\mathcal{N}^{n \times n}$ denote the cone of all n-by-n symmetric matrices whose entries are all nonnegative. Express the following doubly nonnegative optimization*

$$\begin{cases} \min_{X \in \mathcal{S}^n} & \langle C, X \rangle \\ s.t. & \langle A_i, X \rangle = b_i \ (i = 1, \dots, m_1), \\ & X \in \mathcal{S}_+^n \cap \mathcal{N}^{n \times n}, \end{cases} \tag{1.6.8}$$

*as a linear conic optimization problem, in the primal form, with the cone $\mathbb{R}_+^{\binom{n+1}{2}} \times \mathcal{S}_+^n$. Generate random $A_i, C$ and solve it with the software* SeDuMi.

**Exercise 1.6.4.** *For given $N$ points $u_1, \dots, u_N \in \mathbb{R}^n$, find the smallest ellipsoid to cover all of them. That is, find a positive definite matrix $P_{++}^n$ and a center point $c \in \mathbb{R}^n$, such that*

$$\|(u_i - c)^T P^{-1}(u_i - c)\|_2 \le 1$$

*for all points $U_i$ and the trace of $P$ is minimum. Formulate this question as a linear conic optimization problem, in the dual form, with the Cartesian product of psd cones. Generate random points $u_i$ and then solve it with the software* SeDuMi.

# 1.7 ▪ Overview of Moment and Polynomial Optimization

A general optimization problem is

$$\begin{cases} \min & f(x) \\ s.t. & x \in K, \end{cases} \tag{1.7.1}$$

for a given continuous function $f$ and a Borel set $K \subseteq \mathbb{R}^n$. Recall that Borel sets are the sets that can be constructed from open or closed sets by repeatedly taking countable unions and intersections. Borel measures are the ones that are defined on Borel sets. See [86] for detailed introduction. The main goal is to find the minimum value $f_{min}$ of the optimization (1.7.1) and to get one or several global minimizers, if they exist. The question is more challenging if the function $f$ and/or the set $K$ are nonconvex.

An important fact is that the optimization problem (1.7.1) is equivalent to a linear convex one, no matter if (1.7.1) is convex or not. The support of a Borel measure $\mu$ on $\mathbb{R}^n$ is the smallest closed set $S$ such that $\mu(\mathbb{R}^n \backslash S) = 0$, and it is denoted as $\text{supp}(\mu)$. Let $\mathscr{B}(K)$ denote the set of all Borel measures whose supports are contained in $K$. Then (1.7.1) is equivalent to the linear convex optimization

$$\begin{cases} \min & \int f \mathrm{d}\mu \\ s.t. & \mu(K) = 1, \ \mu \in \mathscr{B}(K). \end{cases} \tag{1.7.2}$$

The integral $\int f \mathrm{d}\mu$ is a linear functional in the Borel measure variable $\mu$. The problem (1.7.2) is a linear optimization problem with a measure variable. If a measure $\mu^*$ is an optimizer of (1.7.2), then every point in $\text{supp}(\mu^*)$ is a minimizer of (1.7.1). We leave the proof as an exercise. It is also interesting to observe that a scalar $\gamma$ is a lower bound of $f$ on $K$ if and only if

$$f(x) \geq \gamma \quad \text{for all} \quad x \in K.$$

Therefore, the minimum value $f_{min}$ equals the maximum lower bound of $f$ on $K$. Let $\mathscr{N}(K)$ denote the cone of all functions that are nonnegative on $K$. Then $f_{min}$ is also equal to the optimal value of the maximization problem

$$\begin{cases} \max & \gamma \\ s.t. & f - \gamma \in \mathscr{N}(K). \end{cases} \tag{1.7.3}$$

We remark that the minimization problem (1.7.2) and the maximization problem (1.7.3) are primal-dual pairs. To see this, for every Borel measure $\mu \in \mathscr{B}(K)$ with $\mu(K) = 1$ and for every lower bound $\gamma$ of $f$ on $K$, it always holds that

$$\int f \mathrm{d}\mu - \gamma = \int (f - \gamma) \mathrm{d}\mu \geq 0.$$

If the optimization (1.7.1) achieves its minimum value, both (1.7.2) and (1.7.3) achieve the same optimal value. There is no duality gap between them.

To illustrate the main picture of moment and polynomial optimization, we consider the case that the feasible set $K = [-1, 1] \subseteq \mathbb{R}^1$ and the objective $f$ is a univariate polynomial of degree four

$$f = f_0 + f_1 x^1 + f_2 x^2 + f_3 x^3 + f_4 x^4.$$

For a Borel measure $\mu$ with the support contained in $[-1, 1]$, one has

$$\int f \mathrm{d}\mu = f_0 y_0 + f_1 y_1 + f_2 y_2 + f_3 y_3 + f_4 y_4,$$

where $y_i = \int x^i \mathrm{d}\mu$ is the $i$th degree moment for the measure $\mu$. Note that $y_0 = 1$ since $\mu(K) = 1$. For the above $f$ and $K$, the linear optimization (1.7.2) can be equivalently expressed as

$$\begin{cases} \min & f_0 y_0 + f_1 y_1 + f_2 y_2 + f_3 y_3 + f_4 y_4 \\ s.t. & y_0 = 1, \ y \in \mathscr{R}_4(K), \end{cases} \tag{1.7.4}$$

where $y = (y_0, y_1, y_2, y_3, y_4)$ and $\mathscr{R}_4(K)$ is the cone

$$\mathscr{R}_4(K) := \left\{ y \in \mathbb{R}^5 \ \middle|\ y = \int [x]_4 \mathrm{d}\mu \text{ for some } \mu \in \mathscr{B}(K) \right\}.$$

In the above, the notation $[x]_4$ stands for the monomial vector

$$\begin{bmatrix} 1 & x & x^2 & x^3 & x^4 \end{bmatrix}^T.$$

The set $\mathscr{R}_4(K)$ is called the cone of degree-4 moment vectors, whose representing measures are supported in $K$. As shown in §3.3, it holds that $y \in \mathscr{R}_4(K)$ if and only if $y$ satisfies two linear matrix inequalities:

$$\begin{bmatrix} y_0 & y_1 & y_2 \\ y_1 & y_2 & y_3 \\ y_2 & y_3 & y_4 \end{bmatrix} \succeq 0, \quad \begin{bmatrix} y_0 - y_2 & y_1 - y_3 \\ y_1 - y_3 & y_2 - y_4 \end{bmatrix} \succeq 0.$$

Similarly, for the above $f$ and $K$, the optimization (1.7.3) can be equivalently expressed as

$$\begin{cases} \max & \gamma \\ s.t. & f - \gamma \in \mathscr{P}_4(K). \end{cases} \tag{1.7.5}$$

In the above, $\mathscr{P}_4(K)$ is the cone of polynomials that have degrees up to 4 and that are non-negative on the set $K$. As we will see in §3.2, the constraint $f - \gamma \in \mathscr{P}_4(K)$ is equivalent to

$$f - \gamma = \begin{bmatrix} 1 \\ x \\ x^2 \end{bmatrix}^T X_0 \begin{bmatrix} 1 \\ x \\ x^2 \end{bmatrix} + (1 - x^2) \begin{bmatrix} 1 \\ x \end{bmatrix}^T X_1 \begin{bmatrix} 1 \\ x \end{bmatrix}$$

for two psd matrices $X_0 \in \mathcal{S}_+^3$, $X_1 \in \mathcal{S}_+^2$. The problem (1.7.4) is a linear optimization problem about moments, while (1.7.5) is a linear optimization problem about nonnegative polynomials. We refer to Chapter 3 for more details about univariate nonnegative polynomials and moment cones.

When $f$ is a multivariate polynomial in $x \in \mathbb{R}^n$ and $K$ is a semialgebraic set in $\mathbb{R}^n$, there are similar versions of (1.7.4) and (1.7.5). However, for the multivariate case, it is generally much harder to get a computationally clean description for the cones $\mathscr{R}_d(K)$ and $\mathscr{P}_d(K)$. A good thing is that these cones can be efficiently approximated by using psd cones. Consequently, polynomial and moment optimization problems can be solved efficiently by semidefinite programming relaxations. They are called Moment-SOS relaxations in this book. The classical books and surveys about Moment-SOS relaxations can be found in [31, 166, 167, 171, 172, 191, 295]. The major motivation of this book is to show how to use Moment-SOS relaxations to solve various moment and polynomial optimization problems.

Moment-SOS relaxations for solving polynomial optimization problems can be programmed conveniently with the software GloptiPoly [124], which calls a linear conic optimization solver (e.g., SeDuMi, SDPT3, MOSEK). The following are some examples for showing how to do this.

- (univariate polynomial optimization) Consider minimizing the polynomial function $x - x^2 - x^3 + x^4$ over the interval $K = [-3, 2]$. The GloptiPoly coding is

```
mpol('x', 1);
f = x-x^2-x^3+x^4;
K = [x+3>=0,2-x>=0];
POP = msdp(min(f),K, 3);
[status, val] = msol(POP);
```

Using the software SeDuMi, we get that $val = -0.6196\cdots$. The syntax double(x) returns the minimizer $-0.6404\cdots$.

- (unconstrained optimization) Consider minimizing the polynomial function $x_1^4 + x_2^4 + x_1^3 - x_1x_2^2 + x_1x_2 - x_2$ in the space $\mathbb{R}^2$. The GloptiPoly coding is

```
mpol('x', 2);
f = x(1)^4+x(2)^4+x(1)^3-x(1)*x(2)^2 + x(1)*x(2)-x(2);
POP = msdp(min(f));
[status, val] = msol(POP);
```

Using the software SeDuMi, we get that $val = -0.7669\cdots$. The syntax double(x) returns the minimizer $(-0.8361\cdots, 0.5947\cdots)$.

- (constrained optimization) Consider minimizing the polynomial function $x_1^3 - 2x_2^3 - 3x_1^2x_2 - 5x_1x_2$ over the set $K$ given by the constraints $-2 \le x_1 \le -1$, $1 \le x_2 \le 2$, and $x_1^2 + x_2^2 \le 5$. The GloptiPoly coding is

```
mpol('x', 2);
f = x(1)^3-2*x(2)^3-3*x(1)^2*x(2)-5*x(1)*x(2);
K = [x(1)+2>=0,-1-x(1)>=0,x(2)-1>=0,2-x(2)>=0,5-x'*x>=0];
POP = msdp(min(f),K);
[status, val] = msol(POP);
```

Using the software SeDuMi, we get that $val = -13.0000\cdots$. The syntax double(x) returns the minimizer $(-1.0000\cdots, 2.0000\cdots)$.

More polynomial optimization examples are in exercises.

## 1.7.1 ▪ Exercises

**Exercise 1.7.1.** *Solve the following optimization problems with the software* GloptiPoly:

(i) $\begin{cases} \min\limits_{x\in\mathbb{R}^1} & 2x - 3x^2 - 4x^3 + x^4 \\ s.t. & x \in [-3, -2]. \end{cases}$

(ii) $\begin{cases} \min\limits_{x\in\mathbb{R}^1} & x^7 - x^6 - x^5 - x^4 - x^3 - x^2 \\ s.t. & x \in [0, \infty). \end{cases}$

(iii) $\begin{cases} \min\limits_{x\in\mathbb{R}^1} & x^8 + x^7 - x^6 + x^3 - x^2 + x \\ s.t. & x \in (-\infty, 0]. \end{cases}$

(iv) $\begin{cases} \min & x_1^6 + x_2^4 - x_1^3x_2 - x_1^2 - x_2^2 + x_1x_2 \\ s.t. & x \in \mathbb{R}^2. \end{cases}$

$$(v) \quad \begin{cases} \min_{x \in \mathbb{R}^3} & x_1^3 + x_2^3 + x_3^3 - x_1^2 x_2 - x_1 x_2^2 + x_1 x_2 x_3 \\ s.t. & x_1^2 + x_2^2 + x_3^2 = 1. \end{cases}$$

$$(vi) \quad \begin{cases} \min_{x \in \mathbb{R}^4} & x_1 x_2 x_3 x_4 + x_1 x_2 x_3 + x_2 x_3 x_4 + x_1 x_2 + x_3 x_4 \\ s.t. & x_1^2 + x_2^2 + x_3^2 + x_4^2 \le 1. \end{cases}$$

**Exercise 1.7.2.** *Suppose $K$ is a Borel set. If a Borel measure $\mu^*$ is a minimizer of the optimization problem (1.7.2), show that every point in the support of $\mu^*$ is a global minimizer of (1.7.1).*

# Chapter 2

# Introductions to Algebra, Polynomials, and Moments

*Moment and polynomial optimization theory and methods require some basic knowledge in the areas of complex and real algebraic geometry, computational algebra, Positivstellensätze, and truncated moment problems. This chapter introduces such basic knowledge.*

## 2.1 ▪ Rings, Ideals, and Varieties

Throughout the book, the symbol $\mathbb{F}$ denotes either the real field $\mathbb{R}$ or the complex field $\mathbb{C}$. The symbol $\mathbb{N}$ stands for the set of nonnegative integers. This section reviews basic knowledge in classical algebraic geometry and computational algebra (see the books [59, 105, 113, 303] for more details). Denote by

$$\mathbb{F}[x] := \mathbb{F}[x_1, \ldots, x_n]$$

the ring of polynomials in $x := (x_1, \ldots, x_n)$ and with coefficients in the field $\mathbb{F}$. If a polynomial is homogeneous, i.e., all its terms have the same degree, then it is called a *form*. For a degree $d$, $\mathbb{F}[x]_d$ denotes the set of polynomials in $\mathbb{F}[x]$ with degrees at most $d$, while $\mathbb{F}[x]_d^{\mathrm{hom}}$ denotes the set of forms of degrees equal to $d$. The ring $\mathbb{F}[x]$ is generated by the set $\{1, x_1, \ldots, x_n\}$, i.e., every element in $\mathbb{F}[x]$ is a linear combination of finitely many products of powers of $1, x_1, \ldots, x_n$, with coefficients in $\mathbb{F}$. For a set $G$ of polynomials, let $\mathbb{F}[G]$ denote the set of all polynomials that can be expressed as

$$\sum_{\alpha = (\alpha_1, \ldots, \alpha_k) \in \mathbb{N}^k} c_\alpha \cdot g_1^{\alpha_1} \cdots g_k^{\alpha_k}, \quad c_\alpha \in \mathbb{F},$$

where all $g_1, \ldots, g_k \in G$ and only finitely many coefficients $c_\alpha$ are nonzero. The polynomials in $G$ are called *generators* and $G$ is called a *generator set* for $\mathbb{F}[G]$. If $\mathbb{F}[G] = \mathbb{F}[x]$, then $G$ is a generator set and the polynomials in $G$ are generators for $\mathbb{F}[x]$. Clearly, $\{1, x_1, \ldots, x_n\}$ is a generator set and $1, x_1, \ldots, x_n$ are generators for $\mathbb{F}[x]$.

### 2.1.1 ▪ Elementary algebraic geometry

A subset $I \subseteq \mathbb{F}[x]$ is an *ideal* of $\mathbb{F}[x]$ if $f \cdot h \in I$ for all $h \in I$, $f \in \mathbb{F}[x]$, and $p + q \in I$ for all $p, q \in I$. An ideal $I \subseteq \mathbb{C}[x]$ is said to be *prime* if $I \neq \mathbb{C}[x]$ and $a, b \in \mathbb{C}[x], ab \in I$ imply that $a \in I$ or $b \in I$. For a tuple $h = (h_1, \ldots, h_m)$ of polynomials in $\mathbb{F}[x]$, the notation $\mathrm{Ideal}[h] := \mathrm{Ideal}[h_1, \ldots, h_m]$ denotes the smallest ideal containing all $h_i$, or equivalently,

$$\mathrm{Ideal}[h] = h_1 \cdot \mathbb{F}[x] + \cdots + h_m \cdot \mathbb{F}[x].$$

The set Ideal[$h$] is called the ideal generated by $h$. Every ideal of $\mathbb{F}[x]$ is generated by finitely many polynomials, i.e., every ideal is *finitely generated*.

**Theorem 2.1.1.** *(Hilbert Basis Theorem [59]) For every ideal $I \subseteq \mathbb{F}[x]$, there exist finitely many polynomials $g_1, \ldots, g_m \in I$ such that $I = \mathrm{Ideal}[g_1, \ldots, g_m]$.*

For an ideal $I \subseteq \mathbb{F}[x]$, denote its complex zero set

$$V_{\mathbb{C}}(I) := \{x \in \mathbb{C}^n : p(x) = 0 \quad \forall\, p \in I\}. \tag{2.1.1}$$

In optimization, we are often interested in the real zero set

$$V_{\mathbb{R}}(I) := \{x \in \mathbb{R}^n : p(x) = 0 \quad \forall\, p \in I\}. \tag{2.1.2}$$

A set like $V_{\mathbb{C}}(I)$ is called a *variety* (or *affine variety* or *algebraic variety* in $\mathbb{C}^n$). Every subset $T \subseteq \mathbb{C}^n$ is contained in a variety in $\mathbb{C}^n$. The smallest algebraic variety containing $T$ is called the *Zariski* closure of $T$, for which we denote $\mathrm{Zar}(T)$. In the Zariski topology on $\mathbb{C}^n$, closed sets are algebraic varieties, and open sets are complements of varieties. A set $S \subseteq V$ is called a proper subvariety of $V$ if $S \neq V$ and $S$ itself is also a variety. A nonempty variety $V \subseteq \mathbb{C}^n$ is said to be *irreducible* if there are no proper subvarieties $V_1 \neq V_2$ of $V$ such that $V = V_1 \cup V_2$. Every variety is a union of finitely many irreducible varieties. *Real algebraic varieties* in $\mathbb{R}^n$ and the Zariski topology are similarly defined. In the space $\mathbb{F}^n$, a property $\mathcal{P}$ is said to hold *generically* on an irreducible variety $T \subseteq \mathbb{F}^n$ if $\mathcal{P}$ holds in a Zariski open subset of $T$. We refer to [16, 34, 59] for the Zariski topology.

For a set $S \subseteq \mathbb{C}^n$, the *vanishing ideal* $I(S)$ is the ideal

$$I(S) = \{h \in \mathbb{C}[x] : h(u) = 0 \,\forall\, u \in S\}. \tag{2.1.3}$$

It is interesting to note that a variety $V \subseteq \mathbb{C}^n$ is irreducible if and only if its vanishing ideal $I(V)$ is prime. Clearly, if $S = V_{\mathbb{C}}(I)$ and $p \in I$, then $p \in I(S)$. The relationship between $V_{\mathbb{C}}(I)$ and $I(S)$ is given by *Hilbert's Nullstellensatz*.

**Theorem 2.1.2.** *(Hilbert's Nullstellensatz [59]) Let $I \subseteq \mathbb{C}[x]$ be an ideal.*

- *(Weak Nullstellensatz) If $V_{\mathbb{C}}(I) = \emptyset$, then $1 \in I$.*

- *(Strong Nullstellensatz) For $f \in \mathbb{C}[x]$, if $f(u) = 0$ for all $u \in V_{\mathbb{C}}(I)$, then $f^k \in I$ for some power $k \geq 1$.*

The *radical* of an ideal $I \subseteq \mathbb{C}[x]$ is the set

$$\sqrt{I} := \{f \in \mathbb{C}[x] : f^k \in I \text{ for some } k \in \mathbb{N}\}. \tag{2.1.4}$$

The set $\sqrt{I}$ itself is also an ideal (see exercise). Clearly, $I \subseteq \sqrt{I}$. The $I$ is called a *radical ideal* if $\sqrt{I} = I$.

In many applications, we need to consider homogeneous polynomial equations. This requires us to define projective spaces. Let $\mathbb{P}^n$ denote the $n$-dimensional complex projective space, consisting of lines through the origin in $\mathbb{C}^{n+1}$. A point $\tilde{x} \in \mathbb{P}^n$ is represented by the set of nonzero multiples of $(x_0, x_1, \ldots, x_n)$, where the $x_i$'s are not all zeros. For convenience, we write a point in $\mathbb{P}^n$ as

$$\tilde{x} := (x_0, x_1, \ldots, x_n).$$

A set $U$ in $\mathbb{P}^n$ is called a *projective variety* (or projective algebraic variety) if there are homogeneous polynomials $p_1, \ldots, p_r \in \mathbb{C}[x_0, x_1, \ldots, x_n]$ such that

$$U = \{\tilde{x} \in \mathbb{P}^n : p_i(x_0, x_1, \ldots, x_n) = 0, i = 1, \ldots, r\}.$$

The Zariski topology on $\mathbb{P}^n$ is similarly defined: closed sets are projective algebraic varieties and open sets are complements of projective algebraic varieties. For a nonzero form $h \in \mathbb{C}[x_0, \ldots, x_n]$, the set

$$\mathcal{H} = \{\tilde{x} \in \mathbb{P}^n : h(x_0, x_1, \ldots, x_n) = 0\}$$

is called a *hypersurface* in $\mathbb{P}^n$. If $h$ has degree one, then $\mathcal{H}$ is called a *hyperplane*. An ideal $I \subseteq \mathbb{C}[x_0, \ldots, x_n]$ is *homogeneous* if $I$ is generated by a set of homogeneous polynomials. A Zariski open subset $\mathcal{Q}$ of a projective variety $\mathcal{U}$ is called a *quasi-projective variety*, or equivalently, a quasi-projective variety is a locally closed subset of $\mathbb{P}^n$ in the Zariski topology. The irreducibility of a projective variety is similarly defined. Each projective variety is a union of finitely many irreducible ones.

The dimension of an irreducible affine (resp., projective) variety $V$ is the length $k$ of the longest chain of distinct irreducible affine (resp., projective) varieties,

$$V = V_0 \supsetneq V_1 \supsetneq \cdots \supsetneq V_k. \tag{2.1.5}$$

If $I_0$ is the vanishing ideal of $V$, then the dimension of $V$ is also equal to the maximum $k$ such that

$$I_0 \subsetneq I_1 \subsetneq \cdots \subsetneq I_k, \tag{2.1.6}$$

where each $I_i$ is a prime ideal of $\mathbb{C}[x]$. For an irreducible projective variety $V$, its dimension can be equivalently defined as the largest integer $k$ such that every set of $k$ generic hyperplanes have a common intersection point with $V$. For a general variety $V$, its *dimension* is defined to be the maximum dimension of its irreducible components. The dimension of $V$ is denoted as $\dim(V)$. The gap $n - \dim(V)$ is called the *codimension* of $V$. When $V$ is a real algebraic variety of $\mathbb{R}^n$, its dimension is similarly defined (see §2.2.1).

## 2.1.2 ▪ Quotient rings and Gröbner bases

For an ideal $I \subseteq \mathbb{C}[x]$, a polynomial function $f$ on the variety $V_{\mathbb{C}}(I)$ is invariant if $f$ is replaced by $f + h$ for every $h \in I$. This observation leads to the concept of *quotient ring*. For a given ideal $I$, we define the equivalence relation $\sim$ on $\mathbb{C}[x]$ such that $p \sim q$ if $p - q \in I$. If $p \sim q$, we also write that

$$p \equiv q \quad \mathrm{mod} \quad I.$$

The set of all polynomials equivalent to $p$ is called the *equivalent class* of $p$ and it is denoted as $[p]$. Equivalently, $[p] = [q]$ if and only if $p \sim q$. The set of all equivalent classes in $\mathbb{C}[x]$ is called the quotient ring (or coordinate ring) of $I$ and it is denoted as $\mathbb{C}[x]/I$. For the quotient ring $\mathbb{C}[x]/I$, the addition and multiplication are defined as

$$[p] + [q] := [p + q], \quad [p] \cdot [q] := [pq].$$

They are well-defined for equivalent classes. Under these operations, the set $\mathbb{C}[x]/I$ is a ring. It is called the quotient ring of the ideal $I$.

To determine the complex variety $V_{\mathbb{C}}(I)$, we often need a basis for the quotient space $\mathbb{C}[x]/I$. This often requires a Gröbner basis for $I$, which is based on an ordering on monomials (see [59]). A monomial can be written as

$$x^\alpha := x_1^{\alpha_1} \cdots x_n^{\alpha_n}$$

for the power $\alpha := (\alpha_1, \ldots, \alpha_n)$. Its degree is

$$|\alpha| := \alpha_1 + \cdots + \alpha_n.$$

A monomial ordering is a relation $\succ$ on monomials $x^\alpha$ ($\alpha \in \mathbb{N}^n$) satisfying the following: (i) $\succ$ is a total ordering on monomials (i.e., we must have $x^\alpha \succ x^\beta$ or $x^\beta \succ x^\alpha$ or $x^\alpha = x^\beta$ for any two monomials $x^\alpha, x^\beta$); (ii) if $x^\alpha \succ x^\beta$, then $x^\alpha \cdot x^\gamma \succ x^\beta \cdot x^\gamma$ for every monomial $x^\gamma$; (iii) the relation $\succ$ is well ordered (i.e., every nonempty set of monomials has the smallest one under $\succ$). Frequently used monomial orderings are the following:

- *Lexicographic order*: $x^\alpha \succ_{lex} x^\beta$ if and only if the most left nonzero entry of $\alpha - \beta$ is positive.

- *Graded lexicographic order*: $x^\alpha \succ_{grlex} x^\beta$ if and only if $|\alpha| > |\beta|$ or $|\alpha| = |\beta|$ but $\alpha \succ_{lex} \beta$.

- *Graded reverse lexicographic order*: $x^\alpha \succ_{grevlex} x^\beta$ if and only if $|\alpha| > |\beta|$ or $|\alpha| = |\beta|$ but the most right nonzero entry of $\alpha - \beta$ is negative.

Under a given monomial ordering $\succ$, the *leading monomial* of a polynomial $f$ is the monomial that is the biggest; it is denoted as $\mathrm{LM}(f)$. The *leading coefficient* of $f$ is the coefficient of $\mathrm{LM}(f)$, denoted as $\mathrm{LC}(f)$. The product $\mathrm{LC}(f) \cdot \mathrm{LM}(f)$ is called the *leading term* of $f$, denoted as $\mathrm{LT}(f)$. For a set $G \subseteq \mathbb{C}[x]$, the set of its leading monomials is

$$\mathrm{LM}(G) = \{\mathrm{LM}(f) : f \in G\}.$$

The set $\mathrm{LT}(G)$ is similarly defined. For an ideal $I$, the ideal generated by $\mathrm{LT}(I)$ is called the *leading term ideal* or *initial ideal* of $I$, for which we denote $\mathrm{Ideal}[\mathrm{LT}(I)]$. A monomial not belonging to $\mathrm{Ideal}[\mathrm{LT}(I)]$ is called a *standard monomial* of $I$. For a given monomial ordering $\succ$, a finite subset $G \subseteq I$ is called a *Gröbner basis* of $I$ if

$$I = \mathrm{Ideal}[G], \quad \mathrm{Ideal}[\mathrm{LT}(I)] = \mathrm{Ideal}[\mathrm{LT}(G)].$$

For two nonzero polynomials $f, g$, let $x^\gamma$ be the least common multiple of their leading monomials $\mathrm{LM}(f), \mathrm{LM}(g)$. Then their *S-polynomial* is

$$S(f, g) := \frac{x^\gamma}{\mathrm{LT}(f)} f - \frac{x^\gamma}{\mathrm{LT}(g)} g. \tag{2.1.7}$$

Gröbner bases can be detected and obtained by computing $S$-polynomials. Suppose $G = \{g_1, \ldots, g_m\}$ is a basis for an ideal $I$. The set $G$ is a Gröbner basis of $I$ under a monomial ordering $\succ$ if and only if for all distinct $g_i, g_j$ in $G$, the remainder of $S(g_i, g_j)$, under the division by $G$ with the ordering $\succ$, is zero (see [59]). If $G$ is not a Gröbner basis, we can add all such nonzero remainders to $G$. After this is repeated finitely many times, one can get a Gröbner basis for $I$. This is Buchberger's algorithm for computing Gröbner bases. A Gröbner basis $G$ is said to be *reduced* if for every $p \in G$, $\mathrm{LC}(p) = 1$ and no monomial of $p$ belongs to $\mathrm{Ideal}[\mathrm{LT}(G \backslash \{p\})]$. Under a monomial ordering $\succ$, a nonzero ideal has a unique reduced Gröbner basis. We refer to [59] for more details about Gröbner bases.

**Example 2.1.3.** *([59, §2.7]) Consider the ideal* $I = \mathrm{Ideal}[f_1, f_2]$ *with*

$$f_1 = x_1^3 - 2x_1 x_2, \quad f_2 = x_1^2 x_2 - 2x_2^2 + x_1.$$

*Select the graded lexicographic ordering* $\succ_{grlex}$. *We have* $S(f_1, f_2) = -x_1^2$. *The remainder of* $-x_1^2$ *under the division by* $f_1, f_2$ *is still* $-x_1^2$, *which does not belong to* $\mathrm{Ideal}[LT(f_1), LT(f_2)]$.

*So $G_1 = \{f_1, f_2\}$ is not a Gröbner basis for $I$. After applying Buchberger's algorithm, we get the Gröbner basis*

$$G_2 = \{f_1, \ f_2, \ f_3 := -x_1^2, \ f_4 := -2x_1x_2, \ f_5 := -2x_2^2 + x_1\}.$$

*It is not a reduced Gröbner basis. The reduce one is*

$$G_3 = \{f_1, \ f_2, \ \tilde{f}_3 := x_1^2, \ \tilde{f}_4 := x_1x_2, \ \tilde{f}_5 := x_2^2 - \frac{1}{2}x_1\}.$$

The dimension of an ideal $I \subseteq \mathbb{C}[x_1, \ldots, x_n]$ is the dimension of its complex variety $V_{\mathbb{C}}(I)$. It can be determined by using the Hilbert polynomial. Recall that $\mathbb{C}[x_1, \ldots, x_n]_d$ is the space of all polynomials in $(x_1, \ldots, x_n)$ of degrees at most $d$. Denote

$$I_{\leq d} := I \cap \mathbb{C}[x_1, \ldots, x_n]_d.$$

For each integer $s \geq 0$, define the dimension

$$\mathrm{HF}(s) := \dim \mathbb{C}[x_1, \ldots, x_n]_s - \dim I_{\leq s}.$$

The $\mathrm{HF}(\cdot)$ is called the Hilbert function of $I$, defined on $\mathbb{N}$. There exists a univariate polynomial $h(s)$ such that $\mathrm{HF}(s) = h(s)$ for all $s$ sufficiently large. The polynomial $h(s)$ is called the *Hilbert polynomial* of $I$, for which we denote as $\mathrm{HP}(s)$. The dimension of $I$, which is also the dimension of $V_{\mathbb{C}}(I)$, is the degree of the Hilbert polynomial $h(s)$. In particular, an ideal $I \subseteq \mathbb{C}[x]$ is called *zero dimensional* if the complex variety $V_{\mathbb{C}}(I)$ is a finite set. Note that $I$ being zero dimensional is stronger than the real variety $V_{\mathbb{R}}(I)$ being a finite set. For instance, the ideal $I = \langle x_1^2 + x_2^2 \rangle$ is not zero dimensional while the real variety $V_{\mathbb{R}}(I) = \{(0,0)\}$ is a single point. The following is a characterization for zero dimensional ideals.

**Theorem 2.1.4.** *([59, 309]) The following are equivalent for an ideal $I \subseteq \mathbb{C}[x]$:*

*(i) The ideal $I$ is zero dimensional.*

*(ii) The quotient ring $\mathbb{C}[x]/I$ is a finitely dimensional vector space.*

*(iii) If $G$ is a Gröbner basis of $I$, then for each $i = 1, \ldots, n$, there exists an integer $m_i \in \mathbb{N}$ such that $x_i^{m_i}$ is the leading monomial of some $g \in G$.*

The dimension of the coordinate ring $\mathbb{C}[x]/I$ is defined to be the dimension of the ideal $I$. Dimensions for projective varieties can be similarly defined by Hilbert functions. We refer to [59, 105, 309] for more details about quotient rings and related computation.

### 2.1.3 ▪ Solving polynomial equations

For given polynomials $f_1, \ldots, f_m \in \mathbb{C}[x]$, let $I := \mathrm{Ideal}[f_1, \ldots, f_m]$ be the ideal generated by them. Consider the polynomial system

$$f_1(x) = \cdots = f_m(x) = 0. \tag{2.1.8}$$

Assume that $I$ is zero dimensional, i.e., (2.1.8) has finitely many complex solutions, or equivalently, the quotient space $\mathbb{C}[x]/I$ is finitely dimensional. The set of all complex solutions to

(2.1.8) is the variety $V_\mathbb{C}(I)$. The number of its complex solutions (counting their multiplicities) is equal to the dimension of the quotient space

$$D := \dim \mathbb{C}[x]/I.$$

We often need a basis $B = \{[b_1], \ldots, [b_D]\}$ for the quotient space $\mathbb{C}[x]/I$:

$$\mathbb{C}[x]/I = \text{span}\{[b_1], \ldots, [b_D]\}.$$

To compute $B$, a Gröbner basis of $I$ is generally required with respect to a monomial ordering $\succ$. Suppose $G$ is the reduced Gröbner basis for the given monomial ordering. The leading monomials of $G$ generate the initial ideal $\text{Ideal}[\text{LT}(I)]$. The monomials not belonging to $\text{Ideal}[\text{LT}(I)]$ are *standard monomials* of $I$. The set of standard monomials gives a basis of $\mathbb{C}[x]/I$.

The multiplicity of a point $u \in V_\mathbb{C}(I)$ for an ideal $I$ is defined as follows (see [60] for more details). Denote by $\mathbb{C}[x]_{<u>}$ the ring of *rational functions* $p/q$, with $p, q \in \mathbb{C}[x]$ and $q(u) \neq 0$. The product $I\mathbb{C}[x]_{<u>}$ is also an ideal of the ring $\mathbb{C}[x]_{<u>}$. The quotient space

$$\mathbb{C}[x]_{<u>}/I\mathbb{C}[x]_{<u>}$$

is a vector space over $\mathbb{C}$. When $I$ is zero dimensional, the space $\mathbb{C}[x]_{<u>}/I\mathbb{C}[x]_{<u>}$ is finitely dimensional. Its dimension is defined to be the *multiplicity* of the point $u$. Equivalently, the multiplicity of $u$ is the following dimension:

$$\dim \mathbb{C}[x]_{<u>}/I\mathbb{C}[x]_{<u>}.$$

**Proposition 2.1.5.** *[309, Prop. 2.1] The complex variety $V_\mathbb{C}(I)$ is finite if and only if the basis set $B$ is finite, and the cardinality of $|B|$ equals $|V_\mathbb{C}(I)|$, counting multiplicities.*

For each $x_i$, define the multiplication mapping

$$\mathcal{M}_{x_i} : \mathbb{C}[x]/I \to \mathbb{C}[x]/I, \quad [p] \mapsto [x_i p]. \tag{2.1.9}$$

It can be shown that $\mathcal{M}_{x_i}$ is a linear mapping. Let $M_{x_i}$ denote the representing matrix of $\mathcal{M}_{x_i}$, under the basis $B$. It is called the *companion matrix* or *multiplication matrix* of the ideal $I$ with respect to $x_i$. The companion matrices $M_{x_1}, \ldots, M_{x_n}$ commute with each other, so they share common eigenvectors. The variety of $I$ can be determined by eigenvalues of companion matrices. This is summarized by Stickelberger's Theorem (see [309, Theorem 2.6] or [171, Theorem 2.9]).

**Theorem 2.1.6.** *(Stickelberger's Theorem) Let $I \subseteq \mathbb{C}[x]$ be a zero dimensional ideal and let $M_{x_1}, \ldots, M_{x_n}$ be the companion matrices, then*

$$V_\mathbb{C}(I) = \{(\lambda_1, \ldots, \lambda_n) : \exists v \in \mathbb{C}^D \backslash \{0\}, M_{x_i} v = \lambda_i v, i = 1, \ldots, n\}, \tag{2.1.10}$$

*where $D = \dim \mathbb{C}[x]/I$.*

With Stickelberger's Theorem, one can further show that a zero dimensional ideal $I$ is radical if and only if the companion matrices $M_{x_1}, \ldots, M_{x_n}$ are simultaneously diagonalizable, which occurs if and only if the cardinality $|V_\mathbb{C}(I)| = D$ (see [309, §2.3] for details).

**Example 2.1.7.** *([309, §2.3]) Consider the polynomial equations:*

$$x_1^2 + \tfrac{1}{5}x_2 - \tfrac{1}{5}x_3 + \tfrac{2}{25} = 0, \qquad x_2^2 - \tfrac{1}{5}x_1 + \tfrac{1}{5}x_3 + \tfrac{2}{25} = 0,$$
$$x_3^2 + \tfrac{1}{5}x_1 - \tfrac{1}{5}x_2 + \tfrac{2}{25} = 0, \quad x_1x_2 + x_2x_3 + x_3x_1 + \tfrac{1}{25} = 0.$$

*Let I be the ideal generated by the above polynomials. Then,*

$$\{[1], [x_1], [x_2], [x_3], [x_1 x_3], [x_2 x_3]\}$$

*is a basis of the quotient space $\mathbb{C}[x]/I$ (see the exercise). The companion matrices are*

$$M_{x_1} = \begin{bmatrix} 0 & \frac{-2}{25} & \frac{-1}{25} & 0 & \frac{-2}{125} & 0 \\ 1 & 0 & 0 & 0 & \frac{-1}{25} & \frac{1}{25} \\ 0 & \frac{-1}{5} & 0 & 0 & \frac{1}{25} & \frac{1}{25} \\ 0 & \frac{1}{5} & 0 & 0 & \frac{-2}{25} & \frac{1}{25} \\ 0 & 0 & -1 & 1 & 0 & 0 \\ 0 & 0 & -1 & 0 & \frac{-1}{5} & 0 \end{bmatrix}, \quad M_{x_2} = \begin{bmatrix} 0 & \frac{-1}{25} & \frac{-2}{25} & 0 & 0 & \frac{2}{125} \\ 0 & 0 & \frac{1}{5} & 0 & \frac{1}{25} & \frac{1}{25} \\ 1 & 0 & 0 & 0 & \frac{1}{25} & \frac{-1}{25} \\ 0 & 0 & \frac{-1}{5} & 0 & \frac{1}{25} & \frac{-2}{25} \\ 0 & -1 & 0 & 0 & 0 & \frac{1}{5} \\ 0 & -1 & 0 & 1 & 0 & 0 \end{bmatrix},$$

$$M_{x_3} = \begin{bmatrix} 0 & 0 & 0 & \frac{-2}{25} & \frac{1}{125} & \frac{-1}{125} \\ 0 & 0 & 0 & \frac{-1}{5} & \frac{-2}{25} & \frac{1}{25} \\ 0 & 0 & 0 & \frac{1}{5} & \frac{1}{25} & \frac{-2}{25} \\ 1 & 0 & 0 & 0 & \frac{-1}{25} & \frac{-1}{25} \\ 0 & 1 & 0 & 0 & \frac{-1}{5} & \frac{1}{5} \\ 0 & 0 & 1 & 0 & \frac{-1}{5} & \frac{1}{5} \end{bmatrix}.$$

*They can be simultaneously diagonalized by the matrix V equaling*

```
0.0635 + 0.0177i   0.0635 - 0.0177i  -0.0444 - 0.0217i  -0.0444 + 0.0217i  -0.0444 + 0.0217i  -0.0444 - 0.0217i
0.0196 - 0.1738i   0.0196 + 0.1738i   0.0240 - 0.2127i   0.0240 + 0.2127i   0.0291 + 0.1044i   0.0291 - 0.1044i
0.0731 + 0.1589i   0.0731 - 0.1589i   0.0240 - 0.0291i  -0.0291 + 0.1044i  -0.0240 - 0.2127i  -0.0240 + 0.2127i
-0.0238 + 0.0853i  -0.0238 - 0.0853i   0.0895 - 0.1945i   0.0895 + 0.1945i  -0.0895 - 0.1945i  -0.0895 + 0.1945i
0.6807 + 0.0000i   0.6807 + 0.0000i   0.8331 + 0.0000i   0.8331 + 0.0000i   0.1201 + 0.4309i   0.1201 - 0.4309i
0.5825 + 0.3521i   0.5825 - 0.3521i   0.1201 - 0.4309i   0.1201 + 0.4309i   0.8331 + 0.0000i   0.8331 + 0.0000i
```

*where $i = \sqrt{-1}$, i.e., each $V^{-1} M_{x_i} V$ is diagonal. Each column of V gives a vector $v$ in (2.1.10). This gives 6 complex zeros of the above polynomial system:*

```
(-0.1423 + 0.3588i,  -0.1423 + 0.3588i,  -0.1423 + 0.3588i),
(-0.1423 - 0.3588i,  -0.1423 - 0.3588i,  -0.1423 - 0.3588i),
( 0.1423 - 0.3588i,   0.1423 - 0.3588i,   0.1423 - 0.3588i),
( 0.1423 + 0.3588i,   0.1423 + 0.3588i,   0.1423 - 0.3588i),
(-0.0000 + 0.1519i,  -0.0000 + 0.1519i,  -0.0000 + 0.1519i),
( 0.0000 - 0.1519i,   0.0000 - 0.1519i,   0.0000 - 0.1519i).
```

*There is no real solution for this polynomial system.*

To apply Stickelberger's Theorem, one needs to compute common eigenvectors for the companion matrices. A practical method for computing $V_{\mathbb{C}}(I)$ is to apply a generic linear combination of $M_{x_1}, \dots, M_{x_n}$, proposed by Corless, Gianni, and Trager [61]. Choose generic numbers $\xi_1, \dots, \xi_n$ and let

$$M := \xi_1 M_{x_1} + \dots + \xi_n M_{x_n}.$$

Then compute its Schur decomposition (the superscript * denotes the conjugate transpose)

$$Q^* M Q = T := \begin{bmatrix} T_{11} & T_{12} & \cdots & T_{1s} \\ & T_{22} & & T_{2s} \\ & & \ddots & \vdots \\ & & & T_{ss} \end{bmatrix}, \tag{2.1.11}$$

where $Q \in \mathbb{C}^{D \times D}$ is unitary (i.e., $Q^{-1} = Q^*$), $T \in \mathbb{C}^{D \times D}$ is upper triangular, the diagonal of each block $T_{jj}$ is a constant (i.e., $T_{jj}$ has only one eigenvalue), and different $T_{jj}$ has a different diagonal entry. Let

$$\widetilde{M}_{x_i} := Q^* M_{x_i} Q, \quad i = 1, \dots, n.$$

It can be shown that $\widetilde{M}_{x_i}$ is also block upper triangular, with the same pattern as for $T$. So one can write the partition of $\widetilde{M}_{x_i}$ as

$$\widetilde{M}_{x_i} = \begin{bmatrix} M_{11}^{(i)} & M_{12}^{(i)} & \cdots & M_{1s}^{(i)} \\ & M_{22}^{(i)} & & M_{2s}^{(i)} \\ & & \ddots & \vdots \\ & & & M_{ss}^{(i)} \end{bmatrix}.$$

Then, for each $j = 1, \ldots, s$, let

$$u_j := \big(\mathrm{trace}(M_{jj}^{(1)}), \ldots, \mathrm{trace}(M_{jj}^{(n)})\big)/\mathrm{length}(T_{jj}). \tag{2.1.12}$$

The above $u_1, \ldots, u_s$ are the all solutions to (2.1.8). The length of $T_{jj}$ is the multiplicity of $u_j$. We refer to [61] for more details.

**Example 2.1.8.** *[171, Example 2.10] Consider the polynomial equations*

$$\begin{aligned} f_1 &:= x_1^2 + 2x_2^2 - 2x_2 = 0, \\ f_2 &:= x_1 x_2^2 - x_1 x_2 = 0, \\ f_3 &:= x_2^3 - 2x_2^2 + x_2 = 0. \end{aligned}$$

*One can verify that it has two solutions: $(0,0)$ and $(0,1)$. Under the lexicographical monomial ordering $x_1 > x_2$, $\{f_1, f_2, f_3\}$ is a Gröbner basis for the ideal $I = \mathrm{Ideal}[f_1, f_2, f_3]$. The set*

$$B = \big\{[1], [x_2], [x_2^2], [x_1], [x_1 x_2]\big\}$$

*is a basis for $\mathbb{C}[x]/I$. The companion matrices are*

$$M_{x_1} = \begin{bmatrix} 0 & 0 & 0 & 0 & 0 \\ 0 & 0 & 0 & 2 & 2 \\ 0 & 0 & 0 & -2 & -2 \\ 1 & 0 & 0 & 0 & 0 \\ 0 & 1 & 1 & 0 & 0 \end{bmatrix}, \quad M_{x_2} = \begin{bmatrix} 0 & 0 & 0 & 0 & 0 \\ 1 & 0 & -1 & 0 & 0 \\ 0 & 1 & 2 & 0 & 0 \\ 0 & 0 & 0 & 0 & 0 \\ 0 & 0 & 0 & 1 & 1 \end{bmatrix}.$$

*The matrices $M_{x_1}, M_{x_2}$ are not simultaneously diagonalizable. We apply a generic combination of them, say, $M = 3M_{x_1} + 5M_{x_2}$. The Schur decomposition $M = Q^* \cdot T \cdot Q$ is given by*

$$Q = \begin{bmatrix} 0 & -\frac{1}{\sqrt{6}} & -\frac{1}{\sqrt{2}} & 0 & -\frac{1}{\sqrt{3}} \\ 0 & \frac{\sqrt{2}}{\sqrt{3}} & 0 & 0 & -\frac{1}{\sqrt{3}} \\ 0 & -\frac{1}{\sqrt{6}} & \frac{1}{\sqrt{2}} & 0 & -\frac{1}{\sqrt{3}} \\ \frac{1}{\sqrt{2}} & 0 & 0 & \frac{1}{\sqrt{2}} & 0 \\ -\frac{1}{\sqrt{2}} & 0 & 0 & \frac{1}{\sqrt{2}} & 0 \end{bmatrix}, \quad T = \begin{bmatrix} 0 & -\sqrt{3} & -3 & -5 & \frac{\sqrt{3}}{\sqrt{2}} \\ 0 & 0 & -5\sqrt{3} & 6\sqrt{3} & \frac{5}{\sqrt{2}} \\ 0 & 0 & 5 & 0 & -\frac{5\sqrt{3}}{\sqrt{2}} \\ 0 & 0 & 0 & 5 & -\frac{3\sqrt{3}}{\sqrt{2}} \\ 0 & 0 & 0 & 0 & 5 \end{bmatrix}.$$

*The products $Q^* \cdot M_{x_1} \cdot Q$, $Q^* \cdot M_{x_2} \cdot Q$ are respectively*

$$\begin{bmatrix} 0 & -\frac{1}{\sqrt{3}} & -1 & 0 & \frac{1}{\sqrt{6}} \\ 0 & 0 & 0 & 2\sqrt{3} & 0 \\ 0 & 0 & 0 & -2 & 0 \\ 0 & 0 & 0 & 0 & -\frac{\sqrt{3}}{\sqrt{2}} \\ 0 & 0 & 0 & 0 & 0 \end{bmatrix}, \quad \begin{bmatrix} 0 & 0 & 0 & -1 & 0 \\ 0 & 0 & -\sqrt{3} & 0 & \frac{1}{\sqrt{2}} \\ 0 & 0 & 1 & 0 & -\frac{\sqrt{3}}{\sqrt{2}} \\ 0 & 0 & 0 & 1 & 0 \\ 0 & 0 & 0 & 0 & 1 \end{bmatrix}.$$

*The formula (2.1.12) reveals two solutions $(0,0)$ and $(0,1)$ successfully.*

## 2.1.4 ▪ Exercises

**Exercise 2.1.1.** *For an ideal $I \subseteq \mathbb{C}[x]$, show that its radical $\sqrt{I}$ is also an ideal of $\mathbb{C}[x]$.*

**Exercise 2.1.2.** *([59]) Find the reduced Gröbner basis for the following ideals with the given monomial ordering:*

(i) $\mathrm{Ideal}[x_1 x_2^2 - x_1 x_3 + x_2, x_1 x_2 - x_3^2, x_1 - x_2 x_3^4]$ *with the lex order.*

(ii) $\mathrm{Ideal}[x_1^2 - x_2, x_1^3 - x_3]$ *with the grlex order.*

**Exercise 2.1.3.** *([59]) Determine Hilbert polynomials and dimensions for the ideals:*

(i) $\mathrm{Ideal}[x_1 x_3, x_1 x_2 - 1]$;

(ii) $\mathrm{Ideal}[x_3 x_4 - x_2^2, x_1 x_2 - x_3^3]$.

**Exercise 2.1.4.** *For Example 2.1.7, show that*

$$\big\{ [1], [x_1], [x_2], [x_3], [x_1 x_3], [x_2 x_3] \big\}$$

*is a basis of the quotient space $\mathbb{C}[x]/I$.*

**Exercise 2.1.5.** *Use the method in §2.1.3 to find all complex solutions to the following polynomial systems:*

(i) $x_1^3 - x_2 - x_3 - 1 = x_2^3 - x_3 + x_1 + 1 = x_3^3 + x_1 - x_2 + 1 = 0$.

(ii) $(x_1 - x_2)^2 + x_3 + 1 = (x_2 - x_3)^2 - x_1 + 1 = (x_3 - x_1)^2 + x_2 - 1 = 0$.

**Exercise 2.1.6.** *Show that the multiplication map $\mathcal{M}_{x_i}$ as in (2.1.9) is linear and the companion matrices $M_{x_1}, \ldots, M_{x_n}$ commute with each other.*

**Exercise 2.1.7.** *Let $I \subseteq \mathbb{C}[x]$ be an ideal. For a polynomial $\varphi \in \mathbb{C}[x]$, define the multiplication mapping*

$$\mathcal{M}_\varphi : \mathbb{C}[x]/I \to \mathbb{C}[x]/I, \quad [p] \mapsto [\varphi p]. \tag{2.1.13}$$

*For $\varphi, \phi \in \mathbb{C}[x]$, show that $\mathcal{M}_\varphi = \mathcal{M}_\phi$ if and only if $\varphi - \phi \in I$.*

**Exercise 2.1.8.** *Show that the vectors $u_j$ given by (2.1.12) are the solutions to the polynomial system (2.1.8).*

**Exercise 2.1.9.** *([224]) Suppose $I_1, \ldots, I_k$ are ideals of $\mathbb{R}[x]$ such that $V_{\mathbb{C}}(I_i) \cap V_{\mathbb{C}}(I_j) = \emptyset$ for all $i \neq j$. Let $I = I_1 \cap \cdots \cap I_k$. Show that there exist polynomials $a_1, \ldots, a_k \in \mathbb{R}[x]$ such that*

$$a_1^2 + \cdots + a_k^2 - 1 \in I, \quad \text{each } a_i \in \bigcap_{i \neq j \in [k]} I_j.$$

## 2.2 ▪ Algebraic and Semialgebraic Sets

Algebraic and semialgebraic sets are basic concepts in real algebraic geometry (see the books [16, 34]). This section reviews basic knowledge about them.

## 2.2.1 ▪ Algebraic Sets

An *algebraic set* in $\mathbb{R}^n$ is the set of common real zeros of a set of polynomials. For instance, the unit sphere is an algebraic set, which is the real zero set of the polynomial $x_1^2 + \cdots + x_n^2 - 1$. The union of all coordinate lines in $\mathbb{R}^3$ is an algebraic set, since it is the common zero set of monomials $x_i x_j$ ($1 \leq i < j \leq 3$). Moreover, the set of singular $n$-by-$n$ matrices in $\mathbb{R}^{n \times n}$ is an algebraic set, which is the zero set of the determinantal polynomial. For more concrete examples, please see the algebraic sets in Figure 2.1. They are planar curves.

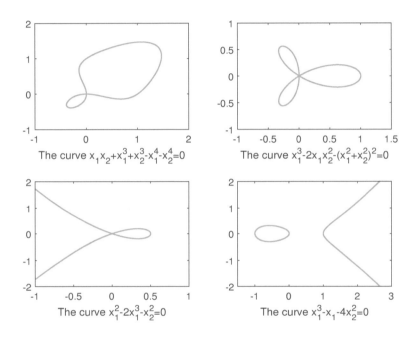

The curve $x_1 x_2 + x_1^3 + x_2^3 - x_1^4 - x_2^4 = 0$

The curve $x_1^3 - 2x_1 x_2^2 - (x_1^2 + x_2^2)^2 = 0$

The curve $x_1^2 - 2x_1^3 - x_2^2 = 0$

The curve $x_1^3 - x_1 - 4x_2^2 = 0$

Figure 2.1: Some algebraic sets that are planar curves.

Intersections and unions of finitely many algebraic sets are again algebraic sets (see exercise). A nonempty algebraic set is said to be *irreducible* if it cannot be written as a union of two distinct proper algebraic subsets; otherwise, it is called *reducible*. A nonempty algebraic set $V$ is irreducible if and only if the vanishing ideal $I(V)$ is a prime ideal. Every algebraic set is the union of finitely many irreducible ones.

The dimension of an irreducible algebraic set is defined as in (2.1.5). For a reducible algebraic set $V \subseteq \mathbb{R}^n$, its dimension $\dim V$ is defined to be the maximum dimension of the irreducible components of $V$. Let

$$R[V] := \mathbb{R}[x]/I(V)$$

be the coordinate ring of $V$. It is also called the ring of polynomial functions on $V$. The dimension $\dim(V)$ is also equal to the maximum number $\ell$ such that

$$I_0 \subsetneqq \cdots \subsetneqq I_\ell \subseteq R[V], \quad \text{each } I_i \text{ is a prime ideal.} \tag{2.2.1}$$

Suppose the vanishing ideal $I(V) = \text{Ideal}[p_1, \ldots, p_k]$. For a point $u \in V$, the tangent space of

$V$ at $u$ is

$$T_u(V) := \bigcap_{j=1}^{k} \nabla p_j(u)^{\perp}.$$

When $V$ is irreducible, dim $T_u(V) \geq \dim(V)$; if the equality holds, then $u$ is called a *nonsingular* point of $V$. When $V$ is reducible, a point $u \in V$ is said to be nonsingular if there is only one irreducible component $V'$ of $V$ that contains $u$ and $u$ is a nonsingular point of $V'$. The algebraic set $V$ is said to be nonsingular if every point in it is nonsingular. We refer to [16, 34] for more properties of algebraic sets.

## 2.2.2 ▪ Semialgebraic sets

More general than algebraic sets are *semialgebraic sets* [16, 34]. A set in $\mathbb{R}^n$ is called semialgebraic if it is given by a finite number of boolean operations (i.e., conjunctions, disjunctions, and negations) of polynomial equations and inequalities. Equivalently, a semialgebraic set can be expressed by a finite number of intersections, unions, and/or complements of sets that are defined like $p(x) = 0$ or $q(x) > 0$. In particular, a set is called a *basic open* semialgebraic set if it can be expressed as $p_1(x) > 0, \ldots, p_r(x) > 0$ for some polynomials $p_1, \ldots, p_r \in \mathbb{R}[x]$; a set is called a *basic closed* semialgebraic set if it can be expressed as $q_1(x) \geq 0, \ldots, q_s(x) \geq 0$ for some polynomials $q_1, \ldots, q_s \in \mathbb{R}[x]$. For instance, the unit ball $B(0,1) := \{x \in \mathbb{R}^n : \|x\| \leq 1\}$ is a basic closed semialgebraic set, while it is not an algebraic set.

A semialgebraic set is a finite union of sets given in the form

$$f_i(x) > 0 \, (\forall \, i \in \mathcal{I}), \quad h_j(x) = 0 \, (\forall \, j \in \mathcal{J})$$

for polynomials $f_i, h_j$. Here $\mathcal{I}, \mathcal{J}$ are finite label sets (possibly empty). Note that a weak inequality like $f(x) \geq 0$ can be expressed as a union of $f(x) > 0$ and $f(x) = 0$. Moreover, every closed semialgebraic set is a union of finitely many basic closed semialgebraic sets [34, Theorem 2.7.2]. That is, if a semialgebraic set $T$ is closed, then $T = \bigcup_{k=1}^{m} T_k$, where each

$$T_k = \{x \in \mathbb{R}^n : g_1^k(x) \geq 0, \ldots, g_{m_k}^k(x) \geq 0\}. \tag{2.2.2}$$

Denote by $\partial T_k$ the *boundary* of $T_k$ in the standard Euclidean topology. For each $u \in \partial T_k$, the active label set

$$I_k(u) := \Big\{ i \in [m_k] : g_i^k(u) = 0 \Big\}$$

is nonempty. The boundary of $T$ is denoted as $\partial T$.

The description of a semialgebraic set by polynomials is usually not unique, and its boundary may have singularities. We say that $u$ is a *nonsingular point* of $\partial T_k$ if $|I_k(u)| = 1$ and $\nabla g_i^k(u) \neq 0$ for $i \in I_k(u)$; otherwise, $u$ is called a *singular* point of $\partial T_k$. A point $u$ on $\partial T_k$ is called a *corner point* of $T_k$ if $|I_k(u)| > 1$. For $u \in \partial T$ and $i \in I_k(u) \neq \emptyset$, we say that $g_i^k$ is *irredundant* at $u$ with respect to $\partial T$ (or just *irredundant* at $u$ if the set $T$ is clear from the context) if there exists a sequence of nonsingular points $\{u_k\}_{k=1}^{\infty} \subseteq V_{\mathbb{R}}(g_i^k) \cap \partial T$ lying on $\partial T_k$ such that $u_k \to u$; otherwise, we say $g_i^k$ is *redundant* at $u$. We say that $g_i^k$ is *nonsingular* at $u$ if $\nabla g_i^k(u) \neq 0$. Geometrically, when $g_i^k$ is nonsingular at $u \in \partial T$, $g_i^k$ being redundant at $u$ means that the inequality $g_i^k(x) \geq 0$ might be unnecessary for defining $T$ locally around $u$.

**Example 2.2.1.** *([31, Chap. 6]) Consider the convex set that is the shaded area of Figure 2.2. It is the union of the following two basic closed semialgebraic sets:*

$$T_1 = \left\{ \begin{array}{l} g_1^1(x) := x_2 \geq 0, \\ g_2^1(x) := 1 - x_2 \geq 0, \\ g_3^1(x) := x_2^4 - x_1^6 \geq 0 \end{array} \right\}, \quad T_2 = \left\{ \begin{array}{l} g_1^2(x) := x_1 \geq 0, \\ g_2^2(x) := 1 - x_2 \geq 0, \\ g_3^2(x) := 10x_2^3 - x_1^5 \geq 0 \end{array} \right\}.$$

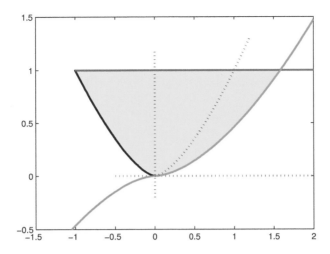

Figure 2.2: The yellow area is the union of $T_1$ and $T_2$ in Example 2.2.1.

*For $T_1$, its corner points are $(-1,1), (0,0), (1,1)$. The $g_3^1$ is irredundant at $(-1,1)$ and $(0,0)$, but redundant at $(1,1)$. The $g_3^1$ is nonsingular at $(-1,1)$ but singular at $(0,0)$. The $g_1^1$ is redundant at $(0,0)$. For $T_2$, its corner points are $(0,0), (0,1), (\sqrt[5]{10}, 1)$. The $g_3^2$ is irredundant at both $(0,0)$ and $(\sqrt[5]{10}, 1)$. It is nonsingular at $(\sqrt[5]{10}, 1)$ but singular at $(0,0)$. The polynomial $g_1^2$ is redundant at $(0,1)$ and $(0,0)$. Both $g_2^1$ and $g_2^2$ are irredundant on the section $x_2 = 1$ of the boundary.*

In optimization, we often have sets that are given as projections of algebraic sets or semial-gebraic sets. It is worthy to note that the projection of an algebraic set may not be algebraic. For instance, consider the hyperbola curve $x_1 x_2 - 1 = 0$. Its projection to the $x_1$-coordinate is the set $x_1 \neq 0$, which is not an algebraic set. However, the projection of an algebraic or semialge-braic set is always semialgebraic. This is a major conclusion of the classical Tarski-Seidenberg theorem (see [16, 34]).

Semialgebraic functions are similarly defined. Let $f : D \to \mathbb{R}$ be a real-valued function, where the domain $D \subseteq \mathbb{R}^n$ is a semialgebraic set. The $f$ is called a *semialgebraic function* if the graph $\{(x, y) : x \in D, f(x) = y\}$ is a semialgebraic set. By the Tarski-Seidenberg theorem, all functions that are given by polynomial equations are semialgebraic. A useful property for semialgebraic functions is *Lojasiewicz's inequality*.

**Theorem 2.2.2.** *(Lojasiewicz's inequality [16, 34]) Let $K$ be a compact semialgebraic set in $\mathbb{R}^n$. If $f, g : K \to \mathbb{R}$ are two continuous semialgebraic functions such that $g \equiv 0$ on $\{x \in K : f(x) = 0\}$, then there exist an integer $N > 0$ and a constant $C > 0$ such that*

$$|g(x)|^N \leq C|f(x)| \quad \text{for all } x \in K.$$

The dimension of a semialgebraic set $S \subseteq \mathbb{R}^n$ can be defined similarly as in (2.2.1). Note that $I(S)$ still denotes the vanishing ideal of $\overline{S}$. The notation $R[S] := \mathbb{R}[x]/I(S)$ denotes the ring of polynomial functions on $S$. The dimension $\dim(S)$ is then defined to be the maximum number $\ell$ such that

$$I_0 \subsetneq \cdots \subsetneq I_\ell \subseteq R[S], \quad \text{each } I_i \text{ is a prime ideal.} \tag{2.2.3}$$

If $S = \cup_{i=1}^m S_m$ is a finite union of semialgebraic sets $S_i$, then

$$\dim(S) = \max\{\dim(S_1), \ldots, \dim(S_m)\}.$$

If $U \subseteq \mathbb{R}^n$, $W \subseteq \mathbb{R}^m$ are two semialgebraic sets, then the Cartesian product $U \times W$ is also a semialgebraic set and

$$\dim(U \times W) = \dim(U) + \dim(W).$$

We refer to [16, 34] for more details about properties of semialgebraic sets.

### 2.2.3 ▪ Exercises

**Exercise 2.2.1.** *Show that the intersection and union of two algebraic sets are also algebraic sets.*

**Exercise 2.2.2.** *Let $T \subseteq \mathbb{R}^n$ be a semialgebraic set. If $T \neq \mathbb{R}^n$ and $T$ has nonempty interior, show that $T$ cannot be an algebraic set.*

**Exercise 2.2.3.** *Show that the set of integer points in $\mathbb{R}^n$ (i.e., the points whose coordinates are all integers) is not semialgebraic.*

**Exercise 2.2.4.** *Let $f : \mathbb{R}^n \to \mathbb{R}^m$ be a semialgebraic function. If $T \subseteq \mathbb{R}^m$ is semialgebraic, show that its preimage $f^{-1}(T) := \{x : f(x) \in T\}$ is also semialgebraic.*

**Exercise 2.2.5.** *([34]) If $A \subseteq \mathbb{R}^n$, $B \subseteq \mathbb{R}^m$ are two algebraic sets, show that the Cartesian product $A \times B$ is also an algebraic set. If $A, B$ are semialgebraic sets, show that $A \times B$ is also semialgebraic. For both cases, show that*

$$\dim(A \times B) = \dim(A) + \dim(B).$$

**Exercise 2.2.6.** *([34]) Let $K \subseteq \mathbb{R}^n$ be a compact semialgebraic set and $Z$ be the real zero set of a polynomial $f \in \mathbb{R}[x]$. Let $dist(x, Z)$ denote the distance of $x$ to $Z$. Use the Lojasiewicz inequality to show that there exist constants $\alpha, C > 0$ such that*

$$dist(x, Z)^\alpha \leq C|f(x)| \quad \text{for all } x \in K.$$

## 2.3 ▪ Resultants and Discriminants

Resultants and discriminants are useful tools for eliminating variables in polynomial systems. We refer to [60, 102, 309] for more detailed introductions.

### 2.3.1 ▪ Resultants

The resultant concerns conditions for a homogeneous polynomial system to have a nonzero complex solution, when the number of equations equals the number of unknowns. Let $f_1, \ldots, f_n$ be forms in $x := (x_1, \ldots, x_n)$. The resultant $\text{Res}(f_1, \ldots, f_n)$ is the polynomial in the coefficients of $f_1, \ldots, f_n$ such that

$$\text{Res}(f_1, \ldots, f_n) = 0 \Leftrightarrow \exists\, u \in \mathbb{C}^n \backslash \{0\} : f_1(u) = \cdots = f_n(u) = 0.$$

The resultant polynomial $\text{Res}(f_1, \ldots, f_n)$ is homogeneous and irreducible. Up to scaling, it has integer coefficients and it is unique up to a sign if its integer coefficients are coprime. If $f_k$ has the degree $d_k$, then $\text{Res}(f_1, \ldots, f_n)$ is homogeneous in each $f_k$ of degree $d_1 \cdots d_{k-1} d_{k+1} \cdots d_n$. Hence, the total degree of $\text{Res}(f_1, \ldots, f_n)$ is

$$d_1 \cdots d_n \cdot \left( d_1^{-1} + \cdots + d_n^{-1} \right).$$

In particular, if each $f_i$ is a linear form, then $\text{Res}(f_1, \ldots, f_n)$ is reduced to the determinant of a matrix.

For the binary case (i.e., $n = 2$), the resultant $\text{Res}(f_1, f_2)$ can be expressed as a determinant (see [309, §4.1]). For instance, if

$$f_1 = a_0 x_1^{d_1} + a_1 x_1^{d_1 - 1} x_2 + \cdots + a_{d_1 - 1} x_1 x_2^{d_1 - 1} + a_{d_1} x_2^{d_1},$$

$$f_2 = b_0 x_1^{d_2} + b_1 x_1^{d_2 - 1} x_2 + \cdots + b_{d_2 - 1} x_1 x_2^{d_2 - 1} + b_{d_2} x_2^{d_2},$$

then

$$\text{Res}(f_1, f_2) = \det \begin{bmatrix} a_0 & a_1 & \cdots & a_{d_1-1} & a_{d_1} & & & \\ & a_0 & a_1 & \cdots & a_{d_1-1} & a_{d_1} & & \\ & & \ddots & \ddots & & \ddots & \ddots & \\ & & & a_0 & a_1 & \cdots & a_{d_1-1} & a_{d_1} \\ b_0 & b_1 & \cdots & b_{d_2-1} & b_{d_2} & & & \\ & b_0 & b_1 & \cdots & b_{d_2-1} & b_{d_2} & & \\ & & \ddots & \ddots & & \ddots & \ddots & \\ & & & b_0 & b_1 & \cdots & b_{d_2-1} & b_{d_2} \end{bmatrix}.$$

In the above matrix, each $a_i$ is repeated $d_2$ times and each $b_i$ is repeated $d_1$ times. So $\text{Res}(f_1, f_2)$ is the determinant of a $(d_1 + d_2)$-by-$(d_1 + d_2)$ matrix and its degree is $d_1 + d_2$. For instance, if $f_1 = a_0 x_1^2 + a_1 x_1 x_2 + a_2 x_2^2$ and $f_2 = b_0 x_1^2 + b_1 x_1 x_2 + b_2 x_2^2$, then

$$\text{Res}(f_1, f_2) = a_2^2 b_0^2 - a_1 a_2 b_0 b_1 + a_0 a_2 b_1^2 + a_1^2 b_0 b_2 - 2a_0 a_2 b_0 b_2 - a_0 a_1 b_1 b_2 + a_0^2 b_2^2.$$

Another interesting case is that $n = 3$ and $d_1 = d_2 = d_3 = 2$. For

$$\begin{aligned} f_1 &= a_0 x_1^2 + a_1 x_1 x_2 + a_2 x_2^2 + + a_3 x_1 x_3 + a_4 x_2 x_3 + a_5 x_3^2, \\ f_2 &= b_0 x_1^2 + b_1 x_1 x_2 + b_2 x_2^2 + + b_3 x_1 x_3 + b_4 x_2 x_3 + b_5 x_3^2, \\ f_3 &= c_0 x_1^2 + c_1 x_1 x_2 + c_2 x_2^2 + + c_3 x_1 x_3 + c_4 x_2 x_3 + c_5 x_3^2, \end{aligned}$$

the determinant of their Jacobian is the polynomial

$$J(x_1, x_2, x_3) := \det \begin{bmatrix} \partial f_1 / \partial x_1 & \partial f_1 / \partial x_2 & \partial f_1 / \partial x_3 \\ \partial f_2 / \partial x_1 & \partial f_2 / \partial x_2 & \partial f_2 / \partial x_3 \\ \partial f_3 / \partial x_1 & \partial f_3 / \partial x_2 & \partial f_3 / \partial x_3 \end{bmatrix}.$$

Write the partial derivatives of $J$ as

$$\begin{aligned} \partial J / \partial x_1 &= u_0 x_1^2 + u_1 x_1 x_2 + u_2 x_2^2 + u_3 x_1 x_3 + u_4 x_2 x_3 + u_5 x_3^2, \\ \partial J / \partial x_2 &= v_0 x_1^2 + v_1 x_1 x_2 + v_2 x_2^2 + v_3 x_1 x_3 + v_4 x_2 x_3 + v_5 x_3^2, \\ \partial J / \partial x_3 &= w_0 x_1^2 + w_1 x_1 x_2 + w_2 x_2^2 + w_3 x_1 x_3 + w_4 x_2 x_3 + w_5 x_3^2. \end{aligned}$$

Then it can be shown that (see [309, Chap. 4])

$$\text{Res}(f_1, f_2, f_3) = \det \begin{bmatrix} a_0 & a_1 & a_2 & a_3 & a_4 & a_5 \\ b_0 & b_1 & b_2 & b_3 & b_4 & b_5 \\ c_0 & c_1 & c_2 & c_3 & c_4 & c_5 \\ u_0 & u_1 & u_2 & u_3 & u_4 & u_5 \\ v_0 & v_1 & v_2 & v_3 & v_4 & v_5 \\ w_0 & w_1 & w_2 & w_3 & w_4 & w_5 \end{bmatrix}. \tag{2.3.1}$$

There are methods for computing resultants for general cases. We refer to [60, Chap. 3], [102, §4, Chap. 3], and [309, Chap. 4].

For nonhomogeneous polynomials, the resultant is defined to be the one for their homogenizations. Let $f_0, f_1, \ldots, f_n$ be nonhomogeneous polynomials in $x := (x_1, \ldots, x_n)$. Let $\tilde{f}_i$ denote the homogenization of $f_i$ ($d_i := \deg(f_i)$):

$$\tilde{f}_i(\tilde{x}) = x_0^{d_i} f_i(x/x_0),$$

where $\tilde{x} := (x_0, x_1, \ldots, x_n)$. The resultant $\text{Res}(f_0, f_1, \ldots, f_n)$ is then defined to be

$$\text{Res}(\tilde{f}_0, \ldots, \tilde{f}_n).$$

If the nonhomogeneous equations

$$f_0(x) = f_1(x) = \cdots = f_n(x) = 0$$

have a solution in $\mathbb{C}^n$, then the homogeneous system

$$\tilde{f}_0(\tilde{x}) = \cdots = \tilde{f}_n(\tilde{x}) = 0$$

has a solution in $\mathbb{P}^n$ (the $n$-dimensional projective space over $\mathbb{C}$). The converse may not be true, because the latter might have a solution at infinity $x_0 = 0$. However, for the univariate case (i.e., $n = 1$), $f_0, f_1$ have a common zero if and only if $\text{Res}(f_0, f_1) = 0$.

## 2.3.2 ▪ Discriminants

The discriminant of a homogeneous polynomial is the resultant of all its partial derivatives. Let $f(x)$ be a form in $x = (x_1, \ldots, x_n)$. Its discriminant $\Delta(f)$ is the polynomial in the coefficients of $f$ such that

$$\Delta(f) = 0 \Leftrightarrow \exists u \in \mathbb{C}^n \backslash \{0\} : \frac{\partial f(u)}{\partial x_1} = \cdots = \frac{\partial f(u)}{\partial x_n} = 0.$$

The discriminant $\Delta(f)$ is homogeneous, is irreducible, and has integer coefficients. It is unique up to a sign if all its integer coefficients are coprime. When $\deg(f) = d$, $\Delta(f)$ has degree $n(d-1)^{n-1}$. For instance, when $n = 2$ and $d = 3$, we have the formula (see [102, Chap. 12])

$$\Delta(ax_1^3 + bx_1^2 x_2 + cx_1 x_2^2 + dx_2^3) = b^2 c^2 - 4ac^3 - 4b^3 d + 18abcd - 27a^2 d^2.$$

When $f$ is nonhomogeneous, its discriminant $\Delta(f)$ is then defined to be the one for its homogenization $\tilde{f}$. Observe that if $u \in \mathbb{C}^n$ is a singular zero point of $f$, i.e., $f(u) = 0$ and $\nabla_x f(u) = 0$, then $\nabla_{\tilde{x}} \tilde{f}(\tilde{u}) = 0$ [3] for $\tilde{u} = (1, u)$. This is because of the identity ($\deg(f) = d$):

$$d \cdot \tilde{f}(\tilde{x}) = x_0 \frac{\partial \tilde{f}(\tilde{x})}{\partial x_0} + x_1 \frac{\partial \tilde{f}(\tilde{x})}{\partial x_1} + \cdots + x_n \frac{\partial \tilde{f}(\tilde{x})}{\partial x_n}. \tag{2.3.2}$$

---

[3]The notation $\nabla_{\tilde{x}} \tilde{f}$ denotes the gradient of $\tilde{f}$ with respect to $\tilde{x} := (x_0, x_1, \ldots, x_n)$.

When $\Delta(f) = 0$, the polynomial $f$ does not necessarily have a singular zero point, because $\nabla_{\tilde{x}}\tilde{f}(\tilde{x}) = 0$ may have a solution at infinity $x_0 = 0$.

Discriminants can also be defined for a group of homogeneous polynomials. Let $f_0(\tilde{x})$, $f_1(\tilde{x}), \dots, f_m(\tilde{x})$ be forms in $\tilde{x} := (x_0, x_1, \dots, x_n)$ of degrees $d_0, d_1, \dots, d_m$ respectively, and $m \leq n$. For convenience, we denote

$$f := (f_0, f_1, \dots, f_m).$$

If every $f_i$ has generic coefficients, the polynomial system

$$f_0(\tilde{x}) = \cdots = f_m(\tilde{x}) = 0 \qquad (2.3.3)$$

has no singular solution in $\mathbb{P}^n$, i.e., for each $\tilde{u} \in \mathbb{P}^n$ satisfying (2.3.3), the Jacobian matrix

$$J_f(\tilde{u}) := \begin{bmatrix} \nabla_{\tilde{x}} f_0(\tilde{u}) & \nabla_{\tilde{x}} f_1(\tilde{u}) & \cdots & \nabla_{\tilde{x}} f_m(\tilde{u}) \end{bmatrix}$$

has full column rank. However, for special $f$, (2.3.3) might have a singular solution. Define the set

$$W(d_0, \dots, d_m) = \left\{ (f_0, \dots, f_m) \;\middle|\; \begin{array}{c} \exists \tilde{u} \in \mathbb{P}^n \quad s.t. \\ f_0(\tilde{u}) = \cdots = f_m(\tilde{u}) = 0, \\ \text{rank } J_f(\tilde{u}) \leq m \end{array} \right\}.$$

When every $d_i = 1$, $W(1, \dots, 1)$ consists of all $(f_0, \dots, f_m)$ such that $f_0, \dots, f_m$ are dependent linear forms. Thus $W(1, \dots, 1)$ consists of all $(n + 1) \times (m + 1)$ matrices whose ranks are at most $m$, which is a determinantal variety of codimension $n + 1 - m$. It is not a hypersurface when $m \leq n - 1$. When every $d_i = d > 1$, $W(d, \dots, d)$ consists of all tuples $(f_0, \dots, f_m)$ such that the multi-homogeneous form in $(\tilde{x}, \tilde{\lambda})$ (here $\tilde{\lambda} = (\lambda_0, \lambda_1, \dots, \lambda_m)$)

$$q(\tilde{x}, \tilde{\lambda}) := \lambda_0 f_0(\tilde{x}) + \lambda_1 f_1(\tilde{x}) + \cdots + \lambda_m f_m(\tilde{x})$$

has a critical point in the projective space $\mathbb{P}^n \times \mathbb{P}^m$. The multi-homogeneous form $q(\tilde{x}, \tilde{\lambda})$ has a critical point in $\mathbb{P}^n \times \mathbb{P}^m$ if and only if its discriminant vanishes (see [102, §2B, Chap. 13]). So $W(d, \dots, d)$ is a hypersurface if $d > 1$.

When the $d_i$'s are not equal and at least one $d_i > 1$, $W(d_0, \dots, d_m)$ is also a hypersurface. This can be implied by Theorem 4.8 of Looijenga [182]. So we assume at least one $d_i > 1$ and hence $W(d_0, \dots, d_m)$ is a hypersurface. Let $\Delta(f_0, \dots, f_m)$ be the minimum degree defining polynomial for $W(d_0, \dots, d_m)$

$$(f_0, \dots, f_m) \in W(d_0, \dots, d_m) \quad \Leftrightarrow \quad \Delta(f_0, \dots, f_m) = 0. \qquad (2.3.4)$$

The $\Delta(f_0, \dots, f_m)$ is called the discriminant of $(f_0, \dots, f_m)$. When $m = 0$, $\Delta(f_0, \dots, f_m)$ becomes the classical discriminant of a single polynomial. So $\Delta(f_0, \dots, f_m)$ is a generalization of $\Delta(f_0)$.

In the following, we give some properties of $\Delta(f_0, \dots, f_m)$ in [219]. For an integer $k \geq 0$, denote by $S_k$ the $k$th complete symmetric polynomial function

$$\mathbf{S}_k(a_1, \dots, a_t) := \sum_{i_1 + \cdots + i_t = k} a_1^{i_1} \cdots a_t^{i_t}. \qquad (2.3.5)$$

**Theorem 2.3.1.** *([219]) Suppose every $d_i > 0$, at least one $d_i > 1$, and $m \leq n$. Then the discriminant polynomial $\Delta(f_0, \dots, f_m)$, in the coefficients of forms $f_i \in \mathbb{C}[\tilde{x}]_{d_i}^{\mathrm{hom}}$, has the following properties:*

(a) For each $k = 0, \ldots, m$, $\Delta(f_0, \ldots, f_m)$ is homogeneous in $f_k$. Moreover,

$$\Delta(f_0, \ldots, f_m) = 0 \quad \text{whenever} \quad f_i = f_j \quad (i \neq j).$$

(b) For each $k = 0, \ldots, m$, the degree of $\Delta(f_0, f_1, \ldots, f_m)$ in $f_k$ is

$$\frac{d_0 d_1 \cdots d_m}{d_k} \cdot \mathbf{S}_{n-m}\left(d_0 - 1, \ldots, \widehat{\widehat{d_k}}, \ldots, d_m - 1\right). \tag{2.3.6}$$

In the above, $\widehat{\widehat{a}}$ means that $a$ is repeated twice.

(c) The total degree of $\Delta(f_0, \ldots, f_m)$ is

$$d_0 \cdots d_m \left(\sum_{k=0}^{m} \frac{1}{d_k} \mathbf{S}_{n-m}\left(d_0 - 1, \ldots, \widehat{\widehat{d_k - 1}}, \ldots, d_m - 1\right)\right). \tag{2.3.7}$$

For the special case that $m = n$, the Jacobian of (2.3.3) must be singular at every solution $\tilde{u} \in \mathbb{P}^n$, because there are $n + 1$ rows and

$$\tilde{u}^T \left[\nabla_{\tilde{x}} f_0(\tilde{u}) \quad \cdots \quad \nabla_{\tilde{x}} f_n(\tilde{u})\right] = \left[d_0 f_0(\tilde{u}) \quad \cdots \quad d_n f_n(\tilde{u})\right] = 0.$$

So, the polynomial system (2.3.3) has a singular solution if and only if its homogenization

$$f_0(\tilde{x}) = \cdots = f_n(\tilde{x}) = 0$$

has a solution in $\mathbb{P}^n$, which is equivalent to the vanishing of $\text{Res}(f_0, \ldots, f_n)$. Therefore,

$$\Delta(f_0, \ldots, f_n) = 0 \iff \text{Res}(f_0, \ldots, f_n) = 0.$$

Observe that $\Delta(f_0, \ldots, f_n)$ and $\text{Res}(f_0, \ldots, f_n)$ have the same degree

$$d_0 \cdots d_n \left(d_0^{-1} + \cdots + d_n^{-1}\right).$$

So $\Delta(f_0, \ldots, f_n)$ is equal to $\text{Res}(f_0, \ldots, f_n)$, up to a constant factor.

When $f_0, \ldots, f_m$ are nonhomogeneous, the discriminant $\Delta(f_0, \ldots, f_m)$ is then defined to be $\Delta(\widehat{f_0}, \ldots, \widehat{f_m})$, where each $\widehat{f_i}$ is the homogenization of $f_i$. Resultants and discriminants are special cases of elimination theory for polynomial systems. The following is a general elimination theorem.

**Theorem 2.3.2.** *(Elimination Theory [114, Theorem 5.7A]) Let $f_1, \ldots, f_r$ be homogeneous polynomials in $x_0, \ldots, x_n$, having indeterminate coefficients $a_{ij}$. Then there is a set of polynomials $g_1, \ldots, g_t$ in the $a_{ij}$, with integer coefficients, which are homogeneous in the coefficients of each $f_i$ separately, with the following property: for any algebraically closed field $\mathbb{F}$, and for any set of special values of the $a_{ij} \in \mathbb{F}$, a necessary and sufficient condition for the $f_i$ to have a common zero different from $(0, \ldots, 0)$ is that the $a_{ij}$ are a common zero of the polynomials $g_j$.*

## 2.3.3 ▪ Exercises

**Exercise 2.3.1.** *For two univariate polynomials $f, g$ in $x$, show that $f, g$ have a common complex zero if and only if $\text{Res}(f, g) = 0$. In particular, show that $f$ has a double complex root if and only if $\Delta(f) = 0$.*

**Exercise 2.3.2.** *Determine the conditions on $a, b, c, d$ such that the univariate polynomial $f(x) = x^4 + ax^3 + bx^2 + cx + d$ has a triple root.*

**Exercise 2.3.3.** *([309]) Let $f, g$ be two univariate polynomials of the same degree $d$. Their Bézoutian is the polynomial*

$$B(x, y) := \frac{f(x)g(y) - f(y)g(x)}{x - y} = \sum_{i,j=0}^{d-1} c_{ij} x^i y^j.$$

*Show that* $\mathrm{Res}(f, g) = \det C$ *for the matrix $C = (c_{ij})$ formed as above.*

**Exercise 2.3.4.** *Determine conditions on $(a, b, c)$ such that the polynomial system*

$$x_1^2 + ax_1x_2 + 1 = x_2^2 + bx_1x_2 + 1 = x_1x_2 + c(x_1 + x_2) + 1 = 0$$

*has a complex solution.*

**Exercise 2.3.5.** *Compute the discriminant of the trivariate ternary form*

$$x_1^3 + x_2^3 + x_3^3 + ax_1^2x_2 + bx_1x_2x_3.$$

**Exercise 2.3.6.** *Determine the range for $(a, b)$ such that the polynomial*

$$1 + a(x_1 + x_2 + x_3) + b(x_1x_2 + x_2x_3 + x_3x_1) + x_1^2 + x_2^2 + x_3^2 \geq 0 \ on \ \mathbb{R}^3$$

*and it has a real zero point.*

**Exercise 2.3.7.** *([309]) The discriminant of a matrix is defined to be the discriminant of its characteristic polynomial. Determine conditions on $(a, b)$ such that the matrix*

$$\begin{bmatrix} 1 & a & b \\ b & 1 & a \\ a & b & 1 \end{bmatrix}$$

*have a double eigenvalue.*

## 2.4 ▪ SOS and Nonnegative Polynomials

For the variable $x := (x_1, \ldots, x_n)$ and a power $\alpha := (\alpha_1, \ldots, \alpha_n) \in \mathbb{N}^n$, recall the notation

$$|\alpha| := \alpha_1 + \cdots + \alpha_n, \quad x^\alpha := x_1^{\alpha_1} \cdots x_n^{\alpha_n}.$$

Denote the power set

$$\mathbb{N}_d^n := \{\alpha \in \mathbb{N}^n : |\alpha| \leq d\}. \tag{2.4.1}$$

The column vector of all monomials in $x$ and of degrees up to $d$ is denoted as

$$[x]_d := \begin{bmatrix} 1 & x_1 & \cdots & x_n & x_1^2 & x_1x_2 & \cdots & x_n^d \end{bmatrix}^T. \tag{2.4.2}$$

The length of the monomial vector $[x]_d$ is $\binom{n+d}{d}$.

### 2.4.1 ▪ Sum of squares

Sum of squares is a basic concept in polynomial optimization.

**Definition 2.4.1.** *A polynomial* $f \in \mathbb{R}[x]$ *is said to be a sum of squares (SOS) if there are polynomials* $p_1, \ldots, p_k \in \mathbb{R}[x]$ *such that* $f = p_1^2 + \cdots + p_k^2$. *The set of all SOS polynomials in* $x$ *is denoted as* $\Sigma[x]$.

The equation $f = p_1^2 + \cdots + p_k^2$ is called an *SOS decomposition*. For instance, the polynomial $1 + x_1^4 + x_2^4 + x_3^4 - 4x_1x_2x_3$ is SOS, with the SOS decomposition

$$1 + x_1^4 + x_2^4 + x_3^4 - 4x_1x_2x_3 = \tfrac{1}{3}\big[(x_1^2 - x_2^2 - 1 + x_3^2)^2 + (x_1^2 + x_2^2 - 1 - x_3^2)^2$$
$$+ (x_1^2 - x_2^2 - x_3^2 + 1)^2 + 2(x_1 - x_2x_3)^2 + 2(x_1x_2 - x_3)^2 + 2(x_1x_3 - x_2)^2\big].$$

The set $\Sigma[x]$ is a cone of $\mathbb{R}[x]$. The intersection $\Sigma[x] \cap \mathbb{R}[x]_{2d}$ is fully dimensional in the space $\mathbb{R}[x]_{2d}$. We denote the truncation

$$\Sigma[x]_{2d} := \Sigma[x] \cap \mathbb{R}[x]_{2d}. \tag{2.4.3}$$

In some occasions, to show the dimension $n$ of $x$, we also use the notation

$$\Sigma[x_1, \ldots, x_n]_{2d} := \Sigma[x]_{2d}, \quad \Sigma_{n,2d} := \Sigma[x_1, \ldots, x_n]_{2d}. \tag{2.4.4}$$

For $f \in \mathbb{R}[x]_{2d}$, if $f = \sum_{i=1}^k p_i^2$ is an SOS decomposition, then we must have $p_i \in \mathbb{R}[x]_d$. Write each $p_i = v_i^T[x]_d$ for a vector $v_i \in \mathbb{R}^{\binom{n+d}{d}}$, then

$$f = [x]_d^T \Big( \sum_{i=1}^k v_i v_i^T \Big)[x]_d.$$

This implies the following well-known result (see [158, 257, 258]).

**Lemma 2.4.2.** *A polynomial* $f \in \mathbb{R}[x]_{2d}$ *is SOS if and only if there exists* $X \in \mathcal{S}^{\binom{n+d}{d}}$ *such that*

$$\boxed{\begin{aligned} f &= [x]_d^T \cdot X \cdot [x]_d, \\ X &\succeq 0. \end{aligned}} \tag{2.4.5}$$

Checking SOS polynomials can be done by solving semidefinite programs. We write $f$ in the standard monomial form (with coefficients $f_\alpha$):

$$f = \sum_{\alpha \in \mathbb{N}_{2d}^n} f_\alpha x^\alpha.$$

Expand the outer product $[x]_d[x]_d^T$ as

$$[x]_d[x]_d^T = \sum_{\alpha \in \mathbb{N}_{2d}^n} A_\alpha x^\alpha, \tag{2.4.6}$$

with real symmetric matrix coefficients $A_\alpha$. By comparing coefficients, one can see that (2.4.5) holds if and only if $f_\alpha = A_\alpha \bullet X$ for every $\alpha \in \mathbb{N}_{2d}^n$. Therefore, $f$ is SOS if and only if there is a solution to the psd matrix system

$$\boxed{\begin{aligned} A_\alpha \bullet X &= f_\alpha \ (\alpha \in \mathbb{N}_{2d}^n), \\ X &\succeq 0. \end{aligned}} \tag{2.4.7}$$

54                    Chapter 2.  Introductions to Algebra, Polynomials, and Moments

For instance, the following polynomial

$$x_1^4 - x_1^2(1 + 2x_1x_2) + x_1^2x_2^2 + 2x_1x_2 + 2$$

is SOS, because it can be expressed as $1 + x_1^2 + (1 - x_1^2 + x_1x_2)^2$. Equivalently, a feasible solution to (2.4.7) is the psd matrix

$$X = \begin{bmatrix} 2 & 0 & 0 & -1 & 1 & 0 \\ 0 & 1 & 0 & 0 & 0 & 0 \\ 0 & 0 & 0 & 0 & 0 & 0 \\ -1 & 0 & 0 & 1 & -1 & 0 \\ 1 & 0 & 0 & -1 & 1 & 0 \\ 0 & 0 & 0 & 0 & 0 & 0 \end{bmatrix}.$$

Next, we characterize the dual cone of $\Sigma_{n,2d}$. The notation $\mathbb{R}^{\mathbb{N}_{2d}^n}$ denotes the space of real vectors that are labeled by $\alpha \in \mathbb{N}_{2d}^n$, i.e., a vector $y \in \mathbb{R}^{\mathbb{N}_{2d}^n}$ can be labeled as

$$y = (y_\alpha)_{\alpha \in \mathbb{N}_{2d}^n}.$$

The vector $y$ is called *truncated multi-sequences* (tms) of degree $2d$. For each integer $k \in [0, d]$, the $k$th order *moment matrix* of $y$ is the $\binom{n+k}{k}$-by-$\binom{n+k}{k}$ symmetric matrix

$$M_k[y] := \left[ y_{\alpha+\beta} \right]_{\alpha \in \mathbb{N}_k^n, \beta \in \mathbb{N}_k^n}. \tag{2.4.8}$$

The rows and columns of $M_k[y]$ are labeled by $\alpha, \beta \in \mathbb{N}_k^n$ respectively, according to the graded lexicographically ordering. Clearly, if $k_1 \leq k_2$, then $M_{k_1}[y]$ is a leading principal submatrix of $M_{k_2}[y]$. For matrices $A_\alpha$ as in (2.4.6), it holds that

$$M_d[y] = \sum_{\alpha \in \mathbb{N}_{2d}^n} A_\alpha y_\alpha.$$

For the univariate case $n = 1$, $M_d[y]$ is the Hankel matrix

$$M_d[y] = \begin{bmatrix} y_0 & y_1 & y_2 & y_3 & \cdots & y_d \\ y_1 & y_2 & y_3 & y_4 & \cdots & y_{d+1} \\ y_2 & y_3 & y_4 & y_5 & \cdots & y_{d+2} \\ y_3 & y_4 & y_5 & y_6 & \cdots & y_{d+3} \\ \vdots & \vdots & \vdots & \ddots & \ddots & \ddots \\ y_d & y_{d+1} & y_{d+2} & y_{d+3} & \cdots & y_{2d} \end{bmatrix}.$$

For the binary case $n = 2$ and $d = 2$, we have

$$M_2[y] = \begin{bmatrix} y_{00} & y_{10} & y_{01} & y_{20} & y_{11} & y_{02} \\ y_{10} & y_{20} & y_{11} & y_{30} & y_{21} & y_{12} \\ y_{01} & y_{11} & y_{02} & y_{21} & y_{12} & y_{03} \\ y_{20} & y_{30} & y_{21} & y_{40} & y_{31} & y_{22} \\ y_{11} & y_{21} & y_{12} & y_{31} & y_{22} & y_{13} \\ y_{02} & y_{12} & y_{03} & y_{22} & y_{13} & y_{04} \end{bmatrix}.$$

For the ternary case $n = 3$ and $d = 2$, we have

$$M_2[y] = \begin{bmatrix} y_{000} & y_{100} & y_{010} & y_{001} & y_{200} & y_{110} & y_{101} & y_{020} & y_{011} & y_{002} \\ y_{100} & y_{200} & y_{110} & y_{101} & y_{300} & y_{210} & y_{201} & y_{120} & y_{111} & y_{102} \\ y_{010} & y_{110} & y_{020} & y_{011} & y_{210} & y_{120} & y_{111} & y_{030} & y_{021} & y_{012} \\ y_{001} & y_{101} & y_{011} & y_{002} & y_{201} & y_{111} & y_{102} & y_{021} & y_{012} & y_{003} \\ y_{200} & y_{300} & y_{210} & y_{201} & y_{400} & y_{310} & y_{301} & y_{220} & y_{211} & y_{202} \\ y_{110} & y_{210} & y_{120} & y_{111} & y_{310} & y_{220} & y_{211} & y_{130} & y_{121} & y_{112} \\ y_{101} & y_{201} & y_{111} & y_{102} & y_{301} & y_{211} & y_{202} & y_{121} & y_{112} & y_{103} \\ y_{020} & y_{120} & y_{030} & y_{021} & y_{220} & y_{130} & y_{121} & y_{040} & y_{031} & y_{022} \\ y_{011} & y_{111} & y_{021} & y_{012} & y_{211} & y_{121} & y_{112} & y_{031} & y_{022} & y_{013} \\ y_{002} & y_{102} & y_{012} & y_{003} & y_{202} & y_{112} & y_{103} & y_{022} & y_{013} & y_{004} \end{bmatrix}.$$

Each vector in $\mathbb{R}^{\mathbb{N}^n_{2d}}$ determines a linear functional on $\mathbb{R}[x]_{2d}$ as follows. Under the standard monomial basis, each polynomial in $\mathbb{R}[x]_{2d}$ is represented by the vector of its coefficients. For the polynomial $f = \sum_{\alpha \in \mathbb{N}^n_{2d}} f_\alpha x^\alpha$, we denote the coefficient vector

$$\mathrm{vec}(f) := (f_\alpha)_{\alpha \in \mathbb{N}^n_{2d}}. \tag{2.4.9}$$

The length of $\mathrm{vec}(f)$ is $\binom{n+2d}{2d}$. A linear functional on $\mathbb{R}[x]_{2d}$ is determined by its values on basis monomials. Therefore, a tms $y \in \mathbb{R}^{\mathbb{N}^n_{2d}}$ determines the linear functional $\ell_y$ acting on $\mathbb{R}[x]_{2d}$ such that

$$\ell_y : \mathbb{R}[x]_{2d} \to \mathbb{R}, \quad x^\alpha \mapsto y_\alpha. \tag{2.4.10}$$

The above is equivalent to

$$\langle f, y \rangle := \ell_y(f) = \sum_{\alpha \in \mathbb{N}^n_{2d}} f_\alpha y_\alpha. \tag{2.4.11}$$

For the above operation $\langle f, y \rangle$, each $y$ gives a linear functional on $\mathbb{R}[x]_{2d}$ and each $f \in \mathbb{R}[x]_{2d}$ gives a linear functional on $\mathbb{R}^{\mathbb{N}^n_{2d}}$. In other words, $\mathbb{R}^{\mathbb{N}^n_{2d}}$ can be viewed as the dual space of $\mathbb{R}[x]_{2d}$, and vice versa. Observe that

$$\langle q^2, y \rangle = \mathrm{vec}(q)^T \cdot M_d[y] \cdot \mathrm{vec}(q) \tag{2.4.12}$$

for all $q \in \mathbb{R}[x]_d$. If the linear functional $\ell_y \geq 0$ on $\Sigma[x]_{2d}$, then $M_d[y] \succeq 0$. This gives the moment cone in $\mathbb{R}^{\mathbb{N}^n_{2d}}$:

$$\mathscr{S}_{n,2d} := \{ y \in \mathbb{R}^{\mathbb{N}^n_{2d}} : M_d[y] \succeq 0 \}. \tag{2.4.13}$$

The cones $\Sigma_{n,2d}$ and $\mathscr{S}_{n,2d}$ are dual to each other. Recall that a cone is *proper* if it is closed, convex, solid, and pointed.

**Proposition 2.4.3.** *Both $\Sigma_{n,2d}$ and $\mathscr{S}_{n,2d}$ are proper cones, and they are dual to each other, i.e.,*

$$(\Sigma_{n,2d})^\star = \mathscr{S}_{n,2d}, \quad (\mathscr{S}_{n,2d})^\star = \Sigma_{n,2d}. \tag{2.4.14}$$

The proof of Proposition 2.4.3 is left as an exercise. In the following, we consider homogeneous SOS polynomials (i.e., SOS forms). Recall the notation

$$\mathbb{R}[\tilde{x}]^{\mathrm{hom}}_{2d} := \mathbb{R}[x_0, x_1, \ldots, x_n]^{\mathrm{hom}}_{2d},$$

which is the space of degree-$2d$ forms in $\tilde{x} := (x_0, x_1, \ldots, x_n)$. Note that a polynomial $f \in \mathbb{R}[x]_{2d}$ is SOS if and only if its homogenization

$$\tilde{f}(\tilde{x}) := x_0^{2d} f(x/x_0)$$

is SOS. The homogenization gives the isomorphism

$$\mathbb{R}[x_0, x_1, \ldots, x_n]_{2d}^{\mathrm{hom}} \cong \mathbb{R}[x_1, \ldots, x_n]_{2d}.$$

For convenience, we also use the notation

$$\Sigma_{n+1,2d}^{\mathrm{hom}} := \Sigma[\tilde{x}]_{2d}^{\mathrm{hom}} := \Sigma[\tilde{x}]_{2d} \cap \mathbb{R}[\tilde{x}]^{\mathrm{hom}}.$$

To characterize the dual cone of $\Sigma_{n+1,2d}^{\mathrm{hom}}$, we need the notation

$$
\begin{array}{rcl}
\overline{\mathbb{N}}_{2d}^{n+1} & := & \{\alpha \in \mathbb{N}^{n+1} : |\alpha| = 2d\}, \\
\mathbb{R}^{\overline{\mathbb{N}}_{2d}^{n+1}} & := & \{z = (z_\alpha) : \alpha \in \overline{\mathbb{N}}_{2d}^{n+1}\}.
\end{array}
\tag{2.4.15}
$$

Each vector $z \in \mathbb{R}^{\overline{\mathbb{N}}_{2d}^{n+1}}$ is called a *homogeneous truncated multi-sequence* (htms) of degree $2d$. Its moment matrix is similarly defined as

$$M_d[z] := \left[z_{\alpha+\beta}\right]_{\alpha,\beta \in \overline{\mathbb{N}}_{2d}^{n+1}}. \tag{2.4.16}$$

The rows and columns of $M_d[z]$ are labeled by power vectors in $\overline{\mathbb{N}}_{2d}^{n+1}$. Similarly, we define the htms cone

$$\mathscr{S}_{n+1,2d}^{\mathrm{hom}} := \left\{z \in \mathbb{R}^{\overline{\mathbb{N}}_{2d}^{n+1}} : M_d[z] \succeq 0\right\}. \tag{2.4.17}$$

The homogeneous version of Proposition 2.4.3 is the following.

**Proposition 2.4.4.** *Both $\Sigma_{n+1,2d}^{\mathrm{hom}}$ and $\mathscr{S}_{n+1,2d}^{\mathrm{hom}}$ are proper cones, and they are dual to each other, i.e.,*

$$(\Sigma_{n+1,2d}^{\mathrm{hom}})^\star = \mathscr{S}_{n+1,2d}^{\mathrm{hom}}, \quad (\mathscr{S}_{n+1,2d}^{\mathrm{hom}})^\star = \Sigma_{n+1,2d}^{\mathrm{hom}}. \tag{2.4.18}$$

The proof for Proposition 2.4.4 is similar to that for Proposition 2.4.3. It is left as an exercise.

## 2.4.2 ▪ Nonnegative polynomials

In polynomial optimization, nonnegative polynomials play an important role. A polynomial $f \in \mathbb{R}[x]$ is said to be *nonnegative* or *positive semidefinite* (psd) if the evaluation $f(u) \geq 0$ for every $u \in \mathbb{R}^n$. Denote the psd (or nonnegative) polynomial cone

$$\mathscr{P}_{n,2d} := \left\{f \in \mathbb{R}[x_1, \ldots, x_n]_{2d} : f \text{ is psd}\right\}. \tag{2.4.19}$$

In the literature about psd polynomials, people often focus on the homogeneous case, because a polynomial is psd if and only if its homogenization is psd. Therefore, to study the cone $\mathscr{P}_{n,2d}$, it is enough to study psd forms. Since the notation $\mathscr{P}_{n+1,2d}$ denotes the cone of psd polynomials in $\mathbb{R}[x_0, x_1, \ldots, x_n]_{2d}$, we denote the homogeneous subcone

$$\mathscr{P}_{n+1,2d}^{\mathrm{hom}} := \mathscr{P}_{n+1,2d} \cap \mathbb{R}[x_0, x_1, \ldots, x_n]_{2d}^{\mathrm{hom}}.$$

Every SOS polynomial is clearly psd, so

$$\Sigma_{n,2d} \subseteq \mathscr{P}_{n,2d}, \quad \Sigma_{n+1,2d}^{\mathrm{hom}} \subseteq \mathscr{P}_{n+1,2d}^{\mathrm{hom}}.$$

The above containment is not an equality, except for three cases. This was fully characterized by Hilbert in 1888.

**Theorem 2.4.5.** *(Hilbert, 1888) The two cones* $\Sigma_{n+1,2d}^{\text{hom}}$ *and* $\mathscr{P}_{n+1,2d}^{\text{hom}}$ *are equal if and only if*

$$n+1=2, \quad or \quad 2d=2, \quad or \quad (n+1,2d)=(3,4).$$

We refer to [283] for the historical work on SOS and nonnegative polynomials. Although it may not be SOS, a nonnegative polynomial is always a sum of squares of rational functions. This is Hilbert's 17th question, which was answered affirmatively by Artin in 1927. In the following, we list some classical nonnegative forms that are not SOS from the work [283]:

- Motzkin's form: $(x_1^2 + x_2^2 - 3x_3^2)x_1^2 x_2^2 + x_3^6$.

- Robinson's form:

$$x_1^6 + x_2^6 + x_3^6 + 3x_1^2 x_2^2 x_3^2 - (x_1^4 x_2^2 + x_1^2 x_2^4 + x_1^4 x_3^2 + x_1^2 x_3^4 + x_2^4 x_3^2 + x_2^2 x_3^4).$$

- Choi's bi-quadratic form:

$$x_1^2 y_1^2 + x_2^2 y_2^2 + x_3^2 y_3^2 + 2(x_1^2 y_2^2 + x_2^2 y_3^2 + x_3^2 y_1^2) - 2x_1 x_2 y_1 y_2 - 2x_1 x_3 y_1 y_3 - 2x_2 x_3 y_2 y_3.$$

- Choi-Lam forms:

$$x_1^2 x_2^2 + x_1^2 x_3^2 + x_2^2 x_3^2 + x_4^4 - 4x_1 x_2 x_3 x_4, \quad x_1^4 x_2^2 + x_2^4 x_3^2 + x_3^4 x_1^2 - 3x_1^2 x_2^2 x_3^2.$$

- Horn's form: $(x_1^2 + \cdots + x_5^2)^2 - 4(x_1^2 x_2^2 + \cdots + x_4^2 x_5^2 + x_5^2 x_1^2)$.

- Lax-Lax form: $\sum\limits_{i=1}^{5} \prod\limits_{i \neq j \in [5]} (x_i - x_j)$.

- Schmüdgen's form:

$$200\big[(x_1^3 - 4x_1 x_3^2)^2 + (x_2^3 - 4x_2 x_3^2)^2\big] + (x_2^2 - x_1^2)x_1(x_1 + 2x_3)(x_1^2 - 2x_1 x_3 + 2x_2^2 - 8x_3^2).$$

Except for the three cases in Theorem 2.4.5, the psd cone $\mathscr{P}_{n+1,2d}^{\text{hom}}$ is strictly bigger than the SOS cone $\Sigma_{n+1,2d}^{\text{hom}}$. Indeed, for each fixed $d > 1$, there are significantly more nonnegative polynomials in $\mathscr{P}_{n,2d}^{\text{hom}}$ than SOS polynomials in $\Sigma_{n+1,2d}^{\text{hom}}$, as the number $n$ of variables goes to infinity. This was shown in [29]. The difference between them can be quantified in terms of $n, d$. Fix $\nu$ to be the rotationally invariant probability measure (i.e., the uniformly distributed probability measure) on the unit sphere $\mathbb{S}^n$. Consider the hyperplane

$$L_{n+1,2d} := \Big\{ f \in \mathbb{R}[\tilde{x}]_{2d}^{\text{hom}} : \int_{\mathbb{S}^n} f(\tilde{x}) \mathrm{d}\nu(\tilde{x}) = 1 \Big\}.$$

To compare the cones $\mathscr{P}_{n+1,2d}^{\text{hom}}$ and $\Sigma_{n+1,2d}^{\text{hom}}$, it is equivalent to compare the intersections

$$\begin{aligned} \mathscr{P}_{n+1,2d}^{c} &:= \mathscr{P}_{n+1,2d}^{\text{hom}} \cap L_{n+1,2d}, \\ \Sigma_{n+1,2d}^{c} &:= \Sigma_{n+1,2d}^{\text{hom}} \cap L_{n+1,2d}. \end{aligned}$$

Denote the dimensions

$$N = \binom{n+2d}{2d} - 1, \quad D = \binom{n+d}{d}.$$

Denote the unit ball in the space $\mathbb{R}[\tilde{x}]_{2d}^{\text{hom}}$ under the $L^2$-norm:

$$B_N := \left\{ f \in \mathbb{R}[\tilde{x}]_{2d}^{\text{hom}} : \int_{\mathbb{S}^n} f(\tilde{x})^2 \mathrm{d}\nu(\tilde{x}) \leq 1 \right\}.$$

The sets $\mathscr{P}_{n+1,2d}^c$ and $\Sigma_{n+1,2d}^c$ are compact convex sets of dimension $N$. For a set $T \subseteq L_{n+1,2d}$, we use $\mathrm{Vol}(T)$ to denote its volume under the Lebesgue measure. For a scalar $t > 0$, the sets $T$ and $tT$ have exactly the same shape, but their volumes are changed by the factor $t^N$. However, the rooting $\mathrm{Vol}(T)^{1/N}$ is invariant for scaling, so it is more appropriate for measuring the volume of $T$. The following is the comparison about the volumes of $\mathscr{P}_{n+1,2d}^c$, $\Sigma_{n+1,2d}^c$ and $B_N$, which was shown in [29].

**Theorem 2.4.6.** *([29]) Let $n, d, N, D$ be as above, then*

$$\left( \frac{\mathrm{Vol}(\mathscr{P}_{n+1,2d}^c)}{\mathrm{Vol}(B_N)} \right)^{\frac{1}{N}} \geq \frac{(n+1)^{-\frac{1}{2}}}{2\sqrt{4d+2}}, \quad \left( \frac{\mathrm{Vol}(\Sigma_{n+1,2d}^c)}{\mathrm{Vol}(B_N)} \right)^{\frac{1}{N}} \leq 2^{4d+1}\sqrt{\frac{6D}{N}}.$$

The ratio between $\left( \mathrm{Vol}(\mathscr{P}_{n+1,2d}^c) \right)^{\frac{1}{N}}$ and $\left( \mathrm{Vol}(\Sigma_{n+1,2d}^c) \right)^{\frac{1}{N}}$ grows asymptotically to at least $O(n^{\frac{d-1}{2}})$ as $n \to \infty$. Therefore, in the space $\mathbb{R}[x_1,\ldots,x_n]_{2d}$ with $d > 1$, there are significantly more nonnegative than SOS polynomials, as $n$ increases.

The dual cone of $\mathscr{P}_{n,2d}$ can be characterized as follows. For a tms $y \in \mathbb{R}^{\mathbb{N}_{2d}^n}$, let $\ell_y$ be the linear functional $\ell_y$ as in (2.4.10). It is nonnegative on the cone $\mathscr{P}_{n,2d}$ if and only if $\langle f, y \rangle \geq 0$ for all $f \in \mathscr{P}_{n,2d}$. Therefore, the dual cone of $\mathscr{P}_{n,2d}$ is

$$\mathscr{P}_{n,2d}^\star = \left\{ y \in \mathbb{R}^{\mathbb{N}_{2d}^n} : \langle f, y \rangle \geq 0 \, \forall f \in \mathscr{P}_{n,2d} \right\}.$$

To describe the dual cone, we consider the conic hull

$$\mathscr{R}_{n,2d} := \left\{ \sum_{i=1}^{k} \lambda_i [v_i]_{2d} : \lambda_i \geq 0, v_i \in \mathbb{R}^n, k \in \mathbb{N} \right\}. \tag{2.4.20}$$

In the above, the vector $[v]_{2d}$ is defined as in (2.4.2). The set $\mathscr{R}_{n,2d}$ is a convex cone but not closed. However, its closure gives the dual cone $\mathscr{P}_{n,2d}^\star$.

**Theorem 2.4.7.** *The nonnegative polynomial cone $\mathscr{P}_{n,2d}$ is proper and its dual cone equals the closure of $\mathscr{R}_{n,2d}$, i.e.,*

$$\mathscr{R}_{n,2d}^\star = \mathscr{P}_{n,2d}, \quad \mathscr{P}_{n,2d}^\star = \mathrm{cl}\left( \mathscr{R}_{n,2d} \right). \tag{2.4.21}$$

We remark that the dual cone of $\mathscr{P}_{n,2d}$ is not $\mathscr{R}_{n,2d}$, since $\mathscr{R}_{n,2d}$ is not closed (see exercise). This issue can be resolved by homogenization. A polynomial in $\mathbb{R}[x]_{2d}$ is uniquely determined by its homogenization in $\mathbb{R}[\tilde{x}]_{2d}^{\text{hom}}$. A linear functional $\ell$ on $\mathbb{R}[\tilde{x}]_{2d}^{\text{hom}}$ is uniquely determined by the tms $z \in \mathbb{R}^{\overline{\mathbb{N}}_{2d}^{n+1}}$ such that

$$\ell(\tilde{x}^\beta) = z_\beta, \quad \beta \in \overline{\mathbb{N}}_{2d}^{n+1}.$$

We can similarly define the operation $\langle \cdot, \cdot \rangle$ such that

$$\langle p, z \rangle := \sum_{\beta \in \overline{\mathbb{N}}_{2d}^{n+1}} p_\alpha z_\alpha \quad \text{for} \quad p = \sum_{\beta \in \overline{\mathbb{N}}_{2d}^{n+1}} p_\beta \tilde{x}^\beta.$$

Under the above $\langle \cdot, \cdot \rangle$, each $z \in \mathbb{R}^{\overline{N}_{2d}^{n+1}}$ gives a linear functional on $\mathbb{R}[\tilde{x}]_{2d}^{\text{hom}}$. By the definition, the dual cone of $\mathscr{P}_{n+1,2d}^{\text{hom}}$ is given as

$$\left(\mathscr{P}_{n+1,2d}^{\text{hom}}\right)^{\star} = \left\{z \in \mathbb{R}^{\overline{N}_{2d}^{n+1}} : \langle p, z \rangle \geq 0 \quad \forall p \in \mathscr{P}_{n+1,2d}^{\text{hom}}\right\}.$$

To characterize the dual cone, we need a set similar to the one in (2.4.20). For convenience, denote the vector of monomials of degree equaling $2d$ and in $\tilde{x} = (x_0, x_1, \ldots, x_n)$:

$$[\tilde{x}]_{2d}^{\text{hom}} := \begin{bmatrix} x_0^{2d} & x_0^{2d-1}x_1 & x_0^{2d-1}x_2 & \cdots & x_n^{2d} \end{bmatrix}. \tag{2.4.22}$$

In short, $[\tilde{x}]_{2d}^{\text{hom}} = x_0^{2d} \cdot [\frac{x}{x_0}]_{2d}$. Then we denote the cone

$$\mathscr{R}_{n+1,2d}^{\text{hom}} := \left\{\sum_{i=1}^{k} \lambda_i [u_i]_{2d}^{\text{hom}} : \lambda_i \geq 0, u_i \in \mathbb{R}^{n+1}, k \in \mathbb{N}\right\}. \tag{2.4.23}$$

Note that $\mathscr{R}_{n+1,2d}^{\text{hom}}$ is a projection of $\mathscr{R}_{n+1,2d}$. The cone $\mathscr{R}_{n+1,2d}$ is not closed, while $\mathscr{R}_{n+1,2d}^{\text{hom}}$ is closed. This is because $\mathscr{R}_{n+1,2d}^{\text{hom}}$ is the conic hull of the compact set

$$T := \{[u]_{2d}^{\text{hom}} : \|u\| = 1\}$$

and the linear functional $\langle \|\tilde{x}\|^{2d}, \cdot \rangle$ is positive on $T$. The closedness of $\mathscr{R}_{n+1,2d}^{\text{hom}}$ is implied by Proposition 1.2.6. Therefore, we have the following duality theorem.

**Theorem 2.4.8.** *Both $\mathscr{P}_{n+1,2d}^{\text{hom}}$ and $\mathscr{R}_{n+1,2d}^{\text{hom}}$ are proper cones and they are dual to each other, i.e.,*

$$\left(\mathscr{P}_{n+1,2d}^{\text{hom}}\right)^{\star} = \mathscr{R}_{n+1,2d}^{\text{hom}}, \quad \left(\mathscr{R}_{n+1,2d}^{\text{hom}}\right)^{\star} = \mathscr{P}_{n+1,2d}^{\text{hom}}. \tag{2.4.24}$$

The proofs for Theorems 2.4.7 and 2.4.8 can be done by a standard argument with the convex set separation theorem. So they are left as exercises. They can also be implied by properties of moment cones in Chapter 8.

The facial structures of the cone $\mathscr{P}_{n+1,2d}^{\text{hom}}$ are discussed in [31, §4.4]. Not every face of $\mathscr{P}_{n+1,2d}^{\text{hom}}$ is exposed. For instance, for $d > 1$, the extreme ray of $x_0^{2d}$ is a face, but it is not exposed. If $\langle p, z \rangle = 0$ is a supporting hyperplane for $\mathscr{P}_{n+1,2d}^{\text{hom}}$ passing through $x_0^{2d}$, then for every $z \in \mathscr{R}_{n+1,2d}^{\text{hom}}$,

$$z = \sum_{i=1}^{k} \lambda_i [u_i]_{2d}^{\text{hom}}, \lambda_i > 0, \langle x_0^{2d}, z \rangle = 0 \quad \Rightarrow \quad u_i = (0, v_i), v_i \in \mathbb{R}^n.$$

This means that for any form $q$ of degree $d-1$, the psd form $x_0^2 q^2$ belongs to the supporting hyperplane $\langle p, z \rangle = 0$, so the extreme ray of $x_0^{2d}$ is not exposed. In fact, for $p \in \mathscr{P}_{n+1,2d}^{\text{hom}}$, the ray of $p$ is an exposed face of $\mathscr{P}_{n+1,2d}^{\text{hom}}$ if and only if every $f \in \mathscr{P}_{n+1,2d}^{\text{hom}}$ with $V_{\mathbb{R}}(p) \subseteq V_{\mathbb{R}}(f)$ is a multiple of $p$.

## 2.4.3 ▪ Uniform denominators

By Artin's affirmative answer to Hilbert's 17th problem, for every psd form $f \in \mathbb{R}[x]_{2d}^{\text{hom}}$, there exists a nonzero SOS form $p \in \mathbb{R}[x]$ such that the product $p^r f$ is SOS for some power $r$. People are interested in what $p$ is. Unfortunately, there does not exist a uniform $p$ that works for all psd forms (see [284]). However, if $f$ is positive definite (i.e., $f(x) > 0$ for all $x \neq 0$), we can choose $p$ to be the quadratic form $x^T x = x_1^2 + \cdots + x_n^2$. Interestingly, if $r$ is big enough, then $(x^T x)^r f$ is a sum of $(2d + 2r)$th power of real linear forms (also see §8.4.3). For a positive definite form $f$, define the ratio

$$\epsilon(f) := \min_{\|x\|=1} f(x) \Big/ \max_{\|x\|=1} f(x).$$

The following is a classical result about the bound on $r$ by Reznick [283].

**Theorem 2.4.9.** *([283]) Let $f \in \mathbb{R}[x]_{2d}^{\mathrm{hom}}$ be a positive definite form and let $\epsilon(f)$ be as above. If the degree*

$$r \geq \frac{2nd(2d-1)}{(4\log 2)\epsilon(f)} - \frac{n+2d}{2},$$

*then $(x^T x)^r f$ is a sum of the $(2d + 2r)$th power of real linear forms. In particular, for such $r$, the product $(x^T x)^r f$ is SOS.*

### 2.4.4 ▪ Exercises

**Exercise 2.4.1.** *Prove Propositions 2.4.3 and 2.4.4.*

**Exercise 2.4.2.** *Find an explicit tms $z \in \mathbb{R}^{\overline{\mathbb{N}}_6^3}$ such that $M_3[z]$ is positive definite but $\langle M(x), z \rangle < 0$, where $M(x)$ is Motzkin's form.*

**Exercise 2.4.3.** *Let $R(x)$ be the Robinson form in $x \in \mathbb{R}^3$. Show that $R(x)$ is not SOS but the product $(x_1^2 + x_2^2 + x_3^2)R(x)$ is SOS. Give an explicit SOS decomposition for the product.*

**Exercise 2.4.4.** *Show that the Lax-Lax form is nonnegative but not SOS.*

**Exercise 2.4.5.** *Consider the cone $\mathscr{R}_{n,2d}$ in (2.4.20). Show that it is solid and pointed but not closed.*

**Exercise 2.4.6.** *Prove Theorems 2.4.7 and 2.4.8.*

**Exercise 2.4.7.** *([31, §4.4]) For each $0 \neq u \in \mathbb{R}^{n+1}$, show that*

$$F(u) := \big\{ f \in \mathscr{P}_{n+1,2d}^{\mathrm{hom}} : f(u) = 0 \big\}$$

*is a face of $\mathscr{P}_{n+1,2d}^{\mathrm{hom}}$ and has codimension $n + 1$.*

**Exercise 2.4.8.** *([31, §4.4]) Let $u, w \in \mathbb{R}^{n+1}$ be nonzero vectors. Let*

$$F_w(u) := \Big\{ f \in \mathscr{P}_{n+1,2d}^{\mathrm{hom}} : f(u) = 0, \, \nabla^2 f(u) \cdot w = 0 \Big\}.$$

*Show that $F_w(u)$ is a face of $\mathscr{P}_{n+1,2d}^{\mathrm{hom}}$ but it is not exposed.*

## 2.5 ▪ Quadratic Modules and Preorderings

Let $g := (g_1, \ldots, g_m)$ be a tuple of polynomials in $\mathbb{R}[x]$. We look for conditions that ensure a polynomial $f \geq 0$ on the semialgebraic set

$$S(g) := \{ x \in \mathbb{R}^n : g_1(x) \geq 0, \ldots, g_m(x) \geq 0 \}. \tag{2.5.1}$$

Clearly, if there are SOS polynomials $\sigma_0, \ldots, \sigma_m$ such that

$$f = \sigma_0 + \sigma_1 g_1 + \cdots + \sigma_m g_m, \tag{2.5.2}$$

then $f \geq 0$ on $S(g)$. This is because for each $u \in S(g)$, all $g_i(u) \geq 0$ and $\sigma_i(u) \geq 0$. So we consider the set of all polynomials $f$ that can be represented as in (2.5.2):

$$\mathrm{QM}[g] := \Sigma[x] + g_1 \cdot \Sigma[x] + \cdots + g_m \cdot \Sigma[x]. \tag{2.5.3}$$

The set $\mathrm{QM}[g]$ is called the *quadratic module* generated by the polynomial tuple $g$. A set $Q \subseteq \mathbb{R}[x]$ is said to be a quadratic module (see [191]) if

$$Q + Q \subseteq Q, \quad p^2 Q \subseteq Q \quad \text{for all} \quad p \in \mathbb{R}[x], \quad 1 \in Q.$$

To ensure $f \geq 0$ on $S(g)$, a more general representation is the preordering. Clearly, if a polynomial $f$ can be expressed as (note $\prod_{i \in \emptyset} g_i = 1$)

$$f = \sum_{I \subseteq \{1,\ldots,m\}} \sigma_I \cdot \prod_{i \in I} g_i \tag{2.5.4}$$

for some SOS polynomials $\sigma_I \in \Sigma[x]$, then we also have $f \geq 0$ on $S(g)$. The representation (2.5.2) is a special case of (2.5.4). This leads us to consider the *preordering* of $g$:

$$\mathrm{Pre}[g] := \sum_{I \subseteq \{1,\ldots,m\}} \left( \prod_{i \in I} g_i \right) \cdot \Sigma[x]. \tag{2.5.5}$$

A set $P \subseteq \mathbb{R}[x]$ is said to be a preordering (see [191]) if

$$P + P \subseteq P, \quad P \cdot P \subseteq P, \quad p^2 \in P \quad \text{for all} \quad p \in \mathbb{R}[x].$$

Clearly, it holds that $\mathrm{QM}[g] \subseteq \mathrm{Pre}[g]$ for every polynomial tuple $g$.

## 2.5.1 ▪ Truncations of $\mathrm{QM}[g]$ and $\mathrm{Pre}[g]$

We discuss how to check whether or not a polynomial $f$ belongs to the quadratic module $\mathrm{QM}[g]$ or the preordering $\mathrm{Pre}[g]$. There are no degree bounds on SOS polynomials $\sigma_i$, $\sigma_I$ in $\mathrm{QM}[g]$ or $\mathrm{Pre}[g]$. In computation, we often work on polynomials with given degrees. The degree-$d$ truncation of the quadratic module $\mathrm{QM}[g]$ is (let $g_0 := 1$)

$$\mathrm{QM}[g]_d := \left\{ \sum_{i=0}^{m} \sigma_i g_i : \sigma_i \in \Sigma[x], \deg(\sigma_i g_i) \leq d \right\}. \tag{2.5.6}$$

In the above, if some $\deg(g_i) > d$, we just simply let the corresponding $\sigma_i = 0$. So the truncation $\mathrm{QM}[g]_d$ is defined for all degrees $d$, even if $d$ is smaller than the degree of some $g_i$. Usually, we are interested in even degrees. Clearly, it holds that the nesting containment

$$\mathrm{QM}[g]_1 \subseteq \mathrm{QM}[g]_2 \subseteq \cdots \subseteq \mathrm{QM}[g] = \bigcup_{d \geq 1} \mathrm{QM}[g]_d.$$

Note that $f \in \mathrm{QM}[g]$ if and only if $f \in \mathrm{QM}[g]_d$ for some $d$.

For a given degree $d$, the membership in the truncation $\mathrm{QM}[g]_d$ can be checked by solving a semidefinite optimization problem. Assume the degree $d = 2k$ is even. We only need to consider

$g_i$'s with $\deg(g_i) \leq 2k$, because otherwise $\sigma_i = 0$ in $\mathrm{QM}[g]_d$. Consider an SOS polynomial $\sigma_i$ such that $\deg(\sigma_i g_i) \leq 2k$. Let $t_i$ be the maximum integer such that $2t_i + \deg(g_i) \leq 2k$, i.e.,

$$t_i = k - \left\lceil \frac{1}{2} \deg(g_i) \right\rceil.$$

The product $g_i[x]_{t_i}[x]_{t_i}^T$ is a symmetric matrix polynomial of length $\binom{n+t_i}{t_i}$. We expand it such that

$$g_i(x)[x]_{t_i}[x]_{t_i}^T = \sum_{\alpha \in \mathbb{N}_{2k}^n} x^\alpha G_\alpha^{(i)} \tag{2.5.7}$$

for symmetric matrices $G_\alpha^{(i)}$. For an SOS polynomial $\sigma_i$, the product $\sigma_i g_i$ can be equivalently expressed as

$$\sigma_i g_i = \left( [x]_{t_i}^T X_i [x]_{t_i} \right) g_i = \left( g_i(x)[x]_{t_i}[x]_{t_i}^T \right) \bullet X_i$$

for a psd matrix $X_i$ of length $\binom{n+t_i}{t_i}$. By the definition, we have $f \in \mathrm{QM}[g]_d$ if and only if there exist SOS polynomials $\sigma_i \in \Sigma[x]_{2t_i}$ such that (note $g_0 = 1$)

$$f = \sigma_0 g_0 + \sigma_1 g_1 + \cdots + \sigma_m g_m.$$

If $f = \sum_{\alpha \in \mathbb{N}_{2k}^n} f_\alpha x^\alpha$, the above is equivalent to the following $\binom{n+2k}{2k}$ equations:

$$f_\alpha = \sum_{i=0}^m G_\alpha^{(i)} \bullet X_i \quad (\alpha \in \mathbb{N}_{2k}^n) \tag{2.5.8}$$

for a psd matrix tuple $(X_0, X_1, \ldots, X_m) \in \mathcal{K}$, where $\mathcal{K}$ is the cone

$$\mathcal{K} := \mathcal{S}_+^{\binom{n+t_0}{t_0}} \times \mathcal{S}_+^{\binom{n+t_1}{t_1}} \times \cdots \times \mathcal{S}_+^{\binom{n+t_m}{t_m}}. \tag{2.5.9}$$

It is the Cartesian product of $m + 1$ standard psd cones. The above observation implies the following proposition.

**Proposition 2.5.1.** *Let $g, \mathcal{K}$ be as in the above. For an even degree $d = 2k$, a polynomial $f \in \mathrm{QM}[g]_d$ if and only if there exists a matrix tuple*

$$X = (X_0, X_1, \ldots, X_m) \in \mathcal{K}$$

*satisfying the linear equations in (2.5.8).*

By the above proposition, we can check memberships in a truncation of $\mathrm{QM}[g]$ by solving a linear conic optimization problem, with the cone $\mathcal{K}$ as in (2.5.9). The same is for the preordering $\mathrm{Pre}[g]$. The degree-$d$ truncation of the preordering $\mathrm{Pre}[g]$ is

$$\mathrm{Pre}[g]_d := \left\{ \sum_{I \subseteq \{1,\ldots,m\}} \sigma_I \cdot \prod_{i \in I} g_i \,\middle|\, \sigma_I \in \Sigma[x], \deg(\sigma_i \prod_{i \in I} g_i) \leq d \right\}. \tag{2.5.10}$$

In the above, for the case $\deg(\prod_{i \in I} g_i) > d$, we just let $\sigma_I = 0$. The truncation $\mathrm{Pre}[g]_d$ is defined for all degrees $d$. It is interesting to note that if $g$ is a tuple of a single polynomial (i.e., $m = 1$), then $\mathrm{QM}[g]_d = \mathrm{Pre}[g]_d$. If we denote by $g'$ the tuple of all possible products of distinct $g_i$'s, i.e.,

$$g' := (g_1, \ldots, g_m, g_1 g_2, g_1 g_3, \ldots, g_1 \cdots g_m),$$

then $\mathrm{QM}[g']_d = \mathrm{Pre}[g]_d$ for every degree $d$. Therefore, the membership in the truncation $\mathrm{Pre}[g]_d$ is the same as in $\mathrm{QM}[g']_d$. It can be checked by solving a semidefinite optimization problem, similar to the case of $\mathrm{QM}[g]_d$.

## 2.5.2 ▪ Localizing matrices and dual cones

Recall that $\mathbb{N}_d^n = \{\alpha \in \mathbb{N}^n : |\alpha| \leq d\}$, the set of monomial powers with degrees at most $d$. The notation $\mathbb{R}^{\mathbb{N}_d^n}$ stands for the space of all real vectors that are labeled by $\alpha \in \mathbb{N}_d^n$. A vector $y \in \mathbb{R}^{\mathbb{N}_d^n}$ is called *truncated multi-sequences* (tms) of degree $d$.

For a polynomial $f \in \mathbb{R}[x]_{2k}$, let $s$ be the maximum integer such that $2s + \deg(f) \leq 2k$, i.e.,

$$s = k - \lceil \tfrac{1}{2} \deg(f) \rceil.$$

The product $f(x)[x]_s[x]_s^T$ is a symmetric matrix polynomial of length $\binom{n+s}{s}$. We can write the expansion as

$$f(x)[x]_s[x]_s^T = \sum_{\alpha \in \mathbb{N}_{2k}^n} x^\alpha F_\alpha$$

for some symmetric matrices $F_\alpha$. For $y \in \mathbb{R}^{\mathbb{N}_{2k}^n}$, define the matrix

$$L_f^{(k)}[y] := \sum_{\alpha \in \mathbb{N}_{2k}^n} y_\alpha F_\alpha. \qquad (2.5.11)$$

It is called the $k$th order *localizing matrix* of $f$ and generated by $y$. The above expression is equivalent to

$$\langle f(x)(v^T[x]_s)^2, y \rangle = v^T(L_f^{(k)}[y])v$$

for every vector $v$ of length $\binom{n+s}{s}$. For the case $\deg f > 2k$, we just simply let

$$L_f^{(k)}[y] = 0.$$

For fixed $f$, the $L_f^{(k)}[y]$ is linear in $y$; for fixed $y$, $L_f^{(k)}[y]$ is linear in $f$ (with required degree). Observe that if $f(u) \geq 0$ and $y = [u]_{2k}$, then

$$L_f^{(k)}[y] = f(u)[u]_s[u]_s^T \succeq 0.$$

Here are some instances of localizing matrices. When $n = 2$, $k = 2$, and $f = 1 - x_1^2 - x_2^2$,

$$L_f^{(2)}[y] = \begin{bmatrix} y_{00} - y_{20} - y_{02} & y_{10} - y_{30} - y_{12} & y_{01} - y_{21} - y_{03} \\ y_{10} - y_{30} - y_{12} & y_{20} - y_{40} - y_{22} & y_{11} - y_{31} - y_{13} \\ y_{01} - y_{21} - y_{03} & y_{11} - y_{31} - y_{13} & y_{02} - y_{22} - y_{04} \end{bmatrix}.$$

When $n = 3$, $k = 3$, and $f = x_1 x_2 - x_3^3$,

$$L_f^{(3)}[y] = \begin{bmatrix} y_{110} - y_{003} & y_{210} - y_{103} & y_{120} - y_{013} & y_{111} - y_{004} \\ y_{210} - y_{103} & y_{310} - y_{203} & y_{220} - y_{113} & y_{211} - y_{104} \\ y_{120} - y_{013} & y_{220} - y_{113} & y_{130} - y_{023} & y_{121} - y_{014} \\ y_{111} - y_{004} & y_{211} - y_{104} & y_{121} - y_{014} & y_{112} - y_{005} \end{bmatrix}.$$

In particular, if $f = 1$ (the constant 1 polynomial), then $L_f^{(k)}[y]$ is reduced to the moment matrix $M_k[y]$, which is given as in (2.4.8).

A localizing matrix can also be viewed as the moment matrix of a displaced tms. For a degree $t \leq d$ and a polynomial

$$p = \sum_{\beta \in \mathbb{N}_t^n} p_\beta x^\beta,$$

define the operation on $y \in \mathbb{R}^{\mathbb{N}_d^n}$ such that

$$p * y := \left( \sum_{\beta \in \mathbb{N}_t^n} p_\beta y_{\alpha+\beta} \right)_{\alpha \in \mathbb{N}_{d-t}^n} . \qquad (2.5.12)$$

Then $p * y$ is a tms of degree $d - t$. For instance, for $y \in \mathbb{R}^{\mathbb{N}_5^3}$ and $p = x_1 x_2 - x_3^3$,

$$p * y = \big( y_{110} - y_{003}, y_{210} - y_{103}, y_{120} - y_{013}, y_{111} - y_{004}, y_{310} - y_{203}, \\ y_{220} - y_{113}, y_{211} - y_{104}, y_{130} - y_{023}, y_{121} - y_{014}, y_{112} - y_{005} \big).$$

In particular, for $p \in \mathbb{R}[x]_{2k}$ and $d_0 = \lceil \frac{1}{2} \deg(p) \rceil$, it holds that

$$L_p^{(k)}[y] = M_{k-d_0}[p * y]. \qquad (2.5.13)$$

For a polynomial tuple $g := (g_1, \ldots, g_m)$, its localizing matrix of order $k$ is then defined to be the block diagonal matrix

$$L_g^{(k)}[y] := \operatorname{diag}\Big( L_{g_1}^{(k)}[y], \ldots, L_{g_m}^{(k)}[y] \Big). \qquad (2.5.14)$$

It can be used to define the convex tms cone

$$\mathscr{S}[g]_{2k} := \Big\{ y \in \mathbb{R}^{\mathbb{N}_{2k}^n} : M_k[y] \succeq 0, \ L_g^{(k)}[y] \succeq 0 \Big\}. \qquad (2.5.15)$$

For a degree $k > 0$, the truncation $\operatorname{QM}[g]_{2k}$ is a convex cone in the space $\mathbb{R}[x]_{2k}$. The cones $\operatorname{QM}[g]_{2k}$ and $\mathscr{S}[g]_{2k}$ are dual to each other.

**Theorem 2.5.2.** *For a polynomial tuple $g$ as above, we have the following:*

*(i) The cone $\mathscr{S}[g]_{2k}$ is closed convex and dual to $\operatorname{QM}[g]_{2k}$, i.e.,*

$$\langle f, y \rangle \ge 0 \quad \forall f \in \operatorname{QM}[g]_{2k}, \quad \forall y \in \mathscr{S}[g]_{2k}. \qquad (2.5.16)$$

*(ii) If the set $S(g)$ has nonempty interior, then both $\mathscr{S}[g]_{2k}$ and $\operatorname{QM}[g]_{2k}$ are proper cones (i.e., they are closed and pointed) and*

$$(\operatorname{QM}[g]_{2k})^\star = \mathscr{S}[g]_{2k}, \quad (\mathscr{S}[g]_{2k})^\star = \operatorname{QM}[g]_{2k}. \qquad (2.5.17)$$

*In the above, the superscript $^\star$ denotes the dual cone.*

The conclusions in Theorem 2.5.2 are well known in moment and polynomial optimization (see [162, 166, 171, 191]). The item (i) follows from the fact that

$$\langle g_i q^2, y \rangle = \operatorname{vec}(q)^T \big( L_{g_i}^{(k)}[y] \big) \operatorname{vec}(q) \qquad (2.5.18)$$

for every $q \in \mathbb{R}[x]$ such that $\deg(g_i q^2) \le 2k$. In the above, $\operatorname{vec}(q)$ denotes the coefficient vector of $q$, as in (2.4.9). So, $\langle g_i q^2, y \rangle \ge 0$ for every $q$ if and only if $L_{g_i}^{(k)}[y] \succeq 0$. Similarly, for every $q \in \mathbb{R}[x]_k$, it also holds that

$$\langle q^2, y \rangle = \operatorname{vec}(q)^T \big( M_k[y] \big) \operatorname{vec}(q). \qquad (2.5.19)$$

Hence, $\langle q^2, y \rangle \ge 0$ for every $q \in \mathbb{R}[x]_k$ if and only if $M_k[y] \succeq 0$. The item (ii) can be implied by Theorem 2.5.4. We leave the proof details for Theorem 2.5.2 to be filled as an exercise.

When the interior of $S(g)$ is empty, the cone $QM[g]_{2d}$ may not be closed. For instance, for $g = (-x^2)$ and for each $\epsilon > 0$,

$$x + \epsilon = (\sqrt{\epsilon} + \frac{x}{2\sqrt{\epsilon}})^2 + (-x^2)\left(\frac{1}{2\sqrt{\epsilon}}\right)^2.$$

So, $x + \epsilon \in QM[g]_2$ for every $\epsilon > 0$. However, we have $x \notin QM[g]$, i.e., there are no SOS polynomials $\sigma_0, \sigma_1$ such that $x = \sigma_0 - x^2\sigma_1$. The proof for this is left as an exercise.

## 2.5.3 ▪ Sums of ideals and quadratic modules

Let $h = (h_1, \ldots, h_t)$ be a tuple of polynomials in $\mathbb{R}[x]$. It gives the ideal

$$\mathrm{Ideal}[h] := h_1 \cdot \mathbb{R}[x] + \cdots + h_t \cdot \mathbb{R}[x].$$

In optimization, we often need to work with the sum $\mathrm{Ideal}[h] + QM[g]$ for a polynomial tuple $g = (g_1, \ldots, g_m)$. This is because if $f \in \mathrm{Ideal}[h] + QM[g]$, then $f \geq 0$ on the set given as

$$h(x) = 0, \ g(x) \geq 0.$$

For a degree $d > 0$, the truncation

$$\mathrm{Ideal}[h]_d := h_1 \cdot \mathbb{R}[x]_{d-\deg(h_1)} + \cdots + h_t \cdot \mathbb{R}[x]_{d-\deg(h_t)} \qquad (2.5.20)$$

is a subspace of $\mathbb{R}[x]_d$. For the case $\deg(h_i) > d$, we just simply let

$$\mathbb{R}[x]_{d-\deg(h_i)} = \{0\}.$$

For each $p_i \in \mathbb{R}[x]_{d-\deg(h_i)}$, the coefficients of the product $h_i \cdot p_i$ are linear in that of $p_i$. Let $H_i^{(d)}$ be the matrix such that

$$\mathrm{vec}(h_i p_i) = H_i^{(d)} \cdot \mathrm{vec}(p_i), \qquad (2.5.21)$$

where $\mathrm{vec}(p_i)$ denotes the coefficient vector of $p_i$, as in (2.4.9). Clearly, a polynomial $f \in \mathrm{Ideal}[h] + QM[g]$ if and only if

$$f \in \mathrm{Ideal}[h]_{2k} + QM[g]_{2k} \quad \text{for some } k \in \mathbb{N}. \qquad (2.5.22)$$

Let $G_\alpha^{(i)}$ be the matrix as in (2.5.7). Let $\mathcal{Q}$ denote the cone

$$\mathcal{Q} := \mathbb{R}^{\binom{n+s_1}{s_1}} \times \cdots \times \mathbb{R}^{\binom{n+s_l}{s_l}} \times \mathcal{S}_+^{\binom{n+t_1}{t_0}} \times \cdots \times \mathcal{S}_+^{\binom{n+t_1}{t_m}},$$

where each $s_i := 2k - \deg(h_i)$. The space $\mathbb{R}^{\binom{n+s_i}{s_i}}$ is a free cone, i.e., the cone of entire space. Observe that (2.5.22) holds if and only if there exists a tuple

$$(\omega_1, \ldots, \omega_l, X_0, X_1, \ldots, X_m) \in \mathcal{Q}$$

satisfying the linear equations

$$f_\alpha = \sum_{j=1}^{l} (H_i^{(2k)} \cdot \omega_i)_\alpha + \sum_{i=0}^{m} G_\alpha^{(i)} \bullet X_i \quad (\alpha \in \mathbb{N}_{2k}^n).$$

Therefore, checking (2.5.22) can be done by solving a linear conic optimization problem with the cone $\mathcal{Q}$.

A vector $z = (z_\alpha)_{\alpha \in \mathbb{N}_d^n}$ gives the linear functional

$$\ell_z : \mathbb{R}[x]_d \to \mathbb{R}, \quad x^\alpha \mapsto z_\alpha.$$

For the matrix $H_i^{(d)}$ as in (2.5.21), we can see that

$$\ell_z(h_i p_i) = z^T \operatorname{vec}(h_i p_i) = z^T H_i^{(d)} \operatorname{vec}(p_i).$$

The *localizing vector* of the polynomial $h_i$ generated by $z$ is

$$\mathscr{V}_{h_i}^{(d)}[z] := (H_i^{(d)})^T z. \tag{2.5.23}$$

For the case $\deg(h_i) > d$, we just let $\mathscr{V}_{h_i}^{(d)}[z] = 0$. It holds that

$$\ell_z(h_i p_i) = \operatorname{vec}(p_i)^T \mathscr{V}_{h_i}^{(d)}[z].$$

We are interested in linear functionals on $\mathbb{R}[x]_d$ that vanish on the subspace $\mathrm{Ideal}[h]_d$. For the polynomial tuple $h = (h_1, \ldots, h_t)$, its localizing vector is then defined to be

$$\mathscr{V}_h^{(d)}[z] := \begin{bmatrix} \mathscr{V}_{h_1}^{(d)}[z] \\ \vdots \\ \mathscr{V}_{h_t}^{(d)}[z] \end{bmatrix}. \tag{2.5.24}$$

It gives the following subspace of $\mathbb{R}^{\mathbb{N}_d^n}$:

$$\mathscr{Z}[h]_d = \left\{ z \in \mathbb{R}^{\mathbb{N}_d^n} : \mathscr{V}_h^{(d)}[z] = 0 \right\}. \tag{2.5.25}$$

Using localizing vectors, we can get a full characterization for the dual cone of $\mathrm{Ideal}[h]_{2k} + \mathrm{QM}[g]_{2k}$.

**Proposition 2.5.3.** *Let $g, h$ be as above. For any even degree $2k > 0$, we have*

$$\left( \mathrm{Ideal}[h]_{2k} + \mathrm{QM}[g]_{2k} \right)^\star = \mathscr{Z}[h]_{2k} \cap \left( \mathrm{QM}[g]_{2k} \right)^\star. \tag{2.5.26}$$

The sum $\mathrm{Ideal}[h] + \mathrm{QM}[g]$ can be expressed in terms of quotient rings. For an ideal $J \subseteq \mathbb{R}[x]$ and for a polynomial $f$, let $[f]_J$ denote the equivalent class of $f$ with respect to $J$, i.e., $p \in [f]$ if and only if $f - p \in J$. The quotient space $\mathbb{R}[x]/J$ consists of equivalent classes of polynomials with respect to $J$. The quotient $\mathbb{R}[x]/J$ is a vector space over $\mathbb{R}$. Note that

$$[f + p]_J = [f]_J + [p]_J \quad \text{for all } f, p \in \mathbb{R}[x].$$

A norm for the quotient space $\mathbb{R}[x]/J$ can be given as

$$\|[f]\|_J := \min \left\{ \| \operatorname{vec}(f - p) \| : p \in J \right\}. \tag{2.5.27}$$

The image of the quadratic module $\mathrm{QM}[g]$ in $\mathbb{R}[x]/J$ is

$$\mathrm{QM}[g]/J := \left\{ [f]_J : f \in \mathrm{QM}[g] \right\}. \tag{2.5.28}$$

For an even degree $2k$, the truncation of $\mathrm{QM}[g]/J$ is

$$\mathrm{QM}[g]_{2k}/J := \left\{ [f]_J : f \in \mathrm{QM}[g]_{2k} \right\}.$$

The quotient $\mathrm{QM}[g]_{2k}/J$ is a finitely dimensional vector space. Under certain conditions, the truncation $\mathrm{QM}[g]_{2k}/J$ is a closed set.

**Theorem 2.5.4.** *Let $J \subseteq \mathbb{R}[x]$ be an ideal and $g$ be a polynomial tuple. If the vanishing ideal $I\big(S(g) \cap V_{\mathbb{R}}(J)\big) \subseteq J$, then the truncation $\mathrm{QM}[g]_{2k}/J$ is closed for every degree $k$.*

**Proof.** Let $g_0 := 1$. For each equivalent class $[f]_J \in \mathrm{QM}[g]_{2k}/J$, there exist SOS polynomials $\sigma_i \in \Sigma[x]_{2k}$, with $\deg(\sigma_i g_i) \leq 2k$, such that

$$f \equiv \sum_{i=0}^{m} \sigma_i g_i \qquad \mathrm{mod} \quad J. \tag{2.5.29}$$

Without loss of generality, assume each $g_i \notin J$, otherwise drop that $g_i$ from the tuple $g$ while the sum $\mathrm{QM}[g]_{2k} + J$ is not changed. For each $i = 0, 1, \ldots, m$, let

$$J_i = \{p \in \mathbb{R}[x]_{k-t_i} : pg_i \in J\}, \quad \text{where } t_i = \left\lceil \frac{1}{2} \deg(g_i) \right\rceil.$$

Each $J_i$ is a subspace of $\mathbb{R}[x]_{k-t_i}$. Let $B_i$ be a monomial basis of the complement space of $J_i$ in $\mathbb{R}[x]_{k-t_i}$. Denote by $[x]_{B_i}$ the column vector of monomials in $B_i$. Then, (2.5.29) holds if and only if there exist psd matrices $X_i$ of dimension $|B_i| \times |B_i|$ such that

$$f \equiv \sum_{i=0}^{m} ([x]_{B_i}^T \cdot X_i \cdot [x]_{B_i}) \cdot g_i \qquad \mathrm{mod} \quad J. \tag{2.5.30}$$

To prove the closedness of $\mathrm{QM}[g]_{2k}/J$, assume there is a sequence $\{f_\ell\}_{\ell=1}^{\infty} \subseteq \mathbb{R}[x]_{2k}$ such that $[f_\ell]_J \in \mathrm{QM}[g]_{2k}/J$ and $[f_\ell]_J \to [\hat{f}]_J$, under the norm as in (2.5.27). We need to show the limit $[\hat{f}]_J \in \mathrm{QM}[g]_{2k}/J$. As in the above, we can express each $f_\ell$ as

$$f_\ell \equiv \sum_{i=0}^{m} ([x]_{B_i}^T X_i^{(\ell)} [x]_{B_i}) \cdot g_i \qquad \mathrm{mod} \quad J,$$

where each $X_i^{(\ell)}$ is a psd matrix of dimension $|B_i| \times |B_i|$. Next we show that for each $i$, the psd matrix sequence $\{X_i^{(\ell)}\}_{\ell=1}^{\infty}$ is bounded. Let $\mu$ be a probability measure whose support equals $S(g) \cap V_{\mathbb{R}}(J)$. Then, for each $i$,

$$\int f_\ell(x) \mathrm{d}\mu(x) \geq \lambda_{\min}(G_i) \|X_i^{(\ell)}\|_F,$$

where $\lambda_{\min}(G_i)$ is the smallest eigenvalue of the matrix

$$G_i := \int g_i \cdot [x]_{B_i}([x]_{B_i})^T \mathrm{d}\mu(x).$$

We claim that $\lambda_{\min}(G_i) > 0$. Since $G_i \succeq 0$, if otherwise $\lambda_{\min}(G_i) = 0$, there exists a vector $q \neq 0$ such that

$$q^T G_i q = \int g_i \cdot (q^T [x]_{B_i})^2 \mathrm{d}\mu(x) = 0.$$

This implies that $g_i \cdot (q^T [x]_{B_i})^2$ vanishes identically on $S(g) \cap V_{\mathbb{R}}(J)$, and so does $g_i \cdot (q^T [x]_{B_i})$. By the assumption $I(S(g) \cap V_{\mathbb{R}}(J)) \subseteq J$, we have that

$$g_i \cdot (q^T [x]_{B_i}) \in J, \quad q^T [x]_{B_i} \in J_i.$$

However, this contradicts that $B_i$ is basis for the complement space of $J_i$ in $\mathbb{R}[x]_{k-t_i}$. Therefore, $\lambda_{\min}(G_i) > 0$ and

$$\|X_i^{(\ell)}\|_F \le \left(\lambda_{\min}(G_i)\right)^{-1} \int f_\ell(x)\mathrm{d}\mu(x),$$

i.e., the sequence $\{X_i^{(\ell)}\}_{\ell=1}^\infty$ is bounded. Without loss of generality, we can further assume that $X_i^{(\ell)} \to \hat{X}_i$ as $\ell \to \infty$ for each $i$. The limit $\hat{X}_i$ is also psd. Hence, we get

$$\hat{f} \equiv \sum_{i=0}^m ([x]_{B_i}^T \hat{X}_i [x]_{B_i}) \cdot g_i \qquad \mod \quad J,$$

which completes the proof.     □

When the semialgebraic set $S(g)$ has empty interior, the truncation $\mathrm{QM}[g]_{2k}$ may not be closed (e.g., this is the case for $g = -x^2$). However, it becomes closed if it is replaced by the quotient of some ideal. In Theorem 2.5.4, if we choose $J = I(S(g))$, then the resulting quotient truncation is closed. Theorem 2.5.4 implies the following result, which was shown by Marshall [189].

**Theorem 2.5.5.** *([189]) For a polynomial tuple $g$, if $J$ is the vanishing ideal of the set $S(g)$, then the quotient truncation $\mathrm{QM}[g]_{2k}/J$ is closed.*

A proof for Theorem 2.5.5 is also given by Laurent [171]. For convenience, we also denote the quotient

$$\mathrm{QM}'[g]_{2k} = \{[f]_{I(S(g))} : f \in \mathrm{QM}[g]_{2k}\}. \tag{2.5.31}$$

It is a closed convex cone of the quotient space $\mathbb{R}[x]/I(S(g))$. The following is the conclusion about the closedness of the sum $\mathrm{Ideal}[h]_{2k} + \mathrm{QM}[g]_{2k}$.

**Theorem 2.5.6.** *Let $g, h$ be two polynomial tuples such that*

$$I(V_\mathbb{R}(h) \cap S(g)) \subseteq \mathrm{Ideal}[h].$$

*For a degree $k$, if $\mathrm{Ideal}[h] \cap \mathbb{R}[x]_{2k} = \mathrm{Ideal}[h]_{2k}$, then $\mathrm{Ideal}[h]_{2k} + \mathrm{QM}[g]_{2k}$ is closed.*

**Proof.** Suppose $\{p_\ell + q_\ell\}_{\ell=1}^\infty \subseteq \mathrm{Ideal}[h]_{2k} + \mathrm{QM}[g]_{2k}$ is sequence such that each $p_\ell \in \mathrm{Ideal}[h]_{2k}, q_\ell \in \mathrm{QM}[g]_{2k}$, and

$$p_\ell + q_\ell \to f \in \mathbb{R}[x]_{2k} \quad \text{as} \quad \ell \to \infty.$$

Let $J := \mathrm{Ideal}[h]$, then $[q_\ell]_J \to [f]_J$, under the norm as in (2.5.27). By Theorem 2.5.4, we get

$$[f]_J \in \mathrm{QM}[g]_{2k}/J.$$

For each $\ell$, write that $q_\ell = a_\ell + b_\ell$ and $f = a + b$, with

$$a_\ell \in \mathrm{QM}[g]_{2k}, \quad b_\ell \in J, \quad a \in \mathrm{QM}[g]_{2k}, \quad b \in J,$$

such that $a_\ell \to a$. Then $p_\ell + b_\ell \to b$. Note that $b_\ell, b \in \mathbb{R}[x]_{2k}$. Since

$$J \cap \mathbb{R}[x]_{2k} = \mathrm{Ideal}[h]_{2k},$$

we get $b_\ell, b \in \mathrm{Ideal}[h]_{2k}$ and hence $f \in \mathrm{Ideal}[h]_{2k} + \mathrm{QM}[g]_{2k}$.     □

If the polynomials in $h$ form a Gröbner basis for $\mathrm{Ideal}[h]$ with respect to a total degree monomial ordering, then for each degree $d \geq \deg(h)$, we have

$$\mathrm{Ideal}[h] \cap \mathbb{R}[x]_d = \mathrm{Ideal}[h]_d.$$

For such a case, the sum $\mathrm{Ideal}[h]_{2k} + \mathrm{QM}[g]_{2k}$ is closed if

$$I(V_\mathbb{R}(h) \cap S(g)) \subseteq \mathrm{Ideal}[h].$$

### 2.5.4 ▪ Exercises

**Exercise 2.5.1.** *Show that* $1 - x^T x \in \mathrm{QM}[x_1, \ldots, x_n, 1 - e^T x]$.

**Exercise 2.5.2.** *Prove that* $x \notin \Sigma[x] - x^2 \Sigma[x]$ *when* $x$ *is univariate.*

**Exercise 2.5.3.** *When* $x$ *is univariate, determine the smallest* $\epsilon > 0$ *such that*

$$1 + x + \epsilon \in \mathrm{QM}[(1 - x^2)^3]_6.$$

**Exercise 2.5.4.** *Let* $M(x)$ *be the Motzkin form. Find the smallest value of* $\epsilon > 0$ *such that*

$$M(x) + \epsilon\|x\|^6 \in \Sigma[x]_6 + \mathrm{Ideal}[\|x\|^2 - 1]_6.$$

**Exercise 2.5.5.** *Determine the set of triples* $(a_1, a_2, a_3) \in \mathbb{R}^3$ *such that*

$$1 + a_1 x_2 x_3 + a_2 x_3 x_1 + a_3 x_1 x_2 \in \mathrm{QM}[1 - x_1^2, 1 - x_2^2, 1 - x_3^2]_2.$$

**Exercise 2.5.6.** *Determine the set of pairs* $(a, b) \in \mathbb{R}^2$ *such that*

$$1 + a x_1(1 + x_1 + x_2) + b x_2(1 + x_1 - x_2) \in \mathrm{QM}[1 - x_1^2 - x_2^2]_2.$$

**Exercise 2.5.7.** *Determine whether or not* $\mathrm{QM}[x_1^2 - 1, x_2^2 - 1, 1 - x_1^2 - x_2^2]_2$ *is closed.*

**Exercise 2.5.8.** *For a tuple* $g$ *of polynomials in* $\mathbb{R}[x]$, *show that* $\mathrm{QM}[g] \cap - \mathrm{QM}[g]$ *is an ideal of* $\mathbb{R}[x]$.

**Exercise 2.5.9.** *Prove Theorem 2.5.2.*

## 2.6 ▪ Positivstellensätze

In polynomial optimization, there are three types of important questions:

- For a set $S \subseteq \mathbb{R}^n$, how do we certify that a polynomial $f \geq 0$ on $S$?

- For a set of real polynomial equations, what are certificates for them to have no real solutions?

- More generally, if a set is given by polynomial equations and/or inequalities, what are certificates for it to be empty?

These questions can be answered by various types of Positivstellensätze. More details about them can be found in the books [16, 34, 269].

For constrained optimization, we often consider polynomials that are nonnegative on a set $S \subseteq \mathbb{R}^n$. For a degree $d > 0$, denote the sets

$$
\begin{aligned}
\mathscr{P}(S) &:= \{f \in \mathbb{R}[x] : f(u) \geq 0 \ \forall \, u \in S\}, \\
\mathscr{P}_d(S) &:= \mathscr{P}(S) \cap \mathbb{R}[x]_d.
\end{aligned}
\tag{2.6.1}
$$

When $S = \mathbb{R}^n$, $\mathscr{P}_d(\mathbb{R}^n)$ is the same as the psd polynomial cone $\mathscr{P}_{n,d}$. More properties of the cone $\mathscr{P}_d(S)$ will be given in §8.1.

## 2.6.1 ▪ Putinar's and Schmüdgen's Positivstellensätze

Let $g := (g_1, \ldots, g_m)$ be a tuple of polynomials in $\mathbb{R}[x]$. It determines the basic closed semialgebraic set

$$
S(g) := \{x \in \mathbb{R}^n : g_1(x) \geq 0, \ldots, g_m(x) \geq 0\}.
\tag{2.6.2}
$$

We are interested in conditions for a polynomial $f \geq 0$ on $S(g)$.

Recall that $\mathrm{QM}[g]$ is the quadratic module of $g$ (see §2.5). Clearly, if $f \in \mathrm{QM}[g]$, then $f \geq 0$ on $S(g)$. However, the converse is not necessarily true. To ensure $f \in \mathrm{QM}[g]$, we need further conditions on $f$ and $g$. In Positivstellensätze, we typically require that $f > 0$ on $S(g)$. The quadratic module $\mathrm{QM}[g]$ is said to be *archimedean* if there exists $p \in \mathrm{QM}[g]$ such that the set

$$
S(p) = \{x \in \mathbb{R}^n : p(x) \geq 0\}
$$

is compact. It is straightforward to verify that if $\mathrm{QM}[g]$ is archimedean, then the set $S(g)$ as in (2.6.2) must be compact. However, when $S(g)$ is compact, the quadratic module $\mathrm{QM}[g]$ is not necessarily archimedean (see Exercise 2.6.2 for such an example). The following theorem gives a typical condition for the membership $f \in \mathrm{QM}[g]$. In the literature, this conclusion is often referenced as *Putinar's Positivstellensatz*.

**Theorem 2.6.1.** *([272]) Suppose* $\mathrm{QM}[g]$ *is archimedean. If a polynomial* $f > 0$ *on* $S(g)$, *then we have* $f \in \mathrm{QM}[g]$.

When the quadratic module $\mathrm{QM}[g]$ is not archimedean but the set $S(g)$ is compact, there is a similar conclusion by using the preordering $\mathrm{Pre}[g]$ (see §2.5). It is also obvious that if $f \in \mathrm{Pre}[g]$, then $f \geq 0$ on the set $S(g)$. This gives another certificate for a polynomial to be nonnegative on a set. Clearly, it holds that the nesting containment

$$
\mathrm{QM}[g] \subseteq \mathrm{Pre}[g] \subseteq \mathscr{P}(S(g)).
$$

The following theorem is often referenced as *Schmüdgen's Positivstellensatz* in the literature.

**Theorem 2.6.2.** *([298]) Let* $g$ *be a polynomial tuple such that the set* $S(g)$ *is compact. If a polynomial* $f > 0$ *on* $S(g)$, *then* $f \in \mathrm{Pre}[g]$.

We would like to remark that if a polynomial $f \geq 0$ on $S(g)$ and some optimality conditions hold at each zero, then we also have $f \in \mathrm{QM}[g]$ or $f \in \mathrm{Pre}[g]$. This is shown in §5.5.1. Beyond Putinar's and Schmüdgen's Positivstellensätze, there exist other types of Positivstellensätze that ensure a polynomial is nonnegative on $S(g)$. We refer to [162, 166, 171, 172, 295] for more results on this topic. Here, we present two of them.

- (**Krein-Krivine-Nudelman Positivstellensatz** [154, 155]) Suppose the set $S(g)$ is compact and $0 \leq g_j \leq 1$ on $S(g)$ for all $j$. Assume that $1, g_1, \ldots, g_m$ generate the algebra $\mathbb{R}[x]$ (i.e., $\mathbb{R}[x] = \mathbb{R}[1, g_1, \ldots, g_m]$). If a polynomial $f > 0$ on $S$, then there exist *finitely* many nonzero scalars $c_{\alpha,\beta}$ such that

$$f = \sum_{\alpha,\beta \in \mathbb{N}^m} c_{\alpha,\beta} \cdot \prod_{j=1}^{m} g_j^{\alpha_j} \cdot \prod_{j=1}^{m} (1 - g_j)^{\beta_j}, \quad c_{\alpha,\beta} \geq 0. \tag{2.6.3}$$

In the above, $\alpha = (\alpha_1, \ldots, \alpha_m)$ and $\beta = (\beta_1, \ldots, \beta_m)$.

- (**Handelman's Positivstellensatz** [112]) Suppose each $g_i$ has degree one and the set $S(g)$ is bounded. If a polynomial $f > 0$ on $S(g)$, then there exist *finitely* many nonzero scalars $c_\beta$ such that ($\beta := (\beta_1, \ldots, \beta_m)$)

$$f = \sum_{\beta \in \mathbb{N}^m} c_\beta \cdot g_1^{\beta_1} \cdots g_m^{\beta_m}, \quad c_\beta \geq 0. \tag{2.6.4}$$

Putinar's and Schmüdgen's Positivstellensätze can be checked by solving a semidefinite optimization problem. Note that $f \in \mathrm{QM}[g]$ (resp., $f \in \mathrm{Pre}[g]$) if and only if $f \in \mathrm{QM}[g]_{2k}$ (resp., $f \in \mathrm{Pre}[g]_{2k}$) for some degree $k$. The membership $f \in \mathrm{QM}[g]$ can be checked by solving the semidefinite optimization problem of (2.5.8)-(2.5.9). It is similar for $f \in \mathrm{Pre}[g]$. On the other hand, Krein-Krivine-Nudelman's and Handelman's Positivstellensätze can be checked by solving linear programs.

## 2.6.2 ▪ Real Nullstellensatz

Let $h := (h_1, \ldots, h_t)$ be a tuple of polynomials in $\mathbb{R}[x]$. It generates the ideal $\mathrm{Ideal}[h] \subseteq \mathbb{R}[x]$. The real variety $V_\mathbb{R}(h)$ consists of real solutions to the polynomial system

$$h_1(x) = \cdots = h_t(x) = 0.$$

Note that $V_\mathbb{R}(h)$ is also equivalently given by the polynomial inequalities

$$h_1(x) \geq 0, \ldots, h_t(x) \geq 0, -h_1(x) \geq 0, \ldots, -h_t(x) \geq 0.$$

Clearly, if $f \in \mathrm{QM}[h, -h]$, then $f \geq 0$ on $V_\mathbb{R}(h)$. Therefore, Theorem 2.6.1 can be similarly applied for checking polynomials nonnegative on $V_\mathbb{R}(h)$. Also note that

$$\mathrm{QM}[h, -h] = \Sigma[x] + \mathrm{Ideal}[h].$$

When polynomial functions are considered on the real variety $V_\mathbb{R}(h)$, we need to consider the quotient space $\mathbb{R}[x]/\mathrm{Ideal}[h]$. For a polynomial $f$ to be SOS in $\mathbb{R}[x]/\mathrm{Ideal}[h]$, it requires that there exists $q \in \mathrm{Ideal}[h]$ such that $f - q \in \Sigma[x]$, or equivalently, $f \in \Sigma[x] + \mathrm{Ideal}[h]$. There are two basic questions regarding $V_\mathbb{R}(h)$: When does a polynomial $p$ vanish identically on $V_\mathbb{R}(h)$ (i.e., $p \equiv 0$ on $V_\mathbb{R}(h)$)? When is the real variety $V_\mathbb{R}(h)$ empty? This is answered by the so-called *Real Nullstellensatz*.

**Theorem 2.6.3.** *(Real Nullstellensatz [34]) Let $h$ be a tuple of polynomials in $\mathbb{R}[x]$. Then, we have the following:*

(i) *A polynomial $f \in \mathbb{R}[x]$ vanishes identically on $V_\mathbb{R}(h)$ if and only if there exist a power $r \geq 1$ and $\sigma \in \Sigma[x]$ such that*

$$f^{2r} + \sigma \in \mathrm{Ideal}[h]. \tag{2.6.5}$$

*(ii) The real variety $V_{\mathbb{R}}(h) = \emptyset$ if and only if it holds that*

$$-1 \in \Sigma[x] + \text{Ideal}[h], \qquad (2.6.6)$$

*which is equivalent to $\mathbb{R}[x] = \Sigma[x] + \text{Ideal}[h]$.*

The membership in the sum $\Sigma[x] + \text{Ideal}[h]$ can be checked by solving a linear conic optimization problem with psd matrices (see §2.5.3). We can do the same thing for Real Nullstellensatz. Note that $f \equiv 0$ on $V_{\mathbb{R}}(h)$ if and only if $-f^{2r} \in \Sigma[x] + \text{Ideal}[h]$ for some power $r \geq 1$. Real Nullstellensatz motivates the notion of real radical ideals.

**Definition 2.6.4.** *([34]) For an ideal $I \subseteq \mathbb{R}[x]$, the following set*

$$\sqrt[\mathbb{R}]{I} := \left\{ p \in \mathbb{R}[x] : -p^{2k} \in I + \Sigma[x], \ k \in \mathbb{N} \right\} \qquad (2.6.7)$$

*is called the real radical of $I$. In particular, if $\sqrt[\mathbb{R}]{I} = I$, then $I$ is said to be real radical (or just real).*

The following are some examples of Real Nullstellensatz and real radicals.

**Example 2.6.5.** *(i) Consider the polynomial $f = x_1 x_2 - 1$ and the ideal*

$$I = \text{Ideal}[(x_1 - 1)^2 + (x_2 - 1)^2].$$

*Clearly, $f \equiv 0$ on $V_{\mathbb{R}}(I)$. This is certified by the identity*

$$f^2 + \left[ (x_1 x_2 - 2x_2 + 1)^2 + 2(x_1 - 1)^2 + 2x_2^2(x_2 - 1)^2 \right]$$
$$= \left[ (x_1 - 1)^2 + (x_2 - 1)^2 \right] (2 + 2x_2^2).$$

*The real radical $\sqrt[\mathbb{R}]{I} = \text{Ideal}[x_1 - 1, x_2 - 1]$.*

*(ii) ([226]) Let $f \in \mathbb{R}[x]$ be such that $f(0) = 0$ and $I = \text{Ideal}[x_1^{2d} + \cdots + x_n^{2d}]$. Clearly, $f \equiv 0$ on $V_{\mathbb{R}}(I)$. One can write $f$ such that*

$$f = x_1 b_1 + \cdots + x_n b_n, \quad b_1, \ldots, b_n \in \mathbb{R}[x].$$

*For $\lambda > 0$ large enough, we have*

$$\lambda(t_1^{2d} + \cdots + t_n^{2d}) - (t_1 + \cdots + t_n)^{2d} \in \Sigma[t_1, \ldots, t_n]_{2d},$$

*because $t_1^{2d} + \cdots + t_n^{2d}$ lies in the interior of $\Sigma[t_1, \ldots, t_n]_{2d}$. By replacing each $t_i$ by $x_i b_i$ in the above, we have*

$$\psi := \lambda((x_1 b_1)^{2d} + \cdots + (x_n b_n)^{2d}) - f^{2d} \in \Sigma[x].$$

*One can check that*

$$\eta := \lambda \left[ \left( \sum_{i=1}^{n} x_i^{2d} \right) \left( \sum_{i=1}^{n} b_i^{2d} \right) - \left( \sum_{i=1}^{n} (x_i b_i)^{2d} \right) \right] \in \Sigma[x],$$

$$f^{2d} + \psi + \eta = \lambda(x_1^{2d} + \cdots + x_n^{2d})(b_1^{2d} + \cdots + b_n^{2d}) \in I.$$

*The sum $\psi + \eta$ is SOS. The real radical $\sqrt[\mathbb{R}]{I} = \text{Ideal}[x_1, x_2, \ldots, x_n]$.*

### 2.6.3 ▪ Positivstellensatz

More general than Real Nullstellensatz is Positivstellensatz. It concerns the solvability of a
system consisting of polynomial inequalities and/or equalities. Let

$$f = (f_i)_{i \in I_1}, \quad g = (g_j)_{j \in I_2}, \quad h = (h_l)_{l \in I_3}$$

be three tuples of finitely many polynomials in $\mathbb{R}[x]$. The label sets $I_1, I_2, I_3$ are finite (may be
possibly empty). Consider the semialgebraic set

$$S = \left\{ x \in \mathbb{R}^n \;\middle|\; \begin{array}{rcl} f_i(x) & \neq & 0\,(\,i \in I_1), \\ g_j(x) & \geq & 0\,(\,j \in I_2), \\ h_l(x) & = & 0\,(\,l \in I_3) \end{array} \right\}. \tag{2.6.8}$$

When does a polynomial $p$ vanish identically on $S$? When is $S$ empty? These questions are
answered in the following theorem, which is often referenced as *Positivstellensatz*.

**Theorem 2.6.6.** *(Positivstellensatz [34, 155, 306]) Let $f, g, h$ be tuples of polynomials as in the
above. Then the set $S$ given as in (2.6.8) is empty if and only if there exist a power $k \geq 1$, a
polynomial $\sigma \in \mathrm{Pre}[g]$, and $\phi \in \mathrm{Ideal}[h]$ such that*

$$\left( \prod_{i \in I_1} f_i \right)^{2k} + \sigma + \phi = 0. \tag{2.6.9}$$

*(The product $\prod\limits_{i \in I_1} f_i = 1$ if the label set $I_1$ is empty.)*

Positivstellensatz implies a polynomial $p \in \mathbb{R}[x]$ vanishes identically on $S$ if and only if

$$\left( p \cdot \prod_{i \in I_1} f_i \right)^{2k} + \sigma + \phi = 0$$

for some power $k \geq 1$, $\sigma \in \mathrm{Pre}[g]$, and $\phi \in \mathrm{Ideal}[h]$. Real Nullstellensatz (i.e., Theorem 2.6.3)
is a special case of Positivstellensatz. Similarly, the conclusion of Positivstellensatz can also be
checked by solving a linear conic optimization problem with psd matrices. Note that the identity
(2.6.9) holds if and only if

$$-\left( \prod_{i \in I_1} f_i \right)^{2k} \in \mathrm{Ideal}[h] + \mathrm{Pre}[g].$$

We refer to §2.5.3 for how to check the above. The following are some examples.

**Example 2.6.7.** *(i) ([309, §7.4]) The set*

$$\left\{ (x_1, x_2) \in \mathbb{R}^2 : x_1 - x_2^2 + 3 \geq 0, \; x_2 + x_1^2 + 2 = 0 \right\} = \emptyset.$$

*The emptiness is certified by the identity*

$$s + 2(x_1 - x_2^2 + 3) - 6(x_2 + x_1^2 + 2) = -1,$$

*where $s = \frac{1}{3} + 2(x_2 + \frac{3}{2})^2 + 6(x_1 - \frac{1}{6})^2$ is SOS.*

*(ii) ([31, §3.4]) The semialgebraic set given as*

$$x_1^2 + x_2^2 - 1 = 0, \ 3x_2 - x_1^2 - 2 \geq 0, \ x_1 - 8x_2^3 \geq 0$$

*is empty. The Positivstellensatz certificate for the emptiness is*

$$\phi \cdot (x_1^2 + x_2^2 - 1) + \sigma + 3 \cdot (3x_2 - x_1^2 - 2) + 1 \cdot (x_1 - 8x_2^3) = -1,$$

*where $\phi = -3x_1^2 + x_1 - 3x_2^2 + 6x_2 - 2$ and*

$$
\begin{aligned}
\sigma &= \frac{5}{43}x_1^2 + \frac{387}{44}\left(x_1 x_2 - \frac{52}{129}x_1\right)^2 + \frac{11}{5}\left(x_1^2 + \frac{1}{22}x_1 x_2 + \frac{5}{11}x_1 - x_2^2\right)^2 \\
&\quad + \frac{1}{20}\left(-x_1^2 + 2x_1 x_2 + x_2^2 + 5x_2\right)^2 + \frac{3}{4}\left(2 - x_1^2 - x_2^2 - x_2\right)^2.
\end{aligned}
$$

*(iii) ([226]) The following set is empty:*

$$\left\{(x_1, x_2) \in \mathbb{R}^2 : x_1^2 + x_2^2 \neq 0, \ x_1^3 \geq 0, x_2^3 \geq 0, -x_1 - x_2 - x_1 x_2 \geq 0\right\}.$$

*The Positivstellensatz certificate is*

$$(x_1^2 + x_2^2)^4 + \sigma_0 + \sigma_1 \cdot x_1^3 + \sigma_2 \cdot x_2^3 + \sigma_{12} \cdot x_1^3 x_2^3 + \sigma_3 \cdot (-x_1 - x_2 - x_1 x_2) = 0,$$

*where the SOS polynomials $\sigma_0, \sigma_1, \sigma_2, \sigma_{12}, \sigma_3$ are*

$$
\begin{aligned}
\sigma_0 &= (x_1^2 - x_2^2)^4 + 6(x_1^4 - x_2^4)^2, \qquad \sigma_{12} = 32(x_1^2 + x_2^2 + x_1^4 + x_2^4 + x_1^6 + x_2^6), \\
\sigma_1 &= 8\big(x_1^6(x_2 + 1/2 + 4x_2^2) + x_1^4(x_1^2/2 + 2x_1 x_2 + 2x_2^2) + x_1^4(2x_2 + 1/2 + 2x_2^2) \\
&\qquad + x_1^2(x_1^2/2 + x_1 x_2 + 4x_2^2) + 4(x_2^4 + x_2^6 + x_2^8)\big), \\
\sigma_2 &= 8\big(x_2^6(x_1 + 1/2 + 4x_1^2) + x_2^4(x_2^2/2 + 2x_2 x_1 + 2x_1^2) + x_2^4(2x_1 + 1/2 + 2x_1^2) \\
&\qquad + x_2^2(x_2^2/2 + x_2 x_1 + 4x_1^2) + 4(x_1^4 + x_1^6 + x_1^8)\big), \\
\sigma_3 &= 8\big((x_1^8 + x_2^8 + x_1^7 + x_2^7 + x_1^6 + x_2^6) + 4x_1^2 x_2^2(x_1^2 + x_2^2 + x_1^4 + x_2^4 + x_1^6 + x_2^6)\big).
\end{aligned}
$$

### 2.6.4 ▪ Quadratic modules for compact sets

For a polynomial tuple $g$, if the quadratic module $\mathrm{QM}[g]$ is archimedean, then the semialgebraic set $S(g)$ given as in (2.6.2) must be compact. When $S(g)$ is compact, the quadratic module $\mathrm{QM}[g]$ is not necessarily archimedean. Such an example is in Exercise 2.6.2. There are special properties for the quadratic module $\mathrm{QM}[g]$ when the set $S(g)$ is compact. This is given in Chapter 7 of Marshall's book [191]. We review these results in this subsection.

Suppose $g = (g_1, \ldots, g_m)$ is a polynomial tuple such that $S(g)$ is compact. When there is a single polynomial (i.e., $m = 1$), the quadratic module $\mathrm{QM}[g]$ is archimedean since $\mathrm{QM}[g] = \mathrm{Pre}[g]$. When there are two polynomials (i.e., $m = 2$), $\mathrm{QM}[g]$ is archimedean (see [270]). When all $g_1, \ldots, g_m$ are linear, the quadratic module $\mathrm{QM}[g]$ is archimedean (see [191, Theorem 7.1.3]). For the one-dimensional case (i.e., $n = 1$), $\mathrm{QM}[g]$ is archimedean (see [191, Theorem 7.1.2]). Indeed, this result can be generalized a bit more. Let

$$J := \mathrm{QM}[g] \cap -\mathrm{QM}[g]. \tag{2.6.10}$$

The intersection $J$ is an ideal (see Exercise 2.5.8). It is called the *support* of $\mathrm{QM}[g]$. If the dimension of the coordinate ring $\frac{\mathbb{R}[x]}{J}$ is less than or equal to 1, then $\mathrm{QM}[g]$ is archimedean. The following is the conclusion.

**Theorem 2.6.8.** *([191, Corollary 7.4.3]) Let $g$ be a polynomial tuple such that $\dim \frac{\mathbb{R}[x]}{J} \leq 1$, where $J$ is the support of $\mathrm{QM}[g]$ as in (2.6.10). If $S(g)$ is compact, then $\mathrm{QM}[g]$ is archimedean.*

The above theorem implies the following result: Let $I$ be an ideal such that $\dim \frac{\mathbb{R}[x]}{I} \leq 1$. For a polynomial tuple $g$ (the set $S(g)$ is not necessarily compact), if the intersection $S(g) \cap V_{\mathbb{R}}(I)$ is compact, then $\mathrm{QM}[g]+I$ is archimedean. The following conclusions are also shown in Marshall's book.

**Theorem 2.6.9.** *([191, Corollary 7.4.2]) Let $g$ be a polynomial tuple such that $\dim \frac{\mathbb{R}[x]}{J} \leq 1$, where $J$ is as in (2.6.10). The set $S(g)$ is not necessarily compact. Then, we have the following:*

(i) *A polynomial $f > 0$ on $S(g)$ if and only if $pf = 1 + q$ for some $p \in \Sigma[x]$ and $q \in \mathrm{QM}[g]$.*

(ii) *A polynomial $f \geq 0$ on $S(g)$ if and only if $pf = f^{2k} + q$ for some $p \in \Sigma[x]$, $q \in \mathrm{QM}[g]$ and $k \in \mathbb{N}$.*

(iii) *A polynomial $f \equiv 0$ on $S(g)$ if and only if $-f^{2k} \in \mathrm{QM}[g]$ for some $k \in \mathbb{N}$.*

(iv) *The set $S(g)$ is empty if and only if $-1 \in \mathrm{QM}[g]$.*

We remark that the above conclusions cannot be generalized to the case that $\dim \frac{\mathbb{R}[x]}{J} \geq 2$. Moreover, for the case $\dim \frac{\mathbb{R}[x]}{J} \leq 1$, we usually do not have $\mathrm{QM}[g] = \mathrm{Pre}[g]$. For more details, refer to [191, Chap. 7].

## 2.6.5 ▪ Exercises

**Exercise 2.6.1.** *Let $g$ be a polynomial tuple. Show that $\mathrm{QM}[g]$ is archimedean if and only if $R - x^T x \in \mathrm{QM}[g]$ for some scalar $R > 0$.*

**Exercise 2.6.2.** *([269]) For each $n \geq 2$ and the polynomial tuple*

$$g = (x_1 - 1/2, \ldots, x_n - 1/2, 1 - x_1 x_2 \cdots x_n),$$

*show that $R - (x_1^2 + \cdots + x_n^2) \notin \mathrm{QM}[g]$ for all real scalars $R$. This shows that $\mathrm{QM}[g]$ is not archimedean while the set $S(g)$ is compact.*

**Exercise 2.6.3.** *Let $g = -(x_1^{2d} + \cdots + x_n^{2d})$. For each $\epsilon > 0$, find a representation of $x_1 f_1(x) + \cdots + x_n f_n(x) + \epsilon$ in $\mathrm{QM}[g]$, where $f_1, \ldots, f_n \in \mathbb{R}[x]$.*

**Exercise 2.6.4.** *Let $f$ be Motzkin's form (see §2.4.2). Determine whether or not $f \in \Sigma[x] + \mathrm{Ideal}[\|x\|^2 - 1]$. If yes, give an explicit representation for the membership; if not, explain why.*

**Exercise 2.6.5.** *Compute the real radical of $\mathrm{Ideal}[h]$, where $h$ is Horn's form (see §2.4.2).*

**Exercise 2.6.6.** *Show that the representation of a polynomial $f$ as in (2.6.3) or (2.6.4) can be checked by solving a hierarchy of linear programs. For $f = 2 - x_1^3 - x_2^2 - x_1 x_2 - x_2$, find an explicit representation as in (2.6.4) with $g = (x_1, x_2, 1 - x_1 - x_2)$.*

**Exercise 2.6.7.** *Give a Real Nullstellensatz certificate for the polynomial $x_1^2 - x_2^2$ to vanish on the real zero set of $x_1^4 x_2^2 + x_1^2 x_2^4 + 1 - 3x_1^2 x_2^2$.*

**Exercise 2.6.8.** *Use Positivstellensatz to show that*

$$\{(x_1, x_2) \in \mathbb{R}^2 : x_1^2 + x_2^2 - 1 = 0, \, x_2^3 - x_1^3 - 2 \geq 0\} = \emptyset.$$

## 2.7 ▪ Truncated Moment Problems

For a degree $d$, let $\mathbb{R}^{\mathbb{N}_d^n}$ denote the space of real vectors $y \in \mathbb{R}^{\mathbb{N}_d^n}$ labeled as $y = (y_\alpha)_{\alpha \in \mathbb{N}_d^n}$. Such $y$ is called a *truncated multi-sequence* (tms). It is said to *admit* a Borel measure $\mu$ (see §1.7) if

$$y_\alpha = \int x^\alpha d\mu \quad \text{for all } \alpha \in \mathbb{N}_d^n. \tag{2.7.1}$$

The above is equivalent to $y = \int [x]_d d\mu$, where $[x]_d$ is the monomial vector as in (2.4.2). If it exists, such a $\mu$ is called a *representing measure* for $y$ and the vector $y$ is called a *truncated moment sequence*. The truncated moment problem (TMP) concerns the existence of a representing measure for a given truncated multi-sequence. In applications, there are often restrictions about supports of representing measures. For a measure $\mu$ on $\mathbb{R}^n$, its support is the smallest closed set $T \subseteq \mathbb{R}^n$ such that $\mu(\mathbb{R}^n \backslash T) = 0$. The support of $\mu$ is denoted as $\text{supp}(\mu)$. The measure $\mu$ is said to be supported in a set $K$ if $\text{supp}(\mu) \subseteq K$. For a point $u \in \mathbb{R}^n$, $\delta_u$ denotes the unit *Dirac* measure at $u$, i.e., its support is the singleton of $u$ and the total measure is one. In optimization, a representing measure $\mu$ for $y$ is often required to be supported in a prescribed set $K \subseteq \mathbb{R}^n$. If this is the case, such $\mu$ is called a *$K$-representing measure* for $y$. Borel measures supported in $K$ are called *$K$-measures*. Denote by $meas(y, K)$ the set of all $K$-measures admitted by $y$. This gives the cone

$$\mathscr{R}_d(K) := \{y \in \mathbb{R}^{\mathbb{N}_d^n} : meas(y, K) \neq \emptyset\}. \tag{2.7.2}$$

When $y = 0$ (the tms of all zero entries), the representing measure for $y$ is just the identically zero measure. A measure, whose support is a finite set, is called *finitely atomic*. A measure $\mu$ is called *$r$-atomic* if its support has the cardinality $r$, i.e., $|\text{supp}(\mu)| = r$. Finitely atomic measures are preferable for solving truncated moment problems.

### 2.7.1 ▪ The truncated $K$-moment problem

For a given set $K \subseteq \mathbb{R}^n$, the truncated $K$-moment problem (TKMP) concerns the following questions: How do we determine if a tms $y$ admits a $K$-representing measure or not? When it does, how do we get a representing measure? When $K = \mathbb{R}^n$, the TKMP is reduced to TMP. In this subsection, we introduce some basic results about TKMPs.

Finitely atomic measures are preferable for solving TKMPs. Indeed, if a tms admits a representing measure, then it always admits a finitely atomic representing measure. This was shown by Bayer and Teichmann [15].

**Theorem 2.7.1.** *([15]) If a tms $y \in \mathbb{R}^{\mathbb{N}_d^n}$ admits a representing measure $\mu$, then it also admits a finitely atomic representing measure $\nu$ such that*

$$\text{supp}(\nu) \subseteq \text{supp}(\mu), \quad |\text{supp}(\nu)| \leq \binom{n+d}{d}.$$

An elegant proof for Theorem 2.7.1 can be found in Laurent's survey [171, §5.2]. This result implies that we only need to consider finitely atomic representing measures for solving TKMPs.

There exists work [121, 229] following this approach for solving truncated moment problems. Theorem 2.7.1 implies that the cone $\mathscr{R}_d(K)$ can also be expressed as the conic hull

$$\mathscr{R}_d(K) = \left\{ \sum_{i=1}^{N} \lambda_i[u_i]_d : \lambda_i \geq 0, \, u_i \in K, \, N \in \mathbb{N} \right\}. \tag{2.7.3}$$

The TKMP is closely related to nonnegative polynomials and Riesz functionals. A tms $y \in \mathbb{R}^{\mathbb{N}_d^n}$ determines the *Riesz functional* $\mathscr{L}_y$ acting on $\mathbb{R}[x]_d$ such that

$$\mathscr{L}_y\left( \sum_{\alpha \in \mathbb{N}_d^n} p_\alpha x^\alpha \right) := \sum_{\alpha \in \mathbb{N}_d^n} p_\alpha y_\alpha. \tag{2.7.4}$$

For convenience of notation, we also write that

$$\langle p, y \rangle := \mathscr{L}_y(p). \tag{2.7.5}$$

Recall that $\mathscr{P}_d(K)$ denotes the cone of polynomials in $\mathbb{R}[x]_d$ that are nonnegative on $K$. The Riesz functional $\mathscr{L}_y$ is called $K$-*positive* if

$$p \in \mathscr{P}_d(K) \quad \Rightarrow \quad \mathscr{L}_y(p) \geq 0.$$

When $K = \mathbb{R}^n$, if $\mathscr{L}_y$ is $K$-positive, then we just simply say that $\mathscr{L}_y$ is positive. Clearly, if $\mathscr{L}_y$ is $K$-positive, then $\mathscr{L}_y$ must also be positive. Typically, it is difficult to check whether or not $\mathscr{L}_y$ is positive (or $K$-positive). However, there is an obvious necessary condition for $\mathscr{L}_y$ to be positive. For $f \in \mathbb{R}[x]_k$ with $2k \leq d$, the function $\mathscr{L}_y(f^2)$ is a quadratic form in the coefficient vector $\mathrm{vec}(f)$ of the polynomial $f$ (see (2.4.9) for $\mathrm{vec}(f)$). It is worthy to note that

$$\mathscr{L}_y(f^2) = \mathrm{vec}(f)^T \left( M_k[y] \right) \mathrm{vec}(f),$$

where $M_k[y]$ is the moment matrix of $y$. If $\mathscr{L}_y$ is positive, then the moment matrix $M_k[y]$ must be psd. Therefore, we get the following conclusion.

**Lemma 2.7.2.** *For a tms $y \in \mathbb{R}^{\mathbb{N}_d^n}$ with $d \geq 2k$, if $\mathscr{L}_y$ is positive, then*

$$M_k[y] \succeq 0.$$

If $M_k[y] \succeq 0$, the Riesz functional $\mathscr{L}_y$ may not be positive, shown in the following.

**Example 2.7.3.** *Consider the tms $y \in \mathbb{R}^{\mathbb{N}_6^2}$ such that*

$$M_3[y] = \begin{bmatrix} 280 & 0 & 0 & 11 & 0 & 34 & 0 & 0 & 0 & 0 \\ 0 & 11 & 0 & 0 & 0 & 0 & 11 & 0 & 12 & 0 \\ 0 & 0 & 34 & 0 & 0 & 0 & 0 & 12 & 0 & 16 \\ 11 & 0 & 0 & 11 & 0 & 12 & 0 & 0 & 0 & 0 \\ 0 & 0 & 0 & 0 & 12 & 0 & 0 & 0 & 0 & 0 \\ 34 & 0 & 0 & 12 & 0 & 16 & 0 & 0 & 0 & 0 \\ 0 & 11 & 0 & 0 & 0 & 0 & 280 & 0 & 34 & 0 \\ 0 & 0 & 12 & 0 & 0 & 0 & 0 & 34 & 0 & 16 \\ 0 & 12 & 0 & 0 & 0 & 0 & 34 & 0 & 16 & 0 \\ 0 & 0 & 16 & 0 & 0 & 0 & 0 & 16 & 0 & 12 \end{bmatrix}.$$

*The moment matrix $M_3[y]$ is positive definite; however, the Riesz functional $\mathscr{L}_y$ is not positive. For instance, for the dehomogenized Motzkin polynomial $p = x_1^4 + x_1^2 + x_2^6 - 3x_1^2 x_2^2$, we have $\mathscr{L}_y(p) = -2 < 0$ but $p$ is nonnegative everywhere. So, $\mathscr{L}_y$ is not positive.*

If a tms $y$ admits a $K$-representing measure $\mu$, i.e., $\text{supp}(\mu) \subseteq K$, then

$$\mathscr{L}_y(p) = \int p\mathrm{d}\mu = \int_K p\mathrm{d}\mu \geq 0$$

for every $p \in \mathscr{P}_d(K)$. This means that $\mathscr{L}_y$ being $K$-positive is a necessary condition for $y$ to admit a $K$-representing measure. Indeed, the converse is also true if $K$ is compact. This is known in Tchakaloff's work [312].

**Theorem 2.7.4.** *([312]) If $K \subseteq \mathbb{R}^n$ is compact, then a tms $y \in \mathbb{R}^{\mathbb{N}_d^n}$ admits a $K$-representing measure if and only if the Riesz functional $\mathscr{L}_y$ is $K$-positive.*

We refer to Laurent's survey [171, §5.2] for a proof for Theorem 2.7.4. When $K$ is not compact, the conclusion of Theorem 2.7.4 may not hold. For instance, consider the tms $y \in \mathbb{R}^{\mathbb{N}_4^1}$ ($n = 1$) such that

$$M_2[y] = \begin{bmatrix} 1 & 1 & 1 \\ 1 & 1 & 1 \\ 1 & 1 & 2 \end{bmatrix}.$$

For $K = \mathbb{R}^1$, the Riesz functional $\mathscr{L}_y$ is $K$-positive. This is because if $p \in \mathbb{R}[x]_4$ and $p \geq 0$ on $\mathbb{R}$, then $p = p_1^2 + p_2^2$ for two polynomials $p_1, p_2$ (see Theorem 3.1.1). Hence,

$$\mathscr{L}_y(p) = \mathscr{L}_y(p_1^2) + \mathscr{L}_y(p_2^2) \geq 0.$$

However, this tms $y$ does not admit a representing measure. The proof is left as an exercise.

The $K$-positivity is related to localizing matrices. If $\mathscr{L}_y$ is $K$-positive, we have seen that $M_k[y] \succeq 0$. In fact, the localizing matrices of defining polynomials for $K$ must also be psd. For $y \in \mathbb{R}^{\mathbb{N}_d^n}$ and $f \in \mathbb{R}[x]_{2k}$ with $d \geq 2k$, we have

$$\mathscr{L}_y(gf^2) = \text{vec}(f)^T \left( L_g^{(k)}[y] \right) \text{vec}(f) \tag{2.7.6}$$

for every polynomial $f \in \mathbb{R}[x]$, with the degree such that

$$2\deg(f) + \deg(g) \leq 2k.$$

Assume $K$ is the basic closed semialgebraic set such that

$$K = \{x \in \mathbb{R}^n : g_1(x) \geq 0, \ldots, g_m(x) \geq 0\} \tag{2.7.7}$$

for some $g_1, \ldots, g_m \in \mathbb{R}[x]$. If a tms $y \in \mathbb{R}^{\mathbb{N}_d^n}$ admits a $K$-representing measure $\mu$, then

$$\text{vec}(f)^T \left( L_{g_i}^{(k)}[y] \right) \text{vec}(f) = \mathscr{L}_y(g_i f^2) = \int_K g_i f^2 \mathrm{d}\mu \geq 0 \tag{2.7.8}$$

for all $f \in \mathbb{R}[x]$ with appropriate degrees. This implies that (note $M_k[y] = L_{g_0}^{(k)}[y]$ for $g_0 = 1$)

$$M_k[y] \succeq 0, \quad L_{g_1}^{(k)}[y] \succeq 0, \quad \ldots, \quad L_{g_m}^{(k)}[y] \succeq 0. \tag{2.7.9}$$

Therefore, we have the following lemma.

**Lemma 2.7.5.** *Let $K$ be the set as in (2.7.7). If a tms $y \in \mathbb{R}^{\mathbb{N}^n_d}$ admits a $K$-representing measure and $d \geq 2k$, then $y$ must satisfy (2.7.9).*

The condition (2.7.9) may not be sufficient for $y$ to admit a $K$-representing measure. For instance, consider the tms $y$ in Example 2.7.3 and $K = \{\|x\| \leq 20\}$. The $y$ does not admit a $K$-representing measure, since it does not admit any representing measure. The set $K$ is given by the polynomial inequality $g_1 := 400 - x_1^2 - x_2^2 \geq 0$. One can check that

$$L_{g_1}^{(3)}[y] = \begin{bmatrix} 111955 & 0 & 0 & 4377 & 0 & 13572 \\ 0 & 4377 & 0 & 0 & 0 & 0 \\ 0 & 0 & 13572 & 0 & 0 & 0 \\ 4377 & 0 & 0 & 4086 & 0 & 4750 \\ 0 & 0 & 0 & 0 & 4750 & 0 \\ 13572 & 0 & 0 & 4750 & 0 & 6372 \end{bmatrix} \succeq 0.$$

When $K$ is not compact, for $y \in \mathscr{R}_d(K)$, a stronger condition than $K$-positivity is that $\mathscr{L}_y$ is *strictly $K$-positive*, i.e.,

$$p \in \mathscr{P}_d(K),\, p \not\equiv 0 \text{ on } K \quad \Rightarrow \quad \mathscr{L}_y(p) > 0.$$

Generally, it is computationally challenging to check whether $\mathscr{L}_y$ is $K$-positive (or strictly $K$-positive) or not. The set $K$ is said to be a *determining set* of degree $d$ if $p \in \mathbb{R}[x]_d$ and $p \equiv 0$ on $K$ imply that $p = 0$ (the constant zero polynomial). Clearly, if $K$ has nonempty interior, then it is a determining set of any degree. Indeed, we have the following theorem.

**Theorem 2.7.6.** *(i) [95] Suppose $K$ is a determining set of degree $d$. For $y \in \mathbb{R}^{\mathbb{N}^n_d}$, if $\mathscr{L}_y$ is strictly $K$-positive, then $y$ admits a $K$-representing measure.*
*(ii) [64, 95] For a tms $y \in \mathbb{R}^{\mathbb{N}^n_{2k}}$, if $n = 1$ and $M_k[y] \succ 0$, or if $n = 2$ and $M_2[y] \succ 0$ ($2k = 4$), then $y$ admits a representing measure on $\mathbb{R}^n$.*

### 2.7.2 ▪ Flat extensions

We introduce a concrete condition that ensures a tms to admit a unique representing measure. Let $g := (g_1, \ldots, g_m)$ be a tuple of polynomials and $K$ be the set as in (2.7.7). Denote the degree $t$

$$t := \max_{1 \leq i \leq m} \left\{ 1, \left\lceil \frac{1}{2} \deg(g_i) \right\rceil \right\}. \tag{2.7.10}$$

If, in addition to (2.7.9), a tms $y \in \mathbb{R}^{\mathbb{N}^n_{2k}}$ satisfies the rank condition ($k \geq t$)

$$\operatorname{rank} M_{k-t}[y] = \operatorname{rank} M_k[y], \tag{2.7.11}$$

then $y$ admits a unique $K$-representing measure, i.e., $y \in \mathscr{R}_{2k}(K)$. This result is due to Curto and Fialkow [63]. The condition (2.7.11) is called *flat extension*.

**Theorem 2.7.7.** *([63]) Let $K$ be the set given as in (2.7.7). If a tms $y \in \mathbb{R}^{\mathbb{N}^n_{2k}}$ satisfies (2.7.9) and (2.7.11), then $y$ admits a unique $K$-representing measure, which is $r$-atomic with $r = \operatorname{rank} M_k[y]$.*

An elegant and expository proof for Theorem 2.7.7 can also be found in [170, 171]. When the flat extension (2.7.11) holds, a numerical algorithm is given in [122] for computing the unique $r$-atomic representing measure. In the following, we briefly introduce this method, which also sketches a proof for Theorem 2.7.7. Suppose the support $\text{supp}(\mu) = \{u_1, \ldots, u_r\}$ and the $r$-atomic measure $\mu$ is

$$\mu = \lambda_1 \delta_{u_1} + \cdots + \lambda_r \delta_{u_r},$$

with scalars $\lambda_1, \ldots, \lambda_r > 0$. Since $\mu$ is a representing measure for $y$,

$$M_k[y] = \lambda_1 [u_1]_k [u_1]_k^T + \cdots + \lambda_r [u_r]_k [u_r]_k^T. \qquad (2.7.12)$$

A polynomial $p \equiv 0$ on $\text{supp}(\mu)$ if and only if $M_k[y] \, \text{vec}(p) = 0$, because $M_k[y] \succeq 0$ and

$$\text{vec}(p)^T \big( M_k[y] \big) \text{vec}(p) = \int p(x)^2 \mathrm{d}\mu(x).$$

To get $\text{supp}(\mu)$, we look for polynomials $p$ from the null space (or kernel) of $M_k[y]$, for which we denote $\ker M_k[y]$. The null space $\ker M_k[y]$ is positively dimensional when (2.7.11) holds. Consider the ideal

$$I := \text{Ideal}[\, \{p \in \mathbb{R}[x]_k : \text{vec}(p) \in \ker M_k[y]\}\, ]. \qquad (2.7.13)$$

Then we construct a basis for the quotient space $\mathbb{R}[x]/I$. Since $M_k[y] \succeq 0$, it has the condensed spectral decomposition

$$M_k[y] = V \Lambda V^T$$

with a positive diagonal matrix $\Lambda \in \mathbb{R}^{r \times r}$ and $V \in \mathbb{R}^{N \times r}$ ($N$ is the length of $M_k[y]$). Since $V$ has orthonormal columns and $\ker M_k[y] = \text{Null}(V^T)$, one can get a basis for $\ker M_k[y]$ from the reduced row echelon form of $V^T$ with row pivoting. There exist a permutation matrix $P$, a lower triangular matrix $L$, and an upper triangular matrix $U$ such that $PV^T = LU$ and $U$ appears in the standard echelon form

$$U = \begin{bmatrix} 1 & * & 0 & 0 & * & \cdots & 0 & * & * & * \\ & & 1 & 0 & * & \cdots & 0 & * & * & * \\ & & & 1 & * & \cdots & 0 & * & * & * \\ & & & & & \cdots & 0 & * & * & * \\ & & & & & \ddots & 0 & * & * & * \\ & & & & & & 1 & * & * & * \end{bmatrix}.$$

In the above, each $*$ denotes a possibly nonzero entry. When $y_0 \neq 0$, the first column of $U$ must be a pivot column. The null space $\ker M_k[y]$ is the same as that of $U$. The columns of $M_k[y]$, as well as that of $U$, are labeled by $\alpha \in \mathbb{N}_k^n$. Since $\text{rank } M_k[y] = \text{rank } U = r$, the echelon form $U$ has $r$ pivot columns, say, they are labeled by the monomial powers $\beta^{(1)}, \ldots, \beta^{(r)}$. Denote the power set

$$B_0 := \{\beta^{(1)}, \ldots, \beta^{(r)}\}.$$

All other columns of $U$ are linear combinations of those labeled by $\beta^{(i)}$. All pivot entries are ones and each $U(:, \beta^{(j)})$ is a canonical unit basis vector. When the flat extension (2.7.11) holds, we have $|\beta^{(i)}| \leq k - 1$, i.e., $B_0 \subseteq \mathbb{N}_{k-1}^n$. For each $\alpha \in \mathbb{N}_k^n \backslash B_0$, we have

$$U(:, \alpha) = \sum_{j=1}^{r} U(j, \alpha) U(:, \beta^{(j)}). \qquad (2.7.14)$$

The relation (2.7.14) is represented by the polynomial

$$p_\alpha(x) := x^\alpha - \sum_{j=1}^{r} U(j, \alpha) x^{\beta^{(j)}}.$$

Consider the new power set (recall that $e_i$ is the canonical unit basis vector such that $x_i = x^{e_i}$)

$$B_1 := (e_1 + B_0) \cup \cdots \cup (e_n + B_0) \backslash B_0. \qquad (2.7.15)$$

Each vector in $B_1$ is the power of the monomial $x_i \cdot x^\beta$ with $\beta \in B_0$.

**Lemma 2.7.8.** *Let $y, I, U, B_0, B_1$ be as above. Suppose the rank condition (2.7.11) holds. Then, we have*

$$\mathbb{R}[x]/I = \text{span} \left\{ [\beta^{(1)}], \ldots, [\beta^{(r)}] \right\}$$

*and the points $u_1, \ldots, u_r$ in $\text{supp}(\mu)$ are precisely the solutions to the equations*

$$p_\alpha(x) := x^\alpha - \sum_{j=1}^{r} U(j, \alpha) x^{\beta_j} = 0 \quad (\alpha \in B_1). \qquad (2.7.16)$$

**Proof.** Since the flat extension (2.7.11) holds, each monomial in $B_0$ has degree at most $k - 1$. In the moment matrix $M_k[y]$, every column labeled by $\alpha$ with $x^\alpha \notin B_0$ is a linear combination of those columns labeled by $\beta$ with $x^\beta \in B_0$, which are linearly independent by the choice of $B_0$. Therefore, the equivalent classes of monomials in $B_0$ are linearly independent in the quotient space $\mathbb{R}[x]/I$, and every monomial of degree $k$ or higher is equivalent to a linear combination of monomials in $B_0$, modulo $I$. Hence, the equivalent classes of monomials in $B_0$ give a basis of $\mathbb{R}[x]/I$.

For each polynomial $p_\alpha$ in (2.7.16), its coefficient vector $\text{vec}(p_\alpha)$ belongs to the null space of $U$, which is the same as the null space of $M_k[y]$. So, each point in $\text{supp}(\mu)$ must be a zero of $p_\alpha$. The zero vector, which is the power of the constant 1 monomial, belongs to $B_0$. By the construction of $B_1$, the ideal $J$ generated by the polynomials $p_\alpha$ ($\alpha \in B_1$) is zero dimensional. The dimension of the quotient space $\mathbb{R}[x]/J$ is at most $r$, so the equations (2.7.16) have at most $r$ complex solutions, by Theorem 2.1.6. On the other hand, we have (see Exercise 2.7.5)

$$|\text{supp}(\mu)| \geq \text{rank } M_k[y] = r.$$

Every point in $\text{supp}(\mu)$ is a solution to (2.7.16), so $\text{supp}(\mu)$ precisely consists of all complex solutions to (2.7.16). Moreover, these solutions are all real and belong to the set $K$. □

The polynomial system (2.7.16) can be solved as computing a Schur decomposition. For each $i = 1, \ldots, n$, the equations in (2.7.16) for $\alpha \in e_i + B_0$ can be written as

$$\begin{bmatrix} U(1, e_i + \beta_1) & U(2, e_i + \beta_1) & \cdots & U(r, e_i + \beta_1) \\ U(1, e_i + \beta_2) & U(2, e_i + \beta_2) & \cdots & U(r, e_i + \beta_2) \\ \vdots & \vdots & \ddots & \vdots \\ U(1, e_i + \beta_r) & U(2, e_i + \beta_r) & \cdots & U(r, e_i + \beta_r) \end{bmatrix} \begin{bmatrix} x^{\beta_1} \\ x^{\beta_2} \\ \vdots \\ x^{\beta_r} \end{bmatrix} = x_i \begin{bmatrix} x^{\beta_1} \\ x^{\beta_2} \\ \vdots \\ x^{\beta_r} \end{bmatrix}.$$

Denote the coefficient matrix in the above left hand side as $U_i$, then we get the equation ($[x]_{B_0}$ denotes the vector of monomials in $B_0$ evaluated at $x$)

$$U_i \cdot [x]_{B_0} = x_i \cdot [x]_{B_0}$$

for $i = 1, \ldots, n$. There are $n$ matrices $U_1, \ldots, U_n$. For each solution $x$ of (2.7.16), the $i$th coordinate $x_i$ is an eigenvalue of $U_i$. Moreover, $U_1, \ldots, U_n$ share the same eigenvector $[x]_{B_0}$.

Therefore, to solve (2.7.16), we need to diagonalize $U_1, \ldots, U_n$ simultaneously. As in §2.1.3 (also see [61, 122]), we apply a generic linear combination

$$U(\xi) := \xi_1 U_1 + \cdots + \xi_n U_n$$

for a randomly selected vector $\xi = (\xi_1, \ldots, \xi_n)$. Then we compute the ordered Schur decomposition (the superscript $*$ denotes the conjugate transpose)

$$U(\xi) = QTQ^*,$$

where $Q = \begin{bmatrix} q_1 & \cdots & q_r \end{bmatrix}$ is orthogonal and $T = (T_{ij})_{1 \leq i,j \leq r}$ is upper triangular, with the eigenvalues of $U(\xi)$ sorted monotonically increasingly along the diagonal of $T$. From the above Schur decomposition, the $r$ distinct solutions $u_1, \ldots, u_r$ to the polynomial system (2.7.16) can be given as

$$u_i = (q_i^* U_1 q_i, \; q_i^* U_2 q_i, \; \ldots, \; q_i^* U_n q_i), \quad i = 1, \ldots, r. \qquad (2.7.17)$$

Once $u_1, \ldots, u_r$ are obtained as above, the positive scalars $\lambda_i$ satisfying (2.7.12) can be found by solving the linear equation

$$\lambda_1 [u_1]_{2k} + \cdots + \lambda_r [u_r]_{2k} = y.$$

We would like to remark that the above computed vectors $u_1, \ldots, u_r$ must belong to the set $K$ and all $\lambda_i > 0$. This is because $y$ satisfies the psd conditions in (2.7.9). We refer to [63, 122, 170] for more details about the above. The following is an example.

**Example 2.7.9.** *Consider the tms $y \in \mathbb{R}^{\mathbb{N}_4^2}$ such that*

$$M_2[y] = \begin{bmatrix} 3 & 1 & 1 & 9 & 8 & 9 \\ 1 & 9 & 8 & 1 & -2 & -2 \\ 1 & 8 & 9 & -2 & -2 & 1 \\ 9 & 1 & -2 & 33 & 26 & 24 \\ 8 & -2 & -2 & 26 & 24 & 26 \\ 9 & -2 & 1 & 24 & 26 & 33 \end{bmatrix}.$$

*Note that* rank $M_1[y] =$ rank $M_2[y] = 3$. *Applying a row reduction to $M_2[y]$, we get*

$$U = \begin{bmatrix} 1 & 0 & 0 & \frac{22}{7} & \frac{20}{7} & \frac{22}{7} \\ 0 & 1 & 0 & \frac{9}{7} & -\frac{2}{7} & -\frac{12}{7} \\ 0 & 0 & 1 & -\frac{12}{7} & -\frac{2}{7} & \frac{9}{7} \end{bmatrix}.$$

*The first three are pivoting columns, so*

$$B_0 = \{(0,0), (1,0), (0,1)\}, \quad B_1 = \{(2,0), (1,1), (0,2)\}.$$

*The matrices $U_i$ are*

$$U_1 = \begin{bmatrix} 0 & \frac{22}{7} & \frac{20}{7} \\ 1 & \frac{9}{7} & -\frac{2}{7} \\ 0 & -\frac{12}{7} & -\frac{2}{7} \end{bmatrix}, \quad U_2 = \begin{bmatrix} 0 & \frac{20}{7} & \frac{22}{7} \\ 0 & -\frac{2}{7} & -\frac{12}{7} \\ 1 & -\frac{2}{7} & \frac{9}{7} \end{bmatrix}.$$

*Their eigenvalues are $-2, 2, 1$. We apply a random linear combination of $U_1, U_2$, say, $U = 3U_1 + 5U_2$. Then, compute the Schur decomposition $U = QTQ^*$ and get (only 4 decimal digits are displayed)*

$$Q = \begin{bmatrix} 0.9045 & -0.1231 & 0.4082 \\ -0.3015 & -0.8616 & 0.4082 \\ -0.3015 & 0.4924 & 0.8165 \end{bmatrix}.$$

*Using the formula (2.7.17), we get the points*

$$u_1 = (-2, -2), \quad u_2 = (2, 1), \quad u_3 = (1, 2).$$

*The positive scalars $\lambda_i$ satisfying (2.7.12) are all ones. Therefore, the above tms $y$ admits the 3-atomic measure $\delta_{(-2,-2)} + \delta_{(2,1)} + \delta_{(1,2)}$.*

### 2.7.3 ▪ Exercises

**Exercise 2.7.1.** *For the tms $y \in \mathbb{R}^{\mathbb{N}^1_4}$ such that*

$$M_2[y] = \begin{bmatrix} 1 & 1 & 1 \\ 1 & 1 & 1 \\ 1 & 1 & 2 \end{bmatrix},$$

*show that it does not admit a representing measure.*

**Exercise 2.7.2.** *As an exposition for Theorem 2.7.6, for the tms $y \in \mathbb{R}^{\mathbb{N}^2_4}$ such that*

$$M_2[y] = \begin{bmatrix} 8 & -4 & 1 & 12 & 1 & 13 \\ -4 & 12 & 1 & -16 & 1 & -5 \\ 1 & 1 & 13 & 1 & -5 & 1 \\ 12 & -16 & 1 & 36 & -5 & 11 \\ 1 & 1 & -5 & -5 & 11 & 7 \\ 13 & -5 & 1 & 11 & 7 & 37 \end{bmatrix} \succ 0,$$

*find a finitely atomic representing measure for $y$.*

**Exercise 2.7.3.** *For the tms $y \in \mathbb{R}^{\mathbb{N}^2_6}$ such that*

$$M_3[y] = \begin{bmatrix} 7 & -2 & -2 & 6 & -3 & 6 & -2 & -1 & -1 & -2 \\ -2 & 6 & -3 & -2 & -1 & -1 & 6 & -3 & 5 & -3 \\ -2 & -3 & 6 & -1 & -1 & -2 & -3 & 5 & -3 & 6 \\ 6 & -2 & -1 & 6 & -3 & 5 & -2 & -1 & -1 & -1 \\ -3 & -1 & -1 & -3 & 5 & -3 & -1 & -1 & -1 & -1 \\ 6 & -1 & -2 & 5 & -3 & 6 & -1 & -1 & -1 & -2 \\ -2 & 6 & -3 & -2 & -1 & -1 & 6 & -3 & 5 & -3 \\ -1 & -3 & 5 & -1 & -1 & -1 & -3 & 5 & -3 & 5 \\ -1 & 5 & -3 & -1 & -1 & -1 & 5 & -3 & 5 & -3 \\ -2 & -3 & 6 & -1 & -1 & -2 & -3 & 5 & -3 & 6 \end{bmatrix},$$

*determine if it admits a representing measure or not. If it does, give a finitely atomic representing measure for it. If it does not, explain why.*

**Exercise 2.7.4.** *In Theorem 2.7.7, if $y$ admits a $r$-atomic representing measure $\mu = \lambda_1 \delta_{u_1} + \cdots + \lambda_r \delta_{u_r}$ as in (2.7.12), show that $u_1, \ldots, u_r \in K$.*

**Exercise 2.7.5.** *For a tms $y = (y_\alpha)_{\alpha \in \mathbb{N}^n_{2k}}$, if it admits a representing measure $\mu$, show that*

$$|\operatorname{supp}(\mu)| \geq \operatorname{rank} M_k[y].$$

**Exercise 2.7.6.** *For $k = 1$ and $y = (y_\alpha)_{\alpha \in \mathbb{N}^n_{2k}}$, if $y_0 > 0$ and $M_1[y] \succeq 0$, show that $y$ admits a representing measure on $\mathbb{R}^n$.*

**Exercise 2.7.7.** *([170]) Let $y = (y_\alpha)_{\alpha \in \mathbb{N}^n}$ be an infinite sequence. Suppose $M_k[y] \succeq 0$ for all $k$ and $r = \operatorname{rank} M_k[y]$ for all $k$ sufficiently large. Let $I$ be the set of all polynomials $p$ such that $\operatorname{vec}(p) \in \ker M_k[y]$ for some $k$. Show the following:*

*(i) The set $I$ is a zero dimensional ideal of $\mathbb{R}[x]$ and is radical.*

*(ii) The cardinality of $V_{\mathbb{C}}(I)$ is $r$ and $V_{\mathbb{C}}(I) \subseteq \mathbb{R}^n$. Moreover, if the condition (2.7.9) holds, show that $V_{\mathbb{C}}(I) \subseteq K$.*

*(iii) There exist positive scalars $\lambda_1, \ldots, \lambda_r$ satisfying (2.7.12) for all $k$.*

*The above conclusions can be used to prove Theorem 2.7.7.*

# Chapter 3

# Univariate Nonnegative Polynomials and Moments

> *This chapter focuses on univariate polynomials. They are nonnegative if and only if they are sum of squares. There exist semidefinite characterizations for univariate polynomials that are nonnegative over intervals. Moreover, univariate truncated moment sequences also have semidefinite characterizations. We introduce basic results about univariate polynomial optimization as well as univariate truncated moment problems. Polynomials in this chapter are univariate, unless specified in the context.*

## 3.1 ▪ Univariate Nonnegative Polynomials

For univariate polynomials, an important property is that they are nonnegative everywhere if and only if they are SOS. Indeed, it is well known that every nonnegative univariate polynomial is a sum of two squares. The following is a classically well-known result.

**Theorem 3.1.1.** *A univariate polynomial $f$ in $x \in \mathbb{R}^1$ is nonnegative on $\mathbb{R}$ if and only if $f$ is a sum of two squares, i.e., $f = p_1^2 + p_2^2$ for some $p_1, p_2 \in \mathbb{R}[x]$.*

The above theorem can be shown by observing patterns for factorizations of nonnegative univariate polynomials. Its proof is left as an exercise. The following polynomials are nonnegative. Nonnegativity can be certified by SOS decompositions.

- The polynomial $3 + 6x + 4x^2 - 2x^3 + x^4$ is nonnegative and has the SOS decomposition
$$3(1+x)^2 + (x^2 - x)^2.$$

- The polynomial $1 - 2x + 2x^4$ is nonnegative and has the SOS decomposition
$$(1 - x - x^2)^2 + (x - x^2)^2.$$

- The polynomial $5 - 2x - 2x^3 + x^4$ is nonnegative and has the SOS decomposition
$$(x^2 - x - 1)^2 + (2 - x)^2.$$

For each $f \in \mathbb{R}[x]_{2d}$, we can write it in the standard form
$$f = f_0 + f_1 x + \cdots + f_{2d} x^{2d}.$$

Recall the notation

$$[x]_d := \begin{bmatrix} 1 & x & \cdots & x^d \end{bmatrix}^T.$$

Then $f = [x]_d^T X [x]_d$ for a real symmetric matrix $X \in \mathcal{S}^{d+1}$. For instance, $X$ can be chosen to be the tridiagonal matrix

$$\begin{bmatrix} f_0 & \frac{f_1}{2} & & & \\ \frac{f_1}{2} & f_2 & \frac{f_3}{2} & & \\ & \ddots & \ddots & \ddots & \\ & & \frac{f_{2d-3}}{2} & f_{2d-2} & \frac{f_{2d-1}}{2} \\ & & & \frac{f_{2d-1}}{2} & f_{2d} \end{bmatrix}.$$

For convenience, we label the entries $X$ as $X_{ij}$, with the range $i, j \in \{0, 1, \ldots, d\}$. A comparison of coefficients gives the equations

$$f_k = \sum_{i+j=k} X_{ij}, \quad k = 0, 1, \ldots, 2d. \tag{3.1.1}$$

The matrix $X$ satisfying the above is called a *Gram matrix* for $f$. When $d > 1$, the matrix $X$ is not uniquely determined by the coefficients of $f$. For instance, for arbitrary value of $t$,

$$X = \begin{bmatrix} 1 & 2 & t \\ 2 & 6-2t & 2 \\ t & 2 & 1 \end{bmatrix}$$

is a Gram matrix for $1 + 4x + 6x^2 + 4x^3 + x^4$. For $t = 0$, $X$ is not psd. For $t = 1$, $X$ is psd and

$$X = \begin{bmatrix} 1 & 2 & 1 \\ 2 & 4 & 2 \\ 1 & 2 & 1 \end{bmatrix} = \begin{bmatrix} 1 \\ 2 \\ 1 \end{bmatrix} \begin{bmatrix} 1 \\ 2 \\ 1 \end{bmatrix}^T,$$

which gives the SOS decomposition

$$1 + 4x + 6x^2 + 4x^3 + x^4 = (1 + 2x + x^2)^2.$$

The following is a special case of Lemma 2.4.2.

**Lemma 3.1.2.** *A polynomial $f \in \mathbb{R}[x]_{2d}$ is SOS if and only if there exists a psd matrix $X$ satisfying (3.1.1).*

For instance, the polynomial $f = 1 + x + x^2 + x^3 + x^4 + x^5 + x^6$ has the Gram matrix which can be parameterized as

$$X = \begin{bmatrix} 1 & \frac{1}{2} & t_1 & t_2 \\ \frac{1}{2} & 1-2t_1 & \frac{1}{2}-t_2 & t_3 \\ t_1 & \frac{1}{2}-t_2 & 1-2t_3 & \frac{1}{2} \\ t_2 & t_3 & \frac{1}{2} & 1 \end{bmatrix}.$$

For the values $t_1 = t_2 = t_3 = 0$, the Gram matrix $X$ is positive definite and

$$X = \begin{bmatrix} 1 & \frac{1}{2} & & \\ \frac{1}{2} & 1 & \frac{1}{2} & \\ & \frac{1}{2} & 1 & \frac{1}{2} \\ & & \frac{1}{2} & 1 \end{bmatrix} = \begin{bmatrix} 1 & & & \\ \frac{1}{2} & \frac{\sqrt{3}}{2} & & \\ & \frac{1}{\sqrt{3}} & \frac{\sqrt{2}}{\sqrt{3}} & \\ & & \frac{\sqrt{3}}{\sqrt{8}} & \frac{\sqrt{5}}{\sqrt{8}} \end{bmatrix} \begin{bmatrix} 1 & \frac{1}{2} & & \\ & \frac{\sqrt{3}}{2} & \frac{1}{\sqrt{3}} & \\ & & \frac{\sqrt{2}}{\sqrt{3}} & \frac{\sqrt{3}}{\sqrt{8}} \\ & & & \frac{\sqrt{5}}{\sqrt{8}} \end{bmatrix}.$$

This gives the SOS decomposition

$$f = (1 + \frac{1}{2}x)^2 + (\frac{\sqrt{3}}{2}x + \frac{1}{\sqrt{3}}x^2)^2 + (\frac{\sqrt{2}}{\sqrt{3}}x^2 + \frac{\sqrt{3}}{\sqrt{8}}x^3)^2 + (\frac{\sqrt{5}}{\sqrt{8}}x^3)^2.$$

### 3.1.1 ▪ The dual cone

Recall that $\Sigma[x]_{2d}$ denotes the cone of SOS polynomials with degrees at most $2d$. For the univariate case, we have seen that $\Sigma_{1,2d} = \mathscr{P}_{1,2d}$, the cone of nonnegative univariate polynomials of degrees at most $2d$. Moreover, $\Sigma[x]_{2d}$ is a proper cone (i.e., it is closed, convex, solid, and pointed; see Definition 1.2.9). It is full-dimensional in the embedding space $\mathbb{R}^{2d+1}$. The following proposition characterizes the dual cone. Recall the moment cone

$$\mathscr{S}_{1,2d} := \{y \in \mathbb{R}^{2d+1} : M_d[y] \succeq 0\}. \tag{3.1.2}$$

For $y := (y_0, \ldots, y_{2d})$, the moment matrix $M_d[y]$ is a Hankel matrix:

$$M_d[y] = \begin{bmatrix} y_0 & y_1 & y_2 & y_3 & \cdots & y_d \\ y_1 & y_2 & y_3 & y_4 & \cdots & y_{d+1} \\ y_2 & y_3 & y_4 & y_5 & \cdots & y_{d+2} \\ y_3 & y_4 & y_5 & y_6 & \cdots & y_{d+3} \\ \vdots & \vdots & \vdots & \vdots & \ddots & \vdots \\ y_d & y_{d+1} & y_{d+2} & y_{d+3} & \cdots & y_{2d} \end{bmatrix}.$$

Recall that the superscript $\star$ denotes the dual cone (see (1.2.13) for the notation).

**Proposition 3.1.3.** *For the univariate case, it holds that*

$$\left(\mathscr{P}_{1,2d}\right)^\star = \left(\Sigma_{1,2d}\right)^\star = \mathscr{S}_{1,2d}. \tag{3.1.3}$$

***Proof.*** A tms $y \in \mathbb{R}^{2d+1}$ belongs to the dual cone $\left(\Sigma_{1,2d}\right)^\star$ if and only if $\langle p^2, y \rangle \geq 0$ for all $p \in \mathbb{R}[x]_d$. Since

$$\langle p^2, y \rangle = \text{vec}(p)^T \left( M_d[y] \right) \text{vec}(p),$$

one can see that $y \in (\Sigma_{1,2d})^\star$ if and only if $M_d[y] \succeq 0$. The conclusion holds since $\mathscr{P}_{1,2d} = \Sigma_{1,2d}$. □

Recall that the dual cone $\left(\mathscr{P}_{1,2d}\right)^\star$ is the closure of the conic hull of the set of vectors $[t]_{2d}$ with $t \in \mathbb{R}$ (see Theorem 2.4.7). Therefore, $M_d[y] \succeq 0$ if and only if $y$ belong to the closure of the set

$$\mathscr{R}_{1,2d} := \Big\{ \sum_{k=1}^N \lambda_k [t_k]_{2d} \Big| \lambda_k \geq 0, \, t_k \in \mathbb{R}^1, \, N \in \mathbb{N} \Big\}. \tag{3.1.4}$$

In particular, if $M_d[y]$ is positive definite, then $y \in \mathscr{R}_{1,2d}$, i.e., $y$ is a positive linear combination of vectors like $[t]_{2d}$. We leave the proof as an exercise. It is worthy to remark that the cone $\mathscr{R}_{1,2d}$ is not closed.

## 3.1.2 ▪ Exercises

**Exercise 3.1.1.** *Prove Theorem 3.1.1.*

**Exercise 3.1.2.** *For a degree $d > 0$, write the polynomial $1 + x + x^2 + x^3 + \cdots + x^{2d}$ explicitly as a sum of two squares.*

**Exercise 3.1.3.** *For a degree $d > 1$, find the range of values for t such that $1 + x + tx^{2d}$ is SOS.*

**Exercise 3.1.4.** *Determine the range of pairs $(a, b)$ such that the polynomial $1 + ax + bx^3 + x^4$ is SOS. Give defining equations for the boundary of such pairs $(a, b)$.*

**Exercise 3.1.5.** *Determine its dual cone $P^\star$ of*

$$P := \left\{ (f_0, f_1, f_2, f_3) : f_0 + f_1 x^3 + f_2 x^5 + f_3 x^6 \geq 0 \ \forall x \in \mathbb{R} \right\}.$$

*Express both $P$ and $P^\star$ in terms of linear matrix inequalities.*

**Exercise 3.1.6.** *Determine the closure of the convex hull of the set*

$$T := \left\{ (x_0, x_1, x_2, x_3, x_4) \ \middle| \ \begin{matrix} x_0 = 1, \ x_i x_j = x_k x_l \\ \text{for all} \quad i + j = k + l \end{matrix} \right\}.$$

*Express* conv $(T)$ *in terms of linear matrix inequalities.*

**Exercise 3.1.7.** *For $y \in \mathbb{R}^{2d+1}$, if the moment matrix $M_d[y] \succ 0$, show that $y \in \mathscr{R}_{1,2d}$, where $\mathscr{R}_{1,2d}$ is the set as in (3.1.4). Show that $\mathscr{R}_{1,2d}$ is not closed.*

## 3.2 ▪ Polynomials Nonnegative over Intervals

We characterize univariate polynomials that are nonnegative over an interval $I \subseteq \mathbb{R}^1$. Recall that $\mathscr{P}(I)$ stands for the cone of polynomials that are nonnegative in $I$. The intersection $\mathscr{P}(I) \cap \mathbb{R}[x]_d$ is denoted as $\mathscr{P}_d(I)$. The main result of this section is that there are semidefinite representations for the cone $\mathscr{P}_d(I)$. Its dual cone can also be given by semidefinite representations.

### 3.2.1 ▪ The case $I = [a, \infty)$

We characterize polynomials that are nonnegative over the interval $I = [a, \infty)$. Clearly, if $f = s_0 + (x - a)s_1$ for SOS polynomials $s_0, s_1$, then $f \geq 0$ on $I$. Interestingly, the converse is also true. Indeed, the polynomials $s_0, s_1$ can be selected to be perfect squares. This was shown by Lukács [184], Markov [187], and Pólya and Szegö (see [263, 267]).

**Theorem 3.2.1.** *([184, 187, 263, 267]) For a degree $d$, let $d_1 := \lfloor \frac{d}{2} \rfloor$ and $d_2 := \lfloor \frac{d-1}{2} \rfloor$. Then, a polynomial $f \in \mathbb{R}[x]_d$ is nonnegative on $I = [a, \infty)$ if and only if there exist $p \in \mathbb{R}[x]_{d_1}$ and $q \in \mathbb{R}[x]_{d_2}$ such that*

$$f = p^2 + (x - a)q^2.$$

**Proof.** Up to a coordinate shifting, we only need to consider the case $a = 0$ and $I = [0, \infty)$. The proof for the "$\Leftarrow$" direction is obvious. For the "$\Rightarrow$" direction, suppose $f \geq 0$ on $I$. Clearly, the leading coefficient of $f$ is positive, unless $f = 0$. We can factorize $f$ as the product of linear factors. We prove the conclusion by induction on the degree of $f$.

When $\deg(f) = 1$ and $f \neq 0$, we can write $f = c(x - u)$. Since $f \geq 0$ on $\mathbb{R}_+$, we must have $c > 0$ and $u \leq 0$. Then $f = (\sqrt{-uc})^2 + x(\sqrt{c})^2$.

Next, consider the case that $\deg(f) = d > 1$ and $f \neq 0$. Suppose the conclusion is true for polynomials of degrees $\leq d - 1$.

- If $f$ has a real root $u > 0$, then the multiplicity of $u$ must be even, because $f \geq 0$ on $\mathbb{R}_+$. So $f = (x - u)^{2s} g(x)$ with $g \in \mathbb{R}[x]$. Note that $g \geq 0$ on $\mathbb{R}_+$ and $\deg(g) < d$. By induction, $g = g_1^2 + x g_2^2$ with $g_1, g_2$ of proper degrees. Then

$$f = ((x - u)^s g_1)^2 + x((x - u)^s g_2)^2.$$

- If $f$ has a real root $u \leq 0$, then $f = (x - u)g(x)$ for a polynomial $g \in \mathbb{R}[x]_{d-1}$. Note that $g \geq 0$ on $\mathbb{R}_+$. By induction, $g = g_1^2 + x g_2^2$ with $g_1, g_2$ of proper degrees. Then we can see that

$$f = (\sqrt{-u} g_1 - x g_2)^2 + x(g_1 + \sqrt{-u} g_2)^2.$$

- If $f$ has no real roots, then it must have a conjugate pair of complex roots, say, $\alpha \pm \beta \sqrt{-1}$, for $\alpha, \beta \in \mathbb{R}$. Then $f = ((x - \alpha)^2 + \beta^2) g(x)$ for a polynomial $g \in \mathbb{R}[x]_{d-2}$. Note that $g \geq 0$ on $\mathbb{R}_+$. By induction, $g = g_1^2 + x g_2^2$ with $g_1, g_2$ of proper degrees. Since

$$(x - \alpha)^2 + \beta^2 = (x - \sqrt{\alpha^2 + \beta^2})^2 + x(\sqrt{2\sqrt{\alpha^2 + \beta^2} - 2\alpha})^2,$$
$$(p_1^2 + x p_2^2)(g_1^2 + x g_2^2) = (p_1 g_1 - x p_2 g_2)^2 + x(p_1 g_2 + p_2 g_1)^2,$$
$$f = ((x - \alpha)^2 + \beta^2)(g_1^2 + x g_2^2),$$

we can get that

$$f = \left( g_1(x - \sqrt{\alpha^2 + \beta^2}) - x g_2 \sqrt{2\sqrt{\alpha^2 + \beta^2} - 2\alpha} \right)^2$$
$$+ x\left( g_2(x - \sqrt{\alpha^2 + \beta^2}) + g_1 \sqrt{2\sqrt{\alpha^2 + \beta^2} - 2\alpha} \right)^2.$$

By induction, the conclusion holds for $f$ of all degrees.    □

In the following, we characterize the cone $\mathscr{P}_d([a, \infty))$ and its dual cone by semidefinite representations. For a tms $y = (y_0, y_1, \ldots, y_d) \in \mathbb{R}^{d+1}$ with $d = 2m$ or $d = 2m + 1$, the symbol $M_m[y]$ stands for the $m$th order moment matrix:

$$M_m[y] := \begin{bmatrix} y_0 & y_1 & \cdots & y_m \\ y_1 & y_2 & \cdots & y_{m+1} \\ \vdots & \vdots & \ddots & \vdots \\ y_m & y_{m+1} & \cdots & y_{2m} \end{bmatrix}. \tag{3.2.1}$$

If $d = 2m + 1$ is odd, the highest order localizing matrix for $x \geq 0$ is

$$N_m[y] := \begin{bmatrix} y_1 & y_2 & \cdots & y_{m+1} \\ y_2 & y_3 & \cdots & y_{m+2} \\ \vdots & \vdots & \ddots & \vdots \\ y_{m+1} & y_{m+2} & \cdots & y_{2m+1} \end{bmatrix}. \tag{3.2.2}$$

If $d = 2m$ is even, then the largest one is

$$
N_{m-1}[y] := \begin{bmatrix}
y_1 & y_2 & \cdots & y_m \\
y_2 & y_3 & \cdots & y_{m+1} \\
\vdots & \vdots & \ddots & \vdots \\
y_m & y_{m+1} & \cdots & y_{2m-1}
\end{bmatrix}. \tag{3.2.3}
$$

The following characterizes the cone $\mathscr{P}_d([0, \infty))$ and its dual.

**Proposition 3.2.2.** *Let $\mathscr{S}_d([a, \infty))$ be the dual cone of $\mathscr{P}_d([a, \infty))$, i.e.,*

$$
\mathscr{S}_d([a, \infty)) := \left\{ y \in \mathbb{R}^{d+1} : \langle f, y \rangle \geq 0 \ \forall f \in \mathscr{P}_d([a, \infty)) \right\}.
$$

*(i) If the degree $d = 2m + 1$ is odd, then*

$$
\begin{aligned}
\mathscr{P}_d([a, \infty)) &= \Sigma[x]_{2m} + (x - a) \cdot \Sigma[x]_{2m}, \\
\mathscr{S}_d([a, \infty)) &= \left\{ y \in \mathbb{R}^{d+1} : M_m[y] \succeq 0, N_m[y] \succeq a M_m[y] \right\}.
\end{aligned} \tag{3.2.4}
$$

*(ii) If the degree $d = 2m$ is even, then*

$$
\begin{aligned}
\mathscr{P}_d([a, \infty)) &= \Sigma[x]_{2m} + (x - a) \cdot \Sigma[x]_{2m-2}, \\
\mathscr{S}_d([a, \infty)) &= \left\{ y \in \mathbb{R}^{d+1} : M_m[y] \succeq 0, N_{m-1}[y] \succeq a M_{m-1}[y] \right\}.
\end{aligned} \tag{3.2.5}
$$

Proposition 3.2.2 follows readily from Theorem 3.2.1. We leave the proof as an exercise.

## 3.2.2 ▪ The case $I = [a, b]$

Suppose $a < b$ are two real numbers. We characterize polynomials that are nonnegative on the interval $[a, b]$. The interval can be defined by two inequalities: $x - a \geq 0$ and $b - x \geq 0$. It can also be defined by the single inequality $(x - a)(b - x) \geq 0$. By Schmüdgen's Positivstellensatz, one can see that

$$
\Sigma[x] + (x - a) \cdot \Sigma[x] + (b - x) \cdot \Sigma[x] + (x - a)(b - x) \cdot \Sigma[x] \subseteq \mathscr{P}([a, b]).
$$

Interestingly, the above containment is actually an equality. A similar result holds like for the case $I = [0, \infty)$, which was due to Markov, Lukács, Pólya, and Szegö (see [184, 187, 263, 267]).

**Theorem 3.2.3.** *([184, 187, 263, 267]) Let $d = 2m$ or $2m + 1$ and $a < b$. For each $f \in \mathscr{P}_d([a, b])$, we have the following:*

*(i) If $d = 2m + 1$ is odd, then $f = (x - a)p^2 + (b - x)q^2$ for some $p, q \in \mathbb{R}[x]_m$.*

*(ii) If the degree $d = 2m$ is even, then $f = p^2 + (x - a)(b - x)q^2$ for some $p \in \mathbb{R}[x]_m$ and $q \in \mathbb{R}[x]_{m-1}$.*

The above theorem can be similarly proved as for Theorem 3.2.1. Its proof is left as an exercise. The dual cone of $\mathscr{P}_d([a, b])$ is

$$
\mathscr{S}_d([a, b]) := \{ y \in \mathbb{R}^{d+1} : \langle f, y \rangle \geq 0 \ \forall f \in \mathscr{P}_d([a, b]) \}.
$$

It can be similarly characterized as in Proposition 3.2.4. Recall notation $M_m[y], N_m[y], N_{m-1}[y]$ as in (3.2.1)-(3.2.3).

**Proposition 3.2.4.** *For $a < b$, the nonnegative polynomial cone $\mathscr{P}_d([a,b])$ and its dual cone $\mathscr{S}_d([a,b])$ can be characterized as follows:*

*(i) If the degree $d = 2m + 1$ is odd, then*

$$
\boxed{
\begin{aligned}
\mathscr{P}_d([a,b]) &= (x-a) \cdot \Sigma[x]_{2m} + (b-x) \cdot \Sigma[x]_{2m}, \\
\mathscr{S}_d([a,b]) &= \left\{ y \in \mathbb{R}^{d+1} : bM_m[y] \succeq N_m[y] \succeq aM_m[y] \right\}.
\end{aligned}
}
\tag{3.2.6}
$$

*(ii) If the degree $d = 2m$ is even, then*

$$
\boxed{
\begin{aligned}
\mathscr{P}_d([a,b]) &= \Sigma[x]_{2m} + (x-a)(b-x) \cdot \Sigma[x]_{2m-2}, \\
\mathscr{S}_d([a,b]) &= \big\{ y \in \mathbb{R}^{d+1} : M_m[y] \succeq 0, \\
&\qquad (a+b)N_{m-1}[y] \succeq abM_{m-1}[y] + M_{m-1}[\hat{y}] \big\},
\end{aligned}
}
\tag{3.2.7}
$$

*where $\hat{y} := (y_2, \ldots, y_{2m})$.*

**Proof.** The conclusions in Theorem 3.2.3 imply the descriptions for the nonnegative polynomial cone $\mathscr{P}_d([a,b])$. We only need to prove the characterization for the dual cone.

(i) For the case $d = 2m + 1$, by the definition, $y \in \mathscr{S}_d([a,b])$ if and only if $\langle f, y \rangle \geq 0$ for all $f \in \mathscr{P}_d([a,b])$. Such $f$ can be expressed as $(x-a)p^2 + (b-x)q^2$ for $p, q \in \mathbb{R}[x]_m$, so we can get the expansion

$$
\begin{aligned}
\langle (x-a)p^2 + (b-x)q^2, y \rangle &= \operatorname{vec}(p)^T \Big( N_m[y] - aM_m[y] \Big) \operatorname{vec}(p) \\
&\quad + \operatorname{vec}(q)^T \Big( bM_m[y] - N_m[y] \Big) \operatorname{vec}(q).
\end{aligned}
$$

The above is nonnegative for all $p, q$ if and only if

$$
N_m[y] - aM_m[y] \succeq 0, \quad bM_m[y] - N_m[y] \succeq 0.
$$

(ii) For the case $d = 2m$, each $f \in \mathscr{S}_d([a,b])$ can be expressed as $p^2 + (x-a)(b-x)q^2$ for $p \in \mathbb{R}[x]_m$ and $q \in \mathbb{R}[x]_{m-1}$. Note that

$$
\begin{aligned}
\langle p^2 + (x-a)(b-x)q^2, y \rangle &= \operatorname{vec}(p)^T (M_m[y]) \operatorname{vec}(p) \\
+ \operatorname{vec}(q)^T \Big( (a+b)N_{m-1}[y] - abM_{m-1}[y] &- M_{m-1}[\hat{y}] \Big) \operatorname{vec}(q),
\end{aligned}
$$

where $\hat{y} = (y_2, \ldots, y_{2m})$. The above is nonnegative for all $p, q$ if and only if

$$
M_m[y] \succeq 0, \quad (a+b)N_{m-1}[y] - abM_{m-1}[y] - M_{m-1}[\hat{y}] \succeq 0.
$$

So the conclusion holds. $\qquad\square$

Note that $M_{m-1}[(y_2, \ldots, y_{2m})]$ is the moment matrix for the truncated tms $(y_2, \ldots, y_{2m})$, which is

$$
\begin{bmatrix}
y_2 & y_3 & \cdots & y_m \\
y_3 & y_4 & \cdots & y_{m+1} \\
\vdots & \vdots & \ddots & \vdots \\
y_m & y_{m+1} & \cdots & y_{2m}
\end{bmatrix}.
$$

We would like to remark that for an compact interval $[a, b]$, the dual cone $\mathscr{S}_d([a, b])$ coincides with $\mathscr{R}_d([a, b])$, the set of all tms in $\mathbb{R}^{d+1}$ that admit representing measures. This is implied by Theorem 2.7.4.

### 3.2.3 ▪ Exercises

**Exercise 3.2.1.** *Determine polynomials $p, q$ for the following:*

(i) $1 + x + x^3 + \cdots + x^9 = p^2 + xq^2$.

(ii) $x(1 - x^8) = xp^2 + (1 - x)q^2$.

(iii) $x(1 - x^9) = p^2 + x(1 - x)q^2$.

**Exercise 3.2.2.** *Prove Proposition 3.2.2 and Theorem 3.2.3.*

**Exercise 3.2.3.** *Like in Proposition 3.2.2, give a similar representation for the cone $\mathscr{P}_d((-\infty, a])$ and its dual, using psd conditions of matrices.*

**Exercise 3.2.4.** *For the union $I = [-2, -1] \cup [0, \infty)$, give characterizations for the cone $\mathscr{P}_d(I)$ and its dual like Propositions 3.2.2 and 3.2.4.*

## 3.3 ▪ The Univariate TMP

A vector $y := (y_0, y_1, \ldots, y_d) \in \mathbb{R}^{d+1}$ is called a *univariate* truncated multi-sequence (tms) of degree $d$. The univariate truncated moment problem (UTMP) concerns whether or not there exists a Borel measure $\mu$, supported on $\mathbb{R}$, such that

$$y_i = \int x^i d\mu \quad (i = 0, 1, \ldots, d).$$

If it exists, the above $\mu$ is called a *representing measure* for $y$, and $y$ is said to admit the measure $\mu$. The support of $\mu$ is denoted as $\mathrm{supp}(\mu)$. The measure $\mu$ is *finitely atomic* if the cardinality $|\mathrm{supp}(\mu)| < \infty$, and it is *$r$-atomic* if $|\mathrm{supp}(\mu)| = r$. If the above measure $\mu$ is $r$-atomic, then there exist $\lambda_1, \ldots, \lambda_r > 0$ and $t_1, \ldots, t_r \in \mathbb{R}$ such that

$$y = \lambda_1[t_1]_d + \cdots + \lambda_r[t_r]_d. \tag{3.3.1}$$

For an interval $I \subseteq (-\infty, +\infty)$, denote by $\mathscr{R}_d(I)$ the cone of all univariate tms $y \in \mathbb{R}^{d+1}$ such that it admits a measure supported in $I$. By Theorem 2.7.1, the cone $\mathscr{R}_d(I)$ can be expressed as

$$\mathscr{R}_d(I) = \left\{ \sum_{i=1}^{N} \lambda_i[t_i]_d : \lambda_i \geq 0, \, t_i \in I, \, N \in \mathbb{N} \right\}.$$

Suppose $d = 2k$ (the even case) or $d = 2k + 1$ (the odd case). The $k$th order moment matrix of $y = (y_0, y_1, \ldots, y_d)$ is

$$M_k[y] := \begin{bmatrix} y_0 & y_1 & \cdots & y_k \\ y_1 & y_2 & \cdots & y_{k+1} \\ \vdots & \vdots & \ddots & \vdots \\ y_k & y_{k+1} & \cdots & y_{2k} \end{bmatrix}. \tag{3.3.2}$$

If $y$ admits a representing measure on $\mathbb{R}$, then $M_k[y] \succeq 0$, while the converse is not necessarily true. For two integers $s \leq t \leq d$, denote the truncation

$$y|_{s:t} := (y_s, y_{s+1}, \ldots, y_t).$$

For a matrix $A$, the notation Range $(A)$ stands for the range space of $A$. The univariate truncated moment problem was well studied by Curto and Fialkow [64]. In the following, we introduce some basic results on UTMPs. The rank of a tms is defined as follows.

**Definition 3.3.1.** *([64]) Let $d = 2k$ or $d = 2k + 1$. For a tms $y \in \mathbb{R}^{d+1}$, if rank $M_k[y] = k + 1$, the rank of $y$ is $k + 1$; if rank $M_k[y] < k + 1$, the rank of $y$ is defined as the smallest $r$ such that*

$$\begin{bmatrix} y_r \\ y_{r+1} \\ \vdots \\ y_{r+k} \end{bmatrix} \in \text{Range} \begin{bmatrix} y_0 & y_1 & \cdots & y_{r-1} \\ y_1 & y_2 & \cdots & y_r \\ \vdots & \vdots & \ddots & \vdots \\ y_k & y_{k+1} & \cdots & y_{r+k-1} \end{bmatrix}.$$

*The rank of $y$ is denoted as* rank$(y)$.

We remark that if $M_k[y] \succeq 0$, then rank$(y) = $ rank $M_k[y]$. The proof is left as an exercise.

### 3.3.1 ▪ The UTMP on $\mathbb{R}^1$

The set of all tms $y \in \mathbb{R}^{d+1}$ that admits a representing measure on $\mathbb{R}$ is characterized in the following.

**Theorem 3.3.2.** *([64, §3]) Let $y := (y_0, \ldots, y_d) \in \mathbb{R}^{d+1}$ with $y_0 > 0$ and $r := $ rank$(y)$.*

*(i) If $d = 2k + 1$ is odd, then $y$ admits a representing measure on $\mathbb{R}$ if and only if $M_k[y] \succeq 0$ and the linear system*

$$\begin{bmatrix} y_0 & y_1 & \cdots & y_k \\ y_1 & y_2 & \cdots & y_{k+1} \\ \vdots & \vdots & \ddots & \vdots \\ y_k & y_{k+1} & \cdots & y_{2k} \end{bmatrix} \begin{bmatrix} p_0 \\ p_1 \\ \vdots \\ p_k \end{bmatrix} = \begin{bmatrix} y_{k+1} \\ y_{k+2} \\ \vdots \\ y_{2k+1} \end{bmatrix} \qquad (3.3.3)$$

*has a solution.*

*(ii) If $d = 2k$ is even, then $y$ admits a representing measure on $\mathbb{R}$ if and only if $M_k[y] \succeq 0$ and the linear system*

$$\begin{bmatrix} y_0 & y_1 & \cdots & y_{r-1} \\ y_1 & y_2 & \cdots & y_r \\ \vdots & \vdots & \ddots & \vdots \\ y_{2k-r} & y_{2k-r+1} & \cdots & y_{2k-1} \end{bmatrix} \begin{bmatrix} p_0 \\ p_1 \\ \vdots \\ p_{r-1} \end{bmatrix} = \begin{bmatrix} y_r \\ y_{r+1} \\ \vdots \\ y_{2k} \end{bmatrix} \qquad (3.3.4)$$

*has a solution.*

*(iii) If $y$ admits a representing measure, then it always admits a $r$-atomic measure. Moreover, if it exists, the representing measure is unique if and only if $M_k[y]$ is singular.*

***Proof.*** (i) For this case, the degree $d = 2k + 1$ is odd.

"$\Rightarrow$": Suppose $\mu$ is a representing measure for $y$, then $y_i = \int x^i \mathrm{d}\mu$ for $i = 0, 1, \ldots, d$. For every $f = f_0 + f_1 x + \cdots + f_k x^k \in \mathbb{R}[x]_k$, we have that

$$\mathrm{vec}(f)^T M_k[y] \,\mathrm{vec}(f) = \sum_{i,j=0}^{k} y_{i+j} f_i f_j = \int f(x)^2 \mathrm{d}\mu \geq 0.$$

This implies that $M_k[y] \succeq 0$. If $M_k[y]$ is positive definite (i.e., $M_k[y] \succ 0$), then (3.3.3) clearly has a solution. If $M_k[y]$ is singular (i.e., rank $M_k[y] \leq k$), there exists a nonzero vector $q \in \ker M_k[y]$. One can select such $q$ whose last nonzero entry is $-1$, so there exists a vector

$$(q_0, \ldots, q_s, -1, 0, \ldots, 0) \in \ker M_k[y] \quad \text{for some } s \leq k - 1.$$

Then consider the polynomial

$$q(x) := q_0 + q_1 x + \cdots + q_s x^s - x^{s+1} \in \mathbb{R}[x]_{s+1}.$$

Note that $s + 1 \leq k$ and

$$\mathrm{vec}(q)^T M_k[y] \,\mathrm{vec}(q) = \int q(x)^2 \mathrm{d}\mu = 0.$$

This means that $\mathrm{supp}(\mu) \subseteq \mathcal{Z}(q)$, which denotes the zero set of $q$. Let $p = q(x) x^{k-s}$, then

$$\int p(x) x^j \mathrm{d}\mu = 0 \, (j = 0, 1, \ldots, k).$$

Because the leading coefficient of $p$ is $-1$, the above means that (3.3.3) also has a solution.

"$\Leftarrow$": First, we consider the case that $M_k[y] \succ 0$ is positive definite. The linear system (3.3.3) must have a solution $(p_0, p_1, \ldots, p_k)$. Let

$$\begin{aligned} y_{2k+2} &:= p_0 y_{k+1} + p_1 y_{k+2} + \cdots + p_k y_{2k+1}, \\ \tilde{y} &:= (y_0, \ldots, y_{2k+1}, y_{2k+2}). \end{aligned}$$

Then $\tilde{y}$ is a tms of degree $2k + 2$. Note that $M_{k+1}[\tilde{y}] \succeq 0$ and

$$(p_0, p_1, \ldots, p_k, -1) \in \ker M_{k+1}[\tilde{y}].$$

Consider the polynomial

$$p(x) = p_0 + p_1 x + \cdots + p_k x^k - x^{k+1}.$$

We show that $p(x)$ has $k + 1$ distinct real zeros.

- Suppose otherwise that $p(x)$ has a non-real zero, say, $\alpha + \sqrt{-1}\beta$ with $\beta \neq 0$. Then one can write that $|p(x)|^2 = |a(x)|^2 + \beta^2 |b(x)|^2$ with $a(x) \in \mathbb{R}[x]_{k+1}$ and $0 \neq b(x) \in \mathbb{R}[x]_k$. Note that

$$\begin{aligned} 0 &= \mathrm{vec}(p)^T M_{k+1}[\tilde{y}] \,\mathrm{vec}(p) \\ &= \mathrm{vec}(a)^T M_{k+1}[\tilde{y}] \,\mathrm{vec}(a) + \beta^2 \,\mathrm{vec}(b)^T M_k[y] \,\mathrm{vec}(b). \end{aligned}$$

Since $M_{k+1}[\tilde{y}] \succeq 0$ and $M_k[y] \succeq 0$, we must get $\mathrm{vec}(b)^T M_k[y] \,\mathrm{vec}(b) = 0$, which contradicts $M_k[y] \succ 0$. So, all the zeros of $p(x)$ must be real.

- Suppose otherwise that $p(x)$ has a repeated real zero, then there exists $q \in \mathbb{R}[x]_{k-1}$ such that $p(x)q(x) = f(x)^2$ and $0 \neq f \in \mathbb{R}[x]_k$. Then

$$0 = \mathscr{L}_y(pq) = \mathscr{L}_y(f^2) = \mathrm{vec}(f)^T M_k[y]\,\mathrm{vec}(f),$$

which also contradicts that $M_k[y] \succ 0$.

There $p(x)$ has $k+1$ distinct real zeros, say, $u_1 < u_2 < \cdots < u_{k+1}$. Then there exist real scalars $\lambda_1, \ldots, \lambda_{k+1} \in \mathbb{R}$ such that

$$\lambda_1 [u_1]_k + \cdots + \lambda_{k+1}[u_{k+1}]_k = y|_{0:k}.$$

This is because $\big([u_1]_k \quad \cdots \quad [u_{k+1}]_k\big)$ is a nonsingular Vandermonde matrix.

The equations $p(u_i) = 0$ and (3.3.3) imply

$$\lambda_1 [u_1]_{2k+1} + \cdots + \lambda_{k+1}[u_{k+1}]_{2k+1} = y.$$

Next, we show that all the scalars $\lambda_i > 0$. By the interpolation, there exists $f \in \mathbb{R}[x]_k$ such that $f(u_1) = 1$ and $f(u_2) = \cdots = f(u_{k+1}) = 0$. Then

$$
\begin{aligned}
\lambda_1 &= \mathrm{vec}(f)^T \Big(\lambda_1 [u_1]_k [u_1]_k^T + \cdots + \lambda_{k+1}[u_{k+1}]_k [u_{k+1}]_k^T\Big)\mathrm{vec}(f) \\
&= \mathrm{vec}(f)^T M_k[y]\,\mathrm{vec}(f) > 0.
\end{aligned}
$$

Similarly, one can show that all $\lambda_i > 0$. Hence $y$ admits a representing measure supported in $\mathbb{R}$, which is $(k+1)$-atomic.

Second, we consider the case that $M_k[y] \succeq 0$ is singular. Let

$$\hat{y} := (y_0, y_1, \ldots, y_{2r-1}).$$

Since $r = \mathrm{rank}(y)$, one can show that $M_{r-1}[\hat{y}] \succ 0$, so there exists

$$\tilde{p} := \begin{bmatrix} p_0 & p_1 & \cdots & p_{r-1} \end{bmatrix}^T$$

such that

$$M_{r-1}[\hat{y}]\tilde{p} = y|_{r:2r-1}.$$

Like in the earlier case, we can show that the polynomial

$$p(x) := p_0 + p_1 x + \cdots + p_{r-1} x^{r-1} - x^r$$

has $r$ distinct real roots $v_1, \ldots, v_r \in \mathbb{R}$. Similarly, there exist $\lambda_1, \ldots, \lambda_r > 0$ such that

$$\lambda_1 [v_1]_{2r-1} + \cdots + \lambda_r [v_r]_{2r-1} = \hat{y}. \qquad (3.3.5)$$

By the hypothesis that the linear system (3.3.3) has a solution, there exists a value of $y_{2k+2}$ such that $M_{k+1}[\tilde{y}] \succeq 0$, with $\tilde{y} = (y_0, \ldots, y_{2k+1}, y_{2k+2})$. Observe that

$$\mathrm{vec}(p)^T M_r[\tilde{y}]\,\mathrm{vec}(p) = \mathrm{vec}(p)^T M_{k+1}[\tilde{y}]\,\mathrm{vec}(p) = 0.$$

Since $M_{k+1}[\tilde{y}] \succeq 0$, we know that $M_{k+1}[\tilde{y}]\,\mathrm{vec}(p) = 0$. One can further get

$$M_{k+1}[\tilde{y}]\,\mathrm{vec}(px^t) = 0$$

for all $t \leq k - r$ (see Exercise 3.3.3). Hence,

$$p_0 y_i + p_1 y_{i+1} + \cdots + p_{r-1} y_{r+i-1} = y_{r+i} \quad \text{for} \quad i = 0, \ldots, 2k + 1 - r.$$

The above and (3.3.5) then imply that

$$\lambda_1[v_1]_{2k+1} + \cdots + \lambda_r[v_r]_{2k+1} = y. \tag{3.3.6}$$

So, $y$ admits a $r$-atomic representing measure supported in $\mathbb{R}$.

(ii) For this case, the degree $d = 2k$ is even.

"$\Rightarrow$": Suppose $y$ admits a representing measure $\mu$, then $M_k[y]$ must be positive semidefinite. If $r = k + 1$, then $M_k[y]$ is positive definite, and then (3.3.3) has a solution for arbitrary value of $y_{2k+1}$, which implies that (3.3.4) also has a solution. If $r < k + 1$, by the definition of rank$(y)$, the $(r + 1)$th column of $M_k[y]$ is a linear combination of its first $r$ columns. So there exist real scalars $p_0, p_1, \ldots, p_{r-1}$ such that

$$p_0 y_j + p_1 y_{j+1} + \cdots + p_{r-1} y_{r+j-1} = y_{r+j}, \quad j = 0, 1, \ldots, k.$$

Let $p = p_0 + p_1 x + \cdots + p_{r-1} x^{r-1} - x^r$. The above implies

$$\mathrm{vec}(p)^T M_k[y]\, \mathrm{vec}(p) = \int p(x)^2 \mathrm{d}\mu = 0.$$

So, the support of $\mu$ is contained in the zero set of $p$, and hence

$$\int p(x) x^j \mathrm{d}\mu = 0, \quad j = 0, 1, \ldots, 2k - r.$$

This means that $(p_0, p_1, \ldots, p_{r-1})$ is a solution to the linear system (3.3.4).

"$\Leftarrow$": Suppose (3.3.4) has a solution $(p_0, \ldots, p_{r-1})$. Let

$$
\begin{aligned}
y_{2k+1} &:= p_0 y_{2k-r+1} + p_1 y_{2k-r+2} + \cdots + p_{r-1} y_{2k}, \\
\tilde{y} &:= (y_0, \ldots, y_{2k}, y_{2k+1}).
\end{aligned}
$$

Note that $r \leq k + 1$ and

$$0 y_{i-k-1} + \cdots + 0 y_{i-r-1} + p_0 y_{i-r} + p_1 y_{i-r+1} + \cdots + p_{r-1} y_{i-1} = y_i$$

for all $i = k+1, k+2, \ldots, 2k+1$. This means that (3.3.3) has a solution for $\tilde{y}$. By the conclusion of item (i), we know $\tilde{y}$ admits a representing measure, so $y$ also admits one.

(iii) When $y$ admits one, the existence of a $r$-atomic representing measure is proved in the item (i) or (ii). If $r = k + 1$, i.e., $M_k[y]$ is positive definite, then $y$ cannot have a unique representing measure. This is because if $d = 2k$ is even, we can give an arbitrary value for $y_{2k+1}$, and then $(y_0, \ldots, y_{2k}, y_{2k+1})$ also admits a representing measure; if $d = 2k + 1$ is odd, then for all $y_{2k+2}$ sufficiently big, the moment matrix $M_{k+1}[\tilde{y}] \succeq 0$ for

$$\tilde{y} = (y_0, \ldots, y_{2k+1}, y_{2k+2}),$$

and $\tilde{y}$ also admits a representing measure. Therefore, for $y$ to admit a unique representing measure, we must have $r \leq k$.

Now we suppose $r \leq k$. For the case that $d = 2k + 1$ is odd, in the proof of item (i), we have shown that $M_{r-1}[y]$ is positive definite and there exist real scalars $p_0, p_1, \ldots, p_{r-1}$ such that

$$p_0 y_j + p_1 y_{j+1} + \cdots + p_{r-1} y_{r+j-1} = y_{r+j} \quad \text{for} \quad j = 0, 1, \ldots, r - 1.$$

We also showed that the polynomial $p = p_0 + p_1 x + \cdots + p_{r-1} x^{r-1} - x^r$ has $r$ distinct real zeros. For each representing measure $\mu$ for $y$, the support supp$(\mu)$ must be contained in the zero set of $p$. Since $|\mathrm{supp}(\mu)| \geq \mathrm{rank}\, M_r[y] = r$, we know supp$(\mu)$ equals the zero set of

$p$, so the representing measure for $y$ must be unique. For the case that $d = 2k$ is even, let $(p_0, p_1, \ldots, p_{r-1})$ be a solution to (3.3.4). Then we can show that

$$\text{vec}(p)^T M_r[y]\,\text{vec}(p) = \int p^2 \mathrm{d}\mu = 0.$$

This means that the support of every representing measure for $y$ is also contained in the zero set of $p$. Since $M_r[y] \succeq 0$, we also know that $p$ has $r$ distinct real zeros. Since $|\,\text{supp}(\mu)| \geq$ rank $M_r[y] = r$, we know $\text{supp}(\mu)$ equals the zero set of $p$ and hence the representing measure for $y$ must be unique. $\square$

Theorem 3.3.2 gives a complete characterization for a tms $y \in \mathbb{R}^{d+1}$ to admit a representing measure on $\mathbb{R}$. We remark that when the degree $d = 2k + 1$ is odd, the linear system (3.3.4) also has a solution, which is implied by the conclusion for the truncation $y|_{0:2k}$. Moreover, if $y$ admits a representing measure, there always exists a $r$-atomic representing measure, with $r = \text{rank}(y)$. From the proof, we can get the following algorithm for constructing a $r$-atomic representing measure.

**Algorithm 3.3.3.** *Let $y \in \mathbb{R}^{d+1}$ be a tms satisfying the conditions in Theorem 3.3.2. Let $r :=$ rank $M_k[y]$, which also equals $\text{rank}(y)$.*

**Step 1** *If $r = k + 1$ (i.e., $M_k[y]$ is positive definite), let $(p_0, p_1, \ldots, p_k)$ be a solution to (3.3.3) (when $d = 2k$ is even, choose an arbitrary value for $y_{2k+1}$).*

**Step 2** *If $r < k + 1$ (i.e., $M_k[y]$ is singular), let $(p_0, p_1, \ldots, p_{r-1})$ be a solution to (3.3.4).*

**Step 3** *Compute the $r$ distinct real zeros, say, $t_1 < \cdots < t_r$, of the polynomial*

$$p(x) := p_0 + p_1 x + \cdots + p_{r-1} x^{r-1} - x^r. \tag{3.3.7}$$

**Step 4** *Compute positive scalars $\lambda_i > 0$ satisfying the decomposition (3.3.1).*

If $y$ admits a representing measure, then the polynomial $p$ in (3.3.7) must have $r$ distinct real zeros and there must exist $\lambda_i > 0$ satisfying (3.3.1). This is shown in the proof of Theorem 3.3.2.

## 3.3.2 ▪ The UTMP on the interval $[a, b]$

This section discusses when a tms $y := (y_0, \ldots, y_d)$ admits a representing measure supported in a compact interval $[a, b]$. Suppose $a < b$ and $d = 2k$ or $d = 2k + 1$. Recall that the $(k-1)$th order moment matrix of $y$ is

$$M_{k-1}[y] = \begin{bmatrix} y_0 & y_1 & \cdots & y_{k-1} \\ y_1 & y_2 & \cdots & y_k \\ \vdots & \vdots & \ddots & \vdots \\ y_{k-1} & y_k & \cdots & y_{2k-2} \end{bmatrix}. \tag{3.3.8}$$

The $(k-1)$th order moment matrix of the truncation $y|_{2:2k}$ is

$$G_k[y] = \begin{bmatrix} y_2 & y_3 & \cdots & y_{k+1} \\ y_3 & y_4 & \cdots & y_{k+2} \\ \vdots & \vdots & \ddots & \vdots \\ y_{k+1} & y_{k+2} & \cdots & y_{2k} \end{bmatrix}. \tag{3.3.9}$$

When $d = 2k + 1$ is odd, the $k$th order localizing matrix for $x \geq 0$ is

$$N_k[y] := \begin{bmatrix} y_1 & y_2 & \cdots & y_{k+1} \\ y_2 & y_3 & \cdots & y_{k+2} \\ \vdots & \vdots & \ddots & \vdots \\ y_{k+1} & y_{k+2} & \cdots & y_{2k+1} \end{bmatrix}. \tag{3.3.10}$$

When $d = 2k$ is even, the $(k-1)$th order localizing matrix for $x \geq 0$ is

$$N_{k-1}[y] := \begin{bmatrix} y_1 & y_2 & \cdots & y_k \\ y_2 & y_3 & \cdots & y_{k+1} \\ \vdots & \vdots & \ddots & \vdots \\ y_k & y_{k+1} & \cdots & y_{2k-1} \end{bmatrix}. \tag{3.3.11}$$

For the compact interval $[a, b]$, there exists a convenient characterization for $y$ to admit a representing measure supported in $[a, b]$, i.e., for $y \in \mathscr{R}_d([a, b])$. The following is a classical result.

**Theorem 3.3.4.** *([64, 154]) Let* $y := (y_0, \ldots, y_d) \in \mathbb{R}^{d+1}$ *with* $y_0 > 0$.

(i) *If* $d = 2k + 1$ *is odd, then* $y$ *admits a representing measure supported in* $[a, b]$ *if and only if* $y$ *satisfies*

$$bM_k[y] \succeq N_k[y] \succeq aM_k[y]. \tag{3.3.12}$$

(ii) *If* $d = 2k$ *is even, then* $y$ *admits a representing measure supported in* $[a, b]$ *if and only if*

$$M_k[y] \succeq 0, \quad (a + b)N_{k-1}[y] \succeq abM_{k-1}[y] + G_k[y]. \tag{3.3.13}$$

(iii) *If it admits one, the tms* $y$ *must admit a* $\mathrm{rank}(y)$-*atomic representing measure supported in* $[a, b]$.

**Proof.** For the compact interval $[a, b]$, the dual cone of $\mathscr{P}_d([a, b])$ is $\mathscr{R}_d([a, b])$, i.e.,

$$\mathscr{R}_d([a, b]) = \left(\mathscr{P}_d([a, b])\right)^\star.$$

This is implied by Theorem 2.7.4.

(i) When $d = 2k + 1$ is even, Theorem 3.2.3 implies

$$\mathscr{P}_d([a, b]) = \{(x - a)p^2 + (b - x)q^2 : p, q \in \mathbb{R}[x]_k\}.$$

By the duality, we have $y \in \mathscr{R}_d([a, b])$ if and only if

$$\mathscr{L}_y((x - a)p^2 + (b - x)q^2) \geq 0 \quad \text{for all } p, q \in \mathbb{R}[x]_k.$$

Note that

$$\mathscr{L}_y((x - a)p^2) = \mathrm{vec}(p)^T \left( N_k[y] - aM_k[y] \right) \mathrm{vec}(p),$$

$$\mathscr{L}_y((b - x)q^2) = \mathrm{vec}(q)^T \left( bM_k[y] - N_k[y] \right) \mathrm{vec}(q).$$

Therefore, $y \in \mathscr{R}_d([a, b])$ if and only if (3.3.12) is satisfied.

(ii) When the degree $d = 2k$ is even, by Theorem 3.2.3, we know

$$\mathscr{P}_d([a, b]) = \{p^2 + (x - a)(b - x)q^2 : p \in \mathbb{R}[x]_k, q \in \mathbb{R}[x]_{k-1}\}.$$

By the duality, we have $y \in \mathscr{R}_d([a,b])$ if and only if

$$\mathscr{L}_y(p^2 + (x-a)(b-x)q^2) \geq 0 \quad \forall p \in \mathbb{R}[x]_k, q \in \mathbb{R}[x]_{k-1}.$$

Note that $\mathscr{L}_y(p^2) = \text{vec}(p)^T M_k[y]\, \text{vec}(p)$ and

$$\mathscr{L}_y((x-a)(b-x)q^2) = \text{vec}(q)^T \Big((a+b)N_{k-1}[y] - abM_{k-1}[y] - G_k[y]\Big)\text{vec}(q).$$

Hence, $y \in \mathscr{R}_d([a,b])$ is equivalent to

$$\text{vec}(p)^T M_k[y]\, \text{vec}(p) \geq 0,$$
$$\text{vec}(q)^T \Big((a+b)N_{k-1}[y] \succeq abM_{k-1}[y] + G_k[y]\Big)\text{vec}(q) \geq 0$$

for all $p, q$. The above are equivalent to (3.3.13).

(iii) This conclusion can be similarly shown as in the proof of Theorem 3.3.2, since one can further show that all points in the support belong to the interval $[a,b]$. The remaining proof details are left as an exercise. □

Theorem 3.3.4 is well known for univariate moment problems [8, 154]. Curto and Fialkow [64] also gave other equivalent conditions for $y \in \mathscr{R}_d([a,b])$ (see [64, Theorems 4.1 and 4.3]). For the special case that $[a,b] = [0,1]$, the condition (3.3.13) reads explicitly as

$$\begin{bmatrix} y_0 & y_1 & \cdots & y_k \\ y_1 & y_2 & \cdots & y_{k+1} \\ \vdots & \vdots & \ddots & \vdots \\ y_k & y_{k+1} & \cdots & y_{2k} \end{bmatrix} \succeq 0, \tag{3.3.14}$$

$$\begin{bmatrix} y_1 & y_2 & \cdots & y_k \\ y_2 & y_3 & \cdots & y_{k+1} \\ \vdots & \vdots & \ddots & \vdots \\ y_k & y_{k+1} & \cdots & y_{2k-1} \end{bmatrix} \succeq \begin{bmatrix} y_2 & y_3 & \cdots & y_{k+1} \\ y_3 & y_4 & \cdots & y_{k+2} \\ \vdots & \vdots & \ddots & \vdots \\ y_{k+1} & y_{k+2} & \cdots & y_{2k} \end{bmatrix}, \tag{3.3.15}$$

and the condition (3.3.12) reads as

$$\begin{bmatrix} y_0 & y_1 & \cdots & y_k \\ y_1 & y_2 & \cdots & y_{k+1} \\ \vdots & \vdots & \ddots & \vdots \\ y_k & y_{k+1} & \cdots & y_{2k} \end{bmatrix} \succeq \begin{bmatrix} y_1 & y_2 & \cdots & y_{k+1} \\ y_2 & y_3 & \cdots & y_{k+2} \\ \vdots & \vdots & \ddots & \vdots \\ y_{k+1} & y_{k+2} & \cdots & y_{2k+1} \end{bmatrix} \succeq 0. \tag{3.3.16}$$

About uniqueness of representing measures, we have the following results.

**Theorem 3.3.5.** *Let $y := (y_0, \ldots, y_d) \in \mathbb{R}^{d+1}$ be a tms with $y_0 > 0$ and $d = 2k$ or $d = 2k+1$. Suppose it admits a representing measure supported in $[a,b]$ ($a < b$) and $r = \text{rank}\, M_k[y]$.*

*(i) If $r \leq k$, then the representing measure for $y$ is unique.*

*(ii) Suppose $r = k+1$ and $d = 2k+1$ is odd. Let $[\xi, \eta]$ be the range for $y_{2k+2}$ such that the tms $\tilde{y} := (y_0, y_1, \ldots, y_{2k+1}, y_{2k+2})$ satisfies*

$$M_{k+1}[\tilde{y}] \succeq 0, \quad (a+b)N_k[\tilde{y}] \succeq abM_k[\tilde{y}] + G_{k+1}[\tilde{y}]. \tag{3.3.17}$$

*Then $y$ admits a unique representing measure supported in $[a,b]$ if and only if $\xi = \eta$.*

*(iii) Suppose $r = k + 1$ and $d = 2k$ is even. Let $[\xi, \eta]$ be the range for $y_{2k+1}$ such that the tms $\hat{y} := (y_0, y_1, \ldots, y_{2k}, y_{2k+1})$ satisfies*

$$bM_k[y] \succeq N_k[\hat{y}] \succeq aM_k[y]. \tag{3.3.18}$$

*Then $y$ admits a unique representing measure supported in $[a, b]$ if and only if $\xi = \eta$.*

**Proof.** (i) When $r \leq k$, the moment matrix $M_k[y]$ is singular. The representing measure for $y$ is unique, which is shown in Theorem 3.3.2.

(ii) Since $y$ admits a representing measure supported in $[a, b]$, there must exist $y_{2k+2}$ satisfying (3.3.17). Indeed, the inequality $M_{k+1}[\tilde{y}] \succeq 0$ gives the range $y_{2k+2} \geq \xi$ and

$$(a + b)N_k[\tilde{y}] \succeq abM_k[\tilde{y}] + G_{k+1}[\tilde{y}]$$

gives the range $y_{2k+2} \leq \eta$. When $\xi < \eta$, the $[a, b]$-representing measure for $y$ is certainly not unique. When $\xi = \eta$, the $[a, b]$-representing measure for $\tilde{y}$ must be unique, because $M_{k+1}[\tilde{y}]$ is singular. Therefore, the $[a, b]$-representing measure for $y$ is also unique if $\xi = \eta$.

(iii) Since $y \in \mathscr{R}_d([a, b])$, there must exist $y_{2k+1}$ satisfying (3.3.18). Note that $N_k[\hat{y}] \succeq aM_k[y]$ gives the range $y_{2k+1} \geq \xi$ and $bM_k[y] \succeq N_k[\hat{y}]$ gives the range $y_{2k+1} \leq \eta$. When $\xi < \eta$, the $[a, b]$-representing measure for $y$ is clearly not unique. When $\xi = \eta$, the $[a, b]$-representing measure for $\hat{y}$ must be unique, because $N_k[\hat{y}] - aM_k[\hat{y}]$ is singular, implied by Theorem 3.3.7. Therefore, the $[a, b]$-representing measure for $y$ is also unique if $\xi = \eta$. □

The following is the algorithm to get an atomic representing measure supported in $[a, b]$.

**Algorithm 3.3.6.** *Let $y := (y_0, y_1, \ldots, y_d)$ be a tms satisfying the conditions in Theorem 3.3.4. Let $k = \lfloor \frac{d}{2} \rfloor$ and $r = \text{rank } M_k[y]$.*

**Step 1** *If $d = 2k + 1$ is odd, solve the linear system*

$$\begin{bmatrix} y_0 & y_1 & \cdots & y_{r-1} \\ y_1 & y_2 & \cdots & y_r \\ \vdots & \vdots & \ddots & \vdots \\ y_{2k-r+1} & y_{2k-r+2} & \cdots & y_{2k} \end{bmatrix} \begin{bmatrix} g_0 \\ g_1 \\ \vdots \\ g_{r-1} \end{bmatrix} = \begin{bmatrix} y_r \\ y_{r+1} \\ \vdots \\ y_{2k+1} \end{bmatrix}. \tag{3.3.19}$$

**Step 2** *If $d = 2k$ and $r \leq k$, solve the linear system*

$$\begin{bmatrix} y_0 & y_1 & \cdots & y_{r-1} \\ y_1 & y_2 & \cdots & y_r \\ \vdots & \vdots & \ddots & \vdots \\ y_{2k-r} & y_{2k-r+1} & \cdots & y_{2k-1} \end{bmatrix} \begin{bmatrix} g_0 \\ g_1 \\ \vdots \\ g_{r-1} \end{bmatrix} = \begin{bmatrix} y_r \\ y_{r+1} \\ \vdots \\ y_{2k} \end{bmatrix}. \tag{3.3.20}$$

**Step 3** *If $d = 2k$ and $r = k + 1$, compute the smallest value of $y_{2k+1}$ satisfying (3.3.12), and then solve the linear system*

$$\begin{bmatrix} y_0 & y_1 & \cdots & y_k \\ y_1 & y_2 & \cdots & y_{k+1} \\ \vdots & \vdots & \ddots & \vdots \\ y_k & y_{k+1} & \cdots & y_{2k} \end{bmatrix} \begin{bmatrix} g_0 \\ g_1 \\ \vdots \\ g_k \end{bmatrix} = \begin{bmatrix} y_{k+1} \\ y_{k+2} \\ \vdots \\ y_{2k+1} \end{bmatrix}. \tag{3.3.21}$$

**Step 4** *Compute the $r$ distinct roots $t_1, \ldots, t_r$ of the polynomial*

$$g(x) := g_0 + g_1 x + \cdots + g_{r-1} x^{r-1} - x^r.$$

*Determine the coefficients $\lambda_1, \ldots, \lambda_r$ satisfying the equation*

$$y = \lambda_1 [t_1]_d + \cdots + \lambda_r [t_r]_d. \tag{3.3.22}$$

If $y$ satisfies the conditions in Theorem 3.3.4, the roots of $g$ must belong to the interval $[a, b]$ and all $\lambda_1 > 0, \ldots, \lambda_r > 0$. We leave the proof as an exercise.

### 3.3.3 ▪ The UTMP over $[a, \infty)$

Now we discuss the case that $I = [a, \infty)$. Theorem 3.2.1 gives a compete characterization for the cone $\mathscr{P}_d([a, \infty))$, which consists of degree-$d$ polynomials that are nonnegative on $[a, \infty)$. The dual cone of $\mathscr{P}_d([a, \infty))$ is the closure of $\mathscr{R}_d([a, \infty))$. However, the moment cone $\mathscr{R}_d([a, \infty))$ is not closed. By Proposition 3.2.2, for the case $d = 2k + 1$, a necessary condition for $y \in \mathscr{R}_d([a, \infty))$ is $M_k[y] \succeq 0$, $N_k[y] \succeq a M_k[y]$; for the case $d = 2k$, the necessary condition is $M_k[y] \succeq 0$, $N_{k-1}[y] \succeq a M_{k-1}[y]$. However, they might not be sufficient for $y \in \mathscr{R}_d([a, \infty))$. The following gives a complete characterization for the cone $\mathscr{R}_d([a, \infty))$, which is an extension of the results in [64, §5] where the case $a = 0$ is focused. Recall the notation

$$y|_{s:t} := (y_s, y_{s+1}, \ldots, y_t) \quad \text{for} \quad s \leq t.$$

The matrices $N_k[y]$, $N_{k-1}[y]$ are defined in (3.3.10)-(3.3.11).

**Theorem 3.3.7.** *Let $y := (y_0, \ldots, y_d) \in \mathbb{R}^{d+1}$ with $y_0 > 0$.*

(i) *If $d = 2k + 1$ is odd, then $y$ admits a representing measure supported in $[a, \infty)$ if and only if $y$ satisfies*

$$\boxed{\begin{aligned} M_k[y] \succeq 0, \; N_k[y] - a M_k[y] \succeq 0, \\ y|_{k+1:2k+1} \in \text{Range}\,(M_k[y]). \end{aligned}} \tag{3.3.23}$$

*Moreover, if the above holds, then $y$ admits a unique $[a, \infty)$-representing measure if and only if $M_k[y]$ is singular, or $M_k[y]$ is nonsingular but $N_k[y] - a M_k[y]$ is singular.*

(ii) *If $d = 2k$ is even, then $y$ admits a representing measure supported in $[a, \infty)$ if and only if*

$$\boxed{\begin{aligned} M_k[y] \succeq 0, \; N_{k-1}[y] - a M_{k-1}[y] \succeq 0, \\ y|_{k+1:2k} - a y|_{k:2k-1} \in \text{Range}\,(N_{k-1}[y] - a M_{k-1}[y]). \end{aligned}} \tag{3.3.24}$$

*Moreover, if the above holds, then $y$ admits a unique $[a, \infty)$-representing measure if and only if $M_k[y]$ is singular.*

(iii) *When it admits one, $y$ must admit a $\text{rank}(y)$-atomic representing measure supported in $[a, +\infty)$.*

**Proof.** (i) For this case, $d = 2k + 1$ is odd.

"⇒": Suppose $y$ admits a representing measure $\mu$ supported in $[a, \infty)$. Then $M_k[y] \succeq 0$ and $y|_{k+1:2k+1} \in \text{Range}\,(M_k[y])$, by Theorem 3.3.2. Since $\text{supp}(\mu) \subseteq [a, \infty)$, we also have

$$N_k[y] - aM_k[y] = \int (x-a)[x]_k[x]_k^T \mathrm{d}\mu \succeq 0.$$

"⇐": Suppose (3.3.23) holds. Theorem 3.3.2(i) and (iii) imply that $y$ admits a representing measure $\mu$ with $r := |\text{supp}(\mu)| \le k+1$, say,

$$y = \lambda_1[u_1]_{2k+1} + \cdots + \lambda_r[u_r]_{2k+1},$$

with all $\lambda_i > 0$. We need to show that all $u_i \ge a$. For each $i$, there exists $f \in \mathbb{R}[x]_k$ such that $f(u_i) = 1$ and $f(u_j) = 0$ for all $j \ne i$, and then

$$0 \le \text{vec}(f)^T \left( N_k[y] - aM_k[y] \right) \text{vec}(f) = \int (x-a)f(x)^2 \mathrm{d}\mu$$
$$= \sum_{j=1}^{r} \lambda_j(u_j - a)f(u_j)^2 = \lambda_i(u_i - a).$$

Therefore, all $u_i \ge a$.

The uniqueness characterization can be shown as follows. When $M_k[y]$ is singular, the representing measure for $y$ must be unique, by Theorem 3.3.2(iii). When both $M_k[y] \succ 0$ and $N_k[y] - aM_k[y] \succ 0$, we can choose large values for $y_{2k+2}, y_{2k+3}$ such that $M_{k+1}[\tilde{y}] \succ 0$, $N_{k+1}[\tilde{y}] - aM_{k+1}[\tilde{y}] \succ 0$, where

$$\tilde{y} := (y_0, \ldots, y_{2k+1}, y_{2k+2}, y_{2k+3}).$$

Each representing measure for $\tilde{y}$ is also one for $y$. Hence, the $[a, \infty)$-representing measure for $y$ cannot be unique. When $M_k[y] \succ 0$ but $N_k[y] - aM_k[y] \succeq 0$ is singular, let

$$\hat{y} := (y_1, y_2, \ldots, y_{2k+1}) - a(y_0, y_1, \ldots, y_{2k}).$$

It is a tms of even order $2k$ and $M_k[\hat{y}] \succeq 0$ is singular. By Theorem 3.3.2(iii), the representing measure for $\hat{y}$ is unique. Observe that each $[a, \infty)$-representing measure for $y$ is also a representing measure for $\hat{y}$, because

$$y = \sum_{j=1}^{r} \lambda_j[u_j]_{2k+1} \quad \Rightarrow \quad \hat{y} = \sum_{j=1}^{r} \lambda_j(u_j - a)[u_j]_{2k}.$$

Therefore, the $[a, \infty)$-representing measure for $y$ is also unique.

(ii) For this case, $d = 2k$ is even.

"⇒": Suppose $y$ admits a measure $\mu$ with $\text{supp}(\mu) \subseteq [a, \infty)$. Clearly,

$$M_k[y] \succeq 0, \quad N_{k-1}[y] - aM_{k-1}[y] = \int (x-a)[x]_{k-1}([x]_{k-1})^T \mathrm{d}\mu \succeq 0.$$

Let $z := y|_{1:2k} - ay|_{0:2k-1}$, then $z = \int (x-a)[x]_{2k-1} \mathrm{d}\mu$. This means that $z$ is a tms of odd degree $2k-1$ and it admits a representing measure on $[a, \infty)$. By the conclusion in part (i), we know $z|_{k:2k-1} \in \text{Range}\,(M_{k-1}[z])$, which is the same as $y|_{k+1:2k} - ay|_{k:2k-1} \in \text{Range}\,(N_{k-1}[y] - aM_{k-1}[y])$.

"⇐": Suppose (3.3.24) holds. Let $z := (z_0, z_1, \ldots, z_{2k-1})$ be the tms of odd degree $2k-1$ such that

$$z = y|_{1:2k} - ay|_{0:2k-1}.$$

Note that
$$M_{k-1}[z] = N_{k-1}[y] - aM_{k-1}[y].$$
The condition (3.3.24) implies
$$z|_{k:2k-1} \in \text{Range}\,(M_{k-1}[z]),$$
so there are scalars $p_0, p_1, \ldots, p_{k-1}$ such that
$$p_0 z_j + p_1 z_{j+1} + \cdots + p_{k-1} z_{j+k-1} = z_{j+k}, \ j = 0, \ldots, k-1. \tag{3.3.25}$$
Let $y_{2k+1} := z_{2k} + a y_{2k}$, where
$$z_{2k} = p_0 z_k + p_1 z_{k+1} + \cdots + p_{k-1} z_{2k-1}. \tag{3.3.26}$$
Let $\tilde{y} := (y_0, \ldots, y_{2k}, y_{2k+1})$ be the tms of degree $2k+1$, then

$$M_k[\tilde{y}] = M_k[y] \succeq 0, \ N_k[\tilde{y}] - aM_k[\tilde{y}] = \left[ \begin{array}{ccc|c} & & & z_k \\ & M_{k-1}[z] & & \vdots \\ & & & z_{2k-1} \\ \hline z_k & \cdots & z_{2k-1} & z_{2k} \end{array} \right].$$

The last column of $N_k[\tilde{y}] - aM_k[\tilde{y}]$ is a linear combination of its first $k$ columns. Since $M_{k-1}[z] \succeq 0$, we know
$$N_k[\tilde{y}] - aM_k[\tilde{y}] \succeq 0.$$
The equations (3.3.25) and (3.3.26) imply that
$$p_0(y_{j+1} - ay_j) + p_1(y_{j+2} - ay_{j+1}) + \cdots + p_{k-1}(y_{j+k} - ay_{j+k-1})$$
$$= y_{j+k+1} - ay_{j+k} \quad \text{for} \quad j = 0, \ldots, k.$$
The above is equivalent to for each $j = 0, \ldots, k$

$$-ap_0 y_j + \sum_{s=0}^{k-2}(p_s - ap_{s+1})y_{j+s+1} + (p_{k-1} + a)y_{j+k} = y_{j+k+1}.$$

This means that $\tilde{y}|_{k+1:2k+1} \in \text{Range}\,(M_k[\tilde{y}])$. Applying the conclusion of item (i) to the odd degree tms $\tilde{y}$, we know that $\tilde{y}$, and hence $y$, admits a representing measure supported in $[a, \infty)$.

The uniqueness characterization can be shown similarly. When $M_k[y]$ is singular, the representing measure for $y$ must be unique, by Theorem 3.3.2(iii). When $M_k[y]$ is positive definite, for all big values of $y_{2k+1}$, the tms
$$\tilde{y} = (y_0, \ldots, y_{2k}, y_{2k+1}) \in \mathbb{R}^{2k+1}$$
satisfies (3.3.23). So $\tilde{y}$ admits a representing measure supported in $[a, \infty)$. Therefore, if $M_k[y]$ is positive definite, the $[a, \infty)$-representing measure cannot be unique. So, if the representing measure for $y$ is unique, then $M_k[y]$ must be singular.

(iii) This can be similarly shown as in the proof of Theorem 3.3.2. The main idea is to show all points in the support belong to the interval $[a, \infty)$. We leave the details as an exercise. □

From the proof, we can get the following algorithm for computing a $r$-atomic representing measure supported in $[a, \infty)$, if it exists.

**Algorithm 3.3.8.** *Let $y := (y_0, y_1, \ldots, y_d)$ be a tms satisfying the conditions in Theorem 3.3.7. Let $k = \lfloor \frac{d}{2} \rfloor$ and $r = \text{rank}\, M_k[y]$.*

**Step 1** *If $d = 2k + 1$ is odd, solve the linear system (3.3.19).*

**Step 2** *If $d = 2k$ and $r \leq k$, solve the linear system (3.3.20).*

**Step 3** *If $d = 2k$ and $r = k + 1$, compute the smallest value of $y_{2k+1}$ satisfying (3.3.23), and then solve the linear system (3.3.21).*

**Step 4** *Compute the $r$ distinct roots $t_1, \ldots, t_r$ of the polynomial*

$$g(x) := g_0 + g_1 x + \cdots + g_{r-1} x^{r-1} - x^r.$$

*Determine the coefficients $\lambda_1, \ldots, \lambda_r$ satisfying the equation*

$$y = \lambda_1 [t_1]_d + \cdots + \lambda_r [t_r]_d. \tag{3.3.27}$$

When $y$ satisfies the conditions in Theorem 3.3.7, the roots of $g$ must belong to the interval $[a, \infty)$ and all $\lambda_1 > 0, \ldots, \lambda_r > 0$. The proof is left as an exercise.

### 3.3.4 ▪ Exercises

**Exercise 3.3.1.** *Find a $r$-atomic representing measure, with $r$ minimum, for the following tms:*

*(i) $y = \int_{-\infty}^{+\infty} \exp\{-x^2\} \cdot [x]_4 \mathrm{d}x$.*

*(ii) $y = \int_{-1}^{1} [x]_5 \mathrm{d}x$.*

*(iii) $y = \int_0^{+\infty} \exp\{-x\} \cdot [x]_3 \mathrm{d}x$.*

**Exercise 3.3.2.** *Let $k = \lfloor \frac{d}{2} \rfloor$. For a tms $y \in \mathbb{R}^{d+1}$, if $M_k[y] \succeq 0$, show that $\mathrm{rank}(y) = \mathrm{rank}\, M_k[y]$.*

**Exercise 3.3.3.** *Let $y \in \mathbb{R}^{2k+1}$ be such that $M_k[y] \succeq 0$. For a polynomial $p \in \mathbb{R}[x]_{k-1}$, if $M_k[y]\,\mathrm{vec}(p) = 0$, show that $M_k[y]\,\mathrm{vec}(pq) = 0$ for all $q \in \mathbb{R}[x]$ with the degree $\deg(pq) \leq k - 1$. Give an example to show the conclusion does not hold if $\deg(pq) = k$.*

**Exercise 3.3.4.** *For $y \in \mathbb{R}^{2k+1}$ such that $M_k[y] \succ 0$, show that the linear system (3.3.3) must have a solution $(p_0, p_1, \ldots, p_k)$. Show that the polynomial $p_0 + p_1 x + \cdots + p_k x^k - x^{k+1}$ has $k + 1$ distinct real zeros.*

**Exercise 3.3.5.** *Prove the item (iii) of Theorems 3.3.4 and 3.3.7.*

**Exercise 3.3.6.** *In Algorithm 3.3.6, for $y$ satisfying the assumptions there, show that the polynomial $g$ there has $r$ distinct roots in $[a, b]$ and all $\lambda_i > 0$ there. In Algorithm 3.3.8, for $y$ satisfying the assumptions there, show that the polynomial $g$ there has $r$ distinct roots in $[a, \infty)$ and all $\lambda_i > 0$ there.*

**Exercise 3.3.7.** *Let $y \in \mathbb{R}^{d+1}$ and $a \in \mathbb{R}$. If $d = 2k + 1$ is odd and $M_k[y] \succeq 0$, $N_k[y] - aM_k[y] \succeq 0$, show that $y \in \mathrm{cl}\,(\mathscr{R}_d([a, \infty)))$. Similarly, if $d = 2k$ is even and $M_k[y] \succeq 0$, $N_{k-1}[y] - aM_{k-1}[y] \succeq 0$, show that $y \in \mathrm{cl}\,(\mathscr{R}_d([a, \infty)))$.*

## 3.4 ▪ Univariate Polynomial Optimization over Intervals

Let $I \subseteq (-\infty, +\infty)$ be an interval. Consider the optimization problem

$$\begin{cases} \min & f(x) := f_0 + f_1 x + \cdots + f_d x^d \\ s.t. & x \in I. \end{cases} \tag{3.4.1}$$

It can be solved by a single semidefinite relaxation. Let $f_{min}$ denote the minimum value of (3.4.1). We discuss in three different cases of $I$.

### 3.4.1 ▪ The case $I = (-\infty, +\infty)$

When $I = (-\infty, +\infty)$, there is no constraint and (3.4.1) reads as

$$\min_{x \in \mathbb{R}^1} \quad f(x). \tag{3.4.2}$$

For it to have a minimizer, we assume the degree of $f$ is $d = 2m$ and its leading coefficient is positive. The minimum value $f_{min}$ of (3.4.2) equals the largest $\gamma$ such that $f - \gamma$ is a nonnegative polynomial, which is equivalent to $f - \gamma \in \Sigma[x]_{2m}$. So we consider the optimization

$$\begin{cases} \max & \gamma \\ s.t. & f - \gamma \in \Sigma[x]_{2m}. \end{cases} \tag{3.4.3}$$

This is a semidefinite program, which can be written as

$$\begin{cases} \max & \gamma \\ s.t. & f - \gamma = [x]_m^T X [x]_m, \\ & X \in \mathcal{S}_+^{m+1}. \end{cases} \tag{3.4.4}$$

For convenience, we label the rows and columns of $X$ by $i, j = 0, 1, \ldots, m$. The above equality constraints can be expressed as

$$\begin{aligned} f_0 - \gamma &= X_{00}, \\ f_k &= \sum_{i+j=k} X_{ij}, \quad k = 1, \ldots, 2m. \end{aligned}$$

The dual optimization problem of (3.4.4) is

$$\begin{cases} \min & f_0 y_0 + f_1 y_1 + \cdots + f_{2m} y_{2m} \\ s.t. & \begin{bmatrix} y_0 & y_1 & y_2 & \cdots & y_m \\ y_1 & y_2 & y_3 & \cdots & y_{m+1} \\ y_2 & y_3 & y_4 & \cdots & y_{m+2} \\ \vdots & \vdots & \vdots & \ddots & \vdots \\ y_m & y_{m+1} & y_{m+2} & \cdots & y_{2m} \end{bmatrix} \succeq 0, \\ & y_0 = 1, \\ & y = (y_0, y_1, \ldots, y_{2m}) \in \mathbb{R}^{2m+1}. \end{cases} \tag{3.4.5}$$

Let $f_{sos}, f_{mom}$ denote the optimal values of (3.4.4) and (3.4.5) respectively. The constraining matrix in (3.4.5) is precisely the moment matrix $M_m[y]$.

**Theorem 3.4.1.** *Suppose $f$ is a polynomial of degree $d = 2m$ with $f_{2m} > 0$. Then the optimization problems (3.4.2), (3.4.4), (3.4.5) all achieve the same optimal value, i.e., $f_{min} = f_{sos} =$*

$f_{mom}$. *Moreover, the optimization (3.4.5) must have a minimizer $y^*$ and it must satisfy the rank condition*

$$\text{rank } M_{m-1}[y^*] = \text{rank } M_m[y^*]. \tag{3.4.6}$$

*Furthermore, the tms $y^* \in \mathbb{R}^{d+1}$ has a unique $r$-atomic representing measure $\mu$, with $r :=$ rank $M_m[y^*]$, and each point in $\text{supp}(\mu)$ is a minimizer of (3.4.2).*

**Proof.** Since the leading coefficient $f_{2m} > 0$, the optimization (3.4.2) must have a finite minimum value and it is achievable. The equality $f_{min} = f_{sos}$ follows from Theorem 3.1.1. The moment relaxation (3.4.5) has a strictly feasible point, e.g., the tms

$$\hat{y} = \int [x]_{2m} \frac{1}{2\pi} \exp\{-\frac{x^2}{2}\} dx$$

is feasible and $M_m[\hat{y}] \succ 0$. By the strong duality theorem (Theorem 1.5.1), $f_{sos} = f_{mom}$ and (3.4.4) achieves its optimal value. For each minimizer $u$ of (3.4.2), we have $f(u) = f_{min} = f_{mom}$, so the tms $[u]_{2m}$ is a minimizer of (3.4.5). Moreover, for each minimizer

$$y^* = (y_0^*, y_1^*, \dots, y_{2m-1}^*, y_{2m}^*)$$

of (3.4.5), the entry $y_{2m}^*$ is the smallest value of $y_{2m}$ such that

$$M_m[(y_0^*, y_1^*, \dots, y_{2m-1}^*, y_{2m})] \succeq 0,$$

because $f_{2m} > 0$. In fact, $y_{2m}^*$ equals the product

$$\begin{bmatrix} y_m^* \\ y_{m+1}^* \\ \vdots \\ y_{2m-1}^* \end{bmatrix}^T \begin{bmatrix} y_0^* & y_1^* & \cdots & y_{m-1}^* \\ y_1^* & y_2^* & \cdots & y_m^* \\ \vdots & \vdots & \ddots & \vdots \\ y_{m-1}^* & y_m^* & \cdots & y_{2m-2}^* \end{bmatrix}^\dagger \begin{bmatrix} y_m^* \\ y_{m+1}^* \\ \vdots \\ y_{2m-1}^* \end{bmatrix}.$$

(The superscript $\dagger$ denotes Moore-Penrose Pseudoinverse.) Let

$$w := (M_{m-1}[y^*])^\dagger \cdot y^*|_{m:2m-1}, \quad b = y^*|_{m:2m-1}.$$

Then $y_{2m}^* = w^T M_{m-1}[y^*] w = w^T b$. Since

$$M_m[y^*] = \begin{bmatrix} M_{m-1}[y^*] & b \\ b^T & y_{2m}^* \end{bmatrix},$$

one can see that

$$\begin{bmatrix} w \\ -1 \end{bmatrix}^T \cdot M_m[y^*] \cdot \begin{bmatrix} w \\ -1 \end{bmatrix} = w^T M_{m-1}[y^*] w - 2w^T b + y_{2m}^* = 0.$$

Since $M_m[y^*] \succeq 0$, the above implies

$$M_m[y^*] \cdot \begin{bmatrix} w \\ -1 \end{bmatrix} = 0.$$

So, the last column of $M_m[y^*]$ is a linear combination of its first $m$ columns, and hence the rank condition (3.4.6) must be satisfied.

Since (3.4.6) holds, Theorem 2.7.7 implies that $y^*$ admits a unique $r$-atomic measure $\mu$, so there are $r$ distinct real points $u_1, \ldots, u_r$ and positive scalars $\lambda_1, \ldots, \lambda_r > 0$ such that

$$y^* = \lambda_1 [u_1]_{2m} + \cdots + \lambda_r [u_r]_{2m}.$$

Note that $\lambda_1 + \cdots + \lambda_r = 1$ and $\text{supp}(\mu) = \{u_1, \ldots, u_r\}$. Since each $[u_i]_{2m}$ is feasible for (3.4.5) and each $f(u_i) \geq f_{min}$, one can see that

$$f_{min} = f_{mom} = \lambda_1 f(u_1) + \cdots + \lambda_r f(u_r) \geq f_{min}.$$

So $f(u_1) = \cdots = f(u_r) = f_{min}$, i.e., each $u_i$ is a minimizer.     $\Box$

Minimizers of the optimization (3.4.2) can be extracted as follows. Let $r := \text{rank } M_m[y^*]$. Solve the linear system

$$
\begin{bmatrix}
y_0^* & y_1^* & \cdots & y_{r-1}^* \\
y_1^* & y_2^* & \cdots & y_r^* \\
\vdots & \vdots & \ddots & \vdots \\
y_{2m-r}^* & y_{2m-r-1}^* & \cdots & y_{2m-1}^*
\end{bmatrix}
\begin{bmatrix}
p_0 \\
p_1 \\
\vdots \\
p_{r-1}
\end{bmatrix}
=
\begin{bmatrix}
y_r^* \\
y_{r+1}^* \\
\vdots \\
y_{2m}^*
\end{bmatrix}.
\tag{3.4.7}
$$

The above equation must have a solution, by Theorem 3.3.2. One can show that the polynomial

$$p(x) := p_0 + p_1 x + \cdots + p_{r-1} x^{r-1} - x^r$$

has $r$ distinct real roots $u_1, \ldots, u_r$. and each of them is a minimizer of (3.4.2). The proof is left as an exercise.

## 3.4.2 ▪ The case $I = [a, b]$

We consider the case that $I = [a, b]$ is a compact interval, with $a < b$. The optimization (3.4.1) becomes

$$
\begin{cases}
\min & f(x) \\
s.t. & a \leq x \leq b.
\end{cases}
\tag{3.4.8}
$$

To have a minimizer, the polynomial $f$ can have arbitrary degrees and coefficients. The minimum value $f_{min}$ of (3.4.8) equals the largest $\gamma$ such that $f - \gamma$ is nonnegative on $[a, b]$, by Theorem 3.2.3, which is equivalent to

$$
f - \gamma =
\begin{cases}
(x - a)p^2 + (b - x)q^2 & \text{if } d = 2m + 1, \\
\quad p \in \mathbb{R}[x]_m, \, q \in \mathbb{R}[x]_m, \\
p^2 + (x - a)(b - x)q^2 & \text{if } d = 2m, \\
\quad p \in \mathbb{R}[x]_m, \, q \in \mathbb{R}[x]_{m-1}.
\end{cases}
$$

When the degree $d = 2m + 1$ is odd, the minimum $f_{min}$ equals the maximum value of the optimization

$$
\begin{cases}
\max & \gamma \\
s.t. & f - \gamma = (x - a)[x]_m^T X_0 [x]_m + (b - x)[x]_m^T X_1 [x]_m, \\
& X_0 \in \mathcal{S}_+^{m+1}, \, X_1 \in \mathcal{S}_+^{m+1}.
\end{cases}
\tag{3.4.9}
$$

Similarly, we label the rows and columns of $X_0, X_1$ by $i, j = 0, 1, \ldots, m$. The above equality constraints can be expressed as

$$
\begin{aligned}
f_0 - \gamma &= -a(X_0)_{00} + b(X_1)_{00}, \\
f_k &= \sum_{i+j=k-1} (X_0)_{ij} - a \sum_{i+j=k} (X_0)_{ij} \\
&\quad + b \sum_{i+j=k} (X_1)_{ij} - \sum_{i+j=k-1} (X_1)_{ij}, \\
& \qquad k = 1, \ldots, 2m + 1.
\end{aligned}
$$

Hence, the dual optimization problem of (3.4.9) is

$$
\begin{cases}
\min & f_0 y_0 + f_1 y_1 + \cdots + f_{2m+1} y_{2m+1} \\
s.t. & b M_m[y] \succeq N_m[y] \succeq a M_m[y], \\
& y_0 = 1, \\
& y = (y_0, y_1, \ldots, y_{2m}) \in \mathbb{R}^{2m+1}.
\end{cases}
\tag{3.4.10}
$$

Denote by $f_{sos}, f_{mom}$ the optimal values of (3.4.9), (3.4.10) respectively.
When the degree $d = 2m$ is even, $f_{min}$ equals the maximum value of

$$
\begin{cases}
\max & \gamma \\
s.t. & f - \gamma = [x]_m^T X_0 [x]_m + (x - a)(b - x)[x]_{m-1}^T X_1 [x]_{m-1}, \\
& X_0 \in \mathcal{S}_+^{m+1}, \ X_1 \in \mathcal{S}_+^m.
\end{cases}
\tag{3.4.11}
$$

The above equality constraints can be expressed as

$$
\begin{aligned}
f_0 - \gamma &= (X_0)_{00} - ab(X_1)_{00}, \\
f_1 &= 2(X_0)_{10} + (a + b)(X_1)_{00} - 2ab(X_1)_{10}, \\
f_k &= \sum_{i+j=k} (X_0)_{ij} - ab \sum_{i+j=k} (X_1)_{ij} \\
&\quad + (a+b) \sum_{i+j=k-1} (X_1)_{ij} - \sum_{i+j=k-2} (X_1)_{ij}, \\
& \qquad k = 2, \ldots, 2m.
\end{aligned}
$$

Therefore, the dual optimization of (3.4.11) is

$$
\begin{cases}
\min & f_0 y_0 + f_1 y_1 + \cdots + f_{2m} y_{2m} \\
s.t. & (a + b) N_{m-1}[y] \succeq ab M_{m-1}[y] + G_m[y], \\
& M_m[y] \succeq 0, \ y_0 = 1, \\
& y = (y_0, y_1, \ldots, y_{2m}) \in \mathbb{R}^{2m+1}.
\end{cases}
\tag{3.4.12}
$$

In the above, the matrices $G_m[y], N_{m-1}[y]$ are given as in (3.3.9), (3.3.11). For neatness of notation, we still denote by $f_{sos}, f_{mom}$ the optimal values of (3.4.11), (3.4.12) respectively.

**Theorem 3.4.2.** *Suppose $f$ is a polynomial of degree $d = 2m$ or $d = 2m + 1$ and $I = [a, b]$ is a compact interval with $a < b$. Then, all the optimal values $f_{min}, f_{sos}, f_{mom}$ are achieved in each optimization problem and they are all equal to each other. Let $y^*$ be a minimizer of (3.4.10) when $d = 2m + 1$ or (3.4.12) when $d = 2m$. Then $y^*$ must admit a representing measure $\mu$ supported in $[a, b]$. Moreover, each point in the support of $\mu$ is a minimizer of (3.4.8). If $f$ is not constant on $[a, b]$, then $f$ has at most $m + 1$ minimizers and the representing measure $\mu$ must be $r$-atomic with $r = \operatorname{rank} M_m[y^*] \leq m + 1$.*

**Proof.** Since the interval $[a, b]$ is compact, the minimum value $f_{min}$ is achievable. Hence, $f_{min} = f_{sos}$ follows from Theorem 3.2.3. The achievability of the optimal values $f_{sos}, f_{mom}$ and their equality can be shown similarly as in the proof for Theorem 3.4.1. By Theorem 3.3.4, $y^*$ must admit a representing measure $\mu$ supported in $[a, b]$. One can show that each point in the support of $\mu$ is a minimizer of (3.4.8) (see exercise). If $f$ is not a constant polynomial, then $f$ can have at most $m + 1$ minimizers, which can be shown as follows.

- If $d = 2m + 1$ is odd, then $f(x) - f_{min} = (x - a)p^2 + (b - x)q^2$ for $p, q \in \mathbb{R}[x]_m$. If one of $p, q$ is identically zero, then $f$ has most $(m + 1)$ minimizers, unless $f$ is a constant polynomial. Now assume none of $p, q$ is identically zero. Then each of $p, q$ has at most $m$ zeros and a minimizer is either $a$ or $b$ or a common zero of $p, q$ in $(a, b)$.

  - If $f_{min} < \min(f(a), f(b))$, the minimizers are common zeros of $p, q$ in $(a, b)$.
  - If $f(a) = f_{min} < f(b)$, then $q(a) = 0$, and the minimizers are $a$ and common zeros of $p, q$ in $(a, b)$.
  - If $f(b) = f_{min} < f(a)$, then $p(b) = 0$, and the minimizers are $b$ and common zeros of $p, q$ in $(a, b)$.
  - If $f(a) = f(b) = f_{min}$, then $q(a) = p(b) = 0$, and the minimizers are $a, b$ and common zeros of $p, q$ in $(a, b)$.

  In each of the above four cases, $f$ has at most $m + 1$ minimizers.

- If $d = 2m$ is even, then $f(x) - f_{min} = p^2 + (x - a)(b - x)q^2$ for $p \in \mathbb{R}[x]_m$ and $q \in \mathbb{R}[x]_{m-1}$. If $q = 0$ is identically zero, $f$ has most $m$ minimizers; if $p = 0$ is identically zero, $f$ has most $m + 1$ minimizers; if none of $p, q$ is identically zero, then $p$ have at most $m$ zeros and hence $f$ has at most $m$ minimizers.

Because each point in the support of $\mu$ is a minimizer of (3.4.8), the representing measure $\mu$ for $y^*$ must be $r$-atomic with $r \le m + 1$, say, $\mu = \lambda_1 \delta_{u_1} + \cdots + \lambda_r \delta_{u_r}$, with distinct $u_1, \ldots, u_r \in [a, b]$ and all $\lambda_i > 0$. Then

$$M_m[y^*] = V \cdot \text{diag}[\lambda_1, \ldots, \lambda_r] \cdot V^T,$$

where $V = \begin{bmatrix} [u_1]_m & \cdots & [u_r]_m \end{bmatrix}$. Since the points $u_i$ are all distinct, $V$ is a nonsingular Vandermonde matrix and so $r = \text{rank } M_m[y^*]$. $\quad\square$

We refer to Algorithm 3.3.6 for how to compute the points of the support of the representing measure for $y^*$. The points in the support are all minimizers of (3.4.8).

### 3.4.3 ▪ The case $I = [a, +\infty)$

When $I = [a, +\infty)$ is a half line, the optimization (3.4.1) has a single inequality constraint and it reads as

$$\begin{cases} \min & f(x) \\ s.t. & x \ge a. \end{cases} \tag{3.4.13}$$

To have a minimizer, the polynomial $f$ does not need to have an even degree, but its leading coefficient must be positive. The minimum value $f_{min}$ of (3.4.13) equals the largest $\gamma$ such that $f - \gamma$ is nonnegative on $[a, +\infty)$. By Theorem 3.2.1, this requires that $f = p^2 + (x - a)q^2$ for two polynomials $p, q$ of appropriate degrees. Therefore, (3.4.13) is equivalent to the problem

$$\begin{cases} \max & \gamma \\ s.t. & f - \gamma \in \Sigma[x] + (x - a)\Sigma[x]. \end{cases} \tag{3.4.14}$$

When the degree $d = 2m + 1$ is odd, (3.4.14) can be expressed as

$$\begin{cases} \max & \gamma \\ s.t. & f - \gamma = [x]_m^T X_0 [x]_m + (x - a)[x]_m^T X_1 [x]_m, \\ & X_0 \in \mathcal{S}_+^{m+1}, \; X_1 \in \mathcal{S}_+^{m+1}. \end{cases} \tag{3.4.15}$$

Similarly, we label the rows and columns of $X_0, X_1$ by $i, j = 0, 1, \ldots, m$. The above equality constraints can be expressed as

$$\begin{aligned} f_0 - \gamma &= (X_0)_{00} - a(X_1)_{00}, \\ f_k &= \sum_{i+j=k} \left( (X_0)_{ij} - a(X_1)_{ij} \right) + \sum_{i+j=k-1} (X_1)_{ij}, \\ & \qquad\qquad k = 1, \ldots, 2m, \\ f_{2m+1} &= \sum_{i+j=2m} (X_1)_{ij}. \end{aligned}$$

The dual optimization of (3.4.15) can be shown to be

$$\begin{cases} \max & f_0 y_0 + f_1 y_1 + \cdots + f_{2m+1} y_{2m+1} \\ s.t. & N_m[y] - a M_m[y] \succeq 0, \\ & M_m[y] \succeq 0, \; y_0 = 1, \\ & y = (y_0, y_1, \ldots, y_{2m+1}) \in \mathbb{R}^{2m+2}. \end{cases} \tag{3.4.16}$$

In the above, the matrix $N_m[y]$ is as in (3.3.10). We denote by $f_{sos}, f_{mom}$ the optimal values of (3.4.15), (3.4.16) respectively. The matrix $N_m[y] - a M_m[y]$ in (3.4.16) is a localizing matrix for the constraint $x - a \geq 0$.

When the degree $d = 2m$ is even, (3.4.14) can be expressed as

$$\begin{cases} \max & \gamma \\ s.t. & f - \gamma = [x]_m^T X_0 [x]_m + (x - a)[x]_{m-1}^T X_1 [x]_{m-1}, \\ & X_0 \in \mathcal{S}_+^{m+1}, \; X_1 \in \mathcal{S}_+^m. \end{cases} \tag{3.4.17}$$

One can similarly show the dual optimization of (3.4.17) is

$$\begin{cases} \max & f_0 y_0 + f_1 y_1 + \cdots + f_{2m} y_{2m} \\ s.t. & N_{m-1}[y] - a M_{m-1}[y] \succeq 0, \\ & M_m[y] \succeq 0, \; y_0 = 1, \\ & y = (y_0, y_1, \ldots, y_{2m}) \in \mathbb{R}^{2m+1}. \end{cases} \tag{3.4.18}$$

In the above, the matrix $N_{m-1}[y]$ is given as in (3.3.11). For neatness of notation, we still denote by $f_{sos}, f_{mom}$ the optimal values of (3.4.17)-(3.4.18) respectively. In (3.4.18), we use $N_{m-1}[y]$ instead of $N_m[y]$, because the variable $y$ there is a tms of degree $2m$ instead of $2m + 1$.

**Theorem 3.4.3.** *Suppose $f$ is a polynomial of degree $d = 2m$ or $d = 2m+1$, with $f_d > 0$. Then, all the optimal values $f_{min}, f_{sos}, f_{mom}$ are achievable and they are all equal to each other.*

*(i) For the case $d = 2m + 1$, each minimizer $y^*$ of (3.4.16) must satisfy*

$$\operatorname{rank} \left( N_{m-1}[y^*] - a M_{m-1}[y^*] \right) = \operatorname{rank} \left( N_m[y^*] - a M_m[y^*] \right). \tag{3.4.19}$$

*(ii) For the case $d = 2m$, each minimizer $y^*$ of (3.4.18) must satisfy*

$$\operatorname{rank} M_{m-1}[y^*] = \operatorname{rank} M_m[y^*]. \tag{3.4.20}$$

*Moreover, for either case, the $y^*$ has a unique $r$-atomic representing measure $\mu$ supported in $[a, \infty)$, where $r = \text{rank}\, (N_m[y^*] - aM_m[y^*])$ for the case $d = 2m + 1$ and $r = \text{rank}\, M_m[y^*]$ for the case $d = 2m$, and each point in the support of $\mu$ is a minimizer of (3.4.13).*

**Proof.** The proof is quite similar to that for Theorem 3.4.1, with the usage of Theorems 2.7.7 and 3.3.7. The following is the outline.

For the case $d = 2m + 1$, at each minimizer $y^*$ of (3.4.16), the matrix $N_m[y^*] - aM_m[y^*]$ must be psd and singular, because $f_{2m+1} > 0$. As in the proof for Theorem 3.4.1, we can similarly prove the rank condition (3.4.19), which further implies

$$y^*|_{m+1:2m+1} \in \text{Range}\,(M_m[y^*]).$$

The conclusion about the representing measure for $y^*$ follows from Theorem 3.3.7.

For the case $d = 2m$, at each minimizer $y^*$ of (3.4.18), the moment matrix $M_m[y^*]$ must be psd and singular, because $f_{2m} > 0$. As in the proof for Theorem 3.4.1, we can similarly show that the last column of $M_m[y^*]$ is a linear combination of its first $m$ columns. This also implies

$$y^*|_{m+1:2m} - ay^*|_{m:2m-1} \in \text{Range}\,(N_{m-1}[y^*] - aM_{m-1}[y^*]).$$

The conclusion about the representing measure for $y^*$ also follows from Theorem 3.3.7.

We leave the missing details in the above as an exercise. $\square$

We refer to Algorithm 3.3.8 for how to compute the points of the support of the representing measure for $y^*$. The points in the support are all minimizers of (3.4.13).

### 3.4.4 ▪ Exercises

**Exercise 3.4.1.** *Solve the following optimization problems:*

*(i)* $\displaystyle\min_{x \in \mathbb{R}^1} 2x - 3x^2 - 4x^3 - 5x^4 + x^6.$

*(ii)* $\displaystyle\min_{x \geq 0} -3x - 4x^2 + x^3 - x^4 + x^5.$

*(iii)* $\displaystyle\min_{-1 \leq x \leq 1} x - x^2 + x^3 - x^4 + x^5 - x^6.$

**Exercise 3.4.2.** *Let $y^*$ be a minimizer of (3.4.5) as in Theorem 3.4.1 and $(p_0, p_1, \ldots, p_{r-1})$ be a solution to (3.4.7). Show that the polynomial $p(x) := p_0 + p_1 x + \cdots + p_{r-1} x^{r-1} - x^r$ has $r$ distinct real roots and each of them is a minimizer of (3.4.2).*

**Exercise 3.4.3.** *Determine the value of $t$ such that the minimum value of the polynomial $2x + 3x^2 + tx^3 - x^4 - x^5$ on the interval $[-1, 1]$ is maximum. Also determine the value of $t$ such that its maximum value on $[-1, 1]$ is minimum.*

**Exercise 3.4.4.** *In Theorem 3.4.2, show that each point in the support of $\mu$ is a minimizer of (3.4.8).*

**Exercise 3.4.5.** *Give the missing details in the proof of Theorem 3.4.3.*

## 3.5 ▪ Constrained Univariate Polynomial Optimization

We consider the constrained optimization problem

$$\begin{cases} \min & f(x) \\ s.t. & g_1(x) \geq 0, \ldots, g_m(x) \geq 0 \end{cases} \tag{3.5.1}$$

for univariate polynomials $f, g_1, \ldots, g_m \in \mathbb{R}[x]$. Let $K \subseteq \mathbb{R}^1$ be the feasible set of (3.5.1). If it is not empty, $K$ is a union of disjoint closed intervals, say, $K = I_1 \cup \cdots \cup I_N$. One can equivalently minimize $f$ on each interval $I_i$ separately, by using the methods in §3.4, and then select the best one among them. However, this is not mathematically appealing, because it requires one to compute roots of polynomials $g_i$ and we need to solve several optimization problems separately.

We introduce the Moment-SOS relaxation method for solving (3.5.1). It directly uses the defining polynomials. For the univariate case, the Moment-SOS hierarchy has interesting properties which are different from the multivariate case (see Chapters 4 and 5). Clearly, the minimum value $f_{min}$ of (3.5.1) is the maximum $\gamma$ such that $f(x) - \gamma \geq 0$ on $K$. To ensure the nonnegativity, we consider Schmüdgen's Positivstellensatz

$$f(x) - \gamma = \sum_{i_1, \ldots, i_m \in \{0,1\}} \sigma_{i_1, \ldots, i_m} \cdot g_1^{i_1} \cdots g_m^{i_m},$$

with each $\sigma_{i_1, \ldots, i_m} \in \Sigma[x]$. The set of all possible sums in the above right hand side is the preordering $\mathrm{Pre}[g_1, \ldots, g_m]$. For $k = 1, 2, \ldots$, we get a hierarchy of SOS relaxations

$$\begin{cases} \gamma_k := \max & \gamma \\ s.t. & f - \gamma \in \mathrm{Pre}[g_1, \ldots, g_m]_{2k}. \end{cases} \tag{3.5.2}$$

If we write the objective as

$$f(x) = f_0 + f_1 x + \cdots + f_d x^d,$$

then the dual optimization of (3.5.2) is

$$\begin{cases} \gamma_k' := \min & f_0 y_0 + f_1 y_1 + \cdots + f_d y_d \\ s.t. & L_{p_1 \cdots p_m}^{(k)}[y] \succeq 0, \ p_i \in \{1, g_i\}, \ i = 1, \ldots, m, \\ & y_0 = 1, y = (y_0, y_1, \ldots, y_{2k}) \in \mathbb{R}^{2k+1}. \end{cases} \tag{3.5.3}$$

As $k$ increases, the above gives a hierarchy of Moment-SOS relaxations. We have the following convergence property.

**Theorem 3.5.1.** *For the optimization problem (3.5.1), if the feasible set $K$ is compact, then $\gamma_k \leq \gamma_k' \leq f_{min}$ for all $k$ and $\gamma_k \to f_{min}$ as $k \to \infty$. Moreover, if $K$ has nonempty interior, then (3.5.2) achieves its optimal value if it is bounded below.*

The above conclusion can be implied by Schmüdgen's Positivstellensatz (i.e., Theorem 2.6.2), so its proof is left as an exercise. We would like to remark that we may not have $\gamma_k = f_{min}$ for any $k$, even if $K$ is a compact interval. For instance, when the constraint is $(1 - x^2)^3 \geq 0$ and $f(x) = x$, the minimum $f_{min} = -1$. The feasible set is the compact interval $[-1, 1]$. However, we have $\gamma_k < f_{min}$ for all $k$. Suppose otherwise that there are SOS polynomials $\sigma_0, \sigma_1$ such that

$$f(x) - f_{min} = x + 1 = \sigma_0 + (1 - x^2)^3 \sigma_1.$$

Then $-1$ must be a root of $\sigma_0$. Since $\sigma_0$ is SOS, the multiplicity of $-1$ must be two (or a higher even number). Dividing both sides by $x + 1$, we can get

$$1 = (x+1)s_0 + (1+x)^2(1-x)^3\sigma_1$$

for another SOS polynomial $s_0$. This is impossible, because the right hand side vanishes at $-1$, while the left hand side is constantly equaling one. Indeed, it was shown in [307] that there is a constant $C > 0$ such that

$$|\gamma_k - f_{min}| \geq C \cdot k^{-2}, \quad k = 3, 4, \ldots.$$

For this example, one can also show that $\gamma'_k < f_{min}$ for all $k$ (see exercise). Since $K = [-1, 1]$, if we use the method in §3.4.2, the Moment-SOS relaxation is tight, i.e., $f_{sos} = f_{mom} = -1$. The above shows that the convergence of Moment-SOS relaxations depends on the choice of defining polynomials $g_i$.

## 3.5.1 ▪ Bounded feasible sets

An interesting special case is that the feasible set $K = [a, b]$, a bounded interval. For such a case, the defining polynomials $g_1, \ldots, g_m$ are not necessarily the standard ones like $x - a, b - x$. They can be more complicated ones, e.g., $(1 + x^2)(x - a)$, $(1 - x)^4(b - x)^5$. It is worthy to note that when $K = [a, b]$, there exists a single polynomial $h \in \text{Pre}[g_1, \ldots, g_m]$ such that $K = \{h(x) \geq 0\}$. This is shown by Stengle [307]. Recall that $\mathscr{P}([-1, 1])$ is the cone of polynomials that are nonnegative on $[-1, 1]$. Powers and Reznick characterized polynomials in $\mathscr{P}([-1, 1])$ that belong to the preordering $\text{Pre}[h]$.

**Theorem 3.5.2.** *([267, Corollary 11]) Let $h$ be a polynomial such that $\{h(x) \geq 0\} = [-1, 1]$, with $h(x) = (x+1)^{\nu_1}(1-x)^{\nu_2}h_1(x)$ and $h_1(\pm 1) > 0$. Suppose $p$ is a polynomial nonnegative on $[-1, 1]$ such that the following hold:*

*(i) if $p(-1) = 0$ then $(x + 1)^{\nu_1}$ divides $p$;*

*(ii) if $p(1) = 0$ then $(1 - x)^{\nu_2}$ divides $p$.*

*Then, we have $p \in \text{Pre}[h]$. In particular, the equality $\mathscr{P}([-1, 1]) = \text{Pre}[h]$ occurs if and only if $\nu_1 = \nu_2 = 1$.*

By the above theorem, whether the convergence $\gamma_k \to f_{min}$ is finite or not can be detected by considering the polynomial $\hat{f} = f - f_{min}$ and a polynomial $h \in \text{Pre}[g_1, \ldots, g_m]$ such that $\{h(x) \geq 0\} = [a, b]$. Generally, one wants such $h$ whose degree is the smallest. When $a$ and/or $b$ are minimizers, we need to compare the vanishing order of $\hat{f}$ at $a$ and/or $b$ to that of $h$. If the vanishing order of $\hat{f}$ is less than that of $h$, the convergence rate of $\gamma_k \to f_{min}$ can be very slow. Stengle [307] considered the case that $h = (1 - x^2)^3$ and $f = 1 - x^2$. For each $\epsilon > 0$, it was shown that

$$1 - x^2 + \epsilon = s_0 + (1 - x^2)^3 s_1$$

for SOS polynomials $s_0, s_1$. For the above representation to hold, Stengle [307] showed that

$$\deg(s_0), \deg(s_1) \geq C_1 \epsilon^{-\frac{1}{2}}$$

for a constant $C_1 > 0$ independent of $\epsilon$. Moreover, he also showed the existence of satisfactory $s_0, s_1$ with

$$\deg(s_0), \deg(s_1) \leq C_2 \epsilon^{-\frac{1}{2}}|\log \epsilon|$$

for another constant $C_2 > 0$ independent of $\epsilon$.

When $\gamma_k$ fails to have finite convergence to $f_{min}$, it is generally quite difficult to estimate its convergence rate. This issue is related to degree bound for Schmüdgen's Positivstellensatz. We first consider the special case that there is a single inequality constraint $h(x) \geq 0$ and it defines the set $[-1, 1]$. The following is from [267].

**Theorem 3.5.3.** *([267, Corollary 8]) Suppose $h(x) = (1 - x^2)^r q(x)$, where $r$ is odd, $0 \leq q(x) \leq \beta$ for $|x| \leq 1$, and $q(x) \geq \alpha > 0$ for $|x| \geq 1$. Assume that $p \in \mathbb{R}[x]_n$ is positive on $[-1, 1]$ and it has $d$ real roots $u_j$ with $|u_j| \geq u > 1$. Then, there exist SOS polynomials $\sigma_0, \sigma_1$ such that $p = \sigma_0 + \sigma_1 h$ and the degrees of $\sigma_0, \sigma_1$ are at most*

$$d\left(\frac{r}{2(u-1)}(\beta/\alpha)^{\frac{1}{r}} + \deg(h) - 1\right) + n.$$

The above results can be naturally extended to any bounded interval $[a, b]$. To estimate the convergence rate $\gamma_k \to f_{min}$, one needs to know the smallest absolute value of real roots of $f(x) - f_{min} + \epsilon$ for $\epsilon > 0$.

In the following, we present a general result about the convergence rate of $\gamma_k \to f_{min}$, when the feasible set $K$ is compact. The following result is about general multivariate polynomial optimization, which includes $n = 1$ as a special case.

**Theorem 3.5.4.** *([300, Theorem 4]) Suppose the polynomials $g_1, \ldots, g_m \in \mathbb{R}[x_1, \ldots, x_n]$ are such that the set $K$ is nonempty and compact. Then, there exists an integer $c \geq 1$ such that for all $f \in \mathbb{R}[x_1, \ldots, x_n]_d$ it holds that*

$$|\gamma_k - f_{min}| \leq cd^4 n^{2d} \frac{\|f\|}{\sqrt[c]{2k}}$$

*for all $k \geq \frac{1}{2}cd^c n^{cd}$.*

In the above, the norm $\|f\|$ is defined such that

$$\|f\| = \max_\alpha |f_\alpha| \frac{\alpha_1! \cdots \alpha_n!}{|\alpha|!}$$

for $f$ having the expansion $f(x) = \sum_\alpha f_\alpha x^\alpha$. The constant $c$ in Theorem 3.5.4 is generally hard to estimate.

We remark that for the univariate case, the feasible set $K$ is compact if and only if $\mathrm{QM}[g]$ is archimedean (see §2.6.4 or [191, Theorem 7.1.2]). In the relaxation (3.5.2), if the truncated preordering $\mathrm{Pre}[g_1, \ldots, g_m]_{2k}$ is replaced by the truncated quadratic module $\mathrm{QM}[g_1, \ldots, g_m]_{2k}$, we still have the convergence $\gamma_k \to f_{min}$ when $K$ is compact. An estimation for its convergence rate is given in [211]. We refer to Chapter 5 for the convergence theory of the Moment-SOS hierarchy based on quadratic modules.

### 3.5.2 ▪ Unbounded feasible sets

In Theorem 3.5.1, if the feasible set is not compact, the sequence $\{\gamma_k\}$ might fail to have convergence to $f_{min}$. The feasible set $K$ is a union of intervals. If it is unbounded, $K$ must contain $(-\infty, a]$ or $[a, +\infty)$ for a real number $a$. For such cases, the preordering $\mathrm{Pre}[g_1, \ldots, g_m]$ has special properties. The following is a natural extension of Theorem 13 of [267].

**Theorem 3.5.5.** *Let $g_1, \ldots, g_m$ be polynomials such that $[a, +\infty) \subseteq K$ for some $a \in \mathbb{R}^1$. If there exists a sequence $\{p_k\}_{k=1}^{\infty} \subseteq \mathrm{Pre}[g_1, \ldots, g_m]$ such that $p_k \to p$ (coefficientwise) and $\deg(p_k) = \deg(p)$, then $p \in \mathrm{Pre}[g_1, \ldots, g_m]$.*

**Proof.** The proof is similar to that for Theorem 13 of [267]. Write

$$p_k = \sum_{i_1,\ldots,i_m \in \{0,1\}} s_{k,i_1,\ldots,i_m} \cdot g_1^{i_1} \cdots g_m^{i_m}$$

for SOS polynomials $s_{k,i_1,\ldots,i_m}$. Without loss of generality, we can assume each $g_i$ is not identically zero. Since $[a, +\infty) \subseteq K$, each $g_i$ must have positive leading coefficients. In the above representation of $p_k$, there is no cancellation of coefficients for the highest degree. Therefore, there exist $v > u > a$ and $\delta \in (0,1)$ such that each

$$g_i(x) \geq \delta \quad \forall x \in [u, v].$$

Since $p_k \to p$, there exists $T$ such that

$$T \geq p_k(x) \quad \forall x \in [u, v], \, \forall \, k.$$

We can generally assume $\deg(p_k) = \deg(p) = 2d$ (if the degree is odd, we can consider them as polynomials of an even degree, whose highest degree coefficients are zeros). So, each $s_{k,i_1,\ldots,i_m}$ also has the degree at most $2d$. Therefore,

$$T \geq p_k(x) \geq \delta^m s_{k,i_1,\ldots,i_m}(x) \quad \forall x \in [u, v].$$

Write each $s_{k,i_1,\ldots,i_m}$ as

$$s_{k,i_1,\ldots,i_m} = (c_k)^T [x]_{2d}$$

for a coefficient vector $c_k$. Since $s_{k,i_1,\ldots,i_m}$ is SOS, it must be nonnegative, so

$$T^2 \geq \delta^{2m} s_{k,i_1,\ldots,i_m}(x)^2 \quad \forall x \in [u, v].$$

Integrating the above on $[u, v]$ gives

$$T^2(v - u) \geq \delta^{2m} c_k^T \left( \int_u^v [x]_{2d}([x]_{2d})^T \mathrm{d}x \right) c_k.$$

Since $X = \int_u^v [x]_{2d}[x]_{2d}^T \mathrm{d}x$ is positive definite, the sequence $\{c_k\}_{k=1}^{\infty}$ is bounded. Therefore, we can generally assume that $c_k \to c$ as $k \to \infty$. This means that each $s_{k,i_1,\ldots,i_m}$ converges to a polynomial, say, $s_{i_1,\ldots,i_m}$, which must also be SOS and have the same degree. By the limiting, we can get

$$p = \sum_{i_1,\ldots,i_m \in \{0,1\}} s_{i_1,\ldots,i_m} \cdot g_1^{i_1} \cdots g_m^{i_m},$$

which completes the proof. $\square$

The above theorem implies the following facts.

- For the case that $g_1 = x^3$ ($m = 1$) and $f = x$, the minimum value $f_{min}$ clearly equals $0$. But we have $\gamma_k < -1$ for all $k$. If otherwise $\gamma_k \geq -1$ for any $k$, then $1 + x \in \mathrm{Pre}[x^3]$ by Theorem 3.5.5. But this is impossible, as shown in [267]. Suppose otherwise that

$$1 + x = p + x^3 q$$

for some $p, q \in \Sigma[x]$. Because $p, q$ are SOS, their degrees must be even. The degree of the right hand side is $\max\{\deg(p), 3 + \deg(q)\}$, which cannot equal $1$. For this example, we do not have the convergence $\gamma_k \to f_{min}$.

- For the case that $g_1 = x(1 + x^2)$ ($m = 1$) and $f = x^d$, with $d$ an odd degree. Clearly, $f_{min} = 0$. However, as pointed out in [267], we also do not have the convergence $\gamma_k \to f_{min}$. In fact, there exists $\epsilon > 0$ (depending on $d$) such that $\gamma_k < -\epsilon$ for all $k$. If otherwise such $\epsilon$ does not exist, then by Theorem 3.5.5 we can conclude that $x^d \in \text{Pre}[x(1 + x^2)]$. This is impossible, because otherwise

$$x^d = (x + x^3)p(x) + q(x)$$

for some SOS polynomials $p, q$. The proof for the nonexistence of such $p, q$ is left as an exercise.

One wonders when do we have the convergence $\gamma_k \to f_{min}$ for noncompact $K$? For general cases, there is no neat characterization. However, for the case that $(-\infty, a] \subset K$ or $[a, +\infty) \subset K$, Theorem 3.5.5 implies the following characterization.

**Theorem 3.5.6.** *Let $g_1, \ldots, g_m$ be polynomials such that $[a, +\infty) \subseteq K$ for some $a$. Then the convergence $\gamma_k \to f_{min}$ occurs if and only if $\gamma_k = f_{min}$ for some $k$, which is equivalent to*

$$f - f_{min} \in \text{Pre}[g_1, \ldots, g_m].$$

### 3.5.3 ▪ Exercises

**Exercise 3.5.1.** *Give a polynomial $p \in \text{Pre}[(1 + x)x(1 - x), 1 - x^4]$ such that*

$$\{x \in \mathbb{R}^1 : p(x) \geq 0\} = [0, 1].$$

**Exercise 3.5.2.** *In (3.5.1), if the constraint is $(1 - x^2)^3 \geq 0$ and $f = x$, show that the optimal value $\gamma_k'$ of (3.5.3) is less than $f_{min}$ for all $k$.*

**Exercise 3.5.3.** *Let $d > 0$ be an odd integer. Show that there are no SOS polynomials $p, q$ such that $x^d = (x + x^3)p(x) + q(x)$.*

**Exercise 3.5.4.** *Prove Theorems 3.5.1 and 3.5.2.*

# Part II

# Moment-SOS Hierarchies for Polynomial Optimization Problems

# Chapter 4

# Unconstrained Polynomial Optimization

*A basic question in optimization is to minimize a polynomial over the entire space. The standard sum of squares and moment relaxations are applicable to solve this problem. Homogeneous polynomial optimization with sphere constraints is closely related to unconstrained optimization. This chapter introduces basic results for these topics.*

## 4.1 ▪ Minimizing Polynomials

A basic optimization problem is to minimize a polynomial function over the entire space $\mathbb{R}^n$. For a given polynomial $f$ in the variable $x := (x_1, \ldots, x_n)$, consider the unconstrained optimization problem

$$\min_{x \in \mathbb{R}^n} \quad f(x). \tag{4.1.1}$$

For $f$ to have a finite minimum value over $\mathbb{R}^n$, the degree of $f$ must be even, say, $2d$. Otherwise, $f$ is unbounded below in $\mathbb{R}^n$. The optimization problem (4.1.1) has a broad source of applications. For instance, a frequently appearing application is to solve the nonlinear least squares (NLS) problem

$$f_1(x) \approx 0, \quad \ldots, \quad f_m(x) \approx 0,$$

where $f_1, \ldots, f_m$ are given polynomials. This is equivalent to solving (4.1.1) for the polynomial

$$f = f_1(x)^2 + \cdots + f_m(x)^2.$$

The nonlinear equations $f_i(x) = 0$ have a common real solution if and only if the corresponding optimization (4.1.1) has a minimizer and the minimum value is zero. For instance, the best rank-1 tensor approximation problem [228] is to find vectors $a \in \mathbb{R}^{n_1}, b \in \mathbb{R}^{n_2}, c \in \mathbb{R}^{n_3}$ such that $\mathcal{A} \approx a \otimes b \otimes c$ for a given cubic order tensor $\mathcal{A} \in \mathbb{R}^{n_1 \times n_2 \times n_3}$. This requires one to solve the optimization (4.1.1) for the polynomial $f = \|\mathcal{A} - a \otimes b \otimes c\|^2$ (see (11.1.18) for the tensor norm), with the variable $x = (a, b, c)$. Another application of the optimization (4.1.1) is sensor network localization [216]. It requires one to determine locations of a set of vectors such that their mutual distances match given ones.

The task of solving the unconstrained optimization (4.1.1) is NP-hard [158, 204], when the degree of $f$ is 4 or higher. This can be implied by the complexity of the partition problem. For a given integer vector $a \in \mathbb{Z}^n$, the partition problem is to determine whether or not $a$ can be partitioned into two subvectors such that they have equal sums of entries. This can be equivalently reformulated as solving (4.1.1) for the polynomial

$$f = (a^T x)^2 + (x_1^2 - 1)^2 + \cdots + (x_n^2 - 1)^2.$$

Its minimum value is zero if and only if $a$ can be partitioned into two subvectors of equal sums.

Let $f_{min}$ denote the optimal value of (4.1.1). A feasible point $u$ is said to be a minimizer if $f(u) = f_{min}$. Throughout the book, a minimizer means a global minimizer, unless otherwise its meaning is specified. There are three possibilities:

- The minimum value $f_{min}$ is finite and achievable, i.e., $f(x) \geq f_{min}$ for all $x \in \mathbb{R}^n$ and there exists a point $u \in \mathbb{R}^n$ such that $f(u) = f_{min}$. For such a favorable case, we want to compute both $f_{min}$ and the minimizer $u$. When $f$ is a convex polynomial, if $f_{min}$ is finite, then $f$ must have a minimizer [17].

- The minimum value $f_{min}$ is finite but not achievable, i.e., $f(x) > f_{min}$ for all $x \in \mathbb{R}^n$ but for each $\epsilon > 0$ there exists a point $u \in \mathbb{R}^n$ such that $f(u) < f_{min} + \epsilon$. For instance, the bivariate polynomial $f = x_1^2 + (x_1 x_2 - 1)^2$ has the minimum value $f_{min} = 0$, which is not achievable.

- The minimum value $f_{min}$ is infinite, i.e., the problem (4.1.1) is unbounded below. This means that there exists a sequence $\{u_k\}_{k=1}^{\infty} \subseteq \mathbb{R}^n$ such that $f(u_k) \to -\infty$ as $k \to \infty$. For such a case, we are interested in detecting the unboundedness.

It is NP-hard to decide whether the unconstrained optimization (4.1.1) achieves the minimum value or not [2]. When $f_{min}$ is achievable, it is discussed how to compute $f_{min}$ and obtain minimizers in [210, 225]. When $f_{min}$ is not achievable, the computation of $f_{min}$ is discussed in [132, 302, 315]. When $f_{min} = -\infty$, a detection for the unboundedness is often wanted. This issue is discussed in §8.5.

### 4.1.1 ▪ The standard Moment-SOS relaxation

Suppose $f \in \mathbb{R}[x]_{2d}$. Let $f_{min}$ be the minimum value of (4.1.1). A scalar $\gamma \leq f_{min}$ is called a lower bound of $f$ in $\mathbb{R}^n$. Clearly, $\gamma$ is a lower bound for $f$ if and only if the polynomial $f(x) - \gamma$ is nonnegative everywhere, i.e., $f - \gamma \in \mathscr{P}_{n,2d}$. If the nonnegative polynomial cone $\mathscr{P}_{n,2d}$ is replaced by the SOS cone $\Sigma_{n,2d}$, then we get the classical *SOS relaxation*, which was discussed in [158, 257, 258]. The standard SOS relaxation for solving (4.1.1) is

$$\begin{cases} \max & \gamma \\ s.t. & f - \gamma \in \Sigma_{n,2d}. \end{cases} \quad (4.1.2)$$

The dual cone of $\Sigma_{n,2d}$ is $\mathscr{S}_{n,2d}$, the cone of all tms $y \in \mathbb{R}^{\mathbb{N}_{2d}^n}$ such that the moment matrix $M_d[y] \succeq 0$; see (2.4.13) for the cone $\Sigma_{n,2d}$. Hence, the dual optimization of (4.1.2) is

$$\begin{cases} \min & \langle f, y \rangle \\ s.t. & M_d[y] \succeq 0, \\ & y_0 = 1, \, y \in \mathbb{R}^{\mathbb{N}_{2d}^n}. \end{cases} \quad (4.1.3)$$

The equality $y_0 = 1$ is equivalent to $\langle 1, y \rangle = 1$. The optimization (4.1.3) is called a *moment relaxation* of (4.1.1). The primal-dual pair (4.1.2)-(4.1.3) is called the classical *Moment-SOS relaxation*. The Moment-SOS relaxation can be solved as a semidefinite program.

Let $f_{sos}$, $f_{mom}$ denote the optimal values of (4.1.2), (4.1.3) respectively. Interestingly, we always have $f_{sos} = f_{mom}$, i.e., there is no duality gap between (4.1.2) and (4.1.3). The Moment-SOS relaxation (4.1.2)-(4.1.3) is called *tight* if $f_{sos} = f_{min}$. The following is a classical result [158, 167, 171].

**Proposition 4.1.1.** *Let $f_{min}, f_{sos}, f_{mom}$ be the optimal values as above, then*

$$f_{sos} = f_{mom} \leq f_{min}.$$

*Furthermore, (4.1.2) has an optimizer if $f_{sos} > -\infty$.*

**Proof.** The optimization problems (4.1.2) and (4.1.3) are primal-dual pairs. Note that (4.1.3) is strictly feasible, because the interior of the cone $\mathscr{S}_{n,2d}$ intersects the hyperplane $y_0 = 1$. By the strong duality theorem, $f_{sos} = f_{mom}$ and (4.1.2) must achieve its optimal value if $f_{sos}$ is finite. For every $\gamma$ that is feasible in (4.1.2), we must have $\gamma \leq f_{min}$, so $f_{sos} \leq f_{min}$. □

The Moment-SOS relaxation (4.1.2)-(4.1.3) can be conveniently formulated as a semidefinite program by the software `GloptiPoly` and then it can be solved by an SDP solver, e.g., `SeDuMi`. The following is such an example.

**Example 4.1.2.** *For $f = x_1^4 + x_2^4 + x_3^4 - x_1 x_2 x_3$, the minimum value $f_{min} = -\frac{1}{256}$. The Moment-SOS relaxation (4.1.2)-(4.1.3) can be implemented by the software `GloptiPoly` as follows:*

```
mpol('x', 3);
f = x(1)^4+x(2)^4+x(3)^4-x(1)*x(2)*x(3);
PO = msdp(min(f));
[status, fsos] = msol(PO);
```

*The SOS relaxation (4.1.2) returns the lower bound $f_{sos} = -0.0039\cdots$. Indeed, one has $f_{sos} = -\frac{1}{256}$, which is justified by the SOS decomposition*

$$f + \frac{1}{256} = \frac{1}{3}\left[(x_1^2 - x_2^2 - \frac{1}{16} + x_3^2)^2 + (x_1^2 + x_2^2 - \frac{1}{16} - x_3^2)^2 + (x_1^2 - x_2^2 - x_3^2 + \frac{1}{16})^2 \right.$$
$$\left. + 2(\frac{x_1}{4} - x_2 x_3)^2 + 2(x_1 x_2 - \frac{x_3}{4})^2 + 2(x_1 x_3 - \frac{x_2}{4})^2\right].$$

*Using the syntax `double(x)`, we can get four global minimizers:*

$$\tfrac{1}{4}(1,1,1), \quad \tfrac{1}{4}(-1,-1,1), \quad \tfrac{1}{4}(-1,1,-1), \quad \tfrac{1}{4}(1,-1,-1).$$

The following are more examples of the Moment-SOS relaxation (4.1.2)-(4.1.3) for solving (4.1.1).

- For the dehomogenized Motzkin polynomial $f = x_1^4 + x_1^2 + x_2^6 - 3x_1^2 x_2^2$, $f_{min} = 0$ and the minimizers are 0 and $(\pm1, \pm1)$. The SOS relaxation gives the lower bound (see [258])

$$f_{sos} = -\frac{729}{4096} < f_{min}.$$

It is achieved by the SOS decomposition

$$f + \frac{729}{4096} = (-\frac{9}{8}x_2 + x_2^3)^2 + (\frac{27}{64} + x_1^2 - \frac{3}{2}x_2^2)^2 + \frac{5}{32}x_1^2.$$

- Consider $f = x_1^4 x_2^2 + x_1^2 x_2^4 + 1 - 3x_1^2 x_2^2$, which is another dehomogenization of the Motzkin polynomial. We still have $f_{min} = 0$ and the minimizers are $(\pm1, \pm1)$. However, the relaxation (4.1.2) is infeasible, i.e., the lower bound $f_{sos} = -\infty$. The dual optimization problem (4.1.3) is unbounded below.

We would like to remark that the moment relaxation (4.1.3) does not always have a minimizer, even if the Moment-SOS relaxation is tight. The following is such an example.

**Example 4.1.3.** *For the polynomial* $f = x_1^2 + (x_1x_2 - 1)^2$, *one can see that* $f_{sos} = f_{min} = 0$. *In the SOS relaxation (4.1.2), the optimal value is achieved for* $\gamma = 0$, *since* $f$ *itself is already SOS. However, the moment relaxation (4.1.3) does not have a minimizer. The objective in (4.1.3) is* $y_{20} + y_{22} - 2y_{11} + 1$, *with the psd matrix constraint*

$$
M_2[y] := \begin{bmatrix}
1 & y_{10} & y_{01} & y_{20} & y_{11} & y_{02} \\
y_{10} & y_{20} & y_{11} & y_{30} & y_{21} & y_{12} \\
y_{01} & y_{11} & y_{02} & y_{21} & y_{12} & y_{03} \\
y_{20} & y_{30} & y_{21} & y_{40} & y_{31} & y_{22} \\
y_{11} & y_{21} & y_{12} & y_{31} & y_{22} & y_{13} \\
y_{02} & y_{12} & y_{03} & y_{22} & y_{13} & y_{04}
\end{bmatrix} \succeq 0.
$$

*For each feasible* $y$, *the nonnegativity of principal minors implies*

$$
y_{20} \geq 0, \quad y_{22} \geq 0, \quad y_{22} - 2y_{11} + 1 \geq 0.
$$

*If* $y_{20} > 0$, *then the objective of (4.1.3) is positive. If* $y_{20} = 0$, *then* $y_{11} = 0$ *and hence*

$$
y_{20} + y_{22} - 2y_{11} + 1 \geq 1 > 0.
$$

*Therefore, the objective of (4.1.3) is always positive, which implies that (4.1.3) cannot achieve its minimum value.*

### 4.1.2 ▪ Exercises

**Exercise 4.1.1.** *([199]) For the following polynomial*

$$
f = 4x_1^2 - \frac{21}{10}x_1^4 + \frac{1}{3}x_1^6 + x_1x_2 - 4x_2^2 + 4x_2^4,
$$

*determine the optimal values* $f_{sos}, f_{mom}, f_{min}$.

**Exercise 4.1.2.** *Determine the lower bound* $f_{sos}$ *for the following polynomials:*

(i) $f = x_1^4 + x_2^4 + x_3^4 - x_1x_2 - x_2x_3 - x_1x_3$;

(ii) $f = x_1^6 + x_2^6 + x_1^4 + x_1^2 + x_2^2 - 3x_1^2x_2^2$;

(iii) $f = x_1^8 + x_2^8 + x_1^4x_2^2 + x_1^2x_2^4 - 3x_1^2x_2^2$.

**Exercise 4.1.3.** *Determine the range of* $a$ *such that* $f_{sos} > -\infty$ *for the polynomial*

$$
f = a(x_1^6 + x_2^6 + x_3^6) + x_1^4x_2^2 + x_1^2x_2^4 + x_2^6 - 3x_1^2x_2^2x_3^2.
$$

**Exercise 4.1.4.** *For each case of* $(n, 2d)$ *such that* $\mathscr{P}_{n,2d} \neq \Sigma_{n,2d}$, *show that there always exists* $f \in \mathbb{R}[x]_{2d}$ *such that*

$$
-\infty < f_{sos} = f_{mom} < f_{min}.
$$

**Exercise 4.1.5.** *If the minimum value* $f_{min}$ *is achievable for* $f$ *and the SOS relaxation is tight, show that moment relaxation (4.1.3) also achieves its optimal value.*

**Exercise 4.1.6.** *For a given matrix $A \in \mathbb{R}^{m \times n}$, the best rank-1 problem is to find vectors $a \in \mathbb{R}^m, b \in \mathbb{R}^n$ such that the distance $\|A - ab^T\|_F$ is minimum. Formulate this question as a polynomial optimization problem. Show that the Moment-SOS relaxation is tight.*

## 4.2 ▪ Extraction of Minimizers

Solving the Moment-SOS relaxation (4.1.2)-(4.1.3), one can get the lower bounds $f_{sos}$ and $f_{mom}$. The minimum value $f_{min}$ is usually not known in advance. How can we detect whether the Moment-SOS relaxation is tight or not? In most applications, people also like to get a minimizer of (4.1.1). This section addresses this issue.

An appropriate criterion for checking tightness of the Moment-SOS relaxation and for extracting minimizers is the *flat extension* [63] or *flat truncation* condition [225]. For $f \in \mathbb{R}[x]_{2d}$, assume $y^*$ is a minimizer of the moment relaxation (4.1.3).

First, consider the case that rank $M_d[y^*] = 1$. Then, we must have $f_{mom} = f_{sos} = f_{min}$. The minimizer can be selected as

$$u := \left( (y^*)_{e_1}, (y^*)_{e_2}, \ldots, (y^*)_{e_n} \right). \tag{4.2.1}$$

The moment matrix $M_d[y^*]$ is labeled by monomial powers. Due to the constraint $(y^*)_0 = 1$, the $(1, 1)$ entry of $M_d[y^*]$ is 1. If rank $M_d[y^*] = 1$, then the $\alpha$th column of $M_d[y^*]$ is just $(y^*)_\alpha$ times the first column for each $\alpha \in \mathbb{N}_d^n$. Or equivalently, rank $M_d[y^*] = 1$ means that

$$M_d[y^*] = \begin{bmatrix} 1 \\ y_{e_1}^* \\ \vdots \\ y_{e_n}^* \\ \vdots \end{bmatrix} \begin{bmatrix} 1 & y_{e_1}^* & \cdots & y_{e_n}^* & \cdots \end{bmatrix}.$$

So, for each $i = 1, \ldots, n$,

$$y_{\alpha+e_i}^* = y_\alpha^* y_{e_i}^* \quad \left( \forall \alpha \in \mathbb{N}_d^n \right).$$

The above implies that the first column of $M_d[y^*]$ is just the vector $[u]_d$. The rank one condition further implies that

$$M_d[y^*] = [u]_d [u]_d^T,$$

which is equivalent to $y^* = [u]_{2d}$. Therefore, one can further get

$$f_{min} \le f(u) = \langle f, [u]_{2d} \rangle = \langle f, y^* \rangle = f_{mom} \le f_{min}.$$

Hence $f(u) = f_{min}$ and $u$ must be a minimizer of $f$ in $\mathbb{R}^n$.

For instance, for $f = x_1^3 - x_1^5 + x_1^6 + x_1 x_2 - 3x_2^2 + 5x_2^4$, the syntax for solving the relaxations (4.1.2)-(4.1.3) by software `GloptiPoly` is

```
mpol('x', 2);
f = x(1)^3-x(1)^5+x(1)^6+x(1)*x(2)-3*x(2)^2+5*x(2)^4;
PO = msdp(min(f));
[status, fsos] = msol(PO);
```

It gives the value $f_{mom} = -0.91033037\cdots$. Since rank $M[y^*] = 1$, the SOS relaxation is tight. The syntax `double(x)` gives the minimizer

$$u = (-0.68496865\cdots, 0.59774874\cdots).$$

## 4.2.1 ▪ Flat extension

We consider the case that rank $M_d[y^*] > 1$. This usually happens when the Moment-SOS relaxation is not tight, or when the relaxation is tight but there are more than one minimizer. To get one or several minimizers, we typically assume the *flat extension* condition: there is an integer $t \in [0, d)$ such that

$$\text{rank } M_t[y^*] = \text{rank } M_d[y^*]. \tag{4.2.2}$$

The following is a classical result about extracting minimizers from the Moment-SOS relaxation [63, 122, 170, 225].

**Theorem 4.2.1.** *Suppose $y^*$ is a minimizer of the moment relaxation (4.1.3) and the rank condition (4.2.2) holds. Let $r := \text{rank } M_d[y^*]$. Then there exist $r$ distinct points $u_1, \ldots, u_r \in \mathbb{R}^n$ and positive scalars $\lambda_1, \ldots, \lambda_r > 0$ such that*

$$y^* = \lambda_1 [u_1]_{2d} + \cdots + \lambda_r [u_r]_{2d} \tag{4.2.3}$$

*and each $u_i$ is a minimizer of (4.1.1). Furthermore, the Moment-SOS relaxation is tight, i.e., $f_{min} = f_{sos} = f_{mom}$.*

**Proof.** Since the rank condition (4.2.2) holds, Theorem 2.7.7 implies the existence of distinct points $u_i \in \mathbb{R}^n$ and positive scalars $\lambda_i > 0$ satisfying the decomposition (4.2.3). The condition $(y^*)_0 = 1$ implies $\lambda_1 + \cdots + \lambda_r = 1$. Since each $f(u_i) \geq f_{min}$, we get

$$f_{min} \leq \lambda_1 f(u_1) + \cdots + \lambda_r f(u_r) = \lambda_1 \langle f, [u_1]_{2d} \rangle + \cdots + \lambda_r \langle f, [u_r]_{2d} \rangle$$

$$= \langle f, \lambda_1 [u_1]_{2d} + \cdots + \lambda_r [u_r]_{2d} \rangle = \langle f, y^* \rangle = f_{mom} \leq f_{min}.$$

The above implies $f_{min} = f(u_i)$ for all $i$, i.e., each $u_i$ is a minimizer of $f$.  □

**Example 4.2.2.** *Consider the polynomial $f = x_1^4 + x_2^4 + x_3^4 - x_1 x_2 x_3$ in Example 4.1.2. Solving the Moment-SOS relaxation, we get the optimal moment matrix $M_2[y^*]$:*

```
 1.0000   0.0000  -0.0000   0.0000   0.0625   0.0000  -0.0000   0.0625   0.0000   0.0625
 0.0000   0.0625   0.0000  -0.0000   0.0000  -0.0000   0.0000   0.0000   0.0156   0.0000
-0.0000   0.0000   0.0625   0.0000  -0.0000   0.0000   0.0156  -0.0000   0.0000  -0.0000
 0.0000  -0.0000   0.0000   0.0625   0.0000   0.0156   0.0000   0.0000  -0.0000   0.0000
 0.0625   0.0000  -0.0000   0.0000   0.0039   0.0000  -0.0000   0.0039   0.0000   0.0039
 0.0000  -0.0000   0.0000   0.0156   0.0000   0.0039   0.0000   0.0000  -0.0000   0.0000
-0.0000   0.0000   0.0156   0.0000  -0.0000   0.0000   0.0039  -0.0000   0.0000  -0.0000
 0.0625   0.0000  -0.0000   0.0000   0.0039   0.0000  -0.0000   0.0039   0.0000   0.0039
 0.0000   0.0156   0.0000  -0.0000   0.0000  -0.0000   0.0000   0.0000   0.0039   0.0000
 0.0625   0.0000  -0.0000   0.0000   0.0039   0.0000  -0.0000   0.0039   0.0000   0.0039
```

*One can check that* rank $M_1[y^*] = \text{rank } M_2[y^*] = 4$. *The extracted minimizers are*

$$\tfrac{1}{4}(1, 1, 1), \quad \tfrac{1}{4}(-1, -1, 1), \quad \tfrac{1}{4}(-1, 1, -1), \quad \tfrac{1}{4}(1, -1, -1).$$

*Moreover, $f_{sos} = f_{min} = -\tfrac{1}{256}$.*

## 4.2.2 ▪ Flat truncation

The rank condition (4.2.2) is sufficient for the tightness of the Moment-SOS relaxation (4.1.2)-(4.1.3). However, it may not be necessary. For instance, when $f$ is the nonlinear least squares polynomial

$$(x_1 x_2 - 1)^2 + (x_1 x_3 - 1)^2 + (x_2 x_3 - 1)^2,$$

we have $f_{sos} = f_{min} = 0$ but not every minimizer of (4.1.3) satisfies (4.2.2), as shown in Example 4.2.5. Generally, for the rank condition (4.2.2) to hold, the polynomial $f$ can have only finitely many minimizers. A detailed discussion about this is in [171, §6.6] and Proposition 5.3.6. Even for the case that $f$ has a finite number of minimizers, the condition (4.2.2) may still not hold. However, when the Moment-SOS relaxation is tight, (4.2.2) will eventually be satisfied if we apply higher order SOS relaxations. This will be shown in Theorem 4.2.7.

For an integer $k \geq d$, where $2d$ is the degree of $f$, we consider the $k$th order SOS relaxation

$$\begin{cases} \max & \gamma \\ s.t. & f - \gamma \in \Sigma_{n,2k}. \end{cases} \tag{4.2.4}$$

Similarly, its dual optimization is

$$\begin{cases} \min & \langle f, y \rangle \\ s.t. & M_k[y] \succeq 0, \\ & y_0 = 1, \ y \in \mathbb{R}^{\mathbb{N}^n_{2k}}. \end{cases} \tag{4.2.5}$$

The relaxation (4.2.4) is equivalent to (4.1.2), because $f - \gamma \in \Sigma_{n,2k}$ if and only if $f - \gamma \in \Sigma_{n,2d}$. So the optimal value of (4.2.4) is also equal to $f_{sos}$. One can also show that the optimal value of (4.2.5) is also $f_{mom}$. The proof is left as an exercise. However, (4.2.5) and (4.1.3) may be different for extracting minimizers.

**Definition 4.2.3.** *A minimizer $y^* \in \mathbb{R}^{\mathbb{N}^n_{2k}}$ of (4.2.5) is said to have a flat truncation if there exist integers $t_2 > t_1 \geq 0$ such that*

$$\text{rank } M_{t_1}[y^*] = \text{rank } M_{t_2}[y^*], \quad d \leq t_2 \leq k. \tag{4.2.6}$$

**Theorem 4.2.4.** *Suppose $y^*$ is a minimizer of (4.2.5). If $y^*$ has a flat truncation, i.e., (4.2.6) holds, then there exist $r := \text{rank } M_{t_2}[y^*]$ distinct points $u_1, \ldots, u_r \in \mathbb{R}^n$ and positive scalars $\lambda_1, \ldots, \lambda_r > 0$ such that*

$$y^*|_{2t_2} = \lambda_1 [u_1]_{2t_2} + \cdots + \lambda_r [u_r]_{2t_2} \tag{4.2.7}$$

*and each $u_i$ is a minimizer of (4.1.1). Moreover, the Moment-SOS relaxation is tight, i.e., $f_{sos} = f_{mom} = f_{min}$.*

The proof for Theorem 4.2.4 is quite similar to that for Theorem 4.2.1, so it is left as an exercise. We would like to remark that the condition (4.2.6) is weaker than (4.2.2). This is shown in the following example.

**Example 4.2.5.** *([225]) Let $f$ be the following polynomial:*

$$(x_1 x_2 - 1)^2 + (x_1 x_3 - 1)^2 + (x_2 x_3 - 1)^2.$$

*Clearly, $f_{sos} = f_{min} = 0$. The global minimizers are $\pm(1, 1, 1)$. However, not every minimizer of the moment relaxation (4.1.3) has a flat truncation. For instance, for all $a > 0$, the tms (its entries are listed in the graded lexicographical ordering on $\alpha \in \mathbb{N}^3_4$)*

$$y^*(a) := (1, 0, 0, 0, 1, 1, 1, 1, 1, 1, 0, 0, 0, 0, 0, 0, 0, 0, 0, 0,$$
$$1 + a, 1, 1, 1, 1, 1, 1, 1, 1, 1, 1 + a, 1, 1, 1, 1 + a)$$

*is a minimizer of the moment relaxation (4.1.3). For every $a > 0$, one can verify that*

$$\text{rank } M_0[y^*(a)] = 1, \quad \text{rank } M_1[y^*(a)] = 2, \quad \text{rank } M_2[y^*(a)] = 5.$$

*Thus, $y^*(a)$ does not satisfy the flat extension for $a > 0$. If we solve (4.1.3) by* SeDuMi, *a numerical value 2.7232 of $a$ is returned, and the computed minimizer does not have a flat truncation. For $k = 3$, the optimizer of (4.2.5) returned by* SeDuMi *does not have a flat truncation. However, for $k = 4$, the optimizer of (4.2.5) returned by* SeDuMi *has a flat truncation, and we are able to get two global minimizers $\pm(1, 1, 1)$.*

When $f$ has finitely many minimizers, the condition (4.2.6) is sufficient and necessary for detecting tightness of the Moment-SOS relaxation, but the value of $k$ may be bigger than $d$. Recall that $\mathrm{vec}(p)$ denotes the coefficient vector of a polynomial $p$. We need the following lemma.

**Lemma 4.2.6.** *([160, 171]) Let $y \in \mathbb{R}^{\mathbb{N}^n_{2k}}$ be such that $M_k[y] \succeq 0$ and let $p, q \in \mathbb{R}[x]$. If $\deg(pq) \leq k - 1$ and $\mathrm{vec}(q) \in \ker M_k[y]$, then $\mathrm{vec}(pq) \in \ker M_k[y]$. Moreover, if $\mathrm{vec}(q^\ell) \in \ker M_k[y]$ and $2\lceil \ell/2 \rceil \deg(q) \leq k - 1$ for some power $\ell$, then $\mathrm{vec}(q) \in \ker M_k[y]$.*

**Proof.** First, we prove the conclusion for $\deg(p) = 1$, say, $p = x_i$, such that $\deg(qx_i) \leq k - 1$. Note that $\deg(qx_i^2) \leq k$. Define the function

$$\phi(t) := \langle q^2(1 + tx_i^2)^2, y \rangle = \mathrm{vec}(q(1 + tx_i^2))^T M_k[y] \, \mathrm{vec}(q(1 + tx_i^2)).$$

Since $M_k[y] \succeq 0$, $\phi(t) \geq 0$ for all $t \in \mathbb{R}$. Observe that

$$\phi(0) = \langle q^2, y \rangle = \mathrm{vec}(q)^T M_k[y] \, \mathrm{vec}(q) = 0.$$

So, $\phi'(0) = 0$ and

$$\mathrm{vec}(qx_i)^T M_k[y] \, \mathrm{vec}(qx_i) = \langle q^2 x_i^2, y \rangle = \frac{1}{2}\phi'(0) = 0.$$

Since $M_k[y] \succeq 0$, the above implies $\mathrm{vec}(qx_i) \in \ker M_k[y]$.

Second, we can similarly show $\mathrm{vec}(qx_ix_j) \in \ker M_k[y]$ for all $\deg(qx_ix_j) \leq k - 1$, since we already have $\mathrm{vec}(qx_i) \in \ker M_k[y]$. Repeating this, one can see that $\mathrm{vec}(qx^\alpha) \in \ker M_k[y]$ for all $\deg(qx^\alpha) \leq k-1$. This further implies that $\mathrm{vec}(qp) \in \ker M_k[y]$ for all $\deg(qp) \leq k-1$.

Third, suppose $\mathrm{vec}(q^\ell) \in \ker M_k[y]$ for a power $\ell$ with $2\lceil \ell/2 \rceil \deg(q) \leq k-1$. When $\ell = 2s$ is even, we get

$$\mathrm{vec}(q^s)^T M_k[y] \, \mathrm{vec}(q^s) = \langle q^{2s}, y \rangle = \mathrm{vec}(1)^T M_k[y] \, \mathrm{vec}(q^{2s}) = 0.$$

Since $M_k[y] \succeq 0$, we get $\mathrm{vec}(q^s) \in \ker M_k[y]$. When $\ell = 2s - 1$ is odd, the above result for $p = q$ implies $\mathrm{vec}(q^{\ell+1}) \in \ker M_k[y]$, since $2s \deg(q) \leq k - 1$. So, we also get $\mathrm{vec}(q^{2s}) \in \ker M_k[y]$ and hence $\mathrm{vec}(q^s) \in \ker M_k[y]$. Note that $s < \ell$. Repeating this, one can eventually get $\mathrm{vec}(q) \in \ker M_k[y]$. $\square$

When the Moment-SOS relaxation is tight, the following theorem shows that the flat truncation (4.2.6) must be satisfied if $f$ has finitely many minimizers. This was shown in [225].

**Theorem 4.2.7.** *Let $S$ be the set of minimizers for (4.1.1). If $f_{min} = f_{sos}$ and $S \neq \emptyset$ is a finite set, then the relaxation (4.2.5) achieves its minimum value and every minimizer $y^*$ of (4.2.5) has a flat truncation as in (4.2.6), when the order $k$ is big enough.*

**Proof.** Since $f_{sos} = f_{min}$, the subtraction $\sigma := f - f_{min} \in \Sigma[x]$. Since the set $S \neq \emptyset$, the optimization (4.1.1) has a minimizer, say, $u$. Since $f_{min} = f_{sos}$, the optimal value of the moment relaxation (4.2.5) is also $f_{min}$. Note that (4.2.5) must achieve the minimum value $f_{min}$, since the tms $y = [u]_{2k}$ is feasible and

$$\langle f, [u]_{2k} \rangle = f(u) = f_{min}.$$

The remaining proof can be done in three major steps.

**Step 1** The set $S$ is nonempty and finite, so its vanishing ideal

$$I(S) = \{p \in \mathbb{R}[x] : p(u) = 0 \quad \forall u \in S\}$$

is zero dimensional. Let $\{h_1, \ldots, h_r\}$ be a Gröbner basis of $I(S)$ with respect to a total degree ordering. Clearly, each $h_i$ vanishes on the real algebraic set $f(x) - f_{min} = 0$. By Real Nullstellensatz (see Theorem 2.6.3), there exist $\ell \in \mathbb{N}$, $\varphi \in \mathbb{R}[x]$, and $\phi \in \Sigma[x]$ such that

$$h_i^{2\ell} + \phi = (f - f_{min})\varphi.$$

Applying the Riesz functional $\mathscr{L}_{y^*}$ to the above identity (suppose $2k$ is bigger than the degrees of all appearing polynomials), we get that

$$\langle h_i^{2\ell}, y^* \rangle + \langle \phi, y^* \rangle = \langle (f - f_{min})\varphi, y^* \rangle.$$

Note that $\langle (f - f_{min})\varphi, y^* \rangle = \langle \sigma\varphi, y^* \rangle$, where $\sigma := f - f_{min}$.

**Step 2** Since $\langle f - f_{min}, y^* \rangle = 0$, one can further show that $\langle \sigma\varphi, y^* \rangle = 0$ when $k$ is big enough (this is left as an exercise). So,

$$\langle h_i^{2\ell}, y^* \rangle + \langle \phi, y^* \rangle = 0.$$

Since $h_i^{2\ell}$ and $\phi$ are SOS, we have $\langle \phi, y^* \rangle \geq 0$ and each $\langle h_i^{2\ell}, y^* \rangle \geq 0$. Hence, $\langle h_i^{2\ell}, y^* \rangle = 0$, i.e., $h_i^\ell \in \ker M_k[y^*]$. By Lemma 4.2.6, when $k$ is big enough, we have

$$h_i \in \ker M_k[y^*].$$

**Step 3** It's enough to show the conclusion is true for $t_2 = k - 1$. Since the set $S$ is finite, the quotient space $\mathbb{R}[x]/I(S)$ is finitely dimensional. Let $\{b_1, \ldots, b_L\}$ be the standard basis of $\mathbb{R}[x]/I(S)$. Then, for every $\alpha \in \mathbb{N}^n$, we can write

$$x^\alpha = \eta(\alpha) + \sum_{i=1}^{r} \theta_i h_i, \quad \deg(\theta_i h_i) \leq |\alpha|, \quad \eta(\alpha) \in \text{span}\{b_1, \ldots, b_L\}.$$

Since each $h_i \in \ker M_k[y^*]$, Lemma 4.2.6 implies that

$$\theta_i h_i \in \ker M_k[y^*] \quad \text{if } |\alpha| \leq k - 1.$$

So, $x^\alpha - \eta(\alpha) \in \ker M_k[y^*]$ for all $|\alpha| \leq k - 1$. Let

$$d_b := \max_{1 \leq j \leq L} \deg(b_j).$$

Then, for each $|\alpha| \in [d_b + 1, k - 1]$, the $\alpha$th column of $M_k[y^*]$ is a linear combination of $\beta$th columns of $M_k[y^*]$ with $|\beta| \leq d_b$, so

$$\text{rank } M_{d_b}[y^*] = \text{rank } M_t[y^*], \quad t = d_b + 1, \ldots, k - 1.$$

Hence, for each integer $t_1 \in [d_b, k-2]$, we have

$$\operatorname{rank} M_{t_1}[y^*] = \operatorname{rank} M_{k-1}[y^*].$$

This means that (4.2.6) is satisfied for the above $t_1, t_2$, when $k$ is big enough.     □

Theorem 4.2.7 shows that the flat truncation (4.2.6) must be satisfied when $k$ is big enough. Typically, it is very hard to estimate such value of $k$. The polynomial $f$ has finitely many minimizers when its coefficients are generic; see Theorem 5.5.3 (or [217, Prop. 2.1]). If $y^*$ is an optimizer of (4.1.3) instead of (4.2.5), then $y^*$ may not have a flat truncation. This is shown in Example 4.2.5. In this sense, the relaxation (4.2.5) is stronger than (4.1.3), though they have the same optimal value.

### 4.2.3 ▪ Exercises

**Exercise 4.2.1.** *Find global minimizers of the following polynomials:*

(i) $(x_1 x_2 - x_3)^2 + (x_2 x_3 - x_1)^2 + (x_1 x_3 - x_2)^2$.

(ii) $(x_1^2 - 1)^2 + (x_2^2 - 1)^2 + (x_3^2 - 1)^2 + x_1^2 x_2^2 x_3^2$.

(iii) $x_1^4 x_2^2 + x_2^4 x_3^2 + x_3^4 x_1^2 + (x_1^2 + x_2^2 + x_3^2 - 1)^2$.

**Exercise 4.2.2.** *Show that the optimal value of (4.2.5) is equal to $f_{mom}$.*

**Exercise 4.2.3.** *Prove Theorem 4.2.4.*

**Exercise 4.2.4.** *Let $y \in \mathbb{R}^{\mathbb{N}_{2k}^n}$ be such that $M_k[y] \succeq 0$. If $\langle \sigma, y \rangle = 0$ for some $\sigma \in \Sigma[x]_{2d}$ with $d \leq k$, show that $\langle \sigma p, y \rangle = 0$ for all $p \in \mathbb{R}[x]_{2k-2-2d}$.*

**Exercise 4.2.5.** *In Step 2 of the proof for Theorem 4.2.7, show that $\langle \sigma\varphi, y^* \rangle = 0$ when $k$ is big enough.*

## 4.3 ▪ Higher Order Relaxations

The Moment-SOS relaxation (4.1.2)-(4.1.3) may not solve (4.1.1) globally, i.e., it is possible that $f_{sos} < f_{min}$. This is because not every nonnegative polynomial is SOS. However, by Artin's affirmative answer to Hilbert's 17th problem, for each nonnegative polynomial $p$, there exists an SOS polynomial $\sigma \neq 0$ such that the product $\sigma \cdot p$ is SOS. In particular, if $p$ is a positive definite form, we can choose $\sigma = (x^T x)^k$ (see Theorem 2.4.9). When $p$ is inhomogeneous, we may choose $\sigma = (1 + x^T x)^k$. This motivates to use higher order Moment-SOS relaxations.

For each integer $k \geq 0$, $\gamma$ is a lower bound for $f$ in $\mathbb{R}^n$ if and only if the product $(1 + x^T x)^k (f - \gamma)$ is a nonnegative polynomial, i.e.,

$$(1 + x^T x)^k (f - \gamma) \in \mathscr{P}_{n,2d+2k}.$$

If the cone $\mathscr{P}_{n,2d+2k}$ is changed to $\Sigma_{n,2d+2k}$, we get the $k$th order SOS relaxation

$$\begin{cases} \max & \gamma \\ s.t. & (1 + x^T x)^k (f - \gamma) \in \Sigma_{n,2d+2k}. \end{cases} \tag{4.3.1}$$

Let $f_k$ denote the optimal value of (4.3.1) for the power $k$. Clearly, $f_k \leq f_{min}$ for all $k$. If $\gamma$ is feasible for (4.3.1) with a value of $k$, then $\gamma$ must also be feasible for (4.3.1) with all larger values of $k$. Therefore, the sequence of lower bounds $\{f_k\}$ is monotonically increasing, i.e.,

$$f_0 \leq \cdots \leq f_k \leq \cdots \leq f_{min}.$$

When $k = 0$, the relaxation (4.3.1) is the same as (4.1.2), so $f_0 = f_{sos}$. The dual cone of $\Sigma_{n,2d+2k}$ is the moment cone $\mathscr{S}_{n,2d+2k}$ (see (2.4.13) for the notation). Hence, the dual optimization of (4.3.1) is

$$\begin{cases} \min & \langle (1 + x^T x)^k f, y \rangle \\ s.t. & \langle (1 + x^T x)^k, y \rangle = 1, \\ & M_{k+d}[y] \succeq 0, \\ & y \in \mathbb{R}^{\mathbb{N}^n_{2d+2k}}. \end{cases} \qquad (4.3.2)$$

For each $k$, the optimal value of (4.3.2) is also equal to $f_k$, by the strong duality theorem. The proof is left as an exercise. The primal-dual pair (4.3.1)-(4.3.2) is called the $k$th Moment-SOS relaxation for solving (4.1.1). When $k$ goes to infinity, we get a hierarchy of Moment-SOS relaxations. This hierarchy is said to be *tight* if $f_k = f_{min}$ for some $k$. The following is the asymptotic convergence result.

**Theorem 4.3.1.** *For* $f \in \mathbb{R}[x]_{2d}$, *let* $f^{hom}$ *be the homogeneous part of degree $2d$ for $f$. If $f^{hom}$ is a positive definite form, then $f_{min} > -\infty$ and*

$$\lim_{k \to \infty} f_k = f_{min}.$$

**Proof.** Since the form $f^{hom}$ is positive definite, the polynomial function $f(x) \to +\infty$ as $x \to +\infty$. So $f(x)$ must achieve its minimum value and $f_{min} > -\infty$. Moreover, for all $\epsilon > 0$, the homogeneous polynomial

$$\tilde{f}_\epsilon(x_0, x) := x_0^{2d} f(x/x_0) + (\epsilon - f_{min}) x_0^{2d}$$

is positive definite (this is left as an exercise). By Theorem 2.4.9, when $k$ is sufficiently large, we have $(x_0^2 + x^T x)^k \tilde{f}_\epsilon \in \Sigma^{hom}_{n+1,2d+2k}$. Upon dehomogenization, this is equivalent to

$$(1 + x^T x)^k (f - f_{min} + \epsilon) \in \Sigma_{n,2d+2k}.$$

The optimal value $f_k$ of the relaxation (4.3.1) is greater than or equal to $f_{min} - \epsilon$, for all $k$ big enough. The sequence $\{f_k\}$ is monotonically increasing and all $f_k \leq f_{min}$. Since $\epsilon > 0$ can be arbitrarily small, we must have $f_k \to f_{min}$ as $k \to \infty$. □

When the homogeneous part $f^{hom}$ is not positive definite, the convergence of $f_k \to f_{min}$ is not guaranteed. For instance, consider the form

$$f = x_1^4 \cdot (3x_2)^2 + x_1^2 \cdot (3x_2)^4 + (3x_3)^6 - 3x_1^2 \cdot (3x_2)^2 \cdot (3x_3)^2,$$

which is a scaling of the Motzkin form. For each integer $k > 0$, the product $(1 + x^T x)^k f$ cannot be SOS (see [284]). As a consequence, there is no $\gamma$ such that $(1 + x^T x)^k (f - \gamma)$ is SOS, and hence $f_k = -\infty$ for all $k$. However, $f$ itself is nonnegative and $f_{min} = 0$. The sequence of $f_k$ fails to converge to $f_{min}$.

A computational concern is how to get minimizers for (4.1.1) by solving the Moment-SOS hierarchy of relaxations (4.3.1)-(4.3.2). This can be done as follows. Suppose $y^*$ is a minimizer of the moment relaxation (4.3.2). To extract a minimizer for (4.1.1), we typically need the rank condition: there exists an integer $t \in [0, d+k]$ such that

$$\text{rank } M_t[y^*] = \text{rank } M_{d+k}[y^*]. \tag{4.3.3}$$

When (4.3.3) holds, we must have $f_k = f_{min}$ and can get rank $M_t[y^*]$ minimizers for (4.1.1).

**Theorem 4.3.2.** *Suppose $y^*$ is a minimizer of the moment relaxation (4.3.2) for some power $k$. If the rank condition (4.3.3) holds, then there exist $r := \text{rank } M_t[y^*]$ distinct points $u_1, \dots, u_r \in \mathbb{R}^n$ and scalars $\lambda_1, \dots, \lambda_r > 0$ such that*

$$y^*|_{2d+2k} = \lambda_1[u_1]_{2d+2k} + \cdots + \lambda_r[u_r]_{2d+2k}.$$

*Moreover, each $u_i$ is a minimizer of (4.1.1) and $f_k = f_{mom}$.*

Theorem 4.3.2 can be proved similarly to Theorem 4.2.4. It is left as an exercise. We conclude this section with an example for using the Moment-SOS hierarchy of (4.3.1)-(4.3.2).

**Example 4.3.3.** *Consider the dehomogenization of Motzkin form*

$$f = x_1^2 x_2^2 (x_1^2 + x_2^2) - 3x_1^2 x_2^2.$$

*The standard Moment-SOS relaxation does not give a finite lower bound, i.e., $f_{sos} = f_{mom} = -\infty$, but $f_{min} = -1$. We apply Moment-SOS relaxations (4.3.1)-(4.3.2) with $k = 2$. The syntax by software* GloptiPoly *is*

```
mpol('x', 2);
f = x(1)^2*x(2)^2*(x(1)^2+x(2)^2) -3*x(1)^2*x(2)^2;
k = 2; % give a value for k = 0, 1, 2, ...
PO = msdp(min( ((1+x'*x)^k)*f), [mom( (1+x'*x)^k )==1], 3+k);
[status, fk] = msol(PO);
```

*For $k = 1$, we get that $f_1 = -0.99999999 \cdots$, while the rank condition (4.3.3) is not satisfied. For $k = 2$, we get the same bound $f_2 = -0.99999999 \cdots$ but the rank condition (4.3.3) is satisfied. The syntax* double(x) *gives four minimizers $(\pm 1, \pm 1)$.*

### 4.3.1 ▪ Exercises

**Exercise 4.3.1.** *Apply the Moment-SOS hierarchy of (4.3.1)-(4.3.2) to find minimizers of the dehomogenized Robinson polynomial*

$$x_1^6 + x_2^6 + 1 + 3x_1^2 x_2^2 - (x_1^4 x_2^2 + x_1^2 x_2^4 + x_1^4 + x_2^2 + x_2^4 + x_2^2).$$

**Exercise 4.3.2.** *For $f = (x_1^2 + x_2^2 - 1)x_1^2 x_2^2$, what is the smallest $k$ such that $f_k = f_{min}$, where $f_k$ is the optimal value of (4.3.1)? Determine its minimizers by solving (4.3.2).*

**Exercise 4.3.3.** *Show that the dual optimization of the SOS relaxation (4.3.1) is the moment relaxation (4.3.2) and the strong duality holds, i.e., they have the same optimal value and (4.3.1) achieves the optimal value.*

**Exercise 4.3.4.** *Let $f \in \mathbb{R}[x]_{2d}$ be such that its homogeneous part of degree $2d$ is positive definite. Let $f_{min}$ be the minimum value of $f$ in $\mathbb{R}^n$. For each $\epsilon > 0$, show that the form $\tilde{f}_\epsilon(x_0, x) := x_0^{2d} f(x/x_0) + (\epsilon - f_{min})x_0^{2d}$ is positive definite.*

**Exercise 4.3.5.** *([284]) For the polynomial*

$$f = x_1^4 \cdot (3x_2)^2 + x_1^2 \cdot (3x_2)^4 + (3x_3)^6 - 3x_1^2 \cdot (3x_2)^2 \cdot (3x_3)^2,$$

*show that* $(x^T x)^k f$ *cannot be SOS for any power* $k \geq 0$. *Use this fact to show that* $(1 + x^T x)^k (f - \gamma)$ *cannot be SOS for any* $k, \gamma$.

**Exercise 4.3.6.** *Prove Theorem 4.3.2.*

# 4.4 ▪ Homogeneous Polynomial Optimization

Homogeneous polynomial optimization is to minimize forms with sphere constraints. For a given form $f \in \mathbb{R}[x]^{hom}$, consider the optimization problem

$$\left\{ \begin{array}{ll} \min & f(x) \\ s.t. & \|x\| = 1, \, x \in \mathbb{R}^n. \end{array} \right. \tag{4.4.1}$$

The feasible set is the $n - 1$ dimensional unit sphere

$$\mathbb{S}^{n-1} := \{x \in \mathbb{R}^n : \|x\|_2 = 1\}. \tag{4.4.2}$$

Homogeneous polynomial optimization has broad applications. For instance, the best rank-1 approximation for symmetric tensors can be formulated as optimizing forms over unit spheres [228]; the stability number of a graph can be expressed in terms of the minimum value of a homogeneous optimization problem [68]. More applications can be found in [220].

For convenience of notation, we still let $f_{min}$ denote the minimum value of (4.4.1). When $f(x) = a^T x$ is a linear form, $f_{min} = -\|a\|_2$. When $f(x) = x^T F x$ is a quadratic form with a symmetric matrix $F$, $f_{min}$ equals the minimum eigenvalue of $F$. When $\deg(f) > 2$, it is difficult to compute $f_{min}$. The complexity of solving (4.4.1) is already NP-hard when $f$ is cubic [205]. Moment-SOS relaxations are natural choices for solving (4.4.1) globally, as in the work [220]. We discuss how to solve (4.4.1) for even and odd degrees separately.

## 4.4.1 ▪ The even degree case

For an even degree form $f \in \mathbb{R}[x]_{2d}^{hom}$, the standard SOS relaxation for solving (4.4.1) is

$$\left\{ \begin{array}{ll} \max & \gamma \\ s.t. & f(x) - \gamma \cdot (x^T x)^d \in \Sigma_{n,2d}^{hom}. \end{array} \right. \tag{4.4.3}$$

Let $f_{sos}$ be the optimal value of (4.4.3), then $f_{sos} \leq f_{min}$. The dual cone of $\Sigma_{n,2d}^{hom}$ is $\mathscr{S}_{n,2d}^{hom}$, the set of all $z \in \mathbb{R}^{\overline{\mathbb{N}}_{2d}^n}$ such that $M_d[z] \succeq 0$ (see (2.4.17) for the notation). The dual optimization of (4.4.3) is the homogeneous moment relaxation

$$\left\{ \begin{array}{ll} \min & \langle f, z \rangle \\ s.t. & \langle (x^T x)^d, z \rangle = 1, \\ & M_d[z] \succeq 0, \, z \in \mathbb{R}^{\overline{\mathbb{N}}_{2d}^n}. \end{array} \right. \tag{4.4.4}$$

The above variable $z$ is a homogeneous tms (htms) of degree $2d$. The $M_d[z]$ is the homogeneous moment matrix as in (2.4.16). Both relaxations (4.4.3) and (4.4.4) have the same optimal value, and both of them achieve it. The proof is left as an exercise.

To see explicitly what the moment relaxation (4.4.4) is, we consider the case that $n = 3$ and $d = 2$. The constraint $M_d[z] \succeq 0$ reads as

$$M_2[z] = \begin{bmatrix} z_{400} & z_{310} & z_{301} & z_{220} & z_{211} & z_{202} \\ z_{310} & z_{220} & z_{211} & z_{130} & z_{121} & z_{112} \\ z_{301} & z_{211} & z_{202} & z_{121} & z_{112} & z_{103} \\ z_{220} & z_{130} & z_{121} & z_{040} & z_{031} & z_{022} \\ z_{211} & z_{121} & z_{112} & z_{031} & z_{022} & z_{013} \\ z_{202} & z_{112} & z_{103} & z_{022} & z_{013} & z_{004} \end{bmatrix} \succeq 0.$$

The equality constraint $\langle (x^T x)^d, z \rangle = 1$ is

$$z_{400} + 2z_{220} + 2z_{202} + z_{040} + 2z_{022} + z_{004} = 1.$$

The objective $\langle f, z \rangle$ is the linear function in $z$:

$$f_{400}z_{400} + f_{310}z_{310} + f_{301}z_{301} + f_{220}z_{220} + f_{211}z_{211} + f_{202}z_{202} + f_{130}z_{130} + f_{121}z_{121}$$
$$+ f_{112}z_{112} + f_{103}z_{103} + f_{040}z_{040} + f_{031}z_{031} + f_{022}z_{022} + f_{013}z_{013} + f_{004}z_{004},$$

where each $f_\alpha$ is the coefficient of the monomial $x^\alpha$ in $f$.

Minimizers can be extracted from the moment relaxation (4.4.4). Suppose $z^*$ is a minimizer of (4.4.4). The htms $z^*$ can be viewed as a tms $y^* \in \mathbb{R}^{\mathbb{N}_{2d}^{n-1}}$ such that

$$(y^*)_\alpha = (z^*)_{(2d-|\alpha|,\alpha)}$$

for each $\alpha \in \mathbb{N}_{2d}^{n-1}$. Such $y^*$ is called the dehomogenization of $z^*$. It is interesting to note that $M_d[y^*] = M_d[z^*]$. To extract minimizers, we typically need the rank condition: there exists an integer $t \in [0, d)$ such that

$$\text{rank } M_t[y^*] = \text{rank } M_d[y^*]. \tag{4.4.5}$$

**Proposition 4.4.1.** *Let $y^*, z^*$ be as above. If the condition (4.4.5) holds, then there exist $r := \text{rank } M_d[y^*]$ distinct points $u_1, \ldots, u_r \in \mathbb{R}^{n-1}$ and scalars $c_1, \ldots, c_r > 0$ such that*

$$y^* = c_1[u_1]_{2d} + \cdots + c_r[u_r]_{2d}. \tag{4.4.6}$$

*Let $\tau_i > 0$ be such that each $v_i := \tau_i(1, u_i) \in \mathbb{S}^{n-1}$. Then each $v_i$ is a minimizer of (4.4.1) and $f_{sos} = f_{min}$.*

**Proof.** Since $M_d[y^*] \succeq 0$ and $y^*$ is flat, by Theorem 2.7.7, there exist $r$ distinct points $u_1, \ldots, u_r \in \mathbb{R}^{n-1}$ such that $y^*$ has the above decomposition. For the choice of $\tau_i > 0$, one can see that $[v_i]_{2d}^{hom} = \tau_i^{2d}[u_i]_{2d}$. Hence,

$$z^* = y^* = \lambda_1[v_1]_{2d}^{hom} + \cdots + \lambda_r[v_r]_{2d}^{hom}, \quad \lambda_i = c_i\tau_i^{-2d} > 0.$$

Moreover, we also have

$$\langle f, z^* \rangle = \sum_{i=1}^r \lambda_i \langle f, [v_i]_{2d}^{hom} \rangle = \sum_{i=1}^r \lambda_i f(v_i),$$

$$0 \geq \langle f, z^* \rangle - f_{min}\langle (x^T x)^d, z^* \rangle = \sum_{i=1}^r \lambda_i(f(v_i) - f_{min}).$$

Since $f(v_i) \geq f_{min}$ and $\lambda_i > 0$ for all $i$, we must have $f(v_i) = f_{min}$ for every $i$, i.e., each $v_i$ must be a minimizer of $f$ on $\mathbb{S}^{n-1}$. The above also implies $f_{sos} = f_{min}$, since $\langle f, z^* \rangle = f_{sos}$ and $\lambda_1 + \cdots + \lambda_r = 1$. $\square$

It is possible that $f_{sos} < f_{min}$. For instance, this is the case when $f$ is the Motzkin form. Interestingly, the gap $f_{min} - f_{sos}$ can be bounded in terms of $f_{max} - f_{min}$, where $f_{max}$ is the maximum value of $f$ on $\mathbb{S}^{n-1}$. For a degree $d > 0$, denote the matrix

$$\Theta_d := \int_{\mathbb{S}^{d-1}} [u]_d^{hom} ([u]_d^{hom})^T \mathrm{d}\nu, \tag{4.4.7}$$

where $\nu$ is the uniform probability measure on $\mathbb{S}^{d-1}$. For instance,

$$\Theta_2 = \frac{1}{8} \begin{bmatrix} 3 & 0 & 1 \\ 0 & 2 & 0 \\ 1 & 0 & 3 \end{bmatrix}.$$

The matrix $\Theta_d$ is positive definite. Denote the constant

$$\delta_d := \sqrt{\lambda_{min}(\Theta_d)} > 0. \tag{4.4.8}$$

A list of values for $\delta_d$ is as follows:

| $d$ | 2 | 4 | 6 | 8 |
|---|---|---|---|---|
| $\delta_d$ | 0.5000 | 0.0559 | 0.0039 | 0.0002 |

As in [220], one can show that

$$\delta_{2d} \leq \left( \frac{2^d \cdot d!}{\sqrt{3}} \sqrt{\binom{3d}{d}} \right)^{-1}. \tag{4.4.9}$$

The difference $f_{min} - f_{sos}$ can be estimated as follows.

**Theorem 4.4.2.** *([220]) Let $n \geq 2d$. For every $f \in \mathbb{R}[x]_{2d}^{hom}$, let $f_{min}$ (resp., $f_{max}$) be its minimum (resp., maximum) value on $\mathbb{S}^{n-1}$. If $f_{sos}$ is the optimal value of the relaxation (4.4.3), then*

$$1 \leq \frac{f_{max} - f_{sos}}{f_{max} - f_{min}} \leq \frac{1}{\delta_{2d}} \sqrt{\binom{n}{2d}}, \tag{4.4.10}$$

*where the constant $\delta_{2d}$ is given as in (4.4.8).*

## 4.4.2 ▪ Higher order relaxations

The relaxation (4.4.3) may not be enough for solving (4.4.1) globally. Sometimes, we may need higher order relaxations. Consider the case that $f$ has an even degree $2d$. For an integer $k \geq 0$, consider the relaxation

$$\begin{cases} \max & \gamma \\ \text{s.t.} & (x^T x)^k \cdot (f - \gamma(x^T x)^d) \in \Sigma_{n,2k+2d}^{hom}. \end{cases} \tag{4.4.11}$$

Let $f_k$ be the optimal value of (4.4.11). Using the degree bound in Theorem 2.4.9, one can show that (see [94])

$$\frac{f_{min} - f_k}{f_{max} - f_{min}} \leq \frac{nd(2d-1)}{(2k+n+2d)\ln 2 - nd(2d-1)} \tag{4.4.12}$$

for all powers $k$ such that

$$k > \frac{2nd(2d-1)}{4\ln 2} - \frac{n}{2} - d.$$

Clearly, as $k \to \infty$, we get $f_k \to f_{min}$. The computational expense for solving (4.4.11) grows quickly as $k$ increases.

The dual cone of $\Sigma_{n,2k+2d}^{hom}$ is $\mathscr{S}_{n,2k+2d}^{hom}$, so the dual optimization of (4.4.11) is

$$\begin{cases} \min & \langle (x^T x)^k f, z \rangle \\ s.t. & \langle (x^T x)^{k+d}, z \rangle = 1, \\ & M_{d+k}[z] \succeq 0, \ z \in \mathbb{R}^{\overline{\mathbb{N}}_{2d+2k}^n}. \end{cases} \tag{4.4.13}$$

Suppose $z^*$ is a minimizer of (4.4.13). Let $y^* \in \mathbb{R}^{\mathbb{N}_{2d+2k}^{n-1}}$ be the dehomogenization of $z^*$, i.e., $y^*$ is the tms in $\mathbb{R}^{\mathbb{N}_{2d+2k}^{n-1}}$ such that

$$M_{k+d}[y^*] = M_{k+d}[z^*].$$

To obtain minimizes for (4.4.1), we typically need the rank condition: there exists an integer $t \in [0, d+k)$ such that

$$\text{rank } M_t[y^*] = \text{rank } M_{d+k}[y^*]. \tag{4.4.14}$$

Like Proposition 4.4.1, we similarly have the following proposition. The proof is quite similar, so it is left as an exercise.

**Proposition 4.4.3.** *Let $f_k$ be the optimal value of (4.4.11) for the power $k$. Suppose $z^*$ is a minimizer of (4.4.13) and $y^*$ is the dehomogenization as above. If the rank condition (4.4.14) holds, then there exist $r := \text{rank } M_{d+k}[y^*]$ distinct points $v_1, \ldots, v_r \in \mathbb{S}^{n-1}$ and scalars $\lambda_1, \ldots, \lambda_r > 0$ such that*

$$z^* = \lambda_1 [v_1]_{2d+2k}^{hom} + \cdots + \lambda_r [v_1]_{2d+2k}^{hom}$$

*and each $v_i$ is a minimizer of (4.4.1). Moreover, $f_k = f_{min}$ for all $k$ such that (4.4.14) holds,.*

We remark that the homogeneous moment relaxation (4.4.13) can be equivalently formulated as a nonhomogeneous one. Upon dehomogenizing the htms $z$, the relaxation (4.4.13) can be equivalently written as

$$\begin{cases} \min & \langle (1 + x_2^2 + \cdots + x_n^2)^k \overline{f}, y \rangle \\ s.t. & \langle (1 + x_2^2 + \cdots + x_n^2)^{k+d}, y \rangle = 1, \\ & M_{d+k}[y] \succeq 0, \ y \in \mathbb{R}^{\mathbb{N}_{2d+2k}^{n-1}}, \end{cases} \tag{4.4.15}$$

where $\overline{f} = f(1, x_2, \ldots, x_n)$ is the dehomogenization (i.e., let $x_1 = 1$). Therefore, (4.4.15) can be solved by the software GloptiPoly and SDP solvers. The following is an example for showing how to do this.

**Example 4.4.4.** *Consider the homogeneous polynomial*

$$f = x_1^4 x_2^2 + x_2^4 x_3^2 + x_3^4 x_1^2 - 4 x_1^2 x_2^2 x_3^2.$$

*The dehomogenization is $\overline{f} = x_2^2 + x_2^4 x_3^2 + x_3^4 - 4 x_2^2 x_3^2$. To solve (4.4.15), the syntax in GloptiPoly is*

```
mpol('x', 2);
x_2 = x(1); x_3 = x(2);
barf = x_2^2+x_2^4*x_3^2+x_3^4-4*x_2^2*x_3^2;
k = 1; % give value of k = 0, 1, 2, ...
PO = msdp(min( ((1+x'*x)^k)*barf ), ...
          [mom( (1+x'*x)^(k+3) )==1], 3+k);
[status, fk] = msol(PO);
```

*For $k = 0$, we get the lower bound $f_0 = -0.03725470 \cdots$ and the condition (4.4.14) fails to hold. For $k = 1$, we get the lower bound $f_1 = -0.03703705 \cdots$ and the condition (4.4.14) holds. The syntax* `double(x)` *gives four points $(\pm 1, \pm 1)$. Using the normalization trick as in Proposition 4.4.1, we get four optimizers $\frac{1}{\sqrt{3}}(1, \pm 1, \pm 1)$ on the unit sphere. The minimum value $f_{min} = \frac{-1}{27}$.*

### 4.4.3 ▪ The odd degree case

We discuss the case that $f$ has the odd degree $2d - 1$. The SOS relaxation (4.4.3) cannot be applied directly, because SOS forms must have even degrees. However, it can be equivalently formulated as homogeneous polynomial optimization of an even degree, as shown in [220]. Recall that $f_{max}$ denotes the maximum value of $f$ on $\mathbb{S}^{n-1}$, then

$$f_{max} + f_{min} = 0, \quad f_{min} \le 0 \le f_{max}.$$

Let $\hat{f}(\tilde{x}) := f(x) x_{n+1}$. It is an even degree form in $\tilde{x} := (x_1, \ldots, x_n, x_{n+1})$. Denote the optimal values

$$\hat{f}_{min} := \min_{\|x\|_2^2 + x_{n+1}^2 = 1} f(x) x_{n+1},$$

$$\hat{f}_{max} := \max_{\|x\|_2^2 + x_{n+1}^2 = 1} f(x) x_{n+1}.$$

As in [220], one can show that

$$\boxed{\begin{aligned} f_{min} &= \sqrt{2d-1}\left(1 - \tfrac{1}{2d}\right)^{-d} \hat{f}_{min}, \\ f_{max} &= \sqrt{2d-1}\left(1 - \tfrac{1}{2d}\right)^{-d} \hat{f}_{max}. \end{aligned}} \tag{4.4.16}$$

So (4.4.1) is equivalent to the even degree optimization

$$\min_{\|\tilde{x}\|_2 = 1} f(x) x_{n+1}. \tag{4.4.17}$$

The methods and results in previous subsections are applicable for solving (4.4.17). Suppose $(\hat{v}, \hat{x}_{n+1})$ is a minimizer of (4.4.17). Note that $\|\hat{v}\|^2 + \hat{x}_{n+1}^2 = 1$. If $\hat{v} = 0$ or $\hat{x}_{n+1} = 0$, then $f(\hat{v}) \hat{x}_{n+1} = 0$. This occurs only if $f$ is identically zero. So, we consider the case $0 < |\hat{x}_{n+1}| < 1$. Let

$$\hat{u} := \frac{\text{sign}(\hat{x}_{n+1})}{\sqrt{1 - |\hat{x}_{n+1}|^2}} \hat{v}. \tag{4.4.18}$$

One can show that $\hat{u}$ is a minimizer of $f$ on $\mathbb{S}^{n-1}$. This is left as an exercise. The following is an example for showing how to do this.

**Example 4.4.5.** *Consider the homogeneous polynomial*

$$f = x_1^3 + x_2^3 + x_3^3 - (x_1^2 x_2 + x_1 x_2^2 + x_2^2 x_3 + x_2 x_3^2).$$

*The dehomogenization of the product $f \cdot x_4$ with $x_1 = 1$ is*

$$\overline{f} = \left(1 + x_2^3 + x_3^3 - (x_2 + x_2^2 + x_2^2 x_3 + x_2 x_3^2)\right) x_4.$$

*To minimize the even degree form $f \cdot x_4$ on the unit sphere $x_1^2 + x_2^2 + x_3^2 + x_4^2 = 1$, we solve the dehomogenized moment relaxation (4.4.15). The coding in* `GloptiPoly` *is*

```
mpol('x', 3);
x_2 = x(1); x_3 = x(2); x_4 = x(3);
barf = (1+x_2^3+x_3^3-(x_2+x_2^2+x_2^2*x_3+x_2*x_3^2) )*x_4;
k = 0; % give a value of k = 0, 1, 2, ...
PO = msdp(min(((1+x'*x)^k)*(barf)), ...
                [mom((1+x'*x)^(k+2))==1],2+k);
[status, fk] = msol(PO);
```

*For $k = 0$, we get* fk $= -0.37988542\cdots$ *and the condition (4.4.14) holds. The syntax* double(x) *gives the point*

$$u = (-6.09583967\cdots, 1.00000019\cdots, 3.61261373\cdots).$$

*Using the normalization trick as in Proposition 4.4.1, we know that*

$$v = (0.1383955\cdots, -0.84363714\cdots, 0.13839558\cdots, 0.499969700\cdots)$$

*is a minimizer of $f \cdot x_4$ on the unit sphere $x_1^2 + x_2^2 + x_3^2 + x_4^2 = 1$ and the minimum value*

$$\hat{f}_{min} = -0.37988542\cdots.$$

*As in (4.4.16), the minimum value of $f$ on the unit sphere $x_1^2 + x_2^2 + x_3^2 = 1$ is*

$$f_{min} = -1.16974373\cdots.$$

*As in (4.4.18), the minimizer of $f$ on $x_1^2 + x_2^2 + x_3^2 = 1$ is*

$$\hat{u} = (0.15980219\cdots, -0.97412858\cdots, 0.15980223\cdots).$$

### 4.4.4 ▪ Exercises

**Exercise 4.4.1.** *Use the Moment-SOS relaxations to determine the minimum values and minimizers for the following forms over the unit sphere $\|x\| = 1$:*

(i) $x_1^2 x_2^2 + 2x_1^2 x_3^2 + 3x_2^2 x_3^2 + x_4^4 - 5x_1 x_2 x_3 x_4.$

(ii) $x_1^4 x_2^2 + 3x_2^4 x_3^2 + 5x_3^4 x_1^2 - 7x_1^2 x_2^2 x_3^2.$

(iii) $(2x_1 + 3x_2 - 4x_3)x_1 x_2 + x_3^3.$

(iv) $x_1^3 + x_2^3 + x_3^3 + 6x_1 x_2 x_3 - 2(x_1^2 x_2 + x_1 x_2^2 + x_1^2 x_3 + x_1 x_3^2 + x_2^2 x_3 + x_2 x_3^2).$

**Exercise 4.4.2.** *Show that the relaxations (4.4.11) and (4.4.13) have the same optimal value and it is achieved by both of them.*

**Exercise 4.4.3.** *Prove the inequality (4.4.9).*

**Exercise 4.4.4.** *Prove the relation (4.4.16).*

**Exercise 4.4.5.** *Prove the convergence rate (4.4.12) for the hierarchy of SOS relaxations (4.4.11).*

**Exercise 4.4.6.** *If $(\hat{v}, \hat{t})$ is a minimizer of (4.4.17) with $|\hat{t}| < 1$, show that the vector $\hat{u}$ given in (4.4.18) is a minimizer of the odd degree form $f$ on the unit sphere.*

**Exercise 4.4.7.** *Prove Proposition 4.4.3.*

# Chapter 5

# Constrained Polynomial Optimization

*This chapter discusses polynomial optimization with constraints. The Moment-SOS hierarchy is a typical approach for solving constrained polynomial optimization. We present basic results on its asymptotic and finite convergence, extraction of optimizers, optimality conditions, and related topics.*

## 5.1 ▪ Introduction to Constrained Optimization

A class of constrained optimization problems is

$$\begin{cases} \min & f(x) \\ s.t. & c_i(x) = 0 \, (i \in \mathcal{E}), \\ & c_j(x) \geq 0 \, (j \in \mathcal{I}), \end{cases} \tag{5.1.1}$$

where $f$ and all $c_i, c_j$ are polynomial functions in $x := (x_1, \ldots, x_n) \in \mathbb{R}^n$. The $\mathcal{E}$ and $\mathcal{I}$ are two disjoint finite labeling sets for constraining polynomials. Let $K$ denote the feasible set of (5.1.1). For convenience, we still let $f_{min}$ denote the minimum value of (5.1.1). When the feasible set $K$ is empty, we set $f_{min} := +\infty$ by convention. A feasible point $x^*$ for (5.1.1), i.e., $x^* \in K$, is called a *minimizer* if $f(x^*) \leq f(x)$ for all $x \in K$. In optimization, such $x^*$ is also called a *global minimizer*. In contrast to global minimizers, people may also be interested in local minimizers. A point $u \in K$ is called a *local minimizer* if there exists $\delta > 0$ such that

$$f(u) \leq f(x) \quad \text{for all } x \in K \cap B(u, \delta).$$

In the above, if the strict inequality $<$ holds for all $x \neq u$ in the set, then $u$ is called a *strict* local minimizer. Clearly, every global minimizer is a local minimizer, while the reverse is typically not true. Throughout the book, a minimizer means a global minimizer, unless otherwise its meaning is specified in the context.

The constrained polynomial optimization (5.1.1) is quite broad. When all $f, c_i, c_j$ are linear, (5.1.1) is a linear program (LP); when $f$ is quadratic and all $c_i, c_j$ are linear, (5.1.1) is a quadratic program (QP); when all $f, c_i, c_j$ are quadratic, (5.1.1) is a quadratically constrained quadratic program (QCQP). When $f$ is quadratic and the set $K$ is a ball constraint, the optimization (5.1.1) reduces to the classical trust region subproblem in nonlinear programming [329]. When $f$ is quadratic and there are two quadratic inequality constraints, the optimization (5.1.1) includes the classical Celis-Dennis-Tapia (CDT) subproblem [6]. The max-cut problem can be expressed as optimizing a quadratic form over binary constraints [103]. The stability number of a graph can be expressed as optimizing a quadratic form over the simplex constraint [68, 198]. The best rank-1 tensor approximation problem can be expressed as optimizing a homogeneous polynomial with the sphere constraint [228].

When $K$ is nonempty and compact, the minimum value $f_{min}$ is achievable at a feasible point, i.e., the optimization (5.1.1) has a minimizer. When $K$ is noncompact, the minimum value $f_{min}$ is achievable for the following cases.

- When all $f, c_i, c_j$ are linear and $K \neq \emptyset$, if $f_{min} > -\infty$, then $f_{min}$ is achievable. This is a basic property of linear programming.

- When $f$ is quadratic or cubic and all $c_i, c_j$ are linear, if $f_{min} > -\infty$, then $f_{min}$ is achievable (see [97, 145]).

- When $f$ is a convex (or quasi-convex) polynomial, $c_i$ ($i \in \mathcal{E}$) are linear, and $c_j$ ($j \in \mathcal{I}$) are concave (or quasi-concave) polynomials, if $f_{min} > -\infty$, then $f_{min}$ is achievable (see [17, 254]).

For other cases, $f_{min}$ may not be achievable. For unbounded $K$, it is NP-hard to decide whether or not $f_{min}$ is achievable (see [2]).

Since (5.1.1) is a special case of nonlinear optimization, the classical theory for nonlinear programming is applicable straightforwardly. In the following, suppose $u$ is a local minimizer of (5.1.1). The labeling set of active inequality constraints at $u$ is

$$J(u) := \{j \in \mathcal{I} \mid c_j(u) = 0\}. \tag{5.1.2}$$

The *linear independence constraint qualification condition (LICQC)* is said to hold at $u$ if the gradient set $\{\nabla c_i(u)\}_{i \in \mathcal{E} \cup J(u)}$ is linearly independent, where $\nabla c_i(u)$ denotes the gradient of $c_i$ at $u$. Under the LICQC, there exists a Lagrange multiplier vector

$$\lambda := (\lambda_i)_{i \in \mathcal{E}} \cup (\lambda_j)_{j \in \mathcal{I}}$$

satisfying the Karush-Kuhn-Tucker (KKT) conditions

$$\nabla f(u) = \sum_{i \in \mathcal{E}} \lambda_i \nabla c_i(u) + \sum_{j \in \mathcal{I}} \lambda_j \nabla c_j(u), \tag{5.1.3}$$

$$0 \leq c_j(u) \perp \lambda_j \geq 0 \quad \text{for all } j \in \mathcal{I}. \tag{5.1.4}$$

(For two scalars $a, b$, the notation $a \perp b$ means $a \cdot b = 0$.) The equation (5.1.3) is called the *first order optimality condition* (FOOC) and (5.1.4) is called the *complementarity condition* (CC). If (5.1.4) is strengthened to

$$0 \leq c_j(u) \perp \lambda_j \geq 0, \quad \lambda_j + c_j(u) > 0 \quad \text{for all } j \in \mathcal{I}, \tag{5.1.5}$$

then the *strict complementarity condition (SCC)* is said to hold at $u$. The Lagrange multipliers $\lambda_i$ as in (5.1.3)-(5.1.4) are associated to the *Lagrange function*

$$L(x) := f(x) - \sum_{i \in \mathcal{E}} \lambda_i c_i(x) - \sum_{j \in \mathcal{I}} \lambda_j c_j(x). \tag{5.1.6}$$

Clearly, the FOOC (5.1.3) is equivalent to $\nabla L(u) = 0$. There are also second order optimality conditions. Denote the Hessian of the Lagrangian

$$\nabla^2 L(x) := \nabla^2 f(x) - \sum_{i \in \mathcal{E}} \lambda_i \nabla^2 c_i(x) - \sum_{j \in \mathcal{I}} \lambda_j \nabla^2 c_j(x). \tag{5.1.7}$$

Under the LICQC, if $u$ is a local minimizer, then the *second order necessary condition* (SONC) holds:

$$v^T\left(\nabla^2 L(u)\right)v \geq 0 \quad \text{for all} \quad v \in \bigcap_{i \in \mathcal{E} \cup J(u)} \nabla c_i(u)^\perp. \tag{5.1.8}$$

In the above, the notation $\nabla c_i(u)^\perp$ denotes the orthogonal complement:

$$\nabla c_i(u)^\perp := \left\{v \in \mathbb{R}^n : c_i(u)^T v = 0\right\}.$$

A stronger one than (5.1.8) is the *second order sufficient condition* (SOSC):

$$v^T\left(\nabla^2 L(u)\right)v > 0 \quad \text{for all} \quad 0 \neq v \in \bigcap_{i \in \mathcal{E} \cup J(u)} \nabla c_i(u)^\perp. \tag{5.1.9}$$

The relations among these conditions can be summarized as follows. Under the LICQC, the conditions (5.1.3), (5.1.4), and (5.1.8) are necessary for $u$ to be a local minimizer of (5.1.1). Conversely, if (5.1.3), (5.1.5), and (5.1.9) hold at a feasible point $u$, then $u$ is a strict local minimizer of (5.1.1). We refer to [22, §3.3] for more details about local optimality conditions. A feasible point $u$ satisfying (5.1.3)-(5.1.4) is called a *KKT point*. A point $u$ is called a *critical point* or *stationary point* for the optimization (5.1.1) if $u$ satisfies (5.1.3) and $c_i(u) = \lambda_j c_j(u) = 0$ for all $i \in \mathcal{E}$ and for all $j \in \mathcal{I}$. If a critical or stationary point $u$ further satisfies (5.1.8), then it is called a *second order critical or stationary* point. For unconstrained optimization, a second order critical point is also called an $H$-minimizer (see §12.3).

If a local minimizer or stationary point is targeted, the classical nonlinear optimization methods (see [22, 253, 329]) can be directly applied. When the functions are given by polynomials, people often look for a global minimizer. The Moment-SOS hierarchy of semidefinite programming relaxations, originally proposed by Lasserre [158], is a powerful approach for solving (5.1.1) globally, i.e., to find the minimum value $f_{min}$ and the minimizer(s) if they exist. This chapter introduces the basic theory of Moment-SOS relaxations.

### 5.1.1 ▪ Exercises

**Exercise 5.1.1.** *Assume the constraints in (5.1.1) are the simplex*

$$\Delta_n := \{x \in \mathbb{R}^n : x \geq 0, e^T x = 1\}.$$

*If $u$ is a local minimizer, give a direct proof that there exist Lagrange multipliers such that (5.1.3), (5.1.4), and (5.1.8) hold.*

**Exercise 5.1.2.** *Let $u$ be a feasible point for (5.1.1). If the conditions (5.1.3), (5.1.5), and (5.1.9) hold for some Lagrange multipliers, show that $u$ is a strict local minimizer of (5.1.1).*

**Exercise 5.1.3.** *Show that the SOSC (5.1.9) holds if and only if there exist scalars $\mu_i \geq 0$ $(i \in \mathcal{E} \cup J(u))$ such that*

$$\nabla^2 L(u) + \sum_{i \in \mathcal{E} \cup J(u)} \mu_i \nabla c_i(u) \cdot \nabla c_i(u)^T \succ 0.$$

*Does a similar conclusion hold for the SONC (5.1.8)?*

**Exercise 5.1.4.** *([22, §3.3]) For optimality conditions (5.1.3)-(5.1.4) to hold at a local minimizer $u$, the LICQC can be weakened to the Mangasarian-Fromovitz constraint qualification condition*

*(MFCQC): the gradients $\nabla c_i(u)$ $(i \in \mathcal{E})$ are linearly independent, and there exists $d \in \mathbb{R}^n$ such that*

$$\begin{aligned}
\nabla c_i(u)^T d &= 0 \,(i \in \mathcal{E}), \\
\nabla c_j(u)^T d &> 0 \,(j \in J(u)).
\end{aligned}$$

*Show that the MFCQC holds at $u$ if and only if the set of Lagrange multiplier vectors satisfying (5.1.3)-(5.1.4) is nonempty and compact.*

**Exercise 5.1.5.** *([68, 198]) Let $G = (V, E)$ be a graph, with the vertex set $V$ and the edge set $E$. A subset $T \subseteq V$ is called independent if no two vertices in $T$ are connected by an edge. The maximum cardinality of independent subsets of $V$ is called the stability number of $G$, for which we denote $\alpha(G)$. Let $n = |V|$ and let $A = (a_{ij})$ be the adjacency matrix of $G$, i.e., $a_{ij} = 1$ if $(i, j) \in E$ and $a_{ij} = 0$ if $(i, j) \notin E$. Show that*

$$\frac{1}{\alpha(G)} = \min_{x \in \Delta_n} x^T (I_n + A)x.$$

## 5.2 ▪ The Moment-SOS hierarchy

To solve the polynomial optimization (5.1.1) globally, a natural approach is to apply the Moment-SOS hierarchy of semidefinite relaxations, originally introduced by Lasserre [158]. To distinguish equality and inequality constraints, we denote the constraining polynomial tuples

$$c_{eq} := (c_i)_{i \in \mathcal{E}}, \quad c_{in} := (c_j)_{j \in \mathcal{I}}. \tag{5.2.1}$$

Recall that $\mathrm{Ideal}[c_{eq}]$ is the ideal generated by $c_{eq}$, and $\mathrm{QM}[c_{in}]$ is the quadratic module generated by $c_{in}$ (see §2.5 for the notation). The minimum value of (5.1.1) is $f_{min}$ and the feasible set of (5.1.1) is $K$. The feasible set $K$ can be equivalently given as

$$c_{eq}(x) \geq 0, \ -c_{eq}(x) \geq 0, \ c_{in}(x) \geq 0.$$

A scalar $\gamma$ is a lower bound for $f_{min}$ (i.e., $\gamma \leq f_{min}$) if and only if the subtraction polynomial $f - \gamma$ is nonnegative on $K$, i.e., $f - \gamma \in \mathscr{P}(K)$. In view of Putinar's Positivstellensatz (see Theorem 2.6.1), the constraint $f - \gamma \in \mathscr{P}(K)$ is ensured by the membership

$$f - \gamma \in \mathrm{QM}[c_{eq}, -c_{eq}, c_{in}] = \mathrm{Ideal}[c_{eq}] + \mathrm{QM}[c_{in}].$$

If a truncation of degree $2k$ is used, we get the SOS relaxation

$$\left\{ \begin{array}{ll}
\max & \gamma \\
s.t. & f - \gamma \in \mathrm{Ideal}[c_{eq}]_{2k} + \mathrm{QM}[c_{in}]_{2k}.
\end{array} \right. \tag{5.2.2}$$

For $k = 1, 2, \ldots$, we get a hierarchy of SOS relaxations. The integer $k$ is called the *relaxation order*. The relaxation (5.2.2) is equivalent to a semidefinite program (see §2.5.1). The truncation $\mathrm{Ideal}[c_{eq}]_{2k}$ of the ideal $\mathrm{Ideal}[c_{eq}]$ is a subspace of $\mathbb{R}[x]_{2k}$. Its dual cone is the following subspace (see (2.5.23), (2.5.25) for the notation $\mathscr{V}_h^{(d)}[y]$ and $\mathscr{Z}[h]_{2k}$):

$$\mathscr{Z}[c_{eq}]_{2k} = \left\{ y \in \mathbb{R}^{\mathbb{N}_{2k}^n} : \mathscr{V}_{c_i}^{(2k)}[y] = 0 \,(i \in \mathcal{E}) \right\}.$$

The dual cone of $\mathrm{QM}[c_{in}]_{2k}$ is

$$\mathscr{S}[c_{in}]_{2k} = \left\{ y \in \mathbb{R}^{\mathbb{N}_{2k}^n} \left| \begin{array}{ll} L_{c_i}^{(k)}[y] & \succeq \quad 0 \,(i \in \mathcal{I}), \\ M_k[y] & \succeq \quad 0 \end{array} \right. \right\}.$$

By Proposition 2.5.3, the dual cone of $\text{Ideal}[c_{eq}]_{2k} + \text{QM}[c_{in}]_{2k}$ is the intersection

$$\mathscr{Z}[c_{eq}]_{2k} \cap \mathscr{S}[c_{in}]_{2k}.$$

Therefore, the dual optimization of (5.2.2) is the moment relaxation

$$\begin{cases} \min & \langle f, y \rangle \\ \text{s.t.} & y \in \mathscr{Z}[c_{eq}]_{2k} \cap \mathscr{S}[c_{in}]_{2k}, \\ & \langle 1, y \rangle = 1. \end{cases}$$

Since $\langle 1, y \rangle = y_0$, the above can be equivalently written as

$$\begin{cases} \min & \langle f, y \rangle \\ \text{s.t.} & \mathscr{V}_{c_i}^{(2k)}[y] = 0 \ (i \in \mathcal{E}), \\ & L_{c_j}^{(k)}[y] \succeq 0 \ (j \in \mathcal{I}), \\ & M_k[y] \succeq 0, \\ & y_0 = 1, \ y \in \mathbb{R}^{\mathbb{N}_{2k}^n}. \end{cases} \tag{5.2.3}$$

For $k = 1, 2, \ldots$, the sequence of primal-dual pairs (5.2.2)-(5.2.3) is called the *Moment-SOS hierarchy*. For each $k$, let $f_{sos,k}$ and $f_{mom,k}$ denote the optimal values of (5.2.2), (5.2.3) respectively. Recall that $f_{min}$ is the minimum value of (5.1.1). The following is a basic property for the Moment-SOS hierarchy.

**Proposition 5.2.1.** *For every relaxation order $k$, it holds that*

$$f_{sos,k} \leq f_{mom,k} \leq f_{min}. \tag{5.2.4}$$

*Moreover, both sequences $\{f_{sos,k}\}_{k=1}^{\infty}$ and $\{f_{mom,k}\}_{k=1}^{\infty}$ are monotonically increasing.*

(i) *Suppose all equality constraints are linear or $\mathcal{E} = \emptyset$. If there exists $v \in K$ such that $c_j(v) > 0$ for all $j \in \mathcal{I}$, then $f_{sos,k} = f_{mom,k}$ and the relaxation (5.2.2) achieves its maximum value for every $k$.*

(ii) *Suppose there exists a scalar $R > 0$ such that*

$$R - \|x\|^{2k_0} \in \text{Ideal}[c_{eq}]_{2k_0} + \text{QM}[c_{in}]_{2k_0}$$

*for some $k_0 \in \mathbb{N}$. For each $k \geq k_0$, if the moment relaxation (5.2.3) is feasible, then it achieves its minimum value.*

Proposition 5.2.1 can be shown by using properties of truncated quadratic modules and the strong duality theorem (see Theorem 1.5.1). Its proof is left as an exercise. In the following, we show what the moment relaxation (5.2.3) looks like. The objective $\langle f, y \rangle$ is linear in $y$. Let $d = \deg(f)$. If the objective $f$ has the expansion $f = \sum_{\alpha \in \mathbb{N}_d^n} f_\alpha x^\alpha$, then

$$\langle f, y \rangle = \sum_{\alpha \in \mathbb{N}_d^n} f_\alpha y_\alpha.$$

The moment matrix $M_k[y]$ is defined in §2.4.1. For instance, when $n = 3$, the linear matrix inequality $M_2[y] \succeq 0$ reads as

$$
\begin{bmatrix}
y_{000} & y_{100} & y_{010} & y_{001} & y_{200} & y_{110} & y_{101} & y_{020} & y_{011} & y_{002} \\
y_{100} & y_{200} & y_{110} & y_{101} & y_{300} & y_{210} & y_{201} & y_{120} & y_{111} & y_{102} \\
y_{010} & y_{110} & y_{020} & y_{011} & y_{210} & y_{120} & y_{111} & y_{030} & y_{021} & y_{012} \\
y_{001} & y_{101} & y_{011} & y_{002} & y_{201} & y_{111} & y_{102} & y_{021} & y_{012} & y_{003} \\
y_{200} & y_{300} & y_{210} & y_{201} & y_{400} & y_{310} & y_{301} & y_{220} & y_{211} & y_{202} \\
y_{110} & y_{210} & y_{120} & y_{111} & y_{310} & y_{220} & y_{211} & y_{130} & y_{121} & y_{112} \\
y_{101} & y_{201} & y_{111} & y_{102} & y_{301} & y_{211} & y_{202} & y_{121} & y_{112} & y_{103} \\
y_{020} & y_{120} & y_{030} & y_{021} & y_{220} & y_{130} & y_{121} & y_{040} & y_{031} & y_{022} \\
y_{011} & y_{111} & y_{021} & y_{012} & y_{211} & y_{121} & y_{112} & y_{031} & y_{022} & y_{013} \\
y_{002} & y_{102} & y_{012} & y_{003} & y_{202} & y_{112} & y_{103} & y_{022} & y_{013} & y_{004}
\end{bmatrix} \succeq 0.
$$

It is worthwhile to observe that $M_t[y]$ is a leading principal submatrix of $M_k[y]$ for every $t \leq k$. The following are some expositions for the constraints in (5.2.3) with $k = 2$.

- When $K$ is the unit circle $x_1^2 + x_2^2 = 1$ (i.e., $\mathcal{E} = \{1\}$, $c_1 = x_1^2 + x_2^2 - 1$, and $\mathcal{I} = \emptyset$), the constraint $\mathcal{V}_{c_1}^{(4)}[y] = 0$ gives the equations

$$y_{20} + y_{02} - y_{00} = 0, \; y_{30} + y_{12} - y_{10} = 0, \; y_{21} + y_{03} - y_{01} = 0,$$

$$y_{40} + y_{22} - y_{20} = 0, \; y_{31} + y_{13} - y_{11} = 0, \; y_{22} + y_{04} - y_{02} = 0.$$

- When $K$ is the unit ball $1 - x_1^2 - x_2^2 \geq 0$ (i.e., $\mathcal{I} = \{1\}$, $c_1 = 1 - x_1^2 - x_2^2$, and $\mathcal{E} = \emptyset$), the constraint $L_{c_1}^{(2)}[y] \succeq 0$ reads as

$$
\begin{bmatrix}
y_{00} - y_{20} - y_{02} & y_{10} - y_{30} - y_{12} & y_{01} - y_{21} - y_{03} \\
y_{10} - y_{30} - y_{12} & y_{20} - y_{40} - y_{22} & y_{11} - y_{31} - y_{13} \\
y_{01} - y_{21} - y_{03} & y_{11} - y_{31} - y_{13} & y_{02} - y_{22} - y_{04}
\end{bmatrix} \succeq 0.
$$

- When $K$ is the simplex in $\mathbb{R}^3$ (i.e., $\mathcal{E} = \{0\}$, $\mathcal{I} = \{1, 2, 3\}$, and $c_0 := x_1 + x_2 + x_3 - 1$, $c_1 := x_1$, $c_2 := x_2$, $c_3 := x_3$), the constraint $\mathcal{V}_{c_0}^{(2)}[y] = 0$ gives the equations

$$y_{100} + y_{010} + y_{001} - y_{000} = 0, \; y_{200} + y_{110} + y_{101} - y_{100} = 0,$$

$$y_{110} + y_{020} + y_{011} - y_{010} = 0, \; y_{101} + y_{011} + y_{002} - y_{001} = 0.$$

The psd constraints $L_{c_j}^{(2)}[y] \succeq 0$, $j = 1, 2, 3$, are respectively

$$
\begin{bmatrix}
y_{100} & y_{200} & y_{110} & y_{101} \\
y_{200} & y_{300} & y_{210} & y_{201} \\
y_{110} & y_{210} & y_{120} & y_{111} \\
y_{101} & y_{201} & y_{111} & y_{102}
\end{bmatrix} \succeq 0,
$$

$$
\begin{bmatrix}
y_{010} & y_{110} & y_{020} & y_{011} \\
y_{110} & y_{210} & y_{120} & y_{111} \\
y_{020} & y_{120} & y_{030} & y_{021} \\
y_{011} & y_{111} & y_{021} & y_{012}
\end{bmatrix} \succeq 0,
$$

$$
\begin{bmatrix}
y_{001} & y_{101} & y_{011} & y_{002} \\
y_{101} & y_{201} & y_{111} & y_{102} \\
y_{011} & y_{111} & y_{021} & y_{012} \\
y_{002} & y_{102} & y_{012} & y_{003}
\end{bmatrix} \succeq 0.
$$

- When $K$ is the hypercube $[-1, 1]^3$ (i.e., $\mathcal{E} = \emptyset$, $\mathcal{I} = \{1, 2, 3\}$ and $c_1 := 1 - x_1^2$, $c_2 := 1 - x_2^2$, $c_3 := 1 - x_3^2$), the constraints $L_{c_j}^{(2)}[y] \succeq 0$, $j = 1, 2, 3$, are respectively

$$
\begin{bmatrix}
y_{000} - y_{200} & y_{100} - y_{300} & y_{010} - y_{210} & y_{001} - y_{201} \\
y_{100} - y_{300} & y_{200} - y_{400} & y_{110} - y_{310} & y_{101} - y_{301} \\
y_{010} - y_{210} & y_{110} - y_{310} & y_{020} - y_{220} & y_{011} - y_{211} \\
y_{001} - y_{201} & y_{101} - y_{301} & y_{011} - y_{211} & y_{002} - y_{202}
\end{bmatrix} \succeq 0,
$$

$$
\begin{bmatrix}
y_{000} - y_{020} & y_{100} - y_{120} & y_{010} - y_{030} & y_{001} - y_{021} \\
y_{100} - y_{120} & y_{200} - y_{220} & y_{110} - y_{130} & y_{101} - y_{121} \\
y_{010} - y_{030} & y_{110} - y_{130} & y_{020} - y_{040} & y_{011} - y_{031} \\
y_{001} - y_{021} & y_{101} - y_{121} & y_{011} - y_{031} & y_{002} - y_{022}
\end{bmatrix} \succeq 0,
$$

$$
\begin{bmatrix}
y_{000} - y_{002} & y_{100} - y_{102} & y_{010} - y_{012} & y_{001} - y_{003} \\
y_{100} - y_{102} & y_{200} - y_{202} & y_{110} - y_{112} & y_{101} - y_{103} \\
y_{010} - y_{012} & y_{110} - y_{112} & y_{020} - y_{022} & y_{011} - y_{013} \\
y_{001} - y_{003} & y_{101} - y_{103} & y_{011} - y_{013} & y_{002} - y_{004}
\end{bmatrix} \succeq 0.
$$

The Moment-SOS hierarchy of (5.2.2)-(5.2.3) is said to be *tight* (or *exact*) or to have *finite convergence* if $f_{sos,k} = f_{min}$ for some $k$. If $f_{sos,k} \to f_{min}$ as $k \to \infty$, the Moment-SOS hierarchy is said to have *asymptotic convergence*. If $f_{sos,k}$ does not converge to $f_{min}$, we say that the Moment-SOS hierarchy fails to converge. However, we should keep in mind that the sequence $\{f_{sos,k}\}_{k=1}^{\infty}$ always has a limit, unless $f_{sos,k} = -\infty$ for all $k$. This is because it is a monotonically increasing sequence, as shown in (5.2.4). In view of the inequality (5.2.4), the convergence $f_{sos,k} \to f_{min}$ also implies the convergence of $f_{mom,k} \to f_{min}$. In computational practice, the finite convergence occurs frequently (see §5.4). The following is a classical result about the asymptotic convergence, which requires the archimedeanness.

**Theorem 5.2.2.** *([158]) If* Ideal$[c_{eq}]$+QM$[c_{in}]$ *is archimedean, then the Moment-SOS hierarchy has the asymptotic convergence:*

$$
\lim_{k \to \infty} f_{sos,k} = \lim_{k \to \infty} f_{mom,k} = f_{min}.
$$

The above asymptotic convergence can be implied by Theorem 2.6.1. We remark that if the feasible set $K$ is empty and Ideal$[c_{eq}]$ + QM$[c_{in}]$ is archimedean, then the moment relaxation (5.2.3) must be infeasible when $k$ is large enough. The proof is left as Exercise 5.2.7. When the relaxation order $k$ is fixed, an approximation analysis for the Moment-SOS relaxation is given in the work [227]. The following is an example of applying Moment-SOS relaxations to solve polynomial optimization.

**Example 5.2.3.** *Consider the constrained optimization problem*

$$
\begin{cases}
\min & x_1 x_2 x_3 - x_1 x_2 - x_2 x_3 + x_1^4 + x_2^4 + x_3^4 \\
s.t. & 1 - x_1^2 \geq 0, \ 1 - x_2^2 \geq 0, \ 1 - x_3^2 \geq 0.
\end{cases}
$$

*The Moment-SOS hierarchy of (5.2.2)-(5.2.3) can be implemented by the software* GloptiPoly *as follows:*

```
mpol('x', 3);
f = prod(x)-x(1)*x(2)-x(2)*x(3)+sum(x.^4);
K = [1-x(1)^2 >=0, 1-x(2)^2 >=0, 1-x(3)^2 >= 0];
```

```
k = 2; % give an integer value for k;
PO = msdp(min(f), K, k);
[status, fsos_k] = msol(PO);
```

*For $k = 2$, we get $f_{sos,2} = -0.62439478\cdots$, which equals the minimum value $f_{min}$. The syntax* double(x) *returns the minimizer*

$$(-0.688625974\cdots, -0.77353729\cdots, -0.68862597\cdots).$$

The Moment-SOS hierarchy may fail to have finite convergence, even if the archimedean condition is met. The following is such an example.

**Example 5.2.4.** *Consider the polynomial optimization*

$$\begin{cases} \min & x_1^2 x_2^2 (x_1^2 + x_2^2) + (x_3 - 1)^6 - 3x_1^2 x_2^2 (x_3 - 1)^2 \\ s.t. & x_1^2 + x_2^2 + x_3^2 - \frac{1}{2} \geq 0, \\ & 2 - x_1^2 - x_2^2 - x_3^2 \geq 0. \end{cases}$$

*The objective $f$ is shifted from the Motzkin form. The minimum value $f_{min} = 0$. The Moment-SOS hierarchy does not have finite convergence (see [230] or [292, Prop. 6.1]). This is because the minimizer $(0, 0, 1)$ is a zero of $f$ lying in the interior of the feasible set and $f$ is the Motzkin form in $(x_1, x_2, x_3 - 1)$.*

## 5.2.1 ▪ Quadratic optimization and S-Lemma

When the objective and all constraining polynomials are quadratic, the optimization (5.1.1) is reduced to a QCQP. When there is a single quadratic constraint, the Moment-SOS hierarchy is tight.

**Theorem 5.2.5.** *([95]) Suppose (5.1.1) is a QCQP with a single quadratic constraint. Then, $f_{sos,1} = f_{mom,1} = f_{min}$ if either one of the following holds:*

*(i) $\mathcal{E} = \emptyset$ and there exists $u$ such that $c_j(u) > 0$ for $j \in \mathcal{I}$;*

*(ii) $\mathcal{I} = \emptyset$ and for $i \in \mathcal{E}$, $\deg(c_i) > 1$ and $c_i(u) < 0 < c_i(v)$ for some $u, v$.*

**Proof.** The conclusion for the case $f_{min} = -\infty$ is clearly true. If $f_{min} > -\infty$, the conclusion can be implied by the following Theorem 5.2.6.  □

For QCQPs with a single quadratic constraint, the conclusion of Theorem 5.2.5 is often referenced as the *S-Lemma*. The original one is for a single inequality constraint, due to Yakubovich [323]. We refer to [261] for a survey on the S-Lemma. When (5.1.1) is a QCQP with homogeneous quadratic polynomials (up to adding constants), if there are only two inequality constraints, the first order relaxation is also tight (see [115, 185]). See Exercise 5.2.8 for this conclusion. The following are two general versions of the S-Lemma.

**Theorem 5.2.6.** *Let $q \in \mathbb{R}[x]_2$ be a polynomial with $\deg(q) > 1$.*

*(i) (S-Lemma [323]) Suppose there exists $\xi \in \mathbb{R}^n$ such that $q(\xi) > 0$. For $f \in \mathbb{R}[x]_2$, if $f(x) \geq 0$ for all $q(x) \geq 0$, then there exists a scalar $t \geq 0$ such that*

$$f(x) - tq(x) \in \Sigma[x]_2.$$

(ii) ([95]) *Suppose there exist* $\xi, \zeta \in \mathbb{R}^n$ *such that* $q(\xi) > 0 > q(\zeta)$. *For* $f \in \mathbb{R}[x]_2$, *if* $f(x) \geq 0$ *for all* $q(x) = 0$, *then there exists* $t \in \mathbb{R}$ *such that*

$$f(x) - tq(x) \in \Sigma[x]_2.$$

## 5.2.2 ▪ Schmüdgen type hierarchy

We can get stronger relaxations if Schmüdgen Positivstellensatz (see Theorem 2.6.2) is used. In (5.2.2), if the truncated quadratic module $\mathrm{QM}[c_{in}]_{2k}$ is replaced by the truncated preordering $\mathrm{Pre}[c_{in}]_{2k}$, then we get the stronger relaxation

$$\begin{cases} \max & \gamma \\ s.t. & f - \gamma \in \mathrm{Ideal}[c_{eq}]_{2k} + \mathrm{Pre}[c_{in}]_{2k}. \end{cases} \tag{5.2.5}$$

Similarly, the dual optimization problem of (5.2.5) is the moment relaxation

$$\begin{cases} \min & \langle f, y \rangle \\ s.t. & \mathscr{V}_{c_i}^{(2k)}[y] = 0 \, (i \in \mathcal{E}), \\ & L_{c_J}^{(k)}[y] \succeq 0 \, (J \subseteq \mathcal{I}), \\ & y_0 = 1, \, y \in \mathbb{R}^{\mathbb{N}_{2k}^n}. \end{cases} \tag{5.2.6}$$

In the above, for a subset $J = \{j_1, \ldots, j_l\}$, we denote the product

$$c_J := c_{j_1} \cdots c_{j_l}.$$

In particular, if $J = \emptyset$, then $c_\emptyset = 1$ (the constant one polynomial) and $L_{c_\emptyset}^{(k)}[y]$ becomes the moment matrix $M_k[y]$. Let $f'_{sos,k}$, $f'_{mom,k}$ denote the optimal values of (5.2.5), (5.2.6) respectively for the relaxation order $k$. The same kind of conclusions as in Proposition 5.2.1 also hold for the hierarchy of (5.2.5)-(5.2.6). By the weak duality, it also holds that for all $k$

$$f'_{sos,k} \leq f'_{mom,k} \leq f_{min}. \tag{5.2.7}$$

If $K$ has nonempty interior, then each $f'_{sos,k} = f'_{mom,k}$, i.e., there is no duality gap. Both sequences $\{f'_{sos,k}\}_{k=1}^\infty$ and $\{f'_{mom,k}\}_{k=1}^\infty$ are monotonically increasing. Since $\mathrm{Pre}[c_{in}]_{2k} \supseteq \mathrm{QM}[c_{in}]_{2k}$, the relaxation (5.2.5) is stronger than (5.2.2), so

$$f'_{sos,k} \geq f_{sos,k}, \quad f'_{mom,k} \geq f_{mom,k}.$$

We call the sequence of relaxations (5.2.5)-(5.2.6) the *Schmüdgen* type Moment-SOS hierarchy. In contrast, the sequence of relaxations (5.2.2)-(5.2.3) is called the *Putinar* type Moment-SOS hierarchy. If $f'_{sos,k} = f_{min}$ for some order $k$, the Schmüdgen type Moment-SOS hierarchy is said to be *tight* or to have *finite* convergence. The following is a basic result about the Schmüdgen type Moment-SOS hierarchy.

**Theorem 5.2.7.** *([158]) If the feasible set $K$ of (5.1.1) is compact, then the Schmüdgen type Moment-SOS hierarchy has the asymptotic convergence*

$$\lim_{k \to \infty} f'_{sos,k} = \lim_{k \to \infty} f'_{mom,k} = f_{min}.$$

The above can be implied by Theorem 2.6.2. We remark that if the feasible set $K$ is empty, then the moment relaxation (5.2.6) must be infeasible when $k$ is large enough. The proof is left as Exercise 5.2.7.

## 5.2.3 ▪ Exercises

**Exercise 5.2.1.** *Compute the distance between two sets in* $\mathbb{R}^3$ *that are given by the following two equations:*

$$x_1^4 + x_2^4 + x_3^4 = 1, \quad |x_1|^3 + |x_2|^3 + |x_3|^3 - |x_1 x_2 x_3| = 4.$$

*Also find a pair of points on them achieving the distance.*

**Exercise 5.2.2.** *Consider the following problem: for* $N$ *given points* $u_1, \ldots, u_N \in \mathbb{R}^n$, *find a point* $x$ *on the unit sphere* $\mathbb{S}^{n-1}$ *such that the distance from* $x$ *to the set* $\{u_1, \ldots, u_N\}$ *is maximum. Formulate this problem as a constrained polynomial optimization problem. Solve it by the Moment-SOS hierarchy for* $n = 3$ *and* $N = 10$ *with randomly generated points* $u_i$.

**Exercise 5.2.3.** *For given* $A \in \mathcal{S}_{++}^n$ *and* $b \in \mathbb{R}^n$, *find the biggest* $R$ *such that the set given by* $x^T A x + 2b^T x \leq R$ *is contained inside the unit ball* $x^T x \leq 1$. *Formulate this problem equivalently as a semidefinite program by using the S-Lemma.*

**Exercise 5.2.4.** *In (5.1.1), if the equality constraints give a compact set, show that the Moment-SOS hierarchy always has asymptotic convergence.*

**Exercise 5.2.5.** *When* $K$ *is the nonnegative orthant* $\mathbb{R}_+^n$, *give an example of* $f$ *such that the Moment-SOS hierarchy fails to have the asymptotic convergence.*

**Exercise 5.2.6.** *Prove Proposition 5.2.1 and Theorems 5.2.2 and 5.2.7.*

**Exercise 5.2.7.** *Assume the feasible set* $K$ *is empty. Show that the moment relaxation (5.2.6) is infeasible for all* $k$ *large enough. Moreover, if* $\mathrm{Ideal}[c_{eq}] + \mathrm{QM}[c_{in}]$ *is archimedean, show that the moment relaxation (5.2.3) is infeasible when* $k$ *is large enough.*

**Exercise 5.2.8.** *([115, 185]) For given matrices* $A_0, A_1, A_2 \in \mathcal{S}^n$ *and constants* $b_1, b_2$, *consider the homogeneous quadratic optimization*

$$\begin{cases} \min\limits_{x \in \mathbb{R}^n} & x^T A_0 x \\ s.t. & x^T A_1 x \geq b_1, \\ & x^T A_2 x \geq b_2. \end{cases}$$

*If there are scalars* $\lambda_1 \geq 0, \lambda_2 \geq 0$ *such that* $A_0 - \lambda_1 A_1 - \lambda_2 A_2 \succ 0$, *show that* $f_{sos,1} = f_{mom,1} = f_{min}$ *and the moment relaxation (5.2.3) achieves the minimum. Here,* $f_{min}$ *still denotes the minimum value of the above optimization.*

**Exercise 5.2.9.** *([78]) In (5.1.1), assume* $f$ *is a form of degree* $2d$ *and the constraint is the unit sphere* $x^T x = 1$. *Consider the SOS relaxation*

$$\begin{cases} \vartheta_{r+d} := \min & \gamma \\ s.t. & (x^T x)^r (f - \gamma (x^T x)^d) \in \Sigma[x]_{n, 2d+2r} \end{cases}$$

*for a power* $r \geq 0$. *Show that* $\vartheta_{r+d} = f_{sos,r+d}$, *where* $f_{sos,r+d}$ *is the optimal value of the relaxation (5.2.2) with the order* $k = r + d$.

## 5.3 ▪ **Extraction of Minimizers**

We discuss how to obtain minimizers for the constrained polynomial optimization (5.1.1) from the Moment-SOS hierarchy. The extraction of a minimizer is possible if the moment relaxation (5.2.3) is tight, i.e., $f_{mom,k} = f_{min}$ for some order $k$.

Suppose $y^*$ is an optimizer of (5.2.3). Denote the degrees

$$
\boxed{
\begin{aligned}
d_c &:= \max_{i \in \mathcal{E} \cup \mathcal{I}} \left\{ 1, \lceil \tfrac{1}{2} \deg(c_i) \rceil \right\}, \\
d_0 &:= \max \left\{ d_c, \lceil \tfrac{1}{2} \deg(f) \rceil \right\}.
\end{aligned}
}
\tag{5.3.1}
$$

To extract a minimizer for (5.1.1), one could consider the flat extension

$$
\operatorname{rank} M_{k-d_c}[y^*] = \operatorname{rank} M_k[y^*].
\tag{5.3.2}
$$

As shown in §2.7.2, if (5.3.2) holds, then $y^*$ admits a representing measure supported in $K$. However, the condition (5.3.2) is often not satisfied. This can be shown by Example 5.3.2. In computational practice, a more appropriate criterion is the *flat truncation*: there exists an integer $t \in [d_0, k]$ such that

$$
\operatorname{rank} M_{t-d_c}[y^*] = \operatorname{rank} M_t[y^*].
\tag{5.3.3}
$$

Equivalently, (5.3.3) requires that the truncation

$$
y^*|_{2t} := (y^*_\alpha)_{\alpha \in \mathbb{N}^n_{2t}}
$$

is flat, i.e., $y^*|_{2t}$ is a flat tms of degree $2t$.

**Lemma 5.3.1.** *Suppose $y^*$ is an optimizer of (5.2.3). If (5.3.3) holds, then the relaxation (5.2.3) is tight (i.e., $f_{mom,k} = f_{min}$) and the truncation $y^*|_{2t}$ admits a $r$-atomic measure $\mu$, where $r = \operatorname{rank} M_t[y^*]$, such that each point in $\operatorname{supp}(\mu)$ is a minimizer of (5.1.1).*

**Proof.** Let $z := y^*|_{2t}$ be the truncation, which is a flat tms. By Theorem 2.7.7, there exist $r$ distinct points $u_1, \ldots, u_r \in K$ and positive scalars $\lambda_1, \ldots, \lambda_r$ such that

$$
y^*|_{2t} = \lambda_1 [u_1]_{2t} + \cdots + \lambda_r [u_r]_{2t}.
\tag{5.3.4}
$$

The constraint $y_0 = 1$ implies that $\lambda_1 + \cdots + \lambda_r = 1$, so

$$
\lambda_1 f(u_1) + \cdots + \lambda_r f(u_r) = \langle f, y^*|_{2t} \rangle = \langle f, y^* \rangle = f_{mom,k} \le f_{min}.
$$

Since each $u_i \in K$, we have $f(u_i) \ge f_{min}$, so

$$
\lambda_1 f(u_1) + \cdots + \lambda_r f(u_r) \ge (\lambda_1 + \cdots + \lambda_r) f_{min} = f_{min}.
$$

Therefore, each $f(u_i) = f_{min}$, i.e., each $u_i$ is a minimizer of (5.1.1). Similarly, since each $f(u_i) \ge f_{mom,k}$, we also have $f_{mom,k} = f_{min}$.  $\square$

The decomposition (5.3.4) can be obtained by the method in §2.7.2. We remark that the flat truncation (5.3.3) is more appropriate than (5.3.2) for extracting minimizers, as shown in the following example.

**Example 5.3.2.** *([225]) Consider the univariate polynomial optimization*

$$
\begin{cases}
\min\limits_{x \in \mathbb{R}^1} & x^3 \\
s.t. & x \ge 0,\, 1 - x \ge 0.
\end{cases}
$$

*The feasible set $K = [0, 1]$, the minimum value $f_{min} = 0$, and 0 is the only minimizer. We have $f_{sos,k} = f_{mom,k} = 0$ for all $k \geq 2$ (the proof is left as an exercise), so the relaxation (5.2.3) is tight. The flat extension (5.3.2) is violated for almost all optimizers of (5.2.3). However, the flat truncation (5.3.3) is satisfied for all optimizers of (5.2.3). The constraints in (5.2.3) are*

$$\begin{bmatrix} 1 & y_1 & y_2 \\ y_1 & y_2 & y_3 \\ y_2 & y_3 & y_4 \end{bmatrix} \succeq 0, \quad \begin{bmatrix} y_1 & y_2 \\ y_2 & y_3 \end{bmatrix} \succeq 0, \quad \begin{bmatrix} 1 - y_1 & y_1 - y_2 \\ y_1 - y_2 & y_2 - y_3 \end{bmatrix} \succeq 0.$$

*For every optimizer $y^*$ of (5.2.3), we have $(y^*)_3 = 0$, which then implies $(y^*)_1 = (y^*)_2 = 0$. For $k \geq 2$, one can similarly show that*

$$(y^*)_1 = (y^*)_2 = \cdots = (y^*)_{2k-1} = 0.$$

*Therefore, each optimizer $y^*$ of (5.2.3) is in the form*

$$y^*(a) := (1, 0, \ldots, 0, a)$$

*for some $a \geq 0$. In computation, an optimizer $y^*$ with $a > 0$ is typically obtained. For instance, the numerical value $0.7907 \cdots$ of $a$ is obtained for $k = 2$ when the software SeDuMi [308] is used to solve the relaxation (5.2.3). Clearly, (5.3.2) fails to hold for all $a > 0$, while (5.3.3) is always satisfied with $t = k - 1$. The flat truncation (5.3.3) is more appropriate than (5.3.2) for extracting minimizers.*

In the following, we show that the flat truncation (5.3.3) must be satisfied under some genericity assumptions.

## 5.3.1 ▪ Flat truncation for finite convergence

Lemma 5.3.1 shows that if the flat truncation (5.3.3) holds, then $f_{mom,k} = f_{min}$. Furthermore, if there is no duality gap between (5.2.2) and (5.2.3), then $f_{sos,k} = f_{min}$. In computation, one concerns necessity of the flat truncation (5.3.3) for the equality $f_{mom,k} = f_{min}$. When $f_{mom,k} = f_{min}$, if (5.1.1) has a minimizer, say, $x^*$, then $[x^*]_{2k}$ is a minimizer of (5.2.3) and it is always flat. For such a case, (5.2.3) always has a special minimizer that is flat. However, when (5.2.3) is solved numerically (e.g., by interior point methods), we may not be able to get a minimizer of (5.2.3) in the form $[x^*]_{2k}$. Therefore, it makes more sense to ask the following: If the moment relaxation is tight, does *every* optimizer of (5.2.3) have a flat truncation? Indeed, the answer is affirmative under a genericity assumption. The flat truncation is a sufficient and almost necessary criterion for checking finite convergence of the Moment-SOS hierarchy and for extracting minimizers of (5.1.1). This is a major conclusion in [225]. We make the following genericity assumption.

**Assumption 5.3.3.** *There exists $\rho \in \mathrm{QM}[c_{in}]$ such that for every label subset $J \subseteq \mathcal{I}$ such that $J = J(u)$ for some minimizer $u$ of (5.1.1), the equality constrained optimization*

$$\begin{cases} \min & f(x) \\ s.t. & c_i(x) = 0 \ (i \in \mathcal{E} \cup J) \end{cases} \tag{5.3.5}$$

*has finitely many real critical points that belong to the set*

$$\Theta := \{x \in \mathbb{R}^n : f(x) = f_{min}, c_{eq}(x) = 0, \rho(x) \geq 0\}. \tag{5.3.6}$$

When $f$ and all $c_j$ have generic coefficients, the optimization (5.3.5) has finitely many critical points. This is shown in Theorem 5.5.3 (also see [217]). For such a case, Assumption 5.3.3 is satisfied for the simple choice $\rho = 0$. Interestingly, Assumption 5.3.3 holds for more general cases (see [225]). The following is a major conclusion of this section.

**Theorem 5.3.4.** *([225]) Let $S$ be the set of minimizers of (5.1.1). Suppose $S \neq \emptyset$ and Assumption 5.3.3 holds. If $f - f_{min} \in \mathrm{Ideal}[c_{eq}] + \mathrm{QM}[c_{in}]$, then for all $k$ sufficiently large, the moment relaxation (5.2.3) has minimizers and every minimizer must satisfy (5.3.3) for some integer $t \in [d_0, k]$.*

Theorem 5.3.4 assumes that the Moment-SOS hierarchy is tight and the optimal value of (5.2.2) is achievable for some $k$. Moreover, in view of Lemma 5.3.1, for the moment relaxation (5.2.3) to have a minimizer satisfying (5.3.3), the optimizer set $S$ of (5.1.1) must be nonempty. When the feasible set $K$ has nonempty interior, $f_{sos,k} = f_{mom,k}$ and the optimal value $f_{sos,k}$ of (5.2.2) is achievable for every relaxation order $k$. This is shown in Proposition 5.2.1. When $k$ is big enough, Theorem 5.3.4 concludes that every minimizer $y^*$ of (5.2.3) has a flat truncation. Interestingly, we can show that (5.3.3) holds for $t = k - 1$, which was shown in the proof of Theorem 5.3.4. We remark that Theorem 5.3.4 does not assume that $K$ is compact or $\mathrm{Ideal}[c_{eq}] + \mathrm{QM}[c_{in}]$ is archimedean. To prove Theorem 5.3.4, we need the following lemma.

**Lemma 5.3.5.** *([225]) Let $y \in \mathbb{R}^{\mathbb{N}_{2k}^n}, h \in \mathbb{R}[x]_{2k}$ be such that $L_h^{(k)}[y] \succeq 0$. Suppose $s \in \Sigma[x]$ is such that $\deg(hs) \leq 2k$. Then, $\langle hs, y \rangle \geq 0$. Moreover, if $\langle hs, y \rangle = 0$, then $\langle hs\phi, y \rangle = 0$ for every $\phi \in \mathbb{R}[x]_{2\ell}$ with the degree*

$$\deg(hs) + 2\ell \leq 2k - 2.$$

*Proof.* Write $s = \sum_i s_i^2$, then

$$\langle hs, y \rangle = \sum_i \langle hs_i^2, y \rangle = \sum_i \mathrm{vec}(s_i)^T L_h^{(k)}[y]\, \mathrm{vec}(s_i) \geq 0,$$

since $L_h^{(k)}[y] \succeq 0$. In the following, assume $\langle hs, y \rangle = 0$. First, consider the case that $\phi = p_1 \in \mathbb{R}[x]_1$ such that $\deg(hs) + \deg(p_1^2) \leq 2k - 2$. Note that the function

$$a(t) := \langle hs(1 + tp_1^2)^2, y \rangle$$

is nonnegative for all $t \in \mathbb{R}^1$ and $a(0) = 0$, so

$$\langle hsp_1^2, y \rangle = \frac{1}{2}a'(0) = 0.$$

This implies that $\langle hsp_1^2, y \rangle = 0$. Similarly, we can show that

$$\langle hsp_1^2 p_2^2, y \rangle = 0$$

for all $p_2 \in \mathbb{R}[x]_1$ with $\deg(hs) + \deg(p_1^2 p_2^2) \leq 2k - 2$. Repeating this, we can eventually get

$$\langle hsp^2, y \rangle = 0$$

for all $p \in \mathbb{R}[x]_\ell$. Each $\phi \in \mathbb{R}[x]_{2\ell}$ can be written as $\phi = \sum_j (p_j^2 - q_j^2)$ for some polynomials $p_j, q_j \in \mathbb{R}[x]_\ell$, so

$$\langle hs\phi, y \rangle = \sum_j \left( \langle hsp_j^2, y \rangle - \langle hsq_j^2, y \rangle \right) = 0,$$

which completes the proof.      □

Now we can give the proof for Theorem 5.3.4. It is similar to that of Theorem 4.2.7. By the assumption, $f - f_{min} \in \mathrm{Ideal}[c_{eq}]_{2k_0} + \mathrm{QM}[c_{in}]_{2k_0}$ for some order $k_0$, i.e., there exist polynomials $p_i \in \mathbb{R}[x]$ for $i \in \mathcal{E}$ and $s_j \in \Sigma[x]$ for $j \in \mathcal{I} \cup \{0\}$ such that

$$f - f_{min} = \sum_{i \in \mathcal{E}} p_i c_i + \sum_{j \in \mathcal{I}} s_j c_j + s_0. \tag{5.3.7}$$

Note that $f_{sos,k} = f_{mom,k} = f_{min}$ for all $k \geq k_0$. The moment relaxation (5.2.3) always has a minimizer, e.g., for each minimizer $x^*$, the tms $[x^*]_{2k}$ is one. Let $y^*$ be an arbitrary minimizer of (5.2.3). Clearly, $\langle f, y^* \rangle = f_{min}$. Consider the semialgebraic set $C$ (the polynomial $\rho$ is from Assumption 5.3.3):

$$C = \left\{ x \in \mathbb{R}^n \;\middle|\; \begin{array}{rcl} c_i(x) &=& 0 \, (i \in \mathcal{E}), \\ s_0(x) &=& 0, \\ s_j(x)c_j(x) &=& 0 \, (j \in \mathcal{I}), \\ \rho(x) &\geq& 0 \end{array} \right\}. \tag{5.3.8}$$

We complete the proof in three major steps.

**Step 1** First, we show that $C$ is a finite set. Differentiating both sides of (5.3.7) in $x$ gives the identity

$$\nabla f = \sum_{i \in \mathcal{E}} \Big( p_i \nabla c_i + c_i \nabla p_i \Big) + \sum_{j \in \mathcal{I}} \Big( s_j \nabla c_j + c_j \nabla s_j \Big) + \nabla s_0.$$

For each $u \in C$, $f(u) = f_{min}$ and $\rho(u) \geq 0$. Let

$$J(u) := \{ j \in \mathcal{I} : c_j(u) = 0 \}.$$

For each $j \in \mathcal{I} \setminus J(u)$, $s_j(u) = 0$ and so $\nabla s_j(u) = 0$ (because $s_j$ is SOS). Similarly, $\nabla s_0(u) = 0$, so

$$\nabla f(u) = \sum_{i \in \mathcal{E}} p_i(u) \nabla c_i(u) + \sum_{j \in J(u)} s_j(u) \nabla c_j.$$

Hence, $u$ is a critical point of (5.3.5) with $J = J(u)$, and $u \in \Theta$. By Assumption 5.3.3, $C$ must be a finite set.

**Step 2** Since $C$ is finite, its vanishing ideal $I(C)$ (see (2.1.3) for the definition) is zero dimensional. Let $\{h_1, \ldots, h_r\}$ be a Gröbner basis of $I(C)$ with respect to a total degree ordering. Note that each $h_i \equiv 0$ on $C$. By Theorem 2.6.6, there exist $\ell \in \mathbb{N}$, $\phi_i \in \mathbb{R}[x]$, and $\varphi \in \mathrm{Pre}[\rho] = \mathrm{QM}[\rho]$ (because $\rho$ is a single polynomial) such that

$$h_i^{2\ell} + \varphi + \sum_{i \in \mathcal{E}} \phi_i c_i + \sum_{j \in \mathcal{I}} \phi_j s_j c_j + \phi_0 s_0 = 0.$$

When $2k$ is bigger than the degrees of all appearing polynomials in the above, the Riesz functional $\mathscr{L}_{y^*}$ results in the equation

$$\langle h_i^{2\ell}, y^* \rangle + \langle \varphi, y^* \rangle + \sum_{i \in \mathcal{E}} \langle \phi_i c_i, y^* \rangle + \sum_{j \in \mathcal{I}} \langle \phi_j s_j c_j, y^* \rangle + \langle \phi_0 s_0, y^* \rangle = 0. \tag{5.3.9}$$

The equation (5.3.7) implies that (note $\langle f, y^* \rangle = f_{min}$)

$$0 = \langle f - f_{min}, y^* \rangle = \sum_{i \in \mathcal{E}} \langle p_i c_i, y^* \rangle + \sum_{j \in \mathcal{I}} \langle s_j c_j, y^* \rangle + \langle s_0, y^* \rangle.$$

The equality constraints $\mathcal{V}_{c_i}^{(2k)}[y^*] = 0$ in (5.2.3) imply that $\langle p_i c_i, y^* \rangle = 0$ for all $i \in \mathcal{E}$. The inequality constraints $L_{c_j}^{(k)}[y^*] \succeq 0$ in (5.2.3) imply that $\langle s_j c_j, y^* \rangle \geq 0$ for all $j \in \mathcal{I}$. Since $s_0$ is SOS, we also have $\langle s_0, y^* \rangle \geq 0$. Thus, the above implies that

$$\langle s_j c_j, y^* \rangle = \langle s_0, y^* \rangle = 0.$$

By Lemma 5.3.5, if $2k > 2 + \deg(c_j s_j \phi_j)$, then

$$\langle c_j s_j \phi_j, y^* \rangle = 0$$

for each $j$. So, (5.3.9) implies that

$$\langle h_i^{2\ell}, y^* \rangle + \langle \varphi, y^* \rangle = 0. \tag{5.3.10}$$

Since $h_i^{2\ell}$ is SOS, we know $\langle h_i^{2\ell}, y^* \rangle \geq 0$. Note that

$$\varphi \in \text{QM}[\rho] \subseteq \text{Ideal}[c_{eq}] + \text{QM}[c_{in}].$$

We can also get that $\langle \varphi, y^* \rangle \geq 0$. Then (5.3.10) implies $\langle h_i^{2\ell}, y^* \rangle = 0$, i.e., $h_i^\ell \in \ker M_k[y^*]$. By Lemma 4.2.6, if $k$ is big enough, we have

$$h_i \in \ker M_k[y^*].$$

**Step 3** We show that the condition (5.3.3) is satisfied for $t = k - 1$, i.e., $y^*|_{2k-2}$ is flat. Since $C$ is a finite set, the quotient space $\mathbb{R}[x]/I(C)$ is finitely dimensional. Recall that $\{h_1, \ldots, h_r\}$ is a Gröbner basis of $I(C)$ with respect to a total degree ordering. Let $\{b_1, \ldots, b_L\}$ be a standard basis of $\mathbb{R}[x]/I(C)$. Then, for every $\alpha \in \mathbb{N}^n$, we can express that

$$x^\alpha = \eta(\alpha) + \sum_{i=1}^{r} \theta_i h_i, \quad \deg(\theta_i h_i) \leq |\alpha|, \quad \eta(\alpha) \in \text{span}\{b_1, \ldots, b_L\}.$$

Because every $h_i \in \ker M_k[y^*]$, by Lemma 4.2.6, we have

$$\theta_i h_i \in \ker M_k[y^*] \quad \text{if } |\alpha| \leq k - 1.$$

This implies that $x^\alpha - \eta(\alpha) \in \ker M_k[y^*]$ for all $|\alpha| \leq k - 1$. Let

$$d_b := \max_{j=1,\ldots,L} \deg(b_j).$$

For each $|\alpha| \in [d_b + 1, k - 1]$, the $\alpha$th column of $M_k[y^*]$ is a linear combination of $\beta$th columns of $M_k[y^*]$, with $|\beta| \leq d_b$. This implies that

$$\text{rank } M_{d_b}[y^*] = \text{rank } M_t[y^*], \quad t = d_b + 1, \ldots, k - 1.$$

Hence, if $k - 1 - d_c \geq d_b$, then

$$\text{rank } M_{k-1-d_c}[y^*] = \text{rank } M_{k-1}[y^*].$$

So, (5.3.3) must be satisfied for $t = k - 1$, when $k$ is big enough.

If rank $M_k[y^*]$ is maximum among the optimizers of (5.2.3), for every flat truncation of $y^*$, say, $y^*|_{2t}$, we have rank $M_t(y^*) = |S|$. All minimizers of (5.1.1) can be obtained from $y^*|_{2t}$. However, if the $S$ is infinite and rank $M_k[y^*]$ is maximum among all optimizers of (5.2.3), then $y^*$ cannot have a flat truncation. In short, when $y^*$ has a flat truncation and rank $M_k[y^*]$ is maximum, then all minimizers of (5.1.1) can be extracted as in Lemma 5.3.1. This issue was discussed in [171, §6.6]. We summarize this property as follows.

**Proposition 5.3.6.** *([171]) Let $S$ be the set of minimizers of (5.1.1) and let $y^*$ be a minimizer of the moment relaxation (5.2.3). Assume the* rank $M_k[y^*]$ *is maximum among all optimizers of (5.2.3). Then, we have the following:*

   (i) *If the flat truncation (5.3.3) holds for some* $t \in [d_0, k]$, *then* rank $M_t[y^*] = |S|$, *the cardinality of $S$. Therefore, all minimizers of (5.1.1) can be extracted as in Lemma 5.3.1.*

   (ii) *If the set $S$ is infinite, then the flat truncation (5.3.3) cannot be satisfied for any* $t \in [d_0, k]$.

**Proof.** (i) Let $r = $ rank $M_t[y^*]$. Suppose the truncation $y^*|_{2t}$ has the decomposition as in (5.3.4). By Lemma 5.3.1, the points $u_1, \ldots, u_r$ as in (5.3.4) are minimizers of (5.1.1), so $r \le |S|$. We need to show $r = |S|$. Suppose otherwise $r < |S|$. Then there exists $v \in S$ but $v \notin \{u_1, \ldots, u_r\}$. Let $p_1, \ldots, p_s \in \mathbb{R}[x]_t$ be polynomials such that $\{\text{vec}(p_1), \ldots, \text{vec}(p_s)\}$ is a basis for the null space of $M_t[y^*]$. By Lemma 2.7.8, the points $u_1, \ldots, u_r$ are precisely the common zeros of polynomials $p_1, \ldots, p_s$. Thus, there exists

$$p \in \text{span} \{p_1, \ldots, p_s\}$$

such that $p(v) \ne 0$, say, $p(v) = 1$. Let $z^* := [v]_{2k}$, then

$$\text{vec}(p)^T M_k[y^*] \text{vec}(p) = \text{vec}(p)^T M_t[y^*] \text{vec}(p) = 0,$$
$$\text{vec}(p)^T M_k[z^*] \text{vec}(p) = p(v)^2 = 1.$$

Note that $y^*, z^*, \frac{y^*+z^*}{2}$ are all minimizers of (5.2.3) and

$$M_k[y^*] \succeq 0, \quad M_k[z^*] \succeq 0, \quad M_k[y^* + z^*] \succeq 0.$$

Therefore, one can further see that

$$M_k[y^*] \cdot \text{vec}(p) = 0, \quad M_k[z^*] \cdot \text{vec}(p) \ne 0,$$

so we get

$$\ker M_k[y^* + z^*] = \ker M_k[y^*] \cap \ker M_k[z^*] \subsetneq \ker M_k[y^*].$$

This results in the contradiction

$$\text{rank } M_k[\frac{y^* + z^*}{2}] = \text{rank } M_k[y^* + z^*] > \text{rank } M_k[y^*].$$

Therefore, $r = |S|$ and we could get all minimizers of (5.1.1) from $y^*|_{2t}$.

   (ii) The conclusion essentially follows from the item (i). If (5.3.3) holds for some $t \in [d_0, k]$, then the cardinality of $S$ equals rank $M_t[y^*]$, which is impossible when $|S| = \infty$.  □

   When the primal-dual pair (5.2.2)-(5.2.3) is solved by interior-point methods, a minimizer $y^*$ with rank $M_k[y^*]$ being maximum is typically returned. Therefore, if (5.1.1) has finitely many minimizers, the flat truncation (5.3.3) is usually satisfied, by Theorem 5.3.4, and we can extract all minimizers by Proposition 5.3.6. However, if (5.1.1) has infinitely many minimizers, the flat truncation (5.3.3) is usually not satisfiable.

## 5.3.2 ▪ Flat truncation for asymptotic convergence

Another important issue is to check asymptotic convergence of the Moment-SOS hierarchy. When $\text{Ideal}[c_{eq}] + \text{QM}[c_{in}]$ is archimedean, the asymptotic convergence is guaranteed (see Theorem 5.2.2). If the Moment-SOS hierarchy fails to have finite convergence, the flat truncation (5.3.3) cannot be satisfiable. For instance, this is the case for Example 5.2.4. However, flat truncation can still serve as a suitable criterion for checking asymptotic convergence and for extracting minimizers approximately. When (5.1.1) has only finitely many minimizers, the flat truncation (5.3.3) is asymptotically satisfiable.

Let $S$ be the set of minimizers for (5.1.1). For a degree $t$, denote the convex hull

$$F_{2t} := \text{conv}\left(\{[u]_{2t} : u \in S\}\right). \tag{5.3.11}$$

**Proposition 5.3.7.** *([225]) Let $S$ and $F_{2t}$ be as above. When $0 < |S| < \infty$, if $t \geq d_c + |S| - 1$, then each $z \in F_{2t}$ is flat.*

Since the moment relaxation (5.2.3) may not have a minimizer (see Example 4.1.3), we need to consider a nearly optimal solution of (5.2.3). Let $\{y^{(k)}\}_{k=1}^{\infty}$ be a sequence such that $y^{(k)}$ is feasible for (5.2.3) for each $k$. The sequence $\{y^{(k)}\}_{k=1}^{\infty}$ is said to be *asymptotically optimal* if

$$\lim_{k \to \infty} \langle f, y^{(k)} \rangle = \lim_{k \to \infty} f_{mom,k}.$$

**Theorem 5.3.8.** *([225]) Assume $\text{Ideal}[c_{eq}] + \text{QM}[c_{in}]$ is archimedean and the set $S$ of minimizers for (5.1.1) is nonempty. Suppose $\{y^{(k)}\}_{k=1}^{\infty}$ is asymptotically optimal for (5.2.3). Then, for every fixed $t$, the truncated sequence $\{y^{(k)}|_{2t}\}$ is bounded and all its limit points belong to $F_{2t}$. Furthermore, if $|S| < \infty$ and $t \geq d_c + |S| - 1$, then each accumulation point of $\{y^{(k)}|_{2t}\}_{k=1}^{\infty}$ is flat.*

The proofs for Proposition 5.3.7 and Theorem 5.3.8 are given in [225]. The asymptotic convergence of the Moment-SOS hierarchy can be checked as follows. In numerical computation, the condition (5.3.3) requires one to determine the rank of a matrix numerically. For a given tolerance $\epsilon > 0$, the numerical rank of a matrix $A$ is the number of its singular values (or its eigenvalues if $A$ is symmetric positive semidefinite) that are greater than or equal to $\epsilon$. Suppose $S$ is a finite set and the sequence $\{y^{(k)}\}$ is asymptotically optimal for (5.2.3). For every fixed $t \geq d_c + |S| - 1$, each limit point $w$ of the truncated sequence $\{y^{(k)}|_{2t}\}$ is flat, asserted by Theorem 5.3.8. There exists $k$ sufficiently large such that

$$\left\| M_s[y^{(k)}] - M_s[w] \right\| < \epsilon \quad \text{for all } s \leq t.$$

Therefore, the flat truncation (5.3.3) holds numerically for all such $k$.

## 5.3.3 ▪ Flat truncation for the Schmüdgen type hierarchy

When the Schmüdgen type Moment-SOS hierarchy is applied to solve (5.1.1), the flat truncation (5.3.3) is still an appropriate criterion for checking the convergence. Indeed, it is satisfiable under weaker assumptions. The following is such a conclusion from [225].

**Theorem 5.3.9.** *[225] Suppose the set $S$ of minimizers of (5.1.1) is nonempty and finite. Then, the Schmüdgen type Moment-SOS hierarchy of relaxations (5.2.5)-(5.2.6) has finite convergence if and only if every minimizer of (5.2.6) satisfies the flat truncation (5.3.3), when $k$ is sufficiently large.*

The proof of Theorem 5.3.9 is given in [225]. It is similar to that for Theorem 5.3.4, so it is left as an exercise. Compared with Theorem 5.3.4, Theorem 5.3.9 has stronger properties in the following senses:

(i) It does not require the optimal value of (5.2.5) to be achievable.

(ii) Assumption 5.3.3 is weakened to finiteness of the set $S$.

(iii) If it occurs that $\mathrm{Pre}[c_{in}]_{2k} = \mathrm{QM}[c_{in}]_{2k}$ (e.g., this is the case if (5.1.1) has none or has only a single inequality constraint), then every minimizer of the moment relaxation (5.2.3) must satisfy the flat truncation (5.3.3) for all $k$ big enough, under the weaker assumption that $0 < |S| < \infty$.

## 5.3.4 ▪ Exercises

**Exercise 5.3.1.** *Find minimizers of the optimization*

$$\begin{cases} \min_{x \in \mathbb{R}^4} & x_1 x_2 x_3 + x_2 x_3 x_4 - (x_1^3 + x_2^3 + x_3^3 + x_4^3) \\ s.t. & 1 - x_1^2 - x_2^2 - x_3^2 - x_4^2 \geq 0. \end{cases}$$

**Exercise 5.3.2.** *Find minimizers of the optimization*

$$\begin{cases} \min_{x \in \mathbb{R}^6} & 6(x_1^4 + \cdots + x_6^4) - (x_1^2 + \cdots + x_6^2)^2 \\ s.t. & x_1 x_2 = 1, \ x_3 x_4 \geq 1, \ x_2 x_3 + x_5 x_6 \geq 2. \end{cases}$$

**Exercise 5.3.3.** *([225]) Let $\mathcal{V}$ be the set of real critical points of (5.3.5) on which the value of $f$ equals $f_{min}$. Suppose $S$ is finite and $\mathcal{V}$ is infinite, but $\mathcal{V} \backslash K$ is contained in a compact set $T$ not intersecting $K$. Show that Assumption 5.3.3 holds.*

**Exercise 5.3.4.** *([225]) Let $\mathcal{V}$ be the same as in Exercise 5.3.3. Suppose $\mathcal{V}$ is unbounded but, except for finitely many points, lies on a real algebraic set $T$ not intersecting $K$. If $\mathrm{Ideal}[c_{eq}] + \mathrm{QM}[c_{in}]$ is archimedean, show that Assumption 5.3.3 holds.*

**Exercise 5.3.5.** *([225]) Let $y \in \mathbb{R}^{\mathbb{N}_{2k}^n}, h \in \mathbb{R}[x]_{2k}$ be such that $L_h^{(k)}[y] \succeq 0$.*

(i) *Let $\{p_j\}_{j=1}^\infty \subseteq \mathbb{R}[x]$ be a sequence such that each $\deg(hp_j^2) \leq 2k$ and $L_h^{(k)}[y] \cdot p_j \to 0$ as $j \to \infty$. If $q \in \mathbb{R}[x]$ and each $\deg(p_j q) \leq k - \lceil \deg(h)/2 \rceil - 1$, show that*

$$L_h^{(k)}[y] \cdot \mathrm{vec}(p_j q) \to 0 \quad as \quad j \to \infty.$$

(ii) *Let $\{s_j\}_{j=1}^\infty$ be a sequence of SOS polynomials such that*

$$\deg(hs_j) < 2k \quad for \ all \ j, \qquad \langle hs_j, y \rangle \to 0 \ as \ j \to \infty.$$

*If $\phi \in \mathbb{R}[x]_{2\ell}$ and each $\deg(hs_j) + 2\ell \leq 2k - 2$, show that*

$$\langle hs_j \phi, y \rangle \to 0 \quad as \quad j \to \infty.$$

**Exercise 5.3.6.** *Prove Theorem 5.3.9.*

# 5.4 ▪ Finite Convergence Theory

There is a close relationship between the finite convergence theory of Moment-SOS hierarchy and local optimality conditions. Recall the linear independence constraint qualification condition (LICQC), the first order optimality condition (FOOC), the strict complementarity condition (SCC), and the second order sufficient condition (SOSC) introduced in §5.1. The following is the major conclusion for the finite convergence of the Moment-SOS hierarchy for (5.1.1).

**Theorem 5.4.1.** *([230]) Assume* $\text{Ideal}[c_{eq}] + \text{QM}[c_{in}]$ *is archimedean. If the LICQC, SCC, and SOSC hold at every global minimizer of (5.1.1), then the Moment-SOS hierarchy of (5.2.2)-(5.2.3) has finite convergence, i.e.,*

$$f_{sos,k} = f_{mom,k} = f_{min}$$

*for all k big enough.*

**Proof.** By Theorem 5.4.8, the boundary Hessian condition (see §5.4.2) holds at every global minimizer of (5.1.1), when the LICQC, SCC, and SOSC hold. Then, by Theorem 5.4.7, there exists $\sigma_1 \in \text{QM}[c_{in}]$ such that

$$f - f_{min} \equiv \sigma_1 \quad \mod \quad I(V_{\mathbb{R}}(c_{eq})).$$

The above $I(V_{\mathbb{R}}(c_{eq}))$ is the vanishing ideal for the real zero set of equality constraining polynomials. The polynomial

$$\hat{f} := f - f_{min} - \sigma_1 \equiv 0 \ \text{ on } \ V_{\mathbb{R}}(c_{eq}).$$

By Theorem 2.6.3, there exist $\ell \in \mathbb{N}$ and $\sigma_2 \in \Sigma[x]$ such that

$$\hat{f}^{2\ell} + \sigma_2 \in \text{Ideal}[c_{eq}].$$

Let $a > 0$ be a scalar big enough such that $s(t) := 1 + t + at^{2\ell}$ is an SOS univariate polynomial in $t$ (see Lemma 5.6.2). For $\epsilon > 0$, let

$$\sigma_\epsilon := \epsilon \cdot s\left(\hat{f}/\epsilon\right) + a \cdot \epsilon^{1-2\ell} \cdot \sigma_2 + \sigma_1,$$

$$\phi_\epsilon := f - (f_{min} - \epsilon) - \sigma_\epsilon.$$

Since $\phi_\epsilon = -a \cdot \epsilon^{1-2\ell} \cdot (\hat{f}^{2\ell} + \sigma_2) \in \text{Ideal}[c_{eq}]$, let $k_0 \in \mathbb{N}$ be such that

$$\sigma_\epsilon \in \text{QM}[c_{in}]_{2k_0}, \quad \phi_\epsilon \in \text{Ideal}[c_{eq}]_{2k_0}$$

for all $\epsilon > 0$. So, for every $\epsilon > 0$, $\gamma = f_{min} - \epsilon$ is feasible in (5.2.2) for the order $k_0$. Hence, $f_{sos,k_0} \geq f_{min} - \epsilon$. Since $\epsilon > 0$ can be arbitrary, we get $f_{sos,k_0} \geq f_{min}$. Since $f_{sos,k} \leq f_{min}$ for all $k$ and $f_{sos,k}$ is monotonically increasing, we know $f_{sos,k} = f_{min}$ for all $k \geq k_0$. By the relation (5.2.4), we also get $f_{mom,k} = f_{min}$ for all $k \geq k_0$. This shows the Moment-SOS hierarchy has finite convergence.  □

Another good property of the Moment-SOS hierarchy is that the flat truncation is also satisfiable under the archimedeanness and some local optimality conditions.

**Theorem 5.4.2.** *Assume* $\text{Ideal}[c_{eq}] + \text{QM}[c_{in}]$ *is archimedean and* $\text{Ideal}[c_{eq}]$ *is real radical. Suppose the LICQC, SCC, and SOSC hold at every global minimizer of (5.1.1). Then, when k is big enough, every minimizer* $y^*$ *of the moment relaxation (5.2.3) must satisfy the flat truncation (5.3.3).*

*Proof.* The proof is similar to that for Theorem 5.3.4. For convenience of writing, we can generally assume that
$$f_{min} = 0,$$
up to shifting a constant for the objective. Since the LICQC, SCC, and SOSC hold at every global minimizer of (5.1.1), we have $f \in \text{Ideal}[c_{eq}] + \text{QM}[c_{in}]$, by Theorem 5.5.2. Let
$$Q := \text{Ideal}[c_{eq}, f] + \text{QM}[c_{in}] = \text{QM}[c_{in}, c_{eq}, -c_{eq}, f, -f].$$
Since $\text{Ideal}[c_{eq}] + \text{QM}[c_{in}]$ is archimedean, $Q$ is also archimedean. The intersection
$$J := Q \cap -Q$$
is an ideal (see Exercise 2.5.8). Since the boundary Hessian condition holds at each zero of $f$, the ideal $J$ is zero dimensional, i.e., the coordinate ring $\frac{\mathbb{R}[x]}{J}$ has dimension 0. This is shown in the proof of Theorem 2.3 of [190]. The set $S$ of global minimizers of (5.1.1) can be equivalently given as
$$S = \left\{ x \in \mathbb{R}^n \;\middle|\; \begin{array}{rcl} f(x) & = & 0, \\ c_i(x) & = & 0 \, (i \in \mathcal{E}), \\ c_j(x) & \geq & 0 \, (j \in \mathcal{I}) \end{array} \right\}.$$
Since $S$ is contained in the zero set of $J$, the set $S$ is finite.

The vanishing ideal $I(S)$ is zero dimensional. Let $\{h_1, \ldots, h_r\}$ be a Gröbner basis of $I(S)$ with respect to a total degree ordering. Since $h_i \equiv 0$ on $S$ and $J$ is zero dimensional, by Theorem 2.6.9, we have $-h_i^{2\ell} \in Q$ for some $\ell \in \mathbb{N}$. Equivalently, there exist $\phi_0, \phi_i \in \mathbb{R}[x]$, and $\psi_0, \psi_j \in \Sigma[x]$ such that
$$h_i^{2\ell} + \sum_{i \in \mathcal{E}} \phi_i c_i + \phi_0 f + \sum_{j \in \mathcal{I}} \psi_j c_j + \psi_0 = 0.$$

When $2k$ is bigger than the degrees of all appearing polynomials in the above, we have the equation
$$\langle h_i^{2\ell}, y^* \rangle + \sum_{i \in \mathcal{E}} \langle \phi_i c_i, y^* \rangle + \langle \phi_0 f, y^* \rangle + \sum_{j \in \mathcal{I}} \langle \psi_j c_j, y^* \rangle + \langle \psi_0, y^* \rangle = 0. \qquad (5.4.1)$$

Like in the proof for Theorem 5.3.4, we also have
$$\begin{array}{rcl} \langle \phi_i c_i, y^* \rangle & = & 0 \, (i \in \mathcal{E}), \\ \langle \psi_j c_j, y^* \rangle & = & 0 \, (j \in \mathcal{I}), \\ \langle h_i^{2\ell}, y^* \rangle & \geq & 0, \\ \langle \psi_0, y^* \rangle & \geq & 0. \end{array} \qquad (5.4.2)$$

In the following, we show
$$\langle \phi_0 f, y^* \rangle = 0. \qquad (5.4.3)$$
Since $f \in \text{Ideal}[c_{eq}] + \text{QM}[c_{in}]$, there exist polynomials $p_i \in \mathbb{R}[x]$ for $i \in \mathcal{E}$ and $\sigma_j \in \Sigma[x]$ for $j \in \mathcal{I} \cup \{0\}$ such that
$$f = \sum_{i \in \mathcal{E}} p_i c_i + \sum_{j \in \mathcal{I}} \sigma_j c_j + \sigma_0. \qquad (5.4.4)$$
By Theorem 5.4.1, we have
$$\langle f, y^* \rangle = f_{mom,k} = f_{min} = 0$$

when $k$ is large enough. Due to the constraints in (5.2.3), one can see that

$$\langle p_i c_i, y^* \rangle = \langle \sigma_j c_j, y^* \rangle = \langle \sigma_0, y^* \rangle = 0$$

for all $i \in \mathcal{E}$ and $j \in \mathcal{I}$. By Lemma 5.3.5, for every $q \in \mathbb{R}[x]$, we have

$$\langle p_i c_i q^2, y^* \rangle = \langle \sigma_j c_j q^2, y^* \rangle = \langle \sigma_0 q^2, y^* \rangle = 0$$

when $k$ is high enough. Since $\phi_0 = (\phi_0 + \frac{1}{4})^2 - (\phi_0 - \frac{1}{4})^2$, we get

$$\langle p_i c_i \phi_0, y^* \rangle = \langle \sigma_j c_j \phi_0, y^* \rangle = \langle \sigma_0 \phi_0, y^* \rangle = 0.$$

The above implies that (5.4.3) holds when $k$ is big enough.

The relations (5.4.1), (5.4.2), and (5.4.3) imply that $\langle h_i^{2\ell}, y^* \rangle = 0$. By Lemma 4.2.6, when $k$ is big enough, we also have $h_i \in \ker M_k[y^*]$. Finally, we can also show that the flat truncation (5.3.3) is satisfied for some $t \le k$. The remaining proof is the same as for Step 3 in the proof for Theorem 5.3.4.     □


## 5.4.1 ▪ Some facts about finite convergence

Theorem 5.4.1 shows that the Moment-SOS hierarchy is tight when the LICQC, SCC, and SOSC hold at every global minimizer of (5.1.1), under the archimedean condition. This establishes the relationship between local optimality conditions and the global optimality criterion:

$$f(x) - f_{min} \in \mathrm{Ideal}[c_{eq}] + \mathrm{QM}[c_{in}].$$

We discuss this criterion with more details in §5.5.1. The following is an exposition for Theorem 5.4.1.


**Example 5.4.3.** *Consider the optimization (5.1.1) with*

$$f = x_1^6 + x_2^6 + x_3^6 + 3x_1^2 x_2^2 x_3^2 - x_1^4(x_2^2 + x_3^2) - x_2^4(x_3^2 + x_1^2) - x_3^4(x_1^2 + x_2^2)$$

*and the sphere constraint $x_1^2 + x_2^2 + x_3^2 = 1$. The objective $f$ is the Robinson form, which is nonnegative but not SOS. The minimum value $f_{min} = 0$ and the minimizers are*

$$\frac{1}{\sqrt{3}}(\pm 1, \pm 1, \pm 1), \ \frac{1}{\sqrt{2}}(\pm 1, \pm 1, 0), \ \frac{1}{\sqrt{2}}(\pm 1, 0, \pm 1), \ \frac{1}{\sqrt{2}}(0, \pm 1, \pm 1).$$

*The LICQC holds at every feasible point. There is no inequality constraint, so the SCC also holds. The SOSC (5.1.9) holds for all minimizers. For instance, for $u = \frac{1}{\sqrt{3}}(1, 1, 1)$, the Hessian*

$$\nabla^2 L(u) = \frac{4}{9}\left(3I_3 - uu^T\right)$$

*is positive definite in the orthogonal complement $u^\perp$. Theorem 5.4.1 shows that the Moment-SOS hierarchy has finite convergence. The numerical computation indicates $f_{sos,5} = f_{min}$.*


An interesting question is: What is an upper bound on $k$ such that $f_{sos,k} = f_{min}$? This question is mostly open. Such $k$ certainly depends on $f$ and constraining polynomials. One wonders whether or not there is a uniform upper bound on such $k$ for generic polynomials of given degrees. That is, does there exist $N$, which only depends on the degree of $f$ and constraining polynomials, such that $f_{sos,k} = f_{min}$ for all generic $f$ of the given degree and for all $k \ge N$?

Unfortunately, such a bound $N$ does not exist. This is implied by Scheiderer's result [293] on the non-existence of degree bounds for SOS type representations. Indeed, when $K$ is the unit ball, there is no such $N$ (see [225, §5]).

None of the optimality conditions can be removed in Theorem 5.4.1 to guarantee finite convergence. This can be seen as follows.

- When $f = 3x_1 + 2x_2$ and the constraints are

$$x_1^2 - x_2^2 - (x_1^2 + x_2^2)^2 \geq 0, \quad x_1 \geq 0,$$

  the origin $0$ is the unique minimizer. The LICQC fails at $0$, and the FOOC (5.1.3) fails. The feasible set has nonempty interior, so the relaxation (5.2.2) achieves its optimal value. However, the Moment-SOS hierarchy does not have finite convergence, which is implied by Proposition 5.4.4.

- When $f = x_1x_2 + x_1^3 + x_2^3$ and the constraints are

$$x_1 \geq 0, \quad x_2 \geq 0, \quad 1 - x_1 - x_2 \geq 0,$$

  the origin $0$ is the unique minimizer. The LICQC holds at $0$. The Lagrange multipliers are all zeros. The second order sufficient condition (5.1.9) also holds at $0$. However, the strict complementarity condition fails at $0$. The Moment-SOS hierarchy does not have finite convergence. This is shown in [294, Remark 3.9]. The proof is left as Exercise 5.4.1.

- Consider $f = x_1^4x_2^2 + x_1^2x_2^4 + x_3^6 - 3x_1^2x_2^2x_3^2 + \epsilon(x_1^2 + x_2^2 + x_3^2)^3$ and the constraint is the unit ball $1 - x_1^2 - x_2^2 - x_3^2 \geq 0$. For each $\epsilon > 0$, the origin is the unique minimizer, and the LICQC and strict complementarity hold at $0$. However, the second order sufficient condition fails. When $\epsilon > 0$ is sufficiently small, the Moment-SOS hierarchy does not have finite convergence, as shown in [190, Example 2.4].

Whenever the feasible set $K$ has dimension 3 or higher, there always exists $f$ such that the Moment-SOS hierarchy fails to have finite convergence [292, Prop. 6.1]. For instance, this is the case if $f$ is the Motzkin polynomial and $K$ is the unit ball in $\mathbb{R}^3$ (see [224, Example 5.3]).

Theorem 5.4.1 shows that the Moment-SOS hierarchy has finite convergence under the archimedeanness and aforementioned optimality conditions. It is possible that the finite convergence may still hold even if some of them fail. However, the FOOC (5.1.3) is generally necessary for the finite convergence to occur. This is shown in the following.

**Proposition 5.4.4.** *([230]) Suppose the SOS relaxation (5.2.2) achieves its optimal value. If the FOOC (5.1.3) fails at a minimizer of (5.1.1), then $f_{sos,k} < f_{min}$ for all $k$.*

The proof of Proposition 5.4.4 is left as an exercise. The assumption that (5.2.2) achieves its optimal value cannot be dropped. This can be seen for the univariate optimization problem

$$\begin{cases} \min & x \\ s.t. & -x^2 \geq 0. \end{cases} \tag{5.4.5}$$

The minimizer is $0$ and $f_{min} = 0$. The first order optimality condition (5.1.3) fails, but $f_{sos,k} = f_{min}$ for all $k \geq 1$ (see the exercise). We remark that the SOS relaxation (5.2.2) achieves its optimal value when $K$ has nonempty interior (see Proposition 5.2.1).

## 5.4.2 ▪ The boundary Hessian condition

Suppose $u$ is a local minimizer of (5.1.1). Let $\ell$ be the local dimension of the variety $V_{\mathbb{R}}(c_{eq})$ at $u$ (see §2.2 or [34, §2.8]). We assume the feasible set $K$ can be parameterized near $u$ as follows.

**Condition 5.4.5.** *([192]) (i) The point $u$ on $V_{\mathbb{R}}(c_{eq})$ is nonsingular and there exists a neighborhood $\mathcal{O}$ of $u$ such that $V_{\mathbb{R}}(c_{eq}) \cap \mathcal{O}$ is parameterized by parameters $t_1, \ldots, t_\ell$; (ii) there exist $\nu_1, \ldots, \nu_r \in \mathcal{I}$ such that $t_j = c_{\nu_j}(x)$ for $j = 1, \ldots, r$ on $V_{\mathbb{R}}(c_{eq}) \cap \mathcal{O}$ and $K \cap \mathcal{O}$ is defined by $t_1 \geq 0, \ldots, t_r \geq 0$.*

The following condition was introduced by Marshall [190, 192]. It is called the *boundary Hessian condition* (BHC).

**Condition 5.4.6.** *([190, 192]) Assume the Condition 5.4.5 holds for the feasible set $K$ at $u$. Expand $f$ locally around $u$ as $f = f_0 + f_1 + f_2 + \cdots$, where each $f_i$ is homogeneous of degree $i$ in $t_1, \ldots, t_\ell$. It holds that $f_1 = a_1 t_1 + \cdots + a_r t_r$ for positive scalars $a_1 > 0, \ldots, a_r > 0$, and $f_2(0, \ldots, 0, t_{r+1}, \ldots, t_\ell)$ is positive definite in $(t_{r+1}, \ldots, t_\ell)$.*

We remark that if the BHC holds at every minimizer, then (5.1.1) has only finitely many minimizers (see exercise). Recall that $I(V)$ denotes the vanishing ideal of the set $V$. The following result is due to Marshall.

**Theorem 5.4.7.** *([191, Theorem 9.5.3]) Let $V = V_{\mathbb{R}}(c_{eq})$ and let $f_{min}$ be the minimum value of (5.1.1). If $\mathrm{Ideal}[c_{eq}] + \mathrm{QM}[c_{in}]$ is archimedean and the BHC holds at every minimizer of (5.1.1), then*

$$f - f_{min} \in I(V) + \mathrm{QM}[c_{in}].$$

The above result is also stated in Scheiderer's survey [295, Theorem 3.1.7]. If $\mathrm{Ideal}[c_{eq}]$ is real radical (see §2.6.2), i.e., $I(V) = \mathrm{Ideal}[c_{eq}]$, then we have

$$f - f_{min} \in \mathrm{Ideal}[c_{eq}] + \mathrm{QM}[c_{in}]$$

in Theorem 5.4.7. However, this may not hold if $\mathrm{Ideal}[c_{eq}]$ is not real radical. The proof of Theorem 5.4.1 is mainly based on the following result.

**Theorem 5.4.8.** *([230]) Let $u$ be a local minimizer of (5.1.1). If the LICQC, SCC, and SOSC hold at $u$, then $f$ satisfies the BHC at $u$.*

**Proof.** For convenience of writing, suppose $\mathcal{E} = \{1, \ldots, m\}$ and the labeling set $J(u) = \{j \in \mathcal{I} : c_j(u) = 0\}$ of active inequality constraints is $\{j_1, \ldots, j_r\}$. We can also assume $u = 0$, up to a coordinate shifting. Since the LICQC holds at 0, the gradients

$$\nabla c_1(0), \ldots, \nabla c_m(0), \nabla c_{j_1}(0), \ldots, \nabla c_{j_r}(0)$$

are linearly independent. The origin 0 is a nonsingular point of the real variety $V_{\mathbb{R}}(c_{eq})$. Up to a linear coordinate transformation, one can further assume that

$$
\begin{aligned}
\begin{bmatrix} \nabla c_{j_1}(0) & \cdots & \nabla c_{j_r}(0) \end{bmatrix} &= \begin{bmatrix} I_r \\ 0 \end{bmatrix}, \\
\begin{bmatrix} \nabla c_1(0) & \cdots & \nabla c_m(0) \end{bmatrix} &= \begin{bmatrix} 0 \\ I_m \end{bmatrix}.
\end{aligned}
\tag{5.4.6}
$$

In the above, each 0 stands for a zero block matrix of compatible dimension. Let $\ell := n - m$, the local dimension of $V_{\mathbb{R}}(c_{eq})$ at 0. Define the function

$$\varphi(x) = (\varphi_{\mathrm{I}}(x), \varphi_{\mathrm{II}}(x), \varphi_{\mathrm{III}}(x)) : \mathbb{R}^n \to \mathbb{R}^n,$$

where

$$\varphi_{\mathrm{I}}(x) = \begin{bmatrix} c_{j_1}(x) \\ \vdots \\ c_{j_r}(x) \end{bmatrix}, \quad \varphi_{\mathrm{II}}(x) = \begin{bmatrix} x_{r+1} \\ \vdots \\ x_\ell \end{bmatrix}, \quad \varphi_{\mathrm{III}}(x) = \begin{bmatrix} c_1(x) \\ \vdots \\ c_m(x) \end{bmatrix}. \tag{5.4.7}$$

Clearly, $\varphi(0) = 0$, and the Jacobian of $\varphi$ at 0 is the identity matrix $I_n$. By the implicit function theorem, the equation $t = \varphi(x)$ defines a smooth function

$$x = \varphi^{-1}(t)$$

in a neighborhood $\mathcal{O}$ of 0. So $t = (t_1, \ldots, t_n)$ can serve as a coordinate system for $\mathbb{R}^n$ around 0 and $t = \varphi(x)$. In the $t$-coordinate system and in the neighborhood $\mathcal{O}$, $V_{\mathbb{R}}(c_{eq})$ is defined by linear equations $t_{\ell+1} = \cdots = t_n = 0$, and $K \cap \mathcal{O}$ can be equivalently described as

$$t_1 \geq 0, \ldots, t_r \geq 0, \quad t_{\ell+1} = \cdots = t_n = 0.$$

Let $\lambda_i (i \in \mathcal{E})$ and $\lambda_j (j \in \mathcal{I})$ be the Lagrange multipliers satisfying (5.1.3)-(5.1.4). The Lagrange function is

$$L(x) := f(x) - \sum_{i=1}^m \lambda_i c_i(x) - \sum_{k=1}^r \lambda_{j_k} c_{j_k}(x).$$

Note that $\nabla_x L(0) = 0$. In the $t$-coordinate system, define functions

$$F(t) := f(\varphi^{-1}(t)),$$
$$\widehat{L}(t) := L(\varphi^{-1}(t)) = F(t) - \sum_{i=\ell+1}^n \lambda_{i-\ell} t_i - \sum_{k=1}^r \lambda_{j_k} t_k.$$

Clearly, $\nabla_x L(0) = 0$ implies $\nabla_t \widehat{L}(0) = 0$. So, it holds that

$$\frac{\partial F(0)}{\partial t_k} = \lambda_{j_k}, \; k = 1, \ldots, r,$$
$$\frac{\partial F(0)}{\partial t_k} = 0, \; k = r+1, \ldots, \ell,$$
$$\frac{\partial F(0)}{\partial t_k} = \lambda_{k-\ell}, \; k = \ell+1, \ldots, n.$$

Expand $F(t)$ locally around 0 as

$$F(t) = f_0 + f_1(t) + f_2(t) + f_3(t) + \cdots,$$

where each $f_i$ is a form in $t$ of degree $i$. Clearly, we have

$$f_1(t) = \mu_{j_1} t_1 + \cdots + \mu_{j_r} t_r \quad \text{when} \quad t_{\ell+1} = \cdots = t_n = 0.$$

For $t_{r+1}, \ldots, t_\ell$ near zero, it holds that

$$F(0, \ldots, 0, t_{r+1}, \ldots, t_\ell, 0, \ldots, 0) = \widehat{L}(0, \ldots, 0, t_{r+1}, \ldots, t_\ell, 0, \ldots, 0)$$
$$= L(\varphi^{-1}(0, \ldots, 0, t_{r+1}, \ldots, t_\ell, 0, \ldots, 0)).$$

We write that
$$x(t) := \varphi^{-1}(t) = (\varphi_1^{-1}(t), \ldots, \varphi_n^{-1}(t)).$$

For all $i, j$, we have
$$\frac{\partial^2 \widehat{L}(t)}{\partial t_i \partial t_j} = \sum_{1 \le k, s \le n} \frac{\partial^2 L(x(t))}{\partial x_k \partial x_s} \frac{\partial \varphi_k^{-1}(t)}{\partial t_i} \frac{\partial \varphi_s^{-1}(t)}{\partial t_j} + \sum_{1 \le k \le n} \frac{\partial L(x(t))}{\partial x_k} \frac{\partial^2 \varphi_k^{-1}(t)}{\partial t_i \partial t_j}.$$

Evaluating the above at $x = t = 0$, we get (note $\nabla_x L(0) = 0$)
$$\frac{\partial^2 \widehat{L}(0)}{\partial t_i \partial t_j} = \sum_{1 \le k, s \le n} \frac{\partial^2 L(0)}{\partial x_k \partial x_s} \frac{\partial \varphi_k^{-1}(0)}{\partial t_i} \frac{\partial \varphi_s^{-1}(0)}{\partial t_j}.$$

The Jacobians of $\varphi$ and $\varphi^{-1}$ are both the identity matrix at 0. So, for all $r + 1 \le i, j \le \ell$, we have
$$\left. \frac{\partial^2 f_2}{\partial t_i \partial t_j} \right|_{t=0} = \left. \frac{\partial^2 F}{\partial t_i \partial t_j} \right|_{t=0} = \left. \frac{\partial^2 \widehat{L}}{\partial t_i \partial t_j} \right|_{t=0} = \left. \frac{\partial^2 L}{\partial x_i \partial x_j} \right|_{x=0}. \tag{5.4.8}$$

The strict complementarity condition (5.1.5) implies that
$$\lambda_{j_1} > 0, \ldots, \lambda_{j_r} > 0,$$

i.e., the coefficients of $\lambda_{j_1} t_1 + \cdots + \lambda_{j_r} t_r$ are all positive. The SOSC (5.1.9) implies that the sub-Hessian matrix
$$\left( \frac{\partial^2 L(0)}{\partial x_i \partial x_j} \right)_{r+1 \le i, j \le \ell}$$

is positive definite. By (5.4.8), the quadratic form $f_2$ is positive definite in $(t_{r+1}, \ldots, t_\ell)$. Therefore, $f$ satisfies the BHC at 0. $\quad\Box$

Theorem 5.4.8 concludes that the LICQC, SCC, and SOSC imply the BHC. Typically, to check the boundary Hessian condition by its definition, one needs to construct a local parametrization for the feasible set $K$ and verify some sign conditions, which is usually not practical. However, checking these optimality conditions is generally much more convenient than checking the BHC.

## 5.4.3 ▪ Exercises

**Exercise 5.4.1.** *Show that $x_1 x_2 + x_1^3 + x_2^3 \notin \mathrm{QM}[x_1, x_2, 1 - x_1 - x_2]$.*

**Exercise 5.4.2.** *For the optimization (5.4.5), show that $f_{sos,k} = f_{min}$ for all $k \ge 1$, i.e., all Moment-SOS relaxations are tight.*

**Exercise 5.4.3.** *Show that the Moment-SOS hierarchy fails to have finite convergence for the optimization*
$$\begin{cases} \min\limits_{x \in \mathbb{R}^2} & 3x_1 - 2x_2 \\ s.t. & x_1^2 - x_2^2 - (x_1^2 + x_2^2)^3 \ge 0, \ x_1 \ge 0. \end{cases}$$

**Exercise 5.4.4.** *Give an example to show that the Moment-SOS hierarchy may fail to have finite convergence if the archimeddeanness does not hold, even when all LICQC, SCC, and SOSC hold at every minimizer.*

**Exercise 5.4.5.** *Show that (5.1.1) has only finitely many minimizers if the BHC holds at every minimizer.*

**Exercise 5.4.6.** *Prove Proposition 5.4.4.*

# 5.5 ▪ Properties of Optimality Conditions

Recall the local optimality conditions introduced in §5.1. The relations among them are the following: if the LICQC holds at a local minimizer $u$, then the FOOC, CC, SONC hold; if the FOOC, SCC, and SOSC hold at a feasible point $u$, then $u$ is a strict local minimizer of (5.1.1). In short, the FOOC, SCC, and SOSC are sufficient for local optimality, but they may not be necessary. Interestingly, for *generic* polynomials, they are also necessary conditions for local optimality. Moreover, a remarkable fact is that local optimality conditions are related to global optimality certificates. In this section, we introduce certificates for local and global optimality. Properties of critical points are also discussed.

## 5.5.1 ▪ Local versus global optimality certificates

Local and global optimality certificates are presumably to be very different, except special cases like convex optimization. However, for polynomial optimization, there are similar (though different) certificates for local and global optimality. The following is a natural certificate for local optimality.

**Proposition 5.5.1.** *Let $u$ be a feasible point of (5.1.1). Then, $u$ is a local minimizer of (5.1.1) if and only if there exist a neighborhood $\mathcal{O}$ of $u$, scalars $\mu_i$ $(i \in \mathcal{E} \cup \mathcal{I})$, and a function $q(x)$ on $\mathcal{O}$ such that*

$$
\begin{aligned}
f(x) - f(u) &= \sum_{i\in\mathcal{E}} \mu_i c_i(x) + \sum_{j\in\mathcal{I}} \mu_j c_j(x) + q(x) \text{ on } \mathcal{O}, \\
\mu_j &\geq 0 \ (j \in \mathcal{I}), \\
q(x) &\geq 0 \text{ on } \mathcal{O} \cap K.
\end{aligned}
\tag{5.5.1}
$$

Proposition 5.5.1 is actually quite straightforward. If the certificate (5.5.1) holds, then $f(x) - f(u) \geq 0$ for all $x \in \mathcal{O} \cap K$ and hence $u$ is a local minimizer. Conversely, if $u$ is a local minimizer, then (5.5.1) holds for the choice that all $\mu_i = 0$ and $q = f - f(u)$. The scalars $\mu_i$ in (5.5.1) may not be Lagrange multipliers. In (5.5.1), we typically cannot expect $q \geq 0$ on $\mathcal{O}$. For instance, this is the case for $f = x_1 + x_2$ and for the constraint $x_1 x_2 = 1$, at the local minimizer $u = (1,1)$. We leave the proof as an exercise.

In the following, we give a global optimality certificate. By the definition, a feasible point $u$ is a global minimizer if and only if

$$
f(x) - f(u) \geq 0 \quad \text{for all } x \in K. \tag{5.5.2}
$$

It is typically difficult to check (5.5.2) directly. People usually prefer easily checkable conditions that ensure (5.5.2). When $f$ and all $c_i$ are polynomials, the certificate (5.5.2) is ensured by Putinar's Positivstellensatz:

$$
\begin{aligned}
f(x) - f(u) &= \sum_{i\in\mathcal{E}} \phi_i c_i + \sum_{j\in\mathcal{I}} \phi_j c_j + \sigma, \\
\phi_i &\in \mathbb{R}[x] \ (i \in \mathcal{E}), \\
\sigma, \phi_j &\in \Sigma[x] \ (j \in \mathcal{I}).
\end{aligned}
\tag{5.5.3}
$$

Clearly, if (5.5.3) holds, then $u$ must be a global minimizer of (5.5.2). The global optimality certificate (5.5.3) is equivalent to the membership

$$f(x) - f(u) \in \text{Ideal}[c_{eq}] + \text{QM}[c_{in}].$$

When $u$ is a global minimizer of (5.1.1), the polynomial

$$\hat{f}(x) := f(x) - f(u)$$

is nonnegative on $K$. But it is not strictly positive, since $\hat{f}(u) = 0$. So Theorem 2.6.1 does not imply (5.5.3). The polynomials $\phi_i$ can be viewed as extensions of Lagrange multipliers, while the membership $\phi_j \in \Sigma[x]$ for $j \in \mathcal{I}$ can be viewed as sign conditions for Lagrange multipliers corresponding to inequality constraints. The representation (5.5.3) is a global optimality certificate for the polynomial optimization (5.1.1).

The local optimality certificate (5.5.1) and the global one (5.5.3) have similarities, but they are also quite different. The similarities and differences are the following: each $\mu_j$ ($j \in \mathcal{I}$) is a nonnegative scalar while each $\sigma_j$ ($j \in \mathcal{I}$) is an SOS polynomial; the function $q$ is nonnegative on $\mathcal{O} \cap K$ while $\sigma$ is SOS. Interestingly, the local optimality conditions LICQC, SCC, and SOSC imply the global optimality certificate (5.5.3), when $\text{Ideal}[c_{eq}] + \text{QM}[c_{in}]$ is archimedean. Recall that an ideal $I \subseteq \mathbb{R}[x]$ is said to be *real radical* (or just *real*; see §2.6.2) if every $p \in \mathbb{R}[x]$, which vanishes identically on the real variety $V_{\mathbb{R}}(I)$, belongs to $I$.

**Theorem 5.5.2.** *Assume that* $\text{Ideal}[c_{eq}]$ *is real radical and* $\text{Ideal}[c_{eq}] + \text{QM}[c_{in}]$ *is archimedean. If the LICQC, SCC, and SOSC hold at every global minimizer of (5.1.1), then the global optimality certificate (5.5.3) holds.*

**Proof.** At every global minimizer $u$, the LICQC, SCC, and SOSC imply that the BHC holds at $u$, by Theorem 5.4.8. Let $V := \{x \in \mathbb{R}^n : c_{eq}(x) = 0\}$. By Theorem 5.4.7, we have

$$f(x) - f(u) \in I(V) + \text{QM}[c_{in}].$$

Since $\text{Ideal}[c_{eq}]$ is real radical, i.e., $\text{Ideal}[c_{eq}] = I(V)$, the certificate (5.5.3) holds. $\qquad \square$

The global optimality certificate (5.5.3) may not hold for some special polynomial optimization problems. For instance, this is the case if $f$ is the Motzkin polynomial and $K$ is the unit ball. However, such cases are not generic. Indeed, Theorem 5.5.9 shows that the optimality conditions LICQC, SCC, and SOSC all hold at every local minimizer if the coefficients of input polynomials lie in an open dense set (actually a Zariski open set) in the embedding space. Therefore, the global optimality certificate (5.5.3) holds generically, when $\text{Ideal}[c_{eq}]$ is real radical and $\text{Ideal}[c_{eq}] + \text{QM}[c_{in}]$ is archimedean. Moreover, when $c_{eq}$ consists of less than $n$ polynomials, the ideal $\text{Ideal}[c_{eq}]$ is prime and real radical for generic cases. This issue was also discussed in [132, 327].

## 5.5.2 ▪ Properties of critical points

First, we count the number of complex critical points. Let $d_0, \ldots, d_k$ be positive degrees and $k \leq n$. For polynomials $p_0 \in \mathbb{R}[x]_{d_0}, \ldots, p_k \in \mathbb{R}[x]_{d_k}$, consider the equality constrained optimization

$$\begin{cases} \min & p_0(x) \\ s.t. & p_1(x) = \cdots = p_k(x) = 0. \end{cases} \qquad (5.5.4)$$

The KKT system is

$$\begin{aligned}\nabla p_0(x) &= \lambda_1 \nabla p_1(x) + \cdots + \lambda_k \nabla p_k(x), \\ p_1(x) &= \cdots = p_k(x) = 0.\end{aligned} \tag{5.5.5}$$

Each $(x, \lambda)$ satisfying (5.5.5) is called a *critical pair* and such $x$ is a critical point. Denote the polynomial tuple

$$p := (p_0, p_1, \ldots, p_k).$$

Consider the complex determinantal variety

$$\mathcal{K}(p) := \left\{ x \in \mathbb{C}^n \; \middle| \; \begin{array}{c} \text{rank} \begin{bmatrix} \nabla p_0(x) & \nabla p_1(x) & \cdots & \nabla p_k(x) \end{bmatrix} \leq k, \\ p_1(x) = \cdots = p_k(x) = 0 \end{array} \right\}. \tag{5.5.6}$$

It is called the *critical variety* of the optimization (5.5.4). Let $\mathbf{S}_r$ be the $r$th complete symmetric function in $k$ variables (also see (2.3.5)):

$$\mathbf{S}_r(n_1, n_2, \ldots, n_k) := \sum_{i_1+i_2+\cdots+i_k=r} n_1^{i_1} \cdots n_k^{i_k}. \tag{5.5.7}$$

**Theorem 5.5.3.** *([217]) If the polynomials $p_0, p_1, \ldots, p_k$ are generic, then the number of complex critical points of (5.5.4) is equal to*

$$d_1 \cdot d_2 \cdots d_k \cdot \mathbf{S}_{n-k}(d_0 - 1, d_1 - 1, \ldots, d_k - 1).$$

Second, we give conditions for $\mathcal{K}(p)$ to intersect a hypersurface $q(x) = 0$ for a polynomial $q \in \mathbb{R}[x]_{d_{k+1}}$. When does the intersection

$$\mathcal{K}(p) \cap \{q(x) = 0\} \neq \emptyset? \tag{5.5.8}$$

For generic $p$, the critical point set $\mathcal{K}(p)$ is finite, so it does not intersect $q(x) = 0$ if $q$ is also generic. Consider the homogenization of (5.5.8):

$$\begin{aligned} \text{rank} \begin{bmatrix} \nabla_x \widetilde{p_0}(\tilde{x}) & \nabla_x \widetilde{p_1}(\tilde{x}) & \cdots & \nabla_x \widetilde{p_k}(\tilde{x}) \end{bmatrix} &\leq k, \\ \widetilde{p_1}(\tilde{x}) = \cdots = \widetilde{p_k}(\tilde{x}) = \widetilde{q}(\tilde{x}) &= 0. \end{aligned} \tag{5.5.9}$$

In the above, each $\widetilde{p_i}(\tilde{x})$ is the homogenization of $p_i(x)$ and $\tilde{x} = (x_0, \ldots, x_n)$. When $k < n$, the matrix in (5.5.9) has rank less than or equal to $k$ if and only if all its maximal minors vanish; when $k = n$, the rank condition in (5.5.9) is always satisfied and can be dropped. We have the following proposition.

**Proposition 5.5.4.** *([230]) There exists a nonzero polynomial $\mathscr{R}$, in the coefficients of $p_0 \in \mathbb{R}[x]_{d_0}, \ldots, p_k \in \mathbb{R}[x]_{d_k}, q \in \mathbb{R}[x]_{d_{k+1}}$, such that (5.5.9) has a solution $0 \neq \tilde{x} \in \mathbb{C}^{n+1}$ if and only if $\mathscr{R}(p_0, \ldots, p_k, q) = 0$. In particular, if $\mathscr{R}(p_0, \ldots, p_k, q) \neq 0$, then the intersection in (5.5.8) is empty.*

There are also conditions for the KKT system (5.5.5) to be nonsingular. Denote

$$L_p(x, \lambda) := p_0(x) - \lambda_1 p_1(x) - \cdots - \lambda_k p_k(x).$$

The system (5.5.5) is nonsingular if and only if the square matrix

$$H_p(x, \lambda) := \begin{bmatrix} \nabla_x^2 L_p(x, \lambda) & \begin{bmatrix} \nabla p_1(x) & \cdots & \nabla p_k(x) \end{bmatrix} \\ \begin{bmatrix} \nabla p_1(x) & \cdots & \nabla p_k(x) \end{bmatrix}^T & 0 \end{bmatrix}$$

is nonsingular at every critical pair $(x, \lambda)$. If every $p_i$ is generic, there are only finitely many critical pairs, and (5.5.5) is nonsingular if $\det H_p(x, \lambda)$ is nonzero at each critical pair $(x, \lambda)$. We have the following proposition.

**Proposition 5.5.5.** *([230]) There exists a nonzero polynomial $\mathscr{D}$, in the coefficients of $p_0 \in \mathbb{R}[x]_{d_0}, \ldots, p_k \in \mathbb{R}[x]_{d_k}$, such that if $\mathscr{D}(p_0, \ldots, p_k) \neq 0$, then (5.5.5) is a nonsingular system.*

The following are expositions for Proposition 5.5.5.

- Suppose $p_i = a_i^T x + b_i$ for $i = 0, \ldots, k$, with generic coefficients. When $k < n$, (5.5.4) does not have a critical point; when $k = n$, the matrix $H_p(x, \lambda)$ has full rank if $a_1, \ldots, a_k$ are linearly independent.

- Suppose there are no constraints. If $\deg(p_0) = 1$ and $p_0$ is nonzero, (5.5.4) does not have a critical point. If $\deg(p_0) = 2$ and $p_0 = x^T A x + 2b^T x + c$ with $A = A^T$ and $\det(A) \neq 0$, then $H_p(x, \lambda) = 2A$ is nonsingular. When $\deg(p_0) \geq 3$, by the definition of discriminants (see §2.3), if

$$\Delta \left( \frac{\partial \widetilde{p_0}}{\partial x_1}, \ldots, \frac{\partial \widetilde{p_0}}{\partial x_n} \right) \neq 0,$$

then the KKT system $\nabla p_0(x) = 0$ is nonsingular.

### 5.5.3 ▪ Genericity of local optimality conditions

We discuss the genericity of the local and global optimality conditions. For convenience of notation, assume the labeling sets in (5.1.1) are

$$\mathcal{E} = \{1, \ldots, m_1\}, \quad \mathcal{I} = \{m_1 + 1, \ldots, m_1 + m_2\}.$$

**Condition 5.5.6.** *The polynomials $f \in \mathbb{R}[x]_{d_0}$, $c_i \in \mathbb{R}[x]_{d_i}$ ($i \in \mathcal{E}$), and $c_j \in \mathbb{R}[x]_{d_j}$ ($j \in \mathcal{I}$) with $m_1 \leq n$ satisfy the following inequalities (Res, $\Delta$ are from §2.3, $\mathscr{R}$ is from Proposition 5.5.4, and $\mathscr{D}$ is from Proposition 5.5.5):*

*(a) If $m_1 + m_2 \geq n + 1$, for all $m_1 + 1 \leq j_1 < \cdots < j_{n-m_1+1} \leq m_1 + m_2$,*

$$\mathrm{Res}(c_1, \ldots, c_{m_1}, c_{j_1}, \ldots, c_{j_{n-m_1+1}}) \neq 0.$$

*(b) For all $m_1 + 1 \leq j_1 < \cdots < j_r \leq m_1 + m_2$ with $0 \leq r \leq n - m_1$,*

$$\Delta(c_1, \ldots, c_{m_1}, c_{j_1}, \ldots, c_{j_r}) \neq 0.$$

*(c) For all $m_1 + 1 \leq j_1 < \cdots < j_r \leq m_1 + m_2$ with $0 \leq r \leq n - m_1$,*

$$\mathscr{R}(f, p_1, \ldots, p_k, p_{k+1}) \neq 0,$$

*where $(p_1, \ldots, p_k, p_{k+1})$ is a permutation of $(c_1, \ldots, c_{m_1}, c_{j_1}, \ldots, c_{j_r})$.*

*(d) For all $m_1 + 1 \leq j_1 < \cdots < j_r \leq m_1 + m_2$ with $0 \leq r \leq n - m_1$,*

$$\mathscr{D}(f, c_1, \ldots, c_{m_1}, c_{j_1}, \ldots, c_{j_r}) \neq 0.$$

The genericity of local optimality conditions can be shown as follows. Let $u \in K$ be a critical point of (5.1.1) and let

$$J(u) = \{j \in \mathcal{I} : c_j(u) = 0\}$$

be the labeling set of active inequality constraints. Denote

$$L(x) := f(x) - \sum_{i \in \mathcal{E}} \lambda_i c_i(x) - \sum_{j \in J(u)} \lambda_j c_j(x), \qquad (5.5.10)$$

$$G(x) := \begin{bmatrix} \nabla c_1(x) & \cdots & \nabla c_{m_1}(x) & \nabla c_{j_1}(x) & \cdots & \nabla c_{j_r}(x) \end{bmatrix}^T, \qquad (5.5.11)$$

$$H(x) := \begin{bmatrix} \nabla_x^2 L(x) & G(x)^T \\ G(x) & 0 \end{bmatrix}. \qquad (5.5.12)$$

**Proposition 5.5.7.** *([230]) Let $u, \lambda_i, L(x), G(x), H(x)$ be as above. Condition 5.5.6 has the following properties:*

*(i) Item (a) implies that the cardinality $|J(u)| \leq n - m_1$ at every point $u \in K$.*

*(ii) Item (b) implies that the LICQC holds at every point of $K$.*

*(iii) Item (c) implies that $\lambda_i \neq 0$ for every $i \in \mathcal{E}$ and $\lambda_j \neq 0$ for every $j \in J(u)$.*

*(iv) Item (d) implies that $H(u)$ is nonsingular, i.e., $\det H(u) \neq 0$.*

**Proposition 5.5.8.** *([230]) The following properties hold at every local minimizer of (5.1.1):*

*(i) Item (a) of Condition 5.5.6 implies that at most $n - m_1$ of inequality constraints are active.*

*(ii) Item (b) of Condition 5.5.6 implies the LICQC.*

*(iii) Items (b) and (c) of Condition 5.5.6 imply the SCC.*

*(iv) Items (b) and (d) of Condition 5.5.6 imply the SOSC.*

*Therefore, if Condition 5.5.6 holds, then all LICQC, SCC, and SOSC hold at every local minimizer of (5.1.1).*

**Theorem 5.5.9.** *([230]) Let $d_0, d_i$ $(i \in \mathcal{E} \cup \mathcal{I})$ be positive degrees. Then, there exist finitely many nonzero polynomials $\varphi_1, \ldots, \varphi_L$, which are in the coefficients of $f$ and $c := (c_i)_{i \in \mathcal{E} \cup \mathcal{I}}$, with $f \in \mathbb{R}[x]_{d_0}$ and $c_i \in \mathbb{R}[x]_{d_i}$, such that if*

$$\varphi_1(f, c) \neq 0, \ldots, \varphi_L(f, c) \neq 0,$$

*then all LICQC, SCC, and SOSC hold at every local minimizer of (5.1.1).*

Theorem 5.5.9 concludes that the LICQC, SCC, and SOSC hold at every local minimizer of (5.1.1) if the tuple of input polynomials (with given degrees) does not belong to a union of finitely many hypersurfaces. So, they hold in an open dense set in the space of input polynomials. This implies that the LICQC, SCC, and SOSC can be used as sufficient and necessary conditions for certifying local optimality, for *almost all* polynomial optimization problems.

Assume Ideal$[c_{eq}]$ + QM$[c_{in}]$ is archimedean and Ideal$[c_{eq}]$ is real radical. If the LICQC, SCC, and SOSC hold at every global minimizer of (5.1.1), then the global optimality certificate

(5.5.3) must hold (see Theorem 5.5.2). Theorem 5.5.9 implies that (5.5.3) holds for generic polynomial optimization problems, under the archimedeanness and real radicalness. As in [158], the global optimality certificate (5.5.3) can be viewed as a generalization of KKT conditions. Therefore, both the local optimality certificate (5.5.1) and the global optimality certificate (5.5.3) can be unified in the KKT framework.

Putinar's Positivstellensatz (i.e., Theorem 2.6.1) is generally used for certifying nonnegativity of polynomials on basic closed semialgebraic sets. Recall that $\mathscr{P}_d(K)$ denotes the cone of polynomials in $\mathbb{R}[x]_d$ that are nonnegative on $K$, and $\partial \mathscr{P}_d(K)$ denotes the boundary of $\mathscr{P}_d(K)$. The membership $p \in \text{Ideal}[c_{eq}] + \text{QM}[c_{in}]$ guarantees that $p \geq 0$ on $K$. If $p$ lies in the interior of $\mathscr{P}_d(K)$, then one has $p \in \text{Ideal}[c_{eq}] + \text{QM}[c_{in}]$, implied by Theorem 2.6.1, under the archimedeanness of $\text{Ideal}[c_{eq}] + \text{QM}[c_{in}]$. Moreover, if $\text{Ideal}[c_{eq}]$ is real radical and $p$ is a generic point on the boundary $\partial \mathscr{P}_d(K)$, Theorems 5.5.2 and 5.5.9 imply $p \in \text{Ideal}[c_{eq}] + \text{QM}[c_{in}]$. Therefore, the conclusion of Putinar's Positivstellensatz holds for generic nonnegative polynomials.

The archimedeanness assumption cannot be removed in Theorems 5.4.1 and 5.4.7. For instance, consider the unconstrained optimization

$$\min_{x \in \mathbb{R}^3} \quad x_1^2 x_2^2 (x_1^2 + x_2^2 - 3x_3^2) + x_3^6 + x_1^2 + x_2^2 + x_3^2.$$

The origin 0 is the unique minimizer. The archimedeanness fails, since there is no constraint. The objective $f$ is the sum of the Motzkin polynomial and the positive definite quadratic form $x^T x$. The SOSC holds at the unique minimizer 0. However, $f - \gamma$ cannot be SOS for any scalar $\gamma$. So the global optimality certificate (5.5.3) does not hold for this instance.

The archimedeanness is not a genericity condition. For a degree $d$, let $\mathscr{A}(d)$ denote the set of polynomials $g \in \mathbb{R}[x]_d$ such that $\text{QM}[g]$ is archimedean. The set $\mathscr{A}(d)$ is not dense in $\mathbb{R}[x]_d$. For instance, when $d = 2$, both $\mathscr{A}(d)$ and its complement $\mathbb{R}[x]_d \backslash \mathscr{A}(d)$ have nonempty interior:

- Let $b_1 = x^T x - 1$. Clearly, $b_1 \in \mathbb{R}[x]_2 \backslash \mathscr{A}(2)$. For all $q \in \mathbb{R}[x]_2$ with sufficiently small coefficients, we have $b_1 + q \notin \mathscr{A}(2)$.

- Let $b_2 = 1 - x^T x$. Clearly, $b_2 \in \mathscr{A}(2)$. For all $q \in \mathbb{R}[x]_2$ with sufficiently small coefficients, we have $b_2 + q \in \mathscr{A}(2)$.

## 5.5.4 ▪ Exercises

**Exercise 5.5.1.** *For the equality constrained optimization*

$$\begin{cases} \min_{x \in \mathbb{R}^2} & x_1 + x_2 \\ s.t. & x_1 x_2 = 1, \end{cases}$$

*show that there are no scalars $\mu_i$ such that (5.5.1) holds for $q \geq 0$ on $\mathcal{O}$, for any open set $\mathcal{O}$ containing the local minimizer $(1, 1)$. Determine the set of all scalars $\mu_i$ satisfying (5.5.1) at the point $(1, 1)$.*

**Exercise 5.5.2.** *For the inequality constrained optimization*

$$\begin{cases} \min_{x \in \mathbb{R}^2} & x_1 + x_2 \\ s.t. & x_1 x_2 \geq 1, \end{cases}$$

*determine the set of all scalars $\mu_i$ satisfying the local optimality certificate (5.5.1), at the local minimizer $(1, 1)$.*

**Exercise 5.5.3.** *For the constrained optimization*

$$\begin{cases} \min\limits_{x \in \mathbb{R}^2} & 2x_1 + x_2 \\ \quad s.t. & x_1^2 + x_2^2 \geq 1, \ x_1 \geq 0, \ x_2 \geq 0, \end{cases}$$

*determine the set of all scalars $\mu_i$ satisfying the local optimality certificate (5.5.1) at $u = (0,1)$. Also find a certificate (5.5.3) for the global optimality of $u$.*

**Exercise 5.5.4.** *Find a global optimality certificate (5.5.3) for*

$$\begin{cases} \min\limits_{x \in \mathbb{R}^2} & x_1^3 + x_2^3 \\ \quad s.t. & x_1^2 + x_2^2 \leq 1. \end{cases}$$

**Exercise 5.5.5.** *([230]) Let $u$ be a local minimizer of (5.1.1) with $\lambda_i$ satisfying (5.1.3)-(5.1.4). Let $L(x), G(x), H(x)$ be as in Proposition 5.5.7. If $G(u)$ has full rank, show that the SOSC (5.1.9) holds at $u$ if and only if $\det H(u) \neq 0$.*

## 5.6 ▪ Optimization with Finite Sets

When the feasible set $K$ of the optimization (5.1.1) is finite, there are special properties for Moment-SOS relaxations.

### 5.6.1 ▪ The case of finite real variety

We consider the case that the equality constraints of (5.1.1) give a finite set. Recall that $c_{eq}$ denotes the tuple of equality constraining polynomials, as in (5.2.1). The Moment-SOS hierarchy of (5.2.2)-(5.2.3) has finite convergence when the real variety $V_{\mathbb{R}}(c_{eq})$ is finite. This is a major conclusion in [226].

**Theorem 5.6.1.** *([226]) For the polynomial optimization (5.1.1), if the real variety $V_{\mathbb{R}}(c_{eq})$ is finite, then the Moment-SOS hierarchy of (5.2.2)-(5.2.3) has finite convergence, i.e.,*

$$f_{sos,k} = f_{mom,k} = f_{min}$$

*for all $k$ big enough. Moreover, if $K \neq \emptyset$, then the moment relaxation (5.2.3) has minimizers and each minimizer $y^*$ must satisfy the flat truncation (5.3.3), when $k$ is big enough.*

***Proof.*** When $V_{\mathbb{R}}(c_{eq}) = \emptyset$, the feasible set $K$ is empty, and hence $f_{min} = +\infty$. By Theorem 2.6.3, $-1 \in \text{Ideal}[c_{eq}] + \Sigma[x]$. For all $\gamma > 0$, we have

$$f - \gamma = \left(1 + \frac{f}{4}\right)^2 + (-1)\left(\gamma + \left(1 - \frac{f}{4}\right)^2\right) \in \text{Ideal}[c_{eq}]_{2k} + \text{QM}[c_{in}]_{2k},$$

when $k$ is big enough. So, (5.2.2) is unbounded above, and hence $f_{sos,k} = +\infty$. For this case, the Moment-SOS hierarchy of (5.2.2)-(5.2.3) has finite convergence.

When $V_{\mathbb{R}}(c_{eq}) \neq \emptyset$ is finite, one can write that

$$V_{\mathbb{R}}(c_{eq}) = \{u_1, \ldots, u_D\}$$

for distinct points $u_1, \ldots, u_D \in \mathbb{R}^n$. Let $\varphi_1, \ldots, \varphi_D \in \mathbb{R}[x]$ be polynomials such that

$$\varphi_i(u_j) = 0 \quad \text{for} \quad i \neq j, \qquad \varphi_i(u_i) = 1.$$

For each $u_i$, if $f(u_i) - f_{min} < 0$, then at least one of $c_j(u_i)$ $(j \in \mathcal{I})$ is negative, say, $c_{j_i}(u_i) < 0$. We choose the polynomials $a_1, \ldots, a_D$ as follows:

$$a_i := \begin{cases} (f(u_i) - f_{min}) \cdot \varphi_i(x)^2 & \text{if } f(u_i) - f_{min} \geq 0; \\ \frac{f(u_i) - f_{min}}{c_{j_i}(u_i)} \cdot c_{j_i}(x) \cdot \varphi_i(x)^2 & \text{if } f(u_i) - f_{min} < 0. \end{cases}$$

All of them belong to $\mathrm{QM}[c_{in}]$. Let $a := a_1 + \cdots + a_D$. Then, $a \in \mathrm{QM}[c_{in}]_{2N_1}$ for some degree $N_1 > 0$. Note that the polynomial

$$\hat{f} := f - f_{min} - a$$

vanishes identically on $V_{\mathbb{R}}(c_{eq})$. By Theorem 2.6.3, there exist a power $\ell > 0$ and a polynomial $q \in \Sigma[x]$ such that

$$\hat{f}^{2\ell} + q \in \mathrm{Ideal}[c_{eq}].$$

Apply Lemma 5.6.2 in the below with $p := \hat{f}$, $q$, and $\omega \geq \frac{1}{2\ell}$. Then, there exists $N \geq N_1$ such that for all $\epsilon > 0$, $\hat{f} + \epsilon = \phi_\epsilon + \theta_\epsilon$, with

$$\phi_\epsilon \in \mathrm{Ideal}[c_{eq}]_{2N}, \quad \theta_\epsilon \in \mathrm{QM}[c_{in}]_{2N}.$$

Therefore, we get $f - (f_{min} - \epsilon) = \phi_\epsilon + \sigma_\epsilon$, where

$$\sigma_\epsilon = \theta_\epsilon + a \in \mathrm{QM}[c_{in}]_{2N}$$

for all $\epsilon > 0$. This implies that $\gamma = f_{min} - \epsilon$ is feasible for (5.2.2) for all $\epsilon > 0$, when the relaxation order is $N$. So $f_{sos,N} \geq f_{min}$. Since $f_{sos,k} \leq f_{min}$ for all $k$ and $f_{sos,k}$ is monotonically increasing, we must have $f_{sos,k} = f_{min}$ for all $k \geq N$. The relation (5.2.4) implies that $f_{mom,k} = f_{min}$ for all $k \geq N$.

If the feasible set $K \neq \emptyset$, (5.1.1) must have a minimizer, say $u$, since $K$ is finite. Therefore, (5.2.3) must also have a minimizer, say $[u]_{2k}$. For all big $k$, every minimizer $y^*$ of (5.2.3) must satisfy (5.3.3). This is implied by Exercise 5.6.5. □

The following is a useful lemma, whose proof is left as an exercise.

**Lemma 5.6.2.** *([226]) Let $\ell \geq 1$ be an integer.*

(i) *If $\omega \geq \omega_0 := \frac{1}{2\ell}\left(1 - \frac{1}{2\ell}\right)^{2\ell-1}$, then $s_\omega(t) := 1 + t + \omega t^{2\ell}$ is an SOS polynomial in $t$.*

(ii) *For $p, q \in \mathbb{R}[x]$ and for all $\epsilon > 0$, $\omega \in \mathbb{R}$, we have $p + \epsilon = \phi_\epsilon + \theta_\epsilon$, with*

$$\phi_\epsilon = -\omega \epsilon^{1-2\ell}(p^{2\ell} + q), \quad \theta_\epsilon = \epsilon \cdot s_\omega(p/\epsilon) + \omega \epsilon^{1-2\ell} q.$$

(iii) *In the item (ii), if $\omega \geq \omega_0$, $p^{2\ell} + q \in \mathrm{Ideal}[c_{eq}]$ and $q \in \mathrm{QM}[c_{in}]$, then there exists an integer $N > 0$ such that for all $\epsilon > 0$,*

$$\phi_\epsilon \in \mathrm{Ideal}[c_{eq}]_{2N}, \quad \theta_\epsilon \in \mathrm{QM}[c_{in}]_{2N}.$$

The following are expositions for Theorem 5.6.1.

**Example 5.6.3.** *([226]) (i) Consider the objective $f := x_1 x_2$ and constraints*

$$c_1(x) := (x_1^2 - 1)^2 + (x_2^2 - 1)^2 = 0, \quad c_2(x) := x_1 + x_2 - 1 \geq 0.$$

*Clearly, $V_{\mathbb{R}}(c_1) = \{(\pm 1, \pm 1)\}$, $K = \{(1,1)\}$ and $f_{min} = 1$. Note that*

$$a \ := \ \frac{1}{2}(x_1 + x_2 - 1)(x_1 - x_2)^2 \in \mathrm{QM}[c_2]_4,$$

$$\hat{f} \ := \ f - 1 - a = \frac{1}{2}\left[(x_2^2 - 1)(x_1 - x_2 + 1) - (x_1^2 - 1)(x_1 - x_2 - 1)\right].$$

*Since $\hat{f} \equiv 0$ on $V_{\mathbb{R}}(c_1)$, we get*

$$\hat{f}^2 + q \ = \ \frac{1}{2}((x_1 - x_2)^2 + 1)c_1 \in \mathrm{Ideal}[c_1]_6, \quad where$$

$$q \ = \ \frac{1}{4}\left((x_1^2 - 1)(x_1 - x_2 + 1) + (x_2^2 - 1)(x_1 - x_2 - 1)\right)^2.$$

*For each $\epsilon > 0$, $\phi_\epsilon = -\frac{1}{4\epsilon}(\hat{f}^2 + q) \in \mathrm{Ideal}[c_1]_6$ and*

$$\sigma_\epsilon = \epsilon\left(1 + \frac{\hat{f}}{2\epsilon}\right)^2 + \frac{1}{4\epsilon}q + a \in \mathrm{QM}[c_2]_6.$$

*Then, $f - 1 + \epsilon = \phi_\epsilon + \sigma_\epsilon$ for all $\epsilon > 0$. So, $f_{sos,k} = 1$ for all $k \geq 3$.*
*(ii) Consider $f \in \mathbb{R}[x]$ with $f(0) = 0$ and the constraint*

$$c_1(x) := x_1^{2d} + \cdots + x_n^{2d} = 0.$$

*Clearly, $f_{min} = 0$. There are no inequality constraints. One can write $f$ as*

$$f = x_1 b_1 + \cdots + x_n b_n, \quad b_1, \ldots, b_n \in \mathbb{R}[x].$$

*The form $t_1^{2d} + \cdots + t_n^{2d}$ lies in the interior of the SOS cone $\Sigma[t_1, \ldots, t_n]_{2d}^{\mathrm{hom}}$, so there exists $\lambda > 0$ such that*

$$\lambda(t_1^{2d} + \cdots + t_n^{2d}) - (t_1 + \cdots + t_n)^{2d} \in \Sigma[t_1, \ldots, t_n]_{2d}^{\mathrm{hom}}.$$

*If each $t_i$ is replaced by $x_i b_i$ in the above, we get*

$$\psi \ := \ \lambda((x_1 b_1)^{2d} + \cdots + (x_n b_n)^{2d}) - f^{2d} \in \Sigma[x],$$

$$\eta \ := \ \lambda\left[\left(\sum_{i=1}^{n} x_i^{2d}\right)\left(\sum_{i=1}^{n} b_i^{2d}\right) - \left(\sum_{i=1}^{n} (x_i b_i)^{2d}\right)\right] \in \Sigma[x],$$

$$f^{2d} + \psi + \eta \ = \ \lambda(x_1^{2d} + \cdots + x_n^{2d})(b_1^{2d} + \cdots + b_n^{2d}) \in \mathrm{Ideal}[c_1].$$

*Let $q := \psi + \eta \in \Sigma[x]$, then $f \equiv 0$ on $V_{\mathbb{R}}(c_1)$ and $f^{2d} + q \in \mathrm{Ideal}[c_1]$. Let $r = \deg(f)$. Apply Lemma 5.6.2 with $\omega = \frac{1}{2d}$, $\ell = d$, and $p = f$. For $\epsilon > 0$, let*

$$\phi_\epsilon \ := \ -\frac{1}{2d}\epsilon^{1-2d}(f^{2d} + q) \in \mathrm{Ideal}[c_1]_{2dr},$$

$$\sigma_\epsilon \ := \ \epsilon\left(1 + \frac{f}{\epsilon} + \frac{1}{2d}(\frac{f}{\epsilon})^{2d}\right) + \frac{1}{2d}\epsilon^{1-2d}q \in \Sigma[x]_{2dr}.$$

*Note that $\mathrm{QM}[\emptyset]_{2dr} = \Sigma[x]_{2dr}$. So, $f + \epsilon = \sigma_\epsilon + \phi_\epsilon$ for all $\epsilon > 0$, and hence $f_{sos,k} = 0$ for all $k \geq dr$.*

We remark that the SOS relaxation (5.2.2) might not achieve its optimal value $f_{sos,k}$ for any order $k$, when $V_{\mathbb{R}}(c_{eq})$ is a finite set. For instance, consider the optimization problem

$$\left\{ \begin{array}{ll} \min & x_1 \\ s.t. & x_1^2 + x_2^2 + \cdots + x_n^2 = 0. \end{array} \right.$$

As in Example 5.6.3(ii), we have seen that $f_{sos,k} = 0$ for all $k \geq 1$. However, the polynomial $\varphi = x_1 - (x_1^2 + x_2^2 + \cdots + x_n^2)\phi$ cannot be SOS for any $\phi \in \mathbb{R}[x]$, since $\varphi(0) = 0$ but $\nabla\varphi(0) \neq 0$. For this case, the relaxation (5.2.2) does not have a maximizer for all $k \geq 1$. However, in Theorem 5.6.1, if the ideal $\text{Ideal}[c_{eq}]$ is *real radical*, i.e., $\text{Ideal}[c_{eq}] = I(V_{\mathbb{R}}(c_{eq}))$ (see §2.6.2 for the definition), then (5.2.2) achieves its optimal value for all $k$ big enough.

**Proposition 5.6.4.** *In Theorem 5.6.1, if, in addition, $\text{Ideal}[c_{eq}]$ is real radical, then (5.2.2) achieves its optimal value for all $k$ big enough.*

The proof of Proposition 5.6.4 is left as an exercise. When $V_{\mathbb{R}}(c_{eq})$ is not a finite set, the conclusion of Proposition 5.6.4 also holds under another assumption.

**Proposition 5.6.5.** *([226]) Let $K$ be the feasible set of (5.1.1). Suppose that $f_{min}$ is finite and the Moment-SOS hierarchy of (5.2.2)-(5.2.3) has finite convergence. If $\text{Ideal}[c_{eq}] = I(K)$, then (5.2.2) achieves its optimal value for all $k$ big enough.*

**Proof.** Since the Moment-SOS hierarchy has finite convergence, there is $N_1$ such that $f_{sos,k} = f_{min}$ for all $k \geq N_1$. The condition $I(K) = \text{Ideal}[c_{eq}]$ implies that the quotient set

$$\text{QM}[c_{in}]_{2k} / \text{Ideal}[c_{eq}]$$

is closed for all $k$, by Theorem 2.5.4. Suppose $\{\gamma_i\}_{i=1}^{\infty}$ is a sequence such that each $\gamma_i$ is feasible for (5.2.2) with $k = N_1$ and $\gamma_i \to f_{min}$ as $i \to \infty$. Note that each

$$f - \gamma_i \in \text{QM}[c_{in}]_{2N_1} / \text{Ideal}[c_{eq}]$$

and $f - \gamma_i \to f - f_{min}$. So, $f - f_{min} \in \text{QM}[c_{in}]_{2N_1} / \text{Ideal}[c_{eq}]$, i.e., there exist $\phi^* \in \text{Ideal}[c_{eq}]$ and $\sigma^* \in \text{QM}[c_{in}]_{2N_1}$ such that

$$f - f_{min} = \phi^* + \sigma^*.$$

Let $N_2 \geq N_1$ be such that $\phi^* \in \text{Ideal}[c_{eq}]_{2N_2}$. Then, $\gamma = f_{min}$ is feasible for (5.2.2) for all $k \geq N_2$, i.e., (5.2.2) achieves its optimal value for all $k \geq N_2$. □

## 5.6.2 ▪ The case of finite semialgebraic sets

It is possible that the feasible set $K$ of (5.1.1) is finite while the real variety $V_{\mathbb{R}}(c_{eq})$ is not. To handle this case, a natural trick is to introduce new variables, say, $x_{n+1}, \ldots, x_{n+\ell}$, with $\ell$ the number of inequality constraints. Then, $K$ can be equivalently given by the equations

$$c_i(x) = 0 \ (i \in \mathcal{E}), \quad c_j(x) - x_{n+j}^2 = 0 \ (j \in \mathcal{I}).$$

Clearly, $K$ is a finite set if and only if the above equations have finitely many real solutions in the space $\mathbb{R}^{n+\ell}$. When $K$ is finite, by Theorem 5.6.1, the Moment-SOS hierarchy of (5.2.2)-(5.2.3) has finite convergence if new variables $x_{n+1}, \ldots, x_{n+\ell}$ are used. In computational practice, people often prefer not introducing new variables, because they would make the resulting Moment-SOS relaxations more expensive to solve.

When the semialgebraic set $K$ is finite, the Putinar type Moment-SOS hierarchy of (5.2.2)-(5.2.3) has finite convergence under an additional assumption. Let

$$J := \big(\text{Ideal}[c_{eq}] + \text{QM}[c_{in}]\big) \cap -\big(\text{Ideal}[c_{eq}] + \text{QM}[c_{in}]\big). \tag{5.6.1}$$

It is an ideal (see Exercise 2.5.8) and is called the support of $\text{Ideal}[c_{eq}] + \text{QM}[c_{in}]$. The dimension of the coordinate ring $\frac{\mathbb{R}[x]}{J}$ is the dimension of the ideal $J$.

**Theorem 5.6.6.** *Let $K$ be the feasible set of (5.1.1) and let $J$ be the ideal as in (5.6.1). Assume $K$ is finite and $\dim \frac{\mathbb{R}[x]}{J} \leq 1$. Then, the Moment-SOS hierarchy of (5.2.2)-(5.2.3) has finite convergence, i.e., $f_{sos,k} = f_{mom,k} = f_{min}$ for all $k$ big enough. Moreover, when $K \neq \emptyset$, the moment relaxation (5.2.3) has minimizers and each minimizer $y^*$ must satisfy the flat truncation (5.3.3), when $k$ is big enough.*

**Proof.** For the case $K = \emptyset$ (so $f_{min} = +\infty$), we have

$$-1 \in \text{Ideal}[c_{eq}] + \text{QM}[c_{in}],$$

by Theorem 2.6.9(iv), under the given assumptions. This implies that (5.2.2) is unbounded above for $k$ big enough, so $f_{sos,k} = +\infty$. Hence,

$$f_{sos,k} = f_{mom,k} = f_{min}$$

for all $k$ big enough.

For the case $K \neq \emptyset$, one can write that $K = \{u_1, \ldots, u_D\}$, since it is finite. Let $\varphi_1, \ldots, \varphi_D \in \mathbb{R}[x]$ be the interpolating polynomials such that

$$\varphi_i(u_j) = 0 \, (i \neq j), \quad \varphi_i(u_j) = 1 \, (i = j).$$

Note that $f(u_i) - f_{min} \geq 0$ for all $i$. Let

$$a := \sum_{i=1}^{D} (f(u_i) - f_{min}) \varphi_i^2 \in \Sigma[x].$$

The polynomial $\hat{f} := f - f_{min} - a$ vanishes identically on $K$. By Theorem 2.6.9(iii), there exist integers $\ell, N_1 > 0$ such that

$$q \in \text{QM}[c_{in}]_{2N_1}, \quad \hat{f}^{2\ell} + q \in \text{Ideal}[c_{eq}]_{2N_1}.$$

Applying Lemma 5.6.2 with $\omega \geq \frac{1}{2\ell}$ and $p = \hat{f}$, we get that for all $\epsilon > 0$,

$$
\begin{aligned}
f - (f_{min} - \epsilon) &= p + \epsilon + a = \sigma_\epsilon + \phi_\epsilon, \\
\phi_\epsilon &= -\omega \epsilon^{1-2\ell} (\hat{f}^{2\ell} + q) \in \text{Ideal}[c_{eq}]_{2N_1}, \\
\sigma_\epsilon &= \epsilon(1 + \hat{f}/\epsilon + \omega(\hat{f}/\epsilon)^{2\ell}) + \omega \epsilon^{1-2\ell} q + a.
\end{aligned}
$$

Let $N \geq N_1$ be such that $\sigma_\epsilon \in \text{QM}[c_{in}]_{2N}$ for all $\epsilon > 0$. The above implies that $f_{sos,k} = f_{min}$ for all $k \geq N$. By the relation (5.2.4), we also get $f_{mom,k} = f_{min}$ for all $k$ big enough.

When $K \neq \emptyset$, the moment relaxation (5.2.3) has minimizers, e.g., $[u]_{2k}$ is one for each minimizer $u$ of (5.1.1). The vanishing ideal $I(K)$ is zero dimensional. Let $\{h_1, \ldots, h_r\}$ be a Gröbner basis of $I(K)$ with respect to a total degree ordering. Since $h_i \equiv 0$ on $K$ and

$$\dim \frac{\mathbb{R}[x]}{J} \leq 1,$$

we know $-h_i^{2\ell} \in \text{Ideal}[c_{eq}] + \text{QM}[c_{in}]$ for some $\ell \in \mathbb{N}$, by Theorem 2.6.9(iii). As in the Step 3 of the proof for Theorem 5.3.4, we can similarly show that every minimizer $y^*$ of (5.2.3) satisfies the flat truncation (5.3.3), when $k$ is big enough. $\quad\square$

The Schmüdgen type Moment-SOS hierarchy of relaxations (5.2.5)-(5.2.6) always has finite convergence, when the feasible set $K$ is finite, without the assumption $\dim \frac{\mathbb{R}[x]}{J} \leq 1$. Recall that $f'_{sos,k}, f'_{mom,k}$ are the optimal values of (5.2.5), (5.2.6) respectively, for the relaxation order $k$. The finite convergence of (5.2.5)-(5.2.6) is as follows.

**Theorem 5.6.7.** *([226]) Assume the feasible set $K$ is finite. Then, the Moment-SOS hierarchy of (5.2.5)-(5.2.6) has finite convergence, i.e.,*

$$f'_{sos,k} = f'_{mom,k} = f_{min}$$

*for all $k$ big enough. Furthermore, every minimizer $y^*$ of (5.2.6) satisfies the flat truncation (5.3.3), when $k$ is big enough.*

The following is an exposition for Theorem 5.6.7.

**Example 5.6.8.** *([226]) Consider the optimization (5.1.1) with*

$$f = -x_1^2 - x_2^2, \quad c_1 = x_1^3 \geq 0, \quad c_2 = x_2^3 \geq 0, \quad c_3 = -x_1 - x_2 - x_1 x_2 \geq 0.$$

*We have $K = \{(0,0)\}$ and $f_{min} = 0$. Note that $f \equiv 0$ on $K$. This is confirmed by the identity $f^4 + q = 0$, where*

$$q = \sigma_0 + c_1 \sigma_1 + c_2 \sigma_2 + c_1 c_2 \sigma_{12} + c_3 \sigma_3.$$

*In the above, the SOS polynomials $\sigma_0, \sigma_1, \sigma_2, \sigma_{12}, \sigma_3$ are*

$$\sigma_0 = (x_1^2 - x_2^2)^4 + 6(x_1^4 - x_2^4)^2, \quad \sigma_{12} = 32(x_1^2 + x_2^2 + x_1^4 + x_2^4 + x_1^6 + x_2^6),$$

$$\sigma_1 = 8\Big(x_1^6(x_2 + 1/2 + 4x_2^2) + x_1^4(x_1^2/2 + 2x_1 x_2 + 2x_2^2)$$

$$+ x_1^4(2x_2 + 1/2 + 2x_2^2) + x_1^2(x_1^2/2 + x_1 x_2 + 4x_2^2) + 4(x_2^4 + x_2^6 + x_2^8)\Big),$$

$$\sigma_2 = 8\Big(x_2^6(x_1 + 1/2 + 4x_1^2) + x_2^4(x_2^2/2 + 2x_2 x_1 + 2x_1^2)$$

$$+ x_2^4(2x_1 + 1/2 + 2x_1^2) + x_2^2(x_2^2/2 + x_2 x_1 + 4x_1^2) + 4(x_1^4 + x_1^6 + x_1^8)\Big),$$

$$\sigma_3 = 8\Big((x_1^8 + x_2^8 + x_1^7 + x_2^7 + x_1^6 + x_2^6) + 4x_1^2 x_2^2(x_1^2 + x_2^2 + x_1^4 + x_2^4 + x_1^6 + x_2^6)\Big).$$

*In Lemma 5.6.2, let $\omega = \frac{1}{4}$ and $p = f$. For each $\epsilon > 0$, let*

$$\phi_\epsilon = 0, \quad \sigma_\epsilon := \epsilon\Big(1 + \frac{f}{\epsilon} + \frac{f^4}{4\epsilon^4}\Big) + \frac{1}{4\epsilon^3} q \in \mathrm{QM}[c_{in}]_{12}.$$

*Then, $f + \epsilon = \phi_\epsilon + \sigma_\epsilon$ for all $\epsilon > 0$. So, $f'_{sos,k} = 0$ for all $k \geq 6$.*

### 5.6.3 ▪ Exercises

**Exercise 5.6.1.** *For all polynomials $b_1, \ldots, b_n \in \mathbb{R}[x]$, show that*

$$\Big[\Big(\sum_{i=1}^n x_i^{2d}\Big)\Big(\sum_{i=1}^n b_i^{2d}\Big) - \Big(\sum_{i=1}^n (x_i b_i)^{2d}\Big)\Big] \in \Sigma[x].$$

**Exercise 5.6.2.** *Consider the constrained polynomial optimization*

$$\begin{cases} \min\limits_{x \in \mathbb{R}^n} & x_1 x_2 + x_2 x_3 + \cdots + x_{n-1} x_n \\ s.t. & n(x_1^4 + \cdots + x_n^4) - (x_1^2 + \cdots + x_n^2)^2 = 0. \end{cases}$$

*As in Example 5.6.3, give a direct proof for the finite convergence for the Moment-SOS hierarchy. What is the smallest $k$ such that $f_{sos,k} = f_{min}$?*

**Exercise 5.6.3.** *Consider the polynomial optimization*

$$\begin{cases} \min\limits_{x \in \mathbb{R}^2} & x_1^2 + x_2^2 \\ s.t. & x_1 \geq 1,\ x_2 \geq 1,\ 1 - x_1 x_2 \geq 0. \end{cases}$$

*As in Example 5.6.8, give a direct proof for the finite convergence for the hierarchy of (5.2.5)-(5.2.6). What is the smallest $k$ such that $f'_{sos,k} = f_{min}$? Does the hierarchy of (5.2.2)-(5.2.3) also have finite convergence?*

**Exercise 5.6.4.** *Prove Lemma 5.6.2 and Proposition 5.6.4.*

**Exercise 5.6.5.** *([160, 171]) Suppose the real variety $V_{\mathbb{R}}(c_{eq})$ is finite. Let $y \in \mathbb{R}^{\mathbb{N}_{2k}^n}$ be a tms such that $M_k[y] \succeq 0$ and $y \in \mathcal{Z}[c_{eq}]_{2k}$ (see (2.5.25) for the notation), i.e., $\mathscr{V}_{c_i}^{(2k)}[y] = 0$ for all $i \in \mathcal{E}$. Fix a degree $d$. Show that if $k$ is big enough, then there exists $t \in [d, k]$ such that*

$$\operatorname{rank} M_{t-d}[y] = \operatorname{rank} M_t[y].$$

## 5.7 ▪ Selection of Constraining Polynomials

A set can be equivalently given by different constraints. For instance, all equality constraints can be replaced by a single one, e.g.,

$$c_i(x) = 0\,(i \in \mathcal{E}) \quad \Leftrightarrow \quad \sum_{i \in \mathcal{E}} c_i(x)^2 = 0.$$

We study the convergence property of the Moment-SOS hierarchy for different choices of constraining polynomials.

For the optimization (5.1.1), recall the constraining polynomial tuples $c_{eq}$ and $c_{in}$ as in (5.2.1). Suppose $h = (h_i)_{i \in \mathcal{H}}$ is a new polynomial tuple such that $V_{\mathbb{R}}(h) = V_{\mathbb{R}}(c_{eq})$. Then, (5.1.1) is equivalent to the optimization

$$\begin{cases} \min\limits_{x \in \mathbb{R}^n} & f(x) \\ s.t. & h_i(x) = 0\,(i \in \mathcal{H}), \\ & c_j(x) \geq 0\,(j \in \mathcal{I}). \end{cases} \tag{5.7.1}$$

Similarly, the new hierarchy of SOS relaxation is

$$\begin{cases} \max & \gamma \\ s.t. & f - \gamma \in \operatorname{Ideal}[h]_{2k} + \operatorname{QM}[c_{in}]_{2k}. \end{cases} \tag{5.7.2}$$

The corresponding dual moment relaxation is

$$\begin{cases} \min & \langle f, y \rangle \\ s.t. & \mathscr{V}_{h_i}^{(2k)}[y] = 0\,(i \in \mathcal{H}), \\ & L_{c_j}^{(k)}[y] \succeq 0\,(j \in \mathcal{I}), \\ & M_k[y] \succeq 0, \\ & y_0 = 1,\ y \in \mathbb{R}^{\mathbb{N}_{2k}^n}. \end{cases} \tag{5.7.3}$$

Let $f_{sos,k}^h$, $f_{mom,k}^h$ denote the optimal values of (5.7.2), (5.7.3) respectively. We can similarly show that

$$f_{sos,k}^h \leq f_{mom,k}^h \leq f_{min}.$$

When $\text{Ideal}[h] + \text{QM}[c_{in}]$ is archimedean, both $f_{sos,k}^h$ and $f_{mom,k}^h$ have asymptotic convergence to the minimum value $f_{min}$ of (5.1.1). The finite convergence also holds under some optimality conditions as in §5.4. Recall that $f_{sos,k}$ denotes the optimal value of the relaxation (5.2.2) for the order $k$. We are interested in the following questions:

- If $f_{sos,k}$ has finite convergence to $f_{min}$, does $f_{sos,k}^h$ necessarily have finite convergence to $f_{min}$?

- If $f_{sos,k}$ has no finite convergence to $f_{min}$, is it also necessarily true that $f_{sos,k}^h$ has no finite convergence to $f_{min}$?

There are affirmative answers to the above questions. Recall that $I(K)$ is the vanishing ideal of the set $K$ and $I(V_{\mathbb{R}}(c_{eq}))$ is the vanishing ideal of the real variety $V_{\mathbb{R}}(c_{eq})$. The following theorem answers these questions affirmatively.

**Theorem 5.7.1.** *([226]) Suppose $h$ is a tuple of polynomials such that $V_{\mathbb{R}}(h) = V_{\mathbb{R}}(c_{eq})$. If $I(K) = I(V_{\mathbb{R}}(c_{eq}))$, then $f_{sos,k}$ has finite convergence to $f_{min}$ if and only if $f_{sos,k}^h$ has finite convergence to $f_{min}$.*

The condition $I(K) = I(V_{\mathbb{R}}(c_{eq}))$ requires that a polynomial $p \equiv 0$ on $K$ if and only if $p \equiv 0$ on $V_{\mathbb{R}}(c_{eq})$. This means that the feasible set $K$ and the real variety $V_{\mathbb{R}}(c_{eq})$ have the same Zariski closure. There is a similar version of Theorem 5.7.1 based on the real radical ideal $I(V_{\mathbb{R}}(c_{eq}))$. Suppose $\eta := (\eta_1, \ldots, \eta_t)$ is a tuple of generators for $I(V_{\mathbb{R}}(c_{eq}))$, i.e.,

$$I(V_{\mathbb{R}}(c_{eq})) = \text{Ideal}[\eta].$$

Then (5.1.1) is equivalent to the optimization

$$\begin{cases} \min\limits_{x \in \mathbb{R}^n} & f(x) \\ \text{s.t.} & \eta_i(x) = 0 \, (i = 1, \ldots, t), \\ & c_j(x) \geq 0 \, (j \in \mathcal{I}). \end{cases} \tag{5.7.4}$$

The hierarchy of SOS relaxations for solving (5.7.4) is

$$\begin{cases} \max & \gamma \\ \text{s.t.} & f - \gamma \in \text{Ideal}[\eta]_{2k} + \text{QM}[c_{in}]_{2k}. \end{cases} \tag{5.7.5}$$

Let $f_{sos,k}^\eta$ be the optimal value of (5.7.5) when the relaxation order is $k$. We also have $f_{sos,k}^\eta \leq f_{min}$ for all $k$.

**Theorem 5.7.2.** *([226]) Let $\eta$, $K$ be as above. Suppose $f_{min} > -\infty$ and*

$$I(K) = I(V_{\mathbb{R}}(c_{eq})) = \text{Ideal}[\eta].$$

*Then $f_{sos,k}$ has finite convergence to $f_{min}$ if and only if $f_{sos,k}^\eta$ has finite convergence to $f_{min}$.*

**Proof.** First, assume that $f_{sos,k}^\eta = f_{min}$ for all $k$ big enough. The feasible set of (5.7.4) is $K$ and $\text{Ideal}[\eta] = I(K)$. Applying Proposition 5.6.5 to (5.7.4), we know that (5.7.5) achieves its optimum $f_{min}$, when $k$ is big enough, say, for all $k \geq N_1$. So, $f(x) - f_{min} = p + \sigma_1$ for some polynomials

$$p \in \text{Ideal}[\eta]_{2N_1}, \quad \sigma_1 \in \text{QM}[c_{in}]_{2N_1}.$$

Since $\text{Ideal}[\eta] = I(V_{\mathbb{R}}(c_{eq}))$, $p \equiv 0$ on $V_{\mathbb{R}}(c_{eq})$. By Theorem 2.6.3, there exist a power $\ell > 0$ and a polynomial $q \in \Sigma[x]$ such that

$$p^{2\ell} + q \in \text{Ideal}[c_{eq}].$$

By Lemma 5.6.2, there exists $N_2 > 0$ such that for all $\epsilon > 0$

$$p + \epsilon = \phi_\epsilon + \theta_\epsilon, \quad \phi_\epsilon \in \text{Ideal}[c_{eq}]_{2N_2}, \quad \theta_\epsilon \in \text{QM}[c_{in}]_{2N_2}.$$

For $\sigma_\epsilon := \theta_\epsilon + \sigma_1$ and $N_3 := \max(N_1, N_2)$, we have

$$f - (f_{min} - \epsilon) = \sigma_\epsilon + \phi_\epsilon, \quad \sigma_\epsilon \in \text{QM}[c_{in}]_{2N_3}, \quad \phi_\epsilon \in \text{Ideal}[c_{eq}]_{2N_3}.$$

Hence, $f_{sos,k} = f_{min}$ for all $k \geq N_3$, i.e., $f_{sos,k}$ has finite convergence to $f_{min}$.

Second, assume that $f_{sos,k} = f_{min}$ for all $k \geq M_1$. For every $\epsilon > 0$, there exist $\phi_\epsilon \in \text{Ideal}[c_{eq}]_{2M_1}$ and $\sigma_\epsilon \in \text{QM}[c_{in}]_{2M_1}$ such that

$$f - (f_{min} - \epsilon) = \phi_\epsilon + \sigma_\epsilon.$$

Since $c_i \in I(V_{\mathbb{R}}(c_{eq})) = \text{Ideal}[\eta]$ for each $i \in \mathcal{E}$, there exists $M_2 \geq M_1$ such that

$$\text{Ideal}[c_{eq}]_{2M_1} \subseteq \text{Ideal}[\eta]_{2M_2},$$
$$\text{QM}[c_{in}]_{2M_1} \subseteq \text{QM}[c_{in}]_{2M_2}.$$

Therefore, $f_{sos,k}^\eta \geq f_{min} - \epsilon$ for all $k \geq M_2$ and for all $\epsilon > 0$. Hence, $f_{sos,k}^\eta \geq f_{min}$ for all $k \geq M_2$. Since $f_{sos,k}^\eta \leq f_{min}$ for all $k$, $f_{sos,k}^\eta$ must have finite convergence to $f_{min}$. $\square$

Theorem 5.7.1 can be implied by Theorem 5.7.2. Its proof is left as an exercise. The following is an exposition for Theorem 5.7.1.

**Example 5.7.3.** *([226]) Consider the optimization problem:*

$$\begin{cases} \min\limits_{x \in \mathbb{R}^3} & f(x) := x_1 x_2 x_3 - 2x_3 \\ s.t. & h := (x_1^2 - x_2)^2 + (x_1^3 - x_3)^2 = 0. \end{cases} \quad (5.7.6)$$

*The feasible set is the curve parameterized as $(x_1, x_1^2, x_1^3)$. The minimum $f_{min} = -1$. Let $\sigma_1 := (x_1^3 - 1)^2$ and*

$$p := (x_1^3 - 2)(x_3 - x_1^3) + x_1 x_3 (x_2 - x_1^2).$$

*Then, $f + 1 = p + \sigma_1$. Clearly, $p \equiv 0$ on $V_{\mathbb{R}}(h)$ and $p^2 + q = h\psi$, where*

$$q = \left( x_1 x_3 (x_3 - x_1^3) - (x_1^3 - 2)(x_2 - x_1^2) \right)^2,$$
$$\psi = x_1^2 x_3^2 + (x_1^3 - 2)^2.$$

*For all $\epsilon > 0$, we have $f + 1 + \epsilon = \phi_\epsilon + \sigma_\epsilon$, where*

$$\phi_\epsilon = \frac{-1}{4\epsilon} \psi h \in \text{Ideal}[h]_{12},$$
$$\sigma_\epsilon = \epsilon \left( 1 + \frac{p}{2\epsilon} \right)^2 + \frac{1}{4\epsilon} q + \sigma_1 \in \Sigma[x]_{12}.$$

*So, $f_{sos,k} = -1$ for all $k \geq 6$.*

In optimization, all equality constraints can be replaced by a singe one. In (5.7.1), the equality constraints can be equivalently written as

$$h^{sq}(x) := \sum_{i \in \mathcal{E}} h_i(x)^2 = 0.$$

Theorem 5.7.1 shows that finite convergence of the Moment-SOS hierarchy is preserved when the new equations are used.

We remark that there does not exist a result similar to Theorem 5.7.1 for the case of inequalities. That is, the choice of inequality constraining polynomials might affect finite convergence of the Moment-SOS hierarchy, while the feasible set $K$ is not changed. For instance, consider the optimization

$$\begin{cases} \min_{x \in \mathbb{R}^1} & 1 - x^2 \\ s.t. & 1 - x^2 \geq 0. \end{cases}$$

One can show that $f_{sos,k} = f_{min}$ for all $k \geq 1$. Clearly, the above optimization is equivalent to the following one:

$$\begin{cases} \min_{x \in \mathbb{R}^1} & 1 - x^2 \\ s.t. & (1 - x^2)^3 \geq 0. \end{cases}$$

However, the Moment-SOS hierarchy for this new formulation does not have finite convergence. Indeed, there exists a constant $C > 0$ such that $f_{sos,k} \leq -Ck^{-2}$ for all $k$. This was shown in [307, Theorem 4].

In computation, to describe the constraining set $K$, we usually select constraining polynomials $c_i$ ($i \in \mathcal{E}$), $c_j$ ($j \in \mathcal{I}$) such that the following hold:

(i) The degree $d_c$, as in (5.3.1), is smallest.

(ii) The sum $\mathrm{Ideal}[c_{eq}]_{2d_c} + \mathrm{QM}[c_{in}]_{2d_c}$ is as big as possible.

The item (i) is to make sure that the relaxations (5.2.2)-(5.2.3) exist for the lowest relaxation order $d_c$, and the item (ii) is to make sure that the lower bound $f_{sos,d_c}$ is as big as possible. In particular, if $K$ is compact, say, $x^T x \leq R$ for all $x \in K$, we usually add the inequality $R - x^T x \geq 0$ to the constraints, while the feasibility and optimal value are not changed.

We would like to remark that the sum $\mathrm{Ideal}[c_{eq}]_{2k} + \mathrm{QM}[c_{in}]_{2k}$ is typically not maximum, unless it equals $\mathscr{P}_{2k}(K)$, the cone of polynomials in $\mathbb{R}[x]_{2k}$ that are nonnegative on $K$. This is shown in Exercise 5.7.4.

## 5.7.1 ▪ Exercises

**Exercise 5.7.1.** *The simplex $\Delta_n := \{x \in \mathbb{R}^n : x \geq 0, 1 - e^T x = 0\}$ is contained inside the unit ball $B(0,1)$. Let $h = e^T x - 1$ and $g = (x_1, \dots, x_n)$. For $n \geq 2$, show that for each $k \in \mathbb{N}$*

$$\mathrm{Ideal}[h]_{2k} + \mathrm{QM}[g]_{2k} \subsetneq \mathrm{Ideal}[h]_{2k} + \mathrm{QM}[g, 1 - x^T x]_{2k}.$$

**Exercise 5.7.2.** *If $q \in \mathbb{R}[x]_2$ is nonnegative on the unit ball $B(0,1)$, show that for every $k \in \mathbb{N}$*

$$\mathrm{QM}[1 - x^T x]_{2k} = \mathrm{QM}[1 - x^T x, q]_{2k}.$$

**Exercise 5.7.3.** *For $n \geq 3$, find a polynomial $p \in \mathbb{R}[x]_6$ such that $p \geq 0$ on $B(0,1)$ and $\text{QM}[1 - x^T x]_6 \neq \text{QM}[1 - x^T x, p]_6$.*

**Exercise 5.7.4.** *Among all polynomial tuples $c_{eq}$ and $c_{in}$, which have degrees at most $2k$ and which give the same feasible set $K$ as in (5.1.1), show that $\text{Ideal}[c_{eq}]_{2k} + \text{QM}[c_{in}]_{2k}$ is maximum if and only if*

$$\text{Ideal}[c_{eq}]_{2k} + \text{QM}[c_{in}]_{2k} = \mathscr{P}_{2k}(K).$$

**Exercise 5.7.5.** *Use Theorem 5.7.2 to prove Theorem 5.7.1.*

# Chapter 6

# Tight Moment-SOS Hierarchies

*The classical Moment-SOS hierarchy may not be tight for solving some polynomial optimization problems. This chapter introduces tight relaxation methods. They are based on optimality conditions and Lagrange multiplier expressions. Under certain nonsingularity assumptions on constraining polynomials, we give new Moment-SOS hierarchies that are always tight. Properties of these relaxations are also studied.*

## 6.1 ▪ Properties of Optimality Conditions

Consider the constrained optimization problem

$$
\begin{cases}
\min & f(x) \\
s.t. & c_i(x) = 0 \, (i \in \mathcal{E}), \\
& c_j(x) \geq 0 \, (j \in \mathcal{I}),
\end{cases}
\tag{6.1.1}
$$

where $f$ and all $c_i, c_j$ are polynomials in $x := (x_1, \ldots, x_n) \in \mathbb{R}^n$. Let $K$ denote the feasible set of (6.1.1). The $\mathcal{E}$ and $\mathcal{I}$ are disjoint finite labeling sets for constraining polynomials. The Moment-SOS hierarchy introduced in §5.2 is to solve (6.1.1) globally, i.e., to find the global minimum value $f_{min}$ and global minimizer(s) if they exist. As in (5.2.1), we still use the notation

$$
c_{eq} := (c_i)_{i \in \mathcal{E}}, \quad c_{in} := (c_j)_{j \in \mathcal{I}}.
$$

When the sum $\mathrm{Ideal}[c_{eq}] + \mathrm{QM}[c_{in}]$ is archimedean, the Moment-SOS hierarchy has asymptotic convergence. In addition, under some optimality conditions (i.e., LICQC, SCC, SOSC), it also has finite convergence. These results are shown in §5.2 and §5.4.

When the classical Moment-SOS hierarchy is used to solve (6.1.1), the following issues are of major concern:

- The convergence depends on the archimedeanness of $\mathrm{Ideal}[c_{eq}] + \mathrm{QM}[c_{in}]$, which holds only if the feasible set $K$ is compact. When $K$ is noncompact, how can we get a tight Moment-SOS hierarchy?

- The computational expense of Moment-SOS relaxations grows rapidly as the relaxation order $k$ increases. For a fixed order $k$, how can we get Moment-SOS relaxations that are as strong as possible?

- When some optimality conditions (e.g., SCC, SOSC) fail, how can we get a tight hierarchy of Moment-SOS relaxations?

This chapter introduces a new approach for getting tight Moment-SOS relaxations, with the usage of KKT conditions and Lagrange multipliers.

Let $u$ be a local minimizer of (6.1.1). The labeling set of active inequality constraints is

$$J(u) := \{j \in \mathcal{I} \mid c_j(u) = 0\}. \tag{6.1.2}$$

Under the LICQC at $u$, i.e., the gradients $\nabla c_i(u)$ $(i \in \mathcal{E} \cup J(u))$ are linearly independent, there exist Lagrange multipliers $\lambda_i$ $(i \in \mathcal{E} \cup \mathcal{I})$ satisfying

$$\begin{aligned}
\sum_{i \in \mathcal{E}} \lambda_i \nabla c_i(u) + \sum_{j \in \mathcal{I}} \lambda_j \nabla c_j(u) &= \nabla f(u), \\
c_j(u) \geq 0, \ \lambda_j \geq 0, \ \lambda_j c_j(u) &= 0 \, (j \in \mathcal{I}).
\end{aligned} \tag{6.1.3}$$

Denote the vector of all Lagrange multipliers

$$\lambda := (\lambda_i)_{i \in \mathcal{E} \cup \mathcal{I}}.$$

We consider the set of complex critical pairs

$$\mathcal{K} := \left\{ (x, \lambda) \ \middle| \ \begin{aligned}
\sum_{i \in \mathcal{E}} \lambda_i \nabla c_i(x) + \sum_{j \in \mathcal{I}} \lambda_j \nabla c_j(x) &= \nabla f(x), \\
c_i(x) &= 0 \, (i \in \mathcal{E}), \\
\lambda_j c_j(x) &= 0 \, (j \in \mathcal{I})
\end{aligned} \right\}. \tag{6.1.4}$$

For each $(x, \lambda) \in \mathcal{K}$, the objective value $f(x)$ is called a *critical value* and $x$ is called a *critical point*. The following is a basic property about critical values.

**Theorem 6.1.1.** *([75]) For all $f, c_i, c_j \in \mathbb{C}[x]$ $(i \in \mathcal{E}, \ j \in \mathcal{I})$, the objective $f$ achieves only finitely many values on the critical set $\mathcal{K}$, i.e., the set $\{f(u) : (u, \lambda) \in \mathcal{K}\}$ is finite.*

***Proof.*** For a subset $J \subseteq \mathcal{I}$, denote the sub-critical set

$$\mathcal{K}_J := \{(x, \lambda) \in \mathcal{K} : c_j(x) = 0 \, (j \in J)\}. \tag{6.1.5}$$

Then the critical set $\mathcal{K}$ can be expressed as the finite union

$$\mathcal{K} = \bigcup_{J \subseteq \mathcal{I}} \mathcal{K}_J.$$

We show that $f$ achieves only finitely many values on each $\mathcal{K}_J$. Fix an arbitrary subset $J \subseteq \mathcal{I}$. For convenience of notation, we still denote the vector

$$\lambda := (\lambda_i)_{i \in \mathcal{E}} \cup (\lambda_j)_{j \in J}.$$

Consider the Lagrangian function

$$\mathcal{L}(x, \lambda) = f(x) - \sum_{i \in \mathcal{E}} \lambda_i c_i(x) - \sum_{j \in J} \lambda_j c_j(x).$$

Note that a critical pair $(u, \lambda) \in \mathcal{K}_J$ if and only if

$$\nabla_x \mathcal{L}(u, \lambda) = 0, \ \nabla_\lambda \mathcal{L}(u, \lambda) = 0.$$

The set $\mathcal{K}_J$ is an algebraic variety in the embedding complex space of $(x, \lambda)$, so $\mathcal{K}_J$ consists of finitely many irreducible varieties, say,

$$\mathcal{K}_J = V_1 \cup \cdots \cup V_N,$$

where each variety $V_l$ is irreducible. It is enough to show that $f$ is constant on each $V_l$. Note that $f \equiv \mathcal{L}$ on $\mathcal{K}_J$. Each $V_l$ is connected in the Euclidean topology on $\mathbb{C}^{n+s}$ ($s$ is the dimension of $\lambda$), so it is path-connected (see [305, §4.1.3]). It suffices to show $f(x^{(1)}) = f(x^{(2)})$ for arbitrary two points $(x^{(1)}, \lambda^{(1)})$, $(x^{(2)}, \lambda^{(2)})$ in $V_l$. Since $V_l$ is path-connected, there exists a piecewise-smooth path $\varphi(t) = (x(t), \lambda(t))$ $(0 \le t \le 1)$, contained in $V_l$, such that

$$\varphi(0) = (x^{(1)}, \lambda^{(1)}), \quad \varphi(1) = (x^{(2)}, \lambda^{(2)}).$$

Observe that the polynomial function $\mathcal{L}(x, \lambda)$ has zero gradient on the path $\varphi(t)$ $(0 \le t \le 1)$. Let $h(t) := \mathcal{L}(\varphi(t))$. Then, on smooth points of the path,

$$h'(t) = [\varphi'(t)]^T \nabla_{x,\lambda} \mathcal{L} = 0.$$

By integration, $\mathcal{L}(\varphi(0)) = \mathcal{L}(\varphi(1))$. Since $\mathcal{L}(x, \lambda) = f(x)$ for all $(x, \lambda) \in \mathcal{K}_J$, we get $f(x^{(1)}) = f(x^{(2)})$. Therefore, the objective $f$ is constant on $V_l$. Since $\mathcal{K}$ consists of finitely many irreducible varieties like $V_l$, $f$ achieves only finitely many values on $\mathcal{K}$.          □

We would like to remark that the conclusion of Theorem 6.1.1 holds for all polynomials $f$, $c_i$, $c_j$, even if the LICQC (or any other optimality condition) fails. When there are no inequality constraints and $f$, $c_i$ have generic coefficients, the number of critical points is given in Theorem 5.5.3. If there are inequality constraints, the number of critical points can be obtained by enumerating all possibilities of active labeling sets. For convenience of notation, suppose the labeling sets $\mathcal{E}, \mathcal{I}$ are disjoint and they are given as

$$\mathcal{E} \cup \mathcal{I} = \{1, \ldots, m\}.$$

**Theorem 6.1.2.** *([217]) Let $d_0$ be the degree of $f$ and let $d_i$ be the degree of $c_i$ for each $i \in \mathcal{E} \cup \mathcal{I}$. If the polynomials $f, c_i$ are generic, then the number of complex critical points of (6.1.1) is equal to*

$$\sum_{\mathcal{E} \subseteq \{i_1, \ldots, i_k\} \subseteq \mathcal{E} \cup \mathcal{I}} d_{i_1} \cdot d_{i_2} \cdots d_{i_k} \cdot \mathbf{S}_{n-k}(d_0 - 1, d_{i_1} - 1, \ldots, d_{i_k} - 1),$$

*where $\mathbf{S}_{n-k}$ is the complete symmetric function given as in (5.5.7).*

In (6.1.4), under the LICQC, the Lagrange multipliers $\lambda_i$ are uniquely determined by the gradients $\nabla c_i(u)$ $(i \in \mathcal{E} \cup J(u))$. On the critical set, one can express $\lambda$ as a rational function in $x$. Denote the matrix

$$G(x) := \begin{bmatrix} \nabla c_1(x) & \cdots & \nabla c_m(x) \end{bmatrix}.$$

If $m \le n$ and rank $G(x) = m$, then we get the rational expression

$$\lambda = [G(x)^T G(x)]^{-1} \cdot G(x)^T \cdot \nabla f(x). \tag{6.1.6}$$

The matrix inverse $[G(x)^T G(x)]^{-1}$ is generally expensive for obtaining an explicit formula, because the determinant $\det [G(x)^T G(x)]$ is typically a high degree polynomial. If $m > n$, then $G(x)^T G(x)$ is always singular and the expression (6.1.6) does not exist.

An interesting question is the following: Do there exist polynomials $p_i$ ($i \in \mathcal{E} \cup \mathcal{I}$) such that for every $i \in \mathcal{E} \cup \mathcal{I}$ and for every $(x, \lambda)$ satisfying (6.1.4), it holds that

$$\lambda_i = p_i(x)? \tag{6.1.7}$$

If they exist, then the following hold:

- The KKT system (6.1.4) can be simplified to

$$
\boxed{
\begin{aligned}
\sum_{i \in \mathcal{E}} p_i(x) \nabla c_i(x) + \sum_{j \in \mathcal{I}} p_j(x) \nabla c_j(x) &= \nabla f(x), \\
c_i(x) &= 0 \, (i \in \mathcal{E}), \\
p_j(x) c_j(x) &= 0 \, (j \in \mathcal{I}).
\end{aligned}
}
\tag{6.1.8}
$$

- For each $j \in \mathcal{I}$, the sign condition $\lambda_j \geq 0$ can be posed as

$$p_j(x) \geq 0. \tag{6.1.9}$$

The new systems (6.1.8) and (6.1.9) do not have variables of Lagrange multipliers. This can save a large amount of computational work.

Interestingly and also surprisingly, there generally exist polynomial expressions for Lagrange multipliers as in (6.1.7). Indeed, when the constraining polynomials are nonsingular, they always exist. Moreover, these polynomial expressions can be used to construct tight relaxations for solving (6.1.1). This is a major conclusion in [240]. This chapter introduces these results.

We conclude this section with Fritz-John conditions for optimization (see [22, §3.3] for more detailed introduction). It is well known that the KKT system (6.1.3) may not hold if the LICQC fails at a local minimizer $u$. However, the Fritz-John conditions always hold, even if the LICQC fails. There always exist real scalars $\mu$ and $\lambda_i, \lambda_j$ ($i \in \mathcal{E}, j \in \mathcal{I}$) such that

$$\mu \nabla f(u) = \sum_{i \in \mathcal{E}} \lambda_i \nabla c_i(u) + \sum_{j \in \mathcal{I}} \lambda_j \nabla c_j(u), \tag{6.1.10}$$

$$\mu \in \{0, 1\}, \quad \lambda_j \geq 0, \quad \lambda_j c_j(u) = 0 \ (j \in \mathcal{I}), \tag{6.1.11}$$

$$\mu + \sum_{i \in \mathcal{E}} |\lambda_i| + \sum_{j \in \mathcal{I}} \lambda_j > 0, \tag{6.1.12}$$

$$\forall \epsilon > 0, \ \exists v \in B(u, \epsilon), \ s.t. \ \begin{cases} 0 \neq \lambda_i (i \in \mathcal{E}) \Rightarrow \lambda_i c_i(v) < 0, \\ 0 \neq \lambda_j (j \in \mathcal{I}) \Rightarrow \lambda_j c_j(v) < 0. \end{cases} \tag{6.1.13}$$

The condition (6.1.12) means that not all $\mu, \lambda_i, \lambda_j$ are zeros. We refer to [22, Proposition 3.3.5] for these conclusions. Observe that the KKT system (6.1.3) is the special case of Fritz-John conditions (6.1.10)-(6.1.12) for $\mu = 1$. Each point satisfying (6.1.10)-(6.1.12) is called a *Fritz-John point*. For the case $\mu = 1$, the $\lambda_i$'s in Fritz-John conditions are not necessarily Lagrange multipliers for (6.1.1).

### 6.1.1 ▪ Exercise

**Exercise 6.1.1.** *Determine the number of critical values for the optimization*

$$\begin{cases} \min\limits_{x \in \mathbb{R}^3} & x_1^3 + x_2^3 + x_3^3 \\ s.t. & x_1^2 + x_2^2 + x_3^2 = 1. \end{cases}$$

**Exercise 6.1.2.** *Give polynomial expressions for Lagrange multipliers as in (6.1.7) for the optimization*

$$\begin{cases} \min_{x \in \mathbb{R}^3} & f(x) \\ s.t. & x_1 \geq 0, \, x_1 + x_2 \geq 0, \, x_1 + x_2 + x_3 \geq 0. \end{cases}$$

**Exercise 6.1.3.** *Verify the Fritz-John conditions for the optimization*

$$\begin{cases} \min & x_2 \\ s.t. & x_2^3 - x_1^5 \geq 0, x_1^3 \geq 0 \end{cases}$$

*at the minimizer* $(0,0)$. *Determine the set of all satisfactory* $\mu$ *and* $\lambda_i, \lambda_j$ $(i \in \mathcal{E}, j \in \mathcal{I})$ *for the Fritz-John conditions at* $(0,0)$.

**Exercise 6.1.4.** *([22]) Suppose $u$ is a local minimizer of (6.1.1), satisfying the KKT system (6.1.3) with required Lagrange multipliers* $\lambda_i, \lambda_j$.

(a) *Give an example that for* $\mu = 1$, *the Lagrange multipliers* $\lambda_i, \lambda_j$ *in (6.1.3) may not satisfy the Fritz-John condition (6.1.13).*

(b) *Show that there exist $\mu$ and a different set of* $\hat{\lambda}_i, \hat{\lambda}_j$ $(i \in \mathcal{E}, j \in \mathcal{I})$ *such that* $\mu, \hat{\lambda}_i, \hat{\lambda}_j$ *satisfy all Fritz-John conditions and the gradients*

$$\nabla c_i(u) \, (i \in \mathcal{E} : \hat{\lambda}_i \neq 0), \quad \nabla c_j(u) \, (j \in \mathcal{I} : \hat{\lambda}_j > 0)$$

*are linearly independent.*

## 6.2 ▪ Lagrange Multiplier Expressions

We show how to get polynomial expressions for Lagrange multipliers as in (6.1.7). Recall that critical points for the optimization (6.1.1) are given by the polynomial system (6.1.4).

Suppose there are totally $m$ constraints and we write that

$$\mathcal{E} \cup \mathcal{I} := \{1, \dots, m\}, \quad c := (c_1, \dots, c_m), \quad \lambda := \begin{bmatrix} \lambda_1 & \cdots & \lambda_m \end{bmatrix}^T.$$

The $\lambda$ is the vector of all Lagrange multipliers. If $(x, \lambda)$ is a critical pair, then $\lambda_i c_i(x) = 0$ for every $i \in \mathcal{E} \cup \mathcal{I}$. For convenience, we denote that

$$C(x) := \begin{bmatrix} \nabla c_1(x) & \nabla c_2(x) & \cdots & \nabla c_m(x) \\ c_1(x) & 0 & \cdots & 0 \\ 0 & c_2(x) & \cdots & 0 \\ \vdots & \vdots & \ddots & \vdots \\ 0 & 0 & \cdots & c_m(x) \end{bmatrix}, \quad g(x) := \begin{bmatrix} \nabla f(x) \\ 0 \\ 0 \\ \vdots \\ 0 \end{bmatrix}. \quad (6.2.1)$$

Then, the Lagrange multiplier vector $\lambda$ satisfies the linear equation

$$C(x)\lambda = g(x). \quad (6.2.2)$$

Note that $C(x)$ is a $(m + n)$-by-$m$ matrix polynomial. If there exists a matrix polynomial $L(x) \in \mathbb{R}[x]^{m \times (m+n)}$ such that

$$L(x)C(x) = I_m, \quad (6.2.3)$$

then one can see that
$$\lambda = L(x)C(x)\lambda = L(x)g(x).$$
So, the Lagrange multiplier vector $\lambda$ can be expressed as
$$\lambda = L(x)g(x) = L_1(x)\nabla f(x), \tag{6.2.4}$$
where $L_1(x)$ is the submatrix of $L(x)$ consisting of the first $n$ columns.

When does there exist a matrix polynomial $L(x)$ satisfying (6.2.3)? Observe that (6.2.3) holds only if $C(x)$ has full column rank for every $x \in \mathbb{C}^n$. Interestingly, the existence of such a matrix polynomial $L(x)$ is equivalent to the nonsingularity of constraining polynomials.

**Definition 6.2.1.** *The constraining polynomial tuple* $c = (c_1, \ldots, c_m)$ *is said to be nonsingular if* rank $C(x) = m$ *for all* $x \in \mathbb{C}^n$.

One can show that the polynomial tuple $c$ is nonsingular if and only if for all $u \in \mathbb{C}^n$, the gradients $\nabla c_i(u)$ ($i \in \mathcal{E} \cup J(u)$) are linearly independent, where $J(u)$ is the labeling set as in (6.1.2). An interesting conclusion is that the matrix polynomial $L(x)$ satisfying (6.2.3) exists if and only if $c$ is nonsingular. We need the following proposition.

**Proposition 6.2.2.** *([240]) For a matrix polynomial* $W(x) \in \mathbb{C}[x]^{s \times t}$ *with* $s \geq t$, *we have* rank $W(u) = t$ *for all* $u \in \mathbb{C}^n$ *if and only if there exists* $P(x) \in \mathbb{C}[x]^{t \times s}$ *such that*
$$P(x)W(x) = I_t.$$
*For the case* $W(x) \in \mathbb{R}[x]^{s \times t}$, *one can choose* $P(x) \in \mathbb{R}[x]^{t \times s}$ *in the above.*

We refer to [240] for the proof of Proposition 6.2.2, which uses Hilbert's Nullstellensatz (i.e., Theorem 2.1.2). Clearly, Proposition 6.2.2 implies the following.

**Proposition 6.2.3.** *([240]) The constraining polynomial tuple* $c$ *is nonsingular if and only if there exists* $L(x) \in \mathbb{R}[x]^{m \times (m+n)}$ *satisfying (6.2.3).*

The following are examples of $L(x)$ satisfying (6.2.3).

- The hypercube $[-1, 1]^n$ is defined as
$$1 - x_1^2 \geq 0, 1 - x_2^2 \geq 0, \ldots, 1 - x_n^2 \geq 0.$$
  The tuple $c = (1 - x_1^2, \ldots, 1 - x_n^2)$ and $L(x)C(x) = I_n$ is satisfied for
$$L(x) = \begin{bmatrix} -\frac{1}{2}\text{diag}(x) & I_n \end{bmatrix}. \tag{6.2.5}$$

- Consider the nonnegative portion of the unit sphere
$$x_1 \geq 0, x_2 \geq 0, \ldots, x_n \geq 0, x_1^2 + \cdots + x_n^2 - 1 = 0.$$
  The tuple $c = (x_1, \ldots, x_n, x_1^2 + \cdots + x_n^2 - 1)$ and $L(x)C(x) = I_{n+1}$ is satisfied for
$$L(x) = \begin{bmatrix} I_n - xx^T & x\mathbf{1}_n^T & 2x \\ \frac{1}{2}x^T & -\frac{1}{2}\mathbf{1}_n^T & -1 \end{bmatrix}. \tag{6.2.6}$$

Here, $\mathbf{1}_n$ denotes the all-one vector of length $n$.

- For the set given by $1-x_1^3-x_2^4 \geq 0$, $1-x_3^4-x_4^3 \geq 0$, the tuple $c = (1-x_1^3-x_2^4, 1-x_3^4-x_4^3)$ and $L(x)C(x) = I_2$ is satisfied for

$$L(x) = \begin{bmatrix} -\frac{x_1}{3} & -\frac{x_2}{4} & 0 & 0 & 1 & 0 \\ 0 & 0 & -\frac{x_3}{4} & -\frac{x_4}{3} & 0 & 1 \end{bmatrix}.$$

We would like to remark that if all $c_i$ have generic coefficients, then $c = (c_1, \ldots, c_m)$ is nonsingular. This was shown in [240].

**Theorem 6.2.4.** *([240]) For positive degrees $d_1, \ldots, d_m$, there exists an open dense subset $\mathcal{U}$ of $\mathcal{D} := \mathbb{R}[x]_{d_1} \times \cdots \times \mathbb{R}[x]_{d_m}$ such that every tuple*

$$c = (c_1, \ldots, c_m) \in \mathcal{U}$$

*is nonsingular. Indeed, such $\mathcal{U}$ can be chosen as a Zariski open subset of $\mathcal{D}$, i.e., it is the complement of a proper real variety of $\mathcal{D}$.*

The degree bound for the matrix polynomial $L(x)$ satisfying (6.2.3) is mostly open. We remark that there exist bounds for inverses of bijective polynomial maps (see the work [13]). For a priori degree of $L(x)$, the matrix polynomial $L(x)$ can be found from solving linear equations, which can be obtained by comparing coefficients on both sides of (6.2.3). There exist neat degree bounds for polyhedral constraints.

## 6.2.1 ▪ Polyhedral constraints

Suppose the constraints of (6.1.1) are $Ax - b \geq 0$, for given

$$A = \begin{bmatrix} a_1^T \\ \vdots \\ a_m^T \end{bmatrix} \in \mathbb{R}^{m \times n}, \quad b = \begin{bmatrix} b_1 \\ \vdots \\ b_m \end{bmatrix} \in \mathbb{R}^m.$$

The constraining polynomials $c_j(x) = a_j^T x - b_j$, $j = 1, \ldots, m$. The matrix $C(x)$ in (6.2.1) reads as

$$\begin{bmatrix} a_1 & a_2 & \cdots & a_m \\ a_1^T x - b_1 & 0 & \cdots & 0 \\ 0 & a_2^T x - b_2 & \cdots & 0 \\ \vdots & \vdots & \ddots & \vdots \\ 0 & 0 & \cdots & a_m^T x - b_m \end{bmatrix} \tag{6.2.7}$$

If rank $A = m$, the $L(x)$ in (6.2.3) can be chosen as the constant matrix

$$L(x) = \begin{bmatrix} (AA^T)^{-1}A & 0 \end{bmatrix}.$$

If rank $A < m$, the degree bound for $L(x)$ in (6.2.3) is $m - \text{rank } A$. This is shown in the following theorem.

**Theorem 6.2.5.** *([240]) If the polynomial tuple of $Ax - b$ is nonsingular, there is a matrix polynomial $L(x)$ satisfying (6.2.3) with degree at most $m - \text{rank } A$.*

**Proof.** Let $r := \text{rank } A$. Up to a linear coordinate transformation, we can reduce $x$ to a $r$-dimensional vector variable. Without loss of generality, one can assume that rank $A = n \leq m$.

For a subset $I := \{i_1, \dots, i_{m-n}\} \subseteq [m]$, denote

$$D_I(x) := \begin{bmatrix} c_{i_1}(x) & & \\ & \ddots & \\ & & c_{i_{m-n}}(x) \end{bmatrix}, \quad A_I = \begin{bmatrix} (a_{i_1})^T \\ \vdots \\ (a_{i_{m-n}})^T \end{bmatrix},$$

$$c_I(x) := \prod_{i \in I} c_i(x), \quad E_I(x) := c_I(x) \cdot \begin{bmatrix} \frac{1}{c_{i_1}(x)} & & \\ & \ddots & \\ & & \frac{1}{c_{i_{m-n}}(x)} \end{bmatrix}.$$

For the case that $I = \emptyset$ (the empty set), we set $c_\emptyset(x) = 1$. Let

$$V := \left\{ I \subseteq [m] \;\middle|\; \begin{array}{l} |I| = m - n, \\ \text{rank } A_{[m]\setminus I} = n \end{array} \right\}.$$

In the following, we prove the existence of $L(x)$ satisfying (6.2.3) with the required degree bound in three major steps.

**Step I:**  For each $I \in V$, we construct a matrix polynomial $L_I(x)$ such that

$$L_I(x)C(x) = c_I(x)I_m. \tag{6.2.8}$$

The matrix $L_I := L_I(x)$ satisfying (6.2.8) can be given by the $2 \times 3$ block matrix labeled as follows:

$$\begin{array}{c} \\ I \\ [m]\setminus I \end{array} \begin{array}{ccc} {\scriptstyle [n]} & {\scriptstyle n+I} & {\scriptstyle n+[m]\setminus I} \\ \left[ \begin{array}{ccc} 0 & E_I(x) & 0 \\ c_I(x) \cdot \left(A_{[m]\setminus I}\right)^{-T} & -\left(A_{[m]\setminus I}\right)^{-T}\left(A_I\right)^T E_I(x) & 0 \end{array} \right]. \end{array}$$

Denote by $L_I(\mathcal{J}, \mathcal{K})$ the submatrix whose row labels are from $\mathcal{J}$ and column labels are from $\mathcal{K}$. Then, equivalently, the blocks of $L_I$ are respectively given as

$$L_I(I, [n]) = 0, \quad L_I(I, n + [m]\setminus I) = 0,$$
$$L_I([m]\setminus I, n + [m]\setminus I) = 0, \quad L_I(I, n + I) = E_I(x),$$
$$L_I([m]\setminus I, [n]) = c_I(x)\left(A_{[m]\setminus I}\right)^{-T},$$
$$L_I([m]\setminus I, n + I) = -\left(A_{[m]\setminus I}\right)^{-T}\left(A_I\right)^T E_I(x).$$

For each $I \in V$, $A_{[m]\setminus I}$ is invertible. The superscript $^{-T}$ denotes the inverse of the transpose. Let $G := L_I(x)C(x)$. One can verify that

$$G(I, I) = E_I(x)D_I(x) = c_I(x)I_{m-n}, \quad G(I, [m]\setminus I) = 0,$$

$$G([m]\setminus I, [m]\setminus I) = \begin{bmatrix} c_I(x)\left(A_{[m]\setminus I}\right)^{-T} & -\left(A_{[m]\setminus I}\right)^{-T}\left(A_I\right)^T E_I(x) \end{bmatrix} \begin{bmatrix} \left(A_{[m]\setminus I}\right)^T \\ 0 \end{bmatrix} = c_I(x)I_n,$$

$$G([m]\setminus I, I) = \begin{bmatrix} c_I(x)\left(A_{[m]\setminus I}\right)^{-T} & -\left(A_{[m]\setminus I}\right)^{-T}\left(A_I\right)^T E_I(x) \end{bmatrix} \begin{bmatrix} A_I^T \\ D_I(x) \end{bmatrix} = 0.$$

This shows that the above $L_I(x)$ satisfies (6.2.8).

**Step II:**  We show that there exist real scalars $\nu_I$ satisfying

$$\sum_{I \in V} \nu_I c_I(x) = 1. \tag{6.2.9}$$

This can be shown by induction on $m \geq n$.

- When $m = n$, $V = \emptyset$ and $c_\emptyset(x) = 1$, so (6.2.9) is clearly true.

- When $m > n$, let

$$N := \{i \in [m] \mid \text{rank } A_{[m]\setminus\{i\}} = n\}. \qquad (6.2.10)$$

For each $i \in N$, let $V_i$ be the set of all $I' \subseteq [m]\setminus\{i\}$ such that $|I'| = m - n - 1$ and rank $A_{[m]\setminus(I'\cup\{i\})} = n$. For each $i \in N$, by the assumption, the linear function $A_{m\setminus\{i\}}x - b_{m\setminus\{i\}}$ is nonsingular. By induction, there exist real scalars $\nu_{I'}^{(i)}$ satisfying

$$\sum_{I'\in V_i} \nu_{I'}^{(i)} c_{I'}(x) = 1. \qquad (6.2.11)$$

Since rank $A = n$, we can generally assume that $\{a_1, \ldots, a_n\}$ is linearly independent. So, there exist scalars $\alpha_1, \ldots, \alpha_n$ such that

$$a_m = \alpha_1 a_1 + \cdots + \alpha_n a_n.$$

If all $\alpha_i = 0$, then $a_m = 0$ and $A$ can be replaced by its first $m - 1$ rows, so (6.2.9) is true by induction. We consider the case that at least one $\alpha_i \neq 0$ and we write that

$$\{i : \alpha_i \neq 0\} = \{i_1, \ldots, i_k\}.$$

Then $a_{i_1}, \ldots, a_{i_k}, a_m$ are linearly dependent. For convenience, let $i_{k+1} := m$. Since $Ax - b$ is nonsingular, the linear system

$$c_{i_1}(x) = \cdots = c_{i_k}(x) = c_{i_{k+1}}(x) = 0$$

has no solutions. Hence, there exist real scalars $\mu_1, \ldots, \mu_{k+1}$ such that

$$\mu_1 c_{i_1}(x) + \cdots + \mu_k c_{i_k}(x) + \mu_{k+1} c_{i_{k+1}}(x) = 1.$$

This above can be implied by echelon form for inconsistent linear systems (see Exercise 1.1.1). Note that $i_1, \ldots, i_{k+1} \in N$. For each $j = 1, \ldots, k+1$, by (6.2.11),

$$\sum_{I'\in V_{i_j}} \nu_{I'}^{(i_j)} c_{I'}(x) = 1.$$

Then, we can get

$$1 = \sum_{j=1}^{k+1} \mu_j c_{i_j}(x) = \sum_{j=1}^{k+1} \mu_j \sum_{I'\in V_{i_j}} \nu_{I'}^{(i_j)} c_{i_j}(x) c_{I'}(x)$$

$$= \sum_{\substack{I=I'\cup\{i_j\}, \\ I'\in V_{i_j}, 1\leq j\leq k+1}} \nu_{I'}^{(i_j)} \mu_j c_I(x).$$

Since each $I' \cup \{i_j\} \in V$, (6.2.9) is satisfied for some scalars $\nu_I$.

**Step III:** Let $L_I(x)$ be as in (6.2.8). We construct $L(x)$ as

$$L(x) := \sum_{I\in V} \nu_I L_I(x). \qquad (6.2.12)$$

Clearly, $L(x)$ satisfies (6.2.3) because

$$L(x)C(x) = \sum_{I\in V} \nu_I L_I(x)C(x) = \sum_{I\in V} \nu_I c_I(x)I_m = I_m.$$

Each $L_I(x)$ has degree $\le m - n$, so $L(x)$ has degree at most $m - n$.   $\square$

The following are some examples of Lagrange multiplier expressions for polyhedral constraints.

- For the simplicial set given as

$$x_1 \ge 0, \ldots, x_n \ge 0, 1 - e^T x \ge 0,$$

the tuple $c = (x_1, \ldots, x_n, 1 - e^T x)$ and $L(x)C(x) = I_{n+1}$ is satisfied for

$$L(x) = \begin{bmatrix} 1 - x_1 & -x_2 & \cdots & -x_n & 1 & \cdots & 1 \\ -x_1 & 1 - x_2 & \cdots & -x_n & 1 & \cdots & 1 \\ \vdots & \vdots & \ddots & \vdots & \vdots & \vdots & \vdots \\ -x_1 & -x_2 & \cdots & 1 - x_n & 1 & \cdots & 1 \\ -x_1 & -x_2 & \cdots & -x_n & 1 & \cdots & 1 \end{bmatrix}. \tag{6.2.13}$$

- For the box $[0, 1]^n$, which is given as

$$x_1 \ge 0, \ldots, x_n \ge 0, 1 - x_1 \ge 0, \ldots, 1 - x_n \ge 0,$$

the tuple $c = (x_1, \ldots, x_n, 1 - x_1, \ldots, 1 - x_n)$ and $L(x)C(x) = I_{2n}$ is satisfied for

$$L(x) = \begin{bmatrix} I_n - \operatorname{diag}(x) & I_n & I_n \\ -\operatorname{diag}(x) & I_n & I_n \end{bmatrix}. \tag{6.2.14}$$

- For the polyhedral set

$$1 - x_4 \ge 0, x_4 - x_3 \ge 0, x_3 - x_2 \ge 0, x_2 - x_1 \ge 0, x_1 + 1 \ge 0,$$

the tuple $c = (1 - x_4, x_4 - x_3, x_3 - x_2, x_2 - x_1, x_1 + 1)$ and $L(x)C(x) = I_5$ is satisfied for

$$L(x) = \frac{1}{2} \begin{bmatrix} -x_1 - 1 & -x_2 - 1 & -x_3 - 1 & -x_4 - 1 & 1 & 1 & 1 & 1 & 1 \\ -x_1 - 1 & -x_2 - 1 & -x_3 - 1 & 1 - x_4 & 1 & 1 & 1 & 1 & 1 \\ -x_1 - 1 & -x_2 - 1 & 1 - x_3 & 1 - x_4 & 1 & 1 & 1 & 1 & 1 \\ -x_1 - 1 & 1 - x_2 & 1 - x_3 & 1 - x_4 & 1 & 1 & 1 & 1 & 1 \\ 1 - x_1 & 1 - x_2 & 1 - x_3 & 1 - x_4 & 1 & 1 & 1 & 1 & 1 \end{bmatrix}.$$

- For the polyhedral set

$$1 + x_1 \ge 0, 1 - x_1 \ge 0, 2 - x_1 - x_2 \ge 0, 2 - x_1 + x_2 \ge 0,$$

the tuple $c = (1 + x_1, 1 - x_1, 2 - x_1 - x_2, 2 - x_1 + x_2)$. The matrix polynomial $L(x)$ satisfying $L(x)C(x) = I_4$ is

$$\frac{1}{6} \begin{bmatrix} x_1^2 - 3x_1 + 2 & x_1 x_2 - x_2 & 4 - x_1 & 2 - x_1 & 1 - x_1 & 1 - x_1 \\ 3x_1^2 - 3x_1 - 6 & 3x_2 + 3x_1 x_2 & 6 - 3x_1 & -3x_1 & -3x_1 - 3 & -3x_1 - 3 \\ 1 - x_1^2 & -2x_2 - x_1 x_2 - 3 & x_1 - 1 & x_1 + 1 & x_1 + 2 & x_1 + 2 \\ 1 - x_1^2 & 3 - x_1 x_2 - 2x_2 & x_1 - 1 & x_1 + 1 & x_1 + 2 & x_1 + 2 \end{bmatrix}.$$

## 6.2.2 ▪ Exercises

**Exercise 6.2.1.** *Determine whether or not there is a matrix polynomial $L(x)$ satisfying (6.2.3) for the following polynomial tuples:*

*(i) $c = (1 - x_1^d, \ldots, 1 - x_n^d)$ for a degree $d > 2$.*

*(ii) $c = (1 - x^T x, e^T x)$.*

*(iii) $c = (1 - x^T x, 1 - y^T y, x^T y)$ for two vector variables $x, y \in \mathbb{R}^n$.*

*(iv) $c = (1 + s_1 x_1 + \cdots + s_n x_n : s_i \in \{\pm 1\})$.*

*When it exists, give the satisfactory $L(x)$ explicitly, with the degree as low as possible.*

**Exercise 6.2.2.** *For the simplicial constraint $x_1 \geq 0, \ldots, x_n \geq 0, 1 - e^T x \geq 0$, determine the set of all matrix polynomials $C(x)$ satisfying (6.2.3).*

**Exercise 6.2.3.** *For the polynomial optimization (6.1.1), show that there exists a matrix polynomial $L(x)$ satisfying (6.2.3) if and only if the LICQC holds at every complex critical point.*

**Exercise 6.2.4.** *Determine conditions on $A \in \mathcal{S}^n$ and $b \in \mathbb{R}^n$ such that $1 + 2b^T x + x^T A x$ is singular.*

**Exercise 6.2.5.** *Determine conditions on $A, B \in \mathcal{S}^n$ such that $(1 - x^T A x, 1 - x^T B x)$ is a singular pair.*

**Exercise 6.2.6.** *Determine conditions on $a_1, a_2, a_3, a_4$ such that $1 + a_1 x + a_2 x^2 + a_3 x^3 + a_4 x^4$ is a singular univariate polynomial in $x$.*

**Exercise 6.2.7.** *Determine conditions on vectors $a_1, \ldots, a_m \in \mathbb{C}^n$ and scalars $b_1, \ldots, b_m$ such that $(a_1^T x - b_1, \ldots, a_m^T x - b_m)$ is nonsingular.*

**Exercise 6.2.8.** *([240]) Suppose constraining polynomials are of the type*

$$c_i(x) = \tau_i x_i + q_i(x_{i+1}, \ldots, x_n), \quad i = 1, \ldots, m,$$

*where each $q_i \in \mathbb{R}[x_{i+1}, \ldots, x_n]$ and scalars $\tau_i \neq 0$. Let $T(x)$ be the matrix consisting of first $m$ rows of $[\nabla c_1(x) \cdots \nabla c_m(x)]$. Show that $T(x)$ is invertible and its inverse is also a matrix polynomial. Further show that the vector of Lagrange multipliers can be given as*

$$\lambda = T(x)^{-1} \cdot \left( \nabla f(x) \right)_{1:m}.$$

# 6.3 ▪ A Tight Moment-SOS Hierarchy

This section introduces a tight hierarchy of Moment-SOS relaxations for solving the polynomial optimization (6.1.1). Let $\lambda := (\lambda_i)_{i \in \mathcal{E} \cup \mathcal{I}}$ denote the vector of Lagrange multipliers as in (6.1.3). Recall that $\mathcal{K}$ denotes the set of all critical pairs, i.e., all $(x, \lambda)$ satisfying (6.1.4). The projection

$$\mathcal{K}_c := \{ u \in \mathbb{R}^n : (u, \lambda) \in \mathcal{K} \} \tag{6.3.1}$$

is the set of all real critical points for (6.1.1). We make the following assumption about Lagrange multipliers.

**Assumption 6.3.1.** *([240]) For each $i \in \mathcal{E} \cup \mathcal{I}$, there exists a polynomial $p_i \in \mathbb{R}[x]$ such that $\lambda_i = p_i(x)$ for all $(x, \lambda) \in \mathcal{K}$.*

Assumption 6.3.1 generally holds. Let $C(x), g(x)$ be as in (6.2.1). If there exists a matrix polynomial $L(x)$ satisfying (6.2.3), i.e., $L(x)C(x) = I_m$, then

$$\lambda = L(x)C(x)g(x) = L_1(x)\nabla f(x)$$

as in (6.2.4). Hence, we can select each $p_i$ such that

$$p_i = \big(L_1(x)\nabla f(x)\big)_i. \qquad (6.3.2)$$

When constraining polynomials have generic coefficients, such $L(x)$ always exists (see Theorem 6.2.4). The following are some common examples.

- For the simplex constraint $e^T x - 1 = 0$, $x_1 \geq 0$, $\ldots$, $x_n \geq 0$,

$$\mathcal{E} = \{0\}, \quad \mathcal{I} = [n], \quad c_0(x) = e^T x - 1, \quad c_j(x) = x_j \; (j \in [n]).$$

In view of (6.2.13), the Lagrange multipliers can be expressed as

$$\lambda_0 = x^T \nabla f(x), \quad \lambda_j = f_{x_j} - x^T \nabla f(x), \quad j = 1, \ldots, n. \qquad (6.3.3)$$

The above expressions can also be obtained as follows. The KKT system implies the equation

$$\nabla f(x) = \lambda_0 e + \begin{bmatrix} \lambda_1 \\ \vdots \\ \lambda_n \end{bmatrix}.$$

If we premultiply $x^T$ in the above and use the complementarity condition $x_j \lambda_j = 0$ for each $j$, then

$$x^T \nabla f(x) = \lambda_0 e^T x = \lambda_0 (e^T x - 1) + \lambda_0 = \lambda_0.$$

By the substitution for $\lambda_0$, we can get the expression as in (6.3.3).

- For the hypercube $[-1, 1]^n$, the constraints are

$$c_j := 1 - x_j^2 \geq 0, \quad j = 1, \ldots, n.$$

In view of (6.2.5), one can get that

$$\lambda_j = -\frac{1}{2} x_j f_{x_j}, \quad j = 1, \ldots, n. \qquad (6.3.4)$$

The above expressions can also be obtained as follows. The KKT system implies the equation

$$\nabla f(x) = \begin{bmatrix} -2x_1 \lambda_1 \\ \vdots \\ -2x_n \lambda_n \end{bmatrix}.$$

If we multiply $x_j$ in the $j$th entry of the equation and use the complementarity condition $\lambda_j(1 - x_j^2) = 0$, then

$$x_j f_{x_j} = -2x_j^2 \lambda_j = 2(1 - x_j^2)\lambda_j - 2\lambda_j = -2\lambda_j.$$

This gives the same expression as in (6.3.4).

- For the sphere constraint $1 - x^T x = 0$ or ball constraint $1 - x^T x \geq 0$, $\mathcal{E} \cup \mathcal{I} = \{1\}$, $c = (1 - x^T x)$ and

$$x^T \nabla f(x) = x^T(-2\lambda_1 x) = 2\lambda_1(1 - x^T x) - 2\lambda_1 = -2\lambda_1.$$

Therefore, we can get that

$$\lambda_1 = -\frac{1}{2} x^T \nabla f(x). \tag{6.3.5}$$

For the polynomials $p_i$ as in Assumption 6.3.1, denote the new sets of polynomials

$$\boxed{\begin{aligned} \Phi &:= \left\{ \frac{\partial f}{\partial x_j} - \sum_{i \in \mathcal{E} \cup \mathcal{I}} p_i \frac{\partial c_i}{\partial x_j} \right\}_{j=1}^n \cup \left\{ p_j c_j : j \in \mathcal{I} \right\}, \\ \Psi &:= \left\{ p_j : j \in \mathcal{I} \right\}. \end{aligned}} \tag{6.3.6}$$

If the KKT conditions are posed, the optimization (6.1.1) becomes

$$\begin{cases} \min & f(x) \\ s.t. & c_i(x) = 0 \, (i \in \mathcal{E}), \\ & c_j(x) \geq 0 \, (j \in \mathcal{I}), \\ & \phi(x) = 0 \, (\phi \in \Phi), \\ & \psi(x) \geq 0 \, (\psi \in \Psi). \end{cases} \tag{6.3.7}$$

Let $f_c$ denote the minimum value of (6.3.7). It is equal to the smallest critical value on the feasible set. The Moment-SOS hierarchy can be applied to solve (6.3.7). The $k$th order SOS relaxation is

$$\begin{cases} \max & \gamma \\ s.t. & f - \gamma \in \text{Ideal}[c_{eq}]_{2k} + \text{QM}[c_{in}]_{2k} \\ & \quad + \text{Ideal}[\Phi]_{2k} + \text{QM}[\Psi]_{2k}. \end{cases} \tag{6.3.8}$$

Its dual optimization is the $k$th order moment relaxation

$$\begin{cases} \min & \langle f, y \rangle \\ s.t. & \mathscr{V}_{c_i}^{(2k)}[y] = 0 \, (i \in \mathcal{E}), \\ & L_{c_j}^{(k)}[y] \succeq 0 \, (j \in \mathcal{I}), \\ & \mathscr{V}_{\phi}^{(2k)}[y] = 0 \, (\phi \in \Phi), \\ & L_{\psi}^{(k)}[y] \succeq 0 \, (\psi \in \Psi), \\ & y_0 = 1, \, M_k[y] \succeq 0, \, y \in \mathbb{R}^{\mathbb{N}_{2k}^n}. \end{cases} \tag{6.3.9}$$

For $k = 1, 2, \ldots$, we get a stronger hierarchy of Moment-SOS relaxations. In (6.3.8)-(6.3.9), if we remove the usage of polynomials in $\Phi$ and $\Psi$, then they are the same as the classical Moment-SOS relaxations (5.2.2)-(5.2.3). For convenience, let $f_{sos,k}^c$, $f_{mom,k}^c$ denote the optimal values of (6.3.8), (6.3.9) respectively.

For the polynomial set $\Phi$ as in (6.3.6), under Assumption 6.3.1, the critical point set $\mathcal{K}_c$ can be expressed as

$$\mathcal{K}_c = \left\{ x \in \mathbb{R}^n \, \middle| \, \begin{array}{l} c_i(x) = 0 \quad (i \in \mathcal{E}), \\ \phi(x) = 0 \quad (\phi \in \Phi) \end{array} \right\}.$$

By Theorem 6.1.1, the objective $f$ achieves only finitely many values on $\mathcal{K}_c$, say, they are

$$v_1 < \cdots < v_N. \tag{6.3.10}$$

The value $f_c$ equals one of them, say, for some $t \in \{1, \ldots, N\}$,

$$f_c = v_t.$$

A critical point $u \in \mathcal{K}_c$ may not be feasible for (6.3.7), i.e., $c_j(u) \not\geq 0$ for some $j \in \mathcal{I}$, or $\psi(u) \not\geq 0$ for some $\psi \in \Psi$. If (6.3.7) is infeasible, by convention, we set

$$f_c = +\infty.$$

If the minimum value $f_{min}$ of (6.1.1) is achievable at a KKT point, then $f_c = f_{min}$. This is the case if the feasible set is compact (or if $f$ is coercive[4]) and the LICQC holds. Like classical Moment-SOS relaxations, we also have

$$f_{sos,k}^c \leq f_{mom,k}^c \leq f_c \qquad (6.3.11)$$

for every relaxation order $k$. Moreover, both $f_{sos,k}^c$ and $f_{mom,k}^c$ are monotonically increasing as $k$ increases. If $f_{sos,k}^c = f_c$ for some order $k$, the hierarchy of relaxations (6.3.8)-(6.3.9) is said to be *tight* (or *exact*) or to have finite convergence.

## 6.3.1 ▪ Tightness of the hierarchy

An important property for the hierarchy of relaxations (6.3.8)-(6.3.9) is the tightness. We begin with a general assumption.

**Assumption 6.3.2.** *There exists $\rho \in \mathrm{QM}[c_{in}, \Psi]$ such that if $u \in \mathcal{K}_c$ and $f(u) < f_c$, then $\rho(u) < 0$.*

The membership $\rho \in \mathrm{QM}[c_{in}, \Psi]$ implies that if $u$ is a feasible point for (6.3.7), then $\rho(u) \geq 0$. So Assumption 6.3.2 requires that $\rho(x) = 0$ serves as a hypersurface separating feasible and nonfeasible critical points. Assumption 6.3.2 generally holds (see [240]). The following are some common instances.

- When (6.1.1) has no inequality constraints, $\mathrm{QM}[c_{in}, \Psi] = \Sigma[x]$, Assumption 6.3.2 is satisfied for $\rho := 0$.

- When (6.1.1) has a single inequality constraint, say, $c_1 \geq 0$, Assumption 6.3.2 is satisfied for $\rho := c_1$.

- Let $K^c$ be the feasible set of (6.3.7). Suppose $\mathcal{K}_c \backslash K^c$ is a finite set, say, $\mathcal{K}_c \backslash K^c = \{u_1, \ldots, u_D\}$. Let $\varphi_1, \ldots, \varphi_D$ be real polynomials such that $\varphi_i(u_i) = 1$ and $\varphi_i(u_j) = 0$ for $i \neq j$. For each $i$, there must exist $j_i \in \mathcal{I}$ such that $c_{j_i}(u_i) < 0$. Then Assumption 6.3.2 holds for

$$\rho := \sum_{i=1,\ldots,D} \frac{-1}{c_{j_i}(u_i)} c_{j_i}(x) \varphi_i(x)^2. \qquad (6.3.12)$$

- For each $u \in \mathcal{K}_c$ with $f(u) < f_c$, at least one of the constraints $c_j(x) \geq 0, p_j(x) \geq 0 (j \in \mathcal{I})$ is violated. For each critical value $v_i < f_c$, suppose there exists $q_i \in \{c_j, p_j\}_{j \in \mathcal{I}}$ such that

$$q_i < 0 \quad \text{on} \quad \mathcal{K}_c \cap \{f(x) = v_i\}.$$

Let $\phi_1, \ldots, \phi_N$ be real univariate polynomials such that $\phi_i(v_i) = 1$ and $\phi_i(v_j) = 0$ for $i \neq j$. Then Assumption 6.3.2 holds for

$$\rho := \sum_{i=1,\ldots,t-1} q_i \cdot \big(\phi_i(f(x))\big)^2. \qquad (6.3.13)$$

---

[4]The function $f$ is coercive if the sublevel set $\{x \in \mathbb{R}^n : f(x) \leq \vartheta\}$ is compact for every value $\vartheta$.

The tightness of the hierarchy of relaxations (6.3.8)-(6.3.9) is shown in the following theorem.

**Theorem 6.3.3.** *([240]) Suppose $\mathcal{K}_c \neq \emptyset$ and Assumption 6.3.1 holds. If*

*(i)* $\mathrm{Ideal}[c_{eq}, \Phi] + \mathrm{QM}[c_{in}, \Psi]$ *is archimedean,* **or**

*(ii)* $\mathrm{Ideal}[c_{eq}] + \mathrm{QM}[c_{in}]$ *is archimedean,* **or**

*(iii) Assumption 6.3.2 holds,*

*then $f_{sos,k}^c = f_{mom,k}^c = f_c$ when $k$ is sufficiently large. Therefore, if the minimum value $f_{min}$ of (6.1.1) is achievable at a feasible critical point, then*

$$f_{sos,k}^c = f_{mom,k}^c = f_{min}$$

*for all $k$ big enough if one of the conditions (i)-(iii) holds.*

**Proof.** By Theorem 6.1.1, the objective $f$ achieves only finitely many values on $\mathcal{K}_c$, say, they are $v_1, \ldots, v_N$, ordered as in (6.3.10). Up to the shifting of a constant, we can further assume that $f_c = 0$. Since $f_c$ is among $v_1, \ldots, v_N$, we can assume that

$$v_t = f_c.$$

When either the condition (i) or (ii) holds, the proof is given in [240]. Here we give the proof when the condition (iii) holds.

Suppose Assumption 6.3.2 holds. Let $\varphi_1, \ldots, \varphi_N$ be real univariate polynomials such that $\varphi_i(v_j) = 0$ for $i \neq j$ and $\varphi_i(v_j) = 1$ for $i = j$. Let

$$s_i := (v_i - f_c)\big(\varphi_i(f)\big)^2, \quad i = t, \ldots, N.$$

Then we have $s := s_t + \cdots + s_N \in \Sigma[x]_{2k_1}$ for some degree $k_1 > 0$. Let

$$\hat{f} := f - f_c - s.$$

By Assumption 6.3.2, $\hat{f}(x) \equiv 0$ on the semialgebraic set

$$\mathcal{K}_2 := \left\{ x \in \mathbb{R}^n \,\middle|\, \begin{array}{rcl} c_{eq}(x) & = & 0, \\ \phi(x) & = & 0 \, (\phi \in \Phi), \\ \rho(x) & \geq & 0 \end{array} \right\}.$$

There is only a single inequality for $\mathcal{K}_2$. By Theorem 2.6.6, there exist a power $\ell \in \mathbb{N}$ and $q = b_0 + \rho b_1$ $(b_0, b_1 \in \Sigma[x])$ such that

$$\hat{f}^{2\ell} + q \in \mathrm{Ideal}[c_{eq}, \Phi].$$

By Assumption 6.3.2, $\rho \in \mathrm{QM}[c_{in}, \Psi]$, so we have $q \in \mathrm{QM}[c_{in}, \Psi]$.

For all $\epsilon > 0$ and $\tau > 0$, we have $\hat{f} + \epsilon = \phi_\epsilon + \theta_\epsilon$, where

$$\phi_\epsilon := -\tau \epsilon^{1-2\ell}\big(\hat{f}^{2\ell} + q\big),$$

$$\theta_\epsilon := \epsilon\Big(1 + \hat{f}/\epsilon + \tau(\hat{f}/\epsilon)^{2\ell}\Big) + \tau \epsilon^{1-2\ell} q.$$

By Lemma 5.6.2, for $\tau \geq \frac{1}{2\ell}$, there exists $k_2 \in \mathbb{N}$ such that for all $\epsilon > 0$,

$$\phi_\epsilon \in \mathrm{Ideal}[c_{eq}, \Phi]_{2k_2}, \quad \theta_\epsilon \in \mathrm{QM}[c_{in}, \Psi]_{2k_2}.$$

Hence, we can get

$$f - (f_c - \epsilon) = \phi_\epsilon + \sigma_\epsilon,$$

where $\sigma_\epsilon = \theta_\epsilon + s \in \text{QM}[c_{in}, \Psi]_{2k_2}$ for all $\epsilon > 0$. Note that

$$\text{Ideal}[c_{eq}]_{2k_2} + \text{QM}[c_{in}]_{2k_2} + \text{Ideal}[\Phi]_{2k_2} + \text{QM}[\Psi]_{2k_2}$$

$$= \text{Ideal}[c_{eq}, \Phi]_{2k_2} + \text{QM}[c_{in}, \Psi]_{2k_2}.$$

For all $\epsilon > 0$, $\gamma = f_c - \epsilon$ is feasible in (6.3.8) for the order $k_2$, so $f^c_{sos,k_2} \geq f_c$. Because of the monotonicity of $f^c_{sos,k}$, we have $f^c_{sos,k} = f_c$ for all $k \geq k_2$. The relation (6.3.11) implies $f^c_{mom,k} = f_c$ for all such $k$.  □

For the conclusion of Theorem 6.3.3 to hold, it suffices to have one (not necessarily all) of the conditions (i)-(iii). The condition (ii) is the easiest one for checking, because it only depends on constraining polynomials of (6.1.1). Also note that the condition (ii) implies the condition (i). When $f, c_i, c_j$ have generic coefficients, there are finitely many critical points, i.e., $V_\mathbb{R}(c_{eq}, \Phi)$ is a finite set. The number of critical points is given in Theorem 6.1.2. So the assumptions of Theorem 6.3.3 hold for generic cases.

## 6.3.2 ▪ Detecting tightness and extracting minimizers

The optimal value $f_c$ of (6.3.7) is the smallest objective value on feasible KKT points, and $f_{min}$ is the smallest objective value on feasible points. If $f_{min}$ is achievable at a KKT point, then $f_c = f_{min}$. However, $f_c > f_{min}$ if $f_{min}$ is not achievable at a KKT point. For instance, this is the case for $f = x_1^4 + (x_1 x_2 - 1)^2$ and $K = \mathbb{R}^2$. In Theorem 6.3.3, we have $f^c_{sos,k} = f_c$ for all $k$ big enough, where $f^c_{sos,k}$ is the optimal value of (6.3.8).

In practice, the value $f_c$ or $f_{min}$ is often not known. How can we recognize $f^c_{sos,k} = f_c$ in computational practice? The flat truncation can be used for this purpose. Suppose $y^*$ is a minimizer of (6.3.9) for the relaxation order $k$. Denote the degree

$$d := \max \left\{ \left\lceil \frac{1}{2} \deg(h) \right\rceil : h \in \{c_i\}_{i \in \mathcal{E} \cup \mathcal{I}} \cup \Phi \cup \Psi \right\}. \tag{6.3.14}$$

If the flat truncation holds, there exists $t \in [d, k]$ such that

$$\text{rank } M_t[y^*] = \text{rank } M_{t-d}[y^*], \tag{6.3.15}$$

then $f^c_{sos,k} = f_c$ and we can get $r := \text{rank } M_t[y^*]$ minimizers for (6.3.7). Generally, the condition (6.3.15) is sufficient and necessary for extracting minimizers. When no critical points are feasible for (6.3.7), the infeasibility can also be detected by solving the relaxations (6.3.8)-(6.3.9). The following is the major conclusion for checking convergence and extracting minimizers.

**Theorem 6.3.4.** *([240]) Suppose Lagrange multipliers can be expressed as in Assumption 6.3.1. Then the hierarchy of relaxations (6.3.8)-(6.3.9) has the following properties:*

(i) *If (6.3.9) is infeasible for some order $k$, then the optimization (6.3.7) must be infeasible, i.e., it has no feasible KKT points.*

(ii) *Under Assumption 6.3.2, if (6.3.7) is infeasible, then the relaxation (6.3.9) must be infeasible when $k$ is big enough.*

*In the following, assume (6.3.7) is feasible (i.e., $f_c < +\infty$). Then, the following hold:*

(iii) *If (6.3.15) is satisfied for some $t \in [d, k]$, then $f^c_{mom,k} = f_c$.*

(iv) *If Assumption 6.3.2 holds and (6.3.7) has only finitely many minimizers, then the moment relaxation (6.3.9) has minimizers and each minimizer $y^*$ must satisfy (6.3.15), when $k$ is big enough.*

**Proof.** Under Assumption 6.3.1, each Lagrange multiplier $\lambda_i$ can be expressed as $\lambda_i = p_i(x)$, so $u$ is a KKT point if and only if it is a feasible point for (6.3.7).

(i) For every feasible point $u$ of (6.3.7), the tms $[u]_{2k}$ is feasible for (6.3.9), for all $k$. Therefore, if (6.3.9) is infeasible for some $k$, then (6.3.7) must be infeasible.

(ii) Under Assumption 6.3.2, if (6.3.7) is infeasible, then the set

$$\left\{ x \in \mathbb{R}^n \left| \begin{array}{rcl} c_i(x) &=& 0 \, (i \in \mathcal{E}), \\ \phi(x) &=& 0 \, (\phi \in \Phi), \\ \rho(x) &\geq& 0 \end{array} \right. \right\} = \emptyset.$$

Since $\rho$ is a single polynomial, $\mathrm{QM}[\rho] = \mathrm{Pre}[\rho]$. By Theorem 2.6.6, we have

$$-1 \in \mathrm{Ideal}[c_{eq}, \Phi] + \mathrm{QM}[\rho].$$

By Assumption 6.3.2, it holds that

$$-1 \in \mathrm{Ideal}[c_{eq}] + \mathrm{QM}[c_{in}] + \mathrm{Ideal}[\Phi] + \mathrm{QM}[\Psi].$$

Thus, for all $k$ big enough, (6.3.8) is unbounded above. Hence, (6.3.9) must be infeasible, by the weak duality.

When (6.3.7) is feasible, $f$ achieves only finitely many values on $\mathcal{K}_c$, so (6.3.7) must achieve its optimal value $f_c$.

(iii) The proof is left as an exercise, since it is similar to the proof for earlier results, e.g., Lemma 5.3.1.

(iv) Under Assumption 6.3.2, the optimization (6.3.7) is equivalent to

$$\left\{ \begin{array}{rl} \min & f(x) \\ s.t. & c_i(x) = 0 \, (i \in \mathcal{E}), \\ & \phi(x) = 0 \, (\phi \in \Phi), \\ & \rho(x) \geq 0. \end{array} \right. \qquad (6.3.16)$$

The optimal value of (6.3.16) is also $f_c$. Its $k$th order SOS relaxation is

$$\left\{ \begin{array}{rl} \gamma_k := \max & \gamma \\ s.t. & f - \gamma \in \mathrm{Ideal}[c_{eq}, \Phi]_{2k} + \mathrm{QM}[\rho]_{2k}. \end{array} \right. \qquad (6.3.17)$$

Its dual optimization is the $k$th order moment relaxation

$$\left\{ \begin{array}{rl} \gamma'_k := \min & \langle f, y \rangle \\ s.t. & \mathscr{V}^{(2k)}_{c_i}[y] = 0 \, (i \in \mathcal{E}), \\ & \mathscr{V}^{(2k)}_{\phi}[y] = 0 \, (\phi \in \Phi), \\ & L^{(k)}_{\rho}[y] \succeq 0, \\ & M_k[y] \succeq 0, \\ & y_0 = 1, \, y \in \mathbb{R}^{\mathbb{N}^n_{2k}}. \end{array} \right. \qquad (6.3.18)$$

By repeating the proof for Theorem 6.3.3(iii), one can similarly show that

$$\gamma_k = \gamma'_k = f_c \tag{6.3.19}$$

for all $k$ big enough. Since $\rho \in \text{QM}[c_{in}, \Psi]$, each $y$, which is feasible for (6.3.9), must also be feasible for (6.3.18). So, when $k$ is big enough, each minimizer $y^*$ is also a minimizer of (6.3.18). By the assumption, (6.3.16) also has only finitely many minimizers. By Theorem 5.3.9, the condition (6.3.15) must be satisfied for some $t \in [d, k]$, when $k$ is big enough.  □

If (6.3.7) has infinitely many minimizers, then the flat truncation (6.3.15) typically does not hold. We refer to Proposition 5.3.6 for this fact.

## 6.3.3 ▪ A refined hierarchy based on preorderings

The tightness of the hierarchy of relaxations (6.3.8)-(6.3.9) is shown in Theorem 6.3.3. In addition to Assumption 6.3.1 about Lagrange multiplier expressions, it also requires one (not necessarily all) of the conditions (i), (ii), (iii) there, to guarantee the tightness. Interestingly, if the preordering is used to give Moment-SOS relaxations, we can get a refined hierarchy that is tight without assuming anyone of the conditions (i), (ii), (iii) there. Observe that

$$\text{QM}[c_{in}]_{2k} + \text{QM}[\Psi]_{2k} = \text{QM}[c_{in}, \Psi]_{2k} \subseteq \text{Pre}[c_{in}, \Psi]_{2k}.$$

In (6.3.8), if $\text{QM}[c_{in}]_{2k} + \text{QM}[\Psi]_{2k}$ is replaced by the preordering $\text{Pre}[c_{in}, \Psi]_{2k}$, we can get the stronger relaxation

$$\begin{cases} \max & \gamma \\ \text{s.t.} & f - \gamma \in \text{Ideal}[c_{eq}, \Phi]_{2k} + \text{Pre}[c_{in}, \Psi]_{2k}. \end{cases} \tag{6.3.20}$$

Then its dual optimization is the moment relaxation

$$\begin{cases} \min & \langle f, y \rangle \\ \text{s.t.} & \mathscr{V}_{c_i}^{(2k)}[y] = 0 \, (i \in \mathcal{E}), \\ & \mathscr{V}_{\phi}^{(2k)}[y] = 0 \, (\phi \in \Phi), \\ & L_{\psi_J}^{(k)}[y] \succeq 0 \, (J \subseteq \{c_{in}\} \cup \Psi), \\ & y_0 = 1, \ y \in \mathbb{R}^{\mathbb{N}_{2k}^n}. \end{cases} \tag{6.3.21}$$

In the above, $\psi_J$ denotes the product of polynomials in $J$. In particular, if $J = \emptyset$, then $\psi_J = 1$. Let $f_{sos,k}^o$, $f_{mom,k}^o$ denote the optimal values of (6.3.20), (6.3.21) respectively. The hierarchy of relaxations (6.3.21)-(6.3.20) is always tight, without assuming anyone of the conditions (i)-(iii) in Theorem 6.3.3.

**Theorem 6.3.5.** *([240]) Suppose $\mathcal{K}_c \neq \emptyset$ and Assumption 6.3.1 holds. Then,*

$$f_{sos,k}^o = f_{mom,k}^o = f_c$$

*for all $k$ sufficiently large. Moreover, if (6.3.7) has only finitely many minimizers, then the moment relaxation (6.3.21) has minimizers and every minimizer $y^*$ must satisfy the flat truncation (6.3.15), when $k$ is big enough.*

*Therefore, if the minimum value $f_{min}$ of (6.1.1) is achievable at a feasible critical point, then*

$$f_{sos,k}^o = f_{mom,k}^o = f_{min}$$

*for all $k$ big enough, and when (6.1.1) has only finitely many minimizers, all of them can be obtained by solving (6.3.21).*

**Proof.** The proof is very similar to that of Theorem 6.3.3 for the case of condition (iii). Follow the same argument there. Without Assumption 6.3.2, we still have $\hat{f}(x) \equiv 0$ on the set

$$\mathcal{K}_3 := \left\{ x \in \mathbb{R}^n \;\middle|\; \begin{array}{l} c_{eq}(x) = 0, \; \phi(x) = 0 \, (\phi \in \Phi), \\ c_{in}(x) \geq 0, \; \psi(x) \geq 0 \, (\psi \in \Psi) \end{array} \right\}.$$

By Theorem 2.6.6, there exists an integer $\ell > 0$ and $q \in \mathrm{Pre}[c_{in}, \Psi]$ such that

$$\hat{f}^{2\ell} + q \in \mathrm{Ideal}[c_{eq}, \Phi].$$

The rest of the proof is the same.

When (6.3.7) has only finitely many minimizers, the conclusion about each minimizer $y^*$ of (6.3.21) satisfying (6.3.15) can be implied by Theorem 5.3.9. We leave the missing details in the proof as an exercise.  □

## 6.3.4 ▪ Some examples

To apply the hierarchy of relaxations (6.3.8)-(6.3.9), we first need polynomials $p_i$ as required in Assumption 6.3.1. If there is a matrix polynomial $L(x)$ satisfying (6.2.3), i.e., $L(x)C(x) = I_m$, then one can choose $p_i$ as in (6.3.2). The minimum value $f_{min}$ of (6.1.1) is achievable at a critical point if the feasible set is compact (or if $f$ is coercive) and the LICQC holds. The advantages of relaxations (6.3.8)-(6.3.9) over the classical Moment-SOS relaxations (5.2.2)-(5.2.3) are shown in the following examples, which are from [240].

**Example 6.3.6.** *Consider the optimization problem*

$$\left\{ \begin{array}{ll} \min & x_1 x_2 (10 - x_3) \\ s.t. & x_1 \geq 0, x_2 \geq 0, x_3 \geq 0, \\ & 1 - x_1 - x_2 - x_3 \geq 0. \end{array} \right.$$

*The matrix polynomial $L(x)$ is given as in (6.2.13). Since the feasible set is compact, the minimum $f_{min} = 0$ is achievable at a critical point. The condition (ii) of Theorem 6.3.3 is satisfied. Each feasible point with $x_1 x_2 = 0$ is a global minimizer, so there are infinitely many minimizers. The computational results are shown in Table 6.1. It confirms that $f^c_{sos,k} = f_{min}$ for all $k \geq 3$, up to some numerical errors.*

Table 6.1: Computational results for Example 6.3.6.

| order $k$ | (5.2.2)-(5.2.3) | | (6.3.8)-(6.3.9) | |
|---|---|---|---|---|
| | $f_{mom,k}$ | time | $f^c_{mom,k}$ | time |
| 2 | $-0.0521$ | 0.6841 | $-0.0521$ | 0.1922 |
| 3 | $-0.0026$ | 0.2657 | $-3 \cdot 10^{-8}$ | 0.2285 |
| 4 | $-0.0007$ | 0.6785 | $-6 \cdot 10^{-9}$ | 0.4431 |
| 5 | $-0.0004$ | 1.6105 | $-2 \cdot 10^{-9}$ | 0.9567 |

**Example 6.3.7.** *Consider the optimization problem*

$$
\begin{cases}
\min & x_1^2 + 50x_2^2 \\
\text{s.t.} & x_1^2 - \frac{1}{2} \geq 0, \\
& x_2^2 - 2x_1x_2 - \frac{1}{8} \geq 0, \\
& x_2^2 + 2x_1x_2 - \frac{1}{8} \geq 0.
\end{cases}
$$

*For $L(x)$ to satisfy (6.2.3), the first column of $L(x)$ is*

$$
\begin{bmatrix}
\frac{8x_1^3}{5} + \frac{x_1}{5} \\
\frac{288\,x_2\,x_1^4}{5} - \frac{16\,x_1^3}{5} - \frac{x_2\,x_1^2\,124}{5} + \frac{8\,x_1}{5} - 2\,x_2 \\
-\frac{288\,x_2\,x_1^4}{5} - \frac{16\,x_1^3}{5} + \frac{x_2\,x_1^2\,124}{5} + \frac{8\,x_1}{5} + 2\,x_2
\end{bmatrix}
$$

*and the second column of $L(x)$ is*

$$
\begin{bmatrix}
-\frac{8\,x_1^2\,x_2}{5} + \frac{4\,x_2^3}{5} - \frac{x_2}{10} \\
\frac{288\,x_1^3\,x_2^2}{5} + \frac{16\,x_1^2\,x_2}{5} - \frac{142\,x_1\,x_2^2}{5} - \frac{9\,x_1}{20} - \frac{8\,x_2^3}{5} + \frac{11\,x_2}{5} \\
-\frac{288\,x_1^3\,x_2^2}{5} + \frac{16\,x_1^2\,x_2}{5} + \frac{142\,x_1\,x_2^2}{5} + \frac{9\,x_1}{20} - \frac{8\,x_2^3}{5} + \frac{11\,x_2}{5}
\end{bmatrix}.
$$

*The objective is coercive, so $f_{min}$ is achievable at a critical point. The minimum value*

$$
f_{min} = 56 + \frac{3}{4} + 25\sqrt{5} \approx 112.6517
$$

*and the minimizers are*

$$
\left( \pm\sqrt{\tfrac{1}{2}}, \pm\left(\sqrt{\tfrac{5}{8}} + \sqrt{\tfrac{1}{2}}\right) \right).
$$

*The computational results are shown in Table 6.2. It confirms that $f_{sos,k}^c = f_{min}$ for all $k \geq 4$, up to some numerical errors.*

Table 6.2: Computational results for Example 6.3.7.

| order $k$ | (5.2.2)-(5.2.3) | | (6.3.8)-(6.3.9) | |
|---|---|---|---|---|
| | $f_{mom,k}$ | time | $f_{mom,k}^c$ | time |
| 3 | 6.7535 | 0.4611 | 56.7500 | 0.1309 |
| 4 | 6.9294 | 0.2428 | 112.6517 | 0.2405 |
| 5 | 8.8519 | 0.3376 | 112.6517 | 0.2167 |
| 6 | 16.5971 | 0.4703 | 112.6517 | 0.3788 |
| 7 | 35.4756 | 0.6536 | 112.6517 | 0.4537 |

**Example 6.3.8.** *Consider the optimization problem*

$$
\begin{cases}
\min & x_1^3 + x_2^3 + x_3^3 + 4x_1x_2x_3 - \left(x_1(x_2^2 + x_3^2)\right. \\
& \left. + x_2(x_3^2 + x_1^2) + x_3(x_1^2 + x_2^2)\right) \\
\text{s.t.} & x_1 \geq 0, x_1x_2 - 1 \geq 0, x_2x_3 - 1 \geq 0.
\end{cases}
$$

*The matrix polynomial $L(x)$ satisfying (6.2.3) is*

$$
\begin{bmatrix}
1 - x_1\,x_2 & 0 & 0 & x_2 & x_2 & 0 \\
x_1 & 0 & 0 & -1 & -1 & 0 \\
-x_1 & x_2 & 0 & 1 & 0 & -1
\end{bmatrix}.
$$

*The objective is a variation of Robinson's form [283]. It is a positive definite form over the nonnegative orthant $\mathbb{R}^3_+$, so the minimum value is achievable at a critical point. In computation, we got $f_{min} \approx 0.9492$ and a global minimizer $(0.9071, 1.1024, 0.9071)$. The computational results are shown in Table 6.3. It confirms that $f^c_{sos,k} = f_{min}$ for all $k \geq 3$, up to some numerical errors.*

Table 6.3: Computational results for Example 6.3.8.

| order $k$ | (5.2.2)-(5.2.3) | | (6.3.8)-(6.3.9) | |
|---|---|---|---|---|
| | $f_{mom,k}$ | time | $f^c_{mom,k}$ | time |
| 2 | $-\infty$ | 0.4129 | $-\infty$ | 0.1900 |
| 3 | $-7.8184 \cdot 10^6$ | 0.4641 | 0.9492 | 0.3139 |
| 4 | $-2.0575 \cdot 10^4$ | 0.6499 | 0.9492 | 0.5057 |

**Example 6.3.9.** *Consider the optimization problem (let $x_0 := 1$)*

$$\begin{cases} \min_{x \in \mathbb{R}^4} & x^T x + \sum_{i=0}^4 \prod_{j \neq i} (x_i - x_j) \\ s.t. & x_1^2 - 1 \geq 0, \ x_2^2 - 1 \geq 0, \\ & x_3^2 - 1 \geq 0, \ x_4^2 - 1 \geq 0. \end{cases}$$

*The matrix polynomial $L(x) = \left[ \frac{1}{2} diag(x) \quad -I_4 \right]$ satisfies (6.2.3). The first part of the objective is $x^T x$, while the second part is a nonnegative polynomial [283]. The objective is coercive, so $f_{min}$ is achievable at a critical point. In computation, we got $f_{min} = 4.0000$ and 11 global*

Table 6.4: Computational results for Example 6.3.9.

| order $k$ | (5.2.2)-(5.2.3) | | (6.3.8)-(6.3.9) | |
|---|---|---|---|---|
| | $f_{mom,k}$ | time | $f^c_{mom,k}$ | time |
| 3 | $-\infty$ | 1.1377 | 3.5480 | 1.1765 |
| 4 | $-6.6913 \cdot 10^4$ | 4.7677 | 4.0000 | 3.0761 |
| 5 | $-21.3778$ | 22.9970 | 4.0000 | 10.3354 |

*minimizers:*

$$\begin{array}{llll} (1,1,1,1), & (1,-1,-1,1), & (1,-1,1,-1), & (1,1,-1,-1), \\ (1,-1,-1,-1), & (-1,-1,1,1), & (-1,1,-1,1), & (-1,1,1,-1), \\ (-1,-1,-1,1), & (-1,-1,1,-1), & (-1,1,-1,-1). \end{array}$$

*The computational results are shown in Table 6.4. It confirms that $f^c_{sos,k} = f_{min}$ for all $k \geq 4$, up to some numerical errors.*

## 6.3.5 ▪ Exercises

**Exercise 6.3.1.** *For the following optimization problems, apply the hierarchy of relaxations (6.3.8)-(6.3.9) to solve them. Choose Lagrange multipliers as in (6.3.2). Compute lower bounds $f^c_{mom,k}$ from the lowest $k$ to bigger ones, until $f^c_{mom,k} = f_c$ is met. Decide whether or not $f_c = f_{min}$.*

(i) $\begin{cases} \min\limits_{x\in\mathbb{R}^3} & x_1^4x_2^2 + x_1^2x_2^4 + x_3^6 - 3x_1^2x_2^2x_3^2 + (x_1^4 + x_2^4 + x_3^4) \\ s.t. & x_1^2 + x_2^2 + x_3^2 \geq 1. \end{cases}$

*Verify that $L(x)C(x) = I$ is satisfied for*

$$L(x) = \begin{bmatrix} \frac{1}{2}x_1 & \frac{1}{2}x_2 & \frac{1}{2}x_3 & -1 \end{bmatrix}.$$

(ii) $\begin{cases} \min\limits_{x\in\mathbb{R}^4} & x_1x_2 + x_2x_3 + x_3x_4 - 3x_1x_2x_3x_4 + (x_1^3 + \cdots + x_4^3) \\ s.t. & x_1, x_2, x_3, x_4 \geq 0, 1 - x_1 - x_2 \geq 0, 1 - x_3 - x_4 \geq 0. \end{cases}$

*Verify that $L(x)C(x) = I$ is satisfied for*

$$L(x) = \begin{bmatrix} 1-x_1 & -x_2 & 0 & 0 & 1 & 1 & 0 & 0 & 1 & 0 \\ -x_1 & 1-x_2 & 0 & 0 & 1 & 1 & 0 & 0 & 1 & 0 \\ 0 & 0 & 1-x_3 & -x_4 & 0 & 0 & 1 & 1 & 0 & 1 \\ 0 & 0 & -x_3 & 1-x_4 & 0 & 0 & 1 & 1 & 0 & 1 \\ -x_1 & -x_2 & 0 & 0 & 1 & 1 & 0 & 0 & 1 & 0 \\ 0 & 0 & -x_3 & -x_4 & 0 & 0 & 1 & 1 & 0 & 1 \end{bmatrix}.$$

(iii) $\begin{cases} \min\limits_{x\in\mathbb{R}^3} & x_1^4x_2^2 + x_2^4x_3^2 + x_3^4x_1^2 - 3x_1^2x_2^2x_3^2 + x_2^2 \\ s.t. & x_1 - x_2x_3 \geq 0, -x_2 + x_3^2 \geq 0. \end{cases}$

*Verify that $L(x)C(x) = I$ is satisfied for*

$$L(x) = \begin{bmatrix} 1 & 0 & 0 & 0 & 0 \\ -x_3 & -1 & 0 & 0 & 0 \end{bmatrix}.$$

(iv) $\begin{cases} \min\limits_{x\in\mathbb{R}^4} & x_1^2(x_1 - x_4)^2 + x_2^2(x_2 - x_4)^2 + x_3^2(x_3 - x_4)^2 \\ & \quad + 2x_1x_2x_3(x_1 + x_2 + x_3 - 2x_4) + \sum_{i=1}^3 (x_i - 1)^2 \\ s.t. & x_1 - x_2 \geq 0, \ x_2 - x_3 \geq 0. \end{cases}$

*Verify that $L(x)C(x) = I$ is satisfied for*

$$L(x) = \begin{bmatrix} 1 & 0 & 0 & 0 & 0 & 0 \\ 1 & 1 & 0 & 0 & 0 & 0 \end{bmatrix}.$$

(v) $\begin{cases} \min\limits_{x\in\mathbb{R}^4} & (x_1 + x_2 + x_3 + x_4 + 1)^2 \\ & \quad -4(x_1x_2 + x_2x_3 + x_3x_4 + x_4 + x_1) \\ s.t. & 0 \leq x_1, \ldots, x_4 \leq 1. \end{cases}$

*Use $L(x)$ as in (6.2.14).*

**Exercise 6.3.2.** *Suppose the minimum value $f_{min}$ of (6.1.1) is achievable at a critical point. Let*

$$d_c := \max\left\{ \lceil \tfrac{1}{2}\deg(c_i) \rceil : i \in \mathcal{E} \cup \mathcal{I} \right\}.$$

*Suppose $y^*$ is a minimizer of (6.3.9) and there exists an integer $t \in [d_c, k]$ such that*

$$\text{rank } M_t[y^*] = \text{rank } M_{t-d_c}[y^*]. \tag{6.3.22}$$

*Show that $f_{mom,k}^c = f_{min}$. Furthermore, if $\mu$ is a finitely atomic representing measure for $y^*|_{2t}$, show that every point in $\text{supp}(\mu)$ is a minimizer for (6.1.1).*

**Exercise 6.3.3.** *Give the proof for item (iii) of Theorem 6.3.4 and prove (6.3.19) for all $k$ big enough. If $y$ is feasible for (6.3.9), show that $y$ is also feasible for (6.3.18).*

**Exercise 6.3.4.** *Fill the missing details in the proof of Theorem 6.3.5.*

## 6.4 ▪ Revisiting Unconstrained Optimization

A special case of (6.1.1) is the unconstrained optimization

$$\min_{x \in \mathbb{R}^n} \quad f(x). \tag{6.4.1}$$

If $u$ is a minimizer, then $\nabla f(u) = 0$. This leads us to consider the constrained optimization

$$\begin{cases} \min & f(x) \\ s.t. & \nabla f(x) = 0. \end{cases} \tag{6.4.2}$$

The $k$th order SOS relaxation for solving (6.4.2) is

$$\begin{cases} \max & \gamma \\ s.t. & f - \gamma \in \text{Ideal}[\nabla f]_{2k} + \Sigma[x]_{2k}. \end{cases} \tag{6.4.3}$$

In the above, $\text{Ideal}[\nabla f]$ denotes the ideal generated by the partial derivatives $f_{x_1}, \ldots, f_{x_n}$, where $f_{x_i}$ denotes the partial derivative of $f$ with respect to $x_i$. For each $k$, the relaxation (6.4.3) is stronger than the classical SOS relaxation (4.1.2). Let $f^\nabla_{sos,k}$ denote the optimal value of (6.4.3). Clearly, it is a lower bound for the smallest real critical value $f_c$ of $f$. It holds that

$$\cdots \le f^\nabla_{sos,k} \le f^\nabla_{sos,k+1} \le \cdots \le f_c.$$

Similar to the case of constrained optimization, the dual optimization of (6.4.3) is the $k$th order moment relaxation

$$\begin{cases} \min & \langle f, y \rangle \\ s.t. & \mathscr{V}^{(2k)}_{f_{x_i}}[y] = 0 \ (i = 1, \ldots, n), \\ & M_k[y] \succeq 0, \\ & y_0 = 1, \ y \in \mathbb{R}^{\mathbb{N}^n_{2k}}. \end{cases} \tag{6.4.4}$$

For each $k$, let $f^\nabla_{mom,k}$ denote the optimal value of (6.4.4). By the weak duality, it also holds that

$$f^\nabla_{sos,k} \le f^\nabla_{mom,k} \le f_c.$$

The sequence of $f^\nabla_{mom,k}$ is also monotonically increasing.

One can extract one or several minimizers for (6.4.2) as follows. Suppose $y^*$ is a minimizer of (6.4.4). Let $d := \lceil \frac{1}{2} \deg(f) \rceil$. If there exists an integer $t \in [d, k]$ such that

$$\text{rank } M_{t-d}[y^*] = \text{rank } M_t[y^*], \tag{6.4.5}$$

then $f^\nabla_{mom,k} = f_c$ and we can get rank $M_t[y^*]$ minimizers for (6.4.2). The following are basic properties of (6.4.3)-(6.4.4).

**Proposition 6.4.1.** *For the relaxations (6.4.3)-(6.4.4), we have the following:*

(i) *If the relaxation (6.4.4) is infeasible for some $k$, then $f$ has no real critical points, i.e., (6.4.2) is infeasible.*

(ii) *If $y^*$ is a minimizer of (6.4.4) satisfying (6.4.5), then $f^\nabla_{mom,k} = f_c$ and the truncation $y^*|_{2t}$ has $r$-atomic representing measure $\mu$, where $r = $ rank $M_t[y^*]$. Furthermore, each point $u \in \text{supp}(\mu)$ is a minimizer of (6.4.2).*

(iii) *In the item (ii), if in addition rank $M_t[y^*]$ is maximum among all minimizers of (6.4.4), then $\text{supp}(\mu)$ is the set of all minimizers of (6.4.2).*

Proposition 6.4.1 can be proved similarly as for cases of constrained optimization. So its proof is left as an exercise. The hierarchy of relaxations (6.4.3)-(6.4.4) is always tight. We have the following conclusion.

**Theorem 6.4.2.** *([210, 224, 240]) For the relaxations of (6.4.3)-(6.4.4), we have the following:*

(i) *For all $k$ big enough, $f^{\nabla}_{sos,k} = f^{\nabla}_{mom,k} = f_c$.*

(ii) *If (6.4.2) has only finitely many minimizers, then every minimizer $y^*$ of (6.4.4) must satisfy (6.4.5), when $k$ is big enough.*

**Proof.** (i) Note that (6.4.3)-(6.4.4) are special cases of (6.3.8)-(6.3.9), since (6.4.1) has no constraints. The conclusion is implied by Theorem 6.3.3 (note that Assumption 6.3.2 is always satisfied for polynomial optimization without inequality constraints), or by Theorem 6.3.5.

(ii) By item (i), the hierarchy of relaxations (6.4.3)-(6.4.4) is always tight. Since (6.4.2) has no inequality constraints, the relaxations (6.4.3)-(6.4.4) are the same as the Schmüdgen type Moment-SOS relaxations for (6.4.2). Hence, the conclusion follows from Theorem 5.3.9.  □

As pointed out in [210], it is possible that the minimum value $f_{min}$ of (6.4.1) is not achievable at any critical point. For instance, for $f = x_1^2 + (x_1 x_2 - 1)^2$, we have $f_{min} = 0$ but $f_c = 1$. However, if $f_{min}$ is achievable at a point, then we have the following conclusion.

**Corollary 6.4.3.** *([210]) Suppose the minimum value $f_{min}$ of (6.4.1) is achievable at a critical point. Then, we have the following:*

(i) *In Proposition 6.4.1(ii), each $u_i$ is a minimizer of (6.4.1); in Proposition 6.4.1(iii), the set of all minimizers for (6.4.1) consists of the points $u_1, \dots, u_r$.*

(ii) *For all $k$ big enough, $f^{\nabla}_{sos,k} = f^{\nabla}_{mom,k} = f_{min}$.*

(iii) *If (6.4.1) has only finitely many minimizers, then every minimizer $y^*$ of (6.4.4) must satisfy (6.4.5) for some $t$, when $k$ is big enough.*

The following is an example of applying relaxations (6.4.3)-(6.4.4).

**Example 6.4.4.** *For the polynomial $f = x_1^2 x_2^2(x_1^2 + x_2^2 - 1)$, the gradient $\nabla f(x) = 0$ gives the equality constraints*

$$4x_1^3 x_2^2 + 2x_1 x_2^4 - 2x_1 x_2^2 = 0, \quad 4x_1^2 x_2^3 + 2x_1^4 x_2 - 2x_1^2 x_2 = 0.$$

*The moment relaxation (6.4.4) has equality constraints $\mathscr{V}^{(2k)}_{f_{x_i}}[y] = 0$, $i = 1, 2$. For instance, when $k = 3$, these constraints are*

$$4y_{32} + 2y_{14} - 2y_{12} = 4y_{42} + 2y_{24} - 2y_{22} = 4y_{33} + 2y_{15} - 2y_{13} = 0,$$

$$4y_{23} + 2y_{41} - 2y_{21} = 4y_{33} + 2y_{51} - 2y_{31} = 4y_{24} + 2y_{42} - 2y_{22} = 0.$$

*The minimum value $f_{min} = \frac{-1}{27} = -0.0370370\cdots$. The GloptiPoly syntax for solving the moment relaxation (6.4.4) is*

```
mpol('x',2);
f = x(1)^2*x(2)^2*(x(1)^2+x(2)^2 - 1);
```

```
S = [diff(f, x(1)) == 0, diff(f,x(2)) == 0];
k = 4; % give the relaxation order k
POP = msdp(min(f), S, k);
[sta, f_k] = msol(POP);
```

*For $k = 3$, we get the lower bound $f_3 \approx -16.7349$. For $k = 4$, we get $f_4 \approx -0.0370$. For $k = 10$, the rank condition (6.4.5) is satisfied for $t = 5$, with*

$$\text{rank } M_2[y^*] = \text{rank } M_5[y^*] = 4.$$

*The syntax* `double(x)` *returns four minimizers* $(\pm\sqrt{\frac{1}{3}}, \pm\sqrt{\frac{1}{3}})$.

### 6.4.1 ▪ Exercises

**Exercise 6.4.1.** *For the following polynomials, apply the hierarchy of relaxations (6.4.3)-(6.4.4) to compute $f_c$. Compute lower bounds $f_{mom,k}^\nabla$ from the lowest $k$ to bigger ones, until $f_{mom,k}^\nabla = f_c$. Also determine whether or not $f_c = f_{min}$.*

(i) $x_1^2 x_2^2 + x_2^2 x_3^2 + x_3^2 x_1^2 - 4x_1 x_2 x_3$.

(ii) $x_1^4 x_2^2 + x_1^2 x_2^4 - x_1^2 x_2^2 + 0.01(x_1^2 + x_2^2 - 1)^2$.

(iii) $x_1^4 x_2^2 + 1.01 x_2^4 + x_1^4 - 3x_1^2 x_2^2$.

(iv) $x_1^2(x_1 - 1)^2 + x_2^2(x_2 - 1)^2 + x_3^2(x_3 - 1)^2 + 2x_1 x_2 x_3(x_1 + x_2 + x_3 - 2)$.

(v) $(x_1^2 + x_2^2 + x_3^2 + x_4^2 + 1)^2 - 4(x_1^2 x_2^2 + x_2^2 x_3^2 + x_3^2 x_4^2 + x_4^2 + x_1^2) + 0.01 x_1^2 x_4^2$.

(vi) $\sum_{i=1}^{5} \left( \prod_{j \neq i} (x_i - x_j) + 0.01(x_i^2 - i)^2 \right)$.

**Exercise 6.4.2.** *Prove Proposition 6.4.1.*

## 6.5 ▪ Jacobian Type Relaxations

In §6.3, we have introduced a tight hierarchy of Moment-SOS relaxations for solving (6.1.1). They are based on Lagrange multiplier expressions. There is another way for constructing tight relaxations, without using Lagrange multiplier expressions. It instead uses minors of Jacobians for polynomials in (6.1.1) and the Fritz-John conditions, as introduced in [224].

Consider the constrained optimization

$$\begin{cases} \min & f(x) \\ s.t. & c_i(x) = 0 \, (i \in \mathcal{E}), \\ & c_j(x) > 0 \, (j \in \mathcal{I}), \end{cases} \tag{6.5.1}$$

with only *strict*[5] inequality constraints. When $\mathcal{E} = \emptyset$ (resp., $\mathcal{I} = \emptyset$), there are no equality (resp., strict inequality) constraints. For convenience, we still let $K$ denote the feasible set of (6.5.1) and let $f_{min}$ denote the minimum value. Recall the notation $c_{eq}, c_{in}$ as in (5.2.1). Suppose $u$

---

[5]We do not discuss weakly inequality constraints in this section, because the resulting polynomials induced by minors of Jacobians are more complicated. We refer to [224] for the case of weakly inequality constraints.

is a minimizer of (6.5.1). For strict inequality constraints, their Lagrange multipliers are always zeros. For convenience of notation, suppose

$$\mathcal{E} = \{1, \ldots, m_1\}.$$

Then, $c_{eq} = (c_1, \ldots, c_{m_1})$. By the Fritz-John conditions (see §6.1), there exist scalars $\lambda_0, \lambda_1, \ldots, \lambda_{m_1}$ such that

$$\lambda_0 \nabla f(u) = \lambda_1 \nabla c_1(u) + \cdots + \lambda_{m_1} \nabla c_{m_1}(u) \tag{6.5.2}$$

for $(\lambda_0, \lambda_1, \ldots, \lambda_{m_1}) \neq 0$ and $\lambda_0 \in \{0, 1\}$. This implies that the Jacobian matrix of the vector-valued function $(f, c_1, \ldots, c_{m_1})$ is singular at $u$. Denote the determinantal variety

$$\mathcal{H} := \{x \in \mathbb{R}^n : \operatorname{rank} B(x) \leq m_1\}, \tag{6.5.3}$$

where $B(x)$ is the matrix polynomial

$$B(x) := \begin{bmatrix} \nabla f(x) & \nabla c_1(x) & \cdots & \nabla c_{m_1}(x) \end{bmatrix}. \tag{6.5.4}$$

Therefore, we have $u \in V_\mathbb{R}(c_{eq}) \cap \mathcal{H}$. One can use $\mathcal{H}$ to construct Moment-SOS relaxations. To do this, we need to express the determinantal variety $\mathcal{H}$ in a computationally efficient way. Obviously, a point $x$ belongs to $\mathcal{H}$ if and only if all maximal minors of the matrix $B(x)$ are zeros. The number of maximal minors is typically large. People often prefer the smallest number of defining equations for $\mathcal{H}$.

Throughout this section, we assume $m_1 \leq n$. For the case $m_1 > n$, the real variety $V_\mathbb{R}(c_{eq})$ is generally empty, which makes (6.5.1) infeasible.

### 6.5.1 ▪ Expressions for determinantal varieties

For a positive integer $k \leq n$, let $X = (X_{ij})$ be a $n \times k$ matrix variable with entries $X_{ij}$. For given $t \in [k]$, consider the determinantal variety

$$D_{t-1}^{n,k} := \{X \in \mathbb{C}^{n \times k} : \operatorname{rank} X < t\}. \tag{6.5.5}$$

For a set $I = \{i_1, \ldots, i_k\} \subseteq [n]$, denote by $\det_I(X)$ the $(i_1, \ldots, i_k) \times (1, \ldots, k)$-minor of the matrix $X$, i.e., $\det_I(X)$ is the determinant of the $k$-by-$k$ submatrix of $X$ whose rows are labeled by $i_1, \ldots, i_k$. Obviously,

$$D_{k-1}^{n,k} = \{X \in \mathbb{C}^{n \times k} : \det_I(X) = 0 \ \forall I \subseteq [n], |I| = k\}.$$

There are $\binom{n}{k}$ equations of degree $k$ in the above. Interestingly, the number of defining equations can be much smaller. A set of $nk - t^2 + 1$ equations are enough for defining $D_{t-1}^{n,k}$, as shown in [38, 39]. A general method for getting $nk - t^2 + 1$ defining equations for $D_{t-1}^{n,k}$ is given in [38, Chap. 5]. We introduce this method in the following.

Let $\Gamma(X)$ be the set of all $k$-by-$k$ minors of $X$, whose row labels are *strictly increasing*. For $i_1 < \cdots < i_k$, let

$$[i_1, \ldots, i_k]$$

denote the $(i_1, \ldots, i_k) \times (1, \ldots, k)$-minor of $X$. We define the following partial ordering on $\Gamma(X)$:

$$[i_1, \ldots, i_k] < [j_1, \ldots, j_k] \quad \Leftrightarrow \quad \begin{cases} i_1 \leq j_1, \ldots, i_k \leq j_k, \\ i_1 + \cdots + i_k < j_1 + \cdots + j_k. \end{cases}$$

For each $I = [i_1, \ldots, i_k] \in \Gamma(X)$, define its rank as

$$rk(I) := \max \left\{ \ell : I = I^{(\ell)} > \cdots > I^{(1)}, \quad \text{each } I^{(i)} \in \Gamma(X) \right\}.$$

The maximum minor in $\Gamma(X)$ is $[n-k+1,\ldots,n]$, whose rank is $nk-k^2+1$. For each integer $\ell \in [1, nk-k^2+1]$, define the polynomial

$$\eta_\ell(X) := \sum_{I \in \Gamma(X),\, rk(I)=\ell} \det{}_I(X). \tag{6.5.6}$$

**Lemma 6.5.1.** *([38, Lemma 5.9]) The determinantal variety $D_{k-1}^{n,k}$ can be equivalently given as follows:*

$$D_{k-1}^{n,k} = \left\{ X \in \mathbb{C}^{n \times k} : \eta_\ell(X) = 0,\, \ell = 1,\ldots, nk-k^2+1 \right\}.$$

When $k=2$, $D_1^{n,2}$ can be defined by $2n-3$ polynomials. The minor $[n-1,n]$ has the maximum rank $2n-3$. For each $\ell = 1, 2, \ldots, 2n-3$,

$$\eta_\ell(X) = \sum_{1 \le i_1 < i_2 \le n:\, i_1+i_2=\ell+2} [i_1, i_2].$$

Every 2-by-2 minor of $X$ is a summand of some $\eta_\ell(X)$.

When $k=3$, $D_2^{n,3}$ can be defined by $3n-8$ equations using $\eta_\ell(X)$. For instance, when $n=6$, the smallest minor is $[1,2,3]$ and the biggest one is $[4,5,6]$. Other minors in $\Gamma(X)$ are ordered in the following diagram:

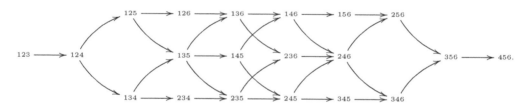

In the above, each arrow points to a bigger minor. Clearly,

$$\eta_1(X) = [1,2,3], \quad \eta_2(X) = [1,2,4], \quad \eta_3(X) = [1,2,5] + [1,3,4],$$

$$\eta_4(X) = [1,2,6]+[1,3,5]+[2,3,4], \quad \eta_5(X) = [1,3,6]+[1,4,5]+[2,3,5],$$

$$\eta_6(X) = [1,4,6]+[2,3,6]+[2,4,5], \quad \eta_7(X) = [1,5,6]+[2,4,6]+[3,4,5],$$

$$\eta_8(X) = [2,5,6]+[3,4,6], \quad \eta_9(X) = [3,5,6], \quad \eta_{10}(X) = [4,5,6].$$

Each above $\eta_\ell(X)$ has degree 3 in the entries of $X$. The summands $[i_1, i_2, i_3]$ in each $\eta_\ell(X)$ have the same sum $i_1 + i_2 + i_3$. For each $\ell = 1, \ldots, 3n-8$,

$$\eta_\ell(X) = \sum_{\substack{1 \le i_1 < i_2 < i_3 \le n, \\ i_1+i_2+i_3=\ell+5}} [i_1, i_2, i_3].$$

When $k > 3$, the determinantal variety $D_{k-1}^{n,k}$ can be defined by $nk-k^2+1$ equations like $\eta_\ell(X) = 0$, where for each $\ell = 1, 2, \ldots, nk-k^2+1$

$$\eta_\ell(X) = \sum_{\substack{1 \le i_1 < \cdots < i_k \le n, \\ i_1+\cdots+i_k=\ell+\binom{k+1}{2}-1}} [i_1, \ldots, i_k].$$

### 6.5.2 ▪ A tight hierarchy of Putinar-Jacobian type

Let $\mathcal{H}$ be the determinantal variety as in (6.5.3). Let

$$N := n(m_1 + 1) - (m_1 + 1)^2 + 1. \tag{6.5.7}$$

For each $\ell = 1, \ldots, N$, define the polynomials,

$$\varphi_\ell(x) := \eta_\ell\big(B(x)\big), \tag{6.5.8}$$

where $B(x)$ is as in (6.5.4) and $\eta_\ell$ is as in (6.5.6). Denote the tuple

$$\varphi := (\varphi_1, \varphi_2, \ldots, \varphi_N). \tag{6.5.9}$$

Every minimizer of (6.5.1) is a Fritz-John point (see §6.1), so it must be a common zero of $\varphi$. This motivates the following optimization problem:

$$\begin{cases} \min & f(x) \\ s.t. & c_i(x) = 0 \, (i \in \mathcal{E}), \\ & \varphi_\ell(x) = 0 \, (\ell \in [N]), \\ & c_j(x) \geq 0 \, (j \in \mathcal{I}). \end{cases} \tag{6.5.10}$$

It is worthy to note that (6.5.10) has weak inequality constraints while (6.5.1) has strict ones. Let $f_{FJ}$ denote the minimum value of (6.5.10) and let $K^{FJ}$ denote its feasible set. If the minimum value of (6.5.1) is achievable and at least one minimizer of (6.5.10) is feasible for (6.5.1), then $f_{min} = f_{FJ}$. If $f_{min}$ is not achievable, then $f_{min} < f_{FJ}$ is possible. Denote the set of Fritz-John points

$$\mathcal{F} := \left\{ x \in \mathbb{R}^n \,\middle|\, \begin{array}{l} c_i(x) = 0 \, (i \in \mathcal{E}), \\ \varphi_\ell(x) = 0 \, (\ell \in [N]) \end{array} \right\}. \tag{6.5.11}$$

The singular locus of $V_{\mathbb{R}}(c_{eq})$ is the set

$$\mathcal{F}_{\text{sig}} := \left\{ x \in \mathbb{R}^n \,\middle|\, \begin{array}{l} c_i(x) = 0 \, (i \in \mathcal{E}), \\ \text{rank} \begin{bmatrix} \nabla c_1(x) & \cdots & \nabla c_{m_1}(x) \end{bmatrix} < m_1 \end{array} \right\}. \tag{6.5.12}$$

It consists of points in $V_{\mathbb{R}}(c_{eq})$ at which the LICQC fails. If $c_{eq}$ is nonsingular, then $\mathcal{F}_{\text{sig}} = \emptyset$ and every Fritz-John point is a critical point (see Exercise 6.5.1). For this case, the objective $f$ achieves only finitely many values on $\mathcal{F}$, implied by Theorem 6.1.1.

For an integer $k > 0$, the $k$th order SOS relaxation for (6.5.10) is

$$\begin{cases} \max & \gamma \\ s.t. & f - \gamma \in \text{Ideal}[c_{eq}, \varphi]_{2k} + \text{QM}[c_{in}]_{2k}. \end{cases} \tag{6.5.13}$$

Its dual optimization is the $k$th order moment relaxation

$$\begin{cases} \min & \langle f, y \rangle \\ s.t. & \mathscr{V}_{c_i}^{(2k)}[y] = 0 \, (i \in \mathcal{E}), \\ & \mathscr{V}_{\varphi_\ell}^{(2k)}[y] = 0 \, (\ell \in [N]), \\ & L_{c_j}^{(k)}[y] \succeq 0 \, (j \in \mathcal{I}), \\ & M_k[y] \succeq 0, \\ & y_0 = 1, y \in \mathbb{R}^{\mathbb{N}_{2k}^n}. \end{cases} \tag{6.5.14}$$

Let $f_{sos,k}^{PJ}$, $f_{mom,k}^{PJ}$ denote the optimal values of (6.5.13), (6.5.14) respectively. By the weak duality,

$$f_{sos,k}^{PJ} \leq f_{mom,k}^{PJ} \leq f_{FJ}.$$

Like in earlier sections, we can use flat truncation to check tightness of (6.5.13)-(6.5.14) and extract minimizers. If (6.5.14) is infeasible for some $k$, then (6.5.10) must be infeasible, i.e., (6.5.1) does not have any Fritz-John points and hence it has no minimizers. When it is feasible, suppose $y^*$ is a minimizer of (6.5.14). Denote the degree

$$d_J := \max_{i \in \mathcal{E}, \ell \in [N], j \in \mathcal{I}} \left\{ \lceil \deg(c_i)/2 \rceil, \lceil \deg(\varphi_\ell)/2 \rceil, \lceil \deg(c_j)/2 \rceil \right\}.$$

If there exists $t \in [d_J, k]$ such that

$$\operatorname{rank} M_t[y^*] = \operatorname{rank} M_{t-d_J}[y^*], \tag{6.5.15}$$

then $f_{mom,k}^{PJ} = f_{FJ}$ and we can extract one or several minimizers. Like before, we have the following proposition.

**Proposition 6.5.2.** *The relaxations (6.5.13)-(6.5.14) have the following properties:*

(i) *If the relaxation (6.5.14) is infeasible for some $k$, then the optimization (6.5.10) is also infeasible and (6.5.1) does not have any feasible Fritz-John points.*

(ii) *If $y^*$ is a minimizer of (6.5.14) and satisfies (6.5.15), then $f_{mom,k}^{PJ} = f_{FJ}$ and the truncation $y^*|_{2t}$ admits a $r$-atomic representing measure $\mu$, where $r = \operatorname{rank} M_t[y^*]$. Furthermore, each point $u \in \operatorname{supp}(\mu)$ is a minimizer of (6.5.10).*

Proposition 6.5.2 can be shown similarly as before, so it is left as an exercise. The hierarchy of relaxations (6.5.13)-(6.5.14) is tight, under a generic assumption.

**Assumption 6.5.3.** *There exists $\rho \in \operatorname{QM}[c_{in}]$ such that if $u \in \mathcal{F}$ and $f(u) < f_{FJ}$, then $\rho(u) < 0$.*

For every $\rho \in \operatorname{QM}[c_{in}]$, we must have $\rho(u) \geq 0$ for all $u \in K^{FJ}$, the feasible set of (6.5.10). The polynomial $\rho$ in Assumption 6.5.3 separates feasible and infeasible Fritz-John points for (6.5.10). It is similar to Assumption 6.3.2 and holds generically. If (6.5.10) has a single inequality constraint (or none), or if $\mathcal{F} \backslash K^{FJ}$ is a finite set, Assumption 6.5.3 must hold (see exercise). The remarks following Assumption 6.3.2 also apply here. The tightness of the hierarchy of relaxations (6.5.13)-(6.5.14) can be shown as follows.

**Theorem 6.5.4.** *Let $f_{sos,k}^{PJ}$, $f_{mom,k}^{PJ}$, $f_{FJ}$, $\mathcal{F}_{sig}$ be as above. Then, we have the following:*

(i) *If $\mathcal{F}$ is a finite set, then $f_{sos,k}^{PJ} = f_{FJ}$ and every optimizer $y^*$ of (6.5.14) must satisfy (6.5.15), when $k$ is big enough.*

(ii) *If $c_{eq}$ is nonsingular and $\operatorname{Ideal}[c_{eq}, \varphi] + \operatorname{QM}[c_{in}]$ is archimedean, then $f_{sos,k}^{PJ} = f_{FJ}$ when $k$ is big enough.*

(iii) *Suppose Assumption 6.5.3 holds. If $f$ achieves only finitely many values on $\mathcal{F}_{sig}$, then $f_{sos,k}^{PJ} = f_{FJ}$ for all $k$ big enough. Furthermore, if in addition (6.5.10) has only finitely many minimizers, then every optimizer $y^*$ of (6.5.14) must satisfy (6.5.15) when $k$ is big enough.*

**Proof.** The item (i) is implied by Theorem 5.6.1, and the item (ii) is implied by Theorem 4.4 of [224]. We prove the item (iii) in the following.

The objective $f$ achieves only finitely many values on $\mathcal{F}\backslash\mathcal{F}_{\text{sig}}$ (see Exercise 6.5.3). By the assumption that $f$ achieves only finitely many values on $\mathcal{F}_{\text{sig}}$, we know $f$ also achieves only finitely many values on $\mathcal{F}$, say, they are ordered as

$$v_1 < v_2 < \cdots < v_D.$$

Let $p_1, \ldots, p_D$ be univariate polynomials such that

$$p_i(v_i) = 1, \quad p_i(v_j) = 0 \; (i \neq j).$$

The value $f_{FJ}$ is among $v_1, \ldots, v_D$, say, $v_s = f_{FJ}$. Let

$$h := (v_{s+1} - f_{FJ})\big(p_{s+1}(f)\big)^2 + \cdots + (v_D - f_{FJ})\big(p_D(f)\big)^2 \in \Sigma[x].$$

Let $\rho$ be the polynomial in Assumption 6.5.3. If $u \in \mathcal{F}$ and $f(u) < f_{FJ}$, then $\rho(u) < 0$. So the polynomial $\hat{f} := f - f_{FJ} - h$ vanishes identically on the semialgebraic set

$$\mathcal{F}_2 := \left\{ x \in \mathbb{R}^n \left| \begin{array}{rcl} c_{eq}(x) & = & 0, \\ \varphi(x) & = & 0, \\ \rho(x) & \geq & 0 \end{array} \right. \right\}.$$

There is a single inequality for $\mathcal{F}_2$ and $\mathrm{QM}[\rho] = \mathrm{Pre}[\rho]$. By Positivstellensatz (see Theorem 2.6.6), there exist a power $\ell$ and a polynomial $q$ such that

$$\hat{f}^{2\ell} + q \in \mathrm{Ideal}[c_{eq}, \varphi], \quad q = b_0 + \rho b_1 \in \mathrm{Pre}[\rho], \quad b_0, b_1 \in \Sigma[x].$$

Note that $\rho \in \mathrm{QM}[c_{in}]$ and $q \in \mathrm{QM}[c_{in}]$.

For all $\epsilon > 0$ and $\omega > 0$, we have $\hat{f} + \epsilon = \phi_\epsilon + \theta_\epsilon$, where

$$\begin{aligned} \phi_\epsilon &:= -\omega\epsilon^{1-2\ell}\big(\hat{f}^{2\ell} + q\big), \\ \theta_\epsilon &:= \epsilon\big(1 + \hat{f}/\epsilon + \omega(\hat{f}/\epsilon)^{2\ell}\big) + \omega\epsilon^{1-2\ell}q. \end{aligned}$$

By Lemma 5.6.2, there exists a power $k_1$ such that

$$\phi_\epsilon \in \mathrm{Ideal}[c_{eq}, \varphi]_{2k_1}, \; \theta_\epsilon \in \mathrm{QM}[c_{in}]_{2k_1}$$

for all $\epsilon > 0$ and for all $\omega \geq \frac{1}{2\ell}$. Hence, we get

$$f - (f_{FJ} - \epsilon) = \phi_\epsilon + \sigma_\epsilon,$$

where $\sigma_\epsilon = \theta_\epsilon + h \in \mathrm{QM}[c_{in}]_{2k_1}$ for all $\epsilon > 0$, when $k_1$ is big enough. This means that $\gamma = f_{FJ} - \epsilon$ is feasible for (6.5.13), so $f_{sos,k_1}^{PJ} \geq f_{FJ}$. On the other hand, $f_{sos,k}^{PJ} \leq f_{FJ}$ for all $k$, so $f_{sos,k_1}^{PJ} = f_{FJ}$. Due to the monotonicity of $f_{sos,k}^{PJ}$, we get $f_{sos,k}^{PJ} = f_{FJ}$ for all $k \geq k_1$.

The proof of satisfaction for the flat truncation (6.5.15) is very similar to that for Theorem 6.3.4. So it is left as an exercise. $\quad\square$

We remark that the assumptions in items (i)-(iii) of Theorem 6.5.4 are all generic. For instance, if $f, c_i, c_j$ are generic polynomials, then $\mathcal{F}$ is a finite set. If $c_{eq}$ is nonsingular, then $\mathcal{F}_{\text{sig}} = \emptyset$; if the singular locus $\mathcal{F}_{\text{sig}}$ is a finite set, then $f$ achieves only finitely many values on $\mathcal{F}_{\text{sig}}$.

Each feasible point of (6.5.10) is a Fritz-John point of (6.5.1). If the minimum value of (6.5.1) is achievable at a minimizer, then it must be achievable at a Fritz-John point and hence the hierarchy of (6.5.13)-(6.5.14) is tight for solving (6.5.1). A minimizer of (6.5.10) is not necessarily a feasible point for (6.5.1), due to the strict inequality constraints. Therefore, we get the following corollary.

**Corollary 6.5.5.** *Under the assumptions of Theorem 6.5.4, if the minimum $f_{min}$ of (6.5.1) is achievable at a feasible point and at least one minimizer of (6.5.10) is feasible for (6.5.1), then $f^{PJ}_{sos,k} = f_{min}$ for all $k$ big enough.*

## 6.5.3 ▪ A tight hierarchy of Schmüdgen-Jacobian type

If the preordering is used, the relaxation (6.5.13) can be strengthened to

$$\begin{cases} \max & \gamma \\ \text{s.t.} & f - \gamma \in \text{Ideal}[c_{eq}, \varphi]_{2k} + \text{Pre}[c_{in}]_{2k}. \end{cases} \tag{6.5.16}$$

Similarly, its dual optimization is the moment relaxation

$$\begin{cases} \min & \langle f, y \rangle \\ \text{s.t.} & \mathscr{V}^{(2k)}_{c_i}[y] = 0 \, (i \in \mathcal{E}), \\ & \mathscr{V}^{(2k)}_{\varphi_\ell}[y] = 0 \, (\ell \in [N]), \\ & L^{(k)}_{c_J}[y] \succeq 0 \, (J \subseteq \mathcal{I}), \\ & y_0 = 1, \, y \in \mathbb{R}^{\mathbb{N}^n_{2k}}. \end{cases} \tag{6.5.17}$$

In the above, the polynomial $c_J$ is the same as in (5.2.6). Let $f^{SJ}_{sos,k}$, $f^{SJ}_{mom,k}$ denote the optimal values of (6.5.16), (6.5.17) respectively. Similarly, $f^{SJ}_{sos,k} \leq f^{SJ}_{mom,k} \leq f_{FJ}$ for all $k$. Since (6.5.16)-(6.5.17) are stronger relaxations than (6.5.13)-(6.5.14), we have

$$f^{PJ}_{sos,k} \leq f^{SJ}_{sos,k}, \quad f^{PJ}_{mom,k} \leq f^{SJ}_{mom,k}.$$

Checking tightness of (6.5.16)-(6.5.17) and extracting minimizers can be done by looking at the flat truncation (6.5.15). We have the same kind of conclusions as in Proposition 6.5.2. Compared to (6.5.13)-(6.5.14), the hierarchy of (6.5.16)-(6.5.17) has stronger convergence properties. The tightness can be guaranteed under weaker assumptions, e.g., Assumption 6.5.3 is not required.

**Theorem 6.5.6.** *Let $f^{SJ}_{sos,k}$, $f^{SJ}_{mom,k}$, $f_{FJ}$, $\mathcal{F}_{sig}$ be as above. If the objective $f$ achieves only finitely many values on the singular locus $\mathcal{F}_{sig}$, then $f^{SJ}_{sos,k} = f_{FJ}$ for all $k$ big enough. Furthermore, if in addition (6.5.10) has only finitely many minimizers, then every optimizer $y^*$ of (6.5.17) must satisfy (6.5.15), when $k$ is big enough.*

*Proof.* The proof is similar to that for item (iii) of Theorem 6.5.4. Since $f$ achieves only finitely many values on $\mathcal{F}_{sig}$, it also achieves finitely many values on the feasible set $K^{FJ}$ of (6.5.10) (see Exercise 6.5.3), say, they are

$$v_1 < v_2 < \cdots < v_D.$$

Let $p_1, \ldots, p_D$ be univariate polynomials such that

$$p_i(v_i) = 1, \quad p_i(v_j) = 0 \, (i \neq j).$$

Clearly, $v_i - f_{FJ} \geq 0$ for all $i$. Then

$$h := (v_1 - f_{FJ})\big(p_1(f)\big)^2 + \cdots + (v_D - f_{FJ})\big(p_D(f)\big)^2 \in \Sigma[x]_{2k_1}$$

for some degree $k_1$. The polynomial $\hat{f} := f - f_{FJ} - h$ vanishes identically on $K^{FJ}$. By Theorem 2.6.6, there exist a power $\ell > 0$ and $q \in \mathrm{Pre}[c_{in}]$ such that

$$\hat{f}^{2\ell} + q \in \mathrm{Ideal}[c_{eq}, \varphi].$$

Applying Lemma 5.6.2, when $k \geq k_1$ is big enough, we have

$$\hat{f} + \epsilon = \phi_\epsilon + \theta_\epsilon,$$

where ($\omega \geq \frac{1}{2\ell}$ is a constant)

$$\begin{aligned}
\phi_\epsilon &:= -\omega\epsilon^{1-2\ell}(\hat{f}^{2\ell} + q) \in \mathrm{Ideal}[c_{eq}, \varphi]_{2k}, \\
\theta_\epsilon &:= \epsilon\left(1 + \hat{f}/\epsilon + \omega(\hat{f}/\epsilon)^{2\ell}\right) + \omega\epsilon^{1-2\ell}q \in \mathrm{Pre}[c_{in}]_{2k}.
\end{aligned}$$

Hence, for all $\epsilon > 0$,

$$f - (f_{FJ} - \epsilon) = \phi_\epsilon + \sigma_\epsilon, \qquad \sigma_\epsilon := \theta_\epsilon + h \in \mathrm{Pre}[c_{in}]_{2k}.$$

This implies that $\gamma = f_{min} - \epsilon$ is feasible in (6.5.16) for every $\epsilon > 0$, when $k$ is big enough. Therefore, $f_{sos,k}^{SJ} \geq f_{FJ}$, and hence $f_{sos,k}^{SJ} = f_{FJ}$, when $k$ is sufficiently large.

Note that (6.5.16)-(6.5.17) are Schmüdgen type Moment-SOS relaxations. When (6.5.10) has only finitely many minimizers, the satisfaction of the flat truncation (6.5.15) for every optimizer of (6.5.17) is implied by Theorem 5.3.9, for $k$ big enough.  □

When (6.5.1) has a single inequality constraint (or none), the hierarchy of relaxations (6.5.13)-(6.5.14) is the same as the hierarchy of (6.5.16)-(6.5.17). Therefore, we have the following conclusion.

**Corollary 6.5.7.** *When (6.5.1) has a single inequality constraint (or none), under the assumptions of Theorem 6.5.6, we have $f_{sos,k}^{SJ} = f_{sos,k}^{PJ} = f_{FJ}$ when $k$ is big enough. Moreover, if in addition (6.5.10) has only finitely many minimizers, then every optimizer $y^*$ of (6.5.17) must satisfy (6.5.15), when $k$ is big enough.*

Compared with Theorem 6.5.4(iii), Theorem 6.5.6 requires weaker assumptions, e.g., Assumption 6.5.3 is not required. It only assumes $f$ achieves finitely many values on $\mathcal{F}_{\mathrm{sig}}$. This is generally satisfiable. See the remarks after the proof of Theorem 6.5.4. If $f$ achieves infinitely many values on $\mathcal{F}_{\mathrm{sig}}$, there is no guarantee for tightness. To see this, consider the rare case that $c_{eq} = 0$ (the constant zero polynomial) and $c_{in} = \emptyset$. Then the singular locus $\mathcal{F}_{\mathrm{sig}} = \mathbb{R}^n$. The relaxation (6.5.16) is equivalent to that $f - \gamma$ is SOS. For such a case, the hierarchy of (6.5.16)-(6.5.17) cannot be tight, unless $f - f_{min}$ is SOS. The following is the relationship between (6.5.16)-(6.5.17) and the optimization (6.5.1).

**Corollary 6.5.8.** *Under the assumptions of Theorem 6.5.6, if the minimum $f_{min}$ of (6.5.1) is achievable at a feasible point and at least one minimizer of (6.5.10) is feasible for (6.5.1), then $f_{sos,k}^{SJ} = f_{min}$ for all $k$ big enough.*

### 6.5.4 ▪ Exercises

**Exercise 6.5.1.** *If $c_{eq}$ is a nonsingular tuple of polynomials, show that $\mathcal{F}_{\mathrm{sig}} = \emptyset$ and every Fritz-John point for (6.5.1) is a critical point.*

**Exercise 6.5.2.** *If (6.5.10) has a single inequality constraint (or none), or if $\mathcal{F}\backslash K^{FJ}$ is a finite set, show that Assumption 6.5.3 must hold.*

**Exercise 6.5.3.** *Show that the objective function $f$ in (6.5.10) achieves only finitely many values on the set $\mathcal{F}\backslash\mathcal{F}_{\text{sig}}$.*

**Exercise 6.5.4.** *Prove Proposition 6.5.2.*

**Exercise 6.5.5.** *Prove the second conclusion of the item (iii) of Theorem 6.5.4. That is, under Assumption 6.5.3, if $f$ achieves only finitely many values on the singular locus $\mathcal{F}_{\text{sig}}$ and (6.5.10) has only finitely many minimizers, show that every optimizer $y^*$ of (6.5.14) must satisfy (6.5.15), when $k$ is big enough.*

**Exercise 6.5.6.** *Prove Corollaries 6.5.5, 6.5.7, and 6.5.8.*

# 6.6 ▪ Some Discussions

This chapter introduced two types of tight hierarchies of Moment-SOS relaxations for solving polynomial optimization. The first one is based on Lagrange multiplier expressions, given in §6.3. The second one is based on Jacobian and Fritz-John conditions, given in §6.5. As a special case of them, the unconstrained optimization is discussed in §6.4. The detection of tightness and extracting minimizers can be done by using flat truncations. In this section, we discuss some remaining issues about them.

## 6.6.1 ▪ Existence of optimizers and singularity

To compute the minimum value $f_{min}$ of (6.1.1) or (6.5.1) by tight hierarchies of Moment-SOS relaxations, it requires that $f_{min}$ is achievable at a critical point or Fritz-John point. When the feasible set is compact, or when the objective is coercive, the minimum value must be achievable at a minimizer. However, if the feasible set is unbounded, the task of checking achievability can be difficult (see [2]).

Consider the cone of polynomials that are bounded below on $K$

$$B_d(K) := \left\{f \in \mathbb{R}[x]_d : \inf_{u \in K} f(u) > -\infty\right\}. \tag{6.6.1}$$

For a polynomial $q$, let $\tilde{q}$ denote its homogenization, i.e.,

$$\tilde{q}(\tilde{x}) = x_0^{\deg(q)} q(x/x_0), \quad \tilde{x} = (x_0, x_1, \ldots, x_n).$$

Moreover, denote by $q^{hom}$ the homogeneous part of $q$ with the highest degree. The homogenization of the set $K$ is

$$K^{hom} := \left\{\tilde{x} \in \mathbb{R}^{n+1} \left| \begin{array}{rcl} \tilde{c}_i(\tilde{x}) & = & 0\,(i \in \mathcal{E}), \\ \tilde{c}_j(\tilde{x}) & \geq & 0\,(j \in \mathcal{I}), \\ x_0 & \geq & 0 \end{array} \right.\right\}. \tag{6.6.2}$$

**Definition 6.6.1.** *([219]) The set $K$ is said to be closed at infinity if*

$$K^{hom} = \text{cl}\left(K^{hom} \cap \{x_0 > 0\}\right).$$

An interesting conclusion is that, under some genericity assumptions, the minimum value of (6.1.1) is achievable if $f$ is bounded below. For generic cases, the tuple of constraining polynomials is nonsingular. The following is a general result about nonsingularity and achievability. We refer to §2.3 for resultants Res and discriminants $\Delta$.

**Theorem 6.6.2.** *([224]) For the optimization (6.1.1), assume $\mathcal{E} = \{1, \ldots, m_1\}$ and $m_1 \leq n$.*

(a) *If, for every subset $\{j_1, \ldots, j_{n-m_1+1}\} \subseteq \mathcal{I}$,*

$$\mathrm{Res}(c_1, \ldots, c_{m_1}, c_{j_1}, \ldots, c_{j_{n-m_1+1}}) \neq 0,$$

*then at most $n - m_1$ of $c_j$ ($j \in \mathcal{I}$) vanish at each feasible point of (6.1.1).*

(b) *If, for every subset $\{j_1, \ldots, j_l\} \subseteq \mathcal{I}$ with $l \leq n - m_1$,*

$$\Delta(c_1, \ldots, c_{m_1}, c_{j_1}, \ldots, c_{j_l}) \neq 0,$$

*then the tuple $(c_1, \ldots, c_{m_1}, c_{j_1}, \ldots, c_{j_l})$ is nonsingular.*

(c) *Suppose $K$ is closed at infinity and $f \in B_d(K)$. If the resultant of every $n$ polynomial of $c_i^{hom}$, $c_j^{hom}$ ($i \in \mathcal{E}$, $j \in \mathcal{I}$ for the case $|\mathcal{E} \cup \mathcal{I}| \geq n$) is nonzero and for every $\{j_1, \ldots, j_l\}$ with $l \leq n - m_1 - 1$,*

$$\Delta(f^{hom}, c_1^{hom}, \ldots, c_{m_1}^{hom}, c_{j_1}^{hom}, \ldots, c_{j_l}^{hom}) \neq 0,$$

*then $f$ achieves its minimum value on $K$.*

(d) *For the case $K = \mathbb{R}^n$, if $f \in B_d(\mathbb{R}^n)$ and $\Delta(f^{hom}) \neq 0$, then $f$ achieves its minimum value in $\mathbb{R}^n$.*

## 6.6.2 ▪ The case of infinitely many minimizers

In §6.3, §6.4, and §6.5, we showed that flat truncation must be satisfied if there are only finitely many minimizers, under some genericity assumptions. However, if there are infinitely many minimizers, the flat truncation is usually not satisfiable (see Proposition 5.3.6).

In the following, we give a practical method for detecting tightness and extracting minimizers when there are infinitely many ones. Suppose a lower bound

$$\vartheta \in \left\{ f_{mom,k}^c, f_{mom,k}^\nabla, f_{mom,k}^{PJ}, f_{mom,k}^{SJ} \right\}$$

is obtained for the optimization problem (6.3.7) or (6.4.2) or (6.5.10) for some relaxation order $k$. The verification of $\vartheta = f_c$ or $\vartheta = f_{FJ}$ can be done as follows. Suppose all the appearing polynomials have degrees at most $2d$. Select a generic matrix $R$ and choose the polynomial $F$ as

$$R \in \mathbb{R}^{\binom{n+d}{d} \times \binom{n+d}{d}}, \quad F = [x]_d^T \cdot R^T R \cdot [x]_d.$$

For convenience, let $\Theta$ denote the set of all equality constraining polynomials and let $\mathcal{G}$ denote the set of all inequality constraining polynomials for the corresponding optimization problem. To check the tightness ($\vartheta = f_c$ or $\vartheta = f_{FJ}$) and to extract minimizers, we consider the following optimization problem:

$$\begin{cases} \min & F(x) \\ s.t. & h(x) = 0 \ (\forall \, h \in \Theta), \\ & f(x) - \vartheta = 0, \\ & g(x) \geq 0 \ (\forall \, g \in \mathcal{G}). \end{cases} \tag{6.6.3}$$

It is important to observe that the lower bound $\vartheta$ is tight for the corresponding optimization problem if and only if the feasible set of (6.6.3) is nonempty. For the generic polynomial $F$ as above, the optimization (6.6.3) has a unique minimizer if it is feasible. To solve (6.6.3), the $k$th order moment relaxation is

$$
\left\{
\begin{array}{ll}
\min & \langle F, y \rangle \\
s.t. & \mathscr{V}_h^{(2k)}[y] = 0 \, (h \in \Theta), \\
& \mathscr{V}_{f-\vartheta}^{(2k)}[y] = 0, \\
& L_g^{(k)}[y] \succeq 0 \, (g \in \mathcal{G}), \\
& M_k[y] \succeq 0, \\
& y_0 = 1, \, y \in \mathbb{R}^{\mathbb{N}_{2k}^n}.
\end{array}
\right.
\tag{6.6.4}
$$

The following is the conclusion about checking tightness and extracting minimizers for the hierarchy of relaxations (6.6.4).

**Theorem 6.6.3.** *Let $f_*$ be the optimal value of the corresponding optimization problem (6.3.7), (6.4.2), or (6.5.10). Let $\vartheta$ be a lower bound for $f_*$ and let $F$ be the generic polynomial as above.*

*(i) If the relaxation (6.6.4) is infeasible for some $k$, we have $\vartheta < f_*$.*

*(ii) Suppose (6.6.4) is feasible and $y^*$ is an optimizer for (6.6.4). If the point*

$$
\hat{u} := \big( (y^*)_{e_1}, \dots, (y^*)_{e_n} \big)
$$

*is feasible for the corresponding optimization problem and $\vartheta = f(\hat{u})$, then $\vartheta = f_*$.*

The proof for Theorem 6.6.3 is quite straightforward, so it is left as an exercise. The convergence of (6.6.4) is shown as in Exercise 6.6.3.

## 6.6.3 ▪ Exercises

**Exercise 6.6.1.** *Prove Theorems 6.6.2 and 6.6.3.*

**Exercise 6.6.2.** *If $F$ is a generic polynomial, show that the minimizer of (6.6.3) is unique if it exists.*

**Exercise 6.6.3.** *Suppose $F$ is a generic polynomial in (6.6.3) and $\vartheta$ equals the optimal value $f_*$ as in Theorem 6.6.3. If $\mathrm{Ideal}[\Theta] + \mathrm{QM}[\mathcal{G}]$ is archimedean, show that the sequence of the point $\hat{u}$, as $k$ increases, converges to a minimizer of the corresponding optimization problem.*

# Part III

# Convexity and Conic Optimization with Moments and Polynomials

# Chapter 7

# Convexity with Polynomials and Sets

*Convexity is an important concept in optimization. This chapter introduces basic results about convex polynomials, convex optimization, semidefinite representations, and spectrahedra and their projections.*

## 7.1 ▪ Convex Polynomials

Recall that a set $\Omega \subseteq \mathbb{R}^n$ is *convex* if $\lambda x + (1 - \lambda)y \in \Omega$ for all $x, y \in \Omega$ and for all $\lambda \in [0, 1]$. Convex sets are frequently appearing in optimization. Basic properties of convex sets are introduced in §1.2.

Let $\mathcal{D} \subseteq \mathbb{R}^n$ be a convex domain with nonempty interior. A function $f : \mathcal{D} \to \mathbb{R}$ is said to be *convex* over $\mathcal{D}$ if

$$f(\lambda x + (1 - \lambda)y) \leq \lambda f(x) + (1 - \lambda)f(y)$$

for all $x, y \in \mathcal{D}$ and for all $\lambda \in [0, 1]$. If $f$ is continuously differentiable in $\mathcal{D}$, then $f$ is convex if and only if

$$f(y) + \nabla f(x)^T (x - y) - f(x) \geq 0 \tag{7.1.1}$$

for all $x, y \in \mathcal{D}$. The inequality (7.1.1) is called the *first order criterion* for convex functions. Furthermore, if $f$ is twice continuously differentiable in $\mathcal{D}$, then $f$ is convex if and only if its Hessian matrix is psd on $\mathcal{D}$, i.e.,

$$\nabla^2 f(x) \succeq 0 \quad \text{for all } x \in \mathcal{D}. \tag{7.1.2}$$

The inequality (7.1.2) is called the *second order criterion* for convex functions.

A polynomial $f \in \mathbb{R}[x]$ is said to be *convex* if the function $f(x)$ is convex over the entire space $\mathbb{R}^n$. Denote the set

$$\mathscr{X}[x] := \{ f \in \mathbb{R}[x] : f \text{ is convex} \}. \tag{7.1.3}$$

Clearly, $\mathscr{X}[x]$ is a convex cone in $\mathbb{R}[x]$. For a degree $d$, denote the truncations

$$\begin{aligned} \mathscr{X}[x]_d &:= \mathscr{X}[x] \cap \mathbb{R}[x]_d, \\ \mathscr{X}[x]_d^{\text{hom}} &:= \mathscr{X}[x] \cap \mathbb{R}[x]_d^{\text{hom}}. \end{aligned} \tag{7.1.4}$$

To determine whether a polynomial $f$ is convex or not, it is equivalent to determining whether or not $\nabla^2 f(x)$ is psd everywhere. For instance, if $f$ is a quadratic polynomial, say, $f = \frac{1}{2}x^T H x + g^T x + b$ for $H \in \mathcal{S}^n$, $g \in \mathbb{R}^n$, $b \in \mathbb{R}^1$, then $f$ is convex if and only if the Hessian $H \succeq 0$ (i.e., $H$ is psd). To ensure that $f$ is convex, we often need convenient conditions other than the first or second order criterion. The SOS-convexity is such one.

217

### 7.1.1 ▪ SOS-convex polynomials

The definition of SOS-convex polynomials was originally introduced in [120], to study semidefinite representations for convex sets. A symmetric matrix polynomial $G \in \mathbb{R}[x]^{n \times n}$ is said to be SOS[6] if $G = RR^T$ for some matrix polynomial $R \in \mathbb{R}[x]^{n \times \ell}$ with $\ell$ columns.

**Definition 7.1.1.** *([120]) A polynomial $f \in \mathbb{R}[x]$ is SOS-convex if its Hessian $\nabla^2 f$ is SOS, i.e., there exists a matrix polynomial $P \in \mathbb{R}[x]^{n \times \ell}$, for some $\ell \in \mathbb{N}$, such that $\nabla^2 f = PP^T$.*

The set of all SOS-convex polynomials is denoted as $\Theta[x]$. For a degree $d > 0$, denote the truncations

$$
\begin{aligned}
\Theta[x]_d &:= \Theta[x] \cap \mathbb{R}[x]_d, \\
\Theta[x]_d^{\mathrm{hom}} &:= \Theta[x] \cap \mathbb{R}[x]_d^{\mathrm{hom}}.
\end{aligned}
\tag{7.1.5}
$$

Clearly, $\Theta[x]_d$ and $\Theta[x]_d^{\mathrm{hom}}$ are closed convex cones. A polynomial $p$ is said to be *SOS-concave* if the negative $-p$ is SOS-convex. For instance, the polynomial $f = \frac{1}{3}(x_1^4 + x_2^4) + x_1^2 x_2^2$ is SOS-convex, since

$$
\nabla^2 f = \begin{bmatrix} 4x_1^2 + 2x_2^2 & 4x_1 x_2 \\ 4x_1 x_2 & 2x_1^2 + 4x_2^2 \end{bmatrix} = PP^T
$$

for the matrix polynomial

$$
P = \begin{bmatrix} 2x_1 & x_2 & x_2 & 0 & 0 \\ 2x_2 & 0 & 0 & x_1 & x_1 \end{bmatrix}.
$$

Every SOS-convex polynomial is clearly convex, so

$$
\Theta[x] \subseteq \mathscr{X}[x], \quad \Theta[x]_d \subseteq \mathscr{X}[x]_d, \ d = 1, 2, \ldots.
$$

One wonders when they are equal. The cases for the equality to hold are the same as Hilbert's characterization on SOS and nonnegative polynomials.

**Theorem 7.1.2.** *([4, 5]) The equality $\Theta[x]_d = \mathscr{X}[x]_d$ holds if and only if*

$$
n = 1, \ \text{or} \ d = 2, \ \text{or} \ (n, d) = (2, 4).
$$

*For the homogeneous case, $\Theta[x]_d^{\mathrm{hom}} = \mathscr{X}[x]_d^{\mathrm{hom}}$ if and only if*

$$
n = 2, \ \text{or} \ d = 2, \ \text{or} \ (n, d) = (3, 4).
$$

Not every convex polynomial is SOS-convex. As given in [3], the following polynomial is convex but not SOS-convex:

$$
\boxed{
\begin{aligned}
&32x_1^8 + 118x_1^6 x_2^2 + 40x_1^6 x_3^2 + 25x_1^4 x_2^4 - 43x_1^4 x_2^2 x_3^2 - 35x_1^4 x_3^4 \\
&+3x_1^2 x_2^4 x_3^2 - 16x_1^2 x_2^2 x_3^4 + 24x_1^2 x_3^6 + 16x_2^8 + 44x_2^6 x_3^2 \\
&+70x_2^4 x_3^4 + 60x_2^2 x_3^6 + 30x_3^8.
\end{aligned}
}
\tag{7.1.6}
$$

Detecting SOS-convex polynomials can be done by solving semidefinite programs (also see §10.2). Note that $f \in \Theta[x]_{2d}$ if and only if

$$
\nabla^2 f = \sum_{j=1}^{\ell} p_j p_j^T,
$$

---

[6]SOS matrix polynomials are also called sums of Hermitian squares (SHS), which are discussed in Chapter 10.

where each $p_j$ is a polynomial column vector of length $n$. One can write $p_j$ such that

$$p_j = (I_n \otimes [x]_{d-1}^T) v_j, \quad v_j \in \mathbb{R}^{n\binom{n+d-1}{d-1}}.$$

Here, $\otimes$ denotes the classical Kronecker product. Then, one can see that

$$\sum_{j=1}^{\ell} p_j p_j^T = (I_n \otimes [x]_{d-1})^T \cdot \left( \sum_{j=1}^{\ell} v_j v_j^T \right) \cdot (I_n \otimes [x]_{d-1}).$$

This implies the following characterization for SOS-convex polynomials.

**Lemma 7.1.3.** *A polynomial $f \in \mathbb{R}[x]_{2d}$ is SOS-convex if and only if there exists a symmetric matrix $X$ of length $n \cdot \binom{n+d-1}{d-1}$ satisfying*

$$
\boxed{
\begin{aligned}
\nabla^2 f &= (I_n \otimes [x]_{d-1})^T \cdot X \cdot (I_n \otimes [x]_{d-1}), \\
X &\succeq 0.
\end{aligned}
}
\tag{7.1.7}
$$

By comparing coefficients, the equality in (7.1.7) gives a set of linear equations about the psd matrix variable $X$. Therefore, testing the membership $f \in \Theta[x]_{2d}$ is equivalent to solving the semidefinite program of (7.1.7).

**Example 7.1.4.** *Consider the homogeneous polynomial*

$$f = (x^{[d]})^T \cdot A \cdot x^{[d]}$$

*for a degree $d > 0$ and a symmetric matrix $A = (a_{ij}) \in \mathcal{S}^n$. The $x^{[d]}$ denotes the column vector such that*

$$x^{[d]} = \begin{bmatrix} x_1^d & \cdots & x_n^d \end{bmatrix}^T.$$

*A direct calculation shows that*

$$
\begin{aligned}
\nabla^2 f &= 2d^2 \cdot diag(x^{[d-1]}) \cdot A \cdot diag(x^{[d-1]}) \\
&+ 2d(d-1) \cdot diag(x^{[d-2]}) \cdot diag(Ax^{[d]}).
\end{aligned}
$$

*For instance, when $n = 3$ and $d = 2$, for*

$$A = \begin{bmatrix} 2 & 1 & 0 \\ 1 & 3 & 1 \\ 0 & 1 & 1 \end{bmatrix},$$

*the polynomial $f$ is SOS-convex. When $A \succeq 0$, the polynomial $f$ is not necessarily SOS-convex. For instance, the matrix*

$$A = \begin{bmatrix} 1 & -1 & 0 \\ -1 & 2 & 1 \\ 0 & 1 & 2 \end{bmatrix}$$

*is positive definite but $f$ is not SOS-convex.*

The following are two useful properties for SOS-convex polynomials.

**Lemma 7.1.5.** *([120]) (i) If a symmetric matrix polynomial $F \in \mathbb{R}[x]^{r \times r}$ is SOS, then for each $u \in \mathbb{R}^n$ the integral $\int_0^1 \int_0^t F(u + s(x - u)) \, ds \, dt$ is an SOS matrix polynomial.*

*(ii) Let $f \in \mathbb{R}[x]$ be an SOS-convex polynomial. If $f(u) = 0$ and $\nabla f(u) = 0$ for some $u \in \mathbb{R}^n$, then $f$ must be SOS.*

The proof for the above is left as an exercise. It is interesting to note that a convex homogeneous polynomial has to be nonnegative everywhere (see exercise). However, not every convex form is SOS, which is shown in [30]. An explicit non-SOS convex form is given in [290].

### 7.1.2 • Jensen's inequality

A useful property of convex functions is *Jensen's inequality*. For a probability measure $\mu$ on $\mathbb{R}^n$, if $f$ is a convex function on $\mathbb{R}^n$ such that $|f|$ is integrable with respect to $\mu$, then

$$f(\mathbb{E}_\mu[x]) \leq \int f(x)\mathrm{d}\mu, \tag{7.1.8}$$

where $\mathbb{E}_\mu[x] = \int x\mathrm{d}\mu$ denotes the expectation of $x$ for the measure $\mu$. This is called Jensen's inequality. In particular, if a tms $y \in \mathbb{R}^{\mathbb{N}_d^n}$ is such that $y = \int [x]_d \mathrm{d}\mu$ and $u = (y_{e_1}, \ldots, y_{e_n})$, then Jensen's inequality reads as

$$f(u) \leq \langle f, y \rangle.$$

The above inequality may not hold if the tms $y$ does not admit a probability measure. However, for SOS-convex polynomials, Jensen's inequality still holds even if $y$ is not a truncated moment sequence but its moment matrix is psd.

**Theorem 7.1.6.** *([164]) Let $y = (y_\alpha) \in \mathbb{R}^{\mathbb{N}_{2k}^n}$ be a tms such that $y_0 = 1$ and $M_k[y] \succeq 0$. If $f \in \mathbb{R}[x]_{2k}$ is SOS-convex (i.e., $f \in \Theta[x]_{2k}$), then*

$$f(y_{e_1}, \ldots, y_{e_n}) \leq \langle f, y \rangle. \tag{7.1.9}$$

**Proof.** Let $u := (y_{e_1}, \ldots, y_{e_n})$. By the integration formula, we have that

$$f(x) = f(u) + \nabla f(u)^T(x - u) + (x - u)^T F(x, u)(x - u),$$

where the matrix

$$F(x, u) := \int_0^1 \int_0^t \nabla^2 f(u + s(x - u))\mathrm{d}s\mathrm{d}t.$$

Since $f$ is SOS-convex, the integration $F(x, u)$ is an SOS-matrix polynomial, by Lemma 7.1.5. Hence, the product $(x - u)^T F(x, u)(x - u)$ is an SOS polynomial, say,

$$(x - u)^T F(x, u)(x - u) = \sum_{i=1}^\ell p_i^2 \quad \text{for some } p_i \in \mathbb{R}[x]_k.$$

Then, one can see that

$$\langle f - f(u) - \nabla f(u)^T(x - u), y \rangle = \langle (x - u)^T F(x, u)(x - u), y \rangle$$
$$= \sum_{i=1}^\ell \langle p_i^2, y \rangle = \sum_{i=1}^\ell \mathrm{vec}(p_i)^T \cdot M_k[y] \cdot \mathrm{vec}(p_i) \geq 0,$$

since $M_k[y] \succeq 0$. Therefore,

$$\langle f, y \rangle \geq \langle f(u) + \nabla f(u)^T(x - u), y \rangle = f(u)$$

because $\langle f(u), y \rangle = f(u)\langle 1, y \rangle = f(u)y_0 = f(u)$ and

$$\langle \nabla f(u)^T (x - u), y \rangle = \nabla f(u)^T (u - u) = 0.$$

So, the inequality (7.1.9) follows. $\quad\square$

Jensen's inequality in Theorem 7.1.6 is useful for getting feasible points for convex sets. It implies the following corollary.

**Corollary 7.1.7.** *For SOS-concave polynomials $c_1, \ldots, c_m$, let*

$$K = \{x \in \mathbb{R}^n : c_1(x) \geq 0, \ldots, c_m(x) \geq 0\}.$$

*If $y \in \mathbb{R}^{\mathbb{N}^n_{2k}}$ is such that $y_0 = 1$, $M_k[y] \succeq 0$ and all $\langle c_i, y \rangle \geq 0$, then $(y_{e_1}, \ldots, y_{e_n}) \in K$.*

### 7.1.3 • Exercises

**Exercise 7.1.1.** *Determine whether or not the following polynomials are SOS-convex:*

(i) $x_1^4 + x_2^4 + x_3^4 + x_4^4 - x_1 x_2 x_3 x_4$.

(ii) $x_1^4 + x_2^4 + x_1^2 x_2 + x_1 x_2^2 + x_1^2 + x_2^2$.

(iii) $2(x_1^6 + x_2^6 + x_3^6) + x_1^3 x_2^3 + x_2^3 x_3^3 + x_3^3 x_1^3$.

(iv) $x_1^6 + x_2^6 + x_3^6 + x_1^2 x_2^4 + x_2^2 x_3^4 + x_3^2 x_1^4$.

**Exercise 7.1.2.** *Show that if a homogeneous polynomial (i.e., a form) is convex, then it must be nonnegative everywhere.*

**Exercise 7.1.3.** *Prove Lemmas 7.1.3 and 7.1.5.*

**Exercise 7.1.4.** *([4]) For $f \in \mathbb{R}[x]$, show that the following memberships are equivalent (let $x := (x_1, \ldots, x_n)$ and $y := (y_1, \ldots, y_n)$):*

(i) $f \in \Theta[x]$.

(ii) $f(x) + f(y) - 2f(\frac{x+y}{2}) \in \Sigma[x, y]$.

(iii) $f(y) - f(x) - \nabla f(x)^T (y - x) \in \Sigma[x, y]$.

**Exercise 7.1.5.** *Prove Corollary 7.1.7. Moreover, show that the conclusion also holds for each $y \in \mathscr{S}[c]_{2k}$ with $y_0 = 1$, where $c = (c_1, \ldots, c_m)$. (See (2.5.15) for the cone $\mathscr{S}[c]_{2k}$.)*

**Exercise 7.1.6.** *([164]) Let $f$ be a univariate convex polynomial and let $y \in \mathbb{R}^{\mathbb{N}^n_{2k}}$ be a tms such that $y_0 = 1$ and $M_k[y] \succeq 0$. Then, for every $g \in \mathbb{R}[x]$ with $\deg(f(g)) \leq 2k$, show that a similar Jensen's inequality holds:*

$$f(\langle g, y \rangle) \leq \langle f(g), y \rangle.$$

**Exercise 7.1.7.** *Let $f(z_1, \ldots, z_r) \in \mathbb{R}[z_1, \ldots, z_r]$ be an SOS-convex polynomial in $(z_1, \ldots, z_r)$ and let $g_1, \ldots, g_r \in \mathbb{R}[x]$ be such that $\deg f(g_1, \ldots, g_r) \leq 2k$. If $y \in \mathbb{R}^{\mathbb{N}^n_{2k}}$ is a tms such that $y_0 = 1$ and $M_d[y] \succeq 0$, show that*

$$f(\langle g_1, y \rangle, \ldots, \langle g_r, y \rangle) \leq \langle f(g_1, \ldots, g_r), y \rangle.$$

## 7.2 ▪ Convex Polynomial Optimization

This section discusses convex polynomial optimization problems and Moment-SOS relaxations for solving them.

### 7.2.1 ▪ Optimality conditions

Let $c := (c_1, \ldots, c_m)$ be a tuple of functions (not necessarily given by polynomials), defined over a convex domain $\mathcal{D} \subseteq \mathbb{R}^n$ with nonempty interior. Consider the set

$$K := \{x \in \mathcal{D} : c_1(x) \geq 0, \ldots, c_m(x) \geq 0\}. \tag{7.2.1}$$

For convenience, assume the closure $\mathrm{cl}\,(K) \subseteq \mathrm{int}\,(\mathcal{D})$. When a function is optimized over $K$, the KKT and Fritz-John conditions (see §5.1 and §6.1) are often used. When the set $K$ is convex, there are special properties for optimality conditions. If every $c_i$ is concave over $\mathcal{D}$, then $K$ is convex. However, if $K$ is convex, the functions $c_i$ are not necessarily concave. For instance, the set $\{x_1^2 x_2 - 1 \geq 0, x_1 \geq 0\}$ is convex, but the function $x_1^2 x_2 - 1$ is not concave. For convex optimization problems, we still need some constraint qualifications for optimality conditions to hold. The following two are typically used ones (see [165]).

**Definition 7.2.1.** *Let $K$ be as in (7.2.1) such that $\mathrm{cl}\,(K) \subseteq \mathrm{int}\,(\mathcal{D})$.*

(i) *The Slater's condition is said to hold for $K$ if there exists $u \in K$ such that $c_i(u) > 0$ for all $i$.*

(ii) *The set $K$ is said to be nondegenerate if for each $j = 1, \ldots, m$,*

$$u \in K, \; c_j(u) = 0 \quad \Rightarrow \quad \nabla c_j(u) \neq 0.$$

*Otherwise, $K$ is called degenerate.*

We remark that if all $c_i$ are concave, then the Slater's condition implies that $K$ is nondegenerate. This is because if $c_i(u) = 0$ and $\nabla c_i(u) = 0$ for some $i$, then $u$ is a global maximizer of $c_i$, whence there is no $u$ such that $c_i(u) > 0$, violating the Slater's condition. The following is a characterization for convexity of the set $K$.

**Lemma 7.2.2.** *([165]) Let $K$ be as in (7.2.1) such that $\mathrm{cl}\,(K) \subseteq \mathrm{int}\,(\mathcal{D})$. Assume all $c_j$ are differentiable, the set $K$ is nondegenerate, and Slater's condition holds. Then, the set $K$ is convex if and only if for each $j = 1, \ldots, m$*

$$x, u \in K, \; c_j(u) = 0 \quad \Rightarrow \quad \langle \nabla c_j(u), x - u \rangle \geq 0. \tag{7.2.2}$$

**Proof.** "⇐": For every point $u$ on the boundary of $K$, there must exist $j$ such that $c_j(u) = 0$. Note that $\nabla c_j(u) \neq 0$, since $K$ is nondegenerate. The condition (7.2.2) implies that the set $K$ is contained in the half space

$$\langle \nabla c_j(u), x - u \rangle \geq 0.$$

This means that $K$ has a supporting hyperplane at each boundary point. Note that $K$ is closed and has nonempty interior, by the Slater's condition. The convexity of $K$ is then implied by Theorem 1.2.3.

"$\Rightarrow$": For each $j$, pick an arbitrary $u \in K$ with $c_j(u) = 0$ and consider the half space

$$H^+ := \{x \in \mathbb{R}^n : \nabla c_j(u)^T (x - u) \geq 0\}.$$

We need to show that $K \subseteq H^+$. If otherwise $K \nsubseteq H^+$, then there exists $v \in K$ such that $\nabla c_j(u)^T (v - u) < 0$. So, for $t > 0$ small enough, we have

$$c_j(u + t(v - u)) < 0.$$

Since $K$ is convex and $u, v \in K$, we must have $c_j(u + t(v - u)) \geq 0$ for $t > 0$ small. This is a contradiction, so $K \subseteq H^+$. $\square$

For an objective function $f$ (not necessarily a polynomial), we consider the optimization problem

$$\begin{cases} \min_{x \in \mathcal{D}} & f(x) \\ s.t. & c_1(x) \geq 0, \ldots, c_m(x) \geq 0. \end{cases} \tag{7.2.3}$$

It is important to remark that if $f$ and all $-c_i$ are convex polynomials and $f$ is bounded below over the feasible set, then the minimum value of (7.2.3) must be achieved at a feasible point (see [17]). When the set $K$ and objective $f$ are both convex, the classical KKT conditions for (7.2.3) hold under some constraint qualifications.

**Theorem 7.2.3.** *([165]) Let $K$ be as in (7.2.1) such that $\mathrm{cl}\,(K) \subseteq \mathrm{int}\,(\mathcal{D})$. Suppose $f$ and all $c_j$ are differentiable. Assume that the set $K$ is convex and nondegenerate, and the Slater's condition holds for $K$. If $f$ is convex, then a point $u$ is a global minimizer of (7.2.3) if and only if $u$ is a KKT point, i.e., there exist Lagrange multipliers $\lambda_1, \ldots, \lambda_m$ such that*

$$\boxed{\begin{aligned} \nabla f(u) &= \lambda_1 \nabla c_1(u) + \cdots + \lambda_m \nabla c_m(u), \\ \lambda_1 \cdot c_1(u) &= \cdots = \lambda_m \cdot c_m(u) = 0, \\ \lambda_1 &\geq 0, \ldots, \lambda_m \geq 0. \end{aligned}} \tag{7.2.4}$$

*Proof.* "$\Leftarrow$": Suppose $u$ is a KKT point, with $\lambda_j$'s being the Lagrange multipliers. Let $J(u) = \{j_1, \ldots, j_r\}$ be the label set of all $j$ with $c_j(u) = 0$. Then $\lambda_j = 0$ for all $j \notin J(u)$. Since $f$ is convex, we have that for all $x \in K$,

$$f(x) - f(u) \geq \nabla f(u)^T (x - u) = \sum_{j \in J(u)} \lambda_j \nabla c_j(u)^T (x - u) \geq 0,$$

which follows from Lemma 7.2.2 and $\lambda_j \geq 0$.

"$\Rightarrow$": Let $u \in K$ be a minimizer of (7.2.3). By Fritz-John optimality conditions (see §6.1), there exist $\lambda_0 \in \{0, 1\}$ and $\lambda_i$ ($i = 1, \ldots, m$) such that

$$\lambda_0 \nabla f(u) = \sum_{i=1}^m \lambda_i \nabla c_i(u), \; \lambda_i c_i(u) = 0, \; \lambda_i \geq 0,$$
$$(\lambda_0, \lambda_1, \ldots, \lambda_m) \neq (0, 0, \ldots, 0).$$

We show that $\lambda_0 \neq 0$ (so $\lambda_0 = 1$). Suppose otherwise $\lambda_0 = 0$, then the label set

$$J := \{j \in [m] : \lambda_j > 0\}$$

is nonempty. Note that $c_j(u) = 0$ for all $j \in J$. By Slater's condition, there exists $v \in K$ such that all $c_i(v) > 0$. There also exists $\epsilon > 0$ such that $c_i(x) > 0$ for all $x \in B(v, \epsilon) \subseteq K$ and for all $i$. The Fritz-John conditions imply that

$$\sum_{j \in J} \lambda_j \nabla c_j(u)^T (x - u) = 0 \quad \forall\, x \in B(v, \epsilon).$$

Together with Lemma 7.2.2, the above implies that $\nabla c_j(u)^T (x - u) = 0$ for all $j \in J$ and for all $x \in B(v, \epsilon)$. This further implies that $\nabla c_j(u) = 0$ for all $j \in J$, which contradicts that $K$ is nondegenerate. Therefore, we must have $\lambda_0 \neq 0$, and hence $\lambda_0 = 1$ and the KKT condition (7.2.4) holds.    $\square$

In Theorem 7.2.3, the set $K$ is assumed to be convex while the constraining functions $c_j$ are *not necessarily concave*. However, if the $c_j$'s are all concave, then the nondegeneracy assumption of $K$ is not required. This leads to the following classical theorem, whose proof is left as an exercise.

**Theorem 7.2.4.** *([22]) Let $K$ be as in (7.2.1) such that $\mathrm{cl}\,(K) \subseteq \mathrm{int}\,(\mathcal{D})$. Suppose $f$ and all $c_j$ are continuously differentiable in $\mathcal{D}$. Assume that all $c_j$ are concave and the Slater's condition holds for $K$. If $u$ is a local minimizer of (7.2.3), then there exist Lagrange multipliers $\lambda_1, \ldots, \lambda_m$ satisfying the KKT condition (7.2.4).*

### 7.2.2 ▪ Moment-SOS relaxations

Now we assume the objective $f$ and all constraining functions $c_i$ in (7.2.3) are polynomials and the convex domain $\mathcal{D} = \mathbb{R}^n$. Denote the degree

$$k_0 = \min\left\{ \left\lceil \frac{1}{2} \deg(f) \right\rceil, \left\lceil \frac{1}{2} \deg(c_1) \right\rceil, \ldots, \left\lceil \frac{1}{2} \deg(c_m) \right\rceil \right\}. \tag{7.2.5}$$

For each $k \geq k_0$, the $k$th order SOS relaxation for solving (7.2.3) is

$$\begin{cases} \max & \gamma \\ s.t. & f - \gamma \in \mathrm{QM}[c]_{2k}. \end{cases} \tag{7.2.6}$$

Its dual optimization is the $k$th order moment relaxation

$$\begin{cases} \min & \langle f, y \rangle \\ s.t. & L_{c_i}^{(k)}[y] \succeq 0 \ (i = 1, \ldots, m), \\ & M_k[y] \succeq 0, \\ & y_0 = 1, \ y \in \mathbb{R}^{\mathbb{N}_{2k}^n}. \end{cases} \tag{7.2.7}$$

Let $f_{sos,k}, f_{mom,k}$ denote the optimal values of (7.2.6), (7.2.7) respectively. Let $f_{min}$ be the minimum value of (7.2.3). Then it holds that

$$f_{sos,k} \leq f_{mom,k} \leq f_{min}.$$

When $f$ and all $-c_i$ are convex polynomials, we have the following results about the Moment-SOS hierarchy.

**Theorem 7.2.5.** *([71, 164]) Assume that $f$ and all $-c_i$ are convex polynomials. Suppose $u$ is a KKT point, with Lagrange multipliers $\lambda_i$'s satisfying (7.2.4). Let $\mathcal{L}(x)$ be the Lagrange function*

$$\mathcal{L}(x) := f(x) - \sum_{i=1}^{m} \lambda_i c_i(x).$$

*If* $\mathrm{QM}[c]$ *is archimedean and* $\nabla^2 \mathcal{L}(u) \succ 0$ *is positive definite, then the Moment-SOS hierarchy of (7.2.6)-(7.2.7) is tight, i.e.,* $f_{sos,k} = f_{mom,k} = f_{min}$ *for all* $k$ *large enough.*

**Proof.** Denote the matrix polynomial

$$F(x) := \int_0^1 \int_0^t \nabla^2 \mathcal{L}(u + s(x - u)) \mathrm{d}s \mathrm{d}t.$$

Since $u$ is a KKT point, we get $\nabla \mathcal{L}(u) = 0$ and

$$\mathcal{L}(x) - \mathcal{L}(u) = (x - u)^T \cdot F(x) \cdot (x - u).$$

Since $f$ and all $-c_i$ are convex, the matrix polynomial $F(x)$ is psd on $K$. Indeed, we further have that $F(x)$ is positive definite on $K$. To see this, suppose otherwise that $F(x)$ is singular at some point $v \in K$, say, $\xi^T F(v)\xi = 0$ for some vector $\xi \neq 0$, then

$$\xi^T F(v)\xi = \int_0^1 \int_0^t \xi^T \left(\nabla^2 \mathcal{L}(u + s(v - u))\right)\xi \mathrm{d}s \mathrm{d}t = 0.$$

Since $\nabla^2 \mathcal{L} \succeq 0$ everywhere, the above implies

$$\xi^T \left(\nabla^2 \mathcal{L}(u + s(v - u))\right)\xi = 0 \quad \text{for all } s \in [0,1].$$

In particular, we get $\xi^T \left(\nabla^2 \mathcal{L}(u)\right)\xi = 0$, which contradicts the assumption that $\nabla^2 \mathcal{L}(u) \succ 0$. Since $F(x) \succ 0$ on $K$ and $\mathrm{QM}[c]$ is archimedean, there exist SOS matrix polynomials $G_0, \ldots, G_m$ such that

$$F = G_0 + c_1 G_1 + \cdots + c_m G_m$$

(see Theorem 10.1.4). Therefore, we get that

$$\mathcal{L}(x) - \mathcal{L}(u) = (x - u)^T G_0(x - u) + \sum_{i=1}^m c_i(x)(x - u)^T G_i(x - u).$$

The above then implies that

$$f(x) - f(u) = (x - u)^T G_0(x - u) + \sum_{i=1}^m c_i(x)\left[\lambda_i + (x - u)^T G_i(x - u)\right].$$

Note that $(x - u)^T G_0(x - u)$ and all $\lambda_i + (x - u)^T G_i(x - u)$ are SOS polynomials, since each $\lambda_i \geq 0$. This means that $\gamma = f(u)$ is feasible for (7.2.6). Since $u$ is a KKT point and $f, -c_i$ are all convex, we know $f(u) = f_{min}$, so $f_{sos,k} \geq f_{min}$ for all $k$ big enough. Since $f_{sos,k} \leq f_{mom,k} \leq f_{min}$, we get $f_{sos,k} = f_{mom,k} = f_{min}$ when $k$ is large enough. $\square$

The detection of the tightness $f_{sos,k} = f_{mom,k} = f_{min}$ and extracting minimizers can be done similarly by checking the flat truncation condition (5.3.3). Its satisfaction is guaranteed in Theorem 5.3.4, under some additional assumptions. However, when $f$ and all $-c_i$ are SOS-convex polynomials, we have stronger convergence properties. For such cases, the first relaxation in the Moment-SOS hierarchy is already tight and a minimizer can be obtained straightforwardly. This is shown as follows.

**Theorem 7.2.6.** *Assume that* $f$ *and all* $-c_i$ *are SOS-convex polynomials. Then, for every* $k \geq k_0$, *where* $k_0$ *is as in (7.2.5), we have the following:*

(i) *The feasible set $K$ of (7.2.3) is empty if and only if the moment relaxation (7.2.7) is infeasible.*

(ii) *Suppose $K \neq \emptyset$. Then (7.2.3) is unbounded below if and only if (7.2.7) is unbounded below.*

(iii) *Suppose $K \neq \emptyset$ and $f$ is bounded below on $K$. Then, $f_{mom,k} = f_{min}$. Moreover, if $f_{min}$ is achievable at a KKT point, or if $K$ has nonempty interior, then $f_{sos,k} = f_{min}$.*

(iv) *If $y^*$ is a minimizer of the moment relaxation (7.2.7), then the point $u := (y_{e_1}^*, \ldots, y_{e_n}^*)$ is a minimizer of (7.2.3).*

**Proof.** (i) If the relaxation (7.2.7) is infeasible for any $k \geq k_0$, then (7.2.3) is clearly infeasible. If the relaxation (7.2.7) is feasible for some $k \geq k_0$, say, $y$ is a feasible point, then we let $u := (y_{e_1}, \ldots, y_{e_n})$. Since all $-c_i$ are SOS-convex, Jensen's inequality (see Theorem 7.1.6) implies that $c_i(u) \geq \langle c_i, y \rangle \geq 0$. The latter inequality is due to that $\langle c_i, y \rangle$ is the $(1,1)$-entry of the localizing matrix $L_{c_i}^{(k)}[y]$. Therefore, if the feasible set $K$ is empty, then the moment relaxation (7.2.7) must also be infeasible for all $k \geq k_0$.

(ii) "$\Rightarrow$": If (7.2.3) is unbounded below, then (7.2.7) must also be unbounded below, since $f_{mom,k} \leq f_{min}$.

"$\Leftarrow$": Suppose (7.2.7) is unbounded below, then $f_{mom,k} = -\infty$. For every $M > 0$, there exists a tms $y$ that is feasible for (7.2.7) and $\langle f, y \rangle \leq -M$. Let $u := (y_{e_1}, \ldots, y_{e_n})$. By Jensen's inequality, we have $u \in K$ and

$$f(u) \leq \langle f, y \rangle \leq -M.$$

Since $M$ can be arbitrarily large, we know $f$ is unbounded below on $K$.

(iii) By the item (ii), we know (7.2.7) is bounded below, i.e., $f_{mom,k} > -\infty$. For each $\epsilon > 0$, there exists a feasible $y$ such that

$$\langle f, y \rangle < f_{mom,k} + \epsilon.$$

By Jensen's inequality, for the point $u := (y_{e_1}, \ldots, y_{e_n})$, we have $u \in K$ and

$$f(u) \leq \langle f, y \rangle < f_{mom,k} + \epsilon.$$

Hence, $f_{min} \leq f(u) \leq f_{mom,k} + \epsilon$ for all $\epsilon > 0$. This means that $f_{min} \leq f_{mom,k}$. Since $f_{mom,k} \leq f_{min}$, we must have $f_{mom,k} = f_{min}$.

Next, suppose that $f_{min}$ is achievable at a KKT point, say, $u$. Let $\lambda_i$'s be the Lagrange multipliers as in (7.2.4). Let $\mathcal{L}(x)$ be the Lagrange function in Theorem 7.2.5. Since $\nabla \mathcal{L}(u) = 0$, we can get

$$\mathcal{L}(x) - \mathcal{L}(u) = (x-u)^T \Big( \int_0^1 \int_0^t \nabla^2 \mathcal{L}(u + s(x-u)) \mathrm{d}s \mathrm{d}t \Big)(x-u).$$

Since $f$ and all $-c_i$ are SOS-convex, the matrix polynomial given by the above integral is SOS, with degree $2k_0 - 2$, by Lemma 7.1.5. As in the proof for Theorem 7.2.5, we can similarly show that $\mathcal{L}(x) - \mathcal{L}(u)$ is an SOS polynomial of degree $2k_0$. This further implies that $f - f_{min} \in \mathrm{QM}[c]_{2k}$ for all $k \geq k_0$. Therefore, $f_{sos,k} = f_{min}$ for all $k \geq k_0$.

Finally, if $K$ has nonempty interior, then $f_{sos,k} = f_{mom,k}$, by Proposition 5.2.1. So, $f_{sos,k} = f_{min}$.

(iv) Since $f$ and all $-c_i$ are SOS-convex, Jensen's inequality implies that

$$c_i(u) \geq \langle c_i, y^* \rangle \geq 0, \ i = 1, \ldots, m,$$
$$f_{min} \leq f(u) \leq \langle f, y^* \rangle = f_{mom,k} = f_{min}.$$

The last equality is implied by the item (iii). So $u$ is a minimizer of (7.2.3).          □

In Theorem 7.2.6(iii), if $f_{min}$ is not achievable at any KKT point and $K$ has empty interior, then we may not have $f_{sos,k} = f_{min}$. For instance, for the SOS-convex polynomial optimization

$$\begin{cases} \min & x \\ s.t. & -x^2 \geq 0, \end{cases}$$

we have $f_{sos,k} = -\infty$ for all $k$ but $f_{min} = 0$.

## 7.2.3 ▪ Exercises

**Exercise 7.2.1.** *Let $K$ be the set as in (7.2.1) with $c = (c_1, \ldots, c_m)$. For the following $c$, determine whether or not $K$ is degenerate.*

(i) $c = (x_1, \ x_1^2 - x_1^3 - x_2^2)$.

(ii) $c = (x_1, \ x_1^2 - x_2^2 - (x_1^2 + x_2^2)^2)$.

(iii) $c = (x_1(x_1^2 + x_2^2) - x_1^4 - x_1^2 x_2^2 - x_2^4)$.

(iv) $c = (2x_1 x_2 x_3 - x_1^2 - x_2^2 - x_3^2 + 1)$.

(v) $c = (x_1^3 - x_1 - x_2^2, \ (x_1 - 1)^2 + x_2^2 - 1)$.

(vi) $c = (x_2 - x_1^3, \ x_1 - x_2^3)$.

**Exercise 7.2.2.** *Let $K$ be the set as in (7.2.1). If the LICQC holds at every point $u \in K$, show that $K$ must be nondegenerate.*

**Exercise 7.2.3.** *If $c_1, \ldots, c_m \in \mathbb{R}[x]_{2d}$ are SOS-concave, show that the set $K$ in (7.2.1) is nonempty if and only if there exists $y \in \mathbb{R}^{\mathbb{N}^n_{2d}}$ satisfying*

$$\langle c_1, y \rangle \geq 0, \ldots, \langle c_m, y \rangle \geq 0, \ M_d[y] \succeq 0, \ y_0 = 1.$$

**Exercise 7.2.4.** *Prove Theorem 7.2.4.*

**Exercise 7.2.5.** *Let $f = f_1^2 + \cdots + f_r^2$ with each $f_i \in \mathbb{R}[x]_d$. For each $y \in \mathbb{R}^{\mathbb{N}^n_{2d}}$ with $M_d[y] \succeq 0$ and $y_0 = 1$, show that*

$$(\langle f_1, y \rangle)^2 + \cdots + (\langle f_r, y \rangle)^2 \leq \langle f, y \rangle.$$

## 7.3 ▪ Spectrahedra and Projections

A basic question in optimization is how to represent convex sets. Polyhedra are typical convex sets. More general ones are given by semidefinite programs. They are spectrahedra and their projections.

## 7.3.1 ▪ Spectrahedra

A set $S \subseteq \mathbb{R}^n$ is called a *spectrahedron* if it is given by a *linear matrix inequality* (LMI), i.e., there exist symmetric matrices $A_0, A_1, \ldots, A_n$ such that

$$S = \{x \in \mathbb{R}^n : A_0 + x_1 A_1 + \cdots + x_n A_n \succeq 0\}. \qquad (7.3.1)$$

If the origin $0$ lies in the interior of $S$, one can generally assume $A_0 \succ 0$ (i.e., it is positive definite). The matrix polynomial

$$A(x) := A_0 + x_1 A_1 + \cdots + x_n A_n$$

is called a *linear pencil*. In particular, if $A_0 = I$, $A(x)$ is called a *monic* linear pencil. Every spectrahedron is convex. For instance, an ellipsoid

$$E = \{x \in \mathbb{R}^n : (x - c)^T P^{-1} (x - c) \leq 1\},$$

for given $P \in \mathcal{S}_{++}^n$ and $c \in \mathbb{R}^n$, is a spectrahedron, since it can be represented by the linear matrix inequality

$$\begin{bmatrix} P & x - c \\ (x - c)^T & 1 \end{bmatrix} \succeq 0.$$

More general spectrahedra than ellipsoids are $k$-ellipses and matrix cubes parameterized by eigenvalues (see [214, 215] and Exercise 7.3.9). Spectrahedra are convex sets of broad applications (see [19, 36, 37, 209]).

There are necessary conditions, beyond the convexity, for sets to be spectrahedra. Suppose $S$ is a spectrahedron given as in (7.3.1). The boundary of $S$ lies on the determinantal hypersurface

$$\det A(x) = 0.$$

Assume $S$ has nonempty interior. Up to shifting an interior point to $0$, one can generally assume $A_0 \succ 0$. Then $S$ is the closure of the connected component of $\det A(x) > 0$ containing the origin. This observation leads to the definition of *algebraic interior* [118]. A set $T \subseteq \mathbb{R}^n$ is called an algebraic interior if it equals the closure of a connected component of the set given by the inequality $p(x) > 0$ for some polynomial $p \in \mathbb{R}[x]$. The $p$ is called a *defining polynomial* of $T$. The defining polynomial for $T$ is not unique. The smallest degree defining polynomial for $T$ is unique, up to a positive scaling of coefficients, and its degree is called the degree of $T$ [118].

**Example 7.3.1.** *([223]) Consider the elliptope given by the LMI*

$$\begin{bmatrix} 1 & x_1 & x_2 \\ x_1 & 1 & x_3 \\ x_2 & x_3 & 1 \end{bmatrix} \succeq 0.$$

*It is an algebraic interior, whose smallest degree defining polynomial is*

$$p := 2x_1 x_2 x_3 - x_1^2 - x_2^2 - x_3^2 + 1.$$

*The polynomial $p$ is the determinant of the above monic linear pencil. The elliptope can be equivalently defined by four inequalities:*

$$p(x) \geq 0, \quad 1 - x_1^2 \geq 0, \quad 1 - x_2^2 \geq 0, \quad 1 - x_3^2 \geq 0.$$

For a set to be a spectrahedron, the convexity is not enough. It needs to satisfy an additional property for defining polynomials. To see this, suppose $S$ is a spectrahedron as in (7.3.1), with

$A(0) = I$. Then $S$ is the algebraic interior $f := \det A(x) > 0$, containing the origin. For a nonzero vector $d = (d_1, \ldots, d_n) \in \mathbb{R}^n$, consider the line $x(t) := td$ passing through the origin. Note that

$$f(x(t)) = \det(I + tB), \quad \text{where} \quad B = d_1 A_1 + \cdots + d_n A_n.$$

Since $B$ is symmetric, the equation $f(x(t)) = 0$ in $t$ has only real roots. Such $f$ is called a Real Zero polynomial.

**Definition 7.3.2.** *([118]) A polynomial $p \in \mathbb{R}[x]$ is called Real Zero (RZ) with respect to a point $u$ with $p(u) > 0$ if for every $0 \neq d \in \mathbb{R}^n$, the univariate polynomial $p(u + td)$ in $t$ has only real zeros. In case of $u = 0$, such $p$ is just simply called RZ.*

For instance, the cubic polynomial $2x_1 x_2 x_3 - x_1^2 - x_2^2 - x_3^2 + 1$ is RZ, because it is the determinant of the monic linear pencil in Example 7.3.1. The polynomial $p := 1 - (x_1^4 + x_2^4)$ is not RZ [118]. For every $0 \neq (d_1, d_2) \in \mathbb{R}^2$, the univariate polynomial in $t$

$$p(td) = \left(1 - t^2(d_1^4 + d_2^4)^{\frac{1}{2}}\right)\left(1 + t^2(d_1^4 + d_2^4)^{\frac{1}{2}}\right)$$

has two non-real zeros.

RZ polynomials are nonhomogeneous versions of hyperbolic polynomials. A homogeneous polynomial $h \in \mathbb{R}[x]^{hom}$ is called *hyperbolic* with respect to a point $u \in \mathbb{R}^n$ with $h(u) > 0$ if for every $0 \neq d \in \mathbb{R}^n$ the univariate polynomial $t \mapsto h(u + td)$ has only real zeros. Clearly, if a form $h(x)$ is hyperbolic with respect to $u = (1, u_2, \ldots, u_n)$, then the dehomogenization $h(1, x_2, \ldots, x_n)$ is RZ with respect to $(u_2, \ldots, u_n)$. The following is a characterization for algebraic interiors that are given by RZ polynomials.

**Theorem 7.3.3.** *([118]) Suppose $T \subseteq \mathbb{R}^n$ is an algebraic interior with the minimum degree polynomial $p$.*

 (i) *If $p$ is RZ with respect to a point $u \in \mathrm{int}\,(T)$, then $T$ must be convex.*

 (ii) *If $p$ is RZ with respect to a point $u \in \mathrm{int}\,(T)$, then $p$ is RZ with respect to every point $v \in \mathrm{int}\,(T)$.*

The relationship between spectrahedra and RZ polynomials is characterized as follows.

**Theorem 7.3.4.** *([118]) If an algebraic interior $T \subseteq \mathbb{R}^n$ is a spectrahedron, then its minimum degree defining polynomial $p$ must be RZ with respect to an interior point of $T$. For the two dimensional case (i.e., $n = 2$), the converse is also true, and $T$ can be represented by a monic LMI whose size equals the degree of $p$.*

Not every convex algebraic interior is a spectrahedron. For instance, the convex set $1 - x_1^4 - x_2^4 \geq 0$ is not a spectrahedron, because $1 - x_1^4 - x_2^4$ is not RZ with respect to the origin.

## 7.3.2 ▪ Projections of spectrahedra

A set $P \subseteq \mathbb{R}^n$ is called the projection of a spectrahedron, or just a *projected spectrahedron*, if there exists a spectrahedron $S \subseteq \mathbb{R}^{n+k}$ such that

$$P = \left\{ x \in \mathbb{R}^n : (x, y) \in S \text{ for some } y \in \mathbb{R}^k \right\}. \tag{7.3.2}$$

In the above, $y$ is called a *lifting variable* and $S$ is called a *lifting spectrahedron*. If it is a projected spectrahedron, then $P$ is called *semidefinite representable* (SDr) and $P$ is said to have a semidefinite representation. Every projected spectrahedron is convex. If $P$ is SDr, then there exist real symmetric matrices $A_i, B_j$ such that (note $x = (x_1, \ldots, x_n)$ and $y = (y_1, \ldots, y_k)$)

$$P = \left\{ x \in \mathbb{R}^n : A_0 + \sum_{i=1}^{n} x_i A_i + \sum_{j=1}^{k} y_j B_j \succeq 0 \text{ for some } y \in \mathbb{R}^k \right\}. \qquad (7.3.3)$$

The LMI in (7.3.3) is called a semidefinite representation for $P$. For instance, the set $1 - (x_1^4 + x_2^4) \geq 0$ is a projected spectrahedron, because it admits the semidefinite representation

$$\mathrm{diag} \left( \begin{bmatrix} 1 + y_1 & y_2 \\ y_2 & 1 - y_1 \end{bmatrix}, \begin{bmatrix} 1 & x_1 \\ x_1 & y_1 \end{bmatrix}, \begin{bmatrix} 1 & x_2 \\ x_2 & y_2 \end{bmatrix} \right) \succeq 0.$$

It has two lifting variables. We have already seen that it is not a spectrahedron, because its defining polynomial is not RZ. More examples of projected spectrahedra can be found in the exercises.

It is clear that projected spectrahedra must be convex and semialgebraic. The semidefinite representability is related to geometric properties of their boundaries. The *curvature* is a useful property. Let $f : \mathbb{R}^n \to \mathbb{R}$ be a smooth function. Consider the hypersurface

$$\mathcal{H} = \{ x \in \mathbb{R}^n : f(x) = 0 \}.$$

The function $f$ is called *nonsingular* at $u \in \mathcal{H}$ if $\nabla f(u) \neq 0$. If $f$ is nonsingular at $u$, the hypersurface $\mathcal{H}$ has *positive curvature* at $u$ if its *second fundamental form* is definite (either positive or negative), i.e., for some $s \in \{-1, 1\}$,

$$s \cdot v^T \nabla^2 f(u) v > 0 \quad \text{for all } 0 \neq v \in \nabla f(u)^\perp. \qquad (7.3.4)$$

Similarly, $\mathcal{H}$ is said to have *nonnegative curvature* at $u$ if the second fundamental form is semi-definite (either positive or negative). The positive (or nonnegative) curvature of $\mathcal{H}$ at $u$ is independent of the choice of its defining functions (see [119, §3]). Geometrically, when $f$ is nonsingular at $u$, $\mathcal{H}$ has positive curvature at $u$ if and only if there exists a neighborhood $\mathcal{O}$ of $u$ such that $\mathcal{H} \cap \mathcal{O}$ is the graph of a convex function whose Hessian is positive definite. For a subset $T \subseteq \mathcal{H}$, the hypersurface $\mathcal{H}$ is said to have positive curvature on $T$ if for all $u \in T$, $f$ is nonsingular and $\mathcal{H}$ has positive curvature at $u$. For instance, the hypersurface $1 - x_1^4 - x_2^4 = 0$ has zero curvature on the points $(\pm 1, 0), (0, \pm 1)$ but it has positive curvature elsewhere.

A function $f$ is said to be *strictly quasi-convex* at $u$ if the condition (7.3.4) holds for $s = 1$. If (7.3.4) holds when $>$ is replaced by $\geq$ for $s = 1$, then $f$ is said to be *quasi-convex*. The quasi-concavity and strict quasi-concavity are similarly defined for $s = -1$. The definitions of quasi-convexity and quasi-concavity here are slightly less demanding than the classical ones in some literature (see [37, §3.4.3]). For instance, the two-dimensional hyperboloid $x_1 x_2 - 1 \geq 0$ has positive curvature everywhere on the boundary. The defining function $x_1 x_2 - 1$ is neither convex nor concave, but it is strictly quasi-concave.

Recall the definition of semialgebraic sets and the irredundancy of defining polynomials as in §2.2. The following are some properties of projected spectrahedra. They give some necessary conditions for semidefinite representability.

**Theorem 7.3.5.** *([119]) If a set $P \subseteq \mathbb{R}^n$ is the projection of a spectrahedron, then $P$ must be convex and satisfy the following:*

(a) *The interior* $\mathrm{int}(P)$ *is a finite union of basic open semialgebraic sets, i.e.,*

$$\mathrm{int}\,(P) = \bigcup_{k=1}^{m} T_k, \quad T_k = \{x \in \mathbb{R}^n : g_1^k(x) > 0, \cdots, g_{m_k}^k(x) > 0\}$$

*for some polynomials* $g_i^k(x)$.

(b) *The closure* $\mathrm{cl}\,(P)$ *of* $P$ *is a finite union of basic closed semialgebraic sets, i.e.,*

$$\mathrm{cl}\,(P) = \bigcup_{k=1}^{m} T_k, \quad T_k = \{x \in \mathbb{R}^n : g_1^k(x) \geq 0, \ldots, g_{m_k}^k(x) \geq 0\}$$

*for some polynomials* $g_i^k(x)$ *(they might be different from those in (a) above).*

(c) *For* $u$ *on the boundary of* $\mathrm{cl}\,(P)$ *and polynomials* $g_i^k$ *from (b), let*

$$I_k(u) := \{i \in [m_k] : g_i^k(u) = 0\}.$$

*For* $i \in I_k(u) \neq \emptyset$, *if* $g_i^k$ *is irredundant and nonsingular at* $u$, *then* $g_i^k$ *is quasi-concave at* $u$.

As shown in Theorem 7.3.5, the boundary of a projected spectrahedron must have nonnegative curvature at smooth points, in addition to that the set must be convex and semialgebraic. However, they are not sufficient for semidefinite representability, as recently shown by Scheiderer [296]. In the item (c) of Theorem 7.3.5, the assumption that $g_i^k$ is irredundant cannot be dropped.

**Example 7.3.6.** *([223]) Consider the set in* $\mathbb{R}^2$ *defined as*

$$g_1 := x_1^3 - x_1 - x_2^2 \geq 0, \quad g_2 := (x_1 - 1)^2 + x_2^2 - 1 \geq 0.$$

*The real zero set of* $g_1$ *is not connected and has two components, so the inequality* $g_2(x) \geq 0$ *cannot be dropped for defining this set. The polynomial* $g_2$ *is redundant at the origin, and it is not quasi-concave at* $0$.

It is generally difficult to get polynomials $g_i^k$ as in Theorem 7.3.5. They can be obtained by eliminating lifting variables (see [310]). The following gives a sufficient condition for a set to be a projected spectrahedron. It requires the boundary to have positive curvature.

**Theorem 7.3.7.** *([119]) Suppose* $P \subseteq \mathbb{R}^n$ *is a compact convex set given as*

$$P = \bigcup_{k=1}^{m} T_k := \{x \in \mathbb{R}^n : g_1^k(x) \geq 0, \ldots, g_{m_k}^k(x) \geq 0\}$$

*for some polynomials* $g_i^k$. *If for every* $u$ *on the boundary of* $P$ *and for every* $g_i^k$ *satisfying* $g_i^k(u) = 0$, $T_k$ *has interior near* $u$ *(i.e., there exists* $\delta > 0$ *such that* $T_k \cap B(u, \delta)$ *has nonempty interior) and* $g_i^k(x)$ *is strictly quasi-concave at* $u$, *then* $P$ *is a projected spectrahedron.*

### 7.3.3 ▪ Moment relaxations

Semidefinite representations for convex semialgebraic sets may be possibly given by their moment relaxations. For instance, the one-dimensional set of

$$1 - x^2 - x^3 - x^4 \geq 0$$

can be equivalently given as (with new variables $y_i$)

$$1 - x^2 - x^3 - x^4 \geq 0, \quad \begin{bmatrix} y_0 & y_1 & y_2 \\ y_1 & y_2 & y_3 \\ y_2 & y_3 & y_4 \end{bmatrix} = \begin{bmatrix} 1 & x & x^2 \\ x & x^2 & x^3 \\ x^2 & x^3 & x^4 \end{bmatrix} \succeq 0.$$

If the high degree terms $x^2, x^3, x^4$ are replaced by the variables $y_2, y_3, y_4$, then this set is contained in the projection of the following spectrahedron:

$$\mathrm{diag}\left(1 - y_2 - y_3 - y_4, \begin{bmatrix} 1 & x & y_2 \\ x & y_2 & y_3 \\ y_2 & y_3 & y_4 \end{bmatrix}\right) \succeq 0.$$

Interestingly, the above projection precisely equals the original set. This is also possible for higher dimensional sets.

Let $g := (g_1, \ldots, g_m)$ be a tuple of polynomials. Consider the set

$$S := \{x \in \mathbb{R}^n : g_1(x) \geq 0, \ldots, g_m(x) \geq 0\}. \tag{7.3.5}$$

Let $d$ be the degree

$$d := \max\left\{\lceil \tfrac{1}{2}\deg(g_1)\rceil, \ldots, \lceil \tfrac{1}{2}\deg(g_m)\rceil\right\}. \tag{7.3.6}$$

Recall the moment matrix $M_d[y]$ as in (2.4.8). It holds that

$$S = \left\{x \in \mathbb{R}^n \;\middle|\; \begin{array}{l} y = [x]_{2d},\; M_d[y] \succeq 0, \\ \langle g_i, y\rangle \geq 0\,(i = 1, \ldots, m) \end{array}\right\}. \tag{7.3.7}$$

If the equality $y = [x]_{2d}$ is ignored, then $S$ is contained in the set

$$R := \left\{x \in \mathbb{R}^n \;\middle|\; \begin{array}{l} x = (y_{e_1}, \ldots, y_{e_n}),\; y_0 = 1, \\ \langle g_i, y\rangle \geq 0\,(i = 1, \ldots, m), \\ M_d[y] \succeq 0,\; y \in \mathbb{R}^{\mathbb{N}^n_{2d}}. \end{array}\right\}. \tag{7.3.8}$$

Clearly, $R$ is a projected spectrahedron. Since $S \subseteq R$, the set $R$ is called a *semidefinite relaxation* or *moment relaxation* for $S$.

**Example 7.3.8.** *For the set*

$$S = \{x \in \mathbb{R}^2 : 1 - x_1^4 - x_1^2 x_2^2 - x_2^4 \geq 0\},$$

*the relaxation $R$ in (7.3.8) is given by*

$$1 - y_{40} - y_{22} - y_{04} \geq 0, \quad \begin{bmatrix} 1 & x_1 & x_2 & y_{20} & y_{11} & y_{02} \\ x_1 & y_{20} & y_{11} & y_{30} & y_{21} & y_{12} \\ x_2 & y_{11} & y_{02} & y_{21} & y_{12} & y_{03} \\ y_{20} & y_{30} & y_{21} & y_{40} & y_{31} & y_{22} \\ y_{11} & y_{21} & y_{12} & y_{31} & y_{22} & y_{13} \\ y_{02} & y_{12} & y_{03} & y_{22} & y_{13} & y_{04} \end{bmatrix} \succeq 0.$$

*There are* 12 *lifting variables* $y_{ij}$.

The set $R$ in (7.3.8) is generally a convex relaxation for $S$. Interestingly, if each $g_i$ is SOS-concave, then they are equal.

**Theorem 7.3.9.** *([120]) Let $S, R$ be as in (7.3.5), (7.3.8) respectively. If each polynomial $g_i$ is SOS-concave, then $R = S$.*

**Proof.** Since $S \subseteq R$, it is enough to show $R \subseteq S$. For all $x$ and $y$ satisfying (7.3.8), Jensen's inequality (see Theorem 7.1.6) implies

$$g_i(x) \geq \langle g_i, y \rangle \geq 0,$$

since each $g_i$ is SOS-concave. This implies that $R \subseteq S$. $\square$

The polynomial in Example 7.3.8 is SOS-concave, because

$$\nabla^2(x_1^4 + x_2^4 + x_1^2 x_2^2) = 4 \begin{bmatrix} x_1 \\ x_2 \end{bmatrix} \begin{bmatrix} x_1 \\ x_2 \end{bmatrix}^T + 2 \begin{bmatrix} 2x_1 \\ & x_1 \end{bmatrix}^2 + 2 \begin{bmatrix} x_2 \\ & 2x_2 \end{bmatrix}^2.$$

By Theorem 7.3.9, the projected spectrahedron $R$ in (7.3.8) equals $S$ precisely. When the polynomials $g_i$ are not SOS-concave, the convex relaxation $R$ in (7.3.8) may not equal $S$. However, higher order moment relaxations can give tighter (or even exact) semidefinite relaxations.

Let $d$ be the degree as in (7.3.6). We refer to §2.5.2 for the localizing matrix $L_{g_i}^{(k)}[y]$ determined by the polynomial $g_i$. For each integer $k \geq d$, consider the semidefinite relaxation

$$S_k = \left\{ x \in \mathbb{R}^n \,\middle|\, \begin{array}{l} x = (y_{e_1}, \ldots, y_{e_n}), \; y_0 = 1, \\ L_{g_i}^{(k)}[y] \succeq 0 \, (i = 1, \ldots, m), \\ M_k[y] \succeq 0, \; y \in \mathbb{R}^{\mathbb{N}_{2k}^n} \end{array} \right\}. \tag{7.3.9}$$

**Example 7.3.10.** *Consider the convex set $S$ given as*

$$g_1 := x_2 - x_1^3 \geq 0, \quad g_2 := x_2 + x_1^3 \geq 0.$$

*The moment relaxation $S_3$ in (7.3.9) is*

$$L_{g_1}^{(3)}[y] = \begin{bmatrix} y_{01} - y_{30} & y_{11} - y_{40} & y_{02} - y_{31} \\ y_{11} - y_{40} & y_{21} - y_{50} & y_{12} - y_{41} \\ y_{02} - y_{31} & y_{12} - y_{41} & y_{03} - y_{32} \end{bmatrix} \succeq 0,$$

$$L_{g_2}^{(3)}[y] = \begin{bmatrix} y_{01} + y_{30} & y_{11} + y_{40} & y_{02} + y_{31} \\ y_{11} + y_{40} & y_{21} + y_{50} & y_{12} + y_{41} \\ y_{02} + y_{31} & y_{12} + y_{41} & y_{03} + y_{32} \end{bmatrix} \succeq 0,$$

*plus the constraint $x_1 = y_{10}$, $x_2 = y_{01}$, $y_{00} = 1$, and $M_3[y] \succeq 0$. Higher order moment relaxations are given similarly.*

The moment relaxations $S_k$ for $S$ satisfy the nesting containment

$$S_d \supseteq S_{d+1} \supseteq \cdots \supseteq S. \tag{7.3.10}$$

Even if $S$ is nonconvex, each $S_k$ is convex and contains the convex hull of $S$ in the nesting pattern as in (7.3.10). Therefore, we also have

$$S_d \supseteq S_{d+1} \supseteq \cdots \supseteq \text{conv}(S). \tag{7.3.11}$$

Indeed, we have the following result.

**Theorem 7.3.11.** *([163]) Let $S$ be the set given as in (7.3.5). If the quadratic module $\mathrm{QM}[g]$ is archimedean, then*

$$\mathrm{conv}\,(S) = \bigcap_{k \geq d} S_k. \qquad (7.3.12)$$

The proof of Theorem 7.3.11 is left as an exercise. It implies that the convex relaxation $S_k$ can approximate $\mathrm{conv}\,(S)$ arbitrarily accurately. People are interested in whether or not $S_k = \mathrm{conv}\,(S)$ for some $k$. This may or may not occur. In fact, the equality $S_k = \mathrm{conv}\,(S)$ occurs *only if* all faces of $\mathrm{conv}\,(S)$ are exposed (see [206]). Interestingly, convex hulls of some quadratically parameterized sets can be represented by moment relaxations, as shown in the reference [222]. In the following, we give some conditions for the equality $S_k = \mathrm{conv}\,(S)$ to hold.

Note that $\mathrm{conv}\,(S) \subseteq S_k$ for all $k$. If $\mathrm{conv}\,(S) \neq S_k$ for some $k$, then there must exist $u \in S_k$ but $u \notin \mathrm{conv}\,(S)$. If $\mathrm{conv}\,(S)$ is closed, there exists a hyperplane, say, $a^T x + b = 0$, which strictly separates $u$ from $\mathrm{conv}\,(S)$, i.e.,

$$a^T u + b < 0, \qquad a^T x + b > 0 \quad \text{on} \quad \mathrm{conv}\,(S).$$

For the tms $z := [u]_{2k} \in \mathscr{S}[g]_{2k}$, one can see

$$\langle a^T x + b, z \rangle < 0.$$

Since $\mathscr{S}[g]_{2k}$ is dual to $\mathrm{QM}[g]_{2k}$, we can see that $a^T x + b \notin \mathrm{QM}[g]_{2k}$. Since $a^T x + b > 0$ on $S$, there is a linear polynomial that is positive on $S$ but it does not belong to $\mathrm{QM}[g]_{2k}$. This argument shows that if every linear polynomial that is positive on $S$ belongs to $\mathrm{QM}[g]_{2k}$, then $\mathrm{conv}\,(S) = S_k$. This observation leads to the condition called *Putinar-Prestel's Bounded Degree Representation (PP-BDR)* of order $k$ (see [163]): for *almost every* $(a, b) \in \mathbb{R}^n \times \mathbb{R}$

$$a^T x + b > 0 \text{ on } S \quad \Rightarrow \quad a^T x + b \in \mathrm{QM}[g]_{2k}.$$

(The words "almost every" mean that the above implication holds for all $(a, b) \in \mathbb{R}^n \times \mathbb{R}$ except a subset whose Lebesgue measure is zero.) The PP-BDR is a useful criterion for justifying the equality $\mathrm{conv}\,(S) = S_k$.

**Theorem 7.3.12.** *([163]) Suppose $S$ is a compact set as in (7.3.5), given by a polynomial tuple $g$. Then, for the convex relaxation $S_k$ in (7.3.9), the equality $\mathrm{conv}\,(S) = S_k$ holds if and only if the PP-BDR of order $k$ holds for the set $S$ and $g$.*

The proof of Theorem 7.3.12 is left as an exercise. Generally, it is quite difficult to check the PP-BDR for given $g$. In practice, we prefer concrete and easily checkable conditions. Indeed, there exist such conditions about the Hessians of the defining polynomials $g_i$.

**Theorem 7.3.13.** *([120]) Let $S$ be given as in (7.3.5) for a polynomial tuple $g := (g_1, \ldots, g_m)$. Assume that $S$ is compact and has nonempty interior, each $g_i$ is concave on $S$, and $\mathrm{QM}[g]$ is archimedean. If, for each $i = 1, \ldots, m$, either $g_i$ is SOS-concave or $-\nabla^2 g_i(u) \succ 0$ for all $u \in \partial S \cap \{g_i(x) = 0\}$, then the PP-BDR of order $k$ holds, when $k$ is sufficiently large. Therefore, $S_k = S$ for all $k$ big enough.*

The proof of Theorem 7.3.13 is given in [120], which can be outlined as follows. Let $a^T x + b = 0$ be a supporting hyperplane for $S$ through a point $u \in S$. The KKT conditions are

$$a = \lambda_1 \nabla g_1(u) + \cdots + \lambda_m \nabla g_m(u),$$
$$\lambda_1 g_1(u) = \cdots = \lambda_m g_m(u) = 0,$$
$$\lambda_1 \geq 0, \ldots, \lambda_m \geq 0.$$

The Lagrange function is

$$\mathcal{L}(x) := a^T x + b - (\lambda_1 g_1(x) + \cdots + \lambda_m g_m(x)).$$

Since $\mathcal{L}(u) = 0$, $\nabla \mathcal{L}(u) = 0$, we have

$$\mathcal{L}(x) = (x - u)^T \left( \sum_{i=1}^m \lambda_i \underbrace{\int_0^1 \int_0^t -\nabla^2 g_i(u + s(x - u)) ds\, dt}_{H_i(x)} \right) (x - u).$$

If each $H_i(x)$ is an SOS matrix polynomial, then $\mathcal{L}(x)$ must be SOS, which further implies $a^T x + b \in \mathrm{QM}[g]$. The condition that $g_i$ is SOS-concave or $-\nabla^2 g_i \succ 0$ on $\partial S \cap \{g_i(x) = 0\}$ ensures that each $H_i(x)$ has a Putinar type SOS representation. The degree bounds of the SOS polynomials can be obtained by the positive definiteness of $H_i(x)$. Hence, the PP-BDR of order $k$ holds, when $k$ is large enough. We refer to [120] for more details.

When $S$ is convex, the defining polynomials $g_i$ are not necessarily concave. However, the boundary of $S$ has nonnegative curvature at smooth points (see Theorem 7.3.5). This implies the defining polynomials are quasi-concave at smooth points, which is related to quasi-concavity or strict quasi-concavity of defining polynomials. Interestingly, if each $g_i$ is strictly quasi-concave, there exists a new set of strictly concave defining polynomials.

**Theorem 7.3.14.** *([120, Prop. 10]) Let $S$ be a compact set as in (7.3.5), given by a polynomial tuple $g := (g_1, \ldots, g_m)$. If each $g_i$ is strictly quasi-concave on $S$, then there is a polynomial $h_i$, which is positive on $S$, such that the product $p_i := g_i h_i$ satisfies $-\nabla^2 p_i \succ 0$ on $S$.*

## 7.3.4 ▪ Exercises

**Exercise 7.3.1.** *Decide whether or not the following polynomials are RZ with respect to the point $(1, \ldots, 1)$:*

(i) $x_1 \cdots x_n - 1/3$.

(ii) $n x_n^2 - (x_1^2 + \cdots + x_{n-1}^2)$.

(iii) $x_1 \cdots x_n \cdot \sum_{i<j} \frac{1}{x_i x_j}$.

**Exercise 7.3.2.** *Give semidefinite representations for the following sets:*

(i) $S = \{(x_1, x_2) : x_1 \geq 0, x_2 \geq 0, x_1^3 + x_2^3 \leq 1\}$.

(ii) $S = \{(x_1, x_2) : x_1^2 - x_2^2 - (x_1^2 + x_2^2)^2 \geq 0, x_1 \geq 0\}$.

(iii) $P = \left\{ (f_0, f_1, f_2, f_3) : f_0 + f_1 x + f_2 x^2 + f_3 x^3 \geq 0 \ \forall x \in [-1, 1] \right\}$.

*(iv)* $Q = \left\{ (A, b, c) \in \mathcal{S}^n \times \mathbb{R}^n \times \mathbb{R} : x^T A x + 2b^T x + c \geq 0 \; \forall \, \|x\|_2 \leq 1 \right\}.$

**Exercise 7.3.3.** *For a quadratic polynomial $q \in \mathbb{R}[x]_2$, show that the set $Q := \{x \in \mathbb{R}^n : q(x) \geq 0\}$ is convex if and only if $Q$ is a spectrahedron.*

**Exercise 7.3.4.** *If a cone $K \subseteq \mathbb{R}^n$ is a projected spectrahedron, show that its dual cone $K^*$ is also a projected spectrahedron.*

**Exercise 7.3.5.** *Let $K = \{x \in \mathbb{R}^n : g(x) \geq 0\}$ be a convex set defined by a smooth function $g : \mathbb{R}^n \to \mathbb{R}$. Let $u$ be a point on the boundary of $K$ with $\nabla g(u) \neq 0$. Show the following:*

 *(i) The Hessian $-\nabla^2 g(u)$ is psd in the tangent space $\nabla g(u)^\perp$, i.e.,*

$$-v^T \nabla^2 g(u) v \geq 0 \quad \text{for all } v \in \nabla g(u)^\perp.$$

 *(ii) The equation $\nabla g(u)^T (x - u) = 0$ gives a supporting hyperplane for $K$ through $u$.*

**Exercise 7.3.6.** *Complete the proofs for Theorems 7.3.11 and 7.3.12.*

**Exercise 7.3.7.** *([19]) For each rational $r \in [0, \frac{1}{n}]$, show that the $n$-dimensional hyperboloid $\{(x, t) \in \mathbb{R}^n_+ \times \mathbb{R} : (x_1 \cdots x_n)^r \geq t\}$ is a projected spectrahedron.*

**Exercise 7.3.8.** *([19]) For a symmetric $X \in \mathcal{S}^n$, denote by $\lambda_i(X)$ the $i$th largest eigenvalue of $X$. Let $s_k(X) := \lambda_1(X) + \cdots + \lambda_k(X)$, the sum of $k$ largest eigenvalues of $X$. The epigraph of the function $s_k(\cdot)$ is the set*

$$G := \{(X, t) \in \mathcal{S}^n \times \mathbb{R} : s_k(X) \leq t\}.$$

*Show that $G$ is a projected spectrahedron. Hint: show that $s_k(X) \leq t$ if and only if there exists $(Z, \tau) \in \mathcal{S}^n \times \mathbb{R}$ such that*

$$\begin{cases} t - k\tau - Tr(Z) & \geq 0, \\ Z & \succeq 0, \\ Z - X + \tau I_n & \succeq 0. \end{cases}$$

**Exercise 7.3.9.** *([214]) For given points $c_1, \ldots, c_k \in \mathbb{R}^n$ and $r > 0$, the set of points $x \in \mathbb{R}^n$ satisfying*

$$\|x - c_1\|_2 + \cdots + \|x - c_k\|_2 = r$$

*is called a $k$-ellipse. It is known in [214] that the region surrounded by a $k$-ellipse is a spectrahedron. Show that the region surrounded by the following 3-ellipse:*

$$\sqrt{x_1^2 + x_2^2} + \sqrt{(x_1 - 1)^2 + x_2^2} + \sqrt{x_1^2 + (x_2 - 1)^2} = 5$$

*is the spectrahedron given by the following LMI:*

$$\begin{bmatrix}
6 - 3x_1 & x_2 & x_2 - 1 & 0 & x_2 & 0 & 0 & 0 \\
x_2 & 6 - x_1 & 0 & x_2 - 1 & 0 & x_2 & 0 & 0 \\
x_2 - 1 & 0 & 6 - x_1 & x_2 & 0 & 0 & x_2 & 0 \\
0 & x_2 - 1 & x_2 & 6 + x_1 & 0 & 0 & 0 & x_2 \\
x_2 & 0 & 0 & 0 & 4 - x_1 & x_2 & x_2 - 1 & 0 \\
0 & x_2 & 0 & 0 & x_2 & 4 + x_1 & 0 & x_2 - 1 \\
0 & 0 & x_2 & 0 & x_2 - 1 & 0 & 4 + x_1 & x_2 \\
0 & 0 & 0 & x_2 & 0 & x_2 - 1 & x_2 & 4 + 3x_1
\end{bmatrix} \succeq 0.$$

# 7.4 ▪ The QM-Convexity

A convex semialgebraic set may be given by polynomials that are only concave over a convex domain $\mathcal{D} \subseteq \mathbb{R}^n$. For instance, the set $\{x_2 - x_1^3 \geq 0, x_1 \geq 0\}$ is convex, while the polynomial $x_2 - x_1^3$ is concave only on $x_1 \geq 0$. There are also convex sets that are described by rational functions that are concave over a convex domain $\mathcal{D}$. If we describe them with polynomials, the concavity may be lost. For instance, the convex set

$$\left\{ x \in \mathbb{R}^2_{++} : 1 - \frac{1}{x_1 x_2} \geq 0 \right\}$$

is given by a rational function that is concave over $\mathbb{R}^2_{++}$. However, this set can also be given as $x_1 x_2 - 1 \geq 0, x_1 \geq 0$, but the polynomial $x_1 x_2 - 1$ is not concave anywhere. There are also some convex sets that are given by polynomials or rational functions having singularities on the boundary. For instance, the set

$$\{ x \in \mathbb{R}^2 : x_1^2 - x_1^3 - x_2^2 \geq 0, x_1 \geq 0 \}$$

is convex, with $0$ a singular point on the boundary.

We study semidefinite representations for these kinds of convex sets in this section. Assume that $\mathcal{D}$ is a convex domain with nonempty interior, given as

$$\mathcal{D} = \{ x \in \mathbb{R}^n : g_1(x) \geq 0, \ldots, g_m(x) \geq 0 \} \tag{7.4.1}$$

for some polynomials $g_1, \ldots, g_m$. For a polynomial or rational function $f : \mathcal{D} \to \mathbb{R}$, denote its suplevel set

$$S_{\mathcal{D}}(f) := \{ x \in \mathcal{D} : f(x) \geq 0 \}. \tag{7.4.2}$$

The *first order Lagrange remainder* for the function $f$ is

$$R_f(x, u) := f(x) - f(u) - \nabla f(u)^T (x - u). \tag{7.4.3}$$

By the Taylor expansion remainder expression formula,

$$R_f(x, u) = \int_0^1 (1 - t) \cdot (x - u)^T \left[ \nabla^2 f(u + t(x - u)) \right] (x - u) \mathrm{d}t. \tag{7.4.4}$$

When $\mathcal{D}$ has nonempty interior, $f$ is convex over $\mathcal{D}$ if and only if $R_f(x, u) \geq 0$ for all $x, u \in \mathcal{D}$. Similarly, $f$ is concave over $\mathcal{D}$ if and only if $-R_f(x, u) \geq 0$ for all $x, u \in \mathcal{D}$. Putinar's Positivstellensatz (see §2.6) can be applied to verify convexity or concavity.

**Definition 7.4.1.** *A polynomial $f$ is said to be QM-convex with respect to $g := (g_1, \ldots, g_m)$ if (let $g_0 := 1$)*

$$R_f(x, u) = \sum_{i=0}^{m} g_i(x) \left( \sum_{j=0}^{m} g_j(u) \sigma_{ij}(x, u) \right) \tag{7.4.5}$$

*for some SOS polynomials $\sigma_{ij} \in \Sigma[x, u]$. Similarly, $f$ is said to be QM-concave if $-f$ is QM-convex.*

A convex domain $\mathcal{D}$ may be given by different polynomial tuples $g$. Typically, we prefer defining polynomials $g_i$ whose descriptions are the most convenient ones. If $g$ is clear in the context for $\mathcal{D}$, we just simple say that $f$ is QM-convex (or QM-concave) over $\mathcal{D}$ if it is QM-convex (or QM-concave) with respect to $g$. For instance, the polynomial $f = x_1^3 + x_1^2 x_2 + x_1 x_2^2 + x_2^3$ is convex over $\mathbb{R}_+^2$. It is also QM-convex with respect to $g := (x_1, x_2)$, because

$$
\begin{aligned}
R_f(x, u) &= \left(\frac{1}{3}u_1 + \frac{1}{6}x_1\right)\left(4(x_1 - u_1)^2 + 2(x_1 + x_2 - u_1 - u_2)^2\right) \\
&\quad + \left(\frac{1}{3}u_2 + \frac{1}{6}x_2\right)\left(4(x_2 - u_2)^2 + 2(x_1 + x_2 - u_1 - u_2)^2\right).
\end{aligned}
$$

The set $S_{\mathcal{D}}(f)$ is semidefinite representable if $f$ and all $g_i$ are QM-concave with respect to $g$. For each $h \in \{f, g_1, \ldots, g_m\}$, suppose the Lagrange remainder $R_h(x, u)$ can be written as

$$
-R_h(x, u) = \sum_{i=0}^{m} g_i(x)\left(\sum_{j=0}^{m} g_j(u)\sigma_{ij}^h(x, u)\right)
$$

for some SOS polynomials $\sigma_{ij}^h(x, u) \in \Sigma[x, u]$. Denote the degree

$$
d := \max_{h \in \{f, g_1, \ldots, g_m\}} \max_{0 \le i, j \le m} \lceil \frac{1}{2} \deg_x(g_i \sigma_{ij}^h) \rceil, \tag{7.4.6}
$$

where $\deg_x(\cdot)$ denotes the degree in the variable $x$. We consider the following moment relaxation for the suplevel set $S_{\mathcal{D}}(f)$:

$$
\Omega(f) = \left\{ x \in \mathbb{R}^n \;\middle|\; 
\begin{array}{l}
x = (y_{e_1}, \ldots, y_{e_n}),\; y \in \mathbb{R}^{\mathbb{N}_{2d}^n}, \\
L_h^{(d)}[y] \succeq 0 \; (h \in \{1, g_1, \ldots, g_m\}), \\
y_0 = 1,\; \langle f, y \rangle \ge 0.
\end{array}
\right\}. \tag{7.4.7}
$$

We have the following conclusion.

**Theorem 7.4.2.** *([221]) Let $\mathcal{D}, S_{\mathcal{D}}(f), \Omega(f)$ be as above. Assume that $S_{\mathcal{D}}(f)$ has nonempty interior. If $f$ and all $g_i$ are QM-concave with respect to $g$, then $S_{\mathcal{D}}(f) = \Omega(f)$.*

### 7.4.1 ▪ The QM-convexity for rational functions

We now consider the case that $f$ is a rational function. The convex domain $\mathcal{D}$ is still given by $g$ as in (7.4.1). Suppose $f$ has the expansion

$$
f = \frac{1}{D(x)} \sum_{\alpha \in \mathbb{N}_{2d}^n} f_\alpha x^\alpha, \tag{7.4.8}
$$

where $D(x)$ is the denominator polynomial. Assume $D(x)$ is positive in the interior $\text{int}(\mathcal{D})$. If $D(u) = 0$ at a point $u$ on the boundary of $\text{cl}(\mathcal{D})$, the value $f(u)$ is then defined as

$$
f(u) := \liminf_{x \in \text{int}(\mathcal{D}), x \to u} f(x).
$$

Typically, we assume that $S_{\mathcal{D}}(f)$ is the closure of

$$
S_{\text{int}(\mathcal{D})}(f) := \{x \in \text{int}(\mathcal{D}) : f(x) \ge 0\}.
$$

The following is the definition of QM-convexity/concavity for rational functions.

**Definition 7.4.3.** *Let $p, q$ be two polynomials that are positive in $int(\mathcal{D})$. A rational function $f$ is said to be $(p, q)$-QM convex with respect to $g$ if (let $g_0 := 1$)*

$$p(x) \cdot q(u) \cdot R_f(x, u) = \sum_{i=0}^{m} g_i(x) \left( \sum_{j=0}^{m} g_j(u) \sigma_{ij}(x, u) \right) \tag{7.4.9}$$

*for some SOS polynomials $\sigma_{ij} \in \Sigma[x, u]$. Similarly, the rational function $f$ is said to be $(p, q)$-QM-concave with respect to $g$ if $-f$ is $(p, q)$-QM-convex with respect to $g$.*

When $f$ is a rational function as in (7.4.8), a natural choice for $(p, q)$ in (7.4.9) is

$$p(x) := D(x), \quad q(u) := D(u)^2. \tag{7.4.10}$$

If the choice $(p, q)$ in (7.4.10) makes the identity (7.4.9) hold, we just simply say that $f$ is QM-convex with respect to $g$ if the denominator $D(x)$ is clear in the context.

**Example 7.4.4.** *([221]) The rational function*

$$f := \frac{x_1^4 + x_1^2 x_2^2 + x_2^4}{x_1^2 + x_2^2}$$

*is QM-convex over $\mathbb{R}^2$, because*

$$R_f(x, u) = \frac{f_1^2 + f_2^2 + \frac{1}{2}(f_3^2 + f_4^2 + f_5^2 + f_6^2) + f_7^2 + f_8^2 + f_9^2}{(x_1^2 + x_2^2)(u_1^2 + u_2^2)^2},$$

*where the polynomials $f_i$ are given as below:*

$$
\begin{aligned}
f_1 &= -u_1 u_2 x_2^2 - u_1 u_2 x_1^2 + u_1 u_2^2 x_2 + u_1^2 u_2 x_1, & f_6 &= -u_2^2 x_2^2 + u_2^3 x_2 - u_1^2 x_1^2 + u_1^3 x_1, \\
f_2 &= -u_1 u_2 x_2^2 + u_1 u_2 x_1^2 + u_1 u_2^2 x_2 - u_1^2 u_2 x_1, & f_7 &= -2u_1 u_2 x_1 x_2 + u_1 u_2^2 x_1 + u_1^2 u_2 x_2, \\
f_3 &= -u_2^2 x_1 x_2 + u_2^3 x_1 - u_1^2 x_1 x_2 + u_1^3 x_2, & f_8 &= u_2^2 x_1^2 - u_1^2 x_2^2, \\
f_4 &= u_2^2 x_1 x_2 - u_2^3 x_1 - u_1^2 x_1 x_2 + u_1^3 x_2, & f_9 &= -u_1 u_2^2 x_1 + u_1^2 u_2 x_2. \\
f_5 &= u_2^2 x_2^2 - u_2^3 x_2 - u_1^2 x_1^2 + u_1^3 x_1,
\end{aligned}
$$

In the following, we study semidefinite representation for the suplevel set $S_{\mathcal{D}}(f)$. Assume that $f$ and all $g_i$ are $(p, q)$-QM-concave with respect to $g$ for given $(p, q)$. For each $h \in \{f, g_1, \ldots, g_m\}$, assume that

$$-p(x)q(u)R_h(x, u) = \sum_{i=0}^{m} g_i(x) \left( \sum_{j=0}^{m} g_j(u) \sigma_{ij}^{h}(x, u) \right)$$

for some SOS polynomials $\sigma_{ij}^{h}(x, u) \in \Sigma[x, u]$. Recall that $d$ is the degree as in (7.4.6). Let $LE(p)$ denote the exponent of the leading monomial of $p$ under the lexicographical ordering $(x_1 > x_2 > \cdots > x_n)$. Denote the power sets

$$
\begin{aligned}
E_1 &:= \left\{ \alpha \in \mathbb{N}^n : |\alpha| + |LE(p)| \leq 2d \right\}, \\
E_2 &:= \left\{ \beta \in \mathbb{N}^n : \beta < LE(p) \right\}.
\end{aligned}
$$

For each $i = 0, 1, \ldots, m$, define matrices $P_\alpha^{(i)}, P_\alpha^{(\nu)}, Q_\alpha^{(i)}, Q_\alpha^{(\nu)}$ such that

$$\frac{g_i(x)}{p(x)} [x]_{d-d_i} [x]_{d-d_i}^T = \sum_{\alpha \in E_1} Q_\alpha^{(i)} x^\alpha + \sum_{\beta \in E_2} P_\alpha^{(i)} \frac{x^\beta}{p(x)}, \tag{7.4.11}$$

where $g_0 = 1$ and each $d_i = \lceil \frac{1}{2} \deg(g_i) \rceil$. For $y = (y_\alpha) \in \mathbb{R}^{E_1}$ and $z = (z_\beta) \in \mathbb{R}^{E_2}$, define linear functions $(i = 0, 1, \ldots, m)$

$$Q_i[y, z] = \sum_{\alpha \in E_1} Q_\alpha^{(i)} y_\alpha + \sum_{\beta \in E_2} P_\beta^{(i)} z_\beta. \tag{7.4.12}$$

Suppose the rational function $f$ is expanded as

$$f = \sum_{\alpha \in E_1} f_\alpha^{(1)} x^\alpha + \sum_{\beta \in E_2} f_\beta^{(2)} \frac{x^\beta}{p(x)},$$

then define vectors $f^{(1)}, f^{(2)}$ such that

$$(f^{(1)})^T y + (f^{(2)})^T z = \sum_{\alpha \in E_1} f_\alpha^{(1)} y_\alpha + \sum_{\beta \in E_2} f_\beta^{(2)} z_\beta.$$

We consider the following semidefinite representation:

$$\Gamma(f) = \left\{ x \in \mathbb{R}^n \;\middle|\; \begin{array}{l} x = (y_{e_1}, \ldots, y_{e_n}), \; y_0 = 1, \; y \in \mathbb{R}^{E_1}, \\ (f^{(1)})^T y + (f^{(2)})^T z \geq 0, \; z \in \mathbb{R}^{E_2}, \\ Q_i[y, z] \succeq 0 \; (i = 0, 1, \ldots, m) \end{array} \right\}. \tag{7.4.13}$$

For a polynomial $h$, let $\mathcal{Z}(h)$ denote its real zero set. We have the following representation theorem. Recall that $\partial T$ denotes the boundary of a set $T$ in the Euclidean topology.

**Theorem 7.4.5.** *([221]) Let $f, D, S_D(f)$ be as above. Assume $f$ and all $g_i$ are $(p, q)$-QM-concave with respect to $g$ and $S_D(f)$ has nonempty interior. Suppose $\dim(\mathcal{Z}(f) \cap \mathcal{Z}(D) \cap \partial S_D(f)) < n - 1$ and $\dim(\mathcal{Z}(q) \cap \partial S_D(f)) < n - 1$. Then, $S_D(f) = \Gamma(f)$.*

**Example 7.4.6.** *([221]) Consider the convex set in $\mathbb{R}^2$ given as*

$$x_1^2 + x_2^2 \geq x_1^4 + x_1^2 x_2^2 + x_2^4.$$

*It equals $S_{\mathbb{R}^2}(f)$ with the rational function $f = 1 - \frac{x_1^4 + x_1^2 x_2^2 + x_2^4}{x_1^2 + x_2^2}$. In Example 7.4.4, we have seen that $f(x)$ is QM-concave, with $p = x_1^2 + x_2^2$, $q = (u_1^2 + u_2^2)^2$. A polynomial division shows*

$$\frac{1}{x_1^2 + x_2^2}[x]_2[x]_2^T = \begin{bmatrix} 0 & 0 & 0 & 1 & 0 & 0 \\ 0 & 1 & 0 & x_1 & x_2 & 0 \\ 0 & 0 & 0 & x_2 & 0 & 0 \\ 1 & x_1 & x_2 & x_1^2 - x_2^2 & x_1 x_2 & x_2^2 \\ 0 & x_2 & 0 & x_1 x_2 & x_2^2 & 0 \\ 0 & 0 & 0 & x_2^2 & 0 & 0 \end{bmatrix}$$

$$+ \frac{1}{p(x)} \begin{bmatrix} 1 & x_1 & x_2 & -x_2^2 & x_1 x_2 & x_2^2 \\ x_1 & -x_2^2 & x_1 x_2 & -x_1 x_2^2 & -x_2^3 & x_1 x_2^2 \\ x_2 & x_1 x_2 & x_2^2 & -x_2^3 & x_1 x_2^2 & x_2^3 \\ -x_2^2 & -x_1 x_2^2 & -x_2^3 & x_2^4 & -x_1 x_2^3 & -x_2^4 \\ x_1 x_2 & -x_2^3 & x_1 x_2^2 & -x_1 x_2^3 & -x_2^4 & x_1 x_2^3 \\ x_2^2 & x_1 x_2^2 & x_2^3 & -x_2^4 & x_1 x_2^3 & x_2^4 \end{bmatrix}.$$

*The set $\Gamma(f)$ as in (7.4.13) is therefore given as*

$$
\begin{bmatrix}
0 & 0 & 0 & 1 & 0 & 0 \\
0 & 1 & 0 & x_1 & x_2 & 0 \\
0 & 0 & 0 & x_2 & 0 & 0 \\
1 & x_1 & x_2 & y_{20} - y_{02} & y_{11} & y_{02} \\
0 & x_2 & 0 & y_{11} & y_{02} & 0 \\
0 & 0 & 0 & y_{02} & 0 & 0
\end{bmatrix}
$$

$$
+ \begin{bmatrix}
z_{00} & z_{10} & z_{01} & -z_{02} & z_{11} & z_{02} \\
z_{10} & -z_{02} & z_{11} & -z_{12} & -z_{03} & z_{12} \\
z_{01} & z_{11} & z_{02} & -z_{03} & z_{12} & z_{03} \\
-z_{02} & -z_{12} & -z_{03} & z_{04} & -z_{13} & -z_{04} \\
z_{11} & -z_{03} & z_{12} & -z_{13} & -z_{04} & z_{13} \\
z_{02} & z_{12} & z_{03} & -z_{04} & z_{13} & z_{04}
\end{bmatrix} \succeq 0,
$$

*plus the scalar inequality $1 \geq y_{20} + z_{04}$. By Theorem 7.4.5, we know $\Gamma(f)$ represents the set given by the inequality $x_1^2 + x_2^2 \geq x_1^4 + x_1^2 x_2^2 + x_2^4$.*

## 7.4.2 ▪ Convex sets with singularities

We consider the case that the boundary of a convex semialgebraic set has singular points. Without loss of generality, assume the origin 0 belongs to the boundary of $S_{\mathcal{D}}(f)$ and it is a singular point of the hypersurface

$$
\mathcal{Z}(f) := \{x \in \mathbb{R}^n : f(x) = 0\},
$$

that is, $f(0) = 0$, $\nabla f(0) = 0$. A natural approach for getting rid of singularity is to use rational functions. Let $p(x)$ be a polynomial that is positive in int $(\mathcal{D})$. We typically assume that $S_{\mathcal{D}}(f)$ equals the closure of the set

$$
\left\{ x \in \text{int}(\mathcal{D}) : \frac{f(x)}{p(x)} \geq 0 \right\}.
$$

Under some conditions, one can get a semidefinite representation for $S_{\mathcal{D}}(f)$ with the usage of $f/p$. For instance, consider the convex set

$$
\{(x_1, x_2) \in \mathbb{R}^2_+ : -x_1^3 + 3x_1 x_2^2 - (x_1^2 + x_2^2)^2 \geq 0\}.
$$

The origin is a singular point on its boundary. If we choose $p(x) = x_2^2$, then it can equivalently given as

$$
\left\{ (x_1, x_2) \in \mathbb{R}^2_+ : 3x_1 \geq 2x_1^2 + x_2^2 + \frac{x_1^3}{x_2^2} + \frac{x_1^4}{x_2^2} \right\}.
$$

A semidefinite representation can be obtained by looking at the above rational inequality (this is left as an exercise).

The classical trick of *perspective transformation* is useful for removing singularity. Generally, we can assume

$$
S_{\mathcal{D}}(f) \subseteq \mathbb{R}_+ \times \mathbb{R}^{n-1}, \qquad \text{int}\,(S_{\mathcal{D}}(f)) \neq \emptyset.
$$

Define the perspective transformation $\mathfrak{p}$ as

$$
\mathfrak{p}(x) = \left( \frac{1}{x_1}, \frac{x_2}{x_1}, \dots, \frac{x_n}{x_1} \right).
$$

The image of $S_{\mathcal{D}}(f)$ under the perspective transformation $\mathfrak{p}$ is

$$
\left\{ \mathfrak{p}(x) : x \in S_{\mathcal{D}}(f) \right\} \subseteq \mathbb{R}_+ \times \mathbb{R}^{n-1},
$$

which is also convex (see §2.3 in [37]). Define new coordinates

$$\tilde{x}_1 = \frac{1}{x_1}, \quad \tilde{x}_2 = \frac{x_2}{x_1}, \quad \ldots, \quad \tilde{x}_n = \frac{x_n}{x_1}. \tag{7.4.14}$$

Denote $\tilde{x} := (\tilde{x}_1, \ldots, \tilde{x}_n)$ and $\tilde{\tilde{x}} := (\tilde{x}_2, \ldots, \tilde{x}_n)$. Suppose $f(x)$ has the expansion

$$f(x) = f_b(x) - f_{b+1}(x) - \cdots - f_d(x),$$

where each $f_k(x)$ is the homogeneous part of degree $k$. Let

$$\tilde{f}(\tilde{x}) := \tilde{f}_0(\tilde{\tilde{x}}) - \frac{\tilde{f}_1(\tilde{\tilde{x}})}{\tilde{x}_1} - \cdots - \frac{\tilde{f}_{d-b}(\tilde{\tilde{x}})}{\tilde{x}_1^{d-b}}, \tag{7.4.15}$$

where $\tilde{f}_i(\tilde{\tilde{x}}) := x_1^{b+i} f_{b+i}(1, \tilde{x}_2, \ldots, \tilde{x}_n)$. Define the new domain

$$\tilde{\mathcal{D}} := \{\tilde{x} \in \mathbb{R}^n : \tilde{g}_1(\tilde{x}) \geq 0, \ldots, \tilde{g}_m(\tilde{x}) \geq 0\},$$

where each $\tilde{g}_i(\tilde{x}) = g_i(x)/x_1^{\deg(g_i)}$. Note that $\tilde{\mathcal{D}}$ is convex if and only if $\mathcal{D}$ is convex. Under the perspective transformation $\mathfrak{p}$, the set $S_\mathcal{D}(f)$ can be equivalently given as

$$S_{\tilde{\mathcal{D}}}(\tilde{f}) = \{\tilde{x} \in \tilde{\mathcal{D}} : \tilde{f}(\tilde{x}) \geq 0\}.$$

Therefore, a semidefinite representation for $S_\mathcal{D}(f)$ can be obtained by looking at $S_{\tilde{\mathcal{D}}}(\tilde{f})$. We refer to [221] for more details about singular convex sets.

### 7.4.3 ▪ Exercises

**Exercise 7.4.1.** *Determine whether or not the rational function $\frac{1}{x_1 x_2}$ is $(p, q)$-QM-convex with respect to $g = (x_1, x_2)$ for $p = x_1 x_2$ and $q = u_1^2 u_2^2$.*

**Exercise 7.4.2.** *Give a sufficient and necessary condition for a matrix $A \in \mathcal{S}^2$ such that the polynomial $x_1^4 + x_2^4 + x^T A x$ is QM-convex with respect to $g = (1 - x_1^2 - x_2^2)$.*

**Exercise 7.4.3.** *For two scalars $a < b$, show that a univariate polynomial $f(x)$ is convex over $[a, b]$ if and only if $f$ is QM-convex with respect to $g = (x - a, b - x)$.*

**Exercise 7.4.4.** *Suppose $\mathcal{D}$ is a convex domain, given by a polynomial tuple $g$ as in (7.4.1). Assume $\text{QM}[g]$ is archimedean. If a polynomial $f$ has positive definite Hessian on $\mathcal{D}$, show that $f$ must be QM-convex with respect to $g$.*

**Exercise 7.4.5.** *Consider the convex set*

$$S = \left\{ (x_1, x_2) \in \mathbb{R}_+ \times \mathbb{R} : x_1^3 - 3x_1 x_2^2 - (x_1^2 + x_2^2)^2 \geq 0 \right\}.$$

*Show that it can be equivalently given as*

$$\left\{ (x_1, x_2) \in \mathbb{R}_{++} \times \mathbb{R} : x_1 - \frac{3x_2^2}{x_1} - \left( \frac{x_1^2 + x_2^2}{x_1} \right)^2 \geq 0 \right\}.$$

*Give a semidefinite representation for $S$ based on the above rational inequality.*

**Exercise 7.4.6.** *Give a semidefinite representation for the convex set*

$$\{(x_1, x_2) \in \mathbb{R}_+^2 \;:\; -x_1^3 + 3x_1 x_2^2 - (x_1^2 + x_2^2)^2 \ge 0\}$$

*by looking at its rational representation*

$$\left\{(x_1, x_2) \in \mathbb{R}_+^2 \;:\; 3x_1 \ge 2x_1^2 + x_2^2 + \frac{x_1^3}{x_2^2} + \frac{x_1^4}{x_2^2}\right\}.$$

**Exercise 7.4.7.** *Construct a tight convex moment relaxation for solving the rational optimization problem*

$$\begin{cases} \min\limits_{x \in \mathbb{R}^2} & \frac{x_1^4 + x_1^2 x_2^2 + x_2^4}{x_1^2 + x_2^2} \\ s.t. & x_1 \ge 1, \; x_2 \ge 1. \end{cases}$$

*Prove the tightness for the constructed relaxation.*

**Exercise 7.4.8.** *([221]) Let $g := (x_1, \ldots, x_n)$. If $f$ is a cubic polynomial that is concave over $\mathbb{R}_+^n$, show that $S_{\mathbb{R}_+^n}(f) = \Omega(f)$, where $\Omega(f)$ is given as in (7.4.7).*

**Exercise 7.4.9.** *Consider the polynomial optimization*

$$\begin{cases} \min\limits_{x \in \mathbb{R}^n} & f(x) \\ s.t. & g_1(x) \ge 0, \ldots, g_m(x) \ge 0. \end{cases} \tag{7.4.16}$$

*Assume its feasible set is convex. If $f$ and all $-g_i$ are QM-convex with respect to $g := (g_1, \ldots, g_m)$, show that the dth order Moment-SOS relaxation is tight for solving (7.4.16), where d is the degree as in (7.4.6).*

# Chapter 8

# Conic Optimization with Moments and Polynomials

*A nonconvex polynomial optimization problem can be equivalently expressed as a linear convex conic optimization problem with moment variables. This chapter considers more general linear conic optimization problems with moments and nonnegative polynomials. We study geometric properties and memberships for these cones. Moment-SOS relaxations can be used to solve these linear conic optimization problems. The infeasibility, unboundedness, and truncated moment problems are also discussed in this chapter.*

## 8.1 ▪ PSD Polynomial and Moment Cones

Many hard nonconvex optimization problems can be equivalently formulated as linear convex conic optimization problems with moments or nonnegative polynomials. Let $K \subseteq \mathbb{R}^n$ be a set. A polynomial $f \in \mathbb{R}[x]$ is said to be *nonnegative* or *positive semidefinite* (psd) on $K$ if the evaluation $f(u) \geq 0$ for all $u \in K$. When $f$ is psd on $K$, we write that

$$f \geq 0 \quad \text{on } K, \qquad \text{or} \quad f|_K \geq 0.$$

When $f$ is nonnegative in the entire space $\mathbb{R}^n$, it is just simply called a nonnegative or psd polynomial. For a degree $d > 0$, the set

$$\mathscr{P}_d(K) := \{f \in \mathbb{R}[x]_d : f|_K \geq 0\} \tag{8.1.1}$$

is the cone of polynomials in $\mathbb{R}[x]_d$ that are nonnegative on $K$. It is called the psd (or nonnegative) polynomial cone of degree $d$. For every set $K$, the psd polynomial cone $\mathscr{P}_d(K)$ is closed and convex.

Note that $f \geq 0$ on $K$ if and only if $f \geq 0$ on $\mathrm{cl}\,(K)$, the closure of $K$. So we usually assume $K$ is closed. A basic question in optimization is how to characterize polynomials that are nonnegative on $K$. When $K = \mathbb{R}^n$, $\mathscr{P}_d(\mathbb{R}^n)$ reduces to the nonnegative or psd polynomial cone $\mathscr{P}_{n,d}$. For each even degree $d \geq 4$, it is NP-hard to check memberships for $\mathscr{P}_d(\mathbb{R}^n)$. When $K$ is the nonnegative orthant $\mathbb{R}^n_+$, $\mathscr{P}_d(\mathbb{R}^n_+)$ is the cone of *copositive* polynomials of degrees at most $d$. For each degree $d > 1$, it is NP-hard to check memberships in $\mathscr{P}_d(\mathbb{R}^n_+)$. For general $K$, it is typically hard to check memberships for the cone $\mathscr{P}_d(K)$. For some special cases of $d$ and $K$, the memberships for $\mathscr{P}_d(K)$ can be expressed in terms of semidefinite programs.

- When $K = \mathbb{R}^n$, if $d = 2$ or $n = 1$ or $(n, d) = (2, 4)$, the cone $\mathscr{P}_d(\mathbb{R}^n)$ coincides with the SOS cone $\Sigma[x]_d$ (see Theorem 2.4.5).

- When $n = 1$ and $K$ is an interval $I$, the cone $\mathscr{P}_d(I)$ consists of univariate polynomials that are nonnegative on $I$ and that have degrees at most $d$. A semidefinite representation for $\mathscr{P}_d(I)$ is given in §3.2.

- Let $q \in \mathbb{R}[x]_2$ be with degree $> 1$. For $K = \{x \in \mathbb{R}^n : q(x) \geq 0\}$, with $q(u) > 0$ for some point $u$, we have

$$\mathscr{P}_2(K) = \Sigma[x]_2 + q \cdot \mathbb{R}_+.$$

For $K = \{x \in \mathbb{R}^n : q(x) = 0\}$, with $q(u) > 0 > q(v)$ for some points $u, v$, we have

$$\mathscr{P}_2(K) = \Sigma[x]_2 + q \cdot \mathbb{R}.$$

These conclusions follow from Theorem 5.2.6.

- When $K = \{x \in \mathbb{R}^n : a_1^T x - b_1 \geq 0, \ldots, a_m^T x - b_m \geq 0\}$ is a nonempty polyhedron and $d = 1$, the cone $\mathscr{P}_1(K)$ can be expressed as

$$\mathscr{P}_1(K) = \mathbb{R}_+ + (a_1^T x - b_1) \cdot \mathbb{R}_+ + \cdots + (a_m^T x - b_m) \cdot \mathbb{R}_+.$$

This is the nonhomogeneous Farkas Lemma (see Theorem 1.3.4).

## 8.1.1 ▪ The moment cone and the dual relationship

In optimization, people often need to deal with the dual cone of $\mathscr{P}_d(K)$. This requires one to consider linear functionals on the space $\mathbb{R}[x]_d$. A linear functional is uniquely determined by its values on basis vectors. Generally, we use the monomial basis for $\mathbb{R}[x]_d$. For a power $\alpha := (\alpha_1, \ldots, \alpha_n)$ and the variable $x := (x_1, \ldots, x_n)$, recall that $x^\alpha := x_1^{\alpha_1} \cdots x_n^{\alpha_n}$. For a linear functional $\ell$ on $\mathbb{R}[x]_d$, let $y_\alpha$ denote its value on $x^\alpha$, i.e., $\ell(x^\alpha) = y_\alpha$. Therefore, $\ell$ is uniquely determined by the *truncated multi-sequence* (tms) of degree $d$ such that

$$y := (y_\alpha)_{\alpha \in \mathbb{N}_d^n}.$$

The space of all such $y$ is denoted as $\mathbb{R}^{\mathbb{N}_d^n}$. Recall the notation

$$\langle p, y \rangle := \ell(p) = \sum_{\alpha \in \mathbb{N}_d^n} p_\alpha y_\alpha \quad \text{for} \quad p = \sum_{\alpha \in \mathbb{N}_d^n} p_\alpha x^\alpha.$$

For fixed $p$, $\langle p, y \rangle$ is a linear functional in $y \in \mathbb{R}^{\mathbb{N}_d^n}$; for fixed $y$, $\langle p, y \rangle$ is a linear functional in $p \in \mathbb{R}[x]_d$. Under the above operation $\langle \cdot, \cdot \rangle$, $\mathbb{R}^{\mathbb{N}_d^n}$ is the dual space of $\mathbb{R}[x]_d$, and vice versa. The dual cone of $\mathscr{P}_d(K)$ is therefore given as

$$\mathscr{P}_d(K)^\star := \left\{ y \in \mathbb{R}^{\mathbb{N}_d^n} : \langle p, y \rangle \geq 0 \text{ for all } p \in \mathscr{P}_d(K) \right\}.$$

A tms $y = (y_\alpha)_{\alpha \in \mathbb{N}_d^n}$ is called a *truncated moment sequence* if there exists a Borel measure $\mu$ on $\mathbb{R}^n$ such that $y_\alpha = \int x^\alpha d\mu$ for every $\alpha \in \mathbb{N}_d^n$. This is equivalent to $y = \int [x]_d d\mu$. The $\mu$ is called a representing measure for $y$. Recall that the support of $\mu$, for which we denote $\operatorname{supp}(\mu)$, is the smallest closed set $S$ such that $\mu(\mathbb{R}^n \backslash S) = 0$. We refer to §2.7 for truncated moment problems. In applications, we often require $\operatorname{supp}(\mu)$ to be contained in a prescribed set $K \subseteq \mathbb{R}^n$. If $\operatorname{supp}(\mu) \subseteq K$, the $\mu$ is called a *K-representing measure* for $y$. Recall that $meas(y, K)$ denotes the set of all $K$-representing measures for $y$. The following is the frequently used moment cone:

$$\mathscr{R}_d(K) := \{y \in \mathbb{R}^{\mathbb{N}_d^n} : meas(y, K) \neq \emptyset\}. \tag{8.1.2}$$

In particular, for $y = 0$, $meas(y, K)$ consists of the identically zero measure on $K$, whose support is empty. When $K$ is not closed, the set $meas(y, K)$ is still well defined, since it is required that $\text{supp}(\mu) \subseteq K$ for $\mu \in meas(y, K)$. When $K = \emptyset$, we just simply let $\mathscr{R}_d(K) = \{0\}$. The cone $\mathscr{R}_d(K)$ is well defined for every set $K \subseteq \mathbb{R}^n$. Clearly, $\mathscr{R}_d(K)$ is a convex cone in $\mathbb{R}^{\mathbb{N}^n_d}$, but it may not be closed for some $K$.

In computation, people often prefer finitely atomic measures (see §2.7). If a tms $y$ of degree $d$ admits the finitely atomic measure

$$\nu := \lambda_1 \delta_{u_1} + \cdots + \lambda_r \delta_{u_r}, \quad \lambda_i > 0, \quad u_i \in K,$$

then one can show that

$$y = \int [x]_d \mathrm{d}\nu = \lambda_1 [u_1]_d + \cdots + \lambda_r [u_r]_d.$$

This means $y \in \text{cone}\,([K]_d)$, which denotes the conic hull of the set

$$[K]_d := \{[u]_d : u \in K\}. \tag{8.1.3}$$

Interestingly, the moment cone $\mathscr{R}_d(K)$ coincides with the conic hull cone $([K]_d)$.

**Theorem 8.1.1.** *([15, 171]) For every set $K \subseteq \mathbb{R}^n$, we have*

$$\mathscr{R}_d(K) = \text{cone}\,([K]_d). \tag{8.1.4}$$

*In particular, for each $y \in \mathscr{R}_d(K)$ with a $K$-representing measure $\mu$, there must exist $\nu \in meas(y, K)$ such that $\text{supp}(\nu) \subseteq \text{supp}(\mu)$ and $|\text{supp}(\nu)| \leq \binom{n+d}{d}$. Moreover, if $K$ is compact, then the cone $\mathscr{R}_d(K)$ is closed.*

***Proof.*** It is obvious that $\mathscr{R}_d(K) \supseteq \text{cone}\,([K]_d)$. We need to show the reverse containment. Pick an arbitrary $y \in \mathscr{R}_d(K)$. If $y = 0$, the conclusion is clearly true, since the representing measure for $0$ is the identically zero measure. We consider the case $y \neq 0$. Let $\mu$ be a $K$-representing measure for $y$ and let $T := \text{supp}(\mu)$ be its support. Note that $0 < \mu(K) = y_0 < \infty$ and $T$ is a closed set in $\mathbb{R}^n$. Let $\mathcal{C} := \text{cone}\,([T]_d)$ be the conic hull of $[T]_d$.

First, we show that $y$ belongs to $\text{cl}\,(\mathcal{C})$, the closure of $\mathcal{C}$. Suppose otherwise that $y \notin \text{cl}\,(\mathcal{C})$. Then, by the strict separating hyperplane theorem, there exists $p \in \mathbb{R}[x]_d$ such that

$$\langle p, y \rangle < \langle p, z \rangle \quad \forall z \in \text{cl}\,(\mathcal{C}).$$

The above implies that $p \geq 0$ on $T$ and $\langle p, y \rangle < 0$. However, this contradicts $y \in \mathscr{R}_d(T)$, since

$$\langle p, y \rangle = \int p(x) \mathrm{d}\mu \geq 0.$$

So, we must have $y \in \text{cl}\,(\mathcal{C})$.

Second, we show that $y \in \text{ri}\,(\text{cl}\,(\mathcal{C}))$, the relative interior of $\text{cl}\,(\mathcal{C})$, following the proof of [171, Theorem 5.9]. It is enough to show that if $H$ is a supporting hyperplane of $\text{cl}\,(\mathcal{C})$ through $y$, then $\text{cl}\,(\mathcal{C})$ must be entirely contained in $H$. To show this, suppose $H = \{w : a^T w = 0\}$ is a supporting hyperplane through $y$ such that

$$a^T w \geq 0 \quad \text{for all } w \in \mathcal{C}, \qquad a^T y = 0.$$

Note that $\int a^T [x]_d \mathrm{d}\mu = a^T y = 0$. Since $\text{supp}(\mu) = T$ and $a^T [x]_d \geq 0$ on $T$, we must have that $a^T [x]_d \equiv 0$ on $T$. This implies that $a^T w = 0$ for all $w \in \text{cl}\,(\mathcal{C})$, so $\text{cl}\,(\mathcal{C}) \subseteq H$. Since $\text{cl}\,(\mathcal{C})$ and $\mathcal{C}$ have the same relative interior (see Exercise 1.2.5), we have

$$y \in \text{ri}\,(\text{cl}\,(\mathcal{C})) = \text{ri}\,(\mathcal{C}) \subseteq \mathcal{C} \subseteq \text{cone}\,([K]_d).$$

The above is true for arbitrary $y \in \mathscr{R}_d(K)$, so the equality (8.1.4) holds.

Third, for each $y \in \mathcal{C}$, we show that there is a finitely atomic representing measure with required cardinality. The dimension of the space $\mathbb{R}^{\mathbb{N}_d^n}$ is $\binom{n+d}{d}$. By Carathéodory's Theorem (see Theorem 1.2.4), we can always choose $\nu \in meas(y,T)$ such that the cardinality of $supp(\nu)$ is at most $\binom{n+d}{d}$.

Last, when $K$ is compact, the set $[K]_d$ is also compact. For the constant one polynomial $p = 1$, we can see that $\langle p, y \rangle > 0$ for all $y \in [K]_d$. By Proposition 1.2.6, the conic hull cone $([K]_d)$ is closed if $K$ is compact. $\square$

Let $\mu$ be a $K$-representing measure for $y$, then

$$\langle p, y \rangle = \int p(x) \mathrm{d}\mu$$

for every $p \in \mathbb{R}[x]_d$. Therefore, $\langle p, y \rangle \geq 0$ for all $p \in \mathscr{P}_d(K)$. This gives the dual relationship between $\mathscr{R}_d(K)$ and $\mathscr{P}_d(K)$. Clearly, both $\mathscr{P}_d(K)$ and $\mathscr{R}_d(K)$ are convex cones. The cone $\mathscr{P}_d(K)$ is always closed, while $\mathscr{R}_d(K)$ is closed when $K$ is compact. The dual cones of $\mathscr{P}_d(K)$, $\mathscr{R}_d(K)$ are respectively given as

$$\mathscr{P}_d(K)^\star := \left\{ y \in \mathbb{R}^{\mathbb{N}_d^n} : \langle p, y \rangle \geq 0 \, \forall \, p \in \mathscr{P}_d(K) \right\},$$

$$\mathscr{R}_d(K)^\star := \left\{ p \in \mathbb{R}[x]_d : \langle p, y \rangle \geq 0 \, \forall \, y \in \mathscr{R}_d(K) \right\}.$$

When $K$ is compact, they are dual cones of each other. The following duality theorem is well known in the literature (see [312] and [171, §5.2]).

**Theorem 8.1.2.** *If $K \subseteq \mathbb{R}^n$ is compact, then it holds that*

$$\mathscr{R}_d(K)^\star = \mathscr{P}_d(K), \quad \mathscr{P}_d(K)^\star = \mathscr{R}_d(K). \tag{8.1.5}$$

**Proof.** We already know that $\mathscr{R}_d(K)$ and $\mathscr{P}_d(K)$ are convex cones and $\mathscr{P}_d(K)$ is closed. Since $K$ is compact, the cone $\mathscr{R}_d(K)$ is closed, by Theorem 8.1.1. By the definition, a polynomial $p \in \mathscr{R}_d(K)^\star$ if and only if $\langle p, [u]_d \rangle \geq 0$ for all $u \in K$, i.e., $p \geq 0$ on $K$. Hence, $\mathscr{R}_d(K)^\star = \mathscr{P}_d(K)$. By the bi-duality theorem (Theorem 1.2.10), we know $\mathscr{P}_d(K)^\star = \mathscr{R}_d(K)$, since $\mathscr{R}_d(K)$ is closed. $\square$

When $K$ is not compact, the dual cone of $\mathscr{P}_d(K)$ is usually not equal to $\mathscr{R}_d(K)$. This is because that the moment cone $\mathscr{R}_d(K)$ is generally not closed for noncompact $K$. However, $\mathscr{P}_d(K)^\star$ always equals the closure of $\mathscr{R}_d(K)$.

**Theorem 8.1.3.** *For every set $K \subseteq \mathbb{R}^n$, the following dual relationship holds:*

$$\mathscr{R}_d(K)^\star = \mathscr{P}_d(K), \quad \mathscr{P}_d(K)^\star = \mathrm{cl}\left(\mathscr{R}_d(K)\right). \tag{8.1.6}$$

The above can be shown by applying the bi-duality theorem to the closure $\mathrm{cl}\left(\mathscr{R}_d(K)\right)$. The proof is left as an exercise.

### 8.1.2 ▪ Some geometric properties

A tms $y \in \mathbb{R}^{\mathbb{N}_d^n}$ determines the *Riesz functional* $\mathscr{L}_y$ on $\mathbb{R}[x]_d$ such that

$$\mathscr{L}_y(p) = \langle p, y \rangle.$$

The functional $\mathscr{L}_y$ is said to be $K$-*positive* (or positive on $\mathscr{P}_d(K)$) if

$$p \in \mathscr{P}_d(K) \quad \Rightarrow \quad \mathscr{L}_y(p) \geq 0.$$

Similarly, it is *strictly $K$-positive* (or strictly positive on $\mathscr{P}_d(K)$) if

$$p \in \mathscr{P}_d(K),\ p|_K \not\equiv 0 \quad \Rightarrow \quad \mathscr{L}_y(p) > 0.$$

(The notation $p|_K$ denotes the restriction of the function $p$ on $K$.) The strict $K$-positivity is related to the vanishing ideal of $K$. Recall that the vanishing ideal of $K$, for which we denote $I(K)$, is the ideal of polynomials that are identically zero on $K$. In computation, we often work on the truncation of degree $d$:

$$I(K)_d := I(K) \cap \mathbb{R}[x]_d. \tag{8.1.7}$$

We refer to §1.2 for interiors (or relative interiors) and affine hulls of convex sets. These properties for the cones $\mathscr{R}_d(K)$ and $\mathscr{P}_d(K)$ are summarized as follows.

**Theorem 8.1.4.** *Let $K \subseteq \mathbb{R}^n$ be a closed set and let $d > 0$ be a degree. Then, we have the following:*

(i) *It holds that $\mathscr{P}_d(K) \cap -\mathscr{P}_d(K) = I(K)_d$. The cone $\mathscr{P}_d(K)$ is pointed if and only if $I(K)_d = \{0\}$. When $K$ is compact, for $f \in \mathbb{R}[x]_d$, it holds that $f \in \mathrm{int}\,(\mathscr{P}_d(K))$ if and only if $f > 0$ on $K$.*

(ii) *The affine hull of $\mathscr{R}_d(K)$ is the subspace*

$$F := \left\{y \in \mathbb{R}^{\mathbb{N}_d^n} : \langle p, y \rangle = 0 \quad \text{for all } p \in I(K)_d \right\}. \tag{8.1.8}$$

*So, if $I(K)_d = \{0\}$, then $F = \mathbb{R}^{\mathbb{N}_d^n}$ is full-dimensional and $\mathscr{R}_d(K)$ has nonempty interior.*

(iii) *For $y \in \mathbb{R}^{\mathbb{N}_d^n}$, we have $y \in \mathrm{ri}\,(\mathscr{R}_d(K))$ if and only if $\mathscr{L}_y$ is strictly $K$-positive. Moreover, when $K$ is compact, we have $y \in \mathscr{R}_d(K)$ if and only if $\mathscr{L}_y$ is $K$-positive.*

***Proof.*** (i) Note that $f \in \mathscr{P}_d(K) \cap -\mathscr{P}_d(K)$ if and only if $f \equiv 0$ on $K$, i.e., $f \in I(K)_d$. Thus, $\mathscr{P}_d(K)$ is pointed if and only if $I(K)_d = \{0\}$.

Suppose $K$ is compact. If $f > 0$ on $K$, then $f + q > 0$ on $K$ for all $q \in \mathbb{R}[x]_d$ with $\|\mathrm{vec}(q)\|$ sufficiently small, so $f \in \mathrm{int}\,(\mathscr{P}_d(K))$. Similarly, if $f \in \mathrm{int}\,(\mathscr{P}_d(K))$, then we have $f - \epsilon \in \mathscr{P}_d(K)$ for some $\epsilon > 0$. This implies that $f \geq \epsilon > 0$ on $K$.

(ii) The cone $\mathscr{R}_d(K)$ can be equivalently expressed as

$$\mathscr{R}_d(K) = \mathrm{cone}\,(\{[u]_d : u \in K\}) = \mathrm{cone}\,([K]_d),$$

by Theorem 8.1.1. The affine hull $\mathrm{aff}\,(\mathscr{R}_d(K))$ is a subspace, since $\mathscr{R}_d(K)$ contains the origin. A linear functional $\ell \equiv 0$ on $\mathscr{R}_d(K)$ if and only if $\ell([u]_d) = 0$ for all $u \in K$. Therefore, a tms $y \in \mathrm{aff}\,(\mathscr{R}_d(K))$ if and only if $\ell(y) = 0$ for every linear functional $\ell$ such that $\ell \equiv 0$ on $\mathscr{R}_d(K)$. Such $\ell$ can be expressed such that

$$\ell([u]_d) = p(u), \quad u \in K,$$

for some $p \in I(K)_d$. Therefore, the affine hull of $\mathscr{R}_d(K)$ is the set $F$ as in (8.1.8).

(iii) Since $K$ is a closed set in $\mathbb{R}^n$, there exists a probability measure $\mu$ on $\mathbb{R}^n$ whose support is equal to $K$[7]. For $q \in \mathscr{P}_d(K)$, note that $q|_K \not\equiv 0$ if and only if $\int q \mathrm{d}\mu > 0$. Let $\xi := \int [x]_d \mathrm{d}\mu$ and

$$P := \left\{ [p] \in \mathscr{P}_d(K)/I(K)_d : \int p \mathrm{d}\mu = 1 \right\},$$

where $\mathscr{P}_d(K)/I(K)_d$ is the set of quotient classes $[p]$ of $p \in \mathscr{P}_d(K)$ over the subspace $I(K)_d$. We use the norm as in (2.5.27) for the quotient space $\mathscr{P}_d(K)/I(K)_d$. Note that $P$ is a compact set, since $\mathrm{supp}(\mu) = K$. For each $y \in F$, the Riesz functional $\mathscr{L}_y$ is strictly $K$-positive if and only if $\mathscr{L}_y(p) > 0$ for all $[p] \in P$. Clearly, $\mathscr{L}_\xi$ is strictly $K$-positive and $\xi \in F$.

"$\Rightarrow$" If $y \in \mathrm{ri}\,(\mathscr{R}_d(K))$, then $w := y - \epsilon\xi \in \mathscr{R}_d(K)$ for some $\epsilon > 0$ and

$$\mathscr{L}_y(p) = \mathscr{L}_w(p) + \epsilon\mathscr{L}_\xi(p) \geq \epsilon\mathscr{L}_\xi(p) > 0$$

for all $p \in P$. This means that $\mathscr{L}_y$ is strictly $K$-positive.

"$\Leftarrow$" Consider an arbitrary $y \in \mathscr{R}_d(K)$ such that $\mathscr{L}_y$ is strictly $K$-positive. Note that $\langle p, y \rangle = 0$ for all $p \in I(K)_d$, so we have $y \in F$. Since $P$ is compact, we have that

$$\epsilon := \min_{[p]\in P} \mathscr{L}_y(p) > 0, \quad \rho := \max_{\substack{[p]\in P, \\ z\in F, \|z\|_2=1}} |\langle p, z \rangle| < \infty.$$

For each $w \in F$ with $\|w - y\| < \frac{\epsilon}{2\rho}$, we have

$$\mathscr{L}_w(p) = \mathscr{L}_y(p) + \mathscr{L}_{w-y}(p) \geq (\epsilon - \|w - y\|\rho) > 0$$

for all $p \in P$, so $\mathscr{L}_w$ is $K$-positive. By Theorem 8.1.3, every $w \in F$, which is sufficiently close to $y$, belongs to the closure $\mathrm{cl}\,(\mathscr{R}_d(K))$. This implies that $y \in \mathrm{ri}\,(\mathscr{R}_d(K))$, since $F$ is the affine hull of $\mathscr{R}_d(K)$.

When $K$ is compact, by Theorem 8.1.2, we know $y \in \mathscr{R}_d(K)$ if and only if $y$ belongs to $\mathscr{P}_d(K)^*$, which is equivalent to $\mathscr{L}_y$ being $K$-positive. $\quad\square$

In the following, we study the boundary of the cone $\mathscr{P}_d(K)$. It plays an important role in optimization. To describe the boundary, we need to use the notion of *discriminant* (see §2.3.2). For a form $f \in \mathbb{R}[x_1, \ldots, x_n]^{\mathrm{hom}}$, its discriminant $\Delta(f)$ is a homogeneous polynomial in the coefficients of $f$ such that $\Delta(f) = 0$ if and only if $f(x)$ has a nonzero complex critical point, i.e., $\nabla f(x) = 0$ has a nonzero solution in $\mathbb{C}^n$. When $f$ is nonhomogeneous, $\Delta(f)$ is then defined to be the discriminant of its homogenization.

When $K = \mathbb{R}^n$, $\mathscr{P}_d(K)$ is the psd polynomial cone $\mathscr{P}_{n,d}$. Assume the degree $d > 0$ is even. If $f$ lies on the boundary $\partial\mathscr{P}_{n,d}$, then the homogenization of $f$ must have a nonzero complex critical point, and hence $\Delta(f) = 0$. This gives the following theorem.

**Theorem 8.1.5.** *([219]) For each $f \in \partial\mathscr{P}_{n,d}$, we must have $\Delta(f) = 0$. Therefore, the boundary $\partial\mathscr{P}_{n,d}$ lies on the discriminantal hypersurface*

$$\mathcal{D}_{n,d} := \{ p \in \mathbb{R}[x]_d : \Delta(p) = 0 \},$$

*which is irreducible of degree $n(d-1)^{n-1}$.*

---

[7]When $K$ is compact, this conclusion is mentioned in [286]. When $K$ is closed but unbounded, it holds that $K = \cup_{k=1}^\infty S_k$, where each $S_k := K \cap B(0,k)$ is compact. Let $\mu_k$ be a probability measure whose support equals $S_k$. Then $\mu := \sum_{k=1}^\infty \frac{\mu_k}{2^k}$ is a probability measure whose support equals $K$.

Next, we consider the case that $K$ is the semialgebraic set

$$K := \left\{ x \in \mathbb{R}^n \;\middle|\; \begin{array}{l} c_i(x) = 0\,(i \in \mathcal{E}), \\ c_j(x) \geq 0\,(j \in \mathcal{I}) \end{array} \right\} \tag{8.1.9}$$

for given polynomials $c_i, c_j$. The $\mathcal{E}, \mathcal{I}$ are finite labeling sets for constraining polynomials. We investigate conditions for the boundary of $\mathscr{P}_d(K)$. If $f \in \partial \mathscr{P}_d(K)$, does there exist $u \in K$ such that $f(u) = 0$? Indeed, this is not always the case. For instance, when $f = x_1 + x_2$ and $K = \{x_1^3 + x_2^3 - 1 = 0\}$, one can check that $f > 0$ on $K$ but $f$ lies on $\partial P_1(K)$. For every $\epsilon > 0$, the polynomial $x_1 + x_2 - \epsilon$ is not nonnegative on $K$, because

$$\inf_{x_1^3 + x_2^3 = 1} (x_1 + x_2) = 0.$$

Such a case occurs only if $K$ is noncompact.

We need more dedicate characterizations for the boundary $\partial \mathscr{P}_d(K)$ when $K$ is noncompact. For the set $K$ as in (8.1.9), its homogenization is

$$\widetilde{K} := \left\{ \tilde{x} \in \mathbb{R}^{n+1} \;\middle|\; \begin{array}{l} \tilde{x} := (x_0, x_1, \dots, x_n), \\ \tilde{c}_i(\tilde{x}) = 0\,(i \in \mathcal{E}), \\ \tilde{c}_j(\tilde{x}) \geq 0\,(j \in \mathcal{I}), \\ x_0 \geq 0 \end{array} \right\}, \tag{8.1.10}$$

where $\tilde{c}_i(\tilde{x})$, $\tilde{c}_j(\tilde{x})$ are respectively the homogenization of $c_i(x)$, $c_j(x)$. For $f \in \mathbb{R}[x]_d$, its homogenization is

$$\tilde{f}(\tilde{x}) := f(x/x_0)x_0^d.$$

(For the case $d > \deg(f)$, we still define $\tilde{f}(\tilde{x})$ as above for convenience of discussion.) If $\tilde{f} \geq 0$ on $\widetilde{K}$, then we must have $f \geq 0$ on $K$. The converse is not necessarily true. Recall that the set $K$ as in (8.1.9) is said to be *closed at infinity* (see Definition 6.6.1) if

$$\widetilde{K} = \text{cl}\left( \widetilde{K} \cap \{x_0 > 0\} \right).$$

For $f \in \mathbb{R}[x]_d$, define two minimum values

$$\begin{array}{ll} \vartheta_K(f) & := \quad \min_{x \in K} f(x), \\ \vartheta_{\widetilde{K}}(f) & := \quad \min_{\tilde{x} \in \widetilde{K}, \|\tilde{x}\|_2 = 1} f(\tfrac{x}{x_0})x_0^d. \end{array} \tag{8.1.11}$$

The interior and boundary of the cone $\mathscr{P}_d(K)$ can be characterized as follows.

**Theorem 8.1.6.** *([219]) Let $K$ be the set as in (8.1.9). If $K$ is bounded, then we have*

$$\begin{array}{ll} \vartheta_K(f) > 0 & \Leftrightarrow \quad f \in \text{int}\,(\mathscr{P}_d(K)), \\ \vartheta_K(f) = 0 & \Leftrightarrow \quad f \in \partial \mathscr{P}_d(K). \end{array} \tag{8.1.12}$$

*If $K$ is unbounded and closed at $\infty$, then we have*

$$\begin{array}{ll} \vartheta_{\widetilde{K}}(f) > 0 & \Leftrightarrow \quad f \in \text{int}\,(\mathscr{P}_d(K)), \\ \vartheta_{\widetilde{K}}(f) = 0 & \Leftrightarrow \quad f \in \partial \mathscr{P}_d(K). \end{array} \tag{8.1.13}$$

We remark that the set $K$ may not be closed at $\infty$, even if $K$ is compact; see Exercise 8.1.8. The equations for the boundary of the cone $\mathscr{P}_d(K)$ are given as follows.

**Theorem 8.1.7.** *([219]) Let $K \neq \emptyset$ be the set as in (8.1.9). If either $K$ is compact or $K$ is closed at $\infty$, then the boundary $\partial \mathscr{P}_d(K)$ lies on the union of the hypersurfaces*

$$\Delta(f, c_{j_1}, \ldots, c_{j_t}) = 0$$

*such that $\mathcal{E} \subseteq \{j_1, \ldots, j_t\} \subseteq \mathcal{E} \cup \mathcal{I}$ and $t \leq n$. The above $\Delta$ is the discriminant defined as in (2.3.4).*

The above theorem can be shown by using discriminants. The proof is left as an exercise. In contrast, an algebraic characterization for the boundary of the moment cone $\mathscr{R}_d(K)$ is mostly unknown, to the best of the author's knowledge.

### 8.1.3 ▪ Exercises

**Exercise 8.1.1.** *Determine the cone of vectors $(c_0, c_1, c_2, c_3)$ such that*

$$c_0 + c_1 x_1^4 x_2^2 + c_2 x_2^4 x_1^2 + c_3 x_1^3 x_2^3 \geq 0 \quad on \quad 1 - x_1^2 - x_2^2 \geq 0.$$

**Exercise 8.1.2.** *For the tms $y = \int_{[-1,1]^2} [x]_4 \mathrm{d}x$ (the Lebesgue integral), find an explicit representation $y = \lambda_1 [u_1]_4 + \cdots + \lambda_r [u_r]_4$ with $\lambda_i > 0$ and distinct points $u_1, \ldots, u_r \in [-1,1]^2$.*

**Exercise 8.1.3.** *For any two sets $K_1, K_2 \subseteq \mathbb{R}^n$, show*

$$\begin{aligned} \mathscr{P}_d(K_1 \cup K_2) &= \mathscr{P}_d(K_1) \cap \mathscr{P}_d(K_2), \\ \mathscr{R}_d(K_1 \cup K_2) &= \mathscr{R}_d(K_1) + \mathscr{R}_d(K_2). \end{aligned}$$

**Exercise 8.1.4.** *For $K = \mathbb{R}^n$ and $d \geq 1$, show that the cone $\mathscr{R}_d(K)$ is not closed.*

**Exercise 8.1.5.** *When $K$ is the unit sphere $\mathbb{S}^{n-1}$, give explicit conditions for $y \in \mathbb{R}^{\mathbb{N}_2^n}$ to lie in the relative interior of $\mathscr{R}_2(K)$.*

**Exercise 8.1.6.** *Show that $x_1 + x_2$ lies on the boundary of $\mathscr{P}_d(K)$ when $K$ is given by the equation $x_1^3 + x_2^3 = 1$.*

**Exercise 8.1.7.** *When $K$ is given by the equation $x_1^2(x_1 - x_2) - 1 = 0$, show that $K$ is not closed at $\infty$. For the polynomial $f := x_1 - x_2 + 1$, show that $f \in \mathscr{P}_1(K)$ but its homogenization $\tilde{f} = x_1 - x_2 + x_0$ is not nonnegative on $\widetilde{K}$.*

**Exercise 8.1.8.** *For the set $K = \{x \in \mathbb{R}^2 : x_1^2(1 - x_1^2 - x_2^2) - x_2^2 = 0\}$, its homogenization is*

$$\widetilde{K} = \left\{ \tilde{x} \in \mathbb{R}^3 : x_1^2(x_0^2 - x_1^2 - x_2^2) - x_0^2 x_2^2 = 0 \right\}.$$

*Show that $K$ is compact but it is not closed at $\infty$.*

**Exercise 8.1.9.** *Let $K \subseteq \mathbb{R}^n$ be a set. Show that $K$ is finite if and only if $\mathscr{P}_d(K)$ is a polyhedron for every $d \in \mathbb{N}$.*

**Exercise 8.1.10.** *Prove Theorems 8.1.3, 8.1.5, 8.1.6, and 8.1.7.*

## 8.2 ▪ Memberships for Polynomial and Moment Cones

We study memberships for the psd polynomial cone $\mathscr{P}_d(K)$ and the moment cone $\mathscr{R}_d(K)$. Assume $K$ is the semialgebraic set

$$K = \left\{ x \in \mathbb{R}^n \;\middle|\; \begin{array}{l} c_i(x) = 0 \,(i \in \mathcal{E}), \\ c_j(x) \geq 0 \,(j \in \mathcal{I}) \end{array} \right\} \tag{8.2.1}$$

for two given polynomial tuples

$$c_{eq} := (c_i)_{i \in \mathcal{E}}, \quad c_{in} := (c_j)_{j \in \mathcal{I}}.$$

If there are no equality (resp., inequality) constraints, we just let $c_{eq} = \emptyset$ (resp., $c_{in} = \emptyset$). Recall that $\mathrm{Ideal}[c_{eq}]$ denotes the ideal of $c_{eq}$ and $\mathrm{QM}[c_{in}]$ denotes the quadratic module of $c_{in}$.

### 8.2.1 ▪ Memberships for $\mathscr{P}_d(K)$

Putinar's Positivstellensatz (i.e., Theorem 2.6.1) gives a natural certificate for memberships in the cone $\mathscr{P}_d(K)$. If a polynomial $f \in \mathbb{R}[x]_d$ is such that

$$f \in \mathrm{Ideal}[c_{eq}] + \mathrm{QM}[c_{in}],$$

then $f(u) \geq 0$ for all $u \in K$, so we have the membership $f \in \mathscr{P}_d(K)$. In computation, we often work with truncations of $\mathrm{Ideal}[c_{eq}]$ and $\mathrm{QM}[c_{in}]$. For each degree $k$, denote the intersection

$$Q_k := \mathbb{R}[x]_d \cap \left( \mathrm{Ideal}[c_{eq}]_{2k} + \mathrm{QM}[c_{in}]_{2k} \right). \tag{8.2.2}$$

Clearly, $Q_k \subseteq \mathscr{P}_d(K)$ for all $k$. When $\mathrm{Ideal}[c_{eq}] + \mathrm{QM}[c_{in}]$ is archimedean, if $f > 0$ on $K$, we have $f \in Q_k$ for some $k$, by Theorem 2.6.1. The approximation quality of $Q_k$ is shown as follows.

**Proposition 8.2.1.** *Let $K$ be as in (8.2.1). If* $\mathrm{Ideal}[c_{eq}] + \mathrm{QM}[c_{in}]$ *is archimedean, then*

$$\mathrm{int}\,(\mathscr{P}_d(K)) \subseteq \bigcup_{k=1}^{\infty} Q_k \subseteq \mathscr{P}_d(K). \tag{8.2.3}$$

*Proof.* Since each $Q_k \subseteq \mathscr{P}_d(K)$, the second containment in (8.2.3) is obvious. We prove the first one. The archimedeanness of $\mathrm{Ideal}[c_{eq}] + \mathrm{QM}[c_{in}]$ implies that $K$ is compact. By Theorem 8.1.4, a polynomial $f \in \mathrm{int}\,(\mathscr{P}_d(K))$ if and only if $f > 0$ on $K$, which then implies $f \in Q_k$ for some $k$, by Theorem 2.6.1. So, all containments in (8.2.3) hold. $\square$

The second containment inequality in (8.2.3) is generally not an equality. For instance, when $n = 3$ and $K = B(0,1)$, the Motzkin polynomial

$$x_1^2 x_2^2 (x_1^2 + x_2^2 - 3x_3^2) + x_3^6 \in \mathscr{P}_d(K)$$

does not belong to $Q_k$ for any $k$ (see [224, Example 5.3]). Proposition 8.2.1 can be used to check memberships for the cone $\mathscr{P}_d(K)$ as follows. In view of (8.2.3), we get a hierarchy of membership certificates:

$$f \in \mathrm{Ideal}[c_{eq}]_{2k} + \mathrm{QM}[c_{in}]_{2k}, \quad k = 1, 2, \ldots. \tag{8.2.4}$$

When $\text{Ideal}[c_{eq}] + \text{QM}[c_{in}]$ is archimedean, every $f \in \text{int}\,(\mathscr{P}_d(K))$ passes the test (8.2.4). Moreover, when $f$ lies on the boundary of $\mathscr{P}_d(K)$, the $f$ may also pass the test (8.2.4), under some genericity conditions (see §5.5). However, there exists a special $f \in \mathscr{P}_d(K)$ that fails (8.2.4) for all $k$. For instance, this is the case if $f$ is the Motzkin polynomial and $K$ is the unit ball.

To get a full check for memberships in the cone $\mathscr{P}_d(K)$, we can use tight relaxation methods in §6.3. Note that $f \in \mathscr{P}_d(K)$ if and only if the minimum value $f_{min}$ of $f$ on $K$ is nonnegative. Assume $f_{min}$ is achievable at a KKT point (this is the case if $K$ is compact and nonsingular). Suppose Assumption 6.3.1 holds for the optimization (6.1.1), which is satisfiable when $(c_{eq}, c_{in})$ is nonsingular. Let $\Phi, \Psi$ be the polynomial tuples as in (6.3.6). We consider the following hierarchy of relaxations:

$$\begin{cases} f_k := \max & \gamma \\ \quad\;\; s.t. & f - \gamma \in \text{Ideal}[c_{eq}, \Phi]_{2k} + \text{QM}[c_{in}, \Psi]_{2k}. \end{cases} \tag{8.2.5}$$

When either $\text{Ideal}[c_{eq}] + \text{QM}[c_{in}]$ or $\text{Ideal}[c_{eq}, \Phi] + \text{QM}[c_{in}, \Psi]$ is archimedean, we have $f_k = f_{min}$ for all $k$ big enough. This is shown in Theorem 6.3.3. Therefore, $f \in \mathscr{P}_d(K)$ if and only if $f_k \geq 0$ for some $k$. We refer to §6.3 for the tightness of the hierarchy of relaxations (8.2.5).

## 8.2.2 ▪ Memberships for $\mathscr{R}_d(K)$

We now consider memberships for the moment cone $\mathscr{R}_d(K)$. For a tms $z \in \mathbb{R}^{\mathbb{N}_l^n}$ with the degree $l \geq d$, recall the truncation

$$z|_d := (z_\alpha)_{\alpha \in \mathbb{N}_d^n}. \tag{8.2.6}$$

If $y = z|_d$, then we call $z$ an *extension* of $y$ and call $y$ a *truncation* of $z$. For an integer $k \geq d/2$, consider the moment relaxation

$$F_k := \left\{ y \in \mathbb{R}^{\mathbb{N}_d^n} \;\middle|\; \begin{array}{l} \exists z \in \mathbb{R}^{\mathbb{N}_{2k}^n}, \, y = z|_d, \\ M_k[z] \succeq 0, \\ \mathscr{V}_{c_i}^{(2k)}[z] = 0 \, (i \in \mathcal{E}), \\ L_{c_j}^{(k)}[z] \succeq 0 \, (j \in \mathcal{I}) \end{array} \right\}. \tag{8.2.7}$$

In the above, $\mathscr{V}_{c_i}^{(2k)}[z]$ is a localizing vector and $L_{c_j}^{(k)}[y]$ is a localizing matrix (see §2.5.2 and §2.5.3 for the notation). Each $F_k$ is an outer approximation for the moment cone $\mathscr{R}_d(K)$. Indeed, it holds the nesting relation

$$F_1 \supseteq \cdots \supseteq F_k \supseteq F_{k+1} \supseteq \cdots \supseteq \mathscr{R}_d(K). \tag{8.2.8}$$

**Proposition 8.2.2.** *Let $K$ be as in (8.2.1). If* $\text{Ideal}[c_{eq}] + \text{QM}[c_{in}]$ *is archimedean, then*

$$\mathscr{R}_d(K) = \bigcap_{k=1}^{\infty} F_k. \tag{8.2.9}$$

**Proof.** Since $\mathscr{R}_d(K) \subseteq F_k$ for all $k$, $\mathscr{R}_d(K)$ is clearly contained in the intersection of all $F_k$. To prove the opposite containment, it is enough to show that each $y \notin \mathscr{R}_d(K)$ does not belong to $F_k$ when $k$ is big enough. Pick such a tms $y$. The archimedeanness of $\text{Ideal}[c_{eq}] + \text{QM}[c_{in}]$ implies that $K$ is compact. By Theorem 8.1.2, we know $y \notin \mathscr{P}_d(K)^*$, so there exists $p_1 \in \mathscr{P}_d(K)$ such that $\langle p_1, y \rangle < 0$. For $\epsilon > 0$ sufficiently small, we have that

$$p_2 := p_1 + \epsilon > 0 \text{ on } K, \quad \langle p_2, y \rangle < 0.$$

Moreover, we also have $p_2 \in \text{Ideal}[c_{eq}]_{2k_1} + \text{QM}[c_{in}]_{2k_1}$ for some $k_1$, by Theorem 2.6.1. We show that $y \notin F_{k_1}$. Suppose otherwise $y \in F_{k_1}$, then $y$ has an extension $z \in \mathbb{R}^{\mathbb{N}^n_{2k}}$ such that

$$M_k[z] \succeq 0, \quad \mathcal{V}^{(2k)}_{c_i}[z] = 0 \, (i \in \mathcal{E}), \quad L^{(k)}_{c_j}[z] \succeq 0 \, (j \in \mathcal{I}).$$

As a consequence, by Proposition 2.5.3, we get $\langle p_2, z \rangle \geq 0$, which contradicts the inequality $\langle p_2, y \rangle < 0$. Therefore, we have $y \notin F_{k_1}$, which completes the proof. $\qquad\square$

Proposition 8.2.2 implies that $F_k$ can approximate $\mathcal{R}_d(K)$ arbitrarily accurately. The method in [229] can be applied to check memberships for the cone $\mathcal{R}_d(K)$. The main idea is to find a flat extension of $y$, if it belongs to $\mathcal{R}_d(K)$. This can be done by solving a hierarchy of moment relaxations. Select a degree $d_1 > \frac{d}{2}$ and choose a generic polynomial $R \in \mathbb{R}[x]_{2d_1}$. Then solve the moment optimization problem

$$\begin{cases} \min & \langle R, z \rangle \\ s.t. & w|_d = y, \, w \in \mathcal{R}_{d_1}(K). \end{cases} \tag{8.2.10}$$

When $R$ has generic coefficients, each optimizer of (8.2.10) is an extreme point of its feasible set (see [229]). For convenience, we denote the degree

$$d_c := \max\left\{ \lceil \tfrac{1}{2}\deg(c_i) \rceil : i \in \mathcal{E} \cup \mathcal{I} \right\}. \tag{8.2.11}$$

For an order $k \geq \max\{d_1, d_c\}$, solve the semidefinite relaxation

$$\begin{cases} \min & \langle R, z \rangle \\ s.t. & \mathcal{V}^{(2k)}_{c_i}[z] = 0 \, (i \in \mathcal{E}), \\ & L^{(k)}_{c_j}[z] \succeq 0 \, (j \in \mathcal{I}), \\ & M_k[z] \succeq 0, \\ & z|_d = y, \, z \in \mathbb{R}^{\mathbb{N}^n_{2k}}. \end{cases} \tag{8.2.12}$$

Its dual problem is the linear conic optimization

$$\begin{cases} \max & \langle p, y \rangle \\ s.t. & R - p \in \text{Ideal}[c_{eq}]_{2k} + \text{QM}[c_{in}]_{2k}, \\ & p \in \mathbb{R}[x]_d. \end{cases} \tag{8.2.13}$$

The following is a basic property of the relaxation (8.2.12).

**Proposition 8.2.3.** ([229]) Let $K, y$ be as above.

(i) If (8.2.12) is infeasible for some $k$, then we have $y \notin \mathcal{R}_d(K)$.

(ii) If $z^*$ is a minimizer of (8.2.12) such that for some $t \in [d_c, k]$

$$\text{rank } M_{t-d_c}[z^*] = \text{rank } M_t[z^*] = r, \tag{8.2.14}$$

then we have $y \in \mathcal{R}_d(K)$ and $y$ admits a $r$-atomic $K$-representing measure.

The proof for Proposition 8.2.3 is left as an exercise. We can check memberships for the cone $\mathcal{R}_d(K)$ as follows. Start with $k := \max\{d_1, d_c\}$. If (8.2.12) is infeasible, then we have $y \notin \mathcal{R}_d(K)$ and stop. If (8.2.12) is feasible, solve it for a minimizer $z^*$ (if it exists). To ensure that (8.2.12) has a minimizer, we can choose a generic $R \in \text{int}\,(\Sigma[x]_{2d_1})$. When $\text{Ideal}[c_{eq}]+\text{QM}[c_{in}]$

is archimedean, (8.2.12) has a minimizer for all $R \in \mathbb{R}[x]_{2d_1}$. If the rank condition (8.2.14) holds for some $t$, then $y$ admits a $r$-atomic $K$-representing measure. If (8.2.14) fails to hold for all $t$, we can increase $k$ by one and solve the semidefinite relaxation (8.2.12) again. Repeating this procedure, we can eventually check memberships for the moment cone $\mathcal{R}_d(K)$. The following is the algorithm.

**Algorithm 8.2.4.** *For $K$ given as in (8.2.1) and for given $y \in \mathbb{R}^{\mathbb{N}_d^n}$, do the following:*

**Step 0** *Choose a generic $R \in \mathrm{int}\,(\Sigma[x]_{2d_1})$ and let $k := \max\{d_1, d_c\}$.*

**Step 1** *If (8.2.12) is infeasible, output that $y \notin \mathcal{R}_d(K)$ and stop. If (8.2.12) is feasible, solve it for a minimizer $z^*$. Let $t := \max\{d_1, d_c\}$.*

**Step 2** *Check the rank condition (8.2.14). If it holds, go to Step 3; otherwise, go to Step 4.*

**Step 3** *Apply the method in §2.7.2 to compute the $r$-atomic $K$-representing measure*

$$\mu = \lambda_1 \delta_{u_1} + \cdots + \lambda_r \delta_{u_r},$$

*where $r = \mathrm{rank}\, M_t[z^*]$, each $u_i \in K$, and $\lambda_i > 0$. Then, stop.*

**Step 4** *If $t < k$, let $t := t+1$ and go to Step 2; otherwise, set $k := k+1$ and go to Step 1.*

In Step 0, one can generate a random square matrix $U$ of length $\binom{n+d_1}{d_1}$ and let $R = [x]_{d_1}^T (U^T U)[x]_{d_1}$. The convergence properties of Algorithm 8.2.4 are summarized in the following (see Theorems 5.3 and 5.5 in [229] for more details).

**Theorem 8.2.5.** *([229]) Let $K$ be as in (8.2.1) such that $\mathrm{Ideal}[c_{eq}] + \mathrm{QM}[c_{in}]$ is archimedean. Suppose $R$ is a generic polynomial in $\Sigma[x]_{2d_1}$.*

  (i) *For each $y \notin \mathcal{R}_d(K)$, the relaxation (8.2.12) must be infeasible when $k$ is big enough.*

  (ii) *For each $y \in \mathcal{R}_d(K)$, the relaxation (8.2.12) has an optimizer $z^{(k)}$ for all $k \geq d_1$. More-over, when $t$ is sufficiently large, the truncated sequence $\{z^{(k)}|_{2t}\}_{k=d_1}^{\infty}$ is bounded and all its accumulation points are flat extensions of $y$.*

We would like to remark that Algorithm 8.2.4 has finite convergence under some general conditions (see [229]). In computational practice, Algorithm 8.2.4 often checks memberships for $\mathcal{R}_d(K)$ within a few loops.

**Example 8.2.6.** *([229]) Consider the tms $y \in \mathbb{R}^{\mathbb{N}_6^2}$ given as*

$(1, 0, 0, 1/3, 0, 1/3, 0, 0, 0, 0, 1/5, 0, 1/9, 0, 1/5, 0, 0, 0, 0, 0, 0, 1/7, 0, 1/15, 0, 1/15, 0, 1/7),$

*where the entries $y_\alpha$ are listed in the graded lexicographical ordering. The tms $y$ has a repre-senting measure supported on $K = [-1, 1]^2$, since each $y_\alpha$ is the mean value of $x^\alpha$ on $[-1, 1]^2$. We have $c_{eq} = \emptyset$ and $c_{in} = (1 - x_1^2, 1 - x_2^2)$. By Algorithm 8.2.4, we get the 10-atomic measure $\sum_{i=1}^{10} \lambda_i \delta_{u_i}$ for $y$, with $u_i$ and $\lambda_i$ given in Table 8.1.*

Table 8.1: The 10-atomic representing measure for Example 8.2.6.

| $i$ | $u_i$ | $\lambda_i$ | $i$ | $u_i$ | $\lambda_i$ |
|---|---|---|---|---|---|
| 1 | $(-0.7983, -0.9666)$ | 0.0318 | 6 | $(-0.8710, -0.2228)$ | 0.0947 |
| 2 | $(-0.2155, -0.6211)$ | 0.1698 | 7 | $(-0.9175, 0.8643)$ | 0.0328 |
| 3 | $(0.5833, -0.8876)$ | 0.0729 | 8 | $(-0.4834, 0.4714)$ | 0.1662 |
| 4 | $(0.3541, 0.0676)$ | 0.2054 | 9 | $(0.9269, -0.3819)$ | 0.0672 |
| 5 | $(0.0841, 0.9294)$ | 0.0717 | 10 | $(0.8153, 0.6919)$ | 0.0874 |

## 8.2.3 ▪ Exercises

**Exercise 8.2.1.** *For the polynomial $f = x^T x + \sum\limits_{i<j<k} x_i x_j x_k$ with $x \in \mathbb{R}^5$, determine the largest $\rho$ such that $f \in \mathscr{P}_3(K)$ for $K = B(0, \rho)$.*

**Exercise 8.2.2.** *Let $y \in \mathbb{R}^{\mathbb{N}_6^3}$ be the tms consisting of moments for the Gaussian distribution, i.e.,*

$$y_\alpha = \int x^\alpha \cdot \frac{1}{(2\pi)^{3/2}} \exp\left(-\frac{1}{2} x^T x\right) \mathrm{d}x \quad (\alpha \in \mathbb{N}_6^3).$$

*Find a value of $\rho$, as small as possible, such that $y \in \mathscr{R}_6(K)$ for $K = B(0, \rho)$.*

**Exercise 8.2.3.** *Let $y \in \mathbb{R}^{\mathbb{N}_4^4}$ be the tms consisting of moments for the Lebesgue measure, i.e.,*

$$y_\alpha = \int_{\|x\|_2 \leq 1} x^\alpha \mathrm{d}x \quad (\alpha \in \mathbb{N}_4^4).$$

*Find a value of $\rho$, as small as possible, such that $y \in \mathscr{R}_4(K)$ for $K = [-\rho, \rho]^4$.*

**Exercise 8.2.4.** *Prove the nesting containment relation (8.2.8).*

**Exercise 8.2.5.** *Prove Proposition 8.2.3.*

**Exercise 8.2.6.** *Let $K = \{x^T A x \leq 1\}$ for a positive definite matrix $A$. For $y \in \mathbb{R}^{\mathbb{N}_2^n}$, show that $y \in \mathscr{R}_2(K)$ if and only if it satisfies*

$$M_1[y] \succeq 0, \quad \langle 1 - x^T A x, y \rangle \geq 0.$$

**Exercise 8.2.7.** *Suppose $K$ is the unit sphere $x^T x = 1$. Let $f_k$ be the optimal value of the $k$th order relaxation (8.2.5). For $f \in \mathbb{R}[x]_d$, show that $f \geq 0$ on $K$ if and only if $f_k \geq 0$ for some $k$.*

**Exercise 8.2.8.** *If $c_{eq}(x) = 0$ has only finitely many real solutions, show that the hierarchy of moment relaxations (8.2.12) has finite convergence for checking $y \in \mathscr{R}_d(K)$.*

## 8.3 ▪ The Linear Conic Optimization Problem

Let $K \subseteq \mathbb{R}^n$ be the semialgebraic set as in (8.2.1), given by two polynomial tuples $c_{eq}$ and $c_{in}$. We show how to solve linear conic optimization problems with the moment cone $\mathscr{R}_d(K)$ and the psd polynomial cone $\mathscr{P}_d(K)$.

A linear functional $\ell$ on $\mathbb{R}^{\mathbb{N}^n_d}$ can be expressed as

$$\ell(y) = \langle f, y \rangle, \quad y \in \mathbb{R}^{\mathbb{N}^n_d},$$

for some polynomial $f \in \mathbb{R}[x]_d$. For instance,

$$y_{10} - 2y_{12} - 3y_{21} + 4y_{04} = \langle x_1 - 2x_1x_2^2 - 3x_1^2x_2 + 4x_2^4, y \rangle.$$

An affine subspace $F$ of $\mathbb{R}^{\mathbb{N}^n_d}$ is determined by linear equations such as

$$\langle f_1, y \rangle = b_1, \quad \ldots, \quad \langle f_m, y \rangle = b_m$$

for some given polynomials $f_1, \ldots, f_m \in \mathbb{R}[x]_d$ and given constants $b_1, \ldots, b_m$. A linear objective function on $\mathbb{R}^{\mathbb{N}^n_d}$ can be written as $\langle f_0, y \rangle$ for some $f_0 \in \mathbb{R}[x]_d$. So, a linear conic optimization problem with the moment cone $\mathscr{R}_d(K)$ can be given as

$$\begin{cases} \vartheta^* := \min & \langle f_0, y \rangle \\ \qquad\quad s.t. & \langle f_i, y \rangle = b_i, \ i = 1, \ldots, m, \\ & y \in \mathscr{R}_d(K), \end{cases} \tag{8.3.1}$$

for some polynomials $f_0, f_1, \ldots, f_m \in \mathbb{R}[x]_d$ and constants $b_1, \ldots, b_m$. When $f_0 = 0$ (the identically zero polynomial), solving the optimization (8.3.1) is equivalent to finding a feasible point in the intersection of the cone $\mathscr{R}_d(K)$ and the affine subspace $F$. Detecting feasibility/infeasibility is an important question in optimization. The problem (8.3.1) is called a *generalized problem of moments* (GPM) in [161].

Since the dual cone of $\mathscr{R}_d(K)$ is $\mathscr{P}_d(K)$, the dual optimization problem of (8.3.1) is

$$\begin{cases} \theta^* := \max & b_1\lambda_1 + \cdots + b_m\lambda_m \\ \qquad\quad s.t. & f_0 - \sum_{i=1}^m \lambda_i f_i \in \mathscr{P}_d(K). \end{cases} \tag{8.3.2}$$

For convenience of notation, we denote

$$\boxed{\begin{aligned} b &:= \begin{bmatrix} b_1 & \cdots & b_m \end{bmatrix}^T, \\ \lambda &:= \begin{bmatrix} \lambda_1 & \cdots & \lambda_m \end{bmatrix}^T, \\ f(\lambda) &:= f_0 - \lambda_1 f_1 - \cdots - \lambda_m f_m. \end{aligned}} \tag{8.3.3}$$

The dual relationship between the optimization (8.3.1) and (8.3.2) can be seen as follows. If $y$ is feasible for (8.3.1) and $\lambda$ is feasible for (8.3.2), then

$$\langle f_0, y \rangle - b^T\lambda = \langle f_0, y \rangle - \lambda_1\langle f_1, y \rangle - \cdots - \lambda_m\langle f_m, y \rangle = \langle f(\lambda), y \rangle \geq 0,$$

since $f(\lambda) \in \mathscr{P}_d(K)$ and $y \in \mathscr{R}_d(K)$. This is the *weak* duality between (8.3.1) and (8.3.2). Under certain conditions, the *strong* duality also holds.

Many hard optimization problems can be formulated in the form (8.3.1). For instance, what is the measure $\mu$ supported in the unit circle $\mathbb{S}^1 := \{x_1^2 + x_2^2 = 1\}$, such that the total mass of $\mathbb{S}^1$ is minimum and its moments satisfy the equations

$$\int x_1^2 x_2^2 \mathrm{d}\mu = \int (x_1^4 + x_2^4)\mathrm{d}\mu = \int (x_1^6 + x_2^6)\mathrm{d}\mu = 1?$$

The mass of $\mathbb{S}^1$ is just the moment $\int 1\mathrm{d}\mu = \langle 1, y \rangle$. This question is equivalent to solving the linear moment optimization

$$\begin{cases} \min & \langle 1, y \rangle \\ s.t. & \langle x_1^2 x_2^2, y \rangle = 1, \\ & \langle x_1^4 + x_2^4, y \rangle = 1, \\ & \langle x_1^6 + x_2^6, y \rangle = 1, \\ & y \in \mathscr{R}_6(\mathbb{S}^1). \end{cases}$$

The corresponding polynomials are

$$f_0 = 1, \quad f_1 = x_1^2 x_2^2, \quad f_2 = x_1^4 + x_2^4, \quad f_3 = x_1^6 + x_2^6.$$

As shown in Example 8.3.6, there is no measure $\mu$ satisfying all the above constraints.

The dual optimization problem (8.3.2) is also quite broad in applications. For instance, the minimum value of a polynomial $f \in \mathbb{R}[x]_d$ on the set $K$ is equal to the maximum value of the optimization

$$\begin{cases} \max & \lambda \\ \text{s.t.} & f - \lambda \cdot 1 \in \mathscr{P}_d(K). \end{cases}$$

Similarly, the minimum value of a rational function $f_0/f_1$ on $K$, where $f_0, f_1 \in \mathbb{R}[x]_d$ and $f_1 > 0$ on $K$, is equal to the maximum value of the optimization

$$\begin{cases} \max & \lambda \\ \text{s.t.} & f_0 - \lambda \cdot f_1 \in \mathscr{P}_d(K). \end{cases}$$

We refer to §12.2 for optimization with rational functions.

## 8.3.1 ▪ The Moment-SOS hierarchy

The cones $\mathscr{R}_d(K)$ and $\mathscr{P}_d(K)$ can be approximated arbitrarily accurately by Moment-SOS relaxations. For the set $K$ given as in (8.2.1), consider the sets $Q_k$ in (8.2.2) and $F_k$ in (8.2.7). Their approximation qualities are shown in Propositions 8.2.1 and 8.2.2. If $\mathscr{R}_d(K)$ is relaxed to the outer approximation $F_k$, then (8.3.1) is relaxed to the moment optimization

$$\begin{cases} \vartheta^{(k)} := \min & \langle f_0, w \rangle \\ \text{s.t.} & \langle f_i, w \rangle = b_i \, (i \in [m]), \\ & \mathscr{V}_{c_i}^{(2k)}[w] = 0 \, (i \in \mathcal{E}), \\ & L_{c_j}^{(k)}[w] \succeq 0 \, (j \in \mathcal{I}), \\ & M_k[w] \succeq 0, \, w \in \mathbb{R}^{\mathbb{N}_{2k}^n}. \end{cases} \quad (8.3.4)$$

For each $w$ that is feasible for (8.3.4), the truncation $y := w|_d \in F_k$. If $\mathscr{P}_d(K)$ is replaced by the inner approximation $Q_k$, then (8.3.2) is restricted to the linear conic optimization

$$\begin{cases} \theta^{(k)} := \max_{\lambda \in \mathbb{R}^m} & b^T \lambda \\ \text{s.t.} & f(\lambda) \in \text{Ideal}[c_{eq}]_{2k} + \text{QM}[c_{in}]_{2k}. \end{cases} \quad (8.3.5)$$

Interestingly, the optimization problems (8.3.4) and (8.3.5) are dual to each other. For every relaxation order $k$, it holds that

$$\vartheta^{(k)} \leq \vartheta^*, \quad \theta^{(k)} \leq \theta^* \leq \vartheta^*.$$

Suppose $w^{*,k}$ is a minimizer of (8.3.4) and $\lambda^{*,k}$ is a maximizer of (8.3.5). If the truncation

$$y^{(k)} := w^{*,k}|_d \in \mathscr{R}_d(K), \quad (8.3.6)$$

then $\vartheta^{(k)} = \vartheta^*$ and $y^{(k)}$ is a minimizer of (8.3.1), i.e., (8.3.4) is a tight relaxation for (8.3.1). For such a case, if $\theta^{(k)} = \vartheta^{(k)}$, then $\theta^{(k)} = \theta^*$ and $\lambda^{*,k}$ is a maximizer of (8.3.2). If the relaxation (8.3.4) is infeasible, the moment optimization (8.3.1) must also be infeasible. Observe that (8.3.5) is a restriction of (8.3.2). If (8.3.5) is unbounded above, then (8.3.2) must also be unbounded above, and hence (8.3.1) must be infeasible, by the weak duality. We leave the proof for these facts as Exercises 8.3.5 and 8.3.6. Summarizing the above, we get the following algorithm.

**Algorithm 8.3.1.** *For given $f_0, f_1, \ldots, f_m \in \mathbb{R}[x]_d$, $b \in \mathbb{R}^m$ and $K$ as in (8.2.1), let $k := \lceil d/2 \rceil$ and do the following:*

**Step 1** *Solve the relaxations (8.3.4)-(8.3.5). If (8.3.5) is unbounded above, then (8.3.2) is also unbounded above, (8.3.1) is infeasible, and stop. If (8.3.4) is unbounded below, then go to Step 5; otherwise, compute a minimizer $w^{*,k}$ for (8.3.4) and a maximizer $\lambda^{*,k}$ for (8.3.5), if they exist. Let $y^{(k)}$ be the truncation as in (8.3.6).*

**Step 2** *If there exists an integer $t \in [d/2, k]$ such that*

$$\operatorname{rank} M_{t-d_c}[w^{*,k}] = \operatorname{rank} M_t[w^{*,k}], \qquad (8.3.7)$$

*where $d_c$ is as in (8.2.11), then $y^{(k)} \in \mathscr{R}_d(K)$ is a minimizer of (8.3.1) and go to Step 4. Otherwise, go to Step 3.*

**Step 3** *Apply Algorithm 8.2.4 to check whether $y^{(k)} \in \mathscr{R}_d(K)$ or not. If $y^{(k)} \in \mathscr{R}_d(K)$, then $y^{(k)}$ is a minimizer of (8.3.1) and go to Step 4; otherwise, go to Step 5.*

**Step 4** *If $\theta^{(k)} = \vartheta^{(k)}$, then $\lambda^{*,k}$ is a maximizer of (8.3.2), and stop.*

**Step 5** *Let $k := k + 1$ and go to Step 1.*

In Step 2, if the rank condition (8.3.7) holds, then we can apply the method in §2.7.2 to get a finitely atomic $K$-representing measure $\nu$ for the truncation $w^{*,k}|_{2t}$. One can see that $\nu$ must also be a $K$-representing measure for $y^{(k)}$. The following are some examples.

**Example 8.3.2.** *Let $K$ be the unit sphere in $\mathbb{R}^3$, i.e., $c_{eq} = (1 - x_1^2 - x_2^2 - x_3^2)$ and $c_{in} = \emptyset$.*
*(i) Consider the optimization (8.3.1) with*

$$f_0 := x_1^6 + x_2^6 + x_3^6, \ f_1 = x_1^2 x_2^4 + x_2^2 x_3^4 + x_3^2 x_1^4,$$
$$f_2 = x_1^3 x_2^3 + x_2^3 x_3^3 + x_3^3 x_1^3, \ f_3 = x_1^5 x_2 + x_2^5 x_3 + x_3^5 x_1, \ b_1 = b_2 = b_3 = 1.$$

*We solve (8.3.1) by Algorithm 8.3.1. For $k = 3$, the rank condition (8.3.7) holds, so we know $y^{(3)} \in \mathscr{R}_d(K)$. It admits the 2-atomic measure*

$$\frac{9}{2}\left(\delta_{\frac{(1,1,1)}{\sqrt{3}}} + \delta_{\frac{(-1,-1,-1)}{\sqrt{3}}}\right).$$

*Moreover, we also get*

$$\lambda^{*,3} \approx (-1.440395, 2.218992, 0.221403).$$

*Since $\vartheta^{(k)} = \theta^{(k)}$, the above $\lambda^{*,3}$ is also a maximizer for (8.3.2).*
*(ii) We look for the minimum value of the moment $\int (x_1^6 + x_2^6 + x_3^6)\mathrm{d}\mu$ for all measures $\mu$ supported in $K$ with moments satisfying the constraints*

$$\int x_1^3 x_2^3 \mathrm{d}\mu = \int x_2^3 x_3^3 \mathrm{d}\mu = \int x_3^3 x_1^3 \mathrm{d}\mu,$$
$$\int x_1^2 x_2^2 x_3^2 \mathrm{d}\mu = 1, \int (x_1^4 x_2^2 + x_2^4 x_3^2 + x_3^4 x_1^2)\mathrm{d}\mu = 3.$$

*This is equivalent to solving the optimization (8.3.1) with*

$$f_0 = x_1^6 + x_2^6 + x_3^6, \ f_1 = x_1^3 x_2^3 - x_2^3 x_3^3, \ f_2 = x_1^3 x_2^3 - x_3^3 x_1^3,$$
$$f_3 = x_1^2 x_2^2 x_3^2, \ f_4 = x_1^4 x_2^2 + x_2^4 x_3^2 + x_3^4 x_1^2, \ b = (0,0,1,3).$$

*We solve it by Algorithm 8.3.1. For $k = 3$, the rank condition (8.3.7) holds, so we have $y^{(3)} \in \mathscr{R}_d(K)$. The tms $y^{*,3}$ admits the atomic measure*

$$\frac{27}{4}\left(\delta_{\frac{(1,1,1)}{\sqrt{3}}} + \delta_{\frac{(-1,1,1)}{\sqrt{3}}} + \delta_{\frac{(1,-1,1)}{\sqrt{3}}} + \delta_{\frac{(1,1,-1)}{\sqrt{3}}}\right).$$

*The minimum objective value is* 3.

**Example 8.3.3.** *Let $K$ be the simplex in $\mathbb{R}^n$, given by $c_{eq} = (1 - e^T x)$ and $c_{in} = (x_1, \ldots, x_n, 1 - \|x\|^2)$. For $n = 6$ and the following quadratic forms*

$$f_0 = (x_1 + \cdots + x_6)^2, \; f_1 = x_1 x_2 - x_2 x_3 + x_3 x_4 - x_5 x_6 + x_6 x_1,$$

*we are interested in the maximum $\lambda$ such that $f_0 - \lambda f_1 \geq 0$ on $K$, i.e., $f(\lambda)$ is a copositive form. This is equivalent to solving (8.3.2) for $b = 1$. We apply Algorithm 8.3.1 to solve it. For $k = 2$, we get that $\lambda^{*,2} = 4$ and $y^{(2)} \in \mathscr{R}_2(K)$, which admits the atomic measure $4\delta_{(1/2,1/4,0,0,0,1/4)}$. Since $\vartheta^{(2)} = \theta^{(2)}$, the maximum $\lambda$ for the above is 4.*

## 8.3.2 ▪ Convergence analysis

Checking the membership $y^{(k)} \in \mathscr{R}_d(K)$ is a termination criterion for Algorithm 8.3.1. If a truncation $w^{*,k}|_{2t}$ is flat for some $t \geq d/2$, then $y^{(k)} \in \mathscr{R}_d(K)$. It is possible that $y^{(k)} \in \mathscr{R}_d(K)$ while $w^{*,k}|_{2t}$ is not flat for any $t$ (see Exercise 8.3.3). For such a case, we can apply Algorithm 8.2.4 to check $y^{(k)} \in \mathscr{R}_d(K)$. The following is the convergence property for Algorithm 8.3.1.

**Theorem 8.3.4.** *([232]) Let $K$ be as in (8.2.1). Assume that $\mathrm{Ideal}[c_{eq}] + \mathrm{QM}[c_{in}]$ is archimedean, (8.3.1) is feasible, and there exists $\hat{\lambda} \in \mathbb{R}^m$ such that $f(\hat{\lambda}) > 0$ on $K$. Then, we have the following:*

(i) *For all $k$ sufficiently large, (8.3.5) has an interior point and (8.3.4) has a minimizer $w^{*,k}$.*

(ii) *The sequence $\{y^{(k)}\}_{k=1}^{\infty}$ is bounded, and each of its accumulation points is a minimizer of (8.3.1). Moreover, $\vartheta^{(k)} \to \vartheta^*$ as $k$ increases.*

(iii) *The maximum value $\theta^{(k)}$ of (8.3.5) converges to the maximum value $\theta^*$ of (8.3.2).*

**Proof.** (i) The archimedeanness of $\mathrm{Ideal}[c_{eq}] + \mathrm{QM}[c_{in}]$ implies that $K$ is compact. So, there exist $\epsilon_0 > 0$ and $\delta > 0$ such that

$$f(\lambda) > \epsilon_0 \quad \text{for all} \quad \lambda \in B(\hat{\lambda}, \delta).$$

By the degree bound for Putinar's Positivstellensatz (see Theorem 6 of [211]), there exists $N_0 > 0$ such that

$$f(\lambda) - \epsilon_0 \in \mathrm{Ideal}[c_{eq}]_{2N_0} + \mathrm{QM}[c_{in}]_{2N_0} \quad \text{for all} \quad \lambda \in B(\hat{\lambda}, \delta).$$

So, (8.3.5) has an interior point for all $k \geq N_0$, and hence the strong duality holds between (8.3.4) and (8.3.5). Since (8.3.1) is feasible, the relaxation (8.3.4) is also feasible and has a minimizer $w^{*,k}$, by the strong duality theorem (see Theorem 1.5.1).

(ii) The boundedness of the sequence $\{y^{(k)}\}$ is implied by the archimedeanness of $\mathrm{Ideal}[c_{eq}] + \mathrm{QM}[c_{in}]$ and $f(\hat{\lambda}) > 0$. This is shown in [232]. For cleanness, we leave the proof as an exercise. We show that every accumulation point of $\{y^{(k)}\}$ is a minimizer of (8.3.1). Let $y^*$ be an arbitrary accumulation point. We show that $y^*$ is a minimizer of (8.3.1). Without loss of generality, one can assume $y^{(k)} \to y^*$ as $k \to \infty$. Since $K$ is compact, say, $K \subseteq B(0, \rho)$ with $\rho < 1$, up to a scaling of coordinates, there exists $k_1 > 0$ such that the sequence of

$$z^{(k)} := w^{*,k}|_{2k - 2k_1}$$

is also bounded. This is implied by the archimedeanness of $\text{Ideal}[c_{eq}] + \text{QM}[c_{in}]$ and $f(\hat{\lambda}) > 0$, similar to the proof for boundedness of $\{y^{(k)}\}$. Each tms $z^{(k)}$ can be extended to a vector in $\mathbb{R}^{\mathbb{N}^n_\infty}$ by adding zero entries to the tailing. The set $\mathbb{R}^{\mathbb{N}^n_\infty}$ can be viewed as a Hilbert space, equipped with the inner product

$$\langle u, v \rangle := \sum_{\alpha \in \mathbb{N}^n} u_\alpha v_\alpha \quad \forall\, u, v \in \mathbb{R}^{\mathbb{N}^n_\infty}.$$

The sequence $\{z^{(k)}\}$ is bounded in $\mathbb{R}^{\mathbb{N}^n_\infty}$. By Alaoglu's Theorem (cf. [58, Theorem V.3.1] or [162, Theorem C.18]), it has a subsequence $\{z^{k_j}\}$ that is convergent in the weak-$*$ topology. That is, there exists $z^* \in \mathbb{R}^{\mathbb{N}^n_\infty}$ such that

$$\langle p, z^{k_j} \rangle \to \langle p, z^* \rangle \quad \text{as } j \to \infty$$

for all $p \in \mathbb{R}^{\mathbb{N}^n_\infty}$. The above implies that for each $\alpha \in \mathbb{N}^n$, we have

$$(z^{k_j})_\alpha \to (z^*)_\alpha. \tag{8.3.8}$$

Since $z^{(k)}|_d = y^{(k)} \to y^*$ as $k \to \infty$, we get $z^*|_d = y^*$. For each $r = 1, 2, \ldots$, if $k_j \geq 2r$, then (due to the constraints in (8.3.4))

$$M_r[z^{(k_j)}] \succeq 0, \quad \mathscr{V}_{c_i}^{(2r)}[z^{(k_j)}] = 0 \,(i \in \mathcal{E}), \quad L_{c_i}^{(r)}[z^{(k_j)}] \succeq 0 \,(j \in \mathcal{I}).$$

The convergence (8.3.8) implies that for all $r = 1, 2, \ldots$

$$M_r[z^*] \succeq 0, \quad \mathscr{V}_{c_i}^{(2r)}[z^*] = 0 \,(i \in \mathcal{E}), \quad L_{c_i}^{(r)}[z^*] \succeq 0 \,(j \in \mathcal{I}).$$

This means that $z^* \in \mathbb{R}^{\mathbb{N}^n_\infty}$ is a full moment sequence whose localizing matrices of all orders are psd. By Lemma 3.2 of [272], $z^*$ admits a $K$-representing measure. Clearly, $\langle f_i, y^* \rangle = b_i$ for all $i = 1, \ldots, m$. So, $y^*$ is feasible for (8.3.1) and we can get

$$\vartheta^* \leq \langle f_0, y^* \rangle.$$

Since (8.3.4) is a relaxation of (8.3.1) and $w^{*,k}$ is a minimizer of (8.3.4), it holds that

$$\vartheta^* \geq \langle f_0, y^{(k)} \rangle, \quad k = 1, 2, \ldots.$$

Hence, we get

$$\vartheta^* \geq \lim_{k \to \infty} \langle f_0, y^{(k)} \rangle = \langle f_0, y^* \rangle.$$

Therefore, $\vartheta^* = \langle f_0, y^* \rangle$ and $y^*$ is a minimizer of (8.3.1).

(iii) For every $\epsilon > 0$, there exists $\lambda_\epsilon$ such that $f(\lambda_\epsilon) \in \mathscr{P}_d(K)$ and

$$\theta^* - \epsilon < b^T \lambda_\epsilon \leq \theta^*.$$

Let $\lambda(\epsilon) := (1 - \epsilon)\lambda_\epsilon + \epsilon\hat{\lambda}$. Since $f(\hat{\lambda}) > 0$ on $K$, we have $f(\lambda(\epsilon)) > 0$ on $K$ and

$$b^T \lambda(\epsilon) = (1 - \epsilon)b^T \lambda_\epsilon + \epsilon b^T \hat{\lambda} > (1 - \epsilon)(\theta^* - \epsilon) + \epsilon b^T \hat{\lambda}.$$

For $k$ big enough, we have $f(\lambda(\epsilon)) \in \text{Ideal}[c_{eq}]_{2k} + \text{QM}[c_{in}]_{2k}$ and hence

$$\theta^{(k)} > (1 - \epsilon)(\theta^* - \epsilon) + \epsilon b^T \hat{\lambda}.$$

Since $\theta^{(k)} \leq \theta^*$ for all $k$, we get $\theta^{(k)} \to \theta^*$ as $k \to \infty$. $\quad\square$

Under certain additional assumptions, Algorithm 8.3.1 also has finite convergence. We refer to [232] for the details.

### 8.3.3 ▪ Infeasibility and unboundedness

For the linear conic optimization problems (8.3.1)-(8.3.2), we discuss how to certify infeasibility or unboundedness. For (8.3.1), consider the affine subspace

$$F := \left\{ y \in \mathbb{R}^{\mathbb{N}_d^n} : \langle f_i, y \rangle = b_i, \ i = 1, \ldots, m \right\}. \tag{8.3.9}$$

We look for a certificate for the infeasibility

$$\mathscr{R}_d(K) \cap F = \emptyset. \tag{8.3.10}$$

We also discuss how to detect whether (8.3.1) is unbounded below or not. Similarly, for (8.3.2), we are interested in knowing whether or not the psd polynomial cone $\mathscr{P}_d(K)$ intersects the affine subspace of polynomials

$$A := f_0 + \operatorname{span}\{f_1, \ldots, f_m\}. \tag{8.3.11}$$

We anticipate a certificate for the infeasibility

$$A \cap \mathscr{P}_d(K) = \emptyset. \tag{8.3.12}$$

Interestingly, the infeasibility of a linear conic optimization problem is usually certified by the unboundedness of its dual optimization. Therefore, the infeasibility and unboundedness are usually considered together.

Algorithm 8.3.1 can also detect the infeasibility and/or unboundedness for (8.3.1) or (8.3.2). These properties are summarized as follows.

**Theorem 8.3.5.** *([232]) Let $K$ be as in (8.2.1) and let $A$, $F$ be as above.*

(i) *If there exist $\lambda := (\lambda_1, \ldots, \lambda_m)$ and $k \in \mathbb{N}$ such that*

$$b^T \lambda < 0, \quad \lambda_1 f_1 + \cdots + \lambda_m f_m \in \operatorname{Ideal}[c_{eq}]_{2k} + \operatorname{QM}[c_{in}]_{2k}, \tag{8.3.13}$$

*then (8.3.4) must be infeasible, and hence (8.3.1) is also infeasible. Furthermore, if in addition (8.3.2) is feasible, then (8.3.2) must be unbounded above.*

(ii) *Suppose $\operatorname{Ideal}[c_{eq}] + \operatorname{QM}[c_{in}]$ is archimedean and there exists*

$$p \in \operatorname{span}\{f_1, \ldots, f_m\}, \quad p > 0 \quad on \quad K.$$

*If (8.3.1) is infeasible, then there must exist $\lambda$ and $k$ satisfying (8.3.13).*

(iii) *If there exists a tms $y$ such that*

$$\langle f_0, y \rangle < 0, \quad \langle f_i, y \rangle = 0 \, (i = 1, \ldots, m), \quad y \in \mathscr{R}_d(K), \tag{8.3.14}$$

*then the optimization (8.3.2) is infeasible and (8.3.1) is unbounded below if it is feasible.*

(iv) *Suppose $\operatorname{span}\{f_1, \ldots, f_m\} \cap \mathscr{P}_d(K) = I(K)_d$. If (8.3.2) is infeasible, then there must exist $y$ satisfying (8.3.14).*

**Proof.** (i) If (8.3.13) holds for some $\lambda, k$, then (8.3.4) must be infeasible, because any feasible $w$ for (8.3.4) would give the contradiction

$$0 > b^T \lambda = \sum_{i=1}^{m} \lambda_i \langle f_i, w \rangle = \left\langle \sum_{i=1}^{m} \lambda_i f_i, y \right\rangle \geq 0.$$

Since (8.3.4) is a relaxation of (8.3.1), we know (8.3.1) is also infeasible. If it is feasible, the optimization (8.3.2) must be unbounded above, since $-\lambda$ is an improving ray.

(ii) Consider the optimization problem

$$\begin{cases} \max & 0 \\ s.t. & \langle f_i, y \rangle = b_i, \ i = 1, \ldots, m, \\ & y \in \mathscr{R}_d(K). \end{cases} \tag{8.3.15}$$

Its dual optimization problem is

$$\begin{cases} \min & b^T \lambda \\ s.t. & \lambda_1 f_1 + \cdots + \lambda_m f_m \in \mathscr{P}_d(K). \end{cases} \tag{8.3.16}$$

The archimedeanness of $\text{Ideal}[c_{eq}] + \text{QM}[c_{in}]$ implies the set $K$ is compact, so $\mathscr{R}_d(K)$ and $\mathscr{P}_d(K)$ are closed convex cones (see Theorem 8.1.2). By the assumption, (8.3.16) has an interior point. So, the strong duality holds. If (8.3.1) is infeasible, the optimization (8.3.15) is also infeasible, so the optimization (8.3.16) must be unbounded below (see Theorem 1.5.1), i.e., there exists $\hat{\lambda} := (\hat{\lambda}, \ldots, \hat{\lambda})$ satisfying

$$b^T \hat{\lambda} < 0, \quad \hat{\lambda}_1 a_1 + \cdots + \hat{\lambda}_m a_m \in \mathscr{P}_d(K).$$

Since there is $p \in \text{span}\{f_1, \ldots, f_m\}$ such that $p > 0$ on $K$, there exists $\bar{\lambda}$ such that $\bar{\lambda}_1 a_1 + \cdots + \bar{\lambda}_m a_m > 0$ on $K$. For $\epsilon > 0$ small, $\lambda := \hat{\lambda} + \epsilon\bar{\lambda}$ satisfies (8.3.13) for some $k$, by Theorem 2.6.1.

(iii) Suppose (8.3.14) holds for some $y \in \mathbb{R}^{\mathbb{N}^n_d}$. The dual optimization (8.3.2) must be infeasible. This is because if it has a feasible point $\lambda$, then it results in the contradiction

$$0 \le f(\lambda), y \rangle = \langle f_0, y \rangle < 0.$$

Since $y$ give a decreasing ray for (8.3.1), so it must be unbounded below if it is feasible.

(iv) One can generally assume $f_1, \ldots, f_m$ are linearly independent in the quotient space $\mathbb{R}[x]_d/I(K)_d$; otherwise, the redundant ones can be removed. If (8.3.2) is infeasible, $A \cap \mathscr{P}_d(K) = \emptyset$. We show the distance between $A$ and $\mathscr{P}_d(K)$ is positive. Suppose otherwise the distance were zero, then there exist two sequences $\{\lambda^{(k)}\}$ and $\{p_k\} \subseteq \mathbb{R}[x]_d$ such that

$$\| \text{vec}(p_k) \| \to 0, \qquad f(\lambda^{(k)}) + p_k \in \mathscr{P}_d(K).$$

The sequence $\{\lambda^{(k)}\}$ must be unbounded. This is because if it were bounded, one can generally assume $\lambda^{(k)} \to \lambda^*$, which could imply the contradiction $f(\lambda^*) \in A \cap \mathscr{P}_d(K)$. One can further assume $\|\lambda^{(k)}\|_2 \to \infty$. Consider the normalized sequence $\{\lambda^{(k)}/\|\lambda^{(k)}\|_2\}$. One can generally assume

$$\lambda^{(k)}/\|\lambda^{(k)}\|_2 \to \lambda^* := (\lambda_1^*, \ldots, \lambda_m^*) \ne 0.$$

Since each $f(\lambda^{(k)}) + p_k \in \mathscr{P}_d(K)$ and $\| \text{vec}(p_k)\| \to 0$,

$$\frac{f(\lambda^{(k)}) + p_k}{\|\lambda^{(k)}\|} \to q := -(\lambda_1^* f_1 + \cdots + \lambda_m^* f_m) \in \mathscr{P}_d(K).$$

Since $f_1, \ldots, f_m$ are linearly independent in $\mathbb{R}[x]_d/I(K)_d$ and $\lambda^* \ne 0$, we know $q|_K \not\equiv 0$ and $q|_K \ge 0$, which contradicts the given assumption that

$$\text{span}\{f_1, \ldots, f_m\} \cap \mathscr{P}_d(K) = I(K)_d.$$

So the distance between $A$ and $\mathscr{P}_d(K)$ is positive, say, $\epsilon > 0$. Hence,

$$\left(\mathscr{P}_d(K) + B(0, \frac{\epsilon}{2})\right) \cap A = \emptyset.$$

By Theorem 1.2.1, there exist $0 \neq y \in \mathbb{R}^{\mathbb{N}_d^n}$ and $\tau \in \mathbb{R}$ such that

$$\langle p, y \rangle \geq \tau \quad \forall p \in \mathscr{P}_d(K) + B(0, \tfrac{\epsilon}{2}), \qquad \langle p, y \rangle \leq \tau \quad \forall p \in A.$$

The first inequality implies $\tau < 0$ and $y \in \mathscr{P}_d(K)^\star$, so $y \in \mathscr{R}_d(K)$. The latter inequality implies $\langle f_0, y \rangle < 0$ and $\langle f_i, y \rangle = 0$ for all $i$. Thus, such $y$ satisfies (8.3.14). $\quad\square$

If $K$ is compact and there exists $p \in \text{span}\{f_1, \dots, f_m\}$ such that $p > 0$ on $K$, then the intersection $A \cap \mathscr{P}_d(K)$ must be nonempty. For the intersection to be empty, no polynomials in $\text{span}\{f_1, \dots, f_m\}$ can be positive on $K$. This is why we need the assumption $\text{span}\{f_1, \dots, f_m\} \cap \mathscr{P}_d(K) = I(K)_d$ for the item (iv) of Theorem 8.3.5. This assumption cannot be removed. For instance, consider that $K = \mathbb{S}^1$ (the unit circle) and

$$A = x_1 x_2 + \text{span}\{x_1^2\}.$$

Clearly, $A \cap \mathscr{P}_2(\mathbb{S}^1) = \emptyset$, i.e., (8.3.12) is infeasible. However, for all $y \in \mathscr{R}_2(\mathbb{S}^1)$, if $\langle x_1^2, y \rangle = 0$, then $\langle x_1 x_2, y \rangle = 0$. This is because of the Cauchy-Schwarz inequality

$$\left| \int x_1 x_2 \mathrm{d}\mu \right| \leq \left( \int x_1^2 \mathrm{d}\mu \right)^{\frac{1}{2}} \cdot \left( \int x_2^2 \mathrm{d}\mu \right)^{\frac{1}{2}}$$

for every measure $\mu$. So, (8.3.14) is infeasible. The following are examples of certifying infeasibility or unboundedness.

**Example 8.3.6.** *Let $K = \mathbb{S}^1$ be the unit circle, given by $c_{eq} = (x_1^2 + x_2^2 - 1)$ and $c_{in} = \emptyset$ as in (8.2.1). One wonders whether there is a measure $\mu$ supported in $\mathbb{S}^1$ satisfying moment relations*

$$\int x_1^2 x_2^2 \mathrm{d}\mu = 1, \quad \int (x_1^4 + x_2^4)\mathrm{d}\mu = 1, \quad \int (x_1^6 + x_2^6)\mathrm{d}\mu = 1.$$

*This is equivalent to checking whether there exists $y \in \mathscr{R}_6(\mathbb{S}^1)$ satisfying*

$$\langle f_1, y \rangle = 1, \quad \langle f_2, y \rangle = 1, \quad \langle f_3, y \rangle = 1$$

*for $f_1 = x_1^2 x_2^2$, $f_2 = x_1^4 + x_2^4$, $f_3 = x_1^6 + x_2^6$. Indeed, such $y$ cannot exist, because (8.3.13) is satisfied for $\lambda = (-3, 1, 1)$: $\lambda_1 + \lambda_2 + \lambda_3 < 0$ and*

$$-3f_1 + f_2 + f_3 = 2(x_1^2 - x_2^2)^2 + (x_1^4 - x_1^2 x_2^2 + x_2^4)c_{eq} \in \text{Ideal}[c_{eq}]_6 + \Sigma[x]_6.$$

*By Theorem 8.3.5, the above measure $\mu$ does not exist. The dual problem (8.3.2) is feasible for all $f_0$, because $f_1 + f_2 + f_3 > 0$ on $\mathbb{S}^1$. So (8.3.2) must be unbounded above.*

The certificate (8.3.14) can be obtained by solving a linear conic optimization problem with the constraints

$$\langle f_0, y \rangle = -1, \quad \langle f_i, y \rangle = 0 \ (i = 1, \dots, m), \quad y \in \mathscr{R}_d(K). \qquad (8.3.17)$$

This can be done by choosing a generic objective and then applying Algorithm 8.3.1 to solve it.

**Example 8.3.7.** *Let $K = \mathbb{S}^2$ be the unit sphere in $\mathbb{R}^3$, with $c_{eq} = (x^T x - 1)$ and $c_{in} = \emptyset$. One wonders whether there exist $\lambda_1, \lambda_2, \lambda_3$ such that*

$$\underbrace{x_1^2 x_2^2 (x_1^2 + x_2^2 - 4x_3^2) + x_3^6}_{f_0} - \lambda_1 \underbrace{x_1^3 x_2^3}_{f_1} - \lambda_2 \underbrace{x_1^3 x_3^3}_{f_2} - \lambda_3 \underbrace{x_2^3 x_3^3}_{f_3} \in \mathscr{P}_6(\mathbb{S}^2).$$

*Indeed, there are no $\lambda_1, \lambda_2, \lambda_3$ satisfying the above. This is because the certificate (8.3.17) is satisfied for the tms $y$ that admits the measure*

$$\frac{27}{4}\left(\delta_{\frac{(1,1,1)}{\sqrt{3}}} + \delta_{\frac{(-1,1,1)}{\sqrt{3}}} + \delta_{\frac{(1,-1,1)}{\sqrt{3}}} + \delta_{\frac{(1,1,-1)}{\sqrt{3}}}\right).$$

*Consequently, the corresponding optimization (8.3.1) must be unbounded below, if it is feasible.*

## 8.3.4 ▪ Exercises

**Exercise 8.3.1.** *Determine whether or not there exists a measure $\mu$ supported in $[-1,1]^3$ such that*

$$\int (x_1^k + x_2^k + x_3^k)\,\mathrm{d}\mu = k, \ k = 1,\ldots,6.$$

**Exercise 8.3.2.** *Let $K$ be the unit ball in $\mathbb{R}^2$. Solve the optimization problem (8.3.1) with*

$$f_0 := x_1^4 x_2^2 + 6x_1^2 x_2^2 + 4x_1 x_2^4 + x_2^6 + x_2^2,$$
$$f_1 := x_1^3 x_2^2 + x_1 x_2^2, \quad f_2 := x_1^2 x_2^4 + x_2^4, \quad b_1 = b_2 = 1.$$

**Exercise 8.3.3.** *Let $K$ be the unit ball in $\mathbb{R}^2$. Find the maximum $\lambda_1 + \lambda_2$ such that*

$$x_1^4 x_2^2 + 6x_1^2 x_2^2 + 4x_1 x_2^4 + x_2^6 + x_2^2 - \lambda_1(x_1^3 x_2^2 + x_1 x_2^2) - \lambda_2(x_1^2 x_2^4 + x_2^4) \in \mathscr{P}_6(K).$$

**Exercise 8.3.4.** *Determine the smallest sum $\lambda_1 + \lambda_2 + \lambda_3 + \lambda_4$ such that*

$$\lambda_1 x_1^4 + \lambda_2 x_2^4 + \lambda_3 x_3^4 + \lambda_4 x_4^4 + x_1^3 x_2 + x_3^3 x_4 - x_2^2 x_4^2 \geq 0 \quad \text{on} \quad \mathbb{S}^3.$$

**Exercise 8.3.5.** *Let $w^{*,k}$ be a minimizer of (8.3.4) and $\lambda^{*,k}$ be a maximizer of (8.3.5). Suppose the truncation $y^{(k)} := w^{*,k}|_d \in \mathscr{R}_d(K)$. Show that $\vartheta^{(k)} = \vartheta^*$ and $y^{(k)}$ is a minimizer of (8.3.1). Moreover, if $\theta^{(k)} = \vartheta^{(k)}$, show that $\theta^{(k)} = \theta^*$ and $\lambda^{*,k}$ is a maximizer of (8.3.2).*

**Exercise 8.3.6.** *If the relaxation (8.3.4) is infeasible, show that (8.3.1) is also infeasible. If (8.3.5) is unbounded above, show that (8.3.2) is unbounded above and (8.3.1) is infeasible.*

**Exercise 8.3.7.** *For given tms $z_0, \ldots, z_m \in \mathbb{R}^{\mathbb{N}_d^n}$ and given constants $\eta_1, \ldots, \eta_m$, show how to express the following optimization*

$$\begin{cases} \max & \eta_1 \lambda_1 + \cdots + \eta_m \lambda_m \\ s.t. & z_0 + \lambda_1 z_1 + \cdots + \lambda_m z_m \in \mathscr{R}_d(K) \end{cases}$$

*equivalently in the form (8.3.1).*

**Exercise 8.3.8.** *For given $z_0, \ldots, z_m \in \mathbb{R}^{\mathbb{N}_d^n}$ and constants $\eta_1, \ldots, \eta_m$, show how to express the following optimization*

$$\begin{cases} \min & \langle p, z_0 \rangle \\ s.t. & \langle p, z_i \rangle = \eta_i, \ i = 1, \ldots, m, \\ & p \in \mathscr{P}_d(K), \end{cases}$$

*equivalently in the form of (8.3.2).*

**Exercise 8.3.9.** *In the proof of Theorem 8.3.4, show that $\{y^{(k)}\}_{k=1}^{\infty}$ is a bounded sequence.*

**Exercise 8.3.10.** *If $c_{eq}(x) = 0$ has only finitely many real solutions, show that Algorithm 8.3.1 must terminate within finitely many loops, with the flat truncation (8.3.7) satisfied.*

# 8.4 ▪ The $\mathcal{A}$-Truncated $K$-Moment Problem

Some moment problems only concern a subset of entries of a tms, i.e., not all entries are appearing. This results in the $\mathcal{A}$-*truncated $K$-moment problem* ($\mathcal{A}$-TKMP), which was studied in [229].

## 8.4.1 ▪ $\mathcal{A}$-truncated moments

For a power set $\mathcal{A} \subseteq \mathbb{N}_d^n$, denote the subspace of polynomials

$$\mathbb{R}[x]_{\mathcal{A}} := \mathrm{span}\left\{x^{\alpha} : \alpha \in \mathcal{A}\right\}.$$

If $\mathcal{A} = \mathbb{N}_d^n$, then $\mathbb{R}[x]_{\mathcal{A}} = \mathbb{R}[x]_d$. The degree of the power set $\mathcal{A}$ is

$$\deg(\mathcal{A}) := \max\{|\alpha| : \alpha \in \mathcal{A}\}.$$

Like the case of $\mathbb{R}[x]_d$, each linear functional $\ell$ on $\mathbb{R}[x]_{\mathcal{A}}$ is determined by the values $\ell(x^{\alpha})$, so $\ell$ is uniquely determined by the vector

$$y := (y_{\alpha})_{\alpha \in \mathcal{A}}.$$

Such $y$ is called an $\mathcal{A}$-*truncated multi-sequence* ($\mathcal{A}$-tms). Similarly, a polynomial in $\mathbb{R}[x]_{\mathcal{A}}$ is called an $\mathcal{A}$-*sparse polynomial* or just $\mathcal{A}$-*polynomial*. Denote by $\mathbb{R}^{\mathcal{A}}$ the space of all such real vectors $y$. The dual space of $\mathbb{R}[x]_{\mathcal{A}}$ is $\mathbb{R}^{\mathcal{A}}$, and vice versa.

For a set $K \subseteq \mathbb{R}^n$, denote the cone of $\mathcal{A}$-sparse psd polynomials

$$\mathscr{P}_{\mathcal{A}}(K) := \{p \in \mathbb{R}[x]_{\mathcal{A}} : p(u) \geq 0 \,\forall\, u \in K\}. \tag{8.4.1}$$

The subspace $\mathbb{R}[x]_{\mathcal{A}}$ is said to be $K$-*full* if there exists $p \in \mathbb{R}[x]_{\mathcal{A}}$ such that $p > 0$ on $K$. Like for TKMPs, an $\mathcal{A}$-tms $y$ is said to admit a $K$-representing measure $\mu$ if $\mathrm{supp}(\mu) \subseteq K$ and

$$y_{\alpha} = \int x^{\alpha} \mathrm{d}\mu \quad \text{for all } \alpha \in \mathcal{A}.$$

The $\mathcal{A}$-truncated $K$-moment problem ($\mathcal{A}$-TKMP) concerns conditions for the existence of $K$-representing measures for given $\mathcal{A}$-tms. For convenience, we still let $meas(y, K)$ denote the set of $K$-representing measures for $y$. Similarly, denote the $\mathcal{A}$-moment cone

$$\mathscr{R}_{\mathcal{A}}(K) := \{y \in \mathbb{R}^{\mathcal{A}} : meas(y, K) \neq \emptyset\}. \tag{8.4.2}$$

When $y = 0$, the representing measure for $y$ is just the identically zero measure, whose support is the empty set. If $\mathcal{A} = \mathbb{N}_d^n$, then $\mathscr{P}_{\mathcal{A}}(K) = \mathscr{P}_d(K)$ and $\mathscr{R}_{\mathcal{A}}(K) = \mathscr{R}_d(K)$. For notational convenience, we denote that

$$\begin{aligned}
[x]_{\mathcal{A}} &:= (x^{\alpha})_{\alpha \in \mathcal{A}}, \\
[K]_{\mathcal{A}} &:= \{[x]_{\mathcal{A}} : x \in K\}.
\end{aligned} \tag{8.4.3}$$

The conic hull of $[K]_{\mathcal{A}}$ is the cone

$$\mathrm{cone}\,([K]_{\mathcal{A}}) := \mathrm{cone}\,(\{[x]_{\mathcal{A}} : x \in K\}).$$

Clearly, both $\mathscr{P}_{\mathcal{A}}(K)$ and $\mathscr{R}_{\mathcal{A}}(K)$ are convex cones. We have the following theorem.

**Theorem 8.4.1.** *([229]) Let $K \subseteq \mathbb{R}^n$ be a closed set and let $\mathcal{A} \subseteq \mathbb{N}^n$ be a finite power set. Then, we have the following:*

(i) *The cone $\mathscr{P}_{\mathcal{A}}(K)$ is closed convex. If $K$ is compact and $\mathbb{R}[x]_{\mathcal{A}}$ is $K$-full, then $\mathscr{P}_{\mathcal{A}}(K)$ is solid and $p \in \mathrm{int}\,(\mathscr{P}_{\mathcal{A}}(K))$ if and only if $p > 0$ on $K$.*

(ii) *Each $y \in \mathscr{R}_{\mathcal{A}}(K)$ admits a finitely atomic representing measure $\nu$ with $\mathrm{supp}(\nu) \subseteq K$ and $|\mathrm{supp}(\nu)| \leq |\mathcal{A}|$. Therefore, this implies that*

$$\mathscr{R}_{\mathcal{A}}(K) = \mathrm{cone}\,([K]_{\mathcal{A}}). \tag{8.4.4}$$

*Moreover, if $K$ is compact and $\mathbb{R}[x]_{\mathcal{A}}$ is $K$-full, then $\mathscr{R}_{\mathcal{A}}(K)$ is closed.*

The proof of Theorem 8.4.1 is quite similar to the case $\mathcal{A} = \mathbb{N}^n_d$ as in Theorem 8.1.1. So it is left as an exercise. When $K$ is compact, the set $[K]_{\mathcal{A}}$ is also compact. If, in addition, $\mathbb{R}[x]_{\mathcal{A}}$ is $K$-full, then there is a linear functional $\ell$ on $\mathbb{R}^{\mathcal{A}}$ such that $\ell > 0$ on $[K]_{\mathcal{A}}$. This implies that the conic hull $\mathrm{cone}\,([K]_{\mathcal{A}})$ is a closed cone, by Proposition 1.2.6. However, when $K$ is compact but $\mathbb{R}[x]_{\mathcal{A}}$ is not $K$-full, the cone $([K]_{\mathcal{A}})$ may not be closed.

An $\mathcal{A}$-tms $y \in \mathbb{R}^{\mathcal{A}}$ determines the Riesz functional $\mathscr{L}_y$ on $\mathbb{R}[x]_{\mathcal{A}}$ such that

$$\mathscr{L}_y(x^\alpha) = y_\alpha \quad \text{for } \alpha \in \mathcal{A}.$$

Like the case $\mathcal{A} = \mathbb{N}^n_d$, the Riesz functional $\mathscr{L}_y$ is said to be *$K$-positive* if

$$p \in \mathscr{P}_{\mathcal{A}}(K) \quad \Rightarrow \quad \mathscr{L}_y(p) \geq 0.$$

Similarly, it is *strictly $K$-positive* if

$$p \in \mathscr{P}_{\mathcal{A}}(K),\, p|_K \not\equiv 0 \quad \Rightarrow \quad \mathscr{L}_y(p) > 0.$$

For each $y \in \mathscr{R}_{\mathcal{A}}(K)$, it is clear that $\mathscr{L}_y$ is $K$-positive on $\mathbb{R}[x]_{\mathcal{A}}$. The following is a useful characterization for the $\mathcal{A}$-moment cone $\mathscr{R}_{\mathcal{A}}(K)$.

**Theorem 8.4.2.** *([96]) Suppose $K$ is compact and $\mathbb{R}[x]_{\mathcal{A}}$ is $K$-full. Then, for $y \in \mathbb{R}^{\mathcal{A}}$, the Riesz functional $\mathscr{L}_y$ is $K$-positive if and only if $y \in \mathscr{R}_{\mathcal{A}}(K)$.*

**Proof.** By Theorem 8.4.1(ii), the cone $\mathscr{R}_{\mathcal{A}}(K)$ is closed. The proof for the direction "if" is obvious. To prove the "only if" direction, suppose otherwise that $\mathscr{L}_y$ is $K$-positive but $y \notin \mathscr{R}_{\mathcal{A}}(K)$. Since $\mathscr{R}_{\mathcal{A}}(K)$ is closed convex, by Theorem 1.2.1, there exists $p \in \mathbb{R}[x]_{\mathcal{A}}$ such that

$$\langle p, y \rangle < \langle p, z \rangle \quad \text{for all } z \in \mathscr{R}_{\mathcal{A}}(K).$$

Since $\mathscr{R}_{\mathcal{A}}(K)$ is a cone, the above implies that $p \in \mathscr{P}_{\mathcal{A}}(K)$ and $\langle p, y \rangle < 0$, which contradicts the $K$-positivity of $\mathscr{L}_y$. Therefore, if $\mathscr{L}_y$ is $K$-positive, then we must have $y \in \mathscr{R}_{\mathcal{A}}(K)$. □

The following is the duality theorem for $\mathscr{R}_{\mathcal{A}}(K)$ and $\mathscr{P}_{\mathcal{A}}(K)$.

**Theorem 8.4.3.** *If $K \subseteq \mathbb{R}^n$ is compact and $\mathbb{R}[x]_{\mathcal{A}}$ is $K$-full, then*

$$\mathscr{R}_{\mathcal{A}}(K)^{\star} = \mathscr{P}_{\mathcal{A}}(K), \quad \mathscr{P}_{\mathcal{A}}(K)^{\star} = \mathscr{R}_{\mathcal{A}}(K). \tag{8.4.5}$$

*For every $K$ and every finite power set $\mathcal{A} \subseteq \mathbb{N}^n$, we have that*

$$\mathscr{R}_{\mathcal{A}}(K)^{\star} = \mathscr{P}_{\mathcal{A}}(K), \quad \mathscr{P}_{\mathcal{A}}(K)^{\star} = \mathrm{cl}\left(\mathscr{R}_{\mathcal{A}}(K)\right). \tag{8.4.6}$$

**Proof.** Note that $\mathscr{R}_{\mathcal{A}}(K) = \mathrm{cone}\left([K]_{\mathcal{A}}\right)$. By the definition, $p \in \mathscr{R}_d(K)^{\star}$ if and only if $\langle p, [u]_{\mathcal{A}} \rangle \geq 0$ for all $u \in K$, which is equivalent to $p \in \mathscr{P}_{\mathcal{A}}(K)$. So the equality $\mathscr{R}_{\mathcal{A}}(K)^{\star} = \mathscr{P}_{\mathcal{A}}(K)$ holds. If $K \subseteq \mathbb{R}^n$ is compact and $\mathbb{R}[x]_{\mathcal{A}}$ is $K$-full, $\mathscr{R}_{\mathcal{A}}(K)$ is a closed convex cone (by Theorem 8.4.1), so we also have $\mathscr{P}_{\mathcal{A}}(K)^{\star} = \mathscr{R}_{\mathcal{A}}(K)$ by the bi-duality Theorem 1.2.10.

For every $K$ and every finite power set $\mathcal{A} \subseteq \mathbb{N}^n$, it holds that

$$\left(\mathscr{R}_{\mathcal{A}}(K)\right)^{\star} = \left(\mathrm{cl}\left(\mathscr{R}_{\mathcal{A}}(K)\right)\right)^{\star}.$$

So, Theorem 1.2.10 implies that $\mathscr{P}_{\mathcal{A}}(K)^{\star} = \mathrm{cl}\left(\mathscr{R}_{\mathcal{A}}(K)\right)$. $\square$

Geometric properties for the cones $\mathscr{R}_{\mathcal{A}}(K)$ and $\mathscr{P}_{\mathcal{A}}(K)$ are summarized as follows. The $\mathcal{A}$-truncation of the vanishing ideal $I(K)$ is

$$I(K)_{\mathcal{A}} := I(K) \cap \mathbb{R}[x]_{\mathcal{A}}.$$

**Theorem 8.4.4.** *For a closed set $K \subseteq \mathbb{R}^n$ and a finite set $\mathcal{A} \subseteq \mathbb{N}^n$, we have the following:*

(i) *It holds that $\mathscr{P}_{\mathcal{A}}(K) \cap -\mathscr{P}_{\mathcal{A}}(K) = I(K)_{\mathcal{A}}$. Therefore, the cone $\mathscr{P}_{\mathcal{A}}(K)$ is pointed if and only if $I(K)_{\mathcal{A}} = \{0\}$.*

(ii) *The affine hull of $\mathscr{R}_{\mathcal{A}}(K)$ is the subspace*

$$F := \{y \in \mathbb{R}^{\mathcal{A}} : \langle p, y \rangle = 0 \,\forall p \in I(K)_{\mathcal{A}}\}.$$

*In particular, if $I(K)_{\mathcal{A}} = \{0\}$, then $\mathscr{R}_{\mathcal{A}}(K)$ is full-dimensional and has nonempty interior.*

(iii) *For $y \in \mathbb{R}^{\mathcal{A}}$, we have $y \in \mathrm{ri}\left(\mathscr{R}_{\mathcal{A}}(K)\right)$ if and only if $\mathscr{L}_y$ is strictly $K$-positive on $\mathbb{R}[x]_{\mathcal{A}}$. Moreover, when $K$ is compact and $\mathbb{R}[x]_{\mathcal{A}}$ is $K$-full, we have $y \in \mathscr{R}_{\mathcal{A}}(K)$ if and only if $\mathscr{L}_y$ is $K$-positive on $\mathbb{R}[x]_{\mathcal{A}}$.*

The proof for Theorem 8.4.4 is quite similar to the case that $\mathcal{A} = \mathbb{N}^n_d$ as in Theorem 8.1.4, so it is left as an exercise. To check memberships for the cone $\mathscr{P}_{\mathcal{A}}(K)$, we need to test whether a polynomial in $\mathbb{R}[x]_{\mathcal{A}}$ is psd on $K$ or not. The same method in §8.2.1 can be applied to do this. To check memberships for the moment cone $\mathscr{R}_{\mathcal{A}}(K)$, Algorithm 8.2.4 can be similarly applied. The only difference is that the optimization (8.2.12) is changed to

$$\begin{cases} \min & \langle R, z \rangle \\ s.t. & z_{\alpha} = y_{\alpha} \, (\alpha \in \mathcal{A}), \\ & \mathscr{V}^{(2k)}_{c_i}[z] = 0 \, (i \in \mathcal{E}), \\ & L^{(k)}_{c_j}[z] \succeq 0 \, (j \in \mathcal{I}), \\ & M_k[z] \succeq 0, \, z \in \mathbb{R}^{\mathbb{N}^n_{2k}}. \end{cases} \tag{8.4.7}$$

We refer to [229] for details of checking memberships for the cone $\mathscr{R}_{\mathcal{A}}(K)$. The following are some examples.

**Example 8.4.5.** *([229]) (i) Consider the $\mathcal{A}$-tms $y$ with $(\alpha, y_{\alpha})$ given as*

| $\alpha$ | $(2,0)$ | $(0,2)$ | $(2,1)$ | $(1,2)$ | $(2,2)$ | $(4,2)$ | $(2,4)$ |
|----------|---------|---------|---------|---------|---------|---------|---------|
| $y_\alpha$ | *1/3* | *1/3* | *0* | *0* | *1/9* | *1/15* | *1/15* |

*The above $y$ admits a measure supported on the square $[-1,1]^2$, since each $y_\alpha$ is the average of $x^\alpha$ on $[-1,1]^2$. The corresponding set $K$ is given by $c_{eq} = \emptyset$ and $c_{in} = (1 - x_1^2, 1 - x_2^2)$. We apply Algorithm 8.2.4, with (8.2.12) replaced by (8.4.7). For the relaxation order $k = 4$, we got the 3-atomic representing measure $\sum_{i=1}^{3} \lambda_i \delta_{u_i}$, with $u_i$ and $\lambda_i$ given as below:*

| $i$ | *1* | *2* | *3* |
|-----|-----|-----|-----|
| $u_i$ | *(−0.8524, 0.8910)* | *(0.2109, 0.8873)* | *(0.6697, −0.3743)* |
| $\lambda_i$ | *0.1231* | *0.2061* | *0.5233* |

*(ii) Consider the $\mathcal{A}$-tms $y$ with $(\alpha, y_\alpha)$ given as below:*

| $\alpha$ | $(4,0,0)$ | $(2,0,2)$ | $(0,2,2)$ | $(4,0,2)$ | $(2,2,2)$ | $(0,0,6)$ |
|----------|-----------|-----------|-----------|-----------|-----------|-----------|
| $y_\alpha$ | *1/5* | *1/15* | *1/15* | *3/105* | *1/105* | *1/7* |

*The above $y$ admits a measure supported on the unit sphere $\mathbb{S}^2$, since each $y_\alpha$ is the average of $x^\alpha$ on $\mathbb{S}^2$. The corresponding set $K$ is given by $c_{eq} = (\|x\|_2^2 - 1)$ and $c_{in} = \emptyset$. Like the above, we got a 2-atomic representing measure $\sum_{i=1}^{2} \lambda_i \delta_{u_i}$ with $u_i, \lambda_i$ given as below:*

| $i$ | *1* | *2* |
|-----|-----|-----|
| $u_i$ | *(0.3434, 0.4542, 0.8221)* | *(0.8999, 0.2550, −0.3539)* |
| $\lambda_i$ | *0.4610* | *0.2952* |

### 8.4.2 ▪ Homogeneous moment problems

When $\mathcal{A}$ consists of all powers whose degrees are the same, the $\mathcal{A}$-TKMP becomes the *homogeneous truncated moment problem* (HTMP). For a degree $d$, recall the notation

$$\overline{\mathbb{N}}_d^n := \{\alpha \in \mathbb{N}^n : |\alpha| = d\}. \tag{8.4.8}$$

Each $y \in \mathbb{R}^{\overline{\mathbb{N}}_d^n}$ is called a *homogeneous truncated multi-sequence* (htms) of degree $d$. The cardinality of $\overline{\mathbb{N}}_d^n$ is $\binom{n+d-1}{d}$. The HTMP is a special case of $\mathcal{A}$-TKMP with $\mathcal{A} = \overline{\mathbb{N}}_d^n$. For HTMPs, the set $K$ is often the unit sphere $\mathbb{S}^{n-1}$. When $\mathcal{A} = \overline{\mathbb{N}}_d^n$, it holds that (see §2.4.2 for the notation)

$$\mathscr{P}_\mathcal{A}(\mathbb{S}^{n-1}) = \mathscr{P}_{n,d}^{\mathrm{hom}}, \qquad \mathscr{R}_\mathcal{A}(\mathbb{S}^{n-1}) = \mathscr{R}_{n,d}^{\mathrm{hom}}.$$

The homogeneous moment cone can be equivalently given as

$$\mathscr{R}_{n,d}^{\mathrm{hom}} := \left\{ \int [x]_d^{\mathrm{hom}} \mathrm{d}\mu \, \middle| \, \begin{array}{l} \mu \text{ is a Borel measure,} \\ \mathrm{supp}(\mu) \subseteq \mathbb{S}^{n-1} \end{array} \right\}. \tag{8.4.9}$$

By Theorem 8.4.1,

$$\mathscr{R}_{n,d}^{\mathrm{hom}} = \mathrm{cone}\left(\{[u]_d^{\mathrm{hom}} : \|u\| = 1\}\right).$$

Equivalently, each $y \in \mathscr{R}_{n,d}^{\mathrm{hom}}$ can be expressed such that

$$y = \sum_{i=1}^{N} \lambda_i [u_i]_d^{\mathrm{hom}}, \quad \lambda_i \geq 0, \quad u_i \in \mathbb{S}^{n-1}.$$

When $d$ is even, the subspace $\mathbb{R}[x]_d^{\mathrm{hom}}$ is $\mathbb{S}^{n-1}$-full, since the form $(x^T x)^{d/2}$ is positive on the unit sphere. Therefore, Theorem 8.4.3 implies the following.

**Theorem 8.4.6.** *When $d$ is even, $\mathscr{P}_{n,d}^{\mathrm{hom}}$ and $\mathscr{R}_{n,d}^{\mathrm{hom}}$ are proper cones (i.e., closed, convex, solid, and pointed), and they are dual to each other, i.e.,*

$$\left(\mathscr{P}_{n,d}^{\mathrm{hom}}\right)^* = \mathscr{R}_{n,d}^{\mathrm{hom}}, \quad \left(\mathscr{R}_{n,d}^{\mathrm{hom}}\right)^* = \mathscr{P}_{n,d}^{\mathrm{hom}}. \tag{8.4.10}$$

When $d$ is odd, the psd polynomial cone $\mathscr{P}_{n,d}^{\mathrm{hom}}$ consists of the identically zero form, i.e., $\mathscr{P}_{n,d}^{\mathrm{hom}} = \{0\}$, and the homogeneous moment cone $\mathscr{R}_{n+1,d}^{\mathrm{hom}}$ equals the entire space $\mathbb{R}^{\overline{\mathbb{N}}_d^n}$ (see Exercise 8.4.2). Recall the full truncated moment cone

$$\mathscr{R}_{n,d} := \left\{ \int [x]_d \mathrm{d}\mu \mid \mu \text{ is a Borel measure on } \mathbb{R}^n \right\}. \tag{8.4.11}$$

The homogeneous moment cone $\mathscr{R}_{n,d}^{\mathrm{hom}}$ is a projection of $\mathscr{R}_{n,d}$. The cone $\mathscr{R}_{n,d}$ is not closed, while $\mathscr{R}_{n,d}^{\mathrm{hom}}$ is. The cone $\mathscr{R}_{n,d}^{\mathrm{hom}}$ is more convenient for homogeneous polynomial optimization, while $\mathscr{R}_{n,d}$ is more convenient for the nonhomogeneous case.

Each $\tilde{y} \in \mathbb{R}^{\overline{\mathbb{N}}_d^n}$ can be viewed as a tms $y \in \mathbb{R}^{\mathbb{N}_d^{n-1}}$ such that

$$y_\alpha = \tilde{y}_\beta \quad \text{for all} \quad \alpha \in \mathbb{N}_d^{n-1} \quad \text{and} \quad \beta = (d - |\alpha|, \alpha). \tag{8.4.12}$$

The tms $y$ is called the *dehomogenization* of $\tilde{y}$, and the htms $\tilde{y}$ is called the *homogenization* of $y$. The relation in (8.4.12) gives an isomorphism from $\mathbb{R}^{\overline{\mathbb{N}}_d^n}$ to $\mathbb{R}^{\mathbb{N}_d^{n-1}}$. Each tms $y \in \mathbb{R}^{\mathbb{N}_d^{n-1}}$ can be viewed as an htms $\tilde{y} \in \mathbb{R}^{\overline{\mathbb{N}}_d^n}$ via (8.4.12), and vice versa. Indeed, we have the following theorem.

**Theorem 8.4.7.** *([96]) Let $\tilde{y} \in \mathbb{R}^{\overline{\mathbb{N}}_d^n}$ and $y \in \mathbb{R}^{\mathbb{N}_d^{n-1}}$ be related as in (8.4.12).*

(i) *If $y \in \mathscr{R}_{n-1,d}$, then $\tilde{y} \in \mathscr{R}_{n,d}^{\mathrm{hom}}$.*

(ii) *If $\tilde{y} \in \mathscr{R}_{n,d}^{\mathrm{hom}}$ and $d$ is even, then $y \in \mathrm{cl}\left(\mathscr{R}_{n-1,d}\right)$.*

Checking memberships for the cone $\mathscr{R}_{n,d}^{\mathrm{hom}}$ can be done as a special case of $\mathscr{R}_{\mathcal{A}}(K)$; see the optimization problem (8.4.7). In particular, for the cases $\Sigma_{n,d}^{\mathrm{hom}} = \mathscr{P}_{n,d}^{\mathrm{hom}}$, the cone $\mathscr{R}_{n,d}^{\mathrm{hom}}$ can be characterized by psd matrices.

**Theorem 8.4.8.** *[96] Let $\tilde{y} \in \mathbb{R}^{\overline{\mathbb{N}}_d^n}$ and $y \in \mathbb{R}^{\mathbb{N}_d^{n-1}}$ be related as in (8.4.12). Suppose $d = 2k$ is even. If $n = 2$ or $d = 2$ or $(n, d) = (3, 4)$, then $\tilde{y} \in \mathscr{R}_{n,d}^{\mathrm{hom}}$ if and only if $M_k[y] \succeq 0$.*

**Proof.** For the given cases of $(n, d)$, $\mathscr{P}_{n,d}^{\mathrm{hom}} = \Sigma_{n,d}^{\mathrm{hom}}$ (see Theorem 2.4.5). By Theorem 8.4.6, we have $\tilde{y} \in \mathscr{R}_{n,d}^{\mathrm{hom}}$ if and only if $\tilde{y} \in (\Sigma_{n,d}^{\mathrm{hom}})^*$, which is equivalent to $M_k[y] \succeq 0$ (see Proposition 2.4.4). $\quad\square$

### 8.4.3 ▪ SOEP polynomials

A similar version of SOS polynomials is the *sum of powers of linear polynomials* (SOP polynomials). Let $d > 0$ be a degree.

**Definition 8.4.9.** *A polynomial $f \in \mathbb{R}[x]_d$ is called a sum of $d$th powers of linear polynomials if there exist $f_1, \ldots, f_r \in \mathbb{R}[x]_1$ such that*

$$f = f_1^d + \cdots + f_r^d. \tag{8.4.13}$$

*Such $f$ is called a SOP polynomial. The set of all SOP polynomials of degree $d$ is denoted as $\Sigma^d[x]$. The subset of homogeneous polynomials of degree $d$ in $\Sigma^d[x]$ is denoted as $\Sigma^d[x]^{\mathrm{hom}}$.*

In particular, when $d$ is even, a SOP polynomial (form) is said to be a *sum of even powers of linear polynomials* (SOEP). SOEP forms are studied in [282]. Observe that a polynomial is SOP if and only if its homogenization is SOP. Therefore, we only discuss the homogeneous case $\Sigma^d[x]^{\mathrm{hom}}$, for convenience. When $d$ is odd, every $f \in \mathbb{R}[x]_d$ has a decomposition as in (8.4.13). We leave the proof as Exercise 8.4.2. So, the case of even $d$ is more interesting. SOEP forms can be checked by solving an HTMP. For $f \in \mathbb{R}[x]_d^{\mathrm{hom}}$, let $y = (y_\alpha) \in \mathbb{R}^{\overline{\mathbb{N}}_d^n}$ be the htms such that

$$f(x) = \sum_{\alpha \in \overline{\mathbb{N}}_d^n} \binom{d}{\alpha} y_\alpha x_1^{\alpha_1} \cdots x_n^{\alpha_n}. \tag{8.4.14}$$

In the above, $\alpha = (\alpha_1, \ldots, \alpha_n)$ and $\binom{d}{\alpha} = \frac{d!}{\alpha_1! \cdots \alpha_n!}$. Clearly, the htms $y$ is uniquely determined by coefficients of $f$, and vice versa. We denote this linear map as

$$y := \Upsilon(f). \tag{8.4.15}$$

In (8.4.13), if we write each linear form $f_i = u_i^T x$ for some $u_i \in \mathbb{R}^n$, then

$$f(x) = \sum_{\alpha \in \overline{\mathbb{N}}_d^n} \binom{d}{\alpha} \left( \sum_{i=1}^r u_i^\alpha \right) x^\alpha. \tag{8.4.16}$$

A comparison for the coefficients of $f$ in (8.4.14) and (8.4.16) gives the following theorem.

**Theorem 8.4.10.** *Let $\Upsilon$ be the linear map as in (8.4.15) and let $d$ be an even degree. Then, a form $f \in \mathbb{R}[x]_d^{\mathrm{hom}}$ is SOEP if and only if $y \in \mathscr{R}_{n,d}^{\mathrm{hom}}$. Therefore,*

$$\mathscr{R}_{n,d}^{\mathrm{hom}} = \Upsilon\big(\Sigma^d[x]^{\mathrm{hom}}\big).$$

The smallest length $r$ for (8.4.13) to hold is called the *width* or *SOEP rank* of $f$ (see [282]). SOEP forms have applications in Waring's problems, quadrature rules, and sphere designs. We refer to [282] for more properties about SOEP forms. They can be checked by solving $\mathcal{A}$-truncated moment problems with $\mathcal{A} = \overline{\mathbb{N}}_d^n$ and $K = \mathbb{S}^{n-1}$. A similar version of Algorithm 8.2.4 can be applied for checking SOEP forms. If $f$ is not SOEP, Algorithm 8.2.4 can give a certificate for that. If $f$ is SOEP, then we can get a SOEP decomposition as in (8.4.13).

**Example 8.4.11.** *([232]) Consider the ternary sextic form*

$$q_\lambda := (x_1^2 + x_2^2 + x_3^2)^3 - \lambda(x_1^6 + x_2^6 + x_3^6).$$

*It is SOEP if and only if $\lambda \leq \frac{2}{3}$ (see [282, p. 146]). For $\lambda = \frac{2}{3}$, by Algorithm 8.2.4, we got the following SOEP decomposition for $q_{\frac{2}{3}}$:*

$$\frac{1}{15}\big((x_1 + x_2)^6 + (x_2 + x_3)^6 + (x_1 + x_3)^6 + (x_1 - x_2)^6 + (x_1 - x_3)^6 + (x_2 - x_3)^6\big)$$
$$+ \frac{1}{60}\big((x_1 + x_2 + x_3)^6 + (-x_1 + x_2 + x_3)^6 + (x_1 - x_2 + x_3)^6 + (x_1 + x_2 - x_3)^6\big).$$

*For $\lambda = \frac{1}{3}$, we got a SOEP decomposition of length 11 for $q_{\frac{1}{3}}$. When $\lambda = 1$, Algorithm 8.2.4 detected that $q_1$ is not SOEP.*

## 8.4.4 ▪ Exercises

**Exercise 8.4.1.** *Give an example of $\mathcal{A}$ and $K$ such that $K$ is compact but $\mathscr{R}_\mathcal{A}(K)$ is not closed.*

**Exercise 8.4.2.** *If $d$ is odd, show that $\mathscr{R}^{\mathrm{hom}}_{n,d} = \mathbb{R}^{\overline{\mathbb{N}}^n_d}$ and $\mathbb{R}[x]_d = \Sigma^d[x]$.*

**Exercise 8.4.3.** *Determine whether or not*

$$(x_1^3 + x_2^3 + x_3^3)^2 \in \Sigma^6[x]^{\mathrm{hom}}.$$

**Exercise 8.4.4.** *Determine the maximum $\lambda$ such that*

$$(x_1^2 + x_2^2 + x_3^2)^3 - \lambda(x_1^3 x_2^3 + x_2^3 x_3^3 + x_3^3 x_1^3) \in \Sigma^6[x]^{\mathrm{hom}}.$$

**Exercise 8.4.5.** *Determine the maximum $\lambda$ such that*

$$x_1^6 + x_2^6 + x_3^6 - \lambda(x_1^2 + x_2^2 + x_3^2)^3 \in \Sigma^6[x]^{\mathrm{hom}}.$$

**Exercise 8.4.6.** *For $y \in \mathbb{R}^{\overline{\mathbb{N}}^n_d}$, define the polynomial*

$$y(x) := \sum_{\alpha := (\alpha_1,\ldots,\alpha_n) \in \overline{\mathbb{N}}^n_d} \frac{d!}{\alpha_1! \cdots \alpha_n!} y_\alpha x^\alpha.$$

*Show that $y(x) \in \Sigma^d[x]^{\mathrm{hom}}$ for every $y \in \mathscr{R}^{\mathrm{hom}}_{n,d}$.*

**Exercise 8.4.7.** *Prove Theorems 8.4.1, 8.4.4, and 8.4.7.*

# 8.5 ▪ Unboundedness in Polynomial Optimization

Detecting unboundedness of an optimization problem is typically hard. When the functions are polynomials, this is doable. We consider the polynomial optimization problem

$$\begin{cases} \min & f(x) \\ s.t. & c_i(x) = 0 \, (i \in \mathcal{E}), \\ & c_j(x) \geq 0 \, (j \in \mathcal{I}), \end{cases} \tag{8.5.1}$$

where $f$ and all $c_i, c_j$ are polynomials in $x := (x_1, \ldots, x_n)$, and $\mathcal{E}$ and $\mathcal{I}$ are two finite labeling sets. Let $K$ denote the feasible set of (8.5.1). The optimization (8.5.1) is unbounded below if there exists a sequence $\{u_k\}^\infty_{k=1} \subseteq K$ such that $f(u_k) \to -\infty$. This section discusses how to detect unboundedness for (8.5.1).

Suppose $d$ is the degree of $f$. Note that (8.5.1) is bounded below if and only if $f$ has a lower bound $\gamma$ on $K$, i.e., there exists $\gamma$ such that $f(x) - \gamma \in \mathscr{P}_d(K)$. So we consider the linear conic optimization

$$\begin{cases} \max & \gamma \\ s.t. & f(x) - \gamma \in \mathscr{P}_d(K). \end{cases} \tag{8.5.2}$$

Let $\widetilde{K}$ be the homogenization of $K$ as in (8.1.10). Note that $x_0 \geq 0$ on $\widetilde{K}$. When $K$ is closed at $\infty$ (see Definition 6.6.1), if $f(x) - \gamma \in \mathscr{P}_d(K)$, then $\widetilde{f}(\widetilde{x}) - \gamma x_0^d \in \mathscr{P}_d(\widetilde{K})$. The converse is always true, even if $K$ is not closed at $\infty$. Since $\widetilde{K}$ is given by homogeneous constraints, one can see that

$$\widetilde{f}(\tilde{x}) - \gamma x_0^d \in \mathscr{P}_d(\widetilde{K}) \quad \Leftrightarrow \quad \widetilde{f}(\tilde{x}) - \gamma x_0^d \in \mathscr{P}_d(\widetilde{K}_1)$$

for the compact set

$$\widetilde{K}_1 := \widetilde{K} \cap \{\|\tilde{x}\| = 1\}.$$

Hence, (8.5.2) is equivalent to

$$\left\{ \begin{array}{ll} \max & \gamma \\ s.t. & \widetilde{f}(\tilde{x}) - \gamma x_0^d \in \mathscr{P}_d(\widetilde{K}_1). \end{array} \right. \tag{8.5.3}$$

The dual problem of (8.5.3) is the moment optimization

$$\left\{ \begin{array}{ll} \min & \langle \widetilde{f}(\tilde{x}), y \rangle \\ s.t. & \langle x_0^d, y \rangle = 1, \\ & y \in \mathscr{R}_d(\widetilde{K}_1). \end{array} \right. \tag{8.5.4}$$

If (8.5.4) is unbounded below, then (8.5.3) is infeasible, which then implies that (8.5.2) is infeasible (if $K$ is closed at infinity) and hence (8.5.1) is unbounded below. When $K \neq \emptyset$, the optimization (8.5.4) is feasible. The linear conic optimization (8.5.4) is unbounded below if it has a decreasing ray $\Delta y$:

$$\langle \widetilde{f}(\tilde{x}), \Delta y \rangle = -1, \quad \langle x_0^d, \Delta y \rangle = 0, \quad \Delta y \in \mathscr{R}_d(\widetilde{K}_1). \tag{8.5.5}$$

If $\nu$ is a $\widetilde{K}_1$-representing measure for $\Delta y$, the constraint $\langle x_0^d, \Delta y \rangle = 0$ implies that

$$\operatorname{supp}(\nu) \subseteq \widetilde{K}_1 \cap \{x_0 = 0\}.$$

So, (8.5.5) is equivalent to

$$\langle f^{\mathrm{hom}}, z \rangle = -1, \quad z \in \mathscr{R}_d(K^\circ), \tag{8.5.6}$$

where $f^{\mathrm{hom}}$ is the homogeneous part of degree $d$ for $f$ and

$$K^\circ := \left\{ x \in \mathbb{R}^n \left| \begin{array}{ll} c_i^{\mathrm{hom}}(x) & = 0 \quad (i \in \mathcal{E}), \\ c_j^{\mathrm{hom}}(x) & \geq 0 \quad (j \in \mathcal{I}), \\ x^T x & = 1 \end{array} \right. \right\}. \tag{8.5.7}$$

In the above, each $c_i^{\mathrm{hom}}$ denotes the homogeneous part of $c_i$ with the highest degree. Let $d_0 := \lceil \frac{1}{2}d \rceil$. To check the feasibility of (8.5.6), we choose a generic polynomial $R \in \operatorname{int}(\Sigma[x]_{2d_0})$ and then solve the linear moment optimization problem

$$\left\{ \begin{array}{ll} \min & \langle R, z \rangle \\ s.t. & \langle f^{\mathrm{hom}}, z \rangle = -1, \\ & z \in \mathscr{R}_{2d_0}(K^\circ). \end{array} \right. \tag{8.5.8}$$

Algorithm 8.3.1 can be applied to solve (8.5.8). The following is a useful property for (8.5.8).

**Theorem 8.5.1.** *Suppose the feasible set $K$ of (8.5.1) is nonempty.*

*(i) Assume $K$ is closed at infinity. If the system (8.5.6) has a feasible solution, then the optimization (8.5.1) is unbounded below.*

*(ii) Assume (8.5.6) has a solution. If $R$ is generic in the interior of $\Sigma[x]_{2d_0}$, then (8.5.8) has a unique optimizer $z^*$ and we have $z^* = \lambda[u]_d$ for some point $u \in K^\circ$ and $\lambda > 0$.*

**Proof.** Since $K \neq \emptyset$, the sets $\widetilde{K}$ and $\widetilde{K}_1$ are both nonempty.

(i) If (8.5.6) has a feasible solution, the moment optimization (8.5.5) has a decreasing ray. Since $\widetilde{K}_1$ is nonempty, (8.5.4) must be unbounded below. Therefore, the optimization (8.5.3) is infeasible, by the weak duality. Suppose otherwise $f$ had a lower bound on $K$, say, $\gamma$, then $f(x) - \gamma \geq 0$ on $K$. This would imply that $\tilde{f}(\tilde{x}) - \gamma x_0^d \geq 0$ for all $\tilde{x} \in \widetilde{K}_1$, since $K$ is closed at infinity. However, this contradicts the infeasibility of (8.5.3). Therefore, $f$ cannot have a lower bound on $K$, i.e., the optimization (8.5.1) is unbounded below.

(ii) When $R \in \text{int}\left(\Sigma[x]_{2d_0}\right)$ is generic, there exists $\epsilon > 0$ such that $R - \epsilon\|[x]_{d_0}\|^2 \in \Sigma[x]_{2d_0}$. Then for all $z \in \mathscr{R}_{2d_0}(K^\circ)$, it holds that

$$\langle R, z \rangle \geq \epsilon\langle\|[x]_{d_0}\|^2, z\rangle \geq \epsilon \cdot \text{trace}(M_{d_0}[z]).$$

Since (8.5.6) is feasible, the optimization (8.5.8) is also feasible, say, $z^{(0)}$ is a feasible point. Then the optimization (8.5.8) is equivalent to

$$\begin{cases} \min & \langle R, z \rangle \\ s.t. & \text{trace}(M_{d_0}[z]) \leq \frac{1}{\epsilon}\langle R, z^{(0)} \rangle, \\ & \langle f^{\text{hom}}, z \rangle = -1, \, z \in \mathscr{R}_{2d_0}(K^\circ). \end{cases} \tag{8.5.9}$$

The feasible set of (8.5.9) is compact, so it has an optimizer, say, $z^*$. Since $R$ is generic, the linear objective $\langle R, z \rangle$ has generic coefficients, so the optimizer $z^*$ is unique. Suppose $z^*$ has the decomposition

$$z^* = \lambda_1[u_1]_d + \cdots + \lambda_r[u_r]_d$$

for distinct points $u_1, \ldots, u_r \in K^\circ$ and positive scalars $\lambda_i > 0$. Consider the following linear program:

$$\begin{cases} \min_{(\tau_1,\ldots,\tau_r)} & \tau_1 R(u_1) + \cdots + \tau_r R(u_r) \\ s.t. & \tau_1 f^{\text{hom}}(u_1) + \cdots + \tau_r f^{\text{hom}}(u_r) = -1, \\ & \tau_1 \geq 0, \ldots, \tau_r \geq 0. \end{cases} \tag{8.5.10}$$

Note that (8.5.8) and (8.5.10) have the same optimal value. The above linear program has an optimizer that is a basic feasible solution. That is, it has a minimizer $(\tau_1^*, \ldots, \tau_r^*)$ but has only one nonzero entry, say, $\tau_1^* > 0$ and $\tau_2^* = \cdots = \tau_r^* = 0$. This implies that $\tau_1^*[u_1]_d$ is also a minimizer of (8.5.8). Since $z^*$ is unique, we have $z^* = \tau_1^*[u_1]_d$. $\square$

The following is an example of applying (8.5.6) to detect unboundedness.

**Example 8.5.2.** *Consider the optimization problem*

$$\begin{cases} \min_{x \in \mathbb{R}^3} & x_1^2 + x_2^2 + x_3^2 + x_1 x_2 x_3 \\ s.t. & x_1^2 x_2^2(x_1^2 + x_2^2) + x_3^6 - 3x_1^2 x_2^2 x_3^2 - 1 = 0. \end{cases} \tag{8.5.11}$$

*The homogeneous part $f^{\text{hom}} = x_1 x_2 x_3$. A feasible solution to (8.5.6) is the tms $z := [(1, 1, -1)]_6$. This optimization is unbounded below.*

The condition that the optimization (8.5.4) has a decreasing ray, i.e., the system (8.5.5) has a feasible solution, is equivalent to that there is a feasible solution to (8.5.6). For this case, (8.5.6) can serve as a certificate for (8.5.1) to be unbounded below. However, the moment optimization

(8.5.4) may not have a decreasing ray. For such a case, the system (8.5.6) may be infeasible. In summary, (8.5.6) is a sufficient condition for unboundedness of (8.5.1), but it may not be necessary. For instance, the optimization

$$\begin{cases} \min & f := x_1 x_2 x_3 + x_1^2 x_2^2 (x_1^2 + x_2^2) + x_3^6 - 3x_1^2 x_2^2 x_3^2 \\ s.t. & x_1^2 + x_2^2 - 2x_3^2 = 0, \ x_1 x_2 \geq 0 \end{cases} \tag{8.5.12}$$

is unbounded below. This is because $f(t, t, -t) = -t^3 \to -\infty$ as $t \to +\infty$, while $(t, t, -t)$ stays feasible for all $t \geq 0$. However, the system (8.5.6) is infeasible. This is because the homogeneous part

$$f^{\mathrm{hom}} = x_1^2 x_2^2 (x_1^2 + x_2^2) + x_3^6 - 3x_1^2 x_2^2 x_3^2$$

is the Motzkin polynomial, so $\langle f^{\mathrm{hom}}, z \rangle \geq 0$ for all $z \in \mathscr{R}_d(K^\circ)$.

When (8.5.6) fails to be feasible, the question of detecting unboundedness of (8.5.1) is mostly open, to the best of the author's knowledge. This is an important question for investigation. Moreover, we would like to remark that the infeasibility of the optimization (8.5.1) can be detected by the infeasibility of the moment relaxations (5.2.3) and (5.2.6). See Exercise 5.2.7 for the conclusion.

## 8.5.1 ▪ Exercises

**Exercise 8.5.1.** *Decide whether or not the following polynomials are unbounded below on the exterior ball set* $x_1^3 + x_2^3 + x_3^3 + x_4^3 \geq 1$:

$$f_1 = x_1^3 + x_2^3 - x_3^3 + x_2^2 x_4^2, \ f_2 = x_2^3 + x_3^3 - x_4^3 + x_4^2 x_1^2,$$
$$f_3 = x_3^3 + x_4^3 - x_1^3 + x_1^2 x_2^2, \ f_4 = x_4^3 + x_1^3 - x_2^3 + x_2^2 x_3^2.$$

*Does there exist a convex combination of* $f_1, f_2, f_3, f_4$ *such that it is bounded below on the above exterior ball?*

**Exercise 8.5.2.** *Decide whether or not the following polynomials are unbounded below on the set of* $x_1 x_2 x_3 \geq 1$, $x_2 x_3 x_4 \geq 1$:

$$\begin{aligned} f_1 &= (x_1 x_2 + x_3 x_4)(x_1 x_4 + x_2 x_3) + x_1^2 + x_2^2 + x_3^2 + x_4^2, \\ f_2 &= x_1^3 x_2^2 + x_2^3 x_3^2 + x_3^3 x_4^2 + x_4^3 x_1^2, \\ f_3 &= x_1^4 - x_2^4 + x_3^4 - x_4^4 + x_1 x_2 x_4 + x_1 x_3 x_4, \\ f_4 &= (x_1 - x_2)(x_3 - x_4)^2 + (x_1 - x_3)(x_2 - x_4)^2 \\ &\quad + (x_1 - x_4)(x_2 - x_3)^2 + x_1 x_2 + x_2 x_3 + x_3 x_4. \end{aligned}$$

*Does there exist a convex combination of* $f_1, f_2, f_3, f_4$ *such that it is bounded below on the set same as above?*

**Exercise 8.5.3.** *In Theorem 8.5.1(i), the assumption that* $K$ *is closed at infinity can be slightly weakened. Suppose* $z = \lambda_1[v_1] + \cdots + \lambda_r[v_r]$, *with all* $\lambda_i > 0$ *and* $v_i \in K^\circ$, *is a feasible point of (8.5.6). If each* $(0, v_i) \in \mathrm{cl}\left(\widetilde{K}_1 \cap \{x_0 > 0\}\right)$, *show that the optimization (8.5.1) is unbounded below.*

# Chapter 9

# Copositive and CP Optimization

*The copositive (COP) cone consists of homogeneous polynomials that are non-negative over the nonnegative orthant $\mathbb{R}_+^n$. Its dual cone is the set of completely positive (CP) truncated moment sequences, whose representing measures are supported in $\mathbb{R}_+^n$. Many hard optimization problems can be equivalently formulated as linear optimization with COP or CP cones. This chapter gives basic properties of COP/CP cones and shows how to solve linear conic optimization problems with them.*

## 9.1 ▪ Copositive Matrices and Polynomials

Recall that $\mathcal{S}^n$ denotes the space of $n$-by-$n$ real symmetric matrices. A matrix $A \in \mathcal{S}^n$ is said to be *copositive* (COP) if the quadratic form $x^T A x \geq 0$ for every nonnegative vector $x := (x_1, \ldots, x_n) \in \mathbb{R}_+^n$. Clearly, if $A$ is positive semidefinite (psd), i.e., $A \in \mathcal{S}_+^n$, then $A$ must be copositive. Moreover, if $A = (A_{ij})$ has only nonnegative entries, i.e., all $A_{ij} \geq 0$, then $A$ is also copositive. Let $\Delta_n$ denote the standard simplex in $\mathbb{R}^n$, i.e.,

$$\Delta_n := \{x \in \mathbb{R}_+^n : e^T x = 1\},$$

where $e$ is the vector of all ones. Then $A$ is COP if and only if $x^T A x \geq 0$ for all $x \in \Delta_n$. The $A$ is said to be *strictly copositive* if $x^T A x > 0$ for all $x \in \Delta_n$.

We use $\mathcal{N}^n$ to denote the set of all nonnegative symmetric matrices of dimension $n \times n$. For two matrices $X, Y$, the inequality $X \geq Y$ means that all entries of $X - Y$ are nonnegative. In particular, $X \geq 0$ means that all entries of $X$ are nonnegative. Denote by $\mathcal{COP}_n$ the cone of all copositive matrices in $\mathcal{S}^n$. A matrix is said to be *nonnegative plus psd* (NPP) if it belongs to $\mathcal{N}^n + \mathcal{S}_+^n$. It clearly holds that the inclusion

$$\mathcal{NPP}_n := \mathcal{N}^n + \mathcal{S}_+^n \subseteq \mathcal{COP}_n.$$

For $n \leq 4$, it was shown in [193] that $\mathcal{NPP}_n = \mathcal{COP}_n$. The equality follows from the equality for their dual cones (see Theorem 9.2.2). For $n \geq 5$, they are not equal anymore. The following are some well-known COP matrices that are not NPP.

- Horn's matrix

$$H := \begin{bmatrix} 1 & -1 & 1 & 1 & -1 \\ -1 & 1 & -1 & 1 & 1 \\ 1 & -1 & 1 & -1 & 1 \\ 1 & 1 & -1 & 1 & -1 \\ -1 & 1 & 1 & -1 & 1 \end{bmatrix} \tag{9.1.1}$$

is copositive but not NPP [110]. A simple proof (see [88]) for the copositivity follows from the representation

$$\begin{aligned} x^T H x &= (x_1 - x_2 + x_3 + x_4 - x_5)^2 + 4x_2 x_4 + 4x_3(x_5 - x_4) \\ &= (x_1 - x_2 + x_3 - x_4 + x_5)^2 + 4x_2 x_5 + 4x_1(x_4 - x_5). \end{aligned}$$

For both cases $x_5 \geq x_4$ and $x_5 < x_4$, one can see $x^T H x \geq 0$.

- The Hoffman-Pereira matrix

$$P := \begin{bmatrix} 1 & -1 & 1 & 0 & 0 & 1 & -1 \\ -1 & 1 & -1 & 1 & 0 & 0 & 1 \\ 1 & -1 & 1 & -1 & 1 & 0 & 0 \\ 0 & 1 & -1 & 1 & -1 & 1 & 0 \\ 0 & 0 & 1 & -1 & 1 & -1 & 1 \\ 1 & 0 & 0 & 1 & -1 & 1 & -1 \\ -1 & 1 & 0 & 0 & 1 & -1 & 1 \end{bmatrix} \tag{9.1.2}$$

is copositive but not NPP (see [129]).

- Let $B$ be the matrix

$$B := \begin{bmatrix} 1 & -\cos\psi_4 & c_{45} & \cos_{23} & -\cos\psi_3 \\ -\cos\psi_4 & 1 & -\cos\psi_5 & c_{15} & c_{34} \\ c_{45} & -\cos\psi_5 & 1 & -\cos\psi_1 & c_{12} \\ c_{23} & \cos_{15} & -\cos\psi_1 & 1 & -\cos\psi_2 \\ -\cos\psi_3 & c_{34} & c_{12} & -\cos\psi_2 & 1 \end{bmatrix}, \tag{9.1.3}$$

where each $\psi_i > 0$, $\sum_{i=1}^{5} \psi_i < \pi$, and $c_{ij} := \cos(\psi_i + \psi_j)$. The matrix $B$ is copositive but not NPP (see [126]).

It is a co-NP-complete task to decide whether or not a matrix is copositive [201]. Geometric properties of copositive cones are studied in [83]. The interior of $\mathcal{COP}_n$ is the set of strictly copositive matrices. The extreme rays of $\mathcal{COP}_n$ are known as follows. For $n \leq 4$, $\mathcal{COP}_n = \mathcal{NPP}_n$, and the extreme rays are given by $F_1 := e_i e_j^T + e_j e_i^T$ for $1 \leq i, j \leq n$, or given by rank-1 psd matrices $F_2 := uu^T$ for $0 \neq u \in \mathbb{R}^n$. This is shown in [110]. When $n = 5$, $\mathcal{COP}_5$ has four types of extreme rays: those given by $F_1$, $F_2$ as above; those given by matrices like $D\Theta H\Theta^T D$, where $H$ is the Horn matrix, $\Theta$ is a permutation matrix and $D$ is a diagonal matrix of positive diagonals; and those given by matrices of the form $DPBP^T D$, where $D, P$ are the same as above and $B$ is the matrix in (9.1.3). This is shown in [126]. For $n > 5$, the extreme rays of the cone $\mathcal{COP}_n$ are not fully characterized yet. For each $n > 1$, $\mathcal{COP}_n$ is not facially exposed, i.e., it has non-exposed faces. A face $F \subsetneq \mathcal{COP}_n$ is maximal[8] if and only if

$$F = \{A \in \mathcal{COP}_n : v^T A v = 0\}$$

for some vector $0 \neq v \in \mathbb{R}_+^n$. In particular, the cone $\mathcal{COP}_n$ has $n$ facets. They are given by the supporting hyperplanes $e_i^T A e_i = 0$, for $i = 1, \ldots, n$.

---

[8]The face $F$ being maximal means that if $G$ is a proper face of $\mathcal{COP}_n$ such that $F \subseteq G$, then $F = G$.

## 9.1.1 ▪ Copositive forms and Pólya's Theorem

The copositivity is similarly defined for homogeneous polynomials. A form $f \in \mathbb{R}[x]^{\text{hom}}$ is said to be *copositive* if the evaluation $f(u) \geq 0$ for all $u \in \mathbb{R}^n_+$. Similarly, $f$ is said to be *strictly copositive* if $f(u) > 0$ for all $0 \neq u \in \mathbb{R}^n_+$. For a degree $d > 0$, denote the copositive polynomial cone

$$\mathcal{COP}_{n,d} := \{f \in \mathbb{R}[x]^{\text{hom}}_d : f \text{ is copositive}\}. \tag{9.1.4}$$

It is clear that $f \in \mathcal{COP}_{n,d}$ if and only if the doubled degree form $f(x^{[2]})$, where

$$x^{[2]} := (x_1^2, \ldots, x_n^2), \tag{9.1.5}$$

is nonnegative everywhere. Therefore, $f(x^{[2]})$ being psd is equivalent to that $f$ is copositive. The following polynomials are copositive:

$$\begin{cases} f_1 := x_1^2 x_2 + x_1 x_2^2 + x_3^3 - 3x_1 x_2 x_3, \\ f_2 := x_1^3 + x_2^3 + x_3^3 - x_1^2 x_2 - x_1 x_2^2 - x_1^2 x_3 \\ \qquad -x_1 x_3^2 - x_2^2 x_3 - x_2 x_3^2 + 3x_1 x_2 x_3, \\ f_3 := x_1^2 x_2 + x_2^2 x_3 + x_3^2 x_1 - 3x_1 x_2 x_3. \end{cases} \tag{9.1.6}$$

This is because the forms $f_1(x^{[2]}), f_2(x^{[2]}), f_3(x^{[2]})$ are respectively the Motzkin, Robinson, and Choi-Lam forms. They are all psd but not SOS (see §2.4.2).

It is generally hard to check memberships for the copositive cone $\mathcal{COP}_{n,d}$. A sufficient condition for $f \in \mathcal{COP}_{n,d}$ is that $f(x^{[2]})$ is SOS. A more convenient condition is that $f$ has only nonnegative coefficients. The nonnegativity of coefficients is more likely possible if $f$ is multiplied with some factors. If for some degree $N > 0$ the product

$$(x_1 + \cdots + x_n)^N \cdot f(x)$$

has only positive coefficients, then $f$ must be strictly copositive. One wonders whether or not this is the case for all strictly copositive forms. Pólya's Theorem affirms this.

**Theorem 9.1.1.** *(Pólya [262]) If a form $f \in \mathbb{R}[x]^{\text{hom}}$ is strictly copositive, then there exists $N \in \mathbb{N}$ such that $(x_1 + \cdots + x_n)^N f$ has only positive coefficients.*

Pólya's Theorem can be proved explicitly by examining coefficients, as in [268]. Recall that $e$ denotes the vector of all ones and its length is determined by the operation on it. A form $f \in \mathbb{R}[x]^{\text{hom}}_d$ can be written as $f = \sum_{|\alpha|=d} f_\alpha x^\alpha$, with $\alpha = (\alpha_1, \ldots, \alpha_n)$. So, we have the expansion

$$(e^T x)^N \cdot f = \sum_{\substack{\gamma = (\gamma_1, \ldots, \gamma_n), \\ |\gamma| = N}} \frac{N!}{\gamma_1! \cdots \gamma_n!} x^\gamma \left( \sum_{|\alpha|=d} f_\alpha x^\alpha \right).$$

For each $\beta := (\beta_1, \ldots, \beta_n)$ with $|\beta| = N + d$, let $F_\beta$ denote the coefficient of $x^\beta$ in the product $(e^T x)^N \cdot f$, then one can express $F_\beta$ as

$$\begin{aligned} F_\beta &= \sum_{|\alpha|=d, \alpha \leq \beta} \frac{N!}{(\beta_1 - \alpha_1)! \cdots (\beta_n - \alpha_n)!} f_\alpha \\ &= \frac{N!(N+d)^d}{\beta_1! \cdots \beta_n!} \sum_{|\alpha|=d, \alpha \leq \beta} f_\alpha \cdot \prod_{l=1}^{n} \frac{\beta_l!}{(\beta_l - \alpha_l)!(N+d)^{\alpha_l}} \\ &= \frac{N!(N+d)^d}{\beta_1! \cdots \beta_n!} \sum_{|\alpha|=d} f_\alpha \cdot \left( \frac{\beta_1}{N+d} \right)^{\alpha_1}_{\frac{1}{N+d}} \cdots \left( \frac{\beta_n}{N+d} \right)^{\alpha_n}_{\frac{1}{N+d}}. \end{aligned}$$

In the above, we use the notation

$$(t)_\epsilon^k := t(t - \epsilon) \cdots (t - (k-1)\epsilon).$$

Therefore, it holds that

$$\frac{\beta_1! \cdots \beta_n!}{N!(N+d)^d} F_\beta = \sum_{|\alpha|=d} f_\alpha \cdot \left(\frac{\beta_1}{N+d}\right)_{\frac{1}{N+d}}^{\alpha_1} \cdots \left(\frac{\beta_n}{N+d}\right)_{\frac{1}{N+d}}^{\alpha_n}.$$

Note that $\left(\frac{\beta_1}{N+d}, \ldots, \frac{\beta_n}{N+d}\right) \in \Delta_n$. For $\epsilon > 0$, denote the polynomial

$$f_\epsilon(x) := \sum_{|\alpha|=d} f_\alpha \cdot (x_1)_\epsilon^{\alpha_1} \cdots (x_n)_\epsilon^{\alpha_n}.$$

Observe that $f_\epsilon \to f$ uniformly on the simplex $\Delta_n$. Therefore, if $N$ is big enough, then all $F_\beta$ must be positive, when $f$ is strictly copositive. Moreover, a degree bound on $N$ can be obtained for $F_\beta$ to be positive. This was done in [268]. Define the norm and the minimum value ($\overline{\mathbb{N}}_d^n$ is the set as in (8.4.8))

$$\begin{aligned}
\|f\| &:= \max_{\alpha \in \overline{\mathbb{N}}_d^n} \frac{\alpha_1! \cdots \alpha_n!}{d!} |f_\alpha|, \\
f_{min} &:= \min_{x \in \Delta_n} f(x).
\end{aligned} \tag{9.1.7}$$

The form $f$ is strictly copositive if and only if $f_{min} > 0$. We have the following degree bound for $N$.

**Theorem 9.1.2.** *([268]) Let $\|f\|$, $f_{min}$ be as in (9.1.7). If a form $f \in \mathbb{R}[x]_d^{\text{hom}}$ is strictly copositive, then for all*

$$N > \frac{d(d-1)}{2} \frac{\|f\|}{f_{min}} - d,$$

*the product $(e^T x)^N f$ has only positive coefficients.*

When $f$ is copositive but not strictly copositive, the conclusion of Pólya's Theorem may not hold. For instance, the quadratic form $q = (x_1 - x_2)^2$ is copositive but not strictly; however, $(e^T x)^N q$ cannot have only positive (or nonnegative) coefficients for any power $N$. Moreover, the degree bound in Theorem 9.1.2 is sharp. For instance, the quadratic form

$$q_\epsilon := x_1^2 - (2 - \epsilon)x_1 x_2 + x_2^2$$

is strictly copositive for all $\epsilon > 0$. As pointed out in [268], all coefficients of $(x_1 + x_2)^N q_\epsilon$ are positive precisely when $N \geq 2\lceil \frac{2}{\epsilon} \rceil - 3$.

Pólya's Theorem can be generalized to representations with SOS polynomials. Clearly, if the product $(x^T x)^N \cdot f(x^{[2]})$ is SOS for some power $N$, then $f$ must be copositive. Theorem 9.1.2 implies the following.

**Theorem 9.1.3.** *If a form $f \in \mathbb{R}[x]_d^{\text{hom}}$ is strictly copositive, then for all*

$$N > \frac{d(d-1)}{2} \frac{\|f\|}{f_{min}} - d,$$

*the product $(x^T x)^N f(x^{[2]}) \in \Sigma[x]_{2N+2d}$.*

## 9.1.2 ▪ Exercises

**Exercise 9.1.1.** *For Horn's matrix $H$ as in (9.1.1), show that $(x^T x) \cdot ((x^{[2]})^T \cdot H \cdot x^{[2]})$ is SOS but $(e^T x) \cdot (x^T H x)$ has negative coefficients.*

**Exercise 9.1.2.** *If $A = e_i e_j^T + e_j e_i^T$ for some $i, j$ or $A = u u^T$ for some $u \neq 0$, show that $A$ gives an extreme ray for $\mathcal{COP}_n$.*

**Exercise 9.1.3.** *For $n \geq 2$, show that $\mathcal{COP}_n$ has non-exposed faces.*

**Exercise 9.1.4.** *If a face $F \subsetneq \mathcal{COP}_n$ is maximal, show that*

$$F = \{A \in \mathcal{COP}_n : v^T A v = 0\}$$

*for some vector $0 \neq v \in \mathbb{R}_+^n$. If $F$ is further a facet, show that $v = e_i$ for some $i = 1, \ldots, n$.*

**Exercise 9.1.5.** *For $q = (x_1 - x_2)^2$, show that there is no $N$ such that $(e^T x)^N f$ has only nonnegative coefficients.*

**Exercise 9.1.6.** *([268]) For $\epsilon > 0$ and $q_\epsilon := x_1^2 - (2 - \epsilon)x_1 x_2 + x_2^2$, show that $(x_1 + x_2)^N q_\epsilon$ cannot have only positive coefficients unless $N \geq 2\lceil \frac{2}{\epsilon} \rceil - 3$.*

**Exercise 9.1.7.** *Prove Theorem 9.1.3. Give an example of $f$ and $N > 0$ such that $(x^T x)^N f(x^{[2]})$ is SOS while $(e^T x)^N f$ has negative coefficients.*

# 9.2 ▪ CP Matrices and Moments

A symmetric matrix $C \in \mathcal{S}^n$ is *completely positive* (CP) if there exist nonnegative vectors $u_1, \ldots, u_r \in \mathbb{R}_+^n$ such that

$$C = u_1 u_1^T + \cdots + u_r u_r^T. \tag{9.2.1}$$

If it exists, (9.2.1) is called a CP decomposition for $C$. The smallest $r$ in (9.2.1) is called the *CP rank* of $C$, for which we denote $\text{rank}_{cp}(C)$. The cone of all $n$-by-$n$ CP matrices is denoted as $\mathcal{CP}_n$. It is a convex cone. The basic property of CP matrices is the duality to copositive matrices. If $A$ is a copositive matrix and $C$ is CP as in (9.2.1), then

$$A \bullet C = u_1^T A u_1 + \cdots + u_r^T A u_r \geq 0.$$

We have the following duality relation, which is a special case of Theorem 9.2.4.

**Theorem 9.2.1.** *Both $\mathcal{COP}_n$ and $\mathcal{CP}_n$ are proper cones and they are dual to each other, i.e.,*

$$\left(\mathcal{COP}_n\right)^* = \mathcal{CP}_n, \quad \left(\mathcal{CP}_n\right)^* = \mathcal{COP}_n. \tag{9.2.2}$$

To be completely positive, the matrix $C$ must be psd and all its entries are nonnegative, i.e., it holds that the inclusion

$$\mathcal{CP}_n \subseteq \mathcal{DNN}_n := \mathcal{S}_+^n \cap \mathcal{N}^n. \tag{9.2.3}$$

A matrix is called *doubly nonnegative* (DNN) if it is psd and nonnegative simultaneously. The set $\mathcal{DNN}_n$ is called the $n$-by-$n$ DNN cone. Interestingly, the inclusion in (9.2.3) is an equality for $n \leq 4$. This result appeared in [193].

**Theorem 9.2.2.** *([193]) For $n \leq 4$, it holds that $\mathcal{CP}_n = \mathcal{DNN}_n$.*

The proof in [193] can be outlined as follows. For each $C \in \mathcal{DNN}_n$, if a diagonal entry $C_{ii} = 0$, then the entire $i$th row and $i$th column must be zero, and hence $C$ can be reduced to a DNN matrix of smaller size. Generally, we only need to consider the case that all diagonal entries $C_{ii} > 0$. Up to a scaling $DCD$ for a positive diagonal matrix $D$, one can further assume that all diagonal entries $C_{ii} = 1$. When $n \leq 2$, a CP decomposition for $C$ can be obtained by the Cholesky factorization. When $n = 3$, note that $0 \leq C_{ij} \leq 1$ for all $i, j$. If all $C_{ij} = 1$, then $C = ee^T$ is a CP decomposition. Suppose at least one $C_{ij} < 1$. It is impossible that

$$\text{both} \quad C_{ii}C_{jk} - C_{ij}C_{ik} < 0 \quad \text{and} \quad C_{jj}C_{ik} - C_{ij}C_{jk} < 0$$

for $k \neq i, j$ (see exercise for the proof). So, we have

$$\text{either} \quad C_{ii}C_{jk} - C_{ij}C_{ik} \geq 0 \quad \text{or} \quad C_{jj}C_{ik} - C_{ij}C_{jk} \geq 0.$$

There exists a permutation matrix $P$ such that $B := P^T CP$ satisfies

$$1 > B_{12} \geq 0 \quad \text{and} \quad B_{23} - B_{12}B_{13} \geq 0.$$

Then $B = PX^T XP^T$ is a CP decomposition, with

$$X := \begin{bmatrix} 1 & B_{12} & B_{13} \\ 0 & \sqrt{1 - B_{12}^2} & \frac{B_{23} - B_{12}B_{13}}{1 - B_{12}^2} \\ 0 & 0 & \frac{\sqrt{\det B}}{1 - B_{12}^2} \end{bmatrix}.$$

When $n = 4$, the proof is more dedicated. It can be done for two cases.
**Case I**: Consider that

$$C_{ij}C_{kk} - C_{ik}C_{jk} \geq 0, \quad \text{or} \quad C_{ij}C_{mm} - C_{im}C_{jm} \geq 0, \quad \text{or}$$
$$C_{km}C_{ii} - C_{ik}C_{im} \geq 0, \quad \text{or} \quad C_{km}C_{jj} - C_{jk}C_{jm} \geq 0$$

for all $(i, j, k, m) \in \{(1, 2, 3, 4), (1, 3, 2, 4), (1, 4, 2, 3)\}$. Then, as shown in [193], there exists a permutation matrix $P$ such that $B := P^T CP$ satisfies that

$$\det \begin{bmatrix} B_{11} & B_{12} \\ B_{31} & B_{32} \end{bmatrix} \geq 0, \quad \det \begin{bmatrix} B_{11} & B_{12} \\ B_{41} & B_{42} \end{bmatrix} \geq 0, \quad \det \begin{bmatrix} B_{11} & B_{13} \\ B_{41} & B_{43} \end{bmatrix} \geq 0.$$

Note that $B^T = B$ and $B_{11} = B_{22} = B_{33} = B_{44} = 1$. The above inequalities and the Schur complement imply that

$$E := \begin{bmatrix} B_{22} & B_{23} & B_{24} \\ B_{32} & B_{33} & B_{34} \\ B_{42} & B_{43} & B_{44} \end{bmatrix} - \begin{bmatrix} B_{21} \\ B_{31} \\ B_{41} \end{bmatrix} \begin{bmatrix} B_{12} \\ B_{13} \\ B_{14} \end{bmatrix}^T \in \mathcal{DNN}_3.$$

By the earlier conclusion for $n = 3$, there exists $Y \geq 0$ such that $E = Y^T Y$. Then $C = PZ^T ZP^T$ is a CP decomposition for the nonnegative matrix

$$Z = \begin{bmatrix} 1 & b \\ 0 & Y \end{bmatrix}, \quad b = \begin{bmatrix} B_{12} & B_{13} & B_{14} \end{bmatrix}.$$

**Case II**: Consider that

$$C_{ij}C_{kk} - C_{ik}C_{jk} < 0, \quad C_{ij}C_{mm} - C_{im}C_{jm} < 0,$$
$$C_{km}C_{ii} - C_{ik}C_{im} < 0, \quad C_{km}C_{jj} - C_{jk}C_{jm} < 0$$

for $(i, j, k, m) = (1, 2, 3, 4)$ or $(1, 3, 2, 4)$ or $(1, 4, 2, 3)$. Without loss of generality, consider the case $(i, j, k, m) = (1, 2, 3, 4)$. Let $R$ be the upper triangular matrix such that

$$R^{-1} = \begin{bmatrix} 1 & -C_{12} & -y_1 & -y_2 \\ 0 & 1 & 0 & 0 \\ 0 & 0 & 1 & 0 \\ 0 & 0 & 0 & 1 \end{bmatrix}.$$

As shown in [193], one can choose values $y_1 \geq 0, y_2 \geq 0$ such that $B := R^{-T} C R^{-1}$ has four zero entries

$$B_{12} = B_{21} = B_{34} = B_{43} = 0$$

and all other entries are nonnegative. Consider the matrix

$$Z := \begin{bmatrix} Z_{11} & 0 & Z_{13} & 0 \\ 0 & Z_{22} & Z_{23} & 0 \\ Z_{31} & 0 & 0 & Z_{34} \\ 0 & Z_{42} & 0 & Z_{44} \end{bmatrix}.$$

Similarly, as shown in [193], there exist values $Z_{ij} \geq 0$ such that $Z^T Z = B$. Then $C = (ZR)^T (ZR)$ is a CP decomposition, since $R \geq 0$.

The inclusion in (9.2.3) is strict for $n \geq 5$, i.e, $\mathcal{CP}_n \neq \mathcal{DNN}_n$ for $n \geq 5$. For instance, the following two matrices

$$\begin{bmatrix} 1 & 0 & 0 & \frac{1}{2} & \frac{1}{2} \\ 0 & 1 & \frac{3}{4} & 0 & \frac{1}{2} \\ 0 & \frac{3}{4} & 1 & \frac{1}{2} & 0 \\ \frac{1}{2} & 0 & \frac{1}{2} & 1 & 0 \\ \frac{1}{2} & \frac{1}{2} & 0 & 0 & 1 \end{bmatrix}, \quad \begin{bmatrix} 1 & 1 & 0 & 0 & 1 \\ 1 & 2 & 1 & 0 & 0 \\ 0 & 1 & 2 & 1 & 0 \\ 0 & 0 & 1 & 2 & 1 \\ 1 & 0 & 0 & 1 & 6 \end{bmatrix} \tag{9.2.4}$$

are DNN but not CP. The first one appeared in [193], while the second one appeared in [21, Example 2.9]. The gap between CP and DNN matrices can also be characterized by graph theory [150]. For a matrix $A \in \mathcal{S}^n$, its graph $G = ([n], E)$ is defined such that $(i, j)$ is an edge if and only if $A_{ij} \neq 0$. The graph $G$ is said to be completely positive (CP) if every DNN matrix, which is associated to $G$, is a CP matrix. Interesting, it is shown in [150] that a graph $G$ is CP if and only if it does not contain an odd cycle whose length is bigger than 4. This fact confirms that $\mathcal{CP}_5 \neq \mathcal{DNN}_5$.

The interior of the cone $\mathcal{CP}_n$ is characterized in [82, 87, 88] in several equivalent ways. An explicit characterization for the interior is as follows.

**Lemma 9.2.3.** *([88]) The interior of $\mathcal{CP}_n$ can be expressed as*

$$\mathrm{int}\,(\mathcal{CP}_n) = \{BB^T : \mathrm{rank}(B) = n,\ B > 0\}.$$

*The above inequality $B > 0$ means that all entries of $B$ are positive.*

Geometric properties of the cone $\mathcal{CP}_n$ are studied in [83]. We list some of them here. By the definition, one can see that

$$\mathcal{CP}_n = \mathrm{cone}\left(\{xx^T : x \in \Delta_n\}\right).$$

The extreme rays of $\mathcal{CP}_n$ are given by rank-1 matrices $uu^T$, with $u \in \mathbb{R}^n_+$. In fact, every extreme ray of $\mathcal{CP}_n$ is exposed [83]. For $n \leq 4$, every face of $\mathcal{CP}_n$ is exposed, because $\mathcal{CP}_n = \mathcal{S}^n_+ \cap \mathcal{N}^n$.

However, for $n \geq 5$, not every face of $\mathcal{CP}_n$ is exposed. For angles $\theta_1, \ldots, \theta_5 > 0$ such that $\theta_1 + \cdots + \theta_5 = \pi$, let $a, b, c, d$ be vectors such that

$$\begin{bmatrix} a & b & c & d \end{bmatrix} = \begin{bmatrix} \sin(\theta_5) & 0 & 0 & \sin(\theta_2) \\ s_{54} & \sin(\theta_1) & 0 & 0 \\ \sin(\theta_4) & s_{15} & \sin(\theta_2) & 0 \\ 0 & \sin(\theta_5) & s_{21} & \sin(\theta_3) \\ 0 & 0 & \sin(\theta_1) & s_{32} \end{bmatrix},$$

where each $s_{ij} := \sin(\theta_i + \theta_j)$. Then

$$F := \text{cone}\left(\{aa^T, bb^T, cc^T, dd^T\}\right)$$

is a non-exposed face of $\mathcal{CP}_5$ (see [330]). For each $n \geq 2$, the cone $\mathcal{CP}_n$ has $\frac{1}{2}n(n-1)$ facets, which are given by the supporting hyperplanes $e_i^T C e_j = 0$ for $i < j$. This is shown in [83, Theorem 6.6].

## 9.2.1 ▪ CP moment cones

We can similarly define completely positive (CP) moment cones. Recall the power set $\overline{\mathbb{N}}_d^n$ as in (8.4.8). A homogeneous truncated multi-sequence (htms) $y \in \mathbb{R}^{\overline{\mathbb{N}}_d^n}$ is said to be *completely positive* (CP) if there exist nonnegative vectors $u_1, \ldots, u_r \in \mathbb{R}_+^n$ such that (see (2.4.22) for the notation $[u]_d^{\text{hom}}$)

$$y = [u_1]_d^{\text{hom}} + \cdots + [u_r]_d^{\text{hom}}.$$

The set of all $y$ that can decomposed as above is denoted as $\mathcal{CP}_{n,d}$. Since $[x]_d^{\text{hom}}$ consists of monomials of degree equaling $d$, the set $\mathcal{CP}_{n,d}$ is the cone generated by vectors $[u]_d^{\text{hom}}$ ($u \in \mathbb{R}_+^n$). Equivalently, we have

$$\mathcal{CP}_{n,d} = \text{cone}\left(\{[u]_d^{\text{hom}} : u \in \Delta_n\}\right). \tag{9.2.5}$$

Each vector in $\mathcal{CP}_{n,d}$ can be viewed as an $\mathcal{A}$-tms with the power set $\mathcal{A} = \overline{\mathbb{N}}_d^n$. Theorems 8.4.1, 8.4.3, and 8.4.6 imply the following conclusion.

**Theorem 9.2.4.** *The cone $\mathcal{CP}_{n,d}$ can be equivalently expressed as*

$$\mathcal{CP}_{n,d} = \left\{ \int [x]_d^{\text{hom}} \mathrm{d}\mu \,\middle|\, \begin{array}{l} \mu \text{ is a Borel measure,} \\ \text{supp}(\mu) \subseteq \Delta_n \end{array} \right\}. \tag{9.2.6}$$

*Moreover, both $\mathcal{CP}_{n,d}$ and $\mathcal{COP}_{n,d}$ are proper cones, and they satisfy the dual relationship (the superscript $^\star$ denotes the dual cone)*

$$\left(\mathcal{CP}_{n,d}\right)^\star = \mathcal{COP}_{n,d}, \quad \left(\mathcal{COP}_{n,d}\right)^\star = \mathcal{CP}_{n,d}. \tag{9.2.7}$$

For $\mathcal{A} = \overline{\mathbb{N}}_d^n$ and $K = \Delta_n$, all properties of the moment cone $\mathcal{R}_\mathcal{A}(K)$ in §8.4.1 are applicable to the CP moment cone $\mathcal{CP}_{n,d}$.

## 9.2.2 ▪ Exercises

**Exercise 9.2.1.** *Show that the dual cone of $\mathcal{DNN}_n$ is $\mathcal{NPP}_n$. Use this to show the copositive matrices in (9.1.1), (9.1.2), and (9.1.3) are not NPP.*

**Exercise 9.2.2.** *([193]) For $B = (B_{ij}) \in \mathcal{DNN}_n$, show that for all $i, j, k$ at most one of the following is negative:*

$$B_{ij}B_{kk} - B_{ik}B_{jk}, \quad B_{ik}B_{jj} - B_{ij}B_{jk}, \quad B_{jk}B_{ii} - B_{ij}B_{kk}.$$

**Exercise 9.2.3.** *Find a CP matrix $C$ such that $C \bullet H = 0$, where $H$ is Horn's matrix as in (9.1.1).*

**Exercise 9.2.4.** *Show that the matrices in (9.2.4) are doubly nonnegative but are not CP.*

**Exercise 9.2.5.** *([83]) Show that every extreme ray of $\mathcal{CP}_n$ is given by a rank-1 matrix $uu^T$, with $0 \neq u \in \mathbb{R}_+^n$. Also show that every extreme ray of $\mathcal{CP}_n$ is an exposed face.*

**Exercise 9.2.6.** *If a matrix $X = (X_{ij}) \in \mathcal{DNN}_n$ is diagonally dominant, i.e., $X_{ii} \geq \sum_{j \neq i} X_{ij}$ for all $i$, show that $X$ must be CP.*

**Exercise 9.2.7.** *For all $i < j$, show that the set $\{C \in \mathcal{CP}_n : e_i^T C e_j = 0\}$ is a facet of the cone $\mathcal{CP}_n$.*

# 9.3 ▪ Relaxations and Memberships

We give semidefinite relaxations for the copositive cone $\mathcal{COP}_{n,d}$ and the CP moment cone $\mathcal{CP}_{n,d}$. Memberships for them can also be checked by using these relaxations.

## 9.3.1 ▪ Pólya type relaxations

Pólya's Theorem affirms that a form $f \in \mathbb{R}[x]^{\text{hom}}$ is strictly copositive if and only if the product $(e^T x)^k f$ has only positive coefficients for some power $k$. This gives a hierarchy of polyhedral approximations for the copositive cone $\mathcal{COP}_{n,d}$. We refer to [68, 78, 258, 268, 283] for classical work about Pólya type approximations.

For a power $\alpha := (\alpha_1, \ldots, \alpha_n)$, we use $\eth^\alpha p$ to denote the coefficient of the monomial $x^\alpha := x_1^{\alpha_1} \cdots x_n^{\alpha_n}$ in the polynomial $p$, i.e.,

$$p = \sum_\alpha p_\alpha x^\alpha \quad \Rightarrow \quad \eth^\alpha p = p_\alpha.$$

For each $k$, denote the polyhedral cone (see (8.4.8) for the notation $\overline{\mathbb{N}}_{d+k}^n$)

$$\mathcal{PH}_{n,d}^{(k)} := \left\{ f \in \mathbb{R}[x]_d^{\text{hom}} : \eth^\nu [(e^T x)^k f] \geq 0 \; \forall \nu \in \overline{\mathbb{N}}_{d+k}^n \right\}. \tag{9.3.1}$$

Clearly, every form in $\mathcal{PH}_{n,d}^{(k)}$ must be copositive. Pólya's Theorem can be generalized in terms of SOS polynomials. Note that $f$ is copositive if and only if $f(x^{[2]})$ is nonnegative over $\mathbb{R}^n$, where $x^{[2]} := (x_1^2, \ldots, x_n^2)$. For a power $k$, if $(x^T x)^k f(x^{[2]})$ is SOS, then $f$ must be copositive. This gives another type of approximations for $\mathcal{COP}_{n,d}$:

$$\mathcal{PS}_{n,d}^{(k)} := \left\{ f \in \mathbb{R}[x]_d^{\text{hom}} : (x^T x)^k f(x^{[2]}) \in \Sigma[x]_{2k+2d} \right\}. \tag{9.3.2}$$

If $(e^T x)^k f$ has only nonnegative coefficients, then $(x^T x)^k f(x^{[2]})$ must be SOS. Therefore, we have the following nesting relation:

$$\begin{array}{ccccccccc}
\mathcal{PH}_{n,d}^{(0)} & \subseteq & \mathcal{PH}_{n,d}^{(1)} & \subseteq & \cdots & \mathcal{PH}_{n,d}^{(k)} & \subseteq & \cdots & \mathcal{COP}_{n,d} \\
\cap & & \cap & & & \cap & & & \| \\
\mathcal{PS}_{n,d}^{(0)} & \subseteq & \mathcal{PS}_{n,d}^{(1)} & \subseteq & \cdots & \mathcal{PS}_{n,d}^{(k)} & \subseteq & \cdots & \mathcal{COP}_{n,d}.
\end{array} \tag{9.3.3}$$

Theorems 9.1.1 and 9.1.3 imply the following.

**Theorem 9.3.1.** *For the above notation of sets, we have*

$$\text{int}\,(\mathcal{COP}[x]_d) \subseteq \bigcup_{k=0}^{\infty} \mathcal{PH}_{n,d}^{(k)} \subseteq \bigcup_{k=0}^{\infty} \mathcal{PS}_{n,d}^{(k)} \subseteq \mathcal{COP}_{n,d}. \tag{9.3.4}$$

We next discuss relaxations for the CP moment cone $\mathcal{CP}_{n,d}$, which is dual to $\mathcal{COP}_{n,d}$. Recall that $\mathcal{CP}_{n,d}$ can be expressed as the conic hull

$$\mathcal{CP}_{n,d} = \text{cone}\,(\{[u]_d^{\text{hom}} : u \in \Delta_n\}).$$

Note the following expansion (note $\theta := (\theta_1, \ldots, \theta_n)$ and $\binom{k}{\theta} := \frac{k!}{\theta_1! \cdots \theta_n!}$):

$$(x_1 + \cdots + x_n)^k = \sum_{|\theta|=k} \binom{k}{\theta} x_1^{\theta_1} \cdots x_n^{\theta_n}.$$

For $f = \sum_{|\alpha|=d} f_\alpha x^\alpha$ and for $\nu = \theta + \alpha$ with $|\theta| = k$, it holds that

$$\eth^\nu[(e^T x)^k f] = \sum_{\theta+\alpha=\nu} \binom{k}{\theta} f_\alpha. \tag{9.3.5}$$

The cone $\mathcal{PH}_{n,d}^{(k)}$ can be expressed in the polyhedral form

$$\mathcal{PH}_{n,d}^{(k)} = \left\{ f \in \mathbb{R}[x]_d^{\text{hom}} : \sum_{\theta+\alpha=\nu} \binom{k}{\theta} f_\alpha \geq 0 \ \forall \nu \in \overline{\mathbb{N}}_{d+k}^n \right\}. \tag{9.3.6}$$

For each $w \in \mathbb{R}^{\overline{\mathbb{N}}_{d+k}^n}$, note that

$$\langle (e^T x)^k f, w \rangle = \sum_{|\nu|=d+k} \eth^\nu[(e^T x)^k f] \cdot w_\nu = \sum_{|\alpha|=d} f_\alpha \sum_{\nu-\theta=\alpha} \binom{k}{\theta} w_\nu. \tag{9.3.7}$$

By the duality for polyhedral cones (see Exercise 1.3.1), one can show that the dual cone of $\mathcal{PH}_{n,d}^{(k)}$ is given as

$$\left(\mathcal{PH}_{n,d}^{(k)}\right)^\star = \left\{ y \in \mathbb{R}^{\overline{\mathbb{N}}_d^n} \;\middle|\; \begin{array}{l} y_\alpha = \sum_{\nu-\theta=\alpha} \binom{k}{\theta} w_\nu, \\[2mm] \alpha \in \overline{\mathbb{N}}_d^n,\, \nu \in \overline{\mathbb{N}}_{d+k}^n,\, w \in \mathbb{R}_+^{\overline{\mathbb{N}}_{d+k}^n} \end{array} \right\}. \tag{9.3.8}$$

The proof for the above is left as an exercise.

The dual cone of $\mathcal{PS}_{n,d}^{(k)}$ can be obtained similarly. For each $k$, $\mathcal{PS}_{n,d}^{(k)}$ is a linear section of the SOS cone $\Sigma[x]_{2k+2d}$. Note that $f \in \mathcal{PS}_{n,d}^{(k)}$ is equivalent to

$$(x^T x)^k f(x^{[2]}) = ([x]_{k+d}^{\text{hom}})^T \cdot X \cdot [x]_{k+d}^{\text{hom}}$$

for some psd matrix $X \succeq 0$. Moreover, observe the expansion

$$(x^T x)^k f(x^{[2]}) = \sum_{|\theta|=k} \sum_{|\alpha|=d} \binom{k}{\theta} \cdot f_\alpha \cdot x^{2\theta+2\alpha}. \tag{9.3.9}$$

Hence, we get $f \in \mathcal{PS}_{n,d}^{(k)}$ if and only if

$$\langle (x^T x)^k f(x^{[2]}), w \rangle \geq 0 \quad \forall w \in \left( \Sigma[x]_{2k+2d} \right)^\star. \tag{9.3.10}$$

In the above, the dual cone $\left( \Sigma[x]_{2k+2d} \right)^\star$ is given as (see Proposition 2.4.4)

$$\left( \Sigma[x]_{2k+2d} \right)^\star = \left\{ w \in \mathbb{R}^{\overline{\mathbb{N}}_{2d+2k}^n} : M_{d+k}[w] \succeq 0 \right\}.$$

The relation (9.3.10) is equivalent to

$$\sum_{\alpha \in \overline{\mathbb{N}}_d^n} f_\alpha \left( \sum_{|\theta|=k} \binom{k}{\theta} w_{2\theta+2\alpha} \right) \geq 0 \quad \text{for all} \quad w \in \left( \Sigma[x]_{2k+2d} \right)^\star. \tag{9.3.11}$$

The above implies that $\mathcal{PS}_{n,d}^{(k)}$ is the dual cone of the set

$$\left\{ y \in \mathbb{R}^{\overline{\mathbb{N}}_d^n} \;\middle|\; \begin{array}{l} \exists\, w \in \left( \Sigma[x]_{2k+2d} \right)^\star \quad s.t. \\ y_\alpha = \sum\limits_{|\theta|=k} \binom{k}{\theta} w_{2\theta+2\alpha} \,\forall\, \alpha \in \overline{\mathbb{N}}_d^n \end{array} \right\}.$$

For each $y \in \left( \mathcal{PS}_{n,d}^{(k)} \right)^\star$, we have $\langle f, y \rangle \geq 0$ for all $f \in \mathcal{PS}_{n,d}^{(k)}$. By the duality for spectrahedral cones (see Exercise 1.5.3), one can show that each $y \in \left( \mathcal{PS}_{n,d}^{(k)} \right)^\star$ can be expressed as

$$\exists\, w \in \left( \Sigma[x]_{2k+2d} \right)^\star \quad s.t. \quad y_\alpha = \sum_{|\theta|=k} \binom{k}{\theta} w_{2\theta+2\alpha} \quad \text{for each } \alpha \in \overline{\mathbb{N}}_d^n.$$

Therefore, we get the dual cone

$$\left( \mathcal{PS}_{n,d}^{(k)} \right)^\star = \left\{ y \in \mathbb{R}^{\overline{\mathbb{N}}_d^n} \;\middle|\; \begin{array}{l} \exists\, w \in \mathbb{R}^{\overline{\mathbb{N}}_{2d+2k}^n} \; s.t. \quad M_{d+k}[w] \succeq 0, \\ y_\alpha = \sum\limits_{|\theta|=k} \binom{k}{\theta} w_{2\theta+2\alpha} \,\forall\, \alpha \in \overline{\mathbb{N}}_d^n \end{array} \right\}. \tag{9.3.12}$$

The details for the above proof are left as an exercise. Theorem 9.3.1 implies the following.

**Theorem 9.3.2.** *For the above notation, we have*

$$\bigcap_{k=0}^{\infty} \left( \mathcal{PH}_{n,d}^{(k)} \right)^\star = \mathcal{CP}_{n,d} = \bigcap_{k=0}^{\infty} \left( \mathcal{PS}_{n,d}^{(k)} \right)^\star. \tag{9.3.13}$$

## 9.3.2 ▪ Relaxations via dehomogenization

A form $f \in \mathbb{R}[x]_d^{\text{hom}}$ is copositive if and only if $f \geq 0$ on the simplex

$$\Delta_n = \{ x \in \mathbb{R}_+^n : e^T x = 1 \}.$$

One can use $e^T x = 1$ to eliminate one variable, say, $x_n = 1 - x_1 - \cdots - x_{n-1}$. For convenience, denote

$$\bar{x} := (x_1, \ldots, x_{n-1}), \quad \bar{n} := n - 1. \tag{9.3.14}$$

Define the dehomogenization map $\varpi$:

$$\varpi(f) := f(x_1, \ldots, x_{\bar{n}}, 1 - e^T \bar{x}). \tag{9.3.15}$$

(Here, the length of the all-one vector $e$ is $n-1$.) The resulting dehomogenization of the simplex $\Delta_n$ is

$$\overline{\Delta}_n := \{\bar{x} \in \mathbb{R}^{n-1} : \bar{x} \geq 0, 1 - e^T \bar{x} \geq 0). \tag{9.3.16}$$

Clearly, $x \in \Delta_n$ if and only if $\bar{x} \in \overline{\Delta}_n$, which is equivalent to $g(\bar{x}) \geq 0$, for the following polynomial tuple:

$$g := (x_1, \ldots, x_{\bar{n}}, 1 - e^T \bar{x}, 1 - \|\bar{x}\|^2). \tag{9.3.17}$$

Therefore, $f$ is copositive if and only if

$$\varpi(f) \in \mathscr{P}_d(\overline{\Delta}_n).$$

The above gives a dehomogenization approach for studying COP cones. Putinar's Positivstellensatz is applicable for constructing approximations for the copositive cone $\mathcal{COP}_{n,d}$. If $\varpi(f) \in \mathrm{QM}[g]$, then $f \geq 0$ on $\Delta_n$. The above is guaranteed if $f > 0$ on $\Delta_n$, by Theorem 2.6.1. For a given degree $k \geq d/2$, this leads to the inner approximation

$$\mathcal{CQ}_{n,d}^{(k)} := \left\{ f \in \mathbb{R}[x]_d^{\mathrm{hom}} : \varpi(f) \in \mathrm{QM}[g]_{2k} \right\}. \tag{9.3.18}$$

The linear map $\varpi$ in (9.3.15) gives an isomorphism from $\mathbb{R}[x]_d^{\mathrm{hom}}$ to $\mathbb{R}[\bar{x}]_d$. Its adjoint map $\varpi^T$ is an isomorphism from $(\mathbb{R}[\bar{x}]_d)^*$ to $(\mathbb{R}[x]_d^{\mathrm{hom}})^*$. Similarly, the inverse adjoint map $\varpi^{-T}$ is an isomorphism from $(\mathbb{R}[x]_d^{\mathrm{hom}})^*$ to $(\mathbb{R}[\bar{x}]_d)^*$. Note the isomorphisms

$$(\mathbb{R}[\bar{x}]_d)^* \cong \mathbb{R}^{\mathbb{N}_d^{n-1}}, \quad (\mathbb{R}[x]_d^{\mathrm{hom}})^* \cong \mathbb{R}^{\overline{\mathbb{N}}_d^n}.$$

For each $y \in \mathbb{R}^{\overline{\mathbb{N}}_d^n}$ and for each $f \in \mathbb{R}[x]_d^{\mathrm{hom}}$, it holds that

$$\langle f, y \rangle = \langle \varpi(f), \varpi^{-T}(y) \rangle. \tag{9.3.19}$$

Clearly, $f \in \mathcal{COP}_{n,d}$ is equivalent to $\varpi(f) \in \mathscr{P}_d(\overline{\Delta}_n)$. The dehomogenization mapping $\varpi$ can also be used to study CP moment cones, which is done in the work [248]. The following lemma gives the relationship between $\mathcal{CP}_{n,d}$ and $\mathscr{R}_d(\overline{\Delta}_n)$, where $\mathscr{R}_d(\overline{\Delta}_n)$ is the degree-$d$ moment cone for the set $\overline{\Delta}_n$.

**Lemma 9.3.3.** *An htms $y \in \mathbb{R}^{\overline{\mathbb{N}}_d^n}$ has the decomposition*

$$y = \lambda_1 [u_1]_d^{\mathrm{hom}} + \cdots + \lambda_r [u_r]_d^{\mathrm{hom}} \quad (\lambda_i \in \mathbb{R}, \, u_i \in \Delta_n)$$

*if and only if*

$$\varpi^{-T}(y) = \lambda_1 [v_1]_d + \cdots + \lambda_r [v_r]_d \quad (\lambda_i \in \mathbb{R}, \, v_i \in \overline{\Delta}_n),$$

*where each $v_i$ is the subvector consisting of first $n-1$ entries of $u_i$. Therefore, we have*

$$y \in \mathcal{CP}_{n,d} \quad \Leftrightarrow \quad \varpi^{-T}(y) \in \mathscr{R}_d(\overline{\Delta}_n). \tag{9.3.20}$$

Lemma 9.3.3 can be shown for straightforward verification, so the proof is left as an exercise. For each $y \in \left( \mathcal{CQ}_{n,d}^{(k)} \right)^\star$, we have $\langle p, y \rangle \geq 0$ for all $p \in \mathcal{CQ}_{n,d}^{(k)}$, so (9.3.19) implies

$$\langle \varpi(p), \varpi^{-T}(y) \rangle \geq 0 \quad \text{for all } p \in \mathcal{CQ}_{n,d}^{(k)}.$$

The above is the same as

$$\langle q,\, \varpi^{-T}(y)\rangle \geq 0 \quad \text{for all } q \in \mathrm{QM}[g]_{2k} \cap \mathbb{R}[\bar{x}]_d.$$

Therefore, the dual cone of $\mathcal{CQ}_{n,d}^{(k)}$ can be expressed as

$$\left(\mathcal{CQ}_{n,d}^{(k)}\right)^{\star} = \left\{ y \in \mathbb{R}^{\overline{\mathbb{N}}_d^n} \ \middle| \ \begin{array}{c} \exists\, w \in \mathbb{R}^{\mathbb{N}_{2k}^{\bar{n}}},\ \varpi^{-T}(y) = w|_d, \\ M_k[w] \succeq 0,\ L_g^{(k)}[w] \succeq 0 \end{array} \right\}. \tag{9.3.21}$$

In the above, the localizing matrix $L_g^{(k)}[w]$ is given as in (2.5.14) for the polynomial tuple $g$ in (9.3.17). We have the following approximation result for both COP and CP cones.

**Theorem 9.3.4.** *Let $\mathcal{CQ}_{n,d}^{(k)}$ be as above, then*

$$\boxed{\begin{array}{c} \mathrm{int}\,(\mathcal{COP}_{n,d}) \subseteq \bigcup_{k \geq \frac{d}{2}} \mathcal{CQ}_{n,d}^{(k)} \subseteq \mathcal{COP}_{n,d}, \\[2mm] \bigcap_{k \geq \frac{d}{2}} \left(\mathcal{CQ}_{n,d}^{(k)}\right)^{\star} = \mathcal{CP}_{n,d}. \end{array}} \tag{9.3.22}$$

The proof for the above theorem can be done by a standard argument with duality theory and Putinar's Positivstellensatz. So, it is left as an exercise. The linear map tm in (11.5.11) gives an isomorphism from the symmetric tensor space $\mathrm{S}^d(\mathbb{R}^n)$ to the homogeneous moment space $\mathbb{R}^{\overline{\mathbb{N}}_d^n}$. Therefore, all the above results can be equivalently expressed for COP and CP tensor cones $\mathcal{COP}_n^{\otimes d}$, $\mathcal{CP}_n^{\otimes d}$ (see §11.5.2 for the details).

### 9.3.3 ▪ Memberships for COP and CP cones

A form $f$ is copositive if and only if $f \geq 0$ on the simplex $\Delta_n$, or equivalently, $\varpi(f) \geq 0$ on $\overline{\Delta}_n$. The interior of the copositive cone $\mathcal{COP}_{n,d}$ is contained in the relaxations $\mathcal{PH}_{n,d}^{(k)}$, $\mathcal{PS}_{n,d}^{(k)}$, $\mathcal{CQ}_{n,d}^{(k)}$ for some power $k$. Observe that $f \in \mathcal{COP}_{n,d}$ if and only if the optimal value of

$$\begin{cases} \min & f(x_1,\dots,x_{\bar{n}}, 1 - e^T\bar{x}) \\ s.t. & x_1 \geq 0,\dots,x_{\bar{n}} \geq 0, 1 - e^T\bar{x} \geq 0 \end{cases} \tag{9.3.23}$$

is nonnegative. The method in §6.3 (also see §8.2.1) can be applied to solve (9.3.23). We refer to [239] for how to check copositive polynomials.

The cone $\mathcal{CP}_{n,d}$ is characterized in Theorem 9.2.4. An htms $y \in \mathbb{R}^{\overline{\mathbb{N}}_d^n}$ belongs to $\mathcal{CP}_{n,d}$ if and only if $y$ admits a representing measure supported in $\Delta_n$. By Lemma 9.3.3, we have $y \in \mathcal{CP}_{n,d}$ if and only if $\varpi^{-T}(y) \in \mathscr{R}_d(\overline{\Delta}_n)$. So, Algorithm 8.2.4 can be applied to check memberships for the cone $\mathscr{R}_d(\overline{\Delta}_n)$. For each $\alpha = (\alpha_1,\dots,\alpha_{n-1}) \in \mathbb{N}_d^{\bar{n}}$, the entry $\left(\varpi^{-T}(y)\right)_\alpha$ can be given as

$$\left(\varpi^{-T}(y)\right)_\alpha = \langle \bar{x}^\alpha, \varpi^{-T}(y)\rangle = \langle \varpi^{-1}(\bar{x}^\alpha), y\rangle = \langle \bar{x}^\alpha(e^Tx)^{d-|\alpha|}, y\rangle.$$

For instance, when $n = 3$ and $d = 3$, the entries $\left(\varpi^{-T}(y)\right)_\alpha$ are given as follows:

$$
\begin{aligned}
\left(\varpi^{-T}(y)\right)_{00} &= y_{300} + 3y_{210} + 3y_{201} + 3y_{120} + 6y_{111} \\
&\quad + 3y_{102} + y_{030} + 3y_{021} + 3y_{012} + y_{003}, \\
\left(\varpi^{-T}(y)\right)_{10} &= y_{300} + 2y_{210} + 2y_{021} + y_{120} + 2y_{111} + y_{102}, \\
\left(\varpi^{-T}(y)\right)_{01} &= y_{210} + 2y_{120} + 2y_{111} + y_{030} + 2y_{021} + y_{012}, \\
\left(\varpi^{-T}(y)\right)_{20} &= y_{300} + y_{210} + y_{201}, \\
\left(\varpi^{-T}(y)\right)_{11} &= y_{210} + y_{120} + y_{111}, \\
\left(\varpi^{-T}(y)\right)_{02} &= y_{120} + y_{030} + y_{021}, \\
\left(\varpi^{-T}(y)\right)_{30} &= y_{300}, \\
\left(\varpi^{-T}(y)\right)_{21} &= y_{210}, \\
\left(\varpi^{-T}(y)\right)_{12} &= y_{120}, \\
\left(\varpi^{-T}(y)\right)_{03} &= y_{030}.
\end{aligned}
$$

The above leads to the following algorithm for checking memberships in $\mathcal{CP}_{n,d}$.

**Algorithm 9.3.5.** *For given $y \in \mathbb{R}^{\overline{\mathbb{N}}_d^n}$, let $d_1 := \lceil \frac{d+1}{2} \rceil$ and do the following:*

**Step 0** *Choose a generic $R \in \Sigma_{n,2d_1}$ and let $k := d_1$.*

**Step 1** *Solve the following moment optimization:*

$$
\begin{cases}
\min & \langle R, w \rangle \\
s.t. & w_\alpha = \langle \bar{x}^\alpha (e^T x)^{d-|\alpha|}, y \rangle \quad \text{for each } \alpha \in \overline{\mathbb{N}}_d^{n-1}, \\
& L_{g_i}^{(k)}[w] \succeq 0 \; (i = 1, \ldots, n+1), \\
& M_k[w] \succeq 0, \; w \in \mathbb{R}^{\mathbb{N}_d^n}.
\end{cases}
\tag{9.3.24}
$$

*If (9.3.24) is infeasible, output that $y$ is not CP and stop. If (9.3.24) is feasible, solve it for a minimizer $w^*$. Let $t := \lceil \frac{d}{2} \rceil$.*

**Step 2** *Check whether* $\operatorname{rank} M_t[w^*] = \operatorname{rank} M_{t-1}[w^*]$ *holds or not. If it does, go to Step 3; otherwise, go to Step 4.*

**Step 3** *Apply the method in §2.7.2 to compute the $r$-atomic measure $\mu = \lambda_1 \delta_{u_1} + \cdots + \lambda_r \delta_{u_r}$ for the truncation $w^*|_{2t}$, where $r = \operatorname{rank} M_t[w^*]$, each $u_i \in \overline{\Delta}_n$, and $\lambda_i > 0$; then output the CP decomposition for $y$ as*

$$
y = \lambda_1 \Big[ \begin{pmatrix} u_1 \\ \tau_1 \end{pmatrix} \Big]_d^{\text{hom}} + \cdots + \lambda_r \Big[ \begin{pmatrix} u_r \\ \tau_r \end{pmatrix} \Big]_d^{\text{hom}},
\tag{9.3.25}
$$

*where each $\tau_i = 1 - e^T u_i$.*

**Step 4** *If $t < k$, set $t := t + 1$ and go to Step 2; otherwise, set $k := k + 1$ and go to Step 1.*

The convergence properties for Algorithm 9.3.5 are the same as that for Algorithm 8.2.4. We refer to §8.2.2 for the details.

### 9.3.4 ▪ Exercises

**Exercise 9.3.1.** *Apply Algorithm 9.3.5 to determine whether or not the following matrices are CP:*

$$\begin{bmatrix} 6 & 4 & 1 & 2 & 2 \\ 4 & 5 & 0 & 1 & 3 \\ 1 & 0 & 3 & 1 & 2 \\ 2 & 1 & 1 & 1 & 1 \\ 2 & 3 & 2 & 1 & 5 \end{bmatrix}, \quad \begin{bmatrix} 2 & 1 & 0 & 0 & 0 \\ 1 & 2 & 1 & 0 & 0 \\ 0 & 1 & 2 & 2 & 2 \\ 0 & 0 & 2 & 3 & 3 \\ 0 & 0 & 2 & 3 & 4 \end{bmatrix}, \quad \begin{bmatrix} 1 & 1 & 2 & 3 & 4 \\ 1 & 1 & 3 & 2 & 3 \\ 2 & 3 & 3 & 3 & 3 \\ 3 & 3 & 3 & 2 & 4 \\ 4 & 3 & 3 & 4 & 5 \end{bmatrix}.$$

**Exercise 9.3.2.** *Give an explicit matrix $A \in \mathcal{PS}_{5,2}^{(1)}$ but $A \notin \mathcal{PH}_{5,2}^{(1)}$.*

**Exercise 9.3.3.** *For the form $f = x_1^2 + x_2^2 + x_3^2 - 1.99 x_1(x_2 + x_3) + 2x_2 x_3$, find the smallest $k$ such that $f \in \mathcal{PH}_{3,2}^{(k)}$.*

**Exercise 9.3.4.** *Let $P$ be the copositive matrix in (9.1.2). Find the smallest $k$ such that $x^T P x \in \mathcal{PS}_{7,2}^{(k)}$.*

**Exercise 9.3.5.** *Show that all $\mathcal{PH}_{n,d}^{(k)}$, $\mathcal{PS}_{n,d}^{(k)}$, $\mathcal{CQ}_{n,d}^{(k)}$ are proper cones.*

**Exercise 9.3.6.** *Prove the expressions (9.3.8) and (9.3.12). (Hint: use the relations (9.3.7) and (9.3.10).)*

**Exercise 9.3.7.** *Prove Lemma 9.3.3. Use it to prove the decomposition (9.3.25) holds in Step 3 of Algorithm 9.3.5.*

**Exercise 9.3.8.** *Prove Theorems 9.3.2 and 9.3.4.*

## 9.4 ▪ Optimization with COP and CP cones

For every linear functional $\ell$ on $\mathbb{R}^{\overline{\mathbb{N}}_d^n}$, there exists a form $p \in \mathbb{R}[x]_d^{\text{hom}}$ such that $\ell(y) = \langle p, y \rangle$ for all $y$. Therefore, a linear conic optimization problem with the cone $\mathcal{CP}_{n,d}$ can be expressed as (see §8.3 for more details)

$$\begin{cases} \min & \langle f_0, y \rangle \\ s.t. & \langle f_i, y \rangle = b_i, \ i = 1, \ldots, m, \\ & y \in \mathcal{CP}_{n,d}, \end{cases} \tag{9.4.1}$$

for given forms $f_0, f_1, \ldots, f_m \in \mathbb{R}[x]_d^{\text{hom}}$. Its dual optimization is

$$\begin{cases} \max & b_1 \lambda_1 + \cdots + b_m \lambda_m \\ s.t. & f_0 - \lambda_1 f_1 - \cdots - \lambda_m f_m \in \mathcal{COP}_{n,d}. \end{cases} \tag{9.4.2}$$

Many hard optimization problems can be formulated in the form (9.4.1) or (9.4.2). For instance, the CP matrix completion problem (see [334]) can be expressed in the form (9.4.1). Suppose a matrix $X = (X_{ij}) \in \mathcal{S}^n$ is partially given, say, the entries $X_{ij} = v_{ij}$ are known for $(i,j) \in \Omega$, for a given label set $\Omega$, but the entries $X_{ij}$ for $(i,j) \notin \Omega$ are not known. One wonders whether or not there exist values $X_{ij}$ ($(i,j) \notin \Omega$) such that $X$ is CP. When this is possible, one also wishes to optimize about unknown entries $X_{ij}$, say, one wants the sum of unknown entries to be minimum. The matrix $X$ can be represented by an htms of degree 2, i.e., $X$ is represented by $y \in \mathbb{R}^{\overline{\mathbb{N}}_2^n}$ such that $y_{e_i+e_j} = X_{ij}$. Note that $\langle x_i x_j, y \rangle = y_{e_i+e_j}$. This CP matrix completion problem can be equivalently expressed as

$$
\begin{cases}
\min & \sum\limits_{(i,j) \notin \Omega} \langle x_i x_j, y \rangle \\
s.t. & \langle x_i x_j, y \rangle = v_{ij} \text{ for each } (i,j) \in \Omega, \\
& y \in \mathcal{CP}_{n,2}.
\end{cases}
$$

Linear conic optimization with COP or CP cones has broad applications. We refer to [35, 88] for surveys about relevant work. In the following, we present some classical applications.

### 9.4.1 ▪ Nonconvex Quadratic Programs

Nonconvex quadratic programs (QPs) can be formulated as linear CP optimization as in [42]. Consider the QP:

$$
\begin{cases}
\vartheta_{qp} := \min & x^T H x + 2h^T x \\
s.t. & f_i^T x = b_i, \ i = 1, \ldots, m, \\
& x \in \mathbb{R}^n_+,
\end{cases}
\tag{9.4.3}
$$

for given $H \in \mathcal{S}^n$, $h, f_i \in \mathbb{R}^n$. For convenience, we denote

$$
G = \begin{bmatrix} 0 & h^T \\ h & H \end{bmatrix}, \quad A_i = \begin{bmatrix} -b_i \\ f_i \end{bmatrix} \begin{bmatrix} -b_i \\ f_i \end{bmatrix}^T, \quad i = 1, \ldots, m.
$$

Then (9.4.3) can be equivalently written as

$$
\begin{cases}
\min & G \bullet Y \\
s.t. & A_i \bullet Y = 0, \ i = 1, \ldots, m, \\
& Y = \begin{bmatrix} 1 & x^T \\ x & xx^T \end{bmatrix} \in \mathcal{CP}_{n+1}.
\end{cases}
\tag{9.4.4}
$$

Interestingly, the above is equivalent to the following CP optimization:

$$
\begin{cases}
\vartheta_{cp} := \min & G \bullet Y \\
s.t. & A_i \bullet Y = 0, \ i = 1, \ldots, m, \\
& Y = \begin{bmatrix} 1 & x^T \\ x & X \end{bmatrix} \in \mathcal{CP}_{n+1}.
\end{cases}
\tag{9.4.5}
$$

It is clear that $\vartheta_{cp} \le \vartheta_{qp}$. The following result appeared in [42].

**Theorem 9.4.1.** ([42]) *The QP (9.4.3) is equivalent to the CP optimization (9.4.5) in the following senses:*

(i) *The QP (9.4.3) is feasible if and only if (9.4.5) is feasible.*

(ii) *They have the same optimal value, i.e., $\vartheta_{qp} = \vartheta_{cp}$.*

(iii) *If $Y$ is a minimizer of (9.4.5) and has the CP decomposition*

$$
Y = \begin{bmatrix} \tau_1 \\ v_1 \end{bmatrix} \begin{bmatrix} \tau_1 \\ v_1 \end{bmatrix}^T + \cdots + \begin{bmatrix} \tau_r \\ v_r \end{bmatrix} \begin{bmatrix} \tau_r \\ v_r \end{bmatrix}^T, \quad \tau_i \ge 0, \ v_i \in \mathbb{R}^n_+,
\tag{9.4.6}
$$

*then for each $\tau_i > 0$, the vector $v_i/\tau_i$ is a minimizer for (9.4.3).*

***Proof.*** (i) If (9.4.3) is feasible, then (9.4.5) must be feasible since (9.4.5) is a relaxation of (9.4.3). If (9.4.5) is feasible, say, $Y$ is a feasible point, then $Y$ has a CP decomposition like (9.4.6). Since the $(1,1)$-entry of $Y$ is 1, at least one of $\tau_i$ is positive, say, $\tau_1 > 0$. Since each $A_i$ is psd, we have

$$0 = \begin{bmatrix} \tau_1 \\ v_1 \end{bmatrix}^T A_i \begin{bmatrix} \tau_1 \\ v_1 \end{bmatrix} = (-b_i\tau_1 + f_i^T v_1)^2 = \tau_1^2(-b_i + f_i^T v_1/\tau_1)^2.$$

This implies that $v_1/\tau_1$ is a feasible point for (9.4.3).

(ii) In the item (i), we already know that (9.4.3) and (9.4.5) are simultaneously feasible or infeasible. When they are infeasible, both optimal values are $+\infty$, so they are equal. So we only need to consider the case that they are both feasible. Since (9.4.5) is a relaxation of (9.4.3), it always holds that $\vartheta_{cp} \leq \vartheta_{qp}$.

First, we consider the case that $\vartheta_{cp} > -\infty$. The relaxation (9.4.5) is bounded below, so there is no decreasing ray. Hence, for every $X \in \mathcal{CP}_n$ satisfying $f_i^T X f_i = 0$ ($i \in [m]$), we must have $H \bullet X \geq 0$. (Otherwise, if such $X$ exists, then (9.4.5) must be unbounded below.) For every $\epsilon > 0$, there exists a feasible point $Y$ such that $G \bullet Y \leq \vartheta_{cp} + \epsilon$. Suppose $Y$ has the CP decomposition like (9.4.6). Without loss of generality, assume $\tau_j > 0$ for $j = 1, \ldots, t$ and $\tau_j = 0$ for $j > t$. Then

$$Y = \tau_1 \begin{bmatrix} 1 \\ u_1 \end{bmatrix} \begin{bmatrix} 1 \\ u_1 \end{bmatrix}^T + \cdots + \tau_t \begin{bmatrix} 1 \\ u_t \end{bmatrix} \begin{bmatrix} 1 \\ u_t \end{bmatrix}^T + \begin{bmatrix} 0 & 0 \\ 0 & U \end{bmatrix}, \tag{9.4.7}$$

where each $u_j = v_j/\tau_j$ and $U = u_{t+1}u_{t+1}^T + \cdots + u_r u_r^T$. The constraints in (9.4.5) imply that each $u_j$ ($j \leq t$) must be a feasible point for (9.4.3) and $\tau_1 + \cdots + \tau_t = 1$. Note that each $f_i^T U f_i = 0$. Moreover, we also have $H \bullet U \geq 0$. This is because if otherwise $H \bullet U < 0$, then it gives a decreasing ray for (9.4.5), which is impossible for the case $\vartheta_{cp} > -\infty$. Let $q$ denote the quadratic objective function in (9.4.3). Without loss of generality, one can assume that

$$q(u_1) = \min\{q(u_1), \ldots, q(u_t)\}.$$

Then, we can see that

$$\epsilon + \vartheta_{cp} \geq G \bullet Y = (\tau_1 q(u_1) + \cdots + \tau_t q(u_t)) + H \bullet U \geq q(u_1) \geq \vartheta_{qp}.$$

Since $\epsilon > 0$ can be arbitrary, we get $\vartheta_{cp} \geq \vartheta_{qp}$ and hence $\vartheta_{cp} = \vartheta_{qp}$.

Second, we consider the case that $\vartheta_{cp} = -\infty$, i.e., (9.4.5) is unbounded below. For every $M > 0$, there exists $Y \in \mathcal{CP}_{n+1}$ that is feasible for (9.4.5) such that $G \bullet Y < -M$. Since it is CP, the matrix $Y$ has a CP decomposition as in (9.4.7) for $\tau_1, \ldots, \tau_t > 0$. We prove $\vartheta_{qp} = \vartheta_{cp}$ separately in the following two cases.

**Case I:** $H \bullet U < 0$. The matrix $W := \begin{bmatrix} 0 & 0 \\ 0 & U \end{bmatrix} \in \mathcal{CP}_{n+1}$ is a decreasing ray for (9.4.5), i.e.,

$$G \bullet W < 0, \quad A_i \bullet W = 0 \, (i \in [m]).$$

Since $U$ is CP, one can write that

$$U = z_1 z_1^T + \cdots + z_r z_r^T, \quad z_1, \ldots, z_r \in \mathbb{R}_+^n.$$

Then, at least one of $z_j^T H z_j$ is negative, say, $z_1^T H z_1 < 0$. Note that $f_i^T z_1 = 0$ for all $i$. Since (9.4.3) is feasible, this implies that (9.4.3) is unbounded below, in the direction of $z_1$, so $\vartheta_{qp} = -\infty$ and hence $\vartheta_{qp} = \vartheta_{cp}$.

**Case II:** $H \bullet U \geq 0$. Note that all $u_1, \ldots, u_t$ are feasible for (9.4.3) and $\tau_1 + \cdots + \tau_t = 1$. Without loss of generality, we can still assume that $q(u_1)$ is the smallest among $q(u_1), \ldots, q(u_t)$, then

$$
\begin{aligned}
-M > G \bullet Y &= (\tau_1 q(u_1) + \cdots + \tau_t q(u_t)) + H \bullet U \\
&\geq (\tau_1 + \cdots + \tau_t) q(u_1) = q(u_1).
\end{aligned}
$$

Hence, there exists a feasible point $u_1$ such that $q(u_1) < -M$. Since $M$ can be arbitrarily large, this implies that $\vartheta_{qp} = -\infty$ and hence $\vartheta_{cp} = \vartheta_{qp}$.

(iii) Suppose $Y$ is a minimizer of (9.4.5) and it has the CP decomposition (9.4.6). Since each $A_i \succeq 0$, we have $\| - b_i \tau_j + f_i^T v_j \|^2 = 0$ for all $i$. The CP decomposition for $Y$ as in (9.4.6) can be written equivalently as (9.4.7), where each $\tau_j > 0$ for $j \leq t$ and $\tau_j = 0$ for $j > t$. In the above, we already know that each $u_j := v_j / \tau_j$ (for $j \leq t$) is a feasible point for (9.4.3) and $H \bullet U \geq 0$. Note the relations

$$
\begin{aligned}
&\vartheta_{cp} = \sum_{j=1}^{t} \tau_j q(u_j) + H \bullet U \geq \sum_{j=1}^{t} \tau_j q(u_j), \\
&\tau_1 + \cdots + \tau_t = 1, \ \tau_1 > 0, \ldots, \tau_t > 0, \\
&q(u_j) \geq \vartheta_{qp} = \vartheta_{cp} \quad \text{for } j = 1, \ldots, t,
\end{aligned}
$$

where $\vartheta_{qp} = \vartheta_{cp}$ follows from the item (ii). The above implies that each $u_j$ is a minimizer for (9.4.3).  $\square$

## 9.4.2 ▪ Some combinatorial optimization problems

The stability number of a graph can be expressed in COP or CP optimization. Let $G = (V, E)$ be a graph, with $n = |V|$. Its *stability number*, for which we denote $\alpha(G)$, is the maximum cardinality of the subset $T$ of $V$ such that any two vertices in $T$ are not connected by an edge. Such a set $T$ is also called an *independent* (or *stable*) set. It was shown by Motzkin and Straus [198] that

$$
\frac{1}{\alpha(G)} = \min_{x \in \Delta_n} x^T (I_n + A_G) x, \tag{9.4.8}
$$

where $A_G$ denotes the adjacency matrix of the graph $G$, i.e., the $(i,j)$th entry equals one if $(i,j)$ is an edge for $G$ and otherwise it is zero. As shown in [68], we can also express the stability number as

$$
\alpha(G) = \min \left\{ \lambda : \lambda(\mathbf{1}_{n \times n} + A_G) - \mathbf{1}_{n \times n} \in \mathcal{COP}_n \right\}. \tag{9.4.9}
$$

Here, $\mathbf{1}_{n \times n}$ denotes the $n$-by-$n$ matrix of all ones. By the duality, it also holds that

$$
\alpha(G) = \max \left\{ X \bullet \mathbf{1}_{n \times n} : \langle I + A_G, X \rangle = 1, X \in \mathcal{CP}_n \right\}. \tag{9.4.10}
$$

The *chromatic number* of a graph $G$, denoted as $\chi(G)$, is the smallest number of colors needed to color the vertices so that no two adjacent ones share the same color. As shown by Gvozdenović and Laurent [109], $\chi(G)$ equals the optimal value of

$$
\left\{
\begin{aligned}
&\max \quad y \\
&s.t. \quad \tfrac{1}{n^2}(t - y)ee^T + z\left( n(I + A_{G_t}) - ee^T \right) \in \mathcal{COP}_{nt}, \\
&\qquad t = 1, \ldots, n, \ y, z \in \mathbb{R}.
\end{aligned}
\right. \tag{9.4.11}
$$

In the above, $e$ is the vector of all ones and $A_{G_t}$ denotes the adjacency matrix of $G_t$, which is the Cartesian product of $K_t$ (the complete graph of $t$ vertices) and $G$. Equivalently, the vertex set of $G_t$ is

$$[t] \times V := \{(i,j) : i \in [t], j \in V\}$$

and $((i_1, j_1), (i_2, j_2))$ is an edge of $G_t$ if and only if $i_1 \neq i_2, j_1 = j_2$ or $i_1 = i_2, (j_1, j_2) \in E$. The cardinality of the vertex set of $G_t$ is $nt$.

The quadratic assignment problem (QAP) is a traditionally hard question in facility localization theory, well known for its computational challenge. Let $\Theta_n$ denote the set of $n$-by-$n$ permutation matrices. For given $n$-by-$n$ matrices $A, B, C$, the QAP is to solve the quadratic optimization

$$\begin{cases} \min & \langle AXB + C, X \rangle \\ s.t. & X \in \Theta_n. \end{cases} \tag{9.4.12}$$

As shown by Povh and Rendl [265], the QAP is equivalent to

$$\begin{cases} \min & \langle B^T \otimes A + \mathrm{diag}(\mathrm{vec}(C)), Y \rangle \\ s.t. & Y^{11} + \cdots + Y^{nn} = I_n, \\ & \langle I_n, Y^{ij} \rangle = \delta_{ij} \, (\forall i, j), \\ & 1_n^T Y 1_n = n^2, \\ & Y = (Y^{ij})_{1 \le i,j \le n} \in \mathcal{CP}_{n^2}. \end{cases} \tag{9.4.13}$$

In the above, $1_n$ denotes the all-one vector of length $n$, the variable $Y$ is a block matrix, and $Y^{ij}$ is the $(i,j)$th $n$-by-$n$ block of $Y$.

The min-cut problem (MCP) can also be equivalently expressed in CP optimization. For a graph $G = (V, E)$ of $n$ vertices, each edge $(i,j)$ is assigned a weight $a_{ij} \ge 0$. The MCP is to partition vertices of $G$ into subsets $S_1, S_2, S_3$ with given cardinalities $n_1, n_2, n_3$, where $n_1 + n_2 + n_3 = n$, such that the total weight of edges between $S_1, S_2$ is minimal. A partition $(S_1, S_2, S_3)$ of $V$ can be represented by the $n \times 3$ matrix $X = (X_{ij})$ such that

$$X_{ij} = \begin{cases} 1, & i \in S_j, \\ 0, & i \notin S_j. \end{cases}$$

The partition $(S_1, S_2, S_3)$ represented by $X$ has the prescribed sizes $n_1, n_2, n_3$ if and only if

$$X^T X = \begin{bmatrix} n_1 & 0 & 0 \\ 0 & n_2 & 0 \\ 0 & 0 & n_3 \end{bmatrix}, \quad X 1_3 = 1_n, \quad X \ge 0.$$

Let $A = (a_{ij})$ be the weight matrix. The cut between $S_1$ and $S_2$ is

$$cut(S_1, S_2) := X(:, 1)^T A X(:, 2) = \frac{1}{2} \langle X, AXB \rangle,$$

where the matrix $B$ is given as

$$B = \begin{bmatrix} 0 & 1 & 0 \\ 1 & 0 & 0 \\ 0 & 0 & 0 \end{bmatrix}.$$

The quadratic optimization formulation for the MCP is

$$\begin{cases} \min & \frac{1}{2} \langle AXB, X \rangle \\ s.t. & X^T X = \begin{bmatrix} n_1 & 0 & 0 \\ 0 & n_2 & 0 \\ 0 & 0 & n_3 \end{bmatrix}, \\ & X 1_3 = 1_n, \, X \ge 0. \end{cases} \tag{9.4.14}$$

Observe that the matrix variable $X$ satisfies the equations

$$
\begin{aligned}
(e_i^T X \mathbf{1}_3)^2 &= 1 \ (1 \le i \le n), \\
(\mathbf{1}_n^T X e_j)(e_i^T X \mathbf{1}_3) &= n_j \ (1 \le i \le n, \, 1 \le j \le 3), \\
(\mathbf{1}_n^T X e_j)(\mathbf{1}_n^T X e_j) &= n_i n_j \ (1 \le i \le j \le 3).
\end{aligned}
$$

For the nonconvex quadratic optimization (9.4.14), a CP relaxation is to consider the matrix variable

$$
Y = \begin{bmatrix} X(:,1) \\ X(:,2) \\ X(:,3) \end{bmatrix} \begin{bmatrix} X(:,1) \\ X(:,2) \\ X(:,3) \end{bmatrix}^T .
$$

It is shown by Povh and Rendl [266] that the MCP can be equivalently expressed as the following CP conic optimization problem:

$$
\begin{cases}
\min & \frac{1}{2}\langle B \otimes A, Y \rangle \\
s.t. & \langle B_{ij} \otimes I_n, Y \rangle = n_i \delta_{ij} \ (1 \le i \le j \le 3), \\
& \langle \mathbf{1}_{3\times 3} \otimes e_i e_i^T, Y \rangle = 1 \ (1 \le i \le n), \\
& \langle V_i \otimes W_j^T, Y \rangle = n_i \ (1 \le i \le 3, 1 \le j \le n), \\
& \langle B_{ij} \otimes \mathbf{1}_{n \times n}, Y \rangle = n_i n_j \ (1 \le i \le j \le 3), \\
& Y \in \mathcal{CP}_{3n}.
\end{cases}
\tag{9.4.15}
$$

In the above, $B_{ij} = \frac{1}{2}(e_i e_j^T + e_j e_i^T) \in \mathbb{R}^{3\times 3}$, $\delta_{ij} = 0$ for $i \ne j$ and $\delta_{ij} = 1$ for $i = j$, and $V_i = e_i \mathbf{1}_3^T \in \mathbb{R}^{3\times 3}$ and $W_j = e_j \mathbf{1}_n^T \in \mathbb{R}^{n \times n}$ for $1 \le i \le 3$ and $1 \le j \le n$. The symbol $\otimes$ denotes the Kronecker product.

### 9.4.3 ▪ A Moment-SOS hierarchy of relaxations

Let $\varpi$ be the linear map as in (9.3.15) and let $g$ be the polynomial tuple as in (9.3.17). For $p \in \mathbb{R}[x]_d^{\mathrm{hom}}$ and $y \in \mathbb{R}^{\overline{\mathbb{N}}_d^n}$, it holds that

$$
\langle p, y \rangle = \langle \varpi(p), \varpi^{-T}(y) \rangle.
$$

In particular, Lemma 9.3.3 asserts that

$$
y \in \mathcal{CP}_{n,d} \iff \varpi^{-T}(y) \in \mathscr{R}_d(\overline{\Delta}_n).
$$

Therefore, the CP conic optimization (9.4.1) is equivalent to

$$
\begin{cases}
\min & \langle \varpi(f_0), z \rangle \\
s.t. & \langle \varpi(f_i), z \rangle = b_i \ (i = 1, \dots, m), \\
& z \in \mathscr{R}_d(\overline{\Delta}_n).
\end{cases}
\tag{9.4.16}
$$

If $z^*$ is an optimizer of (9.4.16), then $y^* := \varpi^T(z^*)$ is an optimizer of (9.4.1), and vice versa. Similarly, the copositive optimization (9.4.2) is equivalent to

$$
\begin{cases}
\max & b_1 \lambda_1 + \cdots + b_m \lambda_m \\
s.t. & \varpi(f_0) - \lambda_1 \varpi(f_1) - \cdots - \lambda_m \varpi(f_m) \in \mathscr{P}_d(\overline{\Delta}_n).
\end{cases}
\tag{9.4.17}
$$

Algorithm 8.3.1 can be applied to solve (9.4.16)-(9.4.17). The $k$th order moment relaxation for (9.4.16) is

$$
\begin{cases}
\min & \langle \varpi(f_0), w \rangle \\
s.t. & \langle \varpi(f_i), w \rangle = b_i, \ i = 1, \dots, m, \\
& L_{g_j}^{(k)}[w] \succeq 0, \ j = 1, \dots, n+1, \\
& M_k[w] \succeq 0, \ w \in \mathbb{R}^{\mathbb{N}_{2k}^{\bar{n}}}.
\end{cases}
\tag{9.4.18}
$$

In the above, the polynomials $g_j$ are given as in (9.3.17), i.e.,

$$(g_1, \ldots, g_{n+1}) = (x_1, \ldots, x_{n-1}, 1 - e^T \bar{x}, 1 - \|\bar{x}\|^2).$$

The $k$th order SOS relaxation for the dual optimization (9.4.17) is

$$\begin{cases} \max & b_1 \lambda_1 + \cdots + b_m \lambda_m \\ \text{s.t.} & \varpi(f_0) - \lambda_1 \varpi(f_1) - \cdots - \lambda_m \varpi(f_m) \in \text{QM}[g]_{2k}. \end{cases} \tag{9.4.19}$$

When $k$ increases, we get the following algorithm.

**Algorithm 9.4.2.** *For given forms $f_0, \ldots, f_m \in \mathbb{R}[x]_d^{\text{hom}}$, let $k := \lceil \frac{d}{2} \rceil$ and do the following:*

**Step 1** *Solve the Moment-SOS relaxation pair (9.4.18)-(9.4.19). If (9.4.18) is infeasible, stop and output that (9.4.1) is infeasible; if (9.4.19) is unbounded above, stop and output that (9.4.2) is unbounded above; otherwise, compute an optimizer $w^*$ for (9.4.18) and a maximizer $\lambda^* = (\lambda_1^*, \ldots, \lambda_m^*)$ for (9.4.19), if they exist.*

**Step 2** *Let $z^* := w^*|_d$. If $z^* \in \mathscr{R}_d(\overline{\Delta})$, then $y^* := \varpi^T(z^*)$ is a minimizer of (9.4.1). Moreover, if in addition $\langle \varpi(f_0), w^* \rangle = b^T \lambda^*$, then $\lambda^*$ is a maximizer of (9.4.1). Otherwise, let $k := k + 1$ and go to Step 1.*

In Step 2, Algorithm 9.3.5 can be used to check whether $z^* \in \mathscr{R}_d(\overline{\Delta})$ or not. The convergence properties are summarized in §8.3.2. For neatness, we do not repeat them here.

## 9.4.4 ▪ Exercises

**Exercise 9.4.1.** *([334]) Decide whether or not each of the following matrices can be completed to a full CP matrix (∗ means the entry is not known.)*

$$\begin{bmatrix} * & 4 & 1 & 2 & 2 \\ 4 & * & 0 & 1 & 3 \\ 1 & 0 & * & 1 & 2 \\ 2 & 1 & 1 & * & 1 \\ 2 & 3 & 2 & 1 & * \end{bmatrix}, \quad \begin{bmatrix} 1 & 1 & * & * & 0 \\ 1 & 1 & 1 & * & * \\ * & 1 & 1 & 1 & * \\ * & * & 1 & 1 & 1 \\ 0 & * & * & 1 & 1 \end{bmatrix}, \quad \begin{bmatrix} 1 & 1 & 2 & * & 4 \\ 1 & 1 & 3 & * & 3 \\ 2 & 3 & 3 & 3 & * \\ * & * & 3 & 2 & * \\ 4 & 3 & * & * & * \end{bmatrix}.$$

**Exercise 9.4.2.** *Find scalars $a, b, c$ such that $a + b + c$ is minimum and*

$$\begin{bmatrix} 3 & 2 & 1 & a & b \\ 2 & 3 & 2 & 1 & c \\ 1 & 2 & 3 & 2 & 1 \\ a & 1 & 2 & 3 & 2 \\ b & c & 1 & 2 & 3 \end{bmatrix} \in \mathcal{COP}_5.$$

**Exercise 9.4.3.** *For the partially given matrix (∗ means the entry is unknown)*

$$C = \begin{bmatrix} * & 1 & 2 & 3 & 4 \\ 1 & * & 1 & 2 & 3 \\ 2 & 1 & * & 1 & 2 \\ 3 & 2 & 1 & * & 1 \\ 4 & 3 & 2 & 1 & * \end{bmatrix},$$

*find the smallest trace of $C$ such that $C$ is a CP matrix.*

**Exercise 9.4.4.** *The smallest CP matrix greater than or equal to a given matrix $B$ is the CP matrix $X$ such that $X - B \in \mathcal{N}^n$ and the sum of all entries of $X$ is the smallest. Formulate a linear conic optimization problem for how to find the smallest CP matrix for given $B$. Find smallest CP matrices for the two in (9.2.4).*

**Exercise 9.4.5.** *Prove that (9.4.9) and (9.4.10) have the same optimal value. For a given graph $G$, show how to get one of its independent sets of maximum cardinality from an optimizer of (9.4.10).*

**Exercise 9.4.6.** *Show how to get a minimum cut for a given graph $G$ from an optimizer of the CP optimization (9.4.15).*

**Exercise 9.4.7.** *Formulate (9.4.11) equivalently as a linear conic optimization problem with the CP cone for getting the chromatic number $\chi(G)$ of a graph $G$.*

**Exercise 9.4.8.** *For given matrices/vectors $A, b, H, h$, consider the QP*

$$\begin{cases} \min & x^T H x + 2h^T x \\ s.t. & Ax \geq b, \, x \in \mathbb{R}^n. \end{cases}$$

*Express the above QP equivalently as a linear CP optimization problem, in the sense of Theorem 9.4.1.*

**Exercise 9.4.9.** *For given polynomials $f_0, f_1, \ldots, f_m \in \mathbb{R}[x]_d$, give an equivalent linear CP conic optimization formulation for the optimization*

$$\begin{cases} \min & f_0(x) \\ s.t. & f_i(x) \geq 0, \, i = 1, \ldots, m, \\ & x \in \mathbb{R}^n_+. \end{cases}$$

# Part IV

# Matrix Polynomials, Tensors, and Special Topics

# Chapter 10

# Matrix Polynomials, Moments, and Optimization

*Many optimization problems have polynomial matrix inequality constraints. This chapter introduces the topics of psd matrix polynomials, matrix Positivstellensätze, matrix-valued moment problems, polynomial matrix inequality constrained optimization, matrix convexity, and certificates for containments of projections of semialgebraic sets given by polynomial matrix inequalities.*

## 10.1 ▪ PSD and SHS matrix polynomials

For $x := (x_1, \ldots, x_n)$ and integers $\ell, m > 0$, denote the $\ell$-by-$m$ matrix polynomial space

$$\mathbb{R}[x]^{\ell \times m} := \Big\{ P = (P_{ij})_{\substack{1 \leq i \leq \ell \\ 1 \leq j \leq m}} : P_{ij} \in \mathbb{R}[x] \Big\}.$$

For convenience, when $m = 1$, we just simply denote $\mathbb{R}[x]^{\ell} := \mathbb{R}[x]^{\ell \times 1}$, the space of $\ell$-dimensional vector polynomials. For a degree $d$, the truncation $\mathbb{R}[x]_d^{\ell \times m}$ is defined similarly like $\mathbb{R}[x]_d$ by requiring each $P_{ij} \in \mathbb{R}[x]_d$. For the case $\ell = m$, a matrix polynomial $P$ is said to be *symmetric* if $P_{ij} = P_{ji}$ for all $i, j$. Denote the $\ell$-by-$\ell$ symmetric matrix polynomial space

$$\mathcal{S}\mathbb{R}[x]^{\ell \times \ell} := \Big\{ P \in \mathbb{R}[x]^{\ell \times \ell} : P_{ij} = P_{ji} \text{ for all } i, j \Big\}.$$

The truncation $\mathcal{S}\mathbb{R}[x]_d^{\ell \times \ell}$ is defined similarly.

In optimization, we often work with psd matrix polynomials. A matrix polynomial $P \in \mathcal{S}\mathbb{R}[x]^{\ell \times \ell}$ is said to be *positive semidefinite* (psd) on a set $T \subseteq \mathbb{R}^n$ if the evaluation $P(u)$ is a psd matrix (i.e., $P(u) \succeq 0$) for every $u \in T$. When $P$ is psd on $T$, we just simply write that

$$P \succeq 0 \text{ on } T, \quad \text{or} \quad P|_T \succeq 0.$$

In particular, $P$ is said to be a psd matrix polynomial if $P \succeq 0$ on $\mathbb{R}^n$. A basic question is how to check whether a given matrix polynomial is psd or not. We can define SOS type matrix polynomials, similar to the case of scalar polynomials. A symmetric matrix polynomial $A \in \mathcal{S}\mathbb{R}[x]^{\ell \times \ell}$ is said to be a *sum of Hermitian squares* (SHS) if

$$A = B_1^T B_1 + \cdots + B_k^T B_k$$

for some matrix polynomials $B_1, \ldots, B_k \in \mathbb{R}[x]^{\ell \times \ell}$ and for a length $k \in \mathbb{N}$. In the literature, SHS matrix polynomials are also called SOS matrix polynomials. Equivalently, $A$ is SHS if and only if $A = B^T B$ for a single nonsquare matrix polynomial $B \in \mathbb{R}[x]^{s \times \ell}$. Moreover, $A$ being SHS is also equivalent to

$$A = p_1 p_1^T + \cdots + p_N p_N^T$$

for column vector polynomials $p_1, \ldots, p_N \in \mathbb{R}[x]^\ell$. The cone of SHS matrix polynomials is denoted as

$$\Sigma[x]^{\ell \times \ell} := \left\{ \sum_{i=1}^N p_i p_i^T : p_i \in \mathbb{R}[x]^\ell, N \in \mathbb{N} \right\}. \tag{10.1.1}$$

For a degree $d$, the truncation $\Sigma[x]_d^{\ell \times \ell}$ is similarly defined. Obviously, if $A$ is SHS, then $A(x) \succeq 0$ for all $x \in \mathbb{R}^n$.

### 10.1.1 ▪ Some classical results

An especially interesting case is $n = 1$, the case of univariate psd matrix polynomials. It is well known that a scalar univariate polynomial is nonnegative everywhere if and only if it is a sum of two squares (see Theorem 3.1.1). Indeed, this conclusion is also true for the matrix case. This result appeared in [53, 134, 146, 288].

**Theorem 10.1.1.** *For $n = 1$, a matrix polynomial $A \in \mathcal{SR}[x]^{\ell \times \ell}$ is psd on $\mathbb{R}^1$ if and only if $A$ is a sum of two Hermitian squares in $\mathcal{SR}[x]^{\ell \times \ell}$, i.e.,*

$$A = B_1^T B_1 + B_2^T B_2 \tag{10.1.2}$$

*for $B_1, B_2 \in \mathbb{R}[x]^{\ell \times \ell}$. Equivalently, every bivariate homogeneous psd matrix polynomial is a sum of two Hermitian squares.*

We remark that the decomposition of $A$ in (10.1.2) is equivalent to

$$A = b_1 b_1^T + \cdots + b_N b_N^T, \quad b_j \in \mathbb{R}[x]^{\ell \times 1}, \tag{10.1.3}$$

for a length $N \leq 2\ell$, which was shown in [53]. The proof as in [53, §7] can be outlined as follows. We apply induction on $\ell$. For $\ell = 1$, the conclusion is clearly true, since every nonnegative univariate polynomial is a sum of two squares. For induction, assume the conclusion is true for $\ell = k - 1$. Then we show it is also true for $\ell = k$. Let $q := A_{11}$, the $(1,1)$-entry of $A$. It must be a psd univariate polynomial in $\mathbb{R}[x]$. Consider the general case that $q$ is not identically zero. (Otherwise, the first column and first row of $A$ are entirely zero, and the conclusion is clearly true by the induction.) Let $a_1$ be the first column vector of $A$, then

$$qA = a_1 a_1^T + \begin{bmatrix} 0 & 0 \\ 0 & C \end{bmatrix}$$

for a symmetric matrix polynomial $C \in \mathcal{SR}[x]^{(k-1) \times (k-1)}$. Since $A$ is psd, $C$ must also be a psd polynomial by Schur's complement. Applying the induction on $C$, we can get that

$$C = c_1 c_1^T + \cdots + c_{N_1} c_{N_1}^T, \quad c_j \in \mathbb{C}[x]^{(k-1) \times 1},$$

for some length $N_1 \leq 2(k-1)$. Since $q$ is a univariate psd polynomial, it can be factorized as a product of perfect quadratic squares like $(\alpha x + \beta)^2$ and irreducible psd quadratics like $(\alpha x + \beta)^2 + \gamma^2$. By considering each factor of $q$, one can further show that (10.1.3) must hold for some choices of real polynomial vectors $b_1, \ldots, b_N$, with a length $N \leq 2k$. This can be implied by Exercise 10.1.10, which is Lemma 7.5 of [53]. In particular, the upper bound for $N$ in (10.1.3) can be improved to $N \leq \ell + 1$, as shown in [33].

For the case $\ell = 2$, every quadratic psd matrix polynomial must be SHS. This was shown by Calderón [43].

**Theorem 10.1.2.** *([43]) If $F \in \mathcal{S}\mathbb{R}[x]^{2\times 2}$ is a quadratic psd matrix polynomial, then $F$ must be SHS.*

**Proof.** It is sufficient to prove the conclusion when $F$ is homogeneous, because a matrix polynomial is SHS if and only if its homogenization is SHS. Let $\mathscr{Q}$ be the set of all homogeneous psd quadratic matrix polynomials in $\mathcal{S}\mathbb{R}[x]^{2\times 2}$. Note that $\mathscr{Q}$ is a closed pointed convex cone.

First, for every $0 \neq F \in \mathscr{Q}$, we prove that there exists $0 \neq f \in \mathcal{S}\mathbb{R}[x]_1^{2\times 1}$ such that $F - ff^T \in \mathscr{Q}$. The conclusion is clearly true if $F$ is positive definite on $\|x\| = 1$, because one can choose $f = \epsilon x_1 e_1$ for $\epsilon > 0$ sufficiently small. So, we consider the case that $F$ is psd but singular on $\|x\| = 1$. Up to a linear coordinate transformation and congruent matrix transformation, one can write that

$$F = \begin{bmatrix} x_1^2 + y^T A y & x_1(b^T y) + y^T B y \\ x_1(b^T y) + y^T B y & y^T C y \end{bmatrix},$$

where $y := \begin{bmatrix} x_2 & \cdots & x_n \end{bmatrix}^T$ for a vector $b \in \mathbb{R}^{n-1}$ and symmetric matrices $A, B, C \in \mathcal{S}^{n-1}$. Since $F$ is psd, we have $A, C \succeq 0$. If $C = 0$, then $B = 0, b = 0$ and hence the conclusion is true for $f = x_1 e_1$. So, we consider that $C \neq 0$. The determinant of $F$ can be expressed as

$$x_1^2 \Big( y^T C y - (b^T y)^2 \Big) - 2x_1 \Big( b^T y \cdot y^T B y \Big) + \Big( y^T A y \cdot y^T C y - (y^T B y)^2 \Big).$$

It is nonnegative for all real $x_1 \in \mathbb{R}^1$, so

$$\Big( y^T C y - (b^T y)^2 \Big) \Big( y^T A y \cdot y^T C y - (y^T B y)^2 \Big) - \Big( b^T y \cdot y^T B y \Big)^2 \geq 0$$

for all $y$. The above can be equivalently written as

$$y^T C y \cdot \Big( y^T A y \cdot \big( y^T C y - (b^T y)^2 \big) - (y^T B y)^2 \Big) \geq 0.$$

Since $C \neq 0$, the above implies

$$y^T A y \cdot \Big( y^T C y - (b^T y)^2 \Big) - (y^T B y)^2 \geq 0.$$

Since $A \succeq 0$ and $F$ is psd, we can get that

$$\begin{bmatrix} y^T A y & y^T B y \\ y^T B y & y^T C y - (b^T y)^2 \end{bmatrix} \succeq 0,$$

$$F = \begin{bmatrix} y^T A y & y^T B y \\ y^T B y & y^T C y - (b^T y)^2 \end{bmatrix} + \begin{bmatrix} x_1 \\ b^T y \end{bmatrix} \begin{bmatrix} x_1 \\ b^T y \end{bmatrix}^T \succeq \begin{bmatrix} x_1 \\ b^T y \end{bmatrix} \begin{bmatrix} x_1 \\ b^T y \end{bmatrix}^T.$$

So the conclusion is true for $f = \begin{bmatrix} x_1 & b^T y \end{bmatrix}^T$.

Second, let $F$ be a matrix polynomial that gives an extreme ray of the closed pointed convex cone $\mathscr{Q}$. Let $0 \neq f \in \mathbb{R}[x]_1^{2\times 1}$ be such that $F - ff^T \in \mathscr{Q}$, as shown above. The equation

$$F = \frac{1}{2}(F + ff^T) + \frac{1}{2}(F - ff^T)$$

implies that $F$ must be a multiple of $ff^T$. This means that $F \in \Sigma[x]_2^{2\times 2}$.

Third, the cone $\mathscr{Q}$ is pointed, so each point in $\mathscr{Q}$ is a sum of points that give extremal rays (see Exercise 1.2.9). Therefore, we get $\mathscr{Q} = \Sigma_2^{2\times2}$.    $\square$

It is known that every ternary quartic psd form is SOS (see Theorem 2.4.5). However, this conclusion cannot be generalized to the matrix case. For $\ell = 2$ and $n = 3$, not every ternary psd matrix form in $\mathbb{R}[x]_4^{2\times2}$ is SHS, as shown in [53]. For instance, the matrix form

$$P = \begin{bmatrix} x_1^4 + x_1^2 x_2^2 & -2x_1^2 x_2 x_3 \\ -2x_1^2 x_2 x_3 & x_1^2 x_2^2 + x_3^4 \end{bmatrix}$$

is psd on $\mathbb{R}^3$ but it is not SHS. If it were otherwise SHS, then

$$\begin{bmatrix} x_3 \\ x_2 \end{bmatrix}^T P \begin{bmatrix} x_3 \\ x_2 \end{bmatrix} = x_1^4 x_3^2 + x_2^4 x_1^2 + x_3^4 x_2^2 - 3x_1^2 x_2^2 x_3^2$$

is SOS, which is not true (see the Choi-Lam form in §2.4.2). For $\ell \geq 3$, not every psd matrix polynomial is SHS, even if the degree is two. For instance, the following ternary quadratic matrix form

$$G := \begin{bmatrix} x_1^2 + 2x_2^2 & -x_1 x_2 & -x_1 x_3 \\ -x_1 x_2 & x_2^2 + 2x_3^2 & -x_2 x_3 \\ -x_1 x_3 & -x_2 x_3 & x_3^2 + 2x_1^2 \end{bmatrix}$$

is psd everywhere but it is not SHS, as shown in [52].

There is a similar conclusion like Artin's affirmative answer to Hilbert's 17th problem. Every psd matrix polynomial is a sum of Hermitian squares of rational matrix polynomials. This result appeared in [104, 127, 271].

**Theorem 10.1.3.** *A matrix polynomial $F \in S\mathbb{R}[x]^{\ell\times\ell}$ is psd on $\mathbb{R}^n$ if and only if $p^2 F$ is SHS for some $0 \neq p \in \mathbb{R}[x]$.*

## 10.1.2 ▪ Matrix Positivstellensätze

In optimization, we are interested in matrix polynomials that are psd on a given set. For a degree $d$ and a set $K \subseteq \mathbb{R}^n$, denote the cone of psd matrix polynomials

$$\mathscr{P}_d(K)^{\ell\times\ell} := \left\{ P \in S\mathbb{R}[x]_d^{\ell\times\ell} : P|_K \succeq 0 \right\}. \tag{10.1.4}$$

In particular, when $K = \mathbb{R}^n$, we denote that

$$\mathscr{P}_{n,d}^{\ell\times\ell} := \mathscr{P}_d(\mathbb{R}^n)^{\ell\times\ell}. \tag{10.1.5}$$

Clearly, $\mathscr{P}_d(K)^{\ell\times\ell}$ is a closed convex cone. A subset $\mathcal{M} \subseteq S\mathbb{R}[x]^{\ell\times\ell}$ is called a *quadratic module* if it satisfies

$$I_\ell \in \mathcal{M}, \quad \mathcal{M} + \mathcal{M} \subseteq \mathcal{M}, \quad P^T \cdot \mathcal{M} \cdot P \subseteq \mathcal{M} \quad (\forall P \in \mathbb{R}[x]^{\ell\times\ell}).$$

For a set $\mathcal{G}$ of symmetric matrix polynomials, its $\ell$-by-$\ell$ quadratic module is

$$\text{QM}[\mathcal{G}]^{\ell\times\ell} := \left\{ \sum_{i=1}^k P_i^T g_i P_i \;\middle|\; \begin{array}{l} k \in \mathbb{N}, \ g_i \in \{1\} \cup \mathcal{G}, \\ g_i \in S\mathbb{R}[x]^{t_i\times t_i}, \ P_i \in \mathbb{R}[x]^{t_i\times\ell} \end{array} \right\}. \tag{10.1.6}$$

We remark that the polynomials $g_i$ in (10.1.6) may repeat. Note that $\Sigma[x]^{\ell\times\ell} \subseteq \text{QM}[\mathcal{G}]^{\ell\times\ell}$ for all $\mathcal{G}$. In particular, when $\mathcal{G} = \emptyset$,

$$\text{QM}[\emptyset]^{\ell\times\ell} = \Sigma[x]^{\ell\times\ell}.$$

For the case $\ell = 1$, we just simply denote that

$$\mathrm{QM}[\mathcal{G}] := \mathrm{QM}[\mathcal{G}]^{1\times 1}. \tag{10.1.7}$$

When $\mathcal{G} = \{G_1, \ldots, G_m\}$, for convenience, we also write that

$$\mathrm{QM}[G_1, \ldots, G_m]^{\ell\times\ell} := \mathrm{QM}[\mathcal{G}]^{\ell\times\ell}. \tag{10.1.8}$$

A quadratic module $\mathcal{M} \subseteq \mathcal{SR}[x]^{\ell\times\ell}$ is said to be *archimedean* if for every $P \in \mathcal{SR}[x]^{\ell\times\ell}$, there exists a scalar $R > 0$ such that $R \cdot I_\ell - P^T P \in \mathcal{M}$. As shown in [146], a quadratic module $\mathcal{M}$ is archimedean if and only if for some $N \in \mathbb{N}$,

$$(N - \|x\|^2)I_\ell \in \mathcal{M}.$$

A matrix polynomial set $\mathcal{G}$ determines the semialgebraic set

$$S_\mathcal{G} := \big\{ x \in \mathbb{R}^n \mid g(x) \succeq 0 \, \forall g \in \mathcal{G} \big\}. \tag{10.1.9}$$

If $\mathrm{QM}[\mathcal{G}]^{\ell\times\ell}$ is archimedean, the set $S_\mathcal{G}$ must be compact. However, when $S_\mathcal{G}$ is compact, it is not necessarily that $\mathrm{QM}[\mathcal{G}]^{\ell\times\ell}$ is archimedean. This is like the case of scalar polynomials. The matrix version of Putinar's Positivstellensatz (see Theorem 2.6.1) is the following.

**Theorem 10.1.4.** *([297]) Let $\mathcal{G}$ be a set of symmetric matrix polynomials such that $\mathrm{QM}[\mathcal{G}]^{\ell\times\ell}$ is archimedean. If $F \in \mathcal{SR}[x]^{\ell\times\ell}$ is positive definite on $S_G$, then $F \in \mathrm{QM}[\mathcal{G}]^{\ell\times\ell}$.*

In some applications, we need a Positivstellensatz for certifying emptiness of sets given by polynomial matrix inequalities. Consider the set

$$T := \left\{ x \in \mathbb{R}^n \, \middle| \, \begin{array}{ccc} g(x) & \succeq & 0 \, (\forall \, g \in \mathcal{G}), \\ F(x) & \succeq & 0 \end{array} \right\}.$$

Clearly, the set $T = \emptyset$ if and only if $F \not\succeq 0$ on $S_\mathcal{G}$. (The notation $F \not\succeq 0$ on $S_\mathcal{G}$ means that $F(u)$ is not psd for every $u \in S_\mathcal{G}$, or equivalently $F(u)$ has at least one negative eigenvalue for every $u \in S_\mathcal{G}$.) The following is a matrix Positivstellensatz, due to Klep and Schweighofer.

**Theorem 10.1.5.** *([146, Corollary 3.6]) Let $\mathcal{G}$ be a set of symmetric matrix polynomials such that $\mathrm{QM}[\mathcal{G}]^{\ell\times\ell}$ is archimedean. For $F \in \mathcal{SR}[x]^{\ell\times\ell}$, we have that $F \not\succeq 0$ on $S_\mathcal{G}$ if and only if*

$$-I_\ell \in \mathrm{QM}[\mathcal{G} \cup \{F\}]^{\ell\times\ell}. \tag{10.1.10}$$

*In particular, the set $S_\mathcal{G} = \emptyset$ if and only if $-I_\ell \in \mathrm{QM}[\mathcal{G}]^{\ell\times\ell}$.*

For instance, it is straightforward to verify that

$$\begin{bmatrix} 0 & x_1 x_2 + 1 \\ x_1 x_2 + 1 & 0 \end{bmatrix} \not\succeq 0 \quad \text{for all} \quad 1 - x_1^2 - x_2^2 \geq 0.$$

The certificate as in (10.1.10) for the above is

$$
\begin{aligned}
-\begin{bmatrix} 1 & 0 \\ 0 & 1 \end{bmatrix} \; = \; & \begin{bmatrix} 1 & -1 \\ 0 & 0 \end{bmatrix} \begin{bmatrix} 0 & x_1 x_2 + 1 \\ x_1 x_2 + 1 & 0 \end{bmatrix} \begin{bmatrix} 1 & 0 \\ -1 & 0 \end{bmatrix} \\
+ & \begin{bmatrix} 0 & 0 \\ 1 & -1 \end{bmatrix} \begin{bmatrix} 0 & x_1 x_2 + 1 \\ x_1 x_2 + 1 & 0 \end{bmatrix} \begin{bmatrix} 0 & 1 \\ 0 & -1 \end{bmatrix} \\
+ & \Big( (x_1 + x_2)^2 + (1 - x_1^2 - x_2^2) \Big) \begin{bmatrix} 1 & 0 \\ 0 & 1 \end{bmatrix}.
\end{aligned}
$$

In Theorem 10.1.5, the archimedeanness of $\mathrm{QM}[\mathcal{G}]^{\ell \times \ell}$ is required. If it is not archimedean, the conclusion may not hold. For instance, consider that

$$G = \begin{bmatrix} -x_1 & 0 & 0 \\ 0 & -x_2 & 0 \\ 0 & 0 & -(x_1 x_2 + 1) \end{bmatrix}.$$

The set $S_G$ is compact (empty) while $\mathrm{QM}[G]^{\ell \times \ell}$ is not archimedean. As pointed out in [146], one can show that $-I_3 \notin \mathrm{QM}[G]^{3 \times 3}$; otherwise, it would result in the contradiction $-1 \in \mathrm{QM}[-x_1, -x_2, -(x_1 x_2 + 1)]$. However, for the univariate case $n = 1$, the archimedeanness of $\mathrm{QM}[G]^{\ell \times \ell}$ in Theorem 10.1.5 can be dropped.

**Theorem 10.1.6.** *([146]) When $n = 1$, for $F \in \mathcal{SR}[x]^{\ell \times \ell}$, we have $F \not\succeq 0$ on $\mathbb{R}$ if and only if $-I_\ell \in \mathrm{QM}[F]^{\ell \times \ell}$.*

For instance, it is obvious that

$$\begin{bmatrix} 0 & x^2 + 1 \\ x^2 + 1 & 0 \end{bmatrix} \not\succeq 0 \quad \text{for all} \quad x \in \mathbb{R}.$$

The corresponding Positivstellensatz certificate is

$$
-\begin{bmatrix} 1 & 0 \\ 0 & 1 \end{bmatrix} = (1 + 2x^2) \begin{bmatrix} 1 & 0 \\ 0 & 1 \end{bmatrix} + \begin{bmatrix} 1 & -1 \\ 0 & 0 \end{bmatrix} \begin{bmatrix} 0 & x^2 + 1 \\ x^2 + 1 & 0 \end{bmatrix} \begin{bmatrix} 1 & 0 \\ -1 & 0 \end{bmatrix}
$$
$$
+ \begin{bmatrix} 0 & 0 \\ 1 & -1 \end{bmatrix} \begin{bmatrix} 0 & x^2 + 1 \\ x^2 + 1 & 0 \end{bmatrix} \begin{bmatrix} 0 & 1 \\ 0 & -1 \end{bmatrix}.
$$

### 10.1.3 ▪ Exercises

**Exercise 10.1.1.** *Give an explicit SHS decomposition for*

$$\begin{bmatrix} x_1^2 - 2x_1 x_2 + 2x_2^2 + x_3^2 & x_2^2 - x_2 x_3 - x_1 x_2 + x_3^2 - x_1 x_3 \\ x_2^2 - x_2 x_3 - x_1 x_2 + x_3^2 - x_1 x_3 & x_1^2 + x_2^2 - 2x_2 x_3 + 2x_3^2 \end{bmatrix}.$$

**Exercise 10.1.2.** *([146]) Show that $-1 \notin \mathrm{QM}[-x_1, -x_2, -(x_1 x_2 + 1)]$. Use this fact to show that $\mathrm{QM}[G]^{\ell \times \ell}$ is not archimedean for*

$$G = \begin{bmatrix} -x_1 & 0 & 0 \\ 0 & -x_2 & 0 \\ 0 & 0 & -x_1 x_2 - 1 \end{bmatrix}.$$

**Exercise 10.1.3.** *For every $\epsilon > 0$, give an explicit SHS type representation for*

$$\begin{bmatrix} -1 & 0 \\ 0 & -1 \end{bmatrix} \in \mathrm{QM}[(2x_1 x_2 + 1 + \epsilon) \begin{bmatrix} 0 & 1 \\ 1 & 0 \end{bmatrix}, 1 - x_1^2 - x_2^2]^{2 \times 2}.$$

**Exercise 10.1.4.** *([52]) Show that the following matrix polynomial is psd everywhere but it is not SHS:*

$$\begin{bmatrix} x_1^2 + 2x_2^2 & -x_1 x_2 & -x_1 x_3 \\ -x_1 x_2 & x_2^2 + 2x_3^2 & -x_2 x_3 \\ -x_1 x_3 & -x_2 x_3 & x_3^2 + 2x_1^2 \end{bmatrix}.$$

**Exercise 10.1.5.** *Find an explicit SHS decomposition for the matrix polynomial*

$$\begin{bmatrix} x^4 + 12 & x^3 & x^2 & x \\ x^3 & x^4 + 12 & x^3 & x^2 \\ x^2 & x^3 & x^4 + 12 & x^3 \\ x & x^2 & x^3 & x^4 + 12 \end{bmatrix}.$$

**Exercise 10.1.6.** *Show that there are no $P_1, \ldots, P_k \in \mathcal{S}\mathbb{R}[x]^{2 \times 2}$ such that*

$$\begin{bmatrix} 1 & x \\ x & x^2 \end{bmatrix} = P_1^2 + \cdots + P_k^2.$$

*This shows that not every psd matrix polynomial is a sum of squares of symmetric matrix polynomials.*

**Exercise 10.1.7.** *Let $\mathcal{G}$ be a set of symmetric matrix polynomials. If the quadratic module $\mathrm{QM}[\mathcal{G}]^{\ell_1 \times \ell_1}$ is archimedean for some integer $\ell_1 \geq 1$, show that $\mathrm{QM}[\mathcal{G}]^{\ell \times \ell}$ is archimedean for all $\ell \in \mathbb{N}$.*

**Exercise 10.1.8.** *If $G = diag(g_1, \ldots, g_\ell)$ is a diagonal matrix polynomial, show that $\mathrm{QM}[G] = \mathrm{QM}[g]$ for $g := (g_1, \ldots, g_\ell)$.*

**Exercise 10.1.9.** *For every $A \in \Sigma[x]^{\ell \times \ell}$, show that the determinant $\det A$ is SOS, i.e., show that $\det A \in \Sigma[x]$. (Hint: use the Cauchy–Binet formula).*

**Exercise 10.1.10.** *[53, Lemma 7.5] Suppose $A \in \mathbb{R}[x]^{\ell \times \ell}$ and $0 \neq q \in \mathbb{R}[x]$ are both psd in the univariate variable $x \in \mathbb{R}^1$. If $qA = C^T C$ for some $C \in \mathbb{R}[x]^{2d \times \ell}$, show that $A = B^T B$ for some $B \in \mathbb{R}[x]^{2d \times \ell}$.*

## 10.2 ▪ Semidefinite Representations

A matrix polynomial $F \in \mathcal{S}\mathbb{R}[x]^{\ell \times \ell}$ is SHS if $F = P^T P$ for some matrix polynomial $P \in \mathbb{R}[x]^{k \times \ell}$ with $k$ rows. Equivalently, $F$ is SHS if

$$F = p_1^T p_1 + \cdots + p_k^T p_k$$

for row vector polynomials $p_i \in \mathbb{R}[x]^{1 \times \ell}$. Like SOS polynomials, SHS matrix polynomials can be characterized by semidefinite programs (see [297]). For a degree $d$, recall that $[x]_d$ denotes the column vector of monomials of degrees up to $d$. The length of $[x]_d$ is $\binom{n+d}{d}$. The following is a basic result.

**Lemma 10.2.1.** *A matrix polynomial $F \in \mathcal{S}\mathbb{R}[x]_{2d}^{\ell \times \ell}$ is SHS if and only if there exists a psd matrix $X$ of length $\ell \cdot \binom{n+d}{d}$ such that*

$$F = (I_\ell \otimes [x]_d)^T \cdot X \cdot (I_\ell \otimes [x]_d), \tag{10.2.1}$$

*where $\otimes$ denotes the classical Kronecker product.*

The above lemma can be shown as follows. A row vector polynomial $p_i \in \mathbb{R}[x]_d^{1 \times \ell}$ can be written as

$$p_i = \begin{bmatrix} p_{i,1}^T [x]_d & p_{i,2}^T [x]_d & \cdots & p_{i,\ell}^T [x]_d \end{bmatrix}$$

for coefficient vectors $p_{i,j} \in \mathbb{R}^{\binom{n+d}{d}}$. Let $\text{vec}(p_i)$ denote the coefficient vector of $p_i$ such that

$$\text{vec}(p_i)^T = \begin{bmatrix} p_{i,1}^T & p_{i,2}^T & \cdots & p_{i,\ell}^T \end{bmatrix}.$$

Then, we get that

$$p_i = \text{vec}(p_i)^T \left( I_\ell \otimes [x]_d \right).$$

The coefficient vector $\text{vec}(p_i)$ has the length $\ell \cdot \binom{n+d}{d}$. Note that

$$p_i^T p_i = \left( I_\ell \otimes [x]_d \right)^T \left[ \text{vec}(p_i) \text{vec}(p_i)^T \right] \left( I_\ell \otimes [x]_d \right).$$

Therefore, $F = p_1^T p_1 + \cdots + p_k^T p_k$ if and only if

$$F = \left( I_\ell \otimes [x]_d \right)^T \left[ \sum_{i=1}^{k} \text{vec}(p_i) \text{vec}(p_i)^T \right] \left( I_\ell \otimes [x]_d \right).$$

The conclusion of Lemma 10.2.1 follows from factorizations of psd matrices.

For instance, the following matrix polynomial is SHS:

$$\begin{bmatrix} 2x_1^2 + x_2^2 & 2x_1 x_2 \\ 2x_1 x_2 & x_1^2 + 2x_2^2 \end{bmatrix} = \begin{bmatrix} \sqrt{2}x_1 & \sqrt{2}x_2 \\ x_2 & 0 \\ 0 & x_1 \end{bmatrix}^T \begin{bmatrix} \sqrt{2}x_1 & \sqrt{2}x_2 \\ x_2 & 0 \\ 0 & x_1 \end{bmatrix}.$$

The above matrix polynomial can also be written as

$$\left( \begin{bmatrix} 1 & 0 \\ 0 & 1 \end{bmatrix} \otimes \begin{bmatrix} x_1 \\ x_2 \end{bmatrix} \right)^T \begin{bmatrix} 2 & 0 & 0 & 2 \\ 0 & 1 & 0 & 0 \\ 0 & 0 & 1 & 0 \\ 2 & 0 & 0 & 2 \end{bmatrix} \left( \begin{bmatrix} 1 & 0 \\ 0 & 1 \end{bmatrix} \otimes \begin{bmatrix} x_1 \\ x_2 \end{bmatrix} \right).$$

The matrix in the above middle is psd.

### 10.2.1 ▪ Quadratic modules

We show how to express quadratic modules by semidefinite programs. Consider a set $\mathcal{G}$ of symmetric matrix polynomials

$$\mathcal{G} = \{G_1, \ldots, G_m\}, \quad G_i \in \mathcal{SR}[x]^{t_i \times t_i}. \tag{10.2.2}$$

The quadratic module in $\mathcal{SR}[x]^{\ell \times \ell}$ generated by $\mathcal{G}$ is

$$\text{QM}[\mathcal{G}]^{\ell \times \ell} := \left\{ \sum_{k=1}^{N} P_k^T B_k P_k \;\middle|\; \begin{array}{l} B_k \in \{1\} \cup \mathcal{G}, \; N \in \mathbb{N}, \\ P_k \in \mathbb{R}[x]^{\text{len}(B_k) \times \ell} \end{array} \right\}. \tag{10.2.3}$$

In the above, $\text{len}(B_k)$ denotes the length of $B_k$. For the special case that $\mathcal{G}$ is empty, $\text{QM}[\emptyset]^{\ell \times \ell} = \Sigma[x]^{\ell \times \ell}$. In computation, we often work with a *truncation* of degree $2d$:

$$\text{QM}[\mathcal{G}]_{2d}^{\ell \times \ell} := \left\{ \sum_{k=1}^{N} P_k^T B_k P_k \;\middle|\; \begin{array}{l} B_k \in \{1\} \cup \mathcal{G}, \; N \in \mathbb{N}, \\ P_k \in \mathbb{R}[x]^{\text{len}(B_k) \times \ell}, \\ \deg(P_k^T B_k P_k) \leq 2d \end{array} \right\}. \tag{10.2.4}$$

When $\ell = 1$, for convenience, we just write that

$$\text{QM}[\mathcal{G}]_{2d} := \text{QM}[\mathcal{G}]_{2d}^{1 \times 1}.$$

In the following, we show how to check memberships for the cone $\mathrm{QM}[\mathcal{G}]_{2d}^{\ell \times \ell}$ by using semi-definite programs. This issue is also discussed in the work [297]. For $B \in \mathcal{SR}[x]^{L \times L}$ and a $t$-by-$t$ block matrix $Z \in \mathbb{R}^{tL \times tL}$ such as

$$
Z = \begin{bmatrix} Z^{11} & Z^{12} & \cdots & Z^{1t} \\ Z^{21} & Z^{22} & \cdots & Z^{2t} \\ \vdots & \vdots & \ddots & \vdots \\ Z^{t1} & Z^{t2} & \cdots & Z^{tt} \end{bmatrix}, \quad Z^{rs} \in \mathbb{R}^{L \times L},
$$

we define the block product ($\bullet$ denotes the Euclidean inner product of matrices)

$$
B \boxdot_t Z := \begin{bmatrix} B \bullet Z^{11} & B \bullet Z^{12} & \cdots & B \bullet Z^{1t} \\ B \bullet Z^{21} & B \bullet Z^{22} & \cdots & B \bullet Z^{2t} \\ \vdots & \vdots & \ddots & \vdots \\ B \bullet Z^{t1} & B \bullet Z^{t2} & \cdots & B \bullet Z^{tt} \end{bmatrix}. \tag{10.2.5}
$$

Equivalently, $B \boxdot_t Z$ is a $t$-by-$t$ matrix such that

$$
B \boxdot_t Z = (B \bullet Z^{rs})_{\substack{r=1,\ldots,t \\ s=1,\ldots,t}}.
$$

The following is a semidefinite representation for $\mathrm{QM}[\mathcal{G}]_{2d}^{\ell \times \ell}$.

**Theorem 10.2.2.** *Let $\mathcal{G}$ be the matrix polynomial set as in (10.2.2). Then a matrix polynomial $F$ belongs to $\mathrm{QM}[\mathcal{G}]_{2d}^{\ell \times \ell}$ if and only if there exist psd matrices $X_0, X_1, \ldots, X_m$ such that*

$$
F = \sum_{i=0}^{m} G_i \boxdot_{t_i} \left( (I_{t_i \ell} \otimes [x]_{d_i})^T X_i (I_{t_i \ell} \otimes [x]_{d_i}) \right), \tag{10.2.6}
$$

*where $G_0 = 1$, $t_0 = 1$, $d_i = d - \lceil \frac{1}{2} \deg(G_i) \rceil$, and*

$$
\mathrm{len}(X_i) = t_i \cdot \ell \cdot \binom{n + d_i}{d_i}.
$$

**Proof.** Let $P_k(s,:)$ denote the $s$th row vector of a matrix polynomial $P_k$. Then, for each $G_i \in \mathcal{SR}[x]^{t_i \times t_i}$, we can write that

$$
P_k^T G_i P_k = \sum_{r,s=1}^{t_i} (G_i)_{rs} \cdot P_k(r,:)^T P_k(s,:)
$$

$$
= \sum_{r,s=1}^{t_i} (G_i)_{rs} \cdot (I_\ell \otimes [x]_{d_i})^T \left[ \mathrm{vec}(P_k(r,:)) \, \mathrm{vec}(P_k(s,:))^T \right] (I_\ell \otimes [x]_{d_i}).
$$

Let $\mathrm{vec}(P_k)$ denote the column vector of coefficient vectors of rows of $P_k$, in the order from the top to the bottom. For matrices

$$
X_i = \sum_{k=1}^{N_i} \mathrm{vec}(P_k) \, \mathrm{vec}(P_k)^T,
$$

we get the decomposition

$$
\sum_{k=1}^{N_i} P_k^T G_i P_k = G_i \boxdot_{t_i} \left( (I_{t_i \ell} \otimes [x]_{d_i})^T X_i (I_{t_i \ell} \otimes [x]_{d_i}) \right).
$$

Observe that $F \in \mathrm{QM}[\mathcal{G}]_{2d}^{\ell \times \ell}$ if and only if $F$ is a sum of matrix polynomials $\sum_{k=1}^{N_i} P_k^T G_i P_k$ for $i = 1, \ldots, m$. The conclusion then follows from factorizations of psd matrices.  □

## 10.2.2 ▪ Block moment and localizing matrices

We now discuss dual cones of $\Sigma[x]_{2d}^{\ell \times \ell}$ and the truncated quadratic module $\mathrm{QM}[\mathcal{G}]_{2d}^{\ell \times \ell}$. To this end, we need to consider linear functionals that are nonnegative on $\mathrm{QM}[\mathcal{G}]_{2d}^{\ell \times \ell}$. Note that $\mathrm{QM}[\mathcal{G}]_{2d}^{\ell \times \ell}$ is a cone in the space $\mathbb{R}[x]_{2d}^{\ell \times \ell}$. We use the standard monomial basis for $\mathbb{R}[x]_{2d}^{\ell \times \ell}$

$$\mathcal{B} := \{ E_{rs} x^\alpha : \alpha \in \mathbb{N}_{2d}^n, \ 1 \leq r, s \leq \ell \},$$

where each $E_{rs} := e_r e_s^T \in \mathbb{R}^{\ell \times \ell}$. A linear functional on $\mathbb{R}[x]_{2d}^{\ell \times \ell}$ is determined by its values on basis vectors. Consider the linear functional

$$\phi : \mathbb{R}[x]_{2d}^{\ell \times \ell} \to \mathbb{R}, \quad E_{rs} x^\alpha \mapsto y_\alpha^{(r,s)} \quad \text{for } r, s \in [\ell]. \tag{10.2.7}$$

Since $\mathrm{QM}[G]_{2d}^{\ell \times \ell}$ has only symmetric matrix polynomials, we need to restrict $\phi$ to the subspace $\mathcal{S}\mathbb{R}[x]_{2d}^{\ell \times \ell}$. This can be done by enforcing the symmetry pattern

$$y_\alpha^{(r,s)} = y_\alpha^{(s,r)}$$

for all $r, s \in [\ell]$. Therefore, the linear functional $\phi$ on $\mathcal{S}\mathbb{R}[x]_{2d}^{\ell \times \ell}$ is uniquely determined by the vector

$$Y := (Y_\alpha)_{\alpha \in \mathbb{N}_{2d}^n},$$

where each $Y_\alpha$ is the $\ell$-by-$\ell$ symmetric matrix

$$Y_\alpha := \begin{bmatrix} y_\alpha^{(1,1)} & \cdots & y_\alpha^{(1,\ell)} \\ \vdots & \ddots & \vdots \\ y_\alpha^{(\ell,1)} & \cdots & y_\alpha^{(\ell,\ell)} \end{bmatrix}.$$

The vector $Y$ can be viewed as a matrix-valued truncated multi-sequence of degree $2d$ in the space $(\mathcal{S}^\ell)^{\mathbb{N}_{2d}^n}$.

First, we characterize the dual cone of $\Sigma[x]_{2d}^{\ell \times \ell}$. Each matrix in $\Sigma[x]_{2d}^{\ell \times \ell}$ can be written as

$$(I_\ell \otimes [x]_d)^T X (I_\ell \otimes [x]_d) = \sum_{r,s=1}^{\ell} \sum_{\alpha,\beta \in \mathbb{N}_d^n} (X^{rs})_{\alpha,\beta} \cdot x^{\alpha+\beta} \cdot E_{rs},$$

where each $X^{rs}$ is the $(r, s)$th block of a psd matrix $X$. So,

$$\phi\Big( (I_\ell \otimes [x]_d)^T X (I_\ell \otimes [x]_d) \Big) = \sum_{r,s=1}^{\ell} \sum_{\alpha,\beta \in \mathbb{N}_d^n} (X^{rs})_{\alpha,\beta} \cdot y_{\alpha+\beta}^{(r,s)}.$$

For each $Y \in (\mathcal{S}^\ell)^{\mathbb{N}_{2d}^n}$, denote the block moment matrix

$$M_d[Y] := (Y_{\alpha+\beta})_{\alpha, \beta \in \mathbb{N}_d^n}. \tag{10.2.8}$$

It has the length $\ell \cdot \binom{n+d}{d}$. If $\phi$ is the linear functional determined by $Y$, then

$$\phi\Big( (I_\ell \otimes [x]_d)^T X (I_\ell \otimes [x]_d) \Big) = X \bullet M_d[Y].$$

So, $\phi \geq 0$ on $\Sigma[x]^{\ell \times \ell}$ if and only if $M_d[Y] \succeq 0$.

Second, we characterize the dual cone of the truncated quadratic module $\mathrm{QM}[G]_{2d}^{\ell \times \ell}$. For this, we need to define localizing matrices. For a scalar polynomial $h$ of degree at most $2d$, let

$$d_h := d - \lceil \tfrac{1}{2} \deg(h) \rceil$$

and write the expansion of $h$ as

$$h = \sum_{|\theta| \leq \deg(h)} h_\theta x^\theta.$$

We similarly define the block localizing matrix

$$L_h^{(d)}[Y] := \Big( \sum_{|\theta| \leq \deg(h)} h_\theta Y_{\theta+\alpha+\beta} \Big)_{\alpha, \beta \in \mathbb{N}_{d_h}^n} \tag{10.2.9}$$

for $Y = (Y_\alpha) \in (\mathcal{S}^\ell)^{\mathbb{N}_{2d}^n}$. Note that $L_h^{(d)}[Y]$ is a $\binom{n+d_h}{d_h}$-by-$\binom{n+d_h}{d_h}$ block matrix and each block is $\ell$-by-$\ell$. For the map $\phi$ as in (10.2.7), one can verify that

$$\phi\Big( h \cdot (I_\ell \otimes [x]_{d_h})^T X (I_\ell \otimes [x]_{d_h}) \Big) = X \bullet L_h^{(d)}[Y]. \tag{10.2.10}$$

In particular, for $h = 1$, we have

$$L_1^{(d)}[Y] = M_d[Y].$$

Third, we define localizing matrices for matrix polynomials. Consider the matrix polynomial

$$H = (h_{ij})_{\substack{i=1,\ldots,t \\ j=1,\ldots,t}} \in \mathcal{SR}[x]^{t \times t}.$$

Suppose the degree of $H$ is at most $2d$. Let

$$d_H := d - \lceil \tfrac{1}{2} \deg(H) \rceil.$$

Write each entry $h_{ij}$ as

$$h_{ij} = \sum_{|\theta| \leq \deg(H)} h_\theta^{ij} x^\theta.$$

The localizing matrix for $h_{ij}$ is then

$$L_{h_{ij}}^{(d)}[Y] := \Big( \sum_{|\theta| \leq \deg(H)} h_\theta^{ij} Y_{\theta+\alpha+\beta} \Big)_{\alpha, \beta \in \mathbb{N}_{d_H}^n}. \tag{10.2.11}$$

Note that all $L_{h_{ij}}^{(d)}[Y]$ have the same size. The *localizing matrix* for the matrix polynomial $H$ is then defined to be the $t \times t$ block matrix

$$L_H^{(d)}[Y] := \Big( L_{h_{ij}}^{(d)}[Y] \Big)_{\substack{i=1,\ldots,t \\ j=1,\ldots,t}}. \tag{10.2.12}$$

The dual cone of $\mathrm{QM}[G]_{2d}^{\ell \times \ell}$ can be characterized as follows.

**Theorem 10.2.3.** *Let $\mathcal{G}$ be the set as in (10.2.2). If each $G_i$ has degree $2d_i \leq 2d$, then the dual cone of $\mathrm{QM}[G]_{2d}^{\ell \times \ell}$ is (let $G_0 = 1$)*

$$\mathscr{S}[\mathcal{G}]_{2d}^{\ell \times \ell} := \Big\{ Y \in (\mathcal{S}^\ell)^{\mathbb{N}_{2d}^n} : L_{G_i}^{(d)}[Y] \succeq 0, \, i = 0, 1, \ldots, m \Big\}. \tag{10.2.13}$$

***Proof.*** For convenience of discussion, let $H = (h_{ij}) \in \mathcal{SR}[x]^{t \times t}$ be the matrix polynomial in the above. Consider a block matrix variable $X$ with the block pattern

$$
X = \begin{bmatrix} X^{(1,1)} & \cdots & X^{(1,\ell)} \\ \vdots & \ddots & \vdots \\ X^{(\ell,1)} & \cdots & X^{(\ell,\ell)} \end{bmatrix},
$$

where each matrix $X^{(r,s)}$, for $r,s \in [\ell]$, is blocked such that

$$
X^{(r,s)} = \begin{bmatrix} X^{(1,r),(1,s)} & \cdots & X^{(1,r),(t,s)} \\ \vdots & \vdots & \vdots \\ X^{(t,r),(1,s)} & \cdots & X^{(t,r),(t,s)} \end{bmatrix}.
$$

Note that every subblock has the dimension

$$
X^{(i,r),(j,s)} \in \mathbb{R}^{\binom{n+d_H}{d_H} \times \binom{n+d_H}{d_H}} \quad \forall\, i,j \in [t], \quad \forall\, r,s \in [\ell].
$$

Observe the following matrix product blocking pattern:

$$
(I_{t\ell} \otimes [x]_{d_H})^T X (I_{t\ell} \otimes [x]_{d_H}) = \left( [x]_{d_H}^T X^{(i,r),(j,s)} [x]_{d_H} \right)_{\substack{i,j=1,\ldots,t \\ r,s=1,\ldots,\ell}}.
$$

The above product is a $t\ell$-by-$t\ell$ matrix.

For a matrix polynomial $H$, the $(r,s)$th entry of the product

$$
H \,\square_t\, \left( (I_{t\ell} \otimes [x]_{d_H})^T X (I_{t\ell} \otimes [x]_{d_H}) \right)
$$

can be written as

$$
H \bullet \left[ [x]_{d_H}^T \cdot X^{(i,r),(j,s)} \cdot [x]_{d_H} \right]_{\substack{i=1,\ldots,t \\ j=1,\ldots,t}} = \sum_{i,j=1}^{t} X^{(i,r),(j,s)} \bullet \left( H_{ij} \cdot [x]_{d_H} [x]_{d_H}^T \right).
$$

This implies

$$
H \,\square_t\, \left( (I_{t\ell} \otimes [x]_{d_H})^T X (I_{t\ell} \otimes [x]_{d_H}) \right)
$$

$$
= \sum_{r,s=1}^{\ell} \left( \sum_{i,j=1}^{t} X^{(i,r),(j,s)} \bullet H_{ij} \cdot [x]_{d_H} [x]_{d_H}^T \right) E_{rs}.
$$

Thus, for the map $\phi$ as in (10.2.7), we can see that

$$
\phi\left( H \,\square_t\, \left( (I_{t\ell} \otimes [x]_{d_H})^T X (I_{t\ell} \otimes [x]_{d_H}) \right) \right)
$$

$$
= \sum_{r,s=1}^{\ell} \sum_{i,j=1}^{t} X^{(i,r),(j,s)} \bullet L_{H_{ij}}^{(d)}[Y] = X \bullet \left( L_{H_{ij}}^{(d)}[Y] \right)_{\substack{i=1,\ldots,t \\ j=1,\ldots,t}} = X \bullet L_H^{(d)}[Y].
$$

The above is nonnegative for all $X \succeq 0$ if and only if $L_H^{(d)}[Y] \succeq 0$. The conclusion follows from Theorem 10.2.2 for $H = G_0, G_1, \ldots, G_m$ respectively. $\qquad\square$

### 10.2.3 ▪ Exercises

**Exercise 10.2.1.** *Use Lemma 10.2.1 to show that the matrix polynomial*

$$G := \begin{bmatrix} x_1^2 + 2x_2^2 & -x_1 x_2 & -x_1 x_3 \\ -x_1 x_2 & x_2^2 + 2x_3^2 & -x_2 x_3 \\ -x_1 x_3 & -x_2 x_3 & x_3^2 + 2x_1^2 \end{bmatrix}$$

*is not SHS. Determine whether $G \in QM[1 - x_1^2 - x_2^2 - x_3^2]^{3\times 3}$ or not.*

**Exercise 10.2.2.** *Find the smallest $R$ such that*

$$\begin{bmatrix} R & x_1 & x_2 \\ x_1 & R & x_3 \\ x_2 & x_3 & R \end{bmatrix} \in QM[\begin{pmatrix} 1 - x_1 & 0 & x_2 \\ 0 & 1 - x_1 & x_3 \\ x_2 & x_3 & 1 + x_1 \end{pmatrix}]_2^{3\times 3}.$$

**Exercise 10.2.3.** *Determine the smallest value of the minimum eigenvalue of*

$$\begin{bmatrix} x^4 & x^3 & x^2 & x \\ x^3 & x^4 & x^3 & x^2 \\ x^2 & x^3 & x^4 & x^3 \\ x & x^2 & x^3 & x^4 \end{bmatrix} \quad for \quad x \in \mathbb{R}^1.$$

**Exercise 10.2.4.** *Determine whether the following holds or not:*

$$\begin{bmatrix} 1 & 0 & x_1 \\ 0 & 1 & x_2 \\ x_1 & x_2 & 1 \end{bmatrix} \in QM[\begin{pmatrix} 1 - x_1 & x_2 \\ x_2 & 1 + x_1 \end{pmatrix}]_2^{3\times 3}.$$

**Exercise 10.2.5.** *For $A, B, C \in \mathcal{S}^n$, show that $A + 2tB + t^2 C \succeq 0$ for all $t \in \mathbb{R}$ if and only if the following 2-by-2 block matrix*

$$\begin{bmatrix} A & B \\ B & C \end{bmatrix} \succeq 0.$$

**Exercise 10.2.6.** *Let $\mathscr{P}_{1,2d}^{\ell \times \ell}$ be the cone of symmetric psd matrix polynomials in $x \in \mathbb{R}^1$ and with degrees at most $2d$. Give a semidefinite representation for $\mathscr{P}_{1,2d}^{\ell \times \ell}$ and its dual cone. Do the same thing for the cone $\mathscr{P}_{n,2}^{2 \times 2}$ of symmetric psd quadratic matrix polynomials in $x \in \mathbb{R}^n$. (Hint: use Theorems 10.1.1 and 10.1.2.)*

## 10.3 ▪ Matrix-Valued TMPs

Let $\Theta$ denote the set of all Borel sets in $\mathbb{R}^n$. A function $\mu : \Theta \to \mathcal{S}^\ell$, given as

$$\mu := (\mu_{ij})_{i,j=1}^\ell,$$

is called a positive $\mathcal{S}^\ell$-valued measure on $\mathbb{R}^n$ if each $\mu_{ij}$ is a signed Borel measure on $\mathbb{R}^n$ and $\mu(T) \succeq 0$ for all $T \in \Theta$. For $\xi \in \mathbb{R}^\ell$, let $\mu_\xi$ denote the induced scalar-valued measure:

$$\mu_\xi : \Theta \to \mathbb{R}, \quad T \mapsto \xi^T(\mu(T))\xi.$$

Then $\mu_\xi$ is a positive Borel measure on $\mathbb{R}^n$. The support of $\mu$ is defined as

$$\text{supp}(\mu) := cl\Big( \bigcup_{\xi \in \mathbb{R}^\ell} \text{supp}(\mu_\xi) \Big). \tag{10.3.1}$$

For a measurable function $f : \mathbb{R}^n \to \mathbb{R}$, the integral $\int f(x)\mathrm{d}\mu$ is the symmetric matrix in $\mathcal{S}^\ell$ such that

$$\int f(x)\mathrm{d}\mu := \Big( \int f(x)\mathrm{d}\mu_{ij} \Big)^\ell_{i,j=1}, \tag{10.3.2}$$

provided all integrals in the right hand side exist. In particular, for $x^\alpha := x_1^{\alpha_1} \cdots x_n^{\alpha_n}$, the integral $\int x^\alpha \mathrm{d}\mu$ is called the $\alpha$th moment of $\mu$, if it exists.

A vector labeled in the form

$$Y := (Y_\alpha)_{\alpha \in \mathbb{N}_d^n}, \quad Y_\alpha \in \mathcal{S}^\ell,$$

is called a *matrix-valued truncated multi-sequence* (mv-tms) of degree $d$. Let $K \subseteq \mathbb{R}^n$ be a closed set. For a given mv-tms $Y$, the matrix valued truncated $K$-moment problem (MV-TKMP) concerns whether or not there exists a positive matrix-valued Borel measure $\mu$ on $\mathbb{R}^n$ such that

$$\text{supp}(\mu) \subseteq K, \quad Y_\alpha = \int x^\alpha \mathrm{d}\mu \quad \forall \alpha \in \mathbb{N}_d^n.$$

If it exists, such $\mu$ is called a $K$-representing measure for $Y$ and $Y$ is said to admit (or have) the measure $\mu$. The MV-TMP was recently studied in [142, 143].

Let $\delta_v$ denote the unit Dirac measure supported at the point $v$. If there exist psd matrices $C_1, \ldots, C_r \in \mathcal{S}_+^\ell$ and distinct points $v_1, \ldots, v_r \in K$ such that

$$\mu = C_1 \delta_{v_1} + \cdots + C_r \delta_{v_r},$$

then $\mu$ is called $r$-atomic.[9] Similarly, a matrix valued measure is called *finitely atomic* if it is $r$-atomic for some $r \in \mathbb{N}$. For convenience, let $\mathscr{B}_\ell(K)$ denote the set of all matrix-valued measures supported in $K$ as in the above, and denote

$$\mathscr{R}_d(K)^{\ell \times \ell} := \left\{ Y \in (\mathcal{S}^\ell)^{\mathbb{N}_d^n} \;\middle|\; \begin{array}{c} \exists \mu \in \mathscr{B}_\ell(K) \quad s.t. \\ Y_\alpha = \int x^\alpha \mathrm{d}\mu \; \forall \alpha \in \mathbb{N}_d^n \end{array} \right\}. \tag{10.3.3}$$

The equations in the above can be equivalently written as

$$Y = \int [x]_d \mathrm{d}\mu.$$

When $Y$ admits a $K$-representing measure, it also admits a finitely atomic $K$-representing measure. This was shown in [143]. The size of its support can be bounded by Carathéodory's Theorem. The dimension of the space $(\mathcal{S}^\ell)^{\mathbb{N}_d^n}$ is $\binom{\ell+1}{2} \cdot \binom{n+d}{d}$. Therefore, we have the following conclusion.

**Theorem 10.3.1.** *([143]) Let $Y \in (\mathcal{S}^\ell)^{\mathbb{N}_d^n}$ be a mv-tms and let $K \subseteq \mathbb{R}^n$ be a closed set. Then $Y$ admits a $K$-representing measure $\mu$ if and only if $Y$ admits a finitely atomic measure $\nu$ supported in $K$ such that*

$$|\text{supp}(\nu)| \leq \binom{\ell+1}{2} \binom{n+d}{d}.$$

---

[9]In some references, the measure $\mu$ is called $\ell$-atomic with $\ell = \sum_{i=1}^r \text{rank } C_i$. For consistency with the scalar case, we call it $r$-atomic since $r$ equals the cardinality of the support.

Theorem 10.3.1 implies that the cone $\mathscr{R}_d(K)^{\ell\times\ell}$ can be equivalently expressed as the conic hull

$$\mathscr{R}_d(K)^{\ell\times\ell} = \text{cone}\left(\left\{[u]_d \otimes X \;\middle|\; \begin{array}{l} u \in K, \\ \|X\| = 1,\, X \in \mathcal{S}_+^\ell \end{array}\right\}\right). \qquad (10.3.4)$$

If $\mathcal{S}^\ell$ is replaced by the space of $\ell$-by-$\ell$ complex Hermitian matrices, the upper bound for the cardinality $|\operatorname{supp}(\nu)|$ is then changed to $\ell^2\binom{n+d}{d}$, as shown in [143].

## 10.3.1 ▪ Properties of MV-TKMPs

A matrix polynomial $P \in \mathcal{S}\mathbb{R}[x]_d^{\ell\times\ell}$ can be written as

$$P = \sum_{\alpha\in\mathbb{N}_d^n} P_\alpha x^\alpha$$

for symmetric matrix coefficients $P_\alpha \in \mathcal{S}^\ell$. The dual space of $\mathcal{S}\mathbb{R}[x]^{\ell\times\ell}$ is $(\mathcal{S}^\ell)^{\mathbb{N}_d^n}$, the space of degree-$d$ mv-tms's whose entries are matrices in $\mathcal{S}^\ell$. For

$$Y := (Y_\alpha) \in (\mathcal{S}^\ell)^{\mathbb{N}_d^n},$$

the Riesz functional $\mathscr{L}_Y$ acts on $\mathcal{S}\mathbb{R}[x]^{\ell\times\ell}$ such that

$$\mathscr{L}_Y(P) := \sum_{\alpha\in\mathbb{N}_d^n} P_\alpha \bullet Y_\alpha.$$

In the above, $\bullet$ denotes the Euclidean inner product of matrices. For consistency of notation, we also denote

$$\langle P, Y\rangle := \mathscr{L}_Y(P).$$

The Riesz functional $\mathscr{L}_Y$ is said to be $K$-positive if

$$P \in \mathscr{P}_d(K)^{\ell\times\ell} \Rightarrow \mathscr{L}_Y(P) \geq 0.$$

If $Y$ admits a $K$-representing measure, then $\mathscr{L}_Y$ must be $K$-positive (see exercises). Similarly, $\mathscr{L}_Y$ is said to be *strictly $K$-positive* if

$$P|_K \not\equiv 0,\; P \in \mathscr{P}_d(K)^{\ell\times\ell} \quad\Rightarrow\quad \mathscr{L}_Y(P) > 0.$$

Recall that $I(K)$ is the vanishing ideal of $K$ and

$$I(K)_d := \mathbb{R}[x]_d \cap I(K).$$

**Theorem 10.3.2.** *Let $K \subseteq \mathbb{R}^n$ be a closed set. Then, we have the following:*

(i) *The cone $\mathscr{P}_d(K)^{\ell\times\ell}$ is closed convex and it holds that*

$$\mathscr{P}_d(K)^{\ell\times\ell} \cap -\mathscr{P}_d(K)^{\ell\times\ell} = \left(I(K)_d\right)^{\ell\times\ell}. \qquad (10.3.5)$$

*So, the cone $\mathscr{P}_d(K)^{\ell\times\ell}$ is pointed if and only if $I(K)_d = \{0\}$.*

(ii) *The cone $\mathscr{R}_d(K)^{\ell\times\ell}$ is convex. When $K$ is compact, $\mathscr{R}_d(K)^{\ell\times\ell}$ is closed and a mv-tms $Y \in (\mathcal{S}^\ell)^{\mathbb{N}_d^n}$ belongs to $\mathscr{R}_d(K)^{\ell\times\ell}$ if and only if $\mathscr{L}_Y$ is $K$-positive.*

*(iii)  The affine hull of $\mathscr{R}_d(K)^{\ell \times \ell}$ is the subspace*

$$\mathcal{F} := \left\{ Y \in (\mathcal{S}^\ell)^{\mathbb{N}^n_d} \,\big|\, \langle P, Y \rangle = 0 \; \forall P \in \left(I(K)_d\right)^{\ell \times \ell} \right\}. \tag{10.3.6}$$

*So, if $I(K)_d = \{0\}$, then $\mathcal{F} = (\mathcal{S}^\ell)^{\mathbb{N}^n_d}$ and $\mathscr{R}_d(K)$ has nonempty interior. Moreover, a mv-tms $Y$ lies in the relative interior of $\mathscr{R}_d(K)^{\ell \times \ell}$ if and only if $\mathscr{L}_Y$ is strictly $K$-positive.*

**Proof.** (i) Obviously, $\mathscr{P}_d(K)^{\ell \times \ell}$ is closed convex. A matrix polynomial $P \in \mathscr{P}_d(K)^{\ell \times \ell} \cap -\mathscr{P}_d(K)^{\ell \times \ell}$ if and only if $P|_K \equiv 0$. So (10.3.5) holds. Consequently, the pointedness of $\mathscr{P}_d(K)^{\ell \times \ell}$ is equivalent to $I(K)_d = \{0\}$.

(ii) The convexity of $\mathscr{R}_d(K)^{\ell \times \ell}$ follows from (10.3.4), since $\mathscr{R}_d(K)^{\ell \times \ell}$ is the conic hull of the set

$$T := \left\{ [u]_d \otimes X : u \in K, \|X\| = 1, X \in \mathcal{S}^\ell_+ \right\}.$$

In the following, assume $K$ is compact. The set $T$ is then also compact and we show that $\mathscr{R}_d(K)^{\ell \times \ell}$ is closed. Let $I_\ell$ be the constant identity matrix polynomial. For every $Y = [u]_d \otimes X \in T$, with $u \in K$ and $\|X\| = 1, X \succeq 0$,

$$\mathscr{L}_Y(I_\ell) = I_\ell \bullet X = \mathrm{trace}(X) > 0.$$

By Proposition 1.2.6, we know $\mathscr{R}_d(K)^{\ell \times \ell}$ is closed if $K$ is compact.

"$\Rightarrow$": Each $Y \in \mathscr{R}_d(K)^{\ell \times \ell}$ can be written as

$$Y = \sum_{i=1}^r [u_i]_d \otimes X_i, \quad u_i \in K, \quad X_i \in \mathcal{S}^\ell_+.$$

For every $P \in \mathscr{P}_d(K)^{\ell \times \ell}$, we have $P(u_i) \succeq 0$, so

$$\mathscr{L}_Y(P) = \sum_{i=1}^r P(u_i) \bullet X_i \geq 0.$$

This means that $\mathscr{L}_Y$ is $K$-positive.

"$\Leftarrow$": For each $W := (W_\alpha) \in (\mathcal{S}^\ell)^{\mathbb{N}^n_d}$ such that $\mathscr{L}_W$ is $K$-positive, we need to show that $W \in \mathscr{R}_d(K)^{\ell \times \ell}$. Since $K$ is compact, $\mathscr{R}_d(K)^{\ell \times \ell}$ is closed convex. Suppose otherwise $W \notin \mathscr{R}_d(K)^{\ell \times \ell}$. Then, by the convex set separating hyperplane theorem, there exists $0 \neq H \in \mathcal{S}\mathbb{R}[x]^{\ell \times \ell}$ such that

$$\langle H, W \rangle < \langle H, Y \rangle \quad \text{for all } Y \in \mathscr{R}_d(K)^{\ell \times \ell}.$$

Since $\mathscr{R}_d(K)^{\ell \times \ell}$ is a cone, the above implies that $\langle H, Y \rangle \geq 0$ for all $Y \in \mathscr{R}_d(K)^{\ell \times \ell}$. So, $\langle H, W \rangle < 0$. One can write that

$$\langle H, Y \rangle = \sum_{\alpha \in \mathbb{N}^n_d} H_\alpha \bullet Y_\alpha, \quad \text{where} \quad Y = (Y_\alpha)_{\alpha \in \mathbb{N}^n_d}.$$

In particular, for $Y := [u]_d \otimes (\xi \xi^T)$ with $u \in K$ and $\xi \in \mathbb{R}^\ell$, we get

$$\langle H, Y \rangle = \xi^T H(u) \xi \geq 0.$$

The above is true for all $u \in K$ and for all $\xi \in \mathbb{R}^\ell$, so

$$H(x) := \sum_{|\alpha| \leq d} H_\alpha x^\alpha \succeq 0 \quad \text{on} \quad K.$$

This means that $H(x) \in \mathscr{P}_d(K)^{\ell \times \ell}$. However, $\langle H, W \rangle < 0$ contradicts that $\mathscr{L}_W$ is $K$-positive, so we must have $W \in \mathscr{R}_d(K)^{\ell \times \ell}$.

(iii) The affine hull aff $\left( \mathscr{R}_d(K)^{\ell \times \ell} \right)$ is a subspace, since it contains the origin. Note that a vector $Y$ belongs to aff $\left( \mathscr{R}_d(K)^{\ell \times \ell} \right)$ if and only if $\psi(Y) = 0$ for every linear functional $\psi$ that vanishes identically on $\mathscr{R}_d(K)^{\ell \times \ell}$. So, (10.3.6) follows from (10.3.4). Note that $K$ is closed (not necessarily compact). Let $\sigma$ be a probability measure on $\mathbb{R}^n$ such that $\text{supp}(\sigma) = K$. Let $Z = (Z_\alpha) \in (\mathcal{S}^\ell)^{\mathbb{N}_d^n}$ be the mv-tms such that

$$Z_\alpha := \left( \int x^\alpha \mathrm{d}\sigma \right) I_\ell \quad \text{for each} \quad \alpha \in \mathbb{N}_d^n.$$

For each $P \in \mathcal{SR}[x]_d^{\ell \otimes \ell}$, we have

$$\mathscr{L}_Z(P) = \int I_\ell \bullet P(x) \mathrm{d}\sigma.$$

If $P \succeq 0$ on $K$ and $P|_K \not\equiv 0$, then the scalarization $I_\ell \bullet P(x)$ is nonnegative on $K$, but it is not identically zero. Since $\text{supp}(\sigma) = K$, we get

$$\mathscr{L}_Z(P) = \int I_\ell \bullet P(x) \mathrm{d}\sigma > 0.$$

This means that $\mathscr{L}_Z$ is strictly $K$-positive. Also note that $Z \in \mathcal{F}$.

"$\Rightarrow$": Suppose $Y$ is a relative interior point of $\mathscr{R}_d(K)^{\ell \times \ell}$. Then $Y - \epsilon Z \in \mathscr{R}_d(K)^{\ell \times \ell}$ for some $\epsilon > 0$. By the item (ii), for every $P \in \mathcal{SR}[x]_d^{\ell \times \ell}$ with $P|_K \succeq 0$ and $P|_K \not\equiv 0$, we have

$$\mathscr{L}_Y(P) = \mathscr{L}_{Y-\epsilon Z}(P) + \epsilon \mathscr{L}_Z(P) \geq \epsilon \mathscr{L}_Z(P) > 0.$$

This shows that $\mathscr{L}_Y$ is strictly $K$-positive.

"$\Leftarrow$": Suppose $Y$ is such that $\mathscr{L}_Y$ is strictly $K$-positive. Let

$$\mathcal{T} := \left\{ P \in \left( \mathscr{P}_d(K)/I(K)_d \right)^{\ell \times \ell} : \int I_\ell \bullet P(x) \mathrm{d}\sigma = 1 \right\}.$$

Under the topology induced by the norm like (2.5.27) for the quotient $\left( \mathscr{P}_d(K)/I(K)_d \right)^{\ell \times \ell}$, the set $\mathcal{T}$ is compact, since $\text{supp}(\sigma) = K$. For $W \in (\mathcal{S}^\ell)^{\mathbb{N}_d^n}$, the Riesz functional $\mathscr{L}_W$ is strictly $K$-positive if and only if it is positive on $\mathcal{T}$. Let

$$\epsilon := \min_{P \in \mathcal{T}} \mathscr{L}_Y(P) > 0.$$

In view of (10.3.4), the strict $K$-positivity of $\mathscr{L}_Y$ implies $Y \in \mathcal{F}$. Let

$$M := \max_{\substack{P \in \mathcal{T}, \\ V \in \mathcal{F}, \|V\| = 1}} |\mathscr{L}_V(P)| \quad < \quad +\infty.$$

For all $W \in \mathcal{F}$ with $\|W - Y\| \leq \frac{\epsilon}{2M}$, we have

$$\mathscr{L}_W(P) = \mathscr{L}_Y(P) + \mathscr{L}_{W-Y}(P) \geq \epsilon - M \cdot \|W - Y\| \geq \frac{1}{2}\epsilon$$

for all $P \in \mathcal{T}$. So, $\mathscr{L}_W$ is $K$-positive and hence $W \in \text{cl}\left( \mathscr{R}_d(K)^{\ell \times \ell} \right)$ for every $W \in \mathcal{F}$ that is sufficiently close to $Y$. This means that $Y$ lies in the relative interior of $\mathscr{R}_d(K)^{\ell \times \ell}$. $\qquad \square$

The psd matrix polynomial cone $\mathscr{P}_d(K)^{\ell\times\ell}$, given as in (10.1.4), is clearly closed and convex. Like before, the superscript $^*$ denotes the dual cone.

**Proposition 10.3.3.** *Let $K \subseteq \mathbb{R}^n$ be a set. If $K$ is compact, then $\mathscr{P}_d(K)^{\ell\times\ell}$ and $\mathscr{R}_d(K)^{\ell\times\ell}$ are dual to each other, i.e.,*

$$\left(\mathscr{R}_d(K)^{\ell\times\ell}\right)^* = \mathscr{P}_d(K)^{\ell\times\ell}, \quad \left(\mathscr{P}_d(K)^{\ell\times\ell}\right)^* = \mathscr{R}_d(K)^{\ell\times\ell}. \tag{10.3.7}$$

*For more general $K$ (not necessarily compact), the first equality in (10.3.7) still holds but the second one becomes*

$$\left(\mathscr{P}_d(K)^{\ell\times\ell}\right)^* = \mathrm{cl}\left(\mathscr{R}_d(K)^{\ell\times\ell}\right). \tag{10.3.8}$$

**Proof.** When $K$ is compact, $\mathscr{P}_d(K)^{\ell\times\ell}$ and $\mathscr{R}_d(K)^{\ell\times\ell}$ are both closed convex cones, by Theorem 10.3.2(ii). In view of (10.3.4), we have $\langle P, Y\rangle \geq 0$ for all $P \in \mathscr{P}_d(K)^{\ell\times\ell}$ and for all $Y \in \mathscr{R}_d(K)^{\ell\times\ell}$. Therefore, one can directly see that

$$\left(\mathscr{R}_d(K)^{\ell\times\ell}\right)^* = \mathscr{P}_d(K)^{\ell\times\ell}.$$

Since $\mathscr{R}_d(K)^{\ell\times\ell}$ is closed for compact $K$, the bi-duality theorem (Theorem 1.2.10) implies that

$$\mathscr{R}_d(K)^{\ell\times\ell} = \left(\mathscr{R}_d(K)^{\ell\times\ell}\right)^{**} = \left(\mathscr{P}_d(K)^{\ell\times\ell}\right)^*.$$

This proves the dual relationship (10.3.7).

When $K$ is not compact, the first equality in (10.3.7) follows from (10.3.4) and (10.3.8) follows from the bi-duality relation:

$$\mathrm{cl}\left(\mathscr{R}_d(K)^{\ell\times\ell}\right) = \left(\left(\mathscr{R}_d(K)^{\ell\times\ell}\right)^*\right)^* = \left(\mathscr{P}_d(K)^{\ell\times\ell}\right)^*,$$

because $\left(\mathrm{cl}\left(\mathscr{R}_d(K)^{\ell\times\ell}\right)\right)^* = \left(\mathscr{R}_d(K)^{\ell\times\ell}\right)^*$.  $\square$

### 10.3.2 ▪ A matrix version of flat extensions

Now we give a concrete condition for a mv-tms $Y = (Y_\alpha) \in \left(\mathcal{S}^\ell\right)^{\mathbb{N}^n_d}$ to admit a $K$-representing measure. Suppose the set $K$ is given as

$$K := \left\{ x \in \mathbb{R}^n \,\middle|\, \begin{array}{l} h_i(x) = 0, \; i = 1, \ldots, s, \\ g(x) \succeq 0 \; \forall g \in \mathcal{G} \end{array} \right\} \tag{10.3.9}$$

for a given tuple $h := (h_1, \ldots, h_s)$ of scalar polynomials and a given set $\mathcal{G}$ of symmetric matrix polynomials. Assume the degree $d = 2t$ is even and $d$ is greater than or equal to the maximum degree of $h$ and $\mathcal{G}$. Recall the definitions of moment and localizing matrices in §10.2.2. If $Y$ belongs to the cone $\mathscr{R}_d(K)^{\ell\times\ell}$, then it must satisfy the following conditions

$$\begin{array}{rcl} \mathscr{V}^{(2t)}_{h_j}[Y] & = & 0 \; (j = 1, \ldots, s), \\ L^{(t)}_g[Y] & \succeq & 0 \; (\forall g \in \{1\} \cup \mathcal{G}). \end{array} \tag{10.3.10}$$

Its proof is left as an exercise. In the above, $\mathscr{V}^{(2t)}_{h_j}[Y]$ is the localizing vector of $Y$ defined such that

$$\mathscr{V}^{(2t)}_{h_j}[Y] := \sum_{\alpha\in\mathbb{N}^n_{2t}} h_\alpha \otimes Y_\alpha, \tag{10.3.11}$$

where the coefficient vectors $h_\alpha$ are such that

$$\mathscr{V}_{h_j}^{(2t)}[z] = \sum_{\alpha \in \mathbb{N}_{2t}^n} h_\alpha z_\alpha.$$

In the above, each $h_\alpha$ is a vector, $\otimes$ denotes the classical Kronecker product, and $\mathscr{V}_{h_j}^{(2t)}[z]$ is the localizing vector defined as in (2.5.23).

The flat extension theorem (Theorem 2.7.7) (for the scalar case) can be generalized to MV-TKMPs. Note that $M_t[Y]$ is a block matrix. Each block of $M_t[Y]$ is a $\ell$-by-$\ell$ symmetric matrix, labeled by $(\alpha, \beta) \in \mathbb{N}_t^n \times \mathbb{N}_t^n$. In determining of representing measures, we consider linearly independent block columns. The maximum number of linearly independent block columns of $M_t[Y]$ is called the *block rank*, for which we denote $\text{rank}_{bl}\, M_t[Y]$. The length of each block of $M_t[Y]$ is $\ell$, so

$$\text{rank}_{bl}\, M_t[Y] = \dim \text{span}\left\{ M_t[Y] \cdot (\lambda \otimes I_\ell) : \lambda \in \mathbb{R}^{\mathbb{N}_t^n} \right\}.$$

The value of $\text{rank}_{bl}\, M_t[Y]$ can be determined by reducing $Y$ to a scalar-valued tms. This is shown as follows.

**Lemma 10.3.4.** *Let $B \in \mathcal{S}_{++}^\ell$ be a positive definite matrix. For a given $Y \in \left(\mathcal{S}^\ell\right)^{\mathbb{N}_d^n}$, let $y = (y_\alpha) \in \mathbb{R}^{\mathbb{N}_{2t}^n}$ be the scalar-valued tms such that*

$$y_\alpha = B \bullet Y_\alpha \quad (\forall \alpha \in \mathbb{N}_{2t}^n). \tag{10.3.12}$$

*If $M_t[Y] \succeq 0$, then for all integer $k \in [0, t]$*

$$\text{rank}\, M_k[y] = \text{rank}_{bl}\, M_k[Y]. \tag{10.3.13}$$

***Proof.*** The column and row blocks of $M_k[Y]$ are labeled by $\alpha, \beta \in \mathbb{N}_k^n$. Note that each $M_k[Y]$ is a leading principal block submatrix of $M_t[Y]$ for each $k \leq t$. For each $\beta \in \mathbb{N}_k^n$, denote the vector of column blocks

$$\begin{aligned} W_\beta &:= (Y_{\alpha+\beta})_{\alpha \in \mathbb{N}_k^n}, \\ w_\beta &:= (y_{\alpha+\beta})_{\alpha \in \mathbb{N}_k^n}. \end{aligned}$$

For a subset $\Lambda \subseteq \mathbb{N}_k^n$, if $\{W_\beta\}_{\beta \in \Lambda}$ is linearly dependent, then $\{w_\beta\}_{\beta \in \Lambda}$ is clearly linearly dependent, because $y_{\alpha+\beta} = B \bullet Y_{\alpha+\beta}$. So,

$$\text{rank}\, M_k[y] \leq \text{rank}_{bl}\, M_k[Y]. \tag{10.3.14}$$

We prove that the reverse inequality of (10.3.14) also holds. Consider an arbitrary linear dependence relation among the columns of $M_k[y]$, say,

$$M_k[y]\lambda = \sum_{\beta \in \mathbb{N}_k^n} \lambda_\beta w_\beta = 0 \quad \text{for some} \quad \lambda = (\lambda_\beta)_{\beta \in \mathbb{N}_k^n}.$$

Then one can see that

$$\lambda^T M_k[y]\lambda = \sum_{\alpha, \beta \in \mathbb{N}_k^n} y_{\alpha+\beta}\lambda_\alpha\lambda_\beta = \sum_{\alpha, \beta \in \mathbb{N}_k^n} B \bullet Y_{\alpha+\beta}\lambda_\alpha\lambda_\beta = B \bullet X = 0,$$

where the matrix $X$ is such that

$$X = \sum_{\alpha, \beta \in \mathbb{N}_k^n} Y_{\alpha+\beta}\lambda_\alpha\lambda_\beta = (\lambda \otimes I_\ell)^T \cdot M_k[Y] \cdot (\lambda \otimes I_\ell) \succeq 0.$$

Since $B$ is positive definite, $B \bullet X = 0$ implies that

$$X = (\lambda \otimes I_\ell)^T M_k[Y](\lambda \otimes I_\ell) = 0.$$

Since $M_k[Y] \succeq 0$, the above implies that

$$M_k[Y](\lambda \otimes I_\ell) = \sum_{\beta \in \mathbb{N}^n_k} \lambda_\beta W_\beta = 0.$$

This means that the same linear dependence relation holds for the block columns of $M_k[Y]$, so

$$\operatorname{rank} M_k[y] \geq \operatorname{rank}_{bl} M_k[Y]. \tag{10.3.15}$$

Hence, the equality (10.3.13) follows from (10.3.14) and (10.3.15). □

In the following, we give a flat extension theorem for MV-TKMPs, like Theorem 2.7.7. For a vector $\xi \in \mathbb{R}^\ell$, denote the scalar-valued tms

$$y(\xi) := (\xi^T Y_\alpha \xi)_{\alpha \in \mathbb{N}^n_{2t}}. \tag{10.3.16}$$

The moment matrix of $y(\xi)$ can be written as

$$M_t[y(\xi)] = (I_N \otimes \xi)^T M_t[Y](I_N \otimes \xi), \tag{10.3.17}$$

where $N = \binom{n+t}{t}$. For the set $K$ as in (10.3.9), denote the degree

$$d_0 = \max\left\{\left\lceil \frac{1}{2}\deg(h)\right\rceil, \left\lceil \frac{1}{2}\deg(\mathcal{G})\right\rceil\right\}.$$

The mv-tms $Y$ is said to be *flat* if

$$\operatorname{rank}_{bl} M_t[Y] = \operatorname{rank}_{bl} M_{t-d_0}[Y]. \tag{10.3.18}$$

**Theorem 10.3.5.** *Let $K$ be the set as in (10.3.9) and let $Y \in \left(\mathcal{S}^\ell\right)^{\mathbb{N}^n_{2t}}$ be a mv-tms satisfying (10.3.10). If $Y$ satisfies (10.3.18), then it admits a $r$-atomic $K$-representing measure with $r = \operatorname{rank}_{bl} M_t[Y]$. Moreover, the representing measure for $Y$ is unique.*

**Proof.** Pick an arbitrary positive definite matrix $B \in \mathcal{S}^\ell_{++}$. Let $y = (y_\alpha)$ be the scalar-valued tms as in (10.3.12). Consider a psd decomposition for $B$:

$$B = b_1 b_1^T + \cdots + b_\ell b_\ell^T, \quad b_i \in \mathbb{R}^\ell.$$

Then, for each $h_j$ and $g \in \{1\} \cup \mathcal{G}$, we have

$$\mathscr{V}^{(2t)}_{h_j}[y] = \sum_{i=1}^\ell (I_{N_1} \otimes b_i^T) \mathscr{V}^{(2t)}_{h_j}[Y] b_i = 0,$$

$$L^{(t)}_g[y] = \sum_{i=1}^\ell (I_N \otimes b_i)^T L^{(t)}_g[Y](I_N \otimes b_i) \succeq 0,$$

where $N_1$ is the length of $\mathscr{V}^{(2t)}_{h_j}[y]$ and $N$ is that of $L^{(t)}_g[y]$. By Lemma 10.3.4, the rank condition (10.3.18) implies

$$r = \operatorname{rank} M_t[y] = \operatorname{rank} M_{t-d_0}[y].$$

Therefore, the scalar-valued tms $y \in \mathbb{R}^{\mathbb{N}^n_{2t}}$ is flat. For each $\alpha \in \mathbb{N}^n_t$, there exists a polynomial $\phi_\alpha \in \mathbb{R}[x]_{t-d_0}$ such that

$$Y_\alpha = \phi_\alpha(Y) := \sum_{\beta \in \mathbb{N}^n_{t-d_0}} \phi_{\alpha,\beta} Y_\beta.$$

The above implies that

$$y_\alpha = \sum_{\beta \in \mathbb{N}^n_{t-d_0}} \phi_{\alpha,\beta} y_\beta.$$

By Lemma 2.7.8, the polynomial system

$$\psi_\alpha(x) := x^\alpha - \sum_{\beta \in \mathbb{N}^n_{t-d_0}} \phi_{\alpha,\beta} x^\beta = 0 \quad (\alpha \in \mathbb{N}^n_t)$$

has $r$ distinct solutions in $K$, say, $v_1, \ldots, v_r \in K$. Since they are common zeros of the polynomials $\psi_\alpha(x)$, the points $v_1, \ldots, v_r$ are independent of the choice of $B$. By Theorem 2.7.7, there exist positive scalars $\lambda_1, \ldots, \lambda_r > 0$ such that

$$M_t[y] = \lambda_1 [v_1]_t [v_1]_t^T + \cdots + \lambda_r [v_r]_t [v_r]_t^T. \tag{10.3.19}$$

Let $V = \left([v_1]_{t-d_0} \quad \cdots \quad [v_r]_{t-d_0}\right)^T$, then

$$M_{t-d_0}[y] = V^T \cdot \mathrm{diag}(\lambda_1, \ldots, \lambda_r) \cdot V.$$

Since rank $M_{t-d_0}[y] = r$, the above implies that

$$r = \mathrm{rank}\ V.$$

Hence, for each $i = 1, \ldots, r$, there exists $p_i \in \mathbb{R}[x]_{t-d_0}$ such that $V \cdot \mathrm{vec}(p_i) = e_i$. These polynomials $p_1, \ldots, p_r$ satisfy the relations

$$p_i(v_j) = 0\ (i \neq j), \quad p_i(v_i) = 1.$$

Therefore, (10.3.19) implies that for each $i$,

$$\lambda_i = \mathrm{vec}(p_i)^T M_t[y] \, \mathrm{vec}(p_i).$$

Hence, it holds that

$$M_t[y] = \sum_{i=1}^r \left( \mathrm{vec}(p_i)^T M_t[y] \, \mathrm{vec}(p_i) \right) \cdot [v_i]_t [v_i]_t^T.$$

For each $\alpha \in \mathbb{N}^n_{2t}$, since $y_\alpha = B \bullet Y_\alpha$, we get

$$B \bullet Y_\alpha = \sum_{i=1}^r \left( \mathrm{vec}(p_i)^T M_t[y] \, \mathrm{vec}(p_i) \right) (v_i)^\alpha.$$

For each $i = 1, \ldots, r$, let

$$C_i := (\mathrm{vec}(p_i) \otimes I_\ell)^T M_t[Y](\mathrm{vec}(p_i) \otimes I_\ell) \succeq 0.$$

Then one can verify that

$$B \bullet C_i = \mathrm{vec}(p_i)^T M_t[y] \, \mathrm{vec}(p_i).$$

So, it holds that for all $\alpha \in \mathbb{N}_{2t}^n$

$$B \bullet \left( Y_\alpha - \sum_{i=1}^{r}(v_i)^\alpha C_i \right) = 0.$$

The above is true for all positive definite $B \succ 0$, so

$$Y_\alpha = \sum_{i=1}^{r} C_i(v_i)^\alpha \quad \text{for all } \alpha \in \mathbb{N}_{2t}^n.$$

Hence, $\mu := \sum_{i=1}^{r} C_i \delta_{v_i}$ is a $r$-atomic $K$-representing matrix-valued measure for $Y$.

Note that every representing measure for $Y$ automatically gives one for $y$. Since $y$ is flat, the representing measure for $y$ is unique, so $\mu$ is the unique $K$-representing measure for $Y$.   □

### 10.3.3 ▪ Exercises

**Exercise 10.3.1.** *Let $Y = (Y_\alpha)_{\alpha \in \mathbb{N}_2^2}$ be given as*

$$Y_\alpha = \int_{[-1,1]^2}(xx^T)x^\alpha \mathrm{d}x$$

*for each $\alpha \in \mathbb{N}_2^2$. The above integral is defined for the Lebesgue measure. Construct an explicit finitely atomic representing measure for $Y$, supported in $[-1,1]^2$.*

**Exercise 10.3.2.** *If a mv-tms $Y \in (\mathcal{S}^\ell)^{\mathbb{N}_d^n}$ admits a $K$-representing measure, show that the Riesz functional $\mathscr{L}_Y$ is $K$-positive, i.e., show that*

$$\mathscr{L}_Y(P) \geq 0 \quad \text{for all} \quad P \in \mathscr{P}_d(K)^{\ell \times \ell}.$$

**Exercise 10.3.3.** *If a mv-tms $Y$ belongs to the cone $\mathscr{R}_d(K)^{\ell \times \ell}$, for the set $K$ as in (10.3.9), show that $Y$ must satisfy (10.3.10).*

**Exercise 10.3.4.** *Let $Y = (Y_\alpha)_{\alpha \in \overline{\mathbb{N}}_{2d}^n}$ be a homogeneous mv-tms, where each $Y_\alpha \in \mathcal{S}^\ell$ and $\overline{\mathbb{N}}_{2d}^n = \{\alpha \in \mathbb{N}^n : |\alpha| = 2d\}$. Suppose the homogeneous moment matrix*

$$M_d[Y] := (Y_{\alpha+\beta})_{\alpha,\beta \in \overline{\mathbb{N}}_d^n}$$

*is psd. For the case $(\ell, d) = (2, 1)$ or $n = 2$, show that there exists a matrix-valued measure $\mu$, supported in the unit sphere $\mathbb{S}^{n-1}$ such that $Y_\alpha = \int x^\alpha \mathrm{d}\mu$ for all $\alpha \in \overline{\mathbb{N}}_{2d}^n$. (Hint: use Theorems 10.1.1 and 10.1.2.)*

## 10.4 ▪ Optimization with PMIs

Let $h := (h_1, \ldots, h_s)$ be a tuple of scalar polynomials and let $\mathcal{G}$ be a set of symmetric matrix polynomials. For a scalar polynomial $f(x) \in \mathbb{R}[x]$, we consider the polynomial matrix inequality (PMI) constrained optimization

$$\begin{cases} \min & f(x) \\ s.t. & h_i(x) = 0 \, (i = 1, \ldots, s), \\ & g(x) \succeq 0 \, (g \in \mathcal{G}). \end{cases} \tag{10.4.1}$$

When each $g \in G$ is diagonal, the above reduces to a standard constrained polynomial optimization problem in the form (5.1.1).

In this section, we give the Moment-SOS hierarchy to solve (10.4.1), which was introduced in [123]. For an order $k$, the $k$th moment relaxation is

$$\begin{cases} \min & \langle f, y \rangle \\ \text{s.t.} & \mathscr{V}_{h_i}^{(2k)}[y] = 0 \ (i = 1, \ldots, s), \\ & L_g^{(k)}[y] \succeq 0 \ (g \in \mathcal{G}), \\ & M_k[y] \succeq 0, \\ & y_0 = 1, \ y \in \mathbb{R}^{\mathbb{N}_{2k}^n}. \end{cases} \qquad (10.4.2)$$

The localizing matrix $L_g^{(k)}[y]$ is defined as in (10.2.9) for the case $\ell = 1$. Similarly, the dual optimization of (10.4.2) is the $k$th order SOS relaxation

$$\begin{cases} \max & \gamma \\ \text{s.t.} & f - \gamma \in \text{Ideal}[h]_{2k} + \text{QM}[\mathcal{G}]_{2k}. \end{cases} \qquad (10.4.3)$$

Let $\vartheta_*$, $\vartheta_k$ denote the optimal values of (10.4.1), (10.4.3) respectively. Like the case of scalar polynomial optimization, it also holds that the monotonicity relation

$$\vartheta_k \leq \vartheta_{k+1} \leq \cdots \leq \vartheta_*.$$

The following are the convergence properties for the Moment-SOS hierarchy of (10.4.2)-(10.4.3).

**Theorem 10.4.1.** *([123]) Assume the sum* $\text{Ideal}[h] + \text{QM}[\mathcal{G}]$ *is archimedean. Then, we have the following:*

(i) *The optimization (10.4.1) is infeasible if and only if the moment relaxation (10.4.2) is infeasible for some $k$.*

(ii) *If (10.4.1) is feasible and bounded below, then*

$$\lim_{k \to \infty} \vartheta_k = \vartheta_*.$$

**Proof.** Let $H := \text{diag}(h_1, \ldots, h_s, -h_1, \ldots, -h_s)$, then (see exercise)

$$\text{QM}[\mathcal{G} \cup \{H\}] = \text{Ideal}[h] + \text{QM}[\mathcal{G}].$$

A point $x$ is feasible for (10.4.1) if and only if $H(x) \succeq 0$ and $g(x) \succeq 0$ for all $g \in \mathcal{G}$.

(i) The "if" direction is obvious. To prove the "only if" direction, suppose (10.4.1) is infeasible. Then $H(x) \not\succeq 0$ for all $x$ satisfying $g(x) \succeq 0$ $(g \in \mathcal{G})$. Then Theorem 10.1.5 implies that $-1 \in \text{QM}[\mathcal{G} \cup \{H\}]$, which then yields

$$-1 \in \text{Ideal}[h]_{2k} + \text{QM}[\mathcal{G}]_{2k}$$

for some $k \in \mathbb{N}$. This means that (10.4.3) is unbounded above, which then implies that (10.4.2) is infeasible.

(ii) If (10.4.1) is feasible and bounded below, the convergence of $\vartheta_k$ to $\vartheta_*$ is implied by Theorem 10.1.4. $\square$

The minimizers for (10.4.1) can be obtained as follows. Let $K$ be the feasible set of (10.4.1) and let

$$d_0 := \max \left\{ \left\lceil \frac{1}{2} \deg(h) \right\rceil, \left\lceil \frac{1}{2} \deg(\mathcal{G}) \right\rceil \right\}.$$

**Theorem 10.4.2.** *Let $d_0$ be as in the above. Suppose $y^*$ is a minimizer of the moment relaxation (10.4.2). If there exists an integer $t \in [d_0, k]$ such that*

$$r := \operatorname{rank} M_t[y^*] = M_{t-d_0}[y^*],$$

*then $\vartheta_k = \vartheta_*$ and for the $r$-atomic representing measure $\mu$ for $y^*|_{2t}$, each point $u \in \operatorname{supp}(\mu)$ is a minimizer of (10.4.1).*

The proof for Theorem 10.4.2 is similar to the scalar case of polynomial optimization, so it is left as an exercise. To illustrate the relaxation (10.4.2), we give some expositions.

**Example 10.4.3.** *Consider the PMI constrained optimization*

$$\begin{cases} \min_{x \in \mathbb{R}^3} & x_2 x_3 - x_3^2 \\ s.t. & \begin{bmatrix} -x_3 - x_1^2 & x_1 x_2 \\ x_1 x_2 & x_3 - x_2^2 \end{bmatrix} \succeq 0. \end{cases}$$

*For the order $k = 1$, the moment relaxation is*

$$\begin{cases} \min & y_{011} - y_{002} \\ s.t. & \begin{bmatrix} -y_{001} - y_{200} & y_{110} \\ y_{110} & y_{001} - y_{020} \end{bmatrix} \succeq 0, \\ & M_1[y] \succeq 0, \ y_{000} = 1, \\ & y \in \mathbb{R}^{\mathbb{N}_2^3}. \end{cases}$$

*For the order $k = 2$, the moment relaxation is*

$$\begin{cases} \min & y_{011} - y_{002} \\ s.t. & M_2[y] \succeq 0, \\ & L_G^{(2)}[y] \succeq 0, \\ & y_{000} = 1, y \in \mathbb{R}^{\mathbb{N}_4^3}, \end{cases}$$

*where $M_2[y]$ is the moment matrix (see §2.4.1) and*

$$L_G^{(2)}[y] = \begin{bmatrix} L_{-x_3 - x_1^2}^{(2)}[y] & L_{x_1 x_2}^{(2)}[y] \\ L_{x_1 x_2}^{(2)}[y] & L_{x_3 - x_2^2}^{(2)}[y] \end{bmatrix}.$$

*In the above, the localizing matrices are*

$$L_{x_1 x_2}^{(2)}[y] = \begin{bmatrix} y_{110} & y_{210} & y_{120} & y_{111} \\ y_{210} & y_{310} & y_{220} & y_{211} \\ y_{120} & y_{220} & y_{130} & y_{121} \\ y_{111} & y_{211} & y_{121} & y_{112} \end{bmatrix},$$

$$L_{x_3 - x_2^2}^{(2)}[y] = \begin{bmatrix} y_{001} - y_{020} & y_{101} - y_{120} & y_{011} - y_{030} & y_{002} - y_{021} \\ y_{101} - y_{120} & y_{201} - y_{220} & y_{111} - y_{130} & y_{102} - y_{121} \\ y_{011} - y_{030} & y_{111} - y_{130} & y_{021} - y_{040} & y_{012} - y_{031} \\ y_{002} - y_{021} & y_{102} - y_{121} & y_{012} - y_{031} & y_{003} - y_{022} \end{bmatrix},$$

$$L_{-x_3 - x_1^2}^{(2)}[y] = \begin{bmatrix} -y_{001} - y_{200} & -y_{101} - y_{300} & -y_{011} - y_{210} & -y_{002} - y_{201} \\ -y_{101} - y_{300} & -y_{111} - y_{310} & -y_{111} - y_{310} & -y_{102} - y_{301} \\ -y_{011} - y_{210} & -y_{111} - y_{310} & -y_{021} - y_{220} & -y_{012} - y_{211} \\ -y_{002} - y_{201} & -y_{102} - y_{301} & -y_{012} - y_{211} & -y_{003} - y_{202} \end{bmatrix}.$$

*The moment relaxation (10.4.2) can be implemented in* **YALMIP** *[181] as follows:*

```
sdpvar x_1 x_2 x_3
f = x_2*x_3-x_3^2;
G = [-x_3-x_1^2   x_1*x_2;  x_1*x_2   x_3-x_2^2 ];
k = 2; % give a value for the relaxation order
[sol,xopt,momentdata] = solvemoment([G>=0],f,[], 2);
relaxvalue(f),
xopt{:},
```

*When $k = 1$, the moment relaxation (10.4.2) is unbounded below and the SOS relaxation (10.4.3) is infeasible. When $k = 2$, the moment relaxation (10.4.2) is tight and $\vartheta_2 = 0$. We also get the minimizer $(0, 0, 0)$.*

**Example 10.4.4.** *Consider the PMI constrained optimization*

$$
\begin{cases}
\min\limits_{x \in \mathbb{R}^3} & -x_1 - x_2 x_3 \\
s.t. & \begin{bmatrix} 1 - 2x_2^2 & x_1 x_2 & x_1 x_3 \\ x_1 x_2 & 1 - 3x_3^2 & x_2 x_3 \\ x_1 x_3 & x_2 x_3 & 1 - 4x_1^2 \end{bmatrix} \succeq 0.
\end{cases}
$$

*For the order $k = 1$, the moment relaxation is*

$$
\begin{cases}
\min & -y_{100} - y_{011} \\
s.t. & \begin{bmatrix} y_{000} - 2y_{020} & y_{110} & y_{101} \\ y_{110} & y_{000} - 3y_{002} & y_{011} \\ y_{101} & y_{011} & y_{000} - 4y_{200} \end{bmatrix} \succeq 0, \\
& y_{000} = 1, \ M_1[y] \succeq 0.
\end{cases}
$$

*For the order $k = 2$, the moment relaxation is*

$$
\begin{cases}
\min & -y_{100} - y_{011} \\
s.t. & M_2[y] \succeq 0, L_G^{(2)}[y] \succeq 0, \\
& y_{000} = 1, \ y \in \mathbb{R}^{\mathbb{N}_4^3},
\end{cases}
$$

*where the block localizing matrix $L_G^{(2)}[y]$ is*

$$
L_G^{(2)}[y] = \begin{bmatrix} L_{1-2x_2^2}^{(2)}[y] & L_{x_1 x_2}^{(2)}[y] & L_{x_1 x_3}^{(2)}[y] \\ L_{x_1 x_2}^{(2)}[y] & L_{1-3x_3^2}^{(2)}[y] & L_{x_2 x_3}^{(2)}[y] \\ L_{x_1 x_3}^{(2)}[y] & L_{x_2 x_3}^{(2)}[y] & L_{1-4x_1^2}^{(2)}[y] \end{bmatrix}.
$$

*In the above, the localizing matrices are respectively*

$$
L_{1-2x_2^2}^{(2)}[y] = \begin{bmatrix} y_{000} - 2y_{020} & y_{100} - 2y_{120} & y_{010} - 2y_{030} & y_{001} - 2y_{021} \\ y_{100} - 2y_{120} & y_{200} - 2y_{220} & y_{110} - 2y_{130} & y_{101} - 2y_{121} \\ y_{010} - 2y_{030} & y_{110} - 2y_{130} & y_{020} - 2y_{040} & y_{011} - 2y_{031} \\ y_{001} - 2y_{021} & y_{101} - 2y_{121} & y_{011} - 2y_{031} & y_{002} - 2y_{022} \end{bmatrix},
$$

$$
L_{1-3x_3^2}^{(2)}[y] = \begin{bmatrix} y_{000} - 3y_{002} & y_{100} - 3y_{102} & y_{010} - 3y_{012} & y_{001} - 3y_{003} \\ y_{100} - 3y_{102} & y_{200} - 3y_{202} & y_{110} - 3y_{112} & y_{101} - 3y_{103} \\ y_{010} - 3y_{012} & y_{110} - 3y_{112} & y_{020} - 3y_{022} & y_{011} - 3y_{013} \\ y_{001} - 3y_{003} & y_{101} - 3y_{103} & y_{011} - 3y_{013} & y_{002} - 3y_{004} \end{bmatrix},
$$

$$
L_{1-4x_1^2}^{(2)}[y] = \begin{bmatrix} y_{000} - 4y_{200} & y_{100} - 4y_{300} & y_{010} - 4y_{210} & y_{001} - 4y_{201} \\ y_{100} - 4y_{300} & y_{200} - 4y_{400} & y_{110} - 4y_{310} & y_{101} - 4y_{301} \\ y_{010} - 4y_{210} & y_{110} - 4y_{310} & y_{020} - 4y_{220} & y_{011} - 4y_{211} \\ y_{001} - 4y_{201} & y_{101} - 4y_{301} & y_{011} - 4y_{211} & y_{002} - 4y_{202} \end{bmatrix},
$$

$$L^{(2)}_{x_1 x_2}[y] = \begin{bmatrix} y_{110} & y_{210} & y_{120} & y_{111} \\ y_{210} & y_{310} & y_{220} & y_{211} \\ y_{120} & y_{220} & y_{130} & y_{121} \\ y_{111} & y_{211} & y_{121} & y_{112} \end{bmatrix}, \quad L^{(2)}_{x_1 x_3}[y] = \begin{bmatrix} y_{101} & y_{201} & y_{111} & y_{102} \\ y_{201} & y_{301} & y_{211} & y_{202} \\ y_{111} & y_{211} & y_{121} & y_{112} \\ y_{102} & y_{202} & y_{112} & y_{103} \end{bmatrix},$$

$$L^{(2)}_{x_2 x_3}[y] = \begin{bmatrix} y_{011} & y_{111} & y_{021} & y_{012} \\ y_{111} & y_{211} & y_{121} & y_{112} \\ y_{021} & y_{121} & y_{031} & y_{022} \\ y_{012} & y_{112} & y_{022} & y_{013} \end{bmatrix}.$$

*The moment relaxation (10.4.2) can be implemented in* YALMIP *[181] as follows:*

```
sdpvar x_1 x_2 x_3
f = -x_1-x_2*x_3;
G = [1-2*x_2^2    x_1*x_2    x_1*x_3; ...
x_1*x_2    1-3*x_3^2    x_2*x_3; ...
x_1*x_3    x_2*x_3    1-4*x_1^2 ];
k = 2; % give a value for the relaxation order
[sol,xopt,momentdata] = solvemoment([G>=0],f,[], k);
relaxvalue(f),
xopt{:},
```

*When $k = 1$, the moment relaxation (10.4.2) gives the lower bound $\vartheta_1 \approx -0.7496$. When $k = 2$, it gives $\vartheta_2 \approx -0.7301$. The relaxation (10.4.2) is tight for $k = 2$ and we get two optimizers*

$$(0.4472, 0.6325, 0.4472), \quad (0.4472, -0.6325, -0.4472).$$

## 10.4.1 ▪ Exercises

**Exercise 10.4.1.** *Solve the following PMI constrained optimization:*

$$(i) \quad \begin{cases} \min\limits_{x \in \mathbb{R}^3} & x_1 x_2 - x_2 x_3 - x_3 x_1 \\ s.t. & \begin{bmatrix} x_1 & x_3^2 & x_2^2 \\ x_3^2 & x_2 & x_1^2 \\ x_2^2 & x_1^2 & x_3 \end{bmatrix} \succeq 0. \end{cases}$$

$$(ii) \quad \begin{cases} \max\limits_{x \in \mathbb{R}^3} & x_1 x_2 + x_2 x_3 + x_3 x_1 \\ s.t. & \begin{bmatrix} 1 + x_1 & x_1^2 & x_2^2 \\ x_1^2 & 1 + x_2 & x_3^2 \\ x_2^2 & x_3^2 & 1 + x_3 \end{bmatrix} \succeq 0. \end{cases}$$

**Exercise 10.4.2.** *Determine the value of $x \in \mathbb{R}^1$ such that the minimum eigenvalue value of the following matrix is maximum:*

$$\begin{bmatrix} 0 & 1 & x^2 & x^4 \\ 1 & x & 1 & x^6 \\ x^2 & 1 & x^3 & 1 \\ x^4 & x^6 & 1 & x^5 \end{bmatrix}.$$

**Exercise 10.4.3.** *For a tuple $h := (h_1, \ldots, h_s)$ of scalar polynomials, let*

$$H := diag(h_1, \ldots, h_s, -h_1, \ldots, -h_s).$$

*Show that* $\mathrm{QM}[H \cup \mathcal{G}] = \mathrm{Ideal}[h] + \mathrm{QM}[\mathcal{G}]$ *for every set $\mathcal{G}$ of symmetric matrix polynomials.*

**Exercise 10.4.4.** *Prove Theorem 10.4.2.*

## 10.5 ▪ Matrix Convexity

A symmetric matrix polynomial $P \in \mathcal{S}\mathbb{R}[x]^{\ell \times \ell}$ gives a matrix-valued function $P(x)$ in $x \in \mathbb{R}^n$. The matrix function $P(x)$ is said to be *convex* over a convex domain $\mathcal{D} \subseteq \mathbb{R}^n$ if for all $u, v \in \mathcal{D}$ and for all $0 \leq \lambda \leq 1$ it holds that

$$P(\lambda u + (1 - \lambda)v) \preceq \lambda P(u) + (1 - \lambda)P(v).$$

(The inequality $X \preceq Y$ means that $Y - X$ is psd.) If $P$ is convex over the entire space $\mathbb{R}^n$, then $P$ is just simply called convex. Assume $\mathcal{D}$ has nonempty interior. The convexity of $P(x)$ over $\mathcal{D}$ is equivalent to

$$\nabla_x^2 \big(\xi^T P(x)\xi\big) \succeq 0 \quad \forall \xi \in \mathbb{R}^\ell, \forall x \in \mathcal{D}.$$

Similarly, if $-P$ is convex over $\mathcal{D}$, then $P$ is called *concave* over $\mathcal{D}$. Clearly, if $P$ is concave, then the super level set of $P(x) \succeq 0$ is convex, while the converse is typically not true.

For $P \in \mathcal{S}\mathbb{R}[x]^{\ell \times \ell}$, its *block Hessian* is the block matrix

$$H := \big(\nabla_x^2 P_{ij}\big)_{1 \leq i,j \leq \ell}. \tag{10.5.1}$$

If $H \succeq 0$ on $\mathcal{D}$, then $P$ is convex over $\mathcal{D}$, because for all $\xi \in \mathbb{R}^\ell$

$$\nabla_x^2(\xi^T P \xi) = \sum_{i,j=1}^m \xi_i \xi_j \nabla_x^2 P_{ij} = (\xi \otimes I_n)^T H (\xi \otimes I_n) \succeq 0. \tag{10.5.2}$$

Note that $P$ being convex is weaker than $H$ being psd. For instance, for the matrix polynomial $G$ in Example 10.5.3, the block Hessian of $-G$ is not psd while $-G$ is convex.

It is typically difficult to check convexity of matrix polynomials. Even for the case of quadratic matrix polynomials, the task of checking convexity is already NP-hard (see [176]). A relatively easier checkable condition is the *SOS-convexity*,is the *SOS-convexity*.

**Definition 10.5.1.** *([218]) Let $P \in \mathcal{S}\mathbb{R}[x]^{\ell \times \ell}$ be a matrix polynomial.*

(i) *The $P$ is SOS-convex if for every $\xi \in \mathbb{R}^\ell$ there exists a matrix polynomial $F(x) \in \mathbb{R}[x]^{t \times \ell}$, for some $t \in \mathbb{N}$, such that*

$$\nabla_x^2(\xi^T P(x)\xi) = F(x)^T F(x).$$

*The coefficients of the above $F(x)$ depend on $\xi$.*

(ii) *The $P$ is uniformly SOS-convex if there exists $F(\xi, x) \in \mathbb{R}[\xi, x]^{t \times \ell}$, for some $t \in \mathbb{N}$, such that*

$$\nabla_x^2(\xi^T P(x)\xi) = F(\xi, x)^T F(\xi, x).$$

*The above $F(\xi, x)$ is a matrix polynomial in $(\xi, x)$ jointly.*

(iii) *If $-P$ is SOS-convex, then $P$ is called SOS-concave. The same is defined for uniform SOS-concavity.*

The SOS-convexity requires that the scalar polynomial $\xi^T P(x)\xi$ is SOS-convex for every fixed $\xi \in \mathbb{R}^\ell$. This may not be computationally convenient to check. However, it is more convenient to check uniform SOS-convexity. Note that the uniform SOS-convexity of $P$ is equivalent to its block Hessian matrix polynomial $H$ being SHS. When $P$ is quadratic, $P$ is convex if and only if it is SOS-convex. The proofs for these are left as exercises.

When $P(x)$ is SOS-convex, it is not necessarily that $P(x)$ is uniformly SOS-convex. A counterexample can be found in Example 10.5.3. The following are some classes of (uniformly) SOS-convex matrix polynomials.

- If $Q(x) = Q_1(x) + Q_2(x) + Q_{2d}(x)$, where $Q_1(x)$ is linear in $x$, $Q_2(x)$ is homogeneous quadratic in $x$, and $Q_{2d}(x)$ is homogeneous of degree $2d$ $(d > 1)$, then for every $\xi \in \mathbb{R}^m$

$$\nabla_x^2(\xi^T Q(x)\xi) \; = \; \nabla_x^2\left(\xi^T Q_2(x)\xi\right) + \nabla_x^2\left(\xi^T Q_{2d}(x)\xi\right).$$

The $Q(x)$ is SOS-convex if and only if

$$\nabla_x^2(\xi^T Q_2(x)\xi) \succeq 0 \quad \text{and} \quad \nabla_x^2(\xi^T Q_{2d}(x)\xi) \succeq 0$$

for all $x$ and for all $\xi$. For the case $n = 2$, $\nabla_x^2(\xi^T Q_{2d}(x)\xi)$ is psd in $x$ if and only if it is SHS (see Theorem 10.1.1).

- For the univariate case $n = 1$, $P(x)$ is convex if and only if

$$H \; := \; \left(P_{ij}''(x)\right)_{\substack{i=1,\dots,\ell \\ j=1,\dots,\ell}} \succeq 0 \quad \text{on} \quad \mathbb{R}^1.$$

This is equivalent to that $H$ is SHS. In this case, the convexity coincides with the uniform SOS-convexity.

### 10.5.1 ▪ Convex sets given by PMIs

A matrix polynomial $G \in \mathcal{S}\mathbb{R}[x]_{2d}^{\ell \times \ell}$ determines the semialgebraic set

$$S \; := \; \{x \in \mathcal{D} : G(x) \succeq 0\}, \tag{10.5.3}$$

for a convex domain $\mathcal{D} \subseteq \mathbb{R}^n$. In this subsection, the domain $\mathcal{D} = \mathbb{R}^n$. Suppose $G$ has the expansion

$$G(x) \; = \; \sum_{\alpha \in \mathbb{N}_{2d}^n} G_\alpha x_1^{\alpha_1} \cdots x_n^{\alpha_n},$$

where each $G_\alpha$ is a symmetric matrix in $\mathcal{S}^\ell$. We are interested in a semidefinite representation for $S$. Define the linear matrix function

$$G[y] \; := \; \sum_{\alpha \in \mathbb{N}_{2d}^n} y_\alpha G_\alpha \tag{10.5.4}$$

in $y = (y_\alpha) \in \mathbb{R}^{\mathbb{N}_{2d}^n}$. The set $S$ can be equivalently expressed as

$$S = \left\{x \in \mathbb{R}^n : y = [x]_{2d}, \; G[y] \succeq 0, \; M_d[y] \succeq 0\right\}.$$

If the equality $y = [x]_{2d}$ is ignored, then the set $S$ as in (10.5.3) is contained in the projection

$$\Gamma \; := \; \left\{ x \in \mathbb{R}^n \;\middle|\; \begin{array}{c} \exists\, y \in \mathbb{R}^{\mathbb{N}_{2d}^n}, \; y_0 = 1, \\ G[y] \succeq 0, \; M_d[y] \succeq 0, \\ x = (y_{e_1}, \dots, y_{e_n}) \end{array} \right\}. \tag{10.5.5}$$

So, $S \subseteq \Gamma$. The equality $S = \Gamma$ holds under certain conditions.

**Theorem 10.5.2.** *([218]) Let $S$ be the set as in (10.5.3) and $\Gamma$ be as in (10.5.5). If $G$ is SOS-concave, then $S = \Gamma$.*

**Proof.** Since $S \subseteq \Gamma$, we only need to prove the reverse containment. Pick an arbitrary $\xi \in \mathbb{R}^\ell$, then the scalar polynomial $g_\xi := \xi^T G(x)\xi$ is SOS-concave, since $G$ is SOS-concave. Each $y$ in (10.5.5) satisfies

$$\langle g_\xi, y \rangle \geq 0, \quad M_d[y] \succeq 0, \quad y_0 = 1.$$

For each $x = (y_{e_1}, \ldots, y_{e_n}) \in \Gamma$, Jensen's inequality (see Theorem 7.1.6) implies that

$$g_\xi(x) \geq \langle g_\xi, y \rangle = \xi^T G[y]\xi \geq 0.$$

The above holds for all $\xi \in \mathbb{R}^\ell$, so $G(x) \succeq 0$. Hence, $S \supseteq \Gamma$ and they are equal. $\square$

The following are some expositions for Theorem 10.5.2.

**Example 10.5.3.** *([218]) Consider the set $S$ given by the PMI*

$$G(x) := \begin{bmatrix} 2 - x_1^2 - 2x_3^2 & 1 + x_1 x_2 & x_1 x_3 \\ 1 + x_1 x_2 & 2 - x_2^2 - 2x_1^2 & 1 + x_2 x_3 \\ x_1 x_3 & 1 + x_2 x_3 & 2 - x_3^2 - 2x_2^2 \end{bmatrix} \succeq 0.$$

*The Hessian $\nabla_x^2(-\xi^T G\xi) \succeq 0$ for all $\xi \in \mathbb{R}^3$, because the bi-quadratic form*

$$\begin{aligned}
\frac{1}{2} z^T \nabla_x^2 (\xi^T G\xi) z &= z_1^2 \xi_1^2 + z_2^2 \xi_2^2 + z_3^2 \xi_3^2 + 2(z_1^2 \xi_2^2 + z_2^2 \xi_3^2 + z_3^2 \xi_1^2) \\
&\quad -2(z_1 z_2 \xi_1 \xi_2 + z_1 z_3 \xi_1 \xi_3 + z_2 z_3 \xi_2 \xi_3)
\end{aligned}$$

*is nonnegative everywhere (it is Choi's bi-quadratic form in §2.4.2). For each fixed $\xi \in \mathbb{R}^3$, the Hessian $\nabla_x^2(-\xi^T G(x)\xi)$ is SHS, so $G(x)$ is SOS-concave. However, if $\xi$ is considered as an indeterminant variable, then $\nabla_x^2(-\xi^T G(x)\xi)$ is not SHS in $\xi$. This is because $\frac{1}{2} z^T \nabla_x^2 (\xi^T G\xi) z$ is not SOS (see §2.4.2). So, $G(x)$ is not uniformly SOS-concave. By Theorem 10.5.2, the set $S$ can be equivalently given as follows (the $y_{ijk}$'s are lifting variables):*

$$\begin{bmatrix} 2 - y_{200} - 2y_{002} & 1 + y_{110} & y_{101} \\ 1 + y_{110} & 2 - y_{020} - 2y_{200} & 1 + y_{011} \\ y_{101} & 1 + y_{011} & 2 - y_{002} - 2y_{020} \end{bmatrix} \succeq 0,$$

$$\begin{bmatrix} 1 & x_1 & x_2 & x_3 \\ x_1 & y_{200} & y_{110} & y_{101} \\ x_2 & y_{110} & y_{020} & y_{011} \\ x_3 & y_{101} & y_{011} & y_{002} \end{bmatrix} \succeq 0.$$

**Example 10.5.4.** *([218]) Consider the matrix polynomial*

$$G := \begin{bmatrix} 2 - 2x_1^4 - 4x_1^2 x_2^2 - 2x_2^4 & 3 - x_1^3 x_2 - x_1 x_2^3 \\ 3 - x_1^3 x_2 - x_1 x_2^3 & 5 - x_1^4 - 4x_1^2 x_2^2 - x_2^4 \end{bmatrix}.$$

*It is uniformly SOS-concave, since* $\nabla_x^2(-\xi^T G\xi) = H_1 + H_2 + H_3 + H_4$, *where*

$$H_1 = 2\begin{bmatrix} 2\xi_1 x_1 + \xi_2 x_2 \\ 2\xi_1 x_2 + \xi_2 x_1 \end{bmatrix} \begin{bmatrix} 2\xi_1 x_1 + \xi_2 x_2 \\ 2\xi_1 x_2 + \xi_2 x_1 \end{bmatrix}^T, \quad H_2 = 8(\xi_1^2 + \xi_2^2)\begin{bmatrix} x_1^2 & x_1 x_2 \\ x_1 x_2 & x_2^2 \end{bmatrix},$$

$$H_3 = 2\begin{bmatrix} \xi_1 x_1 & \xi_2 x_2 & \xi_2 x_1 \\ \xi_2 x_1 & \xi_1 x_2 & \xi_2 x_2 \end{bmatrix} \begin{bmatrix} \xi_1 x_1 & \xi_2 x_2 & \xi_2 x_1 \\ \xi_2 x_1 & \xi_1 x_2 & \xi_2 x_2 \end{bmatrix}^T,$$

$$H_4 = 2\left( ((\xi^T x)^2 + \xi_2^2 x_1^2)\begin{bmatrix} 1 & 0 \\ 0 & 1 \end{bmatrix} + \xi_1^2 \begin{bmatrix} 2x_1^2 + 4x_2^2 & 0 \\ 0 & 3x_1^2 + 3x_2^2 \end{bmatrix} \right).$$

*By Theorem 10.5.2, the set $S$ can be equivalently given as*

$$\begin{bmatrix} 2 - 2y_{40} - 4y_{22} - 2y_{04} & 3 - y_{31} - y_{13} \\ 3 - y_{31} - y_{13} & 5 - y_{40} - 4y_{22} - y_{04} \end{bmatrix} \succeq 0,$$

$$\begin{bmatrix} 1 & x_1 & x_2 & y_{20} & y_{11} & y_{02} \\ x_1 & y_{20} & y_{11} & y_{30} & y_{21} & y_{12} \\ x_2 & y_{11} & y_{02} & y_{21} & y_{12} & y_{03} \\ y_{20} & y_{30} & y_{21} & y_{40} & y_{31} & y_{22} \\ y_{11} & y_{21} & y_{12} & y_{31} & y_{22} & y_{13} \\ y_{02} & y_{12} & y_{03} & y_{22} & y_{13} & y_{04} \end{bmatrix} \succeq 0.$$

### 10.5.2 ▪ Strictly convex matrix polynomials

We consider the case that $G$ is concave on a convex domain $\mathcal{D}$ such that

$$\mathcal{D} := \{x \in \mathbb{R}^n \mid g_1(x) \geq 0, \ldots, g_m(x) \geq 0\}, \tag{10.5.6}$$

given by a polynomial tuple $g := (g_1, \ldots, g_m)$. Denote

$$d_0 := \max_{1 \leq j \leq m} \left\{ \lceil \tfrac{1}{2}\deg(G)\rceil, \lceil \tfrac{1}{2}\deg(g_j)\rceil \right\}.$$

For each integer $k \geq d_0$, the set $S$ as in (10.5.3) can be equivalently expressed as

$$S = \left\{ x \in \mathbb{R}^n \,\middle|\, \begin{array}{l} y = [x]_{2k}, \\ M_k[y] \succeq 0, \; G[y] \succeq 0, \\ L_{g_j}^{(k)}[y] \succeq 0 \,(j = 1, \ldots, m) \end{array} \right\}.$$

If the equality $y = [x]_{2k}$ is ignored, then the set $S$ as in (10.5.3) is contained in the projection

$$\Gamma_k := \left\{ x \in \mathbb{R}^n \,\middle|\, \begin{array}{l} \exists y \in \mathbb{R}^{\mathbb{N}_{2k}^n}, \; y_0 = 1, \\ x = (y_{e_1}, \ldots, y_{e_n}), \\ M_k[y] \succeq 0, \; G[y] \succeq 0, \\ L_{g_j}^{(k)}[y] \succeq 0 \,(j = 1, \ldots, m) \end{array} \right\}. \tag{10.5.7}$$

Clearly, it holds that the nesting containment

$$\Gamma_{d_0} \supseteq \cdots \supseteq \Gamma_k \supseteq \cdots \supseteq S. \tag{10.5.8}$$

We look for conditions ensuring $\Gamma_k = S$ for some $k$. The matrix function $G(x)$ is said to be *strictly concave* over $\mathcal{D}$ if the Hessian $\nabla_x^2(-\xi^T G(x)\xi)$ is positive definite for all $x \in \mathcal{D}$ and for all $\xi \neq 0$. We have the following conclusion.

**Theorem 10.5.5.** *([218, Theorem 3.2]) Let $S$ be the set as in (10.5.3), where $\mathcal{D}$ is given as in (10.5.6). Assume the following:*

(i) $\mathcal{D}$ is a convex domain, $S \subseteq \text{int}\,(\mathcal{D})$, and $G(u) \succ 0$ for some $u \in \mathcal{D}$;

(ii) the quadratic module $\text{QM}[g]$ is archimedean;

(iii) $G(x)$ is concave over $\mathcal{D}$.

If the Hessian $\nabla_x^2(-\xi^T G(v)\xi)$ is positive definite for all $v$ lying on the boundary of $S$ and for all $\xi \neq 0$, then $S = \Gamma_k$, when $k$ is large enough.

### 10.5.3 ▪ Exercises

**Exercise 10.5.1.** Determine whether or not $\begin{bmatrix} 2x_1^2 & 3x_1x_2 \\ 3x_1x_2 & 5x_2^2 \end{bmatrix}$ is SOS-convex or uniformly SOS-convex.

**Exercise 10.5.2.** For $P \in \mathcal{SR}[x]^{\ell \times \ell}$, show that $P(x)$ is uniformly SOS-convex if and only if the block Hessian $H$ in (10.5.1) is SHS. In particular, if $P$ is quadratic, show that $P(x)$ is convex if and only if it is SOS-convex.

**Exercise 10.5.3.** Suppose $A_0(x)$ is linear in $x$, each $f_i(x)$ is an SOS-convex scalar polynomial in $x$, and $A_1, \ldots, A_k$ are symmetric psd matrices. Show that $P(x) := A_0(x) + f_1(x)A_1 + \cdots + f_k(x)A_k$ is uniformly SOS-convex.

**Exercise 10.5.4.** Suppose $P(x) = A(x) + \text{diag}(f_1(x), \ldots, f_m(x))$, where $A(x)$ is linear in $x$ and each $f_i(x)$ is a scalar polynomial. Show the following:

(i) $P(x)$ is matrix convex if and only if every $f_i(x)$ is convex;

(ii) $P(x)$ is uniformly SOS-convex if and only if every $f_i(x)$ is SOS-convex.

**Exercise 10.5.5.** If $A(x)$ is a linear matrix polynomial in $x$ and $Q(x)$ is a quadratic psd matrix polynomial (they have the same dimension), show that $A(x) + Q(x)$ is SOS-convex.

**Exercise 10.5.6.** For $f \in \mathbb{R}[x]_{2d}$ and $G \in \mathcal{SR}[x]_{2d}^{\ell \times \ell}$, consider the PMI constrained optimization

$$\begin{cases} \min & f(x) \\ s.t. & G(x) \succeq 0. \end{cases} \tag{10.5.9}$$

Consider the following moment relaxation:

$$\begin{cases} \min & \langle f, y \rangle \\ s.t. & M_d[y] \succeq 0,\, G[y] \succeq 0, \\ & y_0 = 1,\, y \in \mathbb{R}^{\mathbb{N}_{2d}^n}. \end{cases} \tag{10.5.10}$$

Let $\vartheta_*$, $\vartheta_d$ denote the optimal values of (10.5.9), (10.5.10) respectively. Assume that $f$ is SOS-convex and $G$ is SOS-concave. Show that the above moment relation is tight, i.e., $\vartheta_d = \vartheta_*$. Moreover, if $y^*$ is a minimizer of (10.5.10), show that $u^* := (y_{e_1}^*, \ldots, y_{e_n}^*)$ is a minimizer for (10.5.9).

## 10.6 ▪ A Matrix Positivstellensatz with Lifting Polynomials

An important question is how to verify that the projection of a semialgebraic set is contained in the projection of another one. We consider semialgebraic sets that are given by PMIs. Checking containments requires a matrix Positivstellensatz with lifting polynomials. For convenience, denote the vector variables

$$x := (x_1, \ldots, x_n), \; y := (y_1, \ldots, y_r), \; z := (z_1, \ldots, z_s).$$

For two symmetric matrix polynomials $G \in \mathcal{SR}[x, y]^{\ell \times \ell}$ and $Q \in \mathcal{SR}[x, z]^{t \times t}$, consider the semialgebraic sets

$$P_G := \{x \in \mathbb{R}^n : \exists\, y \in \mathbb{R}^r,\, G(x, y) \succeq 0\},$$
$$P_Q := \{x \in \mathbb{R}^n : \exists\, z \in \mathbb{R}^s,\, Q(x, z) \succeq 0\}.$$

The $y$ and $z$ are lifting variables. We look for conditions ensuring the inclusion $P_G \subseteq P_Q$. Denote the lifting sets

$$S_G := \{(x, y) : G(x, y) \succeq 0\},$$
$$S_Q := \{(x, z) : Q(x, z) \succeq 0\}.$$

The set $P_G$ (resp., $P_Q$) is the projection of $S_G$ (resp., $S_Q$). We separately discuss the cases that $Q$ is linear in $z$ and $Q$ is nonlinear in $z$. The main results in this section are from the work [147].

### 10.6.1 ▪ The linear case

Assume the matrix polynomial $Q(x, z)$ is given as

$$Q(x, z) := Q_0(x) + z_1 Q_1(x) + \cdots + z_s Q_s(x) \tag{10.6.1}$$

for some matrix polynomials $Q_0(x), \ldots, Q_s(x) \in \mathcal{SR}[x]^{t \times t}$. A sufficient condition for the inclusion $P_G \subseteq P_Q$ is that there exist scalar polynomials

$$p_1(x), \ldots, p_s(x)$$

and matrix polynomials $V_0(x, y), \ldots, V_N(x, y)$ such that

$$\boxed{\begin{aligned} &Q_0(x) + p_1(x)Q_1(x) + \cdots + p_s(x)Q_s(x) \\ &= V_0(x, y)^T V_0(x, y) + \sum_{i=1}^{N} V_i(x, y)^T G(x, y) V_i(x, y). \end{aligned}} \tag{10.6.2}$$

The above is equivalent to the membership

$$Q_0 + p_1 Q_1 + \cdots + p_s Q_s \in \mathrm{QM}[G]^{t \times t}.$$

The representation (10.6.2) implies that for every $\hat{x} \in P_G$, there exists $\hat{y} \in \mathbb{R}^r$ such that $G(\hat{x}, \hat{y}) \succeq 0$, so

$$Q(\hat{x}, \hat{z}) \succeq 0 \quad \text{for} \quad \hat{z} = (p_1(\hat{x}), \ldots, p_s(\hat{x})).$$

Therefore, we see that (10.6.2) ensures the inclusion $P_G \subseteq P_Q$. The representation (10.6.2) is called a *lifted* matrix Positivstellensatz. The $p_s, \ldots, p_s$ are called *lifting polynomials* for the inclusion $P_G \subseteq P_Q$.

Interestingly, the representation (10.6.2) is almost a necessary condition for $P_G \subseteq P_Q$. In fact, if $P_G$ is contained in the interior of $P_Q$, i.e.,

$$P_G \subseteq \mathrm{int}\,(P_Q),$$

then there exist polynomials $p_s, \ldots, p_s$ satisfying (10.6.2), when the quadratic module $QM[G]^{t \times t}$ is archimedean. Since $G$ has variables $x$ and $y$, the quadratic module $QM[G]^{t \times t}$ is a subset of $S\mathbb{R}[x, y]^{t \times t}$. We have the following conclusion.

**Theorem 10.6.1.** *([147]) Let $Q(x, z)$ be as in (10.6.1) and let $G \in S\mathbb{R}[x, y]^{\ell \times \ell}$ be such that $QM[G]^{t \times t}$ is archimedean. Assume that for each $x \in P_G$ there exists $z \in \mathbb{R}^s$ with $Q(x, z) \succ 0$. Then, there exists a polynomial tuple $p(x) := (p_1(x), \ldots, p_s(x))$ such that*

$$Q(x, p(x)) \in QM[G]^{t \times t}.$$

The lifted matrix Positivstellensatz as in (10.6.2) is illustrated in the following examples.

**Example 10.6.2.** *([147]) Consider the matrix polynomials*

$$G(x, y) \quad := \quad \begin{bmatrix} 1 - y - x_1^2 & x_1 x_2 \\ x_1 x_2 & y - x_2^2 \end{bmatrix},$$

$$Q(x, z) \quad := \quad \begin{bmatrix} 1 + z\, x_2 & z - 2\, x_1 \\ z - 2\, x_1 & 1 - z\, x_2 \end{bmatrix}.$$

*The projection*

$$P_G = \{(x_1, x_2) \in \mathbb{R}^2 \colon |x_1 + x_2| \le 1, |x_1 - x_2| \le 1\}$$

*is contained in*

$$P_Q = \{(x_1, x_2) \in \mathbb{R}^2 \colon 1 + x_2^2 - 4x_1^2 x_2^2 \ge 0\}.$$

*The quadratic module $QM[G]^{2 \times 2}$ is archimedean, because*

$$
\begin{aligned}
3 - x_1^2 - x_2^2 - 2y^2 \quad = \quad & e_1^T G(x, y) e_1 + e_2^T G(x, y) e_2 \\
& + (1 - y)^2 + x_2^2 (1 - y)^2 + x_1^2 (1 + y)^2 \\
& + e_1^T G(x, y) e_1 (1 + y)^2 + e_2^T G(x, y) e_2 (1 - y)^2.
\end{aligned}
$$

*We select $p_1 = x_1$ in Theorem 10.6.1, then*

$$Q(x, p(x)) = \frac{1}{2} \begin{bmatrix} x_1 + x_2 & -1 \\ -1 & x_1 - x_2 \end{bmatrix}^2 + \frac{1 - x_1^2 - x_2^2}{2} \begin{bmatrix} 1 & 0 \\ 0 & 1 \end{bmatrix}.$$

*The certificate (10.6.2) for $P_G \subseteq P_Q$ is given by*

$$N = 4, \quad V_0 = \frac{1}{\sqrt{2}} \begin{bmatrix} x_1 + x_2 & -1 \\ -1 & x_1 - x_2 \end{bmatrix},$$

$$V_1 = \begin{bmatrix} \frac{1}{\sqrt{2}} & 0 \\ 0 & 0 \end{bmatrix}, \quad V_2 = \begin{bmatrix} 0 & 0 \\ \frac{1}{\sqrt{2}} & 0 \end{bmatrix}, \quad V_3 = \begin{bmatrix} 0 & \frac{1}{\sqrt{2}} \\ 0 & 0 \end{bmatrix}, \quad V_4 = \begin{bmatrix} 0 & 0 \\ 0 & \frac{1}{\sqrt{2}} \end{bmatrix}.$$

When $QM[G]^{t \times t}$ is not archimedean, the conclusion of Theorem 10.6.1 may not hold. The following is such an example.

**Example 10.6.3.** *([147]) Consider the matrix polynomial*

$$G(x, y) := \begin{bmatrix} y^2(1 - x^2) - 1 & 0 \\ 0 & 2 - x^2 \end{bmatrix}.$$

*Both $x$ and $y$ are univariate. The projection $P_G = (-1, 1)$ is bounded but not closed. The intersection $\mathrm{QM}[G] \cap \mathbb{R}[x]$ is archimedean, since $2 - x^2 \in \mathrm{QM}[G] \cap \mathbb{R}[x]$. However, the quadratic module $\mathrm{QM}[G]$ itself is not archimedean, since $S_G$ is unbounded. One can show that $\mathrm{QM}[G] \cap \mathbb{R}[x]$ is generated by the polynomial $2 - x^2$, i.e.,*

$$\mathrm{QM}[G] \cap \mathbb{R}[x] = \mathrm{QM}[2 - x^2].$$

*This is because for every $g \in \mathrm{QM}[G] \cap \mathbb{R}[x]$, we can write that*

$$g = \sigma_0 + \sigma_1 \cdot (y^2(1 - x^2) - 1) + \sigma_2 \cdot (2 - x^2) \tag{10.6.3}$$

*for SOS polynomials $\sigma_0, \sigma_1, \sigma_2 \in \Sigma[x, y]$. Note that $g$ does not depend on $y$. To cancel $y$ on the right hand side of (10.6.3), we must have $\sigma_1 = 0$. Similarly, $\sigma_0$ and $\sigma_2$ cannot depend on $y$, so we get $g \in \mathrm{QM}[2 - x^2]$. For each $\lambda \in (1, 2)$, the polynomial $\lambda - x^2$ is positive on $P_G$ but it achieves negative values on $[-\sqrt{2}, \sqrt{2}]$, so it does not belong to $\mathrm{QM}[G] \cap \mathbb{R}[x]$. The conclusion of Theorem 10.6.1 fails to hold for $Q = \lambda - x^2$.*

### 10.6.2 ▪ The nonlinear case

We now consider that $Q(x, z)$ is polynomial in $z = (z_1, \ldots, z_s)$ such that

$$Q(x, z) := \sum_{\alpha \in \mathbb{N}_{2d}^s} z_1^{\alpha_1} \cdots z_s^{\alpha_s} Q_\alpha(x) \tag{10.6.4}$$

for some matrix polynomials $Q_\alpha(x) \in \mathcal{SR}[x]^{t \times t}$, where $\alpha = (\alpha_1, \ldots, \alpha_s)$. If we parameterize each $z_i$ by a polynomial $p_i(x)$, a generalization of (10.6.2) is

$$Q(x, p(x)) = \sum_{\alpha \in \mathbb{N}_{2d}^s} p_1(x)^{\alpha_1} \cdots p_s(x)^{\alpha_s} Q_\alpha(x) \in \mathrm{QM}[G]^{t \times t}. \tag{10.6.5}$$

However, (10.6.5) is nonlinear in the coefficients of $p = (p_1, \ldots, p_s)$. Generally, the set of $p$ satisfying (10.6.5) cannot be expressed by semidefinite programs.

In computation, we prefer a more tractable representation than (10.6.5). If each product $p_1(x)^{\alpha_1} \cdots p_s(x)^{\alpha_s}$ is replaced by a new polynomial $p_\alpha(x)$, then (10.6.5) becomes

$$\boxed{\begin{array}{c} p_0 = 1, \quad \sum_{\alpha \in \mathbb{N}_{2d}^s} p_\alpha(x) Q_\alpha(x) \\ = V_0(x, y)^T V_0(x, y) + \sum_{i=1}^{N} V_i(x, y)^T G(x, y) V_i(x, y) \end{array}} \tag{10.6.6}$$

for some matrix polynomials $V_i(x, y)$. Generally, the representation (10.6.6) does not guarantee $P_G \subseteq P_Q$. However, the inclusion $P_G \subseteq P_Q$ holds under some additional assumptions. For a polynomial tuple $p := (p_\alpha)_{\alpha \in \mathbb{N}_{2d}^s}$, we can similarly define the moment matrix polynomial

$$M_d[p] := \left[ p_{\alpha+\beta} \right]_{\alpha, \beta \in \mathbb{N}_d^s}. \tag{10.6.7}$$

For given $x$, recall that $Q(x, z)$ is SOS-concave in $z$ if for every fixed $\xi \in \mathbb{R}^t$, the polynomial $\xi^T Q(x, z) \xi$ is SOS-concave in $z$, i.e., its Hessian $\nabla_z^2(-\xi^T Q(x, z)\xi)$ is SHS in $z$ (see Definition 10.5.1). Under some assumptions, the representation (10.6.6) guarantees the inclusion $P_G \subseteq P_Q$.

**Proposition 10.6.4.** *([147]) Let $Q(x,z)$ be as in (10.6.4). Assume $Q(x,z)$ is SOS-concave in $z$ for every given $x \in P_G$. If there exists a polynomial tuple $p$ satisfying (10.6.6) and $M_d[p] \succeq 0$ on $P_G$, then $P_G \subseteq P_Q$.*

**Proof.** For $x := (x_1, \ldots, x_n)$ and $w := (w_\alpha)_{\alpha \in \mathbb{N}_{2d}^s}$, define

$$F(x,w) := \sum_{\alpha \in \mathbb{N}_{2d}^s} w_\alpha Q_\alpha(x).$$

Pick an arbitrary $x \in P_G$. For each $\alpha$, let $w_\alpha = p_\alpha(x)$. Then note that $w_0 = 1$ and

$$F(x,w) \succeq 0, \quad M_d[w] \succeq 0.$$

For arbitrary $\xi \in \mathbb{R}^t$, the polynomial

$$q(z) := \xi^T Q(x,z)\xi$$

is SOS-concave in $z$. Let $u = (w_1, \ldots, w_s)$, then Jensen's inequality (see Theorem 7.1.6) implies

$$\xi^T Q(x,u)\xi = q(u) \geq \langle q, w \rangle = \xi^T F(x,w)\xi \geq 0.$$

Since $\xi$ is arbitrary, we have $Q(x,u) \succeq 0$ and so $x \in P_Q$. This can also be implied by Theorem 10.5.2. Since $x \in P_G$ is arbitrary, we can conclude $P_G \subseteq P_Q$. $\quad\square$

The above shows that (10.6.6) is a certificate for the inclusion $P_G \subseteq P_Q$, when $Q(x,z)$ is SOS-concave in $z$ and $M_d[p] \succeq 0$ on $P_G$. We remark that (10.6.6) can be checked by solving an SDP, because (10.6.6) is equivalent to

$$\sum_{\alpha \in \mathbb{N}_{2d}^n} p_\alpha Q_\alpha \in \mathrm{QM}[G]^{t \times t}.$$

Like the linear case, (10.6.6) is also necessary for $P_G \subseteq P_Q$, under similar conditions as in Theorem 10.6.1 and Proposition 10.6.4.

**Theorem 10.6.5.** *([147]) Let $G(x,y) \in \mathcal{SR}[x,y]^{\ell \times \ell}$ be such that $\mathrm{QM}[G]^{t \times t}$ is archimedean and let $Q(x,z)$ be as in (10.6.4). Assume that for each given $x \in P_G$, the matrix $Q(x,z)$ is SOS-concave in $z$ and $Q(x,z) \succ 0$ for some $z$. Then, there exist polynomials $p_\alpha \in \mathbb{R}[x]$ ($\alpha \in \mathbb{N}_{2d}^s$) such that (10.6.6) holds and the moment matrix polynomial $M_d[p]$ is SHS.*

The following is an example of the lifted matrix Positivstellensatz as in (10.6.6) for Theorem 10.6.5.

**Example 10.6.6.** *([147]) Consider the matrix polynomials*

$$G(x,y) := \begin{bmatrix} x_1 & y & x_1 \\ y & x_2 & x_2 \\ x_1 & x_2 & 1 \end{bmatrix},$$

$$Q(x,z) := \begin{bmatrix} x_1 + 2z_1 - z_1^2 & z_1 z_2 & x_2 \\ z_1 z_2 & x_2 + 2z_2 - z_2^2 & x_1 \\ x_2 & x_1 & 1 \end{bmatrix}.$$

*Note that $P_G = [0,1]^2$ and $\mathrm{QM}[G]^{3\times3}$ is archimedean, because*

$$1 - x_1^2 = \begin{bmatrix} 1 \\ 0 \\ -1 \end{bmatrix} G \begin{bmatrix} 1 \\ 0 \\ -1 \end{bmatrix} + \begin{bmatrix} 1 \\ 0 \\ -x_1 \end{bmatrix} G \begin{bmatrix} 1 \\ 0 \\ -x_1 \end{bmatrix},$$

$$1 - x_2^2 = \begin{bmatrix} 0 \\ 1 \\ -1 \end{bmatrix} G \begin{bmatrix} 0 \\ 1 \\ -1 \end{bmatrix} + \begin{bmatrix} 0 \\ 1 \\ -x_2 \end{bmatrix} G \begin{bmatrix} 0 \\ 1 \\ -x_2 \end{bmatrix}.$$

*Since $1 - x_i = \frac{1}{2}(1 - x_i^2) + \frac{1}{2}(x_i - 1)^2$, we get $1 - x_i \in \mathrm{QM}[G]^{3\times3}$. Similarly, we have $2 - y^2 \in \mathrm{QM}[G]^{3\times3}$ because*

$$2 - y^2 = (1 - x_2) + (1 - x_1)y^2 + \frac{1}{2}\begin{bmatrix} 1 \\ -y \\ 1 \end{bmatrix}^T G \begin{bmatrix} 1 \\ -y \\ 1 \end{bmatrix} + \frac{1}{2}\begin{bmatrix} 1 \\ -y \\ -1 \end{bmatrix}^T G \begin{bmatrix} 1 \\ -y \\ -1 \end{bmatrix}.$$

*The matrix polynomial $Q(x,z)$ is SOS-concave in $z$. The polynomials $p_\alpha$ in Theorem 10.6.5 can be selected as $p_\alpha = x_2^{\alpha_1} x_1^{\alpha_2}$ for $\alpha = (\alpha_1, \alpha_2) \in \mathbb{N}_2^2$, then*

$$M(p) = \begin{bmatrix} 1 & x_2 & x_1 \\ x_2 & x_1^2 & x_2 x_1 \\ x_1 & x_1 x_2 & x_1^2 \end{bmatrix} = \begin{bmatrix} 1 \\ x_2 \\ x_1 \end{bmatrix}\begin{bmatrix} 1 \\ x_2 \\ x_1 \end{bmatrix}^T \succeq 0,$$

$$Q(x, p(x)) = \begin{bmatrix} x_2 \\ x_1 \\ 1 \end{bmatrix}\begin{bmatrix} x_2 \\ x_1 \\ 1 \end{bmatrix}^T + \begin{bmatrix} x_1 + 2(x_2 - x_2^2) & 0 & 0 \\ 0 & x_2 + 2(x_1 - x_1^2) & 0 \\ 0 & 0 & 0 \end{bmatrix}.$$

*Observe that $x_1 = e_1^T G(x,y)e_1$, $x_2 = e_2^T G(x,y)e_2$ and*

$$x_1 - x_1^2 = \begin{bmatrix} 1 \\ 0 \\ -x_1 \end{bmatrix}^T G(x,y) \begin{bmatrix} 1 \\ 0 \\ -x_1 \end{bmatrix}, \quad x_2 - x_2^2 = \begin{bmatrix} 0 \\ 2 \\ -x_2 \end{bmatrix}^T G(x,y) \begin{bmatrix} 0 \\ 2 \\ -x_2 \end{bmatrix}.$$

*The certificate (10.6.6) for $P_G \subseteq P_Q$ is given by*

$$N = 4, \quad V_0(x,y) = \begin{bmatrix} x_2 & x_1 & 1 \end{bmatrix}, \quad V_1 = \begin{bmatrix} 1 & 0 & 0 \\ 0 & 0 & 0 \\ 0 & 0 & 0 \end{bmatrix},$$

$$V_2 = \begin{bmatrix} 1 & 0 & 0 \\ 0 & 0 & 0 \\ -x_1 & 0 & 0 \end{bmatrix}, \quad V_3 = \begin{bmatrix} 0 & 0 & 0 \\ 0 & 1 & 0 \\ 0 & 0 & 0 \end{bmatrix}, \quad V_4 = \begin{bmatrix} 0 & 0 & 0 \\ 0 & 2 & 0 \\ 0 & -x_2 & 0 \end{bmatrix}.$$

We remark that a continuous function

$$\phi : P_G \to S_Q, \quad x \mapsto (x, z(x))$$

may not exist without the convexity assumption on $Q$, i.e., the lifting function $z(x)$ may not be continuous. For such a case, representations like (10.6.5) and (10.6.6) cannot hold.

**Example 10.6.7.** *([147]) Let $S_Q$ be obtained by clockwise rotating the set $\{x^2 - z^2 \geq 1, x^2 \leq 4\}$ by $60°$ about the origin, i.e.,*

$$Q(x, z) := \mathrm{diag}\left(-4 - (-\sqrt{3}x + z)^2 + (x + \sqrt{3}z)^2, \; 16 - (x + \sqrt{3}z)^2\right).$$

*The sets $S_Q$ and $P_Q$ are shown in the following figure.*

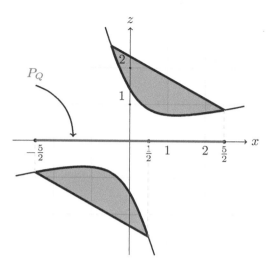

One can see that $P_Q = \left[-\frac{5}{2}, \frac{5}{2}\right]$. *The maximal x-coordinate of a point in the bottom component of $S_Q$ is $\frac{1}{2}$. If $G(x,y)$ is such that $P_G = [-1,1]$, then each point $x$ in $P_G$ can be lifted to a point $(x,z) \in S_Q$ with $Q(x,z) \succ 0$, but there is no continuous function $P_G \to S_Q$. This implies that there are no polynomials $p_i(x)$ satisfying (10.6.5). One can further show that (10.6.6) cannot hold either. The proof for this is left as an exercise.*

## 10.6.3 ▪ Exercises

**Exercise 10.6.1.** *([147]) For the following matrix polynomials*

$$G(x,y) \;:=\; diag\left(\begin{bmatrix} y_1 & x_1 \\ x_1 & 1 \end{bmatrix}, \begin{bmatrix} y_2 & x_2 \\ x_2 & 1 \end{bmatrix}, \begin{bmatrix} 1+y_1 & y_2 \\ y_2 & 1-y_1 \end{bmatrix}\right),$$

$$Q(x,z) \;:=\; \begin{bmatrix} 1 & x_1 & z \\ x_1 & 1 & x_2 \\ z & x_2 & 1 \end{bmatrix},$$

*give an explicit representation as in (10.6.2) to show $P_G \subseteq P_Q$.*

**Exercise 10.6.2.** *([147]) Consider the matrix polynomials*

$$G(x,y) \;:=\; \begin{bmatrix} 1-x_1^2 & x_1+x_2 & x_2^2 \\ x_1+x_2 & 0 & x_1+x_2 \\ x_2^2 & x_1+x_2 & y \end{bmatrix},$$

$$Q(x,z) \;:=\; \begin{bmatrix} 1+2\epsilon+x_2 & x_1^2 \\ x_1^2 & z \end{bmatrix}$$

*for a parameter $\epsilon > 0$. The projection is*

$$P_G = \{(x_1, -x_1) \in \mathbb{R}^2 \colon -1 < x_1 < 1\}.$$

*It is bounded but not closed. Show that the intersection $\mathrm{QM}[G]^{2\times 2} \cap \mathbb{R}[x]^{2\times 2}$ is archimedean but $\mathrm{QM}[G]^{2\times 2}$ itself is not archimedean. Give an explicit representation as in (10.6.2) to show the inclusion $P_G \subseteq P_Q$.*

**Exercise 10.6.3.** *For every fixed* $x = (x_1, x_2)$, *show that the matrix polynomial*

$$Q(x, z) := \begin{bmatrix} x_1 + 2\,z_1 - z_1^2 & z_1 z_2 & x_2 \\ z_1 z_2 & x_2 + 2\,z_2 - z_2^2 & x_1 \\ x_2 & x_1 & 1 \end{bmatrix}$$

*is SOS-concave in* $z = (z_1, z_2)$.

**Exercise 10.6.4.** *In Example 10.6.7, if* $G(x, y)$ *is a symmetric matrix polynomial such that* $[-1, 1] \subseteq P_G \subseteq P_Q$, *show that there are no polynomials* $p_i(x)$ *satisfying (10.6.5) and there are no polynomials* $p_\alpha(x)$ *satisfying (10.6.6).*

# Chapter 11

# Tensor Optimization

*Most tensor optimization problems can be equivalently formulated as homogeneous polynomial optimization with sphere constraints. This chapter introduces basic tensor theory, spectral and nuclear norms, quantum entanglements, tensor positivity, tensor eigenvalues, and singular values. They can be computed by Moment-SOS relaxations.*

## 11.1 ▪ Basic Tensor Theory

Let $\mathbb{F}$ denote the real field $\mathbb{R}$ or the complex field $\mathbb{C}$. For positive integers $n_1, \ldots, n_m$, a tensor $\mathcal{A}$ of dimension $n_1 \times \cdots \times n_m$ can be viewed as a multilinear functional on the Cartesian product space $\mathbb{F}^{n_1} \times \cdots \times \mathbb{F}^{n_m}$. That is,

$$\mathcal{A} : \mathbb{F}^{n_1} \times \cdots \times \mathbb{F}^{n_m} \to \mathbb{F}$$

is a multilinear functional such that

$$\mathcal{A}\Big(\sum_{i_1} \alpha_{i_1} u_{i_1}, \ldots, \sum_{i_m} \alpha_{i_m} u_{i_m}\Big) = \sum_{i_1} \cdots \sum_{i_m} \alpha_{i_1} \cdots \alpha_{i_m} \mathcal{A}(u_{i_1}, \ldots, u_{i_m})$$

for all vectors $u_{i_j} \in \mathbb{F}^{n_j}$ and for all scalars $\alpha_{i_j} \in \mathbb{F}$. The space of all such multilinear functionals is the tensor space $\mathbb{F}^{n_1} \otimes \cdots \otimes \mathbb{F}^{n_m}$, for which we also denote $\mathbb{F}^{n_1 \times \cdots \times n_m}$.

Let $e_1, \ldots, e_{n_k}$ be the canonical basis vectors for $\mathbb{F}^{n_k}$. As a multilinear functional, the tensor $\mathcal{A}$ is uniquely determined by the values $\mathcal{A}(e_{i_1}, \ldots, e_{i_m})$. For convenience, we denote

$$\mathcal{A}_{i_1, \ldots, i_m} := \mathcal{A}(e_{i_1}, \ldots, e_{i_m}).$$

So, $\mathcal{A}$ can be represented by the $m$th order multi-array

$$\mathcal{A} = (\mathcal{A}_{i_1 \ldots i_m}), \quad i_1 \in [n_1], \ldots, i_m \in [n_m].$$

The integer $m$ is called the *order* of $\mathcal{A}$. Tensors of order $m$ are called $m$-tensors. When $m = 3$ (resp., 4), they are called cubic (resp., quartic) tensors.

A tensor can be displayed as a series of matrices. For a 3-tensor $\mathcal{A} = (\mathcal{A}_{i_1 i_2 i_3}) \in \mathbb{F}^{n_1 \times n_2 \times n_3}$, if the third label $i_3$ is fixed, then the array

$$(\mathcal{A}_{i_1 i_2 i_3})_{i_1 \in [n_1], \, i_2 \in [n_2]}$$

is a $n_1 \times n_2$ matrix. Thus, $\mathcal{A}$ can be displayed by $n_3$ matrices of dimension $n_1 \times n_2$. For instance, when $n_1 = n_2 = n_3 = 3$, $\mathcal{A}$ can be displayed as

$$\left[ \begin{array}{ccc|ccc|ccc} \mathcal{A}_{111} & \mathcal{A}_{121} & \mathcal{A}_{131} & \mathcal{A}_{112} & \mathcal{A}_{122} & \mathcal{A}_{132} & \mathcal{A}_{113} & \mathcal{A}_{123} & \mathcal{A}_{133} \\ \mathcal{A}_{211} & \mathcal{A}_{221} & \mathcal{A}_{231} & \mathcal{A}_{212} & \mathcal{A}_{222} & \mathcal{A}_{232} & \mathcal{A}_{213} & \mathcal{A}_{223} & \mathcal{A}_{233} \\ \mathcal{A}_{311} & \mathcal{A}_{321} & \mathcal{A}_{331} & \mathcal{A}_{312} & \mathcal{A}_{322} & \mathcal{A}_{332} & \mathcal{A}_{313} & \mathcal{A}_{323} & \mathcal{A}_{333} \end{array} \right].$$

Similarly, $m$-tensors can also be displayed by listing all such $n_1 \times n_2$ matrices when the last $m - 2$ labels are enumerated.

For vectors $v_i \in \mathbb{F}^{n_i}$, $i = 1, \ldots, m$, their tensor product $\mathcal{T} := v_1 \otimes \cdots \otimes v_m$ is the multilinear functional such that

$$\mathcal{T}(u_1, \ldots, u_m) = (v_1^T u_1) \cdots (v_m^T u_m)$$

for all vectors $u_j \in \mathbb{F}^{n_j}$. Equivalently, the product $v_1 \otimes \cdots \otimes v_m$ is the tensor such that

$$(v_1 \otimes \cdots \otimes v_m)_{i_1 \ldots i_m} = (v_1)_{i_1} \cdots (v_m)_{i_m}$$

for all labels $i_1, \ldots, i_m$ in the range. Tensors of the form $v_1 \otimes \cdots \otimes v_m$ are called rank-1 tensors. Every tensor is a finite sum of rank-1 tensors. For each $\mathcal{A} \in \mathbb{F}^{n_1 \times \cdots \times n_m}$, there exist vectors $u^{k,i} \in \mathbb{F}^{n_i}$, $k = 1, \ldots, r$, such that

$$\mathcal{A} = \sum_{k=1}^{r} u^{k,1} \otimes \cdots \otimes u^{k,m}. \tag{11.1.1}$$

The equation (11.1.1) is called a tensor decomposition of $\mathcal{A}$.

**Definition 11.1.1.** *For a tensor $\mathcal{A} \in \mathbb{F}^{n_1 \times \cdots \times n_m}$, its $\mathbb{F}$-rank is the smallest integer $r$ such that $\mathcal{A}$ is a sum of $r$ rank-1 tensors. The $\mathbb{F}$-rank of $\mathcal{A}$ is denoted as $\mathrm{rank}_\mathbb{F}(\mathcal{A})$. Equivalently,*

$$\mathrm{rank}_\mathbb{F}(\mathcal{A}) = \min \left\{ r : \mathcal{A} = \sum_{k=1}^{r} u^{k,1} \otimes \cdots \otimes u^{k,m}, \ u^{k,i} \in \mathbb{F}^{n_i} \right\}. \tag{11.1.2}$$

When $r$ is the smallest, the decomposition in (11.1.2) is called a $\mathbb{F}$-*rank retaining decomposition* or just $\mathbb{F}$-*rank decomposition*. In some references, the $\mathrm{rank}_\mathbb{F}(\mathcal{A})$ is called the CAN-DECOMP/PARAFAC (CP) rank of $\mathcal{A}$, to be distinguished from other notions of ranks. When the field $\mathbb{F}$ is clear in the context, we just simply write $\mathrm{rank}(\mathcal{A}) := \mathrm{rank}_\mathbb{F}(\mathcal{A})$, and the corresponding decomposition is simply called a *rank retaining decomposition*, or *rank decomposition* or *CP rank decomposition*. When $\mathbb{F} = \mathbb{C}$, $\mathrm{rank}_\mathbb{C}(\mathcal{A})$ is sometimes called the *tensor rank* of $\mathcal{A}$ in some literature. We refer to [57, 152, 156, 179] for more detailed introductions to tensor decompositions and tensor ranks.

The rank of a tensor depends on the field $\mathbb{F}$. For instance, the real tensor

$$\mathcal{A} := e_1 \otimes e_1 \otimes e_1 - e_2 \otimes e_2 \otimes e_1 + e_1 \otimes e_2 \otimes e_2 + e_2 \otimes e_1 \otimes e_2$$

has different real and complex ranks (see [179]). It has $\mathrm{rank}_\mathbb{C}(\mathcal{A}) = 2$, with the complex rank retaining decomposition

$$\mathcal{A} = \frac{1}{2}(z_1 \otimes z_2 \otimes \bar{z}_3 + \bar{z}_1 \otimes \bar{z}_2 \otimes z_3),$$

where $z_1 = z_2 = z_3 = \begin{bmatrix} 1 \\ \sqrt{-1} \end{bmatrix}$ and $\bar{z}_i$ denotes the complex conjugate of $z_i$. However, we have $\mathrm{rank}_\mathbb{R}(\mathcal{A}) = 3$ with the real rank retaining decomposition

$$\mathcal{A} = e_1 \otimes e_1 \otimes (e_1 - e_2) - e_2 \otimes e_2 \otimes (e_1 + e_2) + (e_1 + e_2) \otimes (e_1 + e_2) \otimes e_2.$$

### 11.1.1 ▪ Border, generic, and typical ranks

The border, generic, and typical ranks are basic notions of tensor ranks. More detailed introductions to them can be found in [156, 179]. Here, we give a brief review for them.

**Definition 11.1.2.** *For a tensor $\mathcal{A} \in \mathbb{F}^{n_1 \times \cdots \times n_m}$, its border rank is the smallest $r$ such that $\mathcal{A}$ is the limit of a sequence of tensors whose $\mathbb{F}$-ranks are equal to $r$. The border rank of $\mathcal{A}$ is denoted as $\operatorname{brank}_{\mathbb{F}}(\mathcal{A})$, so*

$$\operatorname{brank}_{\mathbb{F}}(\mathcal{A}) = \min \left\{ r \;\middle|\; \begin{array}{c} \lim_{k \to \infty} \mathcal{A}_k = \mathcal{A}, \; \operatorname{rank}_{\mathbb{F}}(\mathcal{A}_k) = r, \\ \mathcal{A}_k \in \mathbb{F}^{n_1 \times \cdots \times n_m} \end{array} \right\}. \tag{11.1.3}$$

When the field $\mathbb{F}$ is clear in the context, the border rank $\operatorname{brank}_{\mathbb{F}}(\mathcal{A})$ is abbreviated to $\operatorname{brank}(\mathcal{A})$. Clearly, it always holds that

$$\operatorname{brank}_{\mathbb{F}}(\mathcal{A}) \le \operatorname{rank}_{\mathbb{F}}(\mathcal{A}).$$

The strict inequality might hold (see [80, 156, 179]). The following is such a tensor.

**Example 11.1.3.** *([80]) Let $\{x_1, y_1\} \subseteq \mathbb{C}^{n_1}$, $\{x_2, y_2\} \subseteq \mathbb{C}^{n_2}$, $\{x_3, y_3\} \subseteq \mathbb{C}^{n_3}$ be pairs of linearly independent vectors. Then the tensor*

$$\mathcal{A} := x_1 \otimes x_2 \otimes y_3 + x_1 \otimes y_2 \otimes x_3 + y_1 \otimes x_2 \otimes x_3$$

*has the border rank 2, because $\mathcal{A} = \lim_{\epsilon \to 0} \mathcal{A}_\epsilon$ for*

$$\mathcal{A}_\epsilon := \frac{1}{\epsilon} \Big[ (x_1 + \epsilon y_1) \otimes (x_2 + \epsilon y_2) \otimes (x_3 + \epsilon y_3) - x_1 \otimes x_2 \otimes x_3 \Big].$$

*However, $\operatorname{rank}_{\mathbb{C}}(\mathcal{A}) = 3$ (the proof is left as an exercise).*

An integer $r > 0$ is said to be a *typical rank* of $\mathbb{F}^{n_1 \times \cdots \times n_m}$ if there exists a subset $S \subseteq \mathbb{F}^{n_1 \times \cdots \times n_m}$ such that $S$ has nonempty interior (in the Euclidean topology) and $\operatorname{rank}_{\mathbb{F}}(\mathcal{A}) = r$ for all $\mathcal{A} \in S$. For $\mathbb{F} = \mathbb{R}$, the real tensor space $\mathbb{R}^{n_1 \times \cdots \times n_m}$ has several typical ranks when $m > 2$. For $\mathbb{F} = \mathbb{C}$, the complex tensor space $\mathbb{C}^{n_1 \times \cdots \times n_m}$ has a unique typical rank, which is also the maximum of all possible border ranks of tensors in $\mathbb{C}^{n_1 \times \cdots \times n_m}$.

The typical rank of the complex tensor space $\mathbb{C}^{n_1 \times \cdots \times n_m}$ is also called the *generic rank*, for which we denote $\operatorname{grank}(n_1, \ldots, n_m)$. Equivalently,

$$\operatorname{grank}(n_1, \ldots, n_m) = \max \left\{ \operatorname{brank}_{\mathbb{C}}(\mathcal{A}) : \mathcal{A} \in \mathbb{C}^{n_1 \times \cdots \times n_m} \right\}. \tag{11.1.4}$$

There are $n_1 + \cdots + n_m - m + 1$ parameters for parameterizing a rank-1 tensor. By counting dimensions, one can get the inequality

$$\operatorname{grank}(n_1, \ldots, n_m) \ge \left\lceil \frac{n_1 \cdots n_m}{n_1 + \cdots + n_m - m + 1} \right\rceil. \tag{11.1.5}$$

The right hand side of the above is called the *expected rank* of $\mathbb{C}^{n_1 \times \cdots \times n_m}$. The following is a classical result about generic ranks for cubic tensors.

**Theorem 11.1.4.** *([156]) The generic rank of $\mathbb{C}^{3 \times 3 \times 3}$ is 5. When $n \ne 3$, the generic rank of $\mathbb{C}^{n \times n \times n}$ is $\left\lceil \frac{n^3}{3n-2} \right\rceil$.*

We refer to [32, 156] for related work on maximum, typical, and generic ranks for tensors.

## 11.1.2 ▪ Symmetric tensor ranks

When $n_1 = \cdots = n_m$, the tensors in $\mathbb{F}^{n_1 \times \cdots \times n_m}$ are called *hypercubical* (or just *cubical* for convenience). We denote the hypercubical tensor space

$$\mathrm{T}^m(\mathbb{F}^n) := \mathbb{F}^n \otimes \cdots \otimes \mathbb{F}^n,$$

where $\mathbb{F}^n$ appears $m$ times. A hypercubical tensor $\mathcal{A} \in \mathrm{T}^m(\mathbb{F}^n)$ is said to be *symmetric* if its entries are invariant under permutations of labels, i.e.,

$$\mathcal{A}_{i_1 \ldots i_m} = \mathcal{A}_{j_1 \ldots j_m}$$

whenever $(i_1, \ldots, i_m)$ is a permutation of $(j_1, \ldots, j_m)$. The subspace of symmetric tensors in $\mathrm{T}^m(\mathbb{F}^n)$ is denoted as $\mathrm{S}^m(\mathbb{F}^n)$. The dimension of $\mathrm{T}^m(\mathbb{F}^n)$ is $\binom{n+m-1}{m}$. Its proof is left as an exercise.

For symmetric tensors, we are often interested in their symmetric decompositions, i.e., writing them as sums of symmetric rank-1 tensors. A rank-1 tensor $u_1 \otimes \cdots \otimes u_m$ is symmetric if and only if $u_1, \ldots, u_m$ are parallel to each other. For convenience, denote the symmetric tensor power

$$u^{\otimes m} := u \otimes \cdots \otimes u \quad (u \text{ is repeated } m \text{ times}).$$

The following gives a condition for symmetric tensor powers to be linearly independent.

**Lemma 11.1.5.** *([57, 156]) Let $u_1, \ldots, u_r \in \mathbb{C}^n$ be pairwise linearly independent vectors. If $m \geq r - 1$, then the tensors*

$$(u_1)^{\otimes m}, \ldots, (u_r)^{\otimes m}$$

*are linearly independent. Moreover, when equivalent classes of the $u_i$ in $\mathbb{P}^{n-1}$ are not colinear,[10] they are linearly independent if $m \geq r - 2$.*

**Proof.** Suppose there is a linear combination such that

$$\lambda_1 (u_1)^{\otimes m} + \cdots + \lambda_r (u_r)^{\otimes m} = 0. \tag{11.1.6}$$

Up to a linear coordinate transformation, we can generally assume that the first entry $(u_i)_1 \neq 0$ for all $i$. Then, upon scaling, one can further assume that

$$u_i = (1, v_i), \qquad v_i \in \mathbb{C}^{n-1}.$$

Since the $u_i$'s are pairwise linearly independent, the vectors $v_1, \ldots, v_r$ are distinct. The equation (11.1.6) is then equivalent to (see (2.4.2) for the notation $[x]_m$)

$$\lambda_1 [v_1]_m + \cdots + \lambda_r [v_r]_m = 0.$$

If $m \geq r - 1$, there exists a polynomial $p \in \mathbb{R}[x_2, \ldots, x_n]_m$ such that

$$p(v_1) = 1, \quad p(v_2) = \cdots = p(v_r) = 0.$$

Then, we get

$$0 = \mathrm{vec}(p)^T \left( \lambda_1 [v_1]_m + \cdots + \lambda_r [v_r]_m \right) = \lambda_1 p(v_1) + \cdots + \lambda_r p(v_r) = \lambda_1.$$

Similarly, one can show that $\lambda_2 = \cdots = \lambda_r = 0$. So, $(u_1)^{\otimes m}, \ldots, (u_r)^{\otimes m}$ are linearly independent.

---

[10]This means that the points corresponding to equivalent classes of the $u_i$ do not lie in the same line.

Next, suppose the equivalent classes of the $u_i$'s are not colinear, which means that the vectors $v_1, \ldots, v_r$ do not lie in a common line. This is possible only for $r > 2$. We can generally further assume that $v_1$ does not lie in the line passing through $v_{r-1}, v_r$. Hence, there exists a linear function $\ell$ on $\mathbb{C}^{n-1}$ such that it achieves at most $r - 1$ distinct values on the points $v_1, v_2, \ldots, v_r$ and the value $\ell(v_1)$ is different from $\ell(v_2), \ldots, \ell(v_r)$. Therefore, if $m \geq r - 2$, there exists $p \in \mathbb{R}[x_2, \ldots, x_n]_m$ such that $p(v_1) = 1$ and $p(v_i) = 0$ for $i = 2, \ldots, r$. Like in the above, we can also show that $\lambda_1 = 0$. Thus,

$$\lambda_2 [v_2]_m + \cdots + \lambda_r [v_r]_m = 0.$$

Note that $v_2, \ldots, v_r$ are distinct points and $m \geq (r - 1) - 1$. The conclusion in the first part implies that $\lambda_i = 0$ for $i = 2, \ldots, r$. Therefore, $(u_1)^{\otimes m}, \ldots, (u_r)^{\otimes m}$ are linearly independent. □

A basic property of symmetric tensors is that they can be written as sums of symmetric tensor powers. It was shown in [57] that for every $\mathcal{A} \in \mathsf{S}^m(\mathbb{F}^n)$, there exist vectors $u_1, \ldots, u_r \in \mathbb{F}^n$ such that

$$\mathcal{A} = \pm (u_1)^{\otimes m} \pm \cdots \pm (u_r)^{\otimes m}.$$

Because symmetric decompositions always exist, symmetric ranks can be defined for all symmetric tensors.

**Definition 11.1.6.** *For a symmetric tensor $\mathcal{A} \in \mathsf{S}^m(\mathbb{F}^n)$, its symmetric $\mathbb{F}$-rank, for which we denote $\mathrm{srank}_{\mathbb{F}}(\mathcal{A})$, is the smallest integer $r$ such that $\mathcal{A}$ can be written as a sum of $r$ symmetric rank-1 tensors in $\mathsf{S}^m(\mathbb{F}^n)$, i.e.,*

$$\mathrm{srank}_{\mathbb{F}}(\mathcal{A}) = \min \left\{ r : \mathcal{A} = \sum_{i=1}^{r} \lambda_i (u_i)^{\otimes m}, u_i \in \mathbb{F}^n, \lambda_i \in \mathbb{F} \right\}. \qquad (11.1.7)$$

*When the field $\mathbb{F}$ is clear in the context, the symmetric $\mathbb{F}$-rank is just simply called the symmetric rank and is denoted as $\mathrm{srank}(\mathcal{A})$.*

When $r = \mathrm{srank}_{\mathbb{F}}(\mathcal{A})$, the decomposition in (11.1.7) is called a *symmetric $\mathbb{F}$-rank retaining decomposition* or just *symmetric $\mathbb{F}$-rank decomposition* . When $\mathbb{F}$ is clear in the context, the word "$\mathbb{F}$-rank" can be abbreviated to "rank". When the rank $r$ is clear, it is called *symmetric $\mathbb{F}$-decomposition* or just *symmetric decomposition* (if $\mathbb{F}$ is also clear in the context). We refer to the work [235, 237] for how to compute symmetric tensor rank decompositions and approximations.

For instance, consider the tensor $\mathcal{A} \in \mathsf{S}^3(\mathbb{R}^n)$ such that

$$\mathcal{A}_{123} = \mathcal{A}_{132} = \mathcal{A}_{213} = \mathcal{A}_{231} = \mathcal{A}_{312} = \mathcal{A}_{321} = \frac{1}{6}$$

and $\mathcal{A}_{ijk} = 0$ otherwise. Its symmetric rank is 4 (see [280]), achieved for the symmetric rank retaining decomposition

$$\frac{1}{24} \left[ (e_1 + e_2 + e_3)^{\otimes 3} - (e_1 - e_2 + e_3)^{\otimes 3} - (e_1 + e_2 - e_3)^{\otimes 3} + (e_1 - e_2 - e_3)^{\otimes 3} \right].$$

Symmetric ranks also depend on the field. For instance, for the tensor

$$\mathcal{A} = \left[ \begin{array}{cc|cc} -1 & 0 & 0 & 1 \\ 0 & 1 & 1 & 0 \end{array} \right],$$

we have $\mathrm{srank}_{\mathbb{R}}(\mathcal{A}) = 3 > 2 = \mathrm{srank}_{\mathbb{C}}(\mathcal{A})$ (see [57]). Its real and complex symmetric rank retaining decompositions are ($i := \sqrt{-1}$)

$$\mathcal{A} = \frac{1}{2}\begin{bmatrix}1\\1\end{bmatrix}^{\otimes 3} + \frac{1}{2}\begin{bmatrix}1\\-1\end{bmatrix}^{\otimes 3} - 2\begin{bmatrix}1\\0\end{bmatrix}^{\otimes 3} = \frac{i}{2}\begin{bmatrix}-i\\1\end{bmatrix}^{\otimes 3} - \frac{i}{2}\begin{bmatrix}i\\1\end{bmatrix}^{\otimes 3}.$$

For symmetric tensors, there is a symmetric version of border ranks.

**Definition 11.1.7.** *([57]) For $\mathcal{A} \in \mathsf{S}^m(\mathbb{F}^n)$, its symmetric border $\mathbb{F}$-rank, for which we denote $\mathrm{sbrank}_{\mathbb{F}}(\mathcal{A})$, is the smallest $r$ such that it is the limit of a sequence of tensors in $\mathsf{S}^m(\mathbb{F}^n)$ whose symmetric $\mathbb{F}$-ranks are $r$, i.e.,*

$$\mathrm{sbrank}_{\mathbb{F}}(\mathcal{A}) = \min\left\{ r \ \middle| \ \begin{matrix} \mathcal{A}_k \in \mathsf{S}^m(\mathbb{F}^n), \ \lim_{k\to\infty} \mathcal{A}_k = \mathcal{A}, \\ \mathrm{srank}_{\mathbb{F}} \ \mathcal{A}_k = r \end{matrix} \right\}. \qquad (11.1.8)$$

When $\mathbb{F}$ is clear in the context, $\mathrm{sbrank}_{\mathbb{F}}(\mathcal{A})$ is often abbreviated to $\mathrm{sbrank}(\mathcal{A})$. Clearly, it always holds that

$$\mathrm{sbrank}_{\mathbb{F}}(\mathcal{A}) \leq \mathrm{srank}_{\mathbb{F}}(\mathcal{A}). \qquad (11.1.9)$$

The strict inequality may occur in the above (see [57]). For instance, for two linearly independent vectors $x, y \in \mathbb{C}^n$, the tensor

$$\mathcal{A} := x \otimes x \otimes y + x \otimes y \otimes x + y \otimes x \otimes x$$

has the border rank 2, because $\mathcal{A} = \lim_{\epsilon \to 0} \mathcal{A}_\epsilon$ for

$$\mathcal{A}_\epsilon := \frac{1}{\epsilon}\left((x + \epsilon y)^{\otimes 3} - x^{\otimes 3}\right).$$

However, $\mathrm{rank}_{\mathbb{C}}(\mathcal{A}) = 3$, achieved at the symmetric rank retaining decomposition

$$\mathcal{A} = \frac{1}{2}\left((x + y)^{\otimes 3} - (x - y)^{\otimes 3} - 2y^{\otimes 3}\right).$$

Denote by $\mathsf{S}_{r,\circ}^m(\mathbb{C}^n)$ the set of tensors in $\mathsf{S}^m(\mathbb{C}^n)$ whose symmetric ranks equal $r$. Its closure is the set

$$\mathsf{S}_r^m(\mathbb{C}^n) := \overline{\mathsf{S}_{r,\circ}^m(\mathbb{C}^n)}. \qquad (11.1.10)$$

The smallest $r$ such that $\mathsf{S}_r^m(\mathbb{C}^n) = \mathsf{S}^m(\mathbb{C}^n)$ is called the *generic symmetric rank* of $\mathsf{S}^m(\mathbb{C}^n)$. Each rank-1 symmetric tensor $u^{\otimes m}$ is parameterized by $u \in \mathbb{C}^n$. It has $n$ free parameters. By a dimensional counting, for $\mathsf{S}_r^m(\mathbb{C}^n) = \mathsf{S}^m(\mathbb{C}^n)$, it must hold that

$$rn \geq \dim \mathsf{S}^m(\mathbb{C}^n) = \binom{n + m - 1}{m}.$$

The following is a classical result about generic symmetric ranks.

**Theorem 11.1.8.** *([9]) For $m > 2$, a generic tensor $\mathcal{A} \in \mathsf{S}^m(\mathbb{C}^n)$ has the rank*

$$\mathrm{srank}_{\mathbb{C}}(\mathcal{A}) = \left\lceil \frac{1}{n}\binom{n + m - 1}{m} \right\rceil, \qquad (11.1.11)$$

*except for cases $(m, n) \in \{(3, 5), (4, 3), (4, 4), (4, 5)\}$. For these exceptional cases, the rank should be increased by one in the formula.*

For a symmetric tensor $\mathcal{A} \in \mathsf{S}^m(\mathbb{F}^n)$, denote the polynomial

$$\mathcal{A}(x) := \sum_{i_1,\ldots,i_m=1}^{n} \mathcal{A}_{i_1\ldots i_m} x_{i_1} \cdots x_{i_m}. \tag{11.1.12}$$

It is homogeneous of degree $m$ in $x := (x_1, \ldots, x_n)$. This gives a bijection between $\mathsf{S}^m(\mathbb{C}^n)$ and $\mathbb{C}[x]_m^{\mathrm{hom}}$ (see [57, 156, 255]). The tensor $\mathcal{A}$ is uniquely determined by $\mathcal{A}(x)$, and vice versa. For two symmetric tensors $\mathcal{A}, \mathcal{B} \in \mathsf{S}^m(\mathbb{F}^n)$, one can show that $\mathcal{A} = \mathcal{B}$ if and only if $\mathcal{A}(x) = \mathcal{B}(x)$. Interestingly, the symmetric tensor $\mathcal{A}$ has the decomposition

$$\mathcal{A} = \lambda_1 (u_1)^{\otimes m} + \cdots + \lambda_r (u_r)^{\otimes m}$$

if and only if the polynomial $\mathcal{A}(x)$ has the decomposition

$$\mathcal{A}(x) = \lambda_1 (u_1^T x)^{\otimes m} + \cdots + \lambda_r (u_r^T x)^{\otimes m}. \tag{11.1.13}$$

Therefore, a tensor decomposition can be equivalently displayed by showing the decomposition of the polynomial $\mathcal{A}(x)$. The equation (11.1.13) is called a Waring decomposition (see [57]).

For a hypercubical tensor $\mathcal{A} \in \mathsf{T}^m(\mathbb{F}^n)$, its *symmetrization* is the symmetric tensor in $\mathsf{S}^m(\mathbb{F}^n)$, for which we denote $\mathrm{sym}(\mathcal{A})$, such that for all $i_1, \ldots, i_m$

$$\mathrm{sym}(\mathcal{A})_{i_1\ldots i_m} = \frac{1}{m!} \sum_{(j_1,\ldots,j_m)\sim(i_1,\ldots,i_m)} \mathcal{A}_{j_1\ldots j_m}, \tag{11.1.14}$$

where $\sim$ denotes the permutation relationship. For instance, for $\mathcal{A} = a \otimes b \otimes c$, the symmetrization tensor $\mathrm{sym}(\mathcal{A})$ is

$$\frac{1}{6} \big( a \otimes b \otimes c + a \otimes c \otimes b + b \otimes a \otimes c + b \otimes c \otimes a + c \otimes a \otimes b + c \otimes b \otimes a \big).$$

### 11.1.3 ▪ Inner products

Inner products on vector spaces can be naturally extended to their tensor product spaces. For the space $\mathbb{F}^n$, the Euclidean inner product is

$$\langle x, y \rangle = y^H x,$$

where the superscript $^H$ denotes the Hermitian transpose. For two rank-1 tensors $u_1 \otimes \cdots \otimes u_m$, $v_1 \otimes \cdots \otimes v_m$ in $\mathbb{F}^{n_1 \times \cdots \times n_m}$, their inner product is naturally defined as

$$\langle u_1 \otimes \cdots \otimes u_m, v_1 \otimes \cdots \otimes v_m \rangle = \langle u_1, v_1 \rangle \cdots \langle u_m, v_m \rangle. \tag{11.1.15}$$

It can be extended bilinearly to higher rank tensors. The bilinear extension is such that for all rank-1 tensors $\mathcal{U}_i, \mathcal{V}_j$ ($i = 1, \ldots, s, j = 1, \ldots, t$),

$$\Big\langle \sum_{i=1}^{s} \mathcal{U}_i, \sum_{j=1}^{t} \mathcal{V}_j \Big\rangle = \sum_{i=1}^{s} \sum_{j=1}^{t} \langle \mathcal{U}_i, \mathcal{V}_j \rangle. \tag{11.1.16}$$

For two tensors $\mathcal{A} = (\mathcal{A}_{i_1\ldots i_m})$ and $\mathcal{B} = (\mathcal{B}_{i_1\ldots i_m})$, the above induces the *Hilbert-Schmidt inner product*

$$\langle \mathcal{A}, \mathcal{B} \rangle := \sum_{i_1=1}^{n_1} \cdots \sum_{i_m=1}^{n_m} \mathcal{A}_{i_1\ldots i_m} \overline{\mathcal{B}}_{i_1\ldots i_m}. \tag{11.1.17}$$

(The bar ‾ denotes the complex conjugate.) This inner product then induces the *Hilbert-Schmidt norm*

$$\|\mathcal{A}\| := \left( \sum_{i_1=1}^{n_1} \cdots \sum_{i_m=1}^{n_m} |\mathcal{A}_{i_1,\dots,i_m}|^2 \right)^{1/2}. \tag{11.1.18}$$

The tensor space $\mathbb{F}^{n_1 \times \cdots \times n_m}$ owns all properties of inner products. Throughout this chapter, the tensor norm $\|\mathcal{A}\|$ is defined as in (11.1.18), unless its meaning is otherwise specified.

### 11.1.4 ▪ Exercises

**Exercise 11.1.1.** *([179]) For the following real tensor*

$$\mathcal{A} = e_1 \otimes e_1 \otimes e_1 - e_2 \otimes e_2 \otimes e_1 + e_1 \otimes e_2 \otimes e_2 + e_2 \otimes e_1 \otimes e_2,$$

*show that* $\mathrm{rank}_{\mathbb{C}}(\mathcal{A}) = 2 < \mathrm{rank}_{\mathbb{R}}(\mathcal{A}) = 3$.

**Exercise 11.1.2.** *For the tensor $\mathcal{A}$ in Example 11.1.3, show that $\mathrm{rank}_{\mathbb{C}}(\mathcal{A}) = 3$.*

**Exercise 11.1.3.** *([57]) For $\mathbb{F} = \mathbb{R}$ or $\mathbb{C}$, show that the dimension of the symmetric tensor space $\mathrm{S}^m(\mathbb{F}^n)$ is $\binom{n+m-1}{m}$. Use this fact to show that every tensor in $\mathrm{S}^m(\mathbb{F}^n)$ is a sum of at most $\binom{n+m-1}{m}$ rank-1 symmetric tensors.*

**Exercise 11.1.4.** *For two symmetric tensors $\mathcal{A}, \mathcal{B}$, show that $\mathcal{A} = \mathcal{B}$ if and only if $\mathcal{A}(x) = \mathcal{B}(x)$. In particular, show that the symmetric tensor decomposition*

$$\mathcal{A} = \lambda_1 (u_1)^{\otimes m} + \cdots + \lambda_r (u_r)^{\otimes m}$$

*holds if and only if $\mathcal{A}(x)$ has the Waring decomposition:*

$$\mathcal{A}(x) = \lambda_1 (u_1^T x)^m + \cdots + \lambda_r (u_r^T x)^m.$$

**Exercise 11.1.5.** *Show that the inner product (11.1.17) can be implied from (11.1.15) and the bilinear relation (11.1.16).*

**Exercise 11.1.6.** *For the norm given by (11.1.18), show that*

$$\|u_1 \otimes \cdots \otimes u_m\| = \|u_1\| \cdots \|u_m\|.$$

**Exercise 11.1.7.** *If $u_1, \dots, u_m$ are vectors in $\mathbb{C}^n$ that are orthogonal to each other, show that*

$$\|(u_1)^{\otimes m} + \cdots + (u_r)^{\otimes m}\|^2 = \|u_1\|^{2m} + \cdots + \|u_r\|^{2m}.$$

## 11.2 ▪ Tensor Spectral Norms

Let $\mathbb{F} = \mathbb{R}$ or $\mathbb{C}$. For a matrix $A \in \mathbb{F}^{n_1 \times n_2}$, its spectral norm $\|A\|_\sigma$ equals its largest singular value. It can be equivalently expressed as

$$\|A\|_\sigma = \max_{\|x\|=\|y\|=1} |x^T A y| = \max_{x \neq 0, y \neq 0} \frac{|x^T A y|}{\|x\| \cdot \|y\|}.$$

The spectral norm can be similarly defined for tensors. For $\mathcal{A} \in \mathbb{F}^{n_1 \times \cdots \times n_m}$, the value

$$\|\mathcal{A}\|_{\sigma,\mathbb{F}} := \max_{\substack{x^{(i)} \in \mathbb{F}^{n_i}, \|x^{(i)}\|=1, \\ i=1,\ldots,m}} |\langle \mathcal{A}, x^{(1)} \otimes \cdots x^{(m)} \rangle| \qquad (11.2.1)$$

is called the *spectral norm* of $\mathcal{A}$. Equivalently, it can also be given as

$$\|\mathcal{A}\|_{\sigma,\mathbb{F}} = \max_{\substack{0 \neq x^{(i)} \in \mathbb{F}^{n_i}, \\ i=1,\ldots,m}} \frac{|\langle \mathcal{A}, x^{(1)} \otimes \cdots \otimes x^{(m)} \rangle|}{\|x^{(1)}\| \cdots \|x^{(m)}\|}.$$

When $\mathcal{A}$ is real, it has two spectral norms $\|\mathcal{A}\|_{\sigma,\mathbb{R}}$ and $\|\mathcal{A}\|_{\sigma,\mathbb{C}}$. Clearly,

$$\|\mathcal{A}\|_{\sigma,\mathbb{R}} \leq \|\mathcal{A}\|_{\sigma,\mathbb{C}}.$$

It is possible that $\|\mathcal{A}\|_{\sigma,\mathbb{R}} < \|\mathcal{A}\|_{\sigma,\mathbb{C}}$. For instance, for the tensor

$$\mathcal{A} = \frac{1}{2}\big(e_1 \otimes e_1 \otimes e_2 + e_1 \otimes e_2 \otimes e_1 + e_2 \otimes e_1 \otimes e_1 - e_2 \otimes e_2 \otimes e_2\big),$$

we have $\|\mathcal{A}\|_{\sigma,\mathbb{R}} = \frac{1}{2} < \frac{1}{\sqrt{2}} = \|\mathcal{A}\|_{\sigma,\mathbb{C}}$. This is shown in [101]. The spectral norm is closely related to rank-1 tensor approximations.

## 11.2.1 ▪ Rank-1 tensor approximations

For given $\mathcal{A} \in \mathbb{F}^{n_1 \times \cdots \times n_m}$, we look for the rank-1 tensor $\mathcal{B}$ that is closest to $\mathcal{A}$. This requires us to solve the least squares optimization

$$\begin{cases} \min & \|\mathcal{A} - \mathcal{X}\|^2 \\ s.t. & \operatorname{rank}(\mathcal{X}) = 1, \\ & \mathcal{X} \in \mathbb{F}^{n_1 \times \cdots \times n_m}. \end{cases} \qquad (11.2.2)$$

A rank-1 tensor $\mathcal{X}$ can be parameterized as

$$\mathcal{X} := \tau x^{(1)} \otimes \cdots \otimes x^{(m)}, \ \tau \in \mathbb{F}, \ \|x^{(j)}\| = 1.$$

Then, it holds that

$$\begin{aligned} \|\mathcal{A} - \mathcal{X}\|^2 &= \langle \mathcal{A} - \mathcal{X}, \mathcal{A} - \mathcal{X} \rangle = \|\mathcal{A}\|^2 - \langle \mathcal{A}, \mathcal{X} \rangle - \langle \mathcal{X}, \mathcal{A} \rangle + \|\mathcal{X}\|^2 \\ &= \|\mathcal{A}\|^2 - 2\operatorname{Re}(\bar{\tau}\langle \mathcal{A}, x^{(1)} \otimes \cdots \otimes x^{(m)} \rangle) + |\tau|^2. \end{aligned}$$

For fixed $x^{(1)}, \ldots, x^{(m)}$, the minimum of the above is achieved for

$$\tau = \langle \mathcal{A}, x^{(1)} \otimes \cdots \otimes x^{(m)} \rangle,$$

and the minimum value is $\|\mathcal{A}\|^2 - |\tau|^2$. This gives the following theorem.

**Theorem 11.2.1.** *([69]) For given* $\mathcal{A} \in \mathbb{F}^{n_1 \times \cdots \times n_m}$, *the best rank-1 approximation problem (11.2.2) is equivalent to the maximization*

$$\begin{cases} \max & |\langle \mathcal{A}, x^{(1)} \otimes \cdots \otimes x^{(m)} \rangle| \\ s.t. & \|x^{(1)}\| = \cdots = \|x^{(m)}\| = 1, \\ & x^{(1)} \in \mathbb{F}^{n_1}, \ldots, x^{(m)} \in \mathbb{F}^{n_m}. \end{cases} \qquad (11.2.3)$$

*That is, a rank-1 tensor $\lambda \cdot (u_1 \otimes \cdots \otimes u_m)$, with $\lambda \in \mathbb{F}$ and unit length vectors $u_i$, is a best rank-1 approximation for $\mathcal{A}$ if and only if $(u_1, \ldots, u_m)$ is a maximizer of (11.2.3) and $\lambda = \langle \mathcal{A}, x^{(1)} \otimes \cdots \otimes x^{(m)} \rangle$. Moreover, it holds that*

$$\| \mathcal{A} - \lambda \cdot (u_1 \otimes \cdots \otimes u_m) \|^2 = \| \mathcal{A} \|^2 - |\lambda|^2.$$

When the tensor $\mathcal{A}$ is symmetric, we are also interested in a best symmetric rank-1 tensor approximation for (11.2.2). Indeed, when $\mathcal{A}$ is symmetric, the optimization (11.2.2) has a minimizer that is also a symmetric tensor. This result appeared in the work [10, 46, 98, 99, 332]. Equivalently, every symmetric tensor has a best rank-1 approximating tensor that is also symmetric. However, not every best rank-1 approximating tensor is symmetric. In applications, people are often interested in symmetric rank-1 approximations for symmetric tensors.

A symmetric rank-1 tensor can be expressed as $\lambda \cdot x^{\otimes m}$ for a scalar $\lambda \in \mathbb{F}$ and a vector $u \in \mathbb{F}^n$. In the spirit of Theorem 11.2.1, the optimization (11.2.2) is equivalent to the maximization

$$\begin{cases} \max & |\langle \mathcal{A}, x^{\otimes m} \rangle| \\ s.t. & \|x\| = 1, \, x \in \mathbb{F}^n. \end{cases} \tag{11.2.4}$$

If $u$ is a maximizer of (11.2.4) and $\lambda = \langle \mathcal{A}, x^{\otimes m} \rangle$, then $\lambda \cdot u^{\otimes m}$ is the best rank-1 approximating tensor for $\mathcal{A}$.

To find best rank-1 tensor approximations, it is equivalent to solving the maximization (11.2.3). The computational complexity is NP-hard (see [128]). The classical methods for computing rank-1 tensor approximations include alternating least squares (ALS) and higher-order power methods (HOPM). We refer to [69, 149, 331, 332] for these classical methods. These methods typically guarantee stationary points. When the approximating rank is higher than one, we refer to the work [237, 252] for computational methods. In the following, we introduce Moment-SOS relaxation methods for computing best rank-1 approximations.

### 11.2.2 ▪ The case of real symmetric tensors

Consider the case that $\mathbb{F} = \mathbb{R}$ and $\mathcal{A}$ is symmetric. Recall that $x := (x_1, \ldots, x_n)$ and $\mathbb{S}^{n-1}$ stands for the unit sphere in $\mathbb{R}^n$. Denote the following form of degree $m$:

$$f(x) := \langle \mathcal{A}, x^{\otimes m} \rangle.$$

Note that (11.2.4) is equivalent to two optimization problems:

$$\text{(I):} \quad \max_{x \in \mathbb{S}^{n-1}} f(x), \qquad \text{(II):} \quad \min_{x \in \mathbb{S}^{n-1}} f(x). \tag{11.2.5}$$

Suppose $u^+, u^-$ are optimizers of (I), (II) in (11.2.5) respectively. By Theorem 11.2.1, if $|f(u^+)| \geq |f(u^-)|$, then $f(u^+) \cdot (u^+)^{\otimes m}$ is the best real rank-1 approximation; otherwise, $f(u^-) \cdot (u^-)^{\otimes m}$ is the best one. Let $f_{max}$ and $f_{min}$ denote the maximum and minimum values in (11.2.5) respectively. Note that $f_{max} = -f_{min}$ if the degree $m$ is odd.

The KKT system for (11.2.5) gives

$$\text{rank} \begin{bmatrix} \nabla f(x) & x \end{bmatrix} = 1.$$

The above condition can be given by $2n - 3$ equations (see §6.5.1):

$$h_s(x) := \sum_{i+j=s+2} \left( x_j \frac{\partial f}{\partial x_i} - x_i \frac{\partial f}{\partial x_j} \right) = 0, \quad s = 1, \ldots, 2n - 3. \tag{11.2.6}$$

For convenience, denote

$$\begin{aligned}
h_{2n-2} &:= x^T x - 1, \\
h &:= (h_1, \ldots, h_{2n-3}, h_{2n-2}).
\end{aligned} \tag{11.2.7}$$

The degree of $h$ is also $m$. Note that $u$ is a KKT point for (11.2.5) if and only if $h(u) = 0$. Let

$$m_0 := \left\lceil \tfrac{m}{2} \right\rceil.$$

The following is a Moment-SOS algorithm for solving (11.2.5).

**Algorithm 11.2.2.** *([62]) For given $\mathcal{A} \in \mathrm{S}^m(\mathbb{R}^n)$, let $f = \langle \mathcal{A}, x^{\otimes m} \rangle$ and $k := m_0$. Do the following:*

**Step 1** *Solve the following moment relaxation*

$$\begin{cases}
\max & \langle f, y \rangle \\
s.t. & \mathscr{V}_{h_s}^{(2k)}[y] = 0 \, (s = 1, \ldots, 2n - 2), \\
& y_0 = 1, \, M_k[y] \succeq 0, \, y \in \mathbb{R}^{\mathbb{N}_{2k}^n},
\end{cases} \tag{11.2.8}$$

*and get an optimizer $y^*$. Let $\vartheta_k^+$ be the optimal value of the above.*

**Step 2** *If there exists an integer $t \in [m_0, k]$ such that*

$$\mathrm{rank}\, M_{t-1}[y^*] = \mathrm{rank}\, M_t[y^*], \tag{11.2.9}$$

*compute $r_1 := \mathrm{rank}\, M_t[y^*]$ distinct points $u_1, \ldots, u_{r_1} \in \mathbb{S}^{n-1}$ such that*

$$y^*|_{2t} = \lambda_1 [u_1]_{2t} + \cdots + \lambda_{r_1} [u_{r_1}]_{2t}$$

*for positive scalars $\lambda_i > 0$.*

**Step 3** *If $m$ is odd, let $\vartheta_k^- = 0$ and go to Step 5. If $m$ is even, solve the following moment relaxation*

$$\begin{cases}
\min & \langle f, z \rangle \\
s.t. & \mathscr{V}_{h_s}^{(2k)}[z] = 0 \, (s = 1, \ldots, 2n - 2), \\
& z_0 = 1, \, M_k[z] \succeq 0, \, z \in \mathbb{R}^{\mathbb{N}_{2k}^n},
\end{cases} \tag{11.2.10}$$

*and get an optimizer $z^*$. Let $\vartheta_k^-$ be the optimal value of the above.*

**Step 4** *If there exists an integer $l \in [m_0, k]$ such that*

$$\mathrm{rank}\, M_{l-1}[z^*] = \mathrm{rank}\, M_l[z^*], \tag{11.2.11}$$

*compute $r_2 := \mathrm{rank}\, M_l[z^*]$ distinct points $v_1, \ldots, v_{r_2} \in \mathbb{S}^{n-1}$ such that*

$$z^*|_{2l} = \mu_1 [v_1]_{2l} + \cdots + \mu_{r_2} [v_{r_2}]_{2l}$$

*for positive scalars $\mu_i > 0$.*

**Step 5** *If (11.2.9) holds and $|\vartheta_k^+| \geq |\vartheta_k^-|$, then each $f(u_i)(u_i)^{\otimes m}$ $(i \in [r_1])$ is a best real rank-1 approximating tensor, and stop; if (11.2.11) holds and $|\vartheta_k^-| \geq |\vartheta_k^+|$, then each $f(v_j)(v_j)^{\otimes m}$ $(j \in [r_2])$ is a best real rank-1 approximating tensor, and stop.*

**Step 6** *Let $\hat{\vartheta}$ be the one of $\vartheta_k^+, \vartheta_k^-$ whose absolute value is bigger. (If there is a tie, let $\hat{\vartheta} := \vartheta_k^+$.) Choose a generic polynomial $R \in \Sigma[x]_{2m_0}$ and solve the moment relaxation*

$$\begin{cases} \min & \langle R, w \rangle \\ s.t. & \mathscr{V}_{h_s}^{(2k)}[w] = 0 \, (s = 1, \ldots, 2n-2), \\ & \mathscr{V}_{f-\hat{\vartheta}}^{(2k)}[w] = 0, \, w_0 = 1, \\ & M_k[w] \succeq 0, \, w \in \mathbb{R}^{\mathbb{N}_{2k}^n}. \end{cases} \quad (11.2.12)$$

*If (11.2.12) is infeasible, go to Step 7; if it is feasible, solve it for an optimizer $w^*$ and set $\hat{u} := ((w^*)_{e_1}, \ldots, (w^*)_{e_n})$.*

**Step 7** *If $\|\hat{u}\| = 1$ and $f(\hat{u}) = \hat{\vartheta}$, then $f(\hat{u})(\hat{u})^{\otimes m}$ is the best real rank-1 approximating tensor, and stop; otherwise, update $k := k+1$ and go to Step 1.*

We remark that when the order $m$ is odd, there is no need to solve (11.2.10). This is because if $m$ is odd, then $f_{max} = -f_{min}$ and $\mathcal{A}(u)u^{\otimes m}$ must be the best rank-1 approximating tensor for every maximizer $u$ of (11.2.5). If (11.2.9) holds, $\vartheta_k^+ = f_{max}$ and each $u_i$ is a maximizer of (11.2.5). Similarly, if (11.2.11) holds, then $\vartheta_k^- = f_{min}$ and each $v_j$ is a minimizer of (11.2.5). When the maximization (I) in (11.2.5) has a unique maximizer, the rank in (11.2.9) must be one if it holds. The same is true for the minimization (II). The following is the convergence theorem.

**Theorem 11.2.3.** *([62]) Let $f_{max}, f_{min}, \vartheta_k^+, \vartheta_k^-$ be as above.*

*(i) If $k$ is big enough, then*

$$\vartheta_k^+ = f_{max}, \quad \vartheta_k^- = f_{min}, \quad \|\mathcal{A}\|_{\sigma,\mathbb{R}} = \max\{|\vartheta_k^+|, |\vartheta_k^-|\}.$$

*(ii) If (11.2.5) has finitely many maximizers, the flat truncation (11.2.9) must hold when $k$ is sufficiently large. Similarly, if (11.2.5) has finitely many minimizers, the flat truncation (11.2.11) must also hold when $k$ is sufficiently large.*

*(iii) Suppose $R$ is a generic polynomial in $\Sigma[x]_{2m_0}$. Let $u^{(k)}$ be the point $\hat{u}$ produced in Step 6 for the relaxation order $k$. Then the sequence $\{u^{(k)}\}_{k \geq m_0}$ converges to a point $u^*$ such that $\|u^*\| = 1$ and $|f(u^*)| = \|\mathcal{A}\|_{\sigma,\mathbb{R}}$. Consequently, $f(u^*)(u^*)^{\otimes m}$ is the best real rank-1 tensor approximation for $\mathcal{A}$.*

**Proof.** (i) Note that $u$ is a critical point of (11.2.5) if and only if $h(u) = 0$. By Theorem 6.1.1, $f$ achieves only finitely many values on the set of critical points. The optimization (11.2.5) has no inequality constraints. By Theorems 6.5.4 or 6.5.6, we must have $\vartheta_k^+ = f_{max}$ and $\vartheta_k^- = f_{min}$ when $k$ is big enough. The spectral norm $\|\mathcal{A}\|_{\sigma,\mathbb{R}}$ equals the maximum of $|f_{max}|$ and $|f_{min}|$.

(ii) Note that there are no inequality constraints. The conclusion follows from Theorem 6.5.6.

(iii) By the item (i), when $k$ is big enough, we have $f_{max} = \hat{\vartheta}$ or $f_{min} = \hat{\vartheta}$, and $|\hat{\vartheta}| = \|\mathcal{A}\|_{\sigma,\mathbb{R}}$. If $R$ is a generic polynomial in $\Sigma[x]_{2m_0}$, there is a unique minimizer for the following optimization:

$$\begin{cases} \min & R(x) \\ s.t. & h_j(x) = 0 \, (j = 1, \ldots, 2n-2), \\ & f(x) = \hat{\vartheta}. \end{cases}$$

Every feasible point $x$ for the above satisfies $\|x\| = 1$ and $|f(x)| = \|\mathcal{A}\|_{\sigma,\mathbb{R}}$. The conclusion then follows from Theorem 5.3.8. $\quad\square$

## 11.2.3 ▪ The case of real nonsymmetric tensors

Consider the case that $\mathbb{F} = \mathbb{R}$ and $\mathcal{A} \in \mathbb{R}^{n_1 \times \cdots \times n_m}$ is nonsymmetric. Computing the spectral norm needs to solve the maximization (11.2.3). Without loss of generality, one can assume $n_m$ is the biggest dimension, i.e.,

$$n_m \geq n_1, \ldots, n_{m-1}.$$

For convenience of notation, denote

$$m' := m - 1, \quad x' := (x^{(1)}, \ldots, x^{(m')}). \tag{11.2.13}$$

For real vectors $x^{(1)}, \ldots, x^{(m)}$, we can write that

$$\langle \mathcal{A}, x^{(1)} \otimes \cdots \otimes x^{(m)} \rangle = \sum_{j=1}^{n_m} (x^m)_j a_j(x'), \tag{11.2.14}$$

where each $a_j(x')$ is a multilinear form in $(x^{(1)}, \ldots, x^{(m')})$. Let

$$F(x^{(1)}, \ldots, x^{(m')}) := \sum_{j=1}^{n_m} \left| a_j(x') \right|^2,$$

a multi-quadratic form in $(x^{(1)}, \ldots, x^{(m')})$. By the Cauchy-Schwarz inequality,

$$|\langle \mathcal{A}, x^{(1)} \otimes \cdots \otimes x^{(m)} \rangle|^2 \leq F(x^{(1)}, \ldots, x^{(m')}) \cdot \|x^{(m)}\|^2.$$

The equality occurs in the above if and only if $x^{(m)}$ is proportional to the vector

$$\left( a_1(x'), \ldots, a_{n_m}(x') \right).$$

The maximization (11.2.3) is equivalent to

$$\begin{cases} \max & F(x^{(1)}, \ldots, x^{(m')}) \\ \text{s.t.} & \|x^{(j)}\| = 1, \ x^{(j)} \in \mathbb{R}^{n_j}, \\ & j = 1, \ldots, m'. \end{cases} \tag{11.2.15}$$

In the following, we give a Moment-SOS relaxation for solving (11.2.15). Denote the Kronecker product vector

$$b(x') := x^{(1)} \otimes \cdots \otimes x^{(m')}.$$

It has the length $n_1 \cdots n_{m'}$. Denote the label set

$$\Omega := \left\{ (\imath, \jmath) : \imath, \jmath \in [n_1] \times \cdots \times [n_{m'}] \right\}, \tag{11.2.16}$$

where $\imath := (i_1, \ldots, i_{m'})$, $\jmath := (j_1, \ldots, j_{m'})$. Expand the outer product

$$b(x')b(x')^T = \sum_{(\imath,\jmath) \in \Omega} G_{\imath,\jmath} \cdot (x^1)_{i_1}(x^1)_{j_1} \cdots (x^{m'})_{i_{m'}}(x^{m'})_{j_{m'}}$$

for symmetric coefficient matrices $G_{\imath,\jmath}$. Define the linear matrix function

$$G[w] := \sum_{(\imath,\jmath) \in \Omega} G_{\imath,\jmath} \cdot w_{\imath,\jmath} \tag{11.2.17}$$

in the variable

$$w := (w_{i,j})_{(i,j)\in\Omega}.$$

Consider the following expansions:

$$
\begin{aligned}
F(x') &= \sum_{(i,j)\in\Omega} f_{i,j}\cdot (x^{(1)})_{i_1}(x^{(1)})_{j_1}\cdots (x^{(m')})_{i_{m'}}(x^{(m')})_{j_{m'}},\\
\|b(x')\|^2 &= \sum_{(i,j)\in\Omega} c_{i,j}\cdot (x^1)_{i_1}(x^1)_{j_1}\cdots (x^{m'})_{i_{m'}}(x^{m'})_{j_{m'}}.
\end{aligned}
$$

The optimization (11.2.15) is then equivalent to

$$
\left\{
\begin{aligned}
\max_{w\in\mathbb{R}^\Omega} \quad & \sum_{(i,j)\in\Omega} f_{i,j}\cdot w_{i,j},\\
\text{s.t.} \quad & \sum_{(i,j)\in\Omega} c_{i,j}\cdot w_{i,j} = 1,\\
& G[w] = b(x')b(x')^T \succeq 0.
\end{aligned}
\right.
\tag{11.2.18}
$$

Note that if $G[w] = b(x')b(x')^T$, then

$$\mathrm{trace}(G[w]) = \sum_{(i,j)\in\Omega} c_{i,j}\cdot w_{i,j}.$$

If the equality constraint $G[w] = b(x')b(x')^T$ is ignored in (11.2.18), then we get the moment relaxation

$$
\left\{
\begin{aligned}
\max_{w\in\mathbb{R}^\Omega} \quad & \sum_{(i,j)\in\Omega} f_{i,j}\cdot w_{i,j}\\
\text{s.t.} \quad & \mathrm{trace}(G[w]) = 1,\\
& G[w] \succeq 0.
\end{aligned}
\right.
\tag{11.2.19}
$$

The dual optimization of (11.2.19) can be obtained as follows. Let

$$Q := \left\{ b(x')^T Y b(x') : Y \in \mathcal{S}_+^{n_1\cdots n_{m'}} \right\}
\tag{11.2.20}$$

be the cone of sums of squares of multilinear forms in $(x^{(1)},\ldots,x^{(m')})$. The dual optimization of (11.2.19) is the following minimization:

$$
\left\{
\begin{aligned}
\min \quad & \gamma\\
\text{s.t.} \quad & \gamma\cdot\|b(x')\|^2 - F(x') \in Q.
\end{aligned}
\right.
\tag{11.2.21}
$$

Let $F_{max}$, $F_{mom}$ denote the maximum values of (11.2.15), (11.2.19) respectively. Then

$$F_{mom} \ge F_{max} = \|\mathcal{A}\|_{\sigma,\mathbb{R}}^2.$$

When the equality $F_{mom} = F_{max}$ occurs, the relaxation (11.2.19) is said to be tight. The feasible set of (11.2.19) is compact, since $\mathrm{trace}(G[w]) = 1$. So (11.2.19) always has a maximizer, say, $w^*$. If rank $G[w^*] = 1$, then there exist unit length vectors $u_1,\ldots,u_{m'}$ such that

$$G(w^*) = (u_1 \otimes \cdots \otimes u_{m'})(u_1 \otimes \cdots \otimes u_{m'})^T.$$

The above $(u_1,\ldots,u_{m'})$ is a maximizer of (11.2.15). The vectors $u_j$ can be numerically extracted as follows. Let

$$\ell := (\ell_1,\ldots,\ell_{m'}) \in [n_1] \times \cdots \times [n_{m'}]$$

be the label such that

$$w^*_{\ell,\ell} = \max_{(\imath,\imath)\in\Omega} w^*_{\imath,\imath}.$$

For each $j = 1, \ldots, m'$, let

$$\hat{u}_j := \left(w^*_{\hat{\ell}_1,\ell}, w^*_{\hat{\ell}_2,\ell}, \ldots, w^*_{\hat{\ell}_{n_j},\ell}\right), \quad u_j = \hat{u}_j/\|\hat{u}_j\|, \qquad (11.2.22)$$

where $\hat{\ell}_k := \ell + (k - \ell_j) \cdot e_j$ for each $k \in [n_j]$. This gives the following algorithm.

**Algorithm 11.2.4.** *For given $\mathcal{A} \in \mathbb{R}^{n_1 \times \cdots \times n_m}$, do the following:*

**Step 1** *Solve the moment relaxation (11.2.19) for a maximizer $w^*$.*

**Step 2** *Determine $u_1, \ldots, u_{m'}$ as in (11.2.22) and let $u_m := \hat{u}_m/\|\hat{u}_m\|$, where*

$$\hat{u}_m := \left(F_1(u_1, \ldots, u_{m'}), \ldots, F_{n_m}(u_1, \ldots, u_{m'})\right).$$

**Step 3** *Let $\lambda = \mathcal{A}(u_1, \ldots, u_m)$. If $\operatorname{rank} G[w^*] = 1$, then $\lambda u_1 \otimes \cdots \otimes u_m$ is the best real rank-1 approximation for $\mathcal{A}$.*

In Algorithm 11.2.4, if $\operatorname{rank} G[w^*] = 1$, then the output $\lambda \cdot u_1 \otimes \cdots \otimes u_m$ is the best real rank-1 approximation for $\mathcal{A}$. If $\operatorname{rank} G[w^*] > 1$, it may or may not be the best rank-1 approximation. The approximation performance is analyzed in [220, §3]. To find the best rank-1 real approximation, we need to solve the maximization (11.2.15) exactly. When the moment relaxation (11.2.19) is not tight, we refer to tight relaxation methods in §6.3 and §6.5 for solving (11.2.15).

## 11.2.4 ▪ The case of complex tensors

Consider the case $\mathbb{F} = \mathbb{C}$ and $\mathcal{A} \in \mathbb{C}^{n_1 \times \cdots \times n_m}$ is a complex nonsymmetric tensor. We discuss how to solve the maximization (11.2.3) with the field $\mathbb{F} = \mathbb{C}$, to obtain a best rank-1 approximating tensor. Recall that $m' = m - 1$. For each $j = 1, \ldots, n_m$, let $a_j(x^{(1)}, \ldots, x^{(m')})$ be the multilinear form in $(x^{(1)}, \ldots, x^{(m')})$ such that

$$\overline{\langle \mathcal{A}, x^{(1)} \otimes \cdots \otimes x^{(m)} \rangle} = \sum_{j=1}^{n_m} (x^{(m)})_j a_j(x^{(1)}, \ldots, x^{(m')}). \qquad (11.2.23)$$

We similarly denote

$$F(x^{(1)}, \ldots, x^{(m')}) := \sum_{j=1}^{n_m} |a_j(x^{(1)}, \ldots, x^{(m')})|^2.$$

By the Cauchy-Schwarz inequality,

$$|\langle \mathcal{A}, x^{(1)} \otimes \cdots \otimes x^{(m)} \rangle|^2 \leq F(x^{(1)}, \ldots, x^{(m')}) \cdot \|x^{(m)}\|^2.$$

The equality occurs in the above if and only if the conjugate $\overline{x^{(m)}}$ is proportional to

$$\left(a_1(x^{(1)}, \ldots, x^{(m')}), \ldots, a_{n_m}(x^{(1)}, \ldots, x^{(m')})\right).$$

The maximization problem (11.2.3) is then equivalent to

$$\begin{cases} \max & F(x^{(1)}, \ldots, x^{(m')}) \\ \text{s.t.} & \|x^{(i)}\| = 1, \; x^{(i)} \in \mathbb{C}^{n_i}, \\ & i = 1, \ldots, m'. \end{cases} \tag{11.2.24}$$

With the usage of complex Hermitian psd matrix variables, we can similarly construct a Moment-SOS relaxation like (11.2.19) for solving (11.2.24) (see the exercise). It is equivalent to a polynomial optimization problem with real variables. For instance, for each $j = 1, \ldots, m'$, one can write

$$x^{(j)} = x_{\text{re}}^{(j)} + \sqrt{-1} x_{\text{im}}^{(j)}, \quad x_{\text{re}}^{(j)} \in \mathbb{R}^{n_j}, x_{\text{im}}^{(j)} \in \mathbb{R}^{n_j}.$$

The complex sphere constraints become

$$\|x_{\text{re}}^{(j)}\|^2 + \|x_{\text{im}}^{(j)}\|^2 = 1, \quad j = 1, \ldots, m - 1.$$

The methods in §5.2, §6.3, and §6.5 are applicable for solving (11.2.24).

When $\mathcal{A}$ is a symmetric tensor, it has a best rank-1 approximating tensor that is also symmetric. For this case, we can equivalently solve the complex maximization problem

$$\begin{cases} \max & |\langle \mathcal{A}, x^{\otimes m} \rangle| \\ \text{s.t.} & \|x\| = 1, \; x \in \mathbb{C}^n. \end{cases} \tag{11.2.25}$$

Similarly, it is also equivalent to a polynomial optimization problem with real variables. The objective in (11.2.25) is the modulus of a complex polynomial. The Moment-SOS relaxation methods in §5.2, §6.3, and §6.5 can be similarly used to solve (11.2.25).

### 11.2.5 ▪ Exercises

**Exercise 11.2.1.** *For the following tensors, compute the spectral norm* $\|\mathcal{A}\|_{\sigma,\mathbb{F}}$:

 (i) $\mathcal{A} = \frac{1}{\sqrt{2}}(e_1^{\otimes 3} + e_2^{\otimes 3})$ *and* $\mathbb{F} = \mathbb{R}$.

 (ii) $\mathcal{A} = \frac{1}{\sqrt{3}}(e_1^{\otimes 3} + e_2^{\otimes 3} + e_3^{\otimes 3})$ *and* $\mathbb{F} = \mathbb{C}$.

**Exercise 11.2.2.** *([101]) For the following tensor*

$$\mathcal{A} = \frac{1}{2}(e_1 \otimes e_1 \otimes e_2 + e_1 \otimes e_2 \otimes e_1 + e_2 \otimes e_1 \otimes e_1 - e_2 \otimes e_2 \otimes e_2),$$

*show that* $\|\mathcal{A}\|_{\sigma,\mathbb{R}} = \frac{1}{2} < \frac{1}{\sqrt{2}} = \|\mathcal{A}\|_{\sigma,\mathbb{C}}$.

**Exercise 11.2.3.** *For the tensor* $\mathcal{A} \in \mathrm{S}^4(\mathbb{R}^4)$ *such that*

$$\mathcal{A}(x) = 3x_1 x_2 x_3 x_4 + x_1^4 + x_2^4 + x_3^4 + x_4^4,$$

*find the best rank-1 approximating tensors in* $\mathrm{S}^4(\mathbb{R}^4)$ *and* $\mathrm{S}^4(\mathbb{C}^4)$ *respectively.*

**Exercise 11.2.4.** *For the upper triangular tensor* $\mathcal{A} \in \mathrm{T}^3(\mathbb{R}^3)$ *such that*

$$\mathcal{A}_{ijk} = \begin{cases} 1 & \text{if } i \le j \le k, \\ 0 & \text{otherwise}, \end{cases}$$

*find the best rank-1 approximating tensors in* $\mathrm{T}^3(\mathbb{R}^3)$ *and* $\mathrm{T}^3(\mathbb{C}^3)$ *respectively.*

**Exercise 11.2.5.** *In Step 2 of Algorithm 11.2.2, if (11.2.9) holds, show that each point $u_i$ is a maximizer of $f(x)$ over the unit sphere. Similarly, in Step 4, if (11.2.11) holds, show that each $v_j$ is a minimizer of $f(x)$ over the unit sphere.*

**Exercise 11.2.6.** *In Step 5 of Algorithm 11.2.2, if (11.2.9) holds and $|\vartheta_k^+| \geq |\vartheta_k^-|$, show that $|\vartheta_k^+| = \|\mathcal{A}\|_{\sigma,\mathbb{R}}$. Show the same thing if (11.2.11) holds.*

**Exercise 11.2.7.** *If $m = 3$ and one of $n_1, n_2$ is 2, show that the relaxation (11.2.19) is tight for solving the maximization (11.2.15).*

**Exercise 11.2.8.** *Construct a moment relaxation like (11.2.19) for solving (11.2.24), with the usage of complex Hermitian psd matrix variables.*

## 11.3 ▪ Tensor Nuclear Norms

Let $\mathbb{F} = \mathbb{R}$ or $\mathbb{C}$. As introduced in [101], the *nuclear norm* of a tensor $\mathcal{A} \in \mathbb{F}^{n_1 \times \cdots \times n_m}$ is

$$\|\mathcal{A}\|_{*,\mathbb{F}} := \min \left\{ \sum_{i=1}^{r} |\lambda_i| \;\middle|\; \begin{array}{c} \mathcal{A} = \sum_{i=1}^{r} \lambda_i v^{(i,1)} \otimes \cdots \otimes v^{(i,m)}, \\ \|v^{(i,j)}\| = 1, \; v^{(i,j)} \in \mathbb{F}^{n_j}, \; \lambda_i \in \mathbb{F} \end{array} \right\}. \tag{11.3.1}$$

The decomposition of $\mathcal{A}$ in (11.3.1), for which the minimum is achieved, is called a *nuclear decomposition*. The shortest length of nuclear decompositions is called the *nuclear rank*. For every $\mathcal{B} \in \mathbb{F}^{n_1 \times \cdots \times n_m}$ and for every decomposition of $\mathcal{A}$ in (11.3.1), it holds that

$$
\begin{aligned}
|\langle \mathcal{A}, \mathcal{B} \rangle| &= \left| \sum_{i=1}^{r} \lambda_i \langle v^{(i,1)} \otimes \cdots \otimes v^{(i,m)}, \mathcal{B} \rangle \right| \\
&\leq \sum_{i=1}^{r} |\lambda_i| \cdot |\langle \mathcal{B}, v^{(i,1)} \otimes \cdots \otimes v^{(i,m)} \rangle| \\
&\leq \|\mathcal{B}\|_{\sigma,\mathbb{F}} \left( \sum_{i=1}^{r} |\lambda_i| \right).
\end{aligned}
$$

The above implies that

$$|\langle \mathcal{A}, \mathcal{B} \rangle| \leq \|\mathcal{B}\|_{\sigma,\mathbb{F}} \|\mathcal{A}\|_{*,\mathbb{F}}.$$

Similarly, it also holds that

$$|\langle \mathcal{A}, \mathcal{B} \rangle| \leq \|\mathcal{A}\|_{\sigma,\mathbb{F}} \|\mathcal{B}\|_{*,\mathbb{F}}.$$

The above gives the dual relationship between nuclear and spectral norms. It can be shown that the nuclear norm unit ball $\{\mathcal{X} : \|\mathcal{X}\|_{*,\mathbb{F}} \leq 1\}$ is precisely the convex hull of the compact set

$$U := \left\{ \mathcal{X} : \operatorname{rank}_{\mathbb{F}}(\mathcal{X}) = 1, \|\mathcal{X}\| = 1 \right\}.$$

So, the spectral norm $\|\mathcal{A}\|_{\sigma,\mathbb{F}}$ can be equivalently expressed as

$$\|\mathcal{A}\|_{\sigma,\mathbb{F}} = \max_{\mathcal{X} \in U} |\langle \mathcal{A}, \mathcal{X} \rangle|.$$

This fact can further imply that

$$\|\mathcal{A}\|_{\sigma,\mathbb{F}} = \max_{\mathcal{X} \in \mathbb{F}^{n_1 \times \cdots \times n_m}} \left\{ |\langle \mathcal{A}, \mathcal{X} \rangle| : \|\mathcal{X}\|_{*,\mathbb{F}} \leq 1 \right\}. \tag{11.3.2}$$

This implies that $\|\mathcal{A}\|_{\sigma,\mathbb{F}}$ is also the maximum modulus of the linear functional $\langle \mathcal{A}, \mathcal{X} \rangle$ over the nuclear norm unit ball $\|\mathcal{X}\|_{*,\mathbb{F}} \leq 1$. By the dual relation, one can further show that the nuclear norm $\|\mathcal{A}\|_{*,\mathbb{F}}$ equals the maximum modulus of $\langle \mathcal{A}, \mathcal{Y} \rangle$ over the spectral norm unit ball $\|\mathcal{Y}\|_{\sigma,\mathbb{F}} \leq 1$, i.e.,

$$\|\mathcal{A}\|_{*,\mathbb{F}} = \max_{\mathcal{Y} \in \mathbb{F}^{n_1 \times \cdots \times n_m}} \left\{ |\langle \mathcal{A}, \mathcal{Y} \rangle| : \|\mathcal{Y}\|_{\sigma,\mathbb{F}} \leq 1 \right\}. \tag{11.3.3}$$

The proof for the above is left as an exercise. The dual relationship can be summarized as

$$\begin{aligned} \|\mathcal{A}\|_{\sigma,\mathbb{F}} &= \max_{\mathcal{X} \in \mathbb{F}^{n_1 \times \cdots \times n_m}} \left\{ |\langle \mathcal{A}, \mathcal{X} \rangle| : \|\mathcal{X}\|_{*,\mathbb{F}} = 1 \right\}, \\ \|\mathcal{A}\|_{*,\mathbb{F}} &= \max_{\mathcal{Y} \in \mathbb{F}^{n_1 \times \cdots \times n_m}} \left\{ |\langle \mathcal{A}, \mathcal{Y} \rangle| : \|\mathcal{Y}\|_{\sigma,\mathbb{F}} = 1 \right\}. \end{aligned} \tag{11.3.4}$$

Tensor nuclear norms are introduced in [178, 180]. We refer to [101] for more detailed introduction to tensor nuclear norms. When $m > 2$, the complexity of computing nuclear norms is NP-hard [100, 101]. Tensor nuclear norms have broad applications in blind source identification, signal processing, phase transition, and tensor completion (see [45, 311, 328]).

The following is a criterion for checking nuclear decompositions.

**Lemma 11.3.1.** *[101] For a tensor $\mathcal{A} \in \mathbb{F}^{n_1 \times \cdots \times n_m}$, the following*

$$\begin{aligned} \mathcal{A} &= \sum_{i=1}^{r} \lambda_i v^{(i,1)} \otimes \cdots \otimes v^{(i,m)}, \\ v^{(i,j)} &\in \mathbb{F}^{n_j}, \|v^{(i,j)}\| = 1 \end{aligned} \tag{11.3.5}$$

*is a nuclear decomposition for $\mathcal{A}$ if and only if there exists a tensor $\mathcal{B} \in \mathbb{F}^{n_1 \times \cdots \times n_m}$ such that*

$$\begin{aligned} \|\mathcal{B}\|_{\sigma,\mathbb{F}} &= 1, \\ \langle \mathcal{B}, \lambda_i v^{(i,1)} \otimes \cdots \otimes v^{(i,m)} \rangle &= |\lambda_i|, \, i = 1, \ldots, r. \end{aligned} \tag{11.3.6}$$

**Proof.** "$\Leftarrow$": Obviously, $\|\mathcal{A}\|_{*,\mathbb{F}} \leq \sum_{i=1}^{r} |\lambda_i|$ for every decomposition in (11.3.5). If there is $\mathcal{B}$ satisfying (11.3.6), the dual relationship (11.3.4) implies that

$$\|\mathcal{A}\|_{*,\mathbb{F}} \geq |\langle \mathcal{B}, \mathcal{A} \rangle| = |\lambda_1| + \cdots + |\lambda_r|.$$

This shows that (11.3.5) is a nuclear decomposition.

"$\Rightarrow$": Suppose (11.3.5) is a nuclear decomposition. By the dual relationship (11.3.4), there exists a tensor $\mathcal{B}$ such that $\|\mathcal{B}\|_{\sigma,\mathbb{F}} = 1$ and $|\langle \mathcal{B}, \mathcal{A} \rangle| = \|\mathcal{A}\|_{*,\mathbb{F}}$. Up to a unitary scaling for $\mathcal{B}$, one can further assume $\langle \mathcal{B}, \mathcal{A} \rangle = \|\mathcal{A}\|_{*,\mathbb{F}}$, so

$$\begin{aligned} \|\mathcal{A}\|_{*,\mathbb{F}} = \langle \mathcal{B}, \mathcal{A} \rangle &= \sum_{i=1}^{r} \langle \mathcal{B}, \lambda_i v^{(i,1)} \otimes \cdots \otimes v^{(i,m)} \rangle \\ &\leq \sum_{i=1}^{r} |\langle \mathcal{B}, \lambda_i v^{(i,1)} \otimes \cdots \otimes v^{(i,m)} \rangle| \\ &\leq \sum_{i=1}^{r} |\lambda_i| = \|\mathcal{A}\|_{*,\mathbb{F}}. \end{aligned}$$

Since $\|\mathcal{B}\|_{\sigma,\mathbb{F}} = 1$, we have

$$|\langle \mathcal{B}, \lambda_i v^{(i,1)} \otimes \cdots \otimes v^{(i,m)}\rangle| \le |\lambda_i|$$

for each $i = 1, \ldots, m$, so (11.3.6) holds.    □

When $\mathcal{A}$ is a real tensor, it has both real and complex nuclear norms. Clearly, they satisfy the inequality

$$\|\mathcal{A}\|_{*,\mathbb{R}} \ge \|\mathcal{A}\|_{*,\mathbb{C}}.$$

It is possible for the strict inequality to hold. For instance, consider the real tensor

$$\mathcal{T} = \frac{1}{\sqrt{3}}\left(e_1 \otimes e_1 \otimes e_2 + e_1 \otimes e_2 \otimes e_1 + e_2 \otimes e_1 \otimes e_1\right).$$

Applying Lemma 11.3.1, one can show that (see [101])

$$\|\mathcal{T}\|_{*,\mathbb{C}} = \frac{3}{2} < \sqrt{3} = \|\mathcal{T}\|_{*,\mathbb{R}}.$$

We leave the proof as an exercise.

### 11.3.1 ▪ Symmetric tensors

For a symmetric tensor $\mathcal{A} \in \mathbf{S}^m(\mathbb{F}^n)$, its nuclear norm is achievable at a symmetric tensor decomposition, i.e.,

$$\|\mathcal{A}\|_{*,\mathbb{F}} = \min\left\{ \sum_{i=1}^r |\lambda_i| \,\middle|\, \begin{array}{l} \mathcal{A} = \sum_{i=1}^r \lambda_i (v_i)^{\otimes m}, \\ \|v_i\| = 1, v_i \in \mathbb{F}^n, \lambda_i \in \mathbb{F} \end{array} \right\}. \tag{11.3.7}$$

This is shown in [101]. When $\mathcal{A}$ is symmetric, the decomposition achieving the minimum value in (11.3.7) is called a *symmetric nuclear decomposition*.

For an even order $m$, consider symmetric tensors of the form

$$\mathcal{A} := (a_1)^{\otimes m} + \cdots + (a_r)^{\otimes m} \tag{11.3.8}$$

for real vectors $a_1, \ldots, a_r \in \mathbb{R}^n$. This kind of tensors are called *Sum of Even Powers* (SOEP) tensors (also see §8.4.3). One can show that (see [236])

$$\|\mathcal{A}\|_{*,\mathbb{R}} = \|\mathcal{A}\|_{*,\mathbb{C}} = \|a_1\|^m + \cdots + \|a_r\|^m. \tag{11.3.9}$$

Clearly, it holds that

$$\|\mathcal{A}\|_{*,\mathbb{C}} \le \|\mathcal{A}\|_{*,\mathbb{R}} \le \|a_1\|^m + \cdots + \|a_r\|^m.$$

The reverse inequalities of the above also hold. Let $\mathcal{B} \in \mathbf{S}^m(\mathbb{R}^n)$ be the symmetric tensor such that $\langle \mathcal{B}, x^{\otimes m}\rangle = (x^T x)^{\frac{m}{2}}$, then $\|\mathcal{B}\|_{\sigma,\mathbb{C}} = \|\mathcal{B}\|_{\sigma,\mathbb{R}} = 1$. By the duality relation (11.3.4),

$$\|\mathcal{A}\|_{*,\mathbb{C}} \ge \langle \mathcal{B}, \mathcal{A}\rangle = \sum_{i=1}^r \langle \mathcal{B}, (a_i)^{\otimes m}\rangle = \sum_{i=1}^r \|a_i\|^m.$$

Hence, the equalities in (11.3.9) all hold. Moreover, every real symmetric nuclear decomposition of a SOEP tensor must also be in SOEP form. The proof is left as an exercise.

Symmetric tensor nuclear norms can be equivalently expressed in terms of moments. This is done in the work [236, 311]: Recall the following notation of power sets (see (2.4.1) and (8.4.8)):

$$\begin{array}{ll} \mathbb{N}_d^n & := \{\alpha \in \mathbb{N}^n : |\alpha| \le d\}, \\ \overline{\mathbb{N}}_d^n & := \{\alpha \in \mathbb{N}^n : |\alpha| = d\}. \end{array}$$

We show how to compute nuclear norms for symmetric tensors.

## The case of odd order with $\mathbb{F} = \mathbb{R}$

Consider the case that $\mathbb{F} = \mathbb{R}$ and the order $m$ is odd. For each $\mathcal{A} \in \mathrm{S}^m(\mathbb{R}^n)$, one can generally assume $\lambda_i \geq 0$ for the decomposition in (11.3.7), since

$$\lambda_i(v_i)^{\otimes m} = (-\lambda_i)(-v_i)^{\otimes m}.$$

Hence, we have

$$\|\mathcal{A}\|_{*,\mathbb{R}} = \min \left\{ \sum_{i=1}^{r} \lambda_i \, \middle| \, \begin{array}{c} \mathcal{A} = \lambda_1(v_1)^{\otimes m} + \cdots + \lambda_r(v_r)^{\otimes m} \\ \lambda_i \geq 0, \, \|v_i\| = 1, \, v_i \in \mathbb{R}^n \end{array} \right\}. \tag{11.3.10}$$

Every decomposition of $\mathcal{A}$ as in (11.3.10) gives the finitely atomic measure $\mu := \lambda_1 \delta_{v_1} + \cdots + \lambda_r \delta_{v_r}$, which satisfies the equations

$$\mathcal{A} = \int x^{\otimes m} \mathrm{d}\mu, \quad \sum_{i=1}^{r} \lambda_i = \int 1 \mathrm{d}\mu.$$

On the other hand, if there is a Borel measure $\nu$ such that $\mathcal{A} = \int x^{\otimes m} \mathrm{d}\nu$, then there exist scalars $\rho_1, \ldots, \rho_N > 0$ and distinct points $u_1, \ldots, u_N \in \mathbb{S}^{n-1}$ such that

$$\mathcal{A} = \sum_{i=1}^{N} \rho_i u_i^{\otimes m}, \quad \sum_{i=1}^{N} \rho_i = \int 1 \mathrm{d}\nu.$$

This can be implied by Theorem 8.4.1 (see §8.4.2 for details). Hence, the nuclear norm $\|\mathcal{A}\|_{*,\mathbb{R}}$ is equal to the optimal value of

$$\left\{ \begin{array}{ll} \min & \int 1 \mathrm{d}\mu \\ s.t. & \mathcal{A} = \int x^{\otimes m} \mathrm{d}\mu, \end{array} \right. \tag{11.3.11}$$

where the variable $\mu$ is a Borel measure whose support is contained in the unit sphere $\mathbb{S}^{n-1}$ (see [236, 311]).

The optimization problem (11.3.11) can be equivalently expressed in terms of moments. Recall that $x := (x_1, \ldots, x_n)$. For each label $(i_1, \ldots, i_m)$, there is a unique $\alpha = (\alpha_1, \ldots, \alpha_n) \in \overline{\mathbb{N}}_m^n$ such that

$$x^\alpha := x_1^{\alpha_1} \cdots x_n^{\alpha_n} = x_{i_1} \cdots x_{i_m}.$$

The equality constraint in (11.3.11) is the same as

$$\mathcal{A}_{i_1 \ldots i_m} = \int x^\alpha \mathrm{d}\mu \quad (\text{for } x^\alpha = x_{i_1} \cdots x_{i_m}).$$

Denote the moment cone

$$\mathscr{R}_m(\mathbb{S}^{n-1}) := \left\{ \int [x]_m \mathrm{d}\mu \, \middle| \, \begin{array}{c} \mu \text{ is a Borel measure,} \\ \mathrm{supp}(\mu) \subseteq \mathbb{S}^{n-1} \end{array} \right\}. \tag{11.3.12}$$

The cone $\mathscr{R}_m(\mathbb{S}^{n-1})$ can also be expressed as the conic hull

$$\mathscr{R}_m(\mathbb{S}^{n-1}) = \mathrm{cone}\left( \{[u]_m : u \in \mathbb{S}^{n-1}\} \right).$$

The optimization problem (11.3.11) is equivalent to

$$\left\{ \begin{array}{ll} \min & \langle 1, y \rangle \\ s.t. & \langle x^\alpha, y \rangle = \mathcal{A}_{i_1 \ldots i_m} (\text{for each } x^\alpha = x_{i_1} \cdots x_{i_m}), \\ & y \in \mathscr{R}_m(\mathbb{S}^{n-1}). \end{array} \right. \tag{11.3.13}$$

Algorithm 8.3.1 can be applied to solve (11.3.13).

## The case of even order with $\mathbb{F} = \mathbb{R}$

Consider the case that $\mathbb{F} = \mathbb{R}$ and the order $m$ is even. For $\mathcal{A} \in \mathrm{S}^m(\mathbb{R}^n)$, one cannot generally assume $\lambda_i \geq 0$ for the decomposition in (11.3.7). There are two groups of $\lambda_i$'s, one with $\lambda_i > 0$ and the other with $\lambda_i < 0$ (the case $\lambda_i = 0$ does not contribute to the sum). However, we can always decompose $\mathcal{A}$ as

$$
\boxed{
\begin{aligned}
&\mathcal{A} = \sum_{i=1}^{r_1} \lambda_i^+ (v_i^+)^{\otimes m} - \sum_{j=1}^{r_2} \lambda_j^- (v_j^-)^{\otimes m}, \\
&\lambda_i^+ \geq 0, \ \|v_i^+\| = 1, \ e^T v_i^+ \geq 0, \ v_i^+ \in \mathbb{R}^n, \\
&\lambda_j^- \geq 0, \ \|v_j^-\| = 1, \ e^T v_j^- \geq 0, \ v_j^- \in \mathbb{R}^n.
\end{aligned}
}
\tag{11.3.14}
$$

Consider the half unit sphere ($e$ is the vector of all ones)

$$
S^+ := \{x \in \mathbb{R}^n \ : \ \|x\| = 1, e^T x \geq 0\}.
\tag{11.3.15}
$$

The decomposition for $\mathcal{A}$ in (11.3.14) is equivalent to

$$
\mathcal{A} = \int x^{\otimes m} \mathrm{d}\mu^+ - \int x^{\otimes m} \mathrm{d}\mu^-
\tag{11.3.16}
$$

for two Borel measures $\mu^+, \mu^-$ supported in $S^+$. We can similarly show that $\|\mathcal{A}\|_{*,\mathbb{R}}$ is equal to the optimal value of

$$
\begin{cases}
\min & \int 1 \mathrm{d}\mu^+ + \int 1 \mathrm{d}\mu^- \\
s.t. & \mathcal{A} = \int x^{\otimes m} \mathrm{d}\mu^+ - \int x^{\otimes m} \mathrm{d}\mu^-, \\
& \mathrm{supp}(\mu^+), \mathrm{supp}(\mu^-) \subseteq S^+.
\end{cases}
\tag{11.3.17}
$$

We similarly define the moment cone

$$
\mathscr{R}_m(S^+) := \mathrm{cone}\left(\{[u]_m : u \in S^+\}\right).
$$

Then the optimization (11.3.17) is equivalent to

$$
\begin{cases}
\min & (y^+)_0 + (y^-)_0 \\
s.t. & (y^+)_\alpha - (y^-)_\alpha = \mathcal{A}_{i_1 \ldots i_m} \ (\text{for each } x^\alpha = x_{i_1} \cdots x_{i_m}), \\
& y^+, y^- \in \mathscr{R}_m(S^+).
\end{cases}
\tag{11.3.18}
$$

The following algorithm can be applied to solve (11.3.18).

**Algorithm 11.3.2.** *([236]) For given $\mathcal{A} \in \mathrm{S}^m(\mathbb{R}^n)$ with even order $m = 2m_0$, let $k := m_0$ and do the following:*

**Step 1** *Solve the moment relaxation*

$$
\begin{cases}
\min & (z^+)_0 + (z^-)_0 \\
s.t. & (z^+)_\alpha - (z^-)_\alpha = \mathcal{A}_{i_1 \ldots i_m} \ (\text{for each } x^\alpha = x_{i_1} \cdots x_{i_m}), \\
& \mathscr{V}_{x^T x - 1}^{(2k)}[z^+] = 0, \ \mathscr{V}_{x^T x - 1}^{(2k)}[z^-] = 0, \\
& L_{e^T x}^{(k)}[z^+] \succeq 0, \ L_{e^T x}^{(k)}[z^-] \succeq 0, \\
& M_k[z^+] \succeq 0, \ M_k[z^-] \succeq 0, \\
& z^+ \in \mathbb{R}^{\mathbb{N}_{2k}^n}, \ z^- \in \mathbb{R}^{\mathbb{N}_{2k}^n},
\end{cases}
\tag{11.3.19}
$$

*for an optimal pair $(z^{+,*}, z^{-,*})$. Let $y^{+,*} := z^{+,*}|_m$ and $y^{-,*} := z^{-,*}|_m$. Let $\|\mathcal{A}\|_{k*,\mathbb{R}}$ be the optimal value of (11.3.19).*

**Step 2** *If there are integers $t_1, t_2 \in [m_0, k]$ such that*

$$
\begin{aligned}
\operatorname{rank} M_{t_1-1}[z^{+,*}] &= \operatorname{rank} M_{t_1}[z^{+,*}], \\
\operatorname{rank} M_{t_2-1}[z^{-,*}] &= \operatorname{rank} M_{t_2}[z^{-,*}],
\end{aligned}
$$

*then $y^{+,*}, y^{-,*} \in \mathscr{R}_m(S^+)$ and the pair $(z^{+,*}, z^{-,*})$ is an optimizer of (11.3.18); go to Step 3. Otherwise, go to Step 4.*

**Step 3** *Apply the method in §2.7.2 to compute the decompositions*

$$
\begin{aligned}
z^{+,*}|_{2t_1} &= \lambda_1^+ v_1^+|_{2t_1} + \cdots + \lambda_{r_1}^+ v_{r_1}^+|_{2t_1}, \\
z^{-,*}|_{2t_2} &= \lambda_1^- v_1^-|_{2t_2} + \cdots + \lambda_{r_2}^- v_{r_2}^-|_{2t_2}
\end{aligned}
$$

*for points $v_i^+, v_j^- \in S^+$ and scalars $\lambda_i^+, \lambda_j^+ \geq 0$.*

**Step 4** *Apply Algorithm 8.2.4 to check whether $z^{+,*}, z^{-,*} \in \mathscr{R}_m(S^+)$ or not. If one of them does not, let $k := k+1$ and go to Step 1. If both $z^{+,*}, z^{-,*} \in \mathscr{R}_m(S^+)$, then $(z^{+,*}, z^{-,*})$ is an optimal pair for (11.3.18). Apply Algorithm 8.2.4 to compute the decompositions*

$$
\begin{aligned}
y^{+,*} &= \lambda_1^+ [v_1^+]_m + \cdots + \lambda_{r_1}^+ [v_{r_1}^+]_m, \\
y^{-,*} &= \lambda_1^- [v_1^-]_m + \cdots + \lambda_{r_2}^+ [v_{r_2}^-]_m,
\end{aligned}
$$

*with points $v_i^+, v_i^- \in S^+$ and scalars $\lambda_i^+, \lambda_i^- \geq 0$. Then go to Step 5.*

**Step 5** *The decompositions in either Step 3 or Step 4 give the nuclear norm and decomposition*

$$
\begin{aligned}
\|\mathcal{A}\|_{*,\mathbb{R}} &= \sum_{i=1}^{r_1} \lambda_i^+ + \sum_{i=1}^{r_2} \lambda_i^-, \\
\mathcal{A} &= \sum_{i=1}^{r_1} \lambda_i^+ (v_i^+)^{\otimes m} - \sum_{i=1}^{r_2} \lambda_i^- (v_i^-)^{\otimes m}.
\end{aligned}
\tag{11.3.20}
$$

The convergence of Algorithm 11.3.2 was shown in [236]. The results in §8.3.2 can also be used to prove the convergence.

## The case $\mathbb{F} = \mathbb{C}$

Consider the case that $\mathbb{F} = \mathbb{C}$ and the order $m$ is either odd or even. As shown in [101], the nuclear norm of $\mathcal{A} \in S^m(\mathbb{C}^n)$ can be expressed as

$$
\|\mathcal{A}\|_{*,\mathbb{C}} = \min \left\{ \sum_{i=1}^r |\lambda_i| \;\middle|\; \begin{array}{l} \mathcal{A} = \sum_{i=1}^r \lambda_i (w_i)^{\otimes m}, \\ \|w_i\| = 1, w_i \in \mathbb{C}^n \end{array} \right\}.
\tag{11.3.21}
$$

The following gives an equivalent moment reformulation.

**Lemma 11.3.3.** *([236]) The nuclear norm $\|\mathcal{A}\|_{*,\mathbb{C}}$ equals the optimal value of*

$$
\begin{cases}
\min & \sum_{i=1}^r \lambda_i \\
s.t. & \mathcal{A} = \sum_{i=1}^r \lambda_i (u_i + \sqrt{-1} v_i)^{\otimes m}, \\
& \lambda_i \geq 0, \|u_i\|^2 + \|v_i\|^2 = 1, u_i, v_i \in \mathbb{R}^n, \\
& e^T v_i \geq 0, \sin(\frac{2\pi}{m}) e^T u_i - \cos(\frac{2\pi}{m}) e^T v_i \geq 0.
\end{cases}
\tag{11.3.22}
$$

The proof of Lemma 11.3.3 is left as an exercise. The main point for the above is that the decomposition in (11.3.21) is equivalent to

$$\mathcal{A} = \sum_{i=1}^{r} |\lambda_i| (\tau_i w_i)^{\otimes m}$$

for all unitary scalars $\tau_i \in \mathbb{C}$ such that $\lambda_i = |\lambda_i| \tau_i^m$. One can always choose such $\tau_i$ such that the decomposition (11.3.22) holds. A complex vector $z \in \mathbb{C}^n$ can be written as

$$
\begin{aligned}
z &= x^{\mathrm{re}} + \sqrt{-1} x^{\mathrm{im}}, \\
x^{\mathrm{re}} &= (x_1, \ldots, x_n) \in \mathbb{R}^n, \\
x^{\mathrm{im}} &= (x_{n+1}, \ldots, x_{2n}) \in \mathbb{R}^n.
\end{aligned}
$$

In view of Lemma 11.3.3, consider the set

$$
S^c := \left\{ \begin{bmatrix} x^{\mathrm{re}} \\ x^{\mathrm{im}} \end{bmatrix} \middle| \begin{array}{l} \|x^{\mathrm{re}}\|^2 + \|x^{\mathrm{im}}\|^2 = 1, \\ e^T x^{\mathrm{im}} \geq 0, \; x^{\mathrm{re}}, x^{\mathrm{im}} \in \mathbb{R}^n, \\ \sin(\frac{2\pi}{m}) e^T x^{\mathrm{re}} - \cos(\frac{2\pi}{m}) e^T x^{\mathrm{im}} \geq 0 \end{array} \right\}. \tag{11.3.23}
$$

Then, $S^c = \{x : h(x) = 0, g_1(x) \geq 0, g_2(x) \geq 0\}$ for the polynomial tuples

$$
\begin{array}{lll}
h(x) &:= & x^T x - 1, \\
g_1(x) &:= & e^T x^{\mathrm{im}}, \\
g_2(x) &:= & \sin(\frac{2\pi}{m}) e^T x^{\mathrm{re}} - \cos(\frac{2\pi}{m}) e^T x^{\mathrm{im}}.
\end{array} \tag{11.3.24}
$$

The decomposition of $\mathcal{A}$ as in (11.3.22) is equivalent to

$$\mathcal{A} = \int (x^{\mathrm{re}} + \sqrt{-1} x^{\mathrm{im}})^{\otimes m} \mathrm{d}\mu \tag{11.3.25}$$

for the finitely atomic measure

$$\mu := \lambda_1 \delta_{u_1 + \sqrt{-1} v_1} + \cdots + \lambda_r \delta_{u_r + \sqrt{-1} v_r}.$$

Note that $\lambda_1 + \cdots + \lambda_r = \int 1 \mathrm{d}\mu$. By Lemma 11.3.3, $\|\mathcal{A}\|_{*,\mathbb{C}}$ equals the optimal value of

$$
\begin{cases}
\min_{\mu} & \int 1 \mathrm{d}\mu \\
\mathrm{s.t.} & \mathcal{A} = \int (x^{\mathrm{re}} + \sqrt{-1} x^{\mathrm{im}})^{\otimes m} \mathrm{d}\mu, \\
& \mathrm{supp}(\mu) \subseteq S^c.
\end{cases} \tag{11.3.26}
$$

Let $\mathbf{a}^{\mathrm{re}}, \mathbf{a}^{\mathrm{im}} \in \mathbb{R}^{\overline{\mathbb{N}}^n_m}$ be the real-valued tms's such that

$$\mathbf{a}^{\mathrm{re}}_\alpha + \sqrt{-1} \mathbf{a}^{\mathrm{im}}_\alpha = \mathcal{A}_{i_1 \ldots i_m} \quad \text{for each} \quad z^\alpha = z_{i_1} \cdots z_{i_m}, \tag{11.3.27}$$

where $\alpha = (\alpha_1, \ldots, \alpha_n)$ and $z^\alpha = z_1^{\alpha_1} \cdots z_n^{\alpha_n}$. For each $\alpha \in \overline{\mathbb{N}}^n_m$, let

$$R_\alpha, T_\alpha \in \mathbb{R}[x] := \mathbb{R}[x_1, \ldots, x_{2n}]$$

be homogeneous polynomials such that

$$\prod_{i=1}^{n} (x_i + \sqrt{-1} \, x_{n+i})^{\alpha_i} = R_\alpha(x) + \sqrt{-1} \, T_\alpha(x). \tag{11.3.28}$$

Then, they satisfy the moment relation (for every $\alpha \in \overline{\mathbb{N}}_m^n$)

$$\int \prod_{i=1}^n (x_i + \sqrt{-1}\,x_{n+i})^{\alpha_i}\,\mathrm{d}\mu = \int R_\alpha(x)\mathrm{d}\mu + \sqrt{-1}\int T_\alpha(x)\mathrm{d}\mu.$$

Hence, (11.3.26) is equivalent to the optimization

$$\left\{\begin{array}{ll} \min & \langle 1, y\rangle \\ s.t. & \langle R_\alpha, y\rangle = \mathbf{a}_\alpha^{\mathrm{re}} \ (\alpha \in \overline{\mathbb{N}}_m^{2n}), \\ & \langle T_\alpha, y\rangle = \mathbf{a}_\alpha^{\mathrm{im}} \ (\alpha \in \overline{\mathbb{N}}_m^{2n}), \\ & y \in \mathscr{R}_m(S^c). \end{array}\right. \tag{11.3.29}$$

In the above, the set $\mathscr{R}_m(S^c)$ is the moment cone

$$\mathscr{R}_m(S^c) := \operatorname{cone}\left(\{[z]_m : z \in S^c\}\right).$$

Algorithm 8.3.1 can be applied to solve (11.3.29).

## 11.3.2 ▪ Examples of tensor nuclear norms

We give examples for symmetric tensor nuclear norms.

**Example 11.3.4.** *([101]) (i) Consider the tensor in* $\mathrm{S}^3(\mathbb{R}^2)$*:*

$$\mathcal{A} = \frac{1}{\sqrt{3}}\left(e_1 \otimes e_1 \otimes e_2 + e_1 \otimes e_2 \otimes e_1 + e_2 \otimes e_1 \otimes e_1\right).$$

*Applying Lemma 11.3.1, one can show that*

$$\|\mathcal{A}\|_{*,\mathbb{R}} = \sqrt{3} > \|\mathcal{A}\|_{*,\mathbb{C}} = \frac{3}{2}.$$

*The real symmetric nuclear decomposition is*

$$\frac{4}{3\sqrt{3}}\left(\left[\frac{\sqrt{3}e_1}{2} + \frac{e_2}{2}\right]^{\otimes 3} + \left[\frac{-\sqrt{3}e_1}{2} + \frac{e_2}{2}\right]^{\otimes 3} - \frac{e_2^{\otimes 3}}{4}\right),$$

*and the complex one is*

$$\frac{3}{8}\left(\left[\sqrt{\tfrac{2}{3}}e_1 + \sqrt{\tfrac{1}{3}}e_2\right]^{\otimes 3} + \left[-\sqrt{\tfrac{2}{3}}e_1 + \sqrt{\tfrac{1}{3}}e_2\right]^{\otimes 3} \right.$$
$$\left. + \left[\sqrt{-\tfrac{2}{3}}e_1 - \sqrt{\tfrac{1}{3}}e_2\right]^{\otimes 3} + \left[-\sqrt{-\tfrac{2}{3}}e_1 - \sqrt{\tfrac{1}{3}}e_2\right]^{\otimes 3}\right).$$

*(ii) Consider the tensor in* $\mathrm{S}^3(\mathbb{R}^2)$*:*

$$\mathcal{A} = \frac{1}{2}\left(e_1 \otimes e_1 \otimes e_2 + e_1 \otimes e_2 \otimes e_1 + e_2 \otimes e_1 \otimes e_1 - e_2 \otimes e_2 \otimes e_2\right).$$

*We have* $\|\mathcal{A}\|_{*,\mathbb{R}} = 2 > \|\mathcal{A}\|_{*,\mathbb{C}} = \sqrt{2}$*. The real symmetric nuclear decomposition is*

$$\frac{2}{3}\left(\left[\frac{\sqrt{3}e_1}{2} + \frac{e_2}{2}\right]^{\otimes 3} + \left[\frac{-\sqrt{3}e_1}{2} + \frac{e_2}{2}\right]^{\otimes 3} - e_2^{\otimes 3}\right),$$

*while the complex one is*

$$\frac{1}{\sqrt{2}}\left(\left[\frac{-e_2}{\sqrt{2}}+\frac{\sqrt{-1}e_1}{2}\right]^{\otimes 3}+\left[\frac{-e_2}{\sqrt{2}}-\frac{\sqrt{-1}e_1}{2}\right]^{\otimes 3}\right).$$

**Example 11.3.5.** *([236]) (i) Consider the tensor in $S^4(\mathbb{R}^3)$:*

$$\mathcal{A}=(e_1+e_2)^{\otimes 4}+(e_1+e_3)^{\otimes 4}-(e_2+e_3)^{\otimes 4}.$$

*It holds that $\|\mathcal{A}\|_{2*,\mathbb{R}}=\|\mathcal{A}\|_{2*,\mathbb{C}}=12$. The above is simultaneously a real and complex symmetric nuclear decomposition.*
*(ii) Consider the tensor $\mathcal{A}\in S^4(\mathbb{R}^3)$:*

$$\mathcal{A}=(e_1+e_2-e_3)^{\otimes 4}+(e_1-e_2+e_3)^{\otimes 4}+(-e_1+e_2+e_3)^{\otimes 4}-(e)^{\otimes 4}.$$

*We have $\|\mathcal{A}\|_{*,\mathbb{R}}=\|\mathcal{A}\|_{*,\mathbb{C}}=36$. The above simultaneously gives a real and complex symmetric nuclear decomposition.*

A symmetric tensor $\mathcal{A}$ is uniquely determined by the homogeneous polynomial $\mathcal{A}(x)$, given as in (11.1.12). One can equivalently display a symmetric decomposition of $\mathcal{A}$ by showing the Waring decomposition of $\mathcal{A}(x)$, i.e.,

$$\mathcal{A}=\sum_{i=1}^{r}\pm(u_i)^{\otimes m}\Leftrightarrow\mathcal{A}(x)=\sum_{i=1}^{r}\pm(u_i^T x)^m.$$

**Example 11.3.6.** *([236]) (i) Consider the tensor $\mathcal{A}\in S^3(\mathbb{R}^3)$ such that $\mathcal{A}(x)=x_1 x_2 x_3$. The nuclear norms $\|\mathcal{A}\|_{*,\mathbb{R}}=\|\mathcal{A}\|_{*,\mathbb{C}}=\frac{\sqrt{3}}{2}$. The real and complex symmetric nuclear decompositions are simultaneously given by*

$$\mathcal{A}(x)=\frac{1}{24}\Big((-x_1-x_2+x_3)^3+(-x_1+x_2-x_3)^3+(x_1-x_2-x_3)^3$$
$$+(x_1+x_2+x_3)^3\Big).$$

*(ii) Consider the tensor $\mathcal{A}\in S^4(\mathbb{R}^3)$ such that*

$$\mathcal{A}(x)=x_1^2 x_2^2+x_2^2 x_3^2+x_1^2 x_3^2.$$

*The real nuclear norm $\|\mathcal{A}\|_{*,\mathbb{R}}=2$, achieved for the decomposition*

$$\mathcal{A}(x)=\frac{1}{24}\Big((x_1-x_2+x_3)^4+(x_1+x_2-x_3)^4+(-x_1+x_2+x_3)^4$$
$$+(x_1+x_2+x_3)^4\Big)-\frac{1}{6}(x_1^4+x_2^4+x_3^4).$$

*The complex nuclear norm $\|\mathcal{A}\|_{*,\mathbb{C}}=\frac{5}{3}$, achieved for the decomposition*

$$\mathcal{A}(x)=\frac{1}{36}\Big((-x_1+x_2+x_3)^4+(x_1-x_2+x_3)^4+(x_1+x_2-x_3)^4$$
$$+(x_1+x_2+x_3)^4-(x_1-\sqrt{-1}x_3)^4-(x_2-\sqrt{-1}x_3)^4$$
$$-(x_1-\sqrt{-1}x_2)^4-(x_1+\sqrt{-1}x_3)^4$$
$$-(x_2+\sqrt{-1}x_3)^4-(x_1+\sqrt{-1}x_2)^4\Big).$$

*(iii) Consider the tensor $\mathcal{A} \in S^4(\mathbb{R}^3)$ such that*

$$\mathcal{A}(x) = (x_1^2 + x_2^2 + x_3^2)^2.$$

*We have $\|\mathcal{A}\|_{*,\mathbb{R}} = \|\mathcal{A}\|_{*,\mathbb{C}} = 5$. The real and complex symmetric nuclear decompositions are simultaneously given by*

$$
\begin{aligned}
\mathcal{A}(x) &= \frac{1}{12}\Big( (x_1 - x_2 + x_3)^4 + (x_1 + x_2 - x_3)^4 + (-x_1 + x_2 + x_3)^4 \\
&\quad + (x_1 + x_2 + x_3)^4 \Big) + \frac{2}{3}(x_1^4 + x_2^4 + x_3^4).
\end{aligned}
$$

### 11.3.3 ▪ Nonsymmetric tensor nuclear norms

The moment optimization methods in previous subsections can be generalized to compute non-symmetric tensor nuclear norms (see [236]). For cleanness, we only show how to do this for nonsymmetric cubic order tensors with $\mathbb{F} = \mathbb{R}$. For $\mathcal{A} \in \mathbb{R}^{n_1 \times n_2 \times n_3}$, it holds that

$$
\|\mathcal{A}\|_{*,\mathbb{R}} = \min \left\{ \sum_{i=1}^{r} \lambda_i \;\middle|\; 
\begin{array}{l}
\mathcal{A} = \sum_{i=1}^{r} \lambda_i v^{(i,1)} \otimes v^{(i,2)} \otimes v^{(i,3)}, \\
\lambda_i \geq 0, \; \|v^{(i,j)}\| = 1, \; v^{(i,j)} \in \mathbb{R}^{n_j}
\end{array}
\right\}. \tag{11.3.30}
$$

Similarly, one can show that $\|\mathcal{A}\|_{*,\mathbb{R}}$ equals the minimum value of

$$
\left\{
\begin{array}{ll}
\min & \int 1 \mathrm{d}\mu \\
\text{s.t.} & \mathcal{A} = \int x^{(1)} \otimes x^{(2)} \otimes x^{(3)} \mathrm{d}\mu, \\
& \operatorname{supp}(\mu) \subseteq T,
\end{array}
\right. \tag{11.3.31}
$$

where $T$ is the compact set

$$T := \left\{ (x^{(1)}, x^{(2)}, x^{(3)}) : \|x^{(1)}\| = \|x^{(2)}\| = \|x^{(3)}\| = 1 \right\}.$$

Therefore, (11.3.31) is equivalent to the moment optimization

$$
\left\{
\begin{array}{ll}
\min & (y)_0 \\
\text{s.t.} & \langle (x^{(1)})_i (x^{(2)})_j (x^{(3)})_k, y \rangle = \mathcal{A}_{ijk} \; (\text{for all } i, j, k), \\
& y \in \mathscr{R}_3(T).
\end{array}
\right. \tag{11.3.32}
$$

Algorithm 8.3.1 can be applied to solve (11.3.32). We refer to the work [236] for more details.

**Example 11.3.7.** *([236]) Consider the tensor $\mathcal{A} \in \mathbb{R}^{2 \times 2 \times 2}$ such that*

$$\mathcal{A}_{ijk} = i - j - k$$

*for all $i, j, k$ in the range. By solving (11.3.32), we get $\|\mathcal{A}\|_{*,\mathbb{R}} = 6.0000$. The real nuclear decomposition of $\mathcal{A}$ is (only four decimal digits are shown)*

$$
\begin{bmatrix} 1.4363 \\ 0.3140 \end{bmatrix} \otimes \begin{bmatrix} -0.9146 \\ -1.1516 \end{bmatrix} \otimes \begin{bmatrix} 0.9296 \\ 1.1394 \end{bmatrix} + \begin{bmatrix} 0.7375 \\ 1.0525 \end{bmatrix} \otimes \begin{bmatrix} -0.2430 \\ -1.2616 \end{bmatrix} \otimes \begin{bmatrix} 0.5250 \\ 1.1727 \end{bmatrix}
$$

$$
+ \begin{bmatrix} 0.5484 \\ 0.6978 \end{bmatrix} \otimes \begin{bmatrix} 0.8846 \\ 0.0732 \end{bmatrix} \otimes \begin{bmatrix} 0.6501 \\ -0.6040 \end{bmatrix}.
$$

For symmetric tensors, a nuclear decomposition is not necessarily symmetric. The following is such an example.

**Example 11.3.8.** *([236]) Consider the symmetric tensor* $\mathcal{A} \in \mathrm{S}^3(\mathbb{R}^2)$ *such that*

$$\mathcal{A}_{ijk} = i + j + k.$$

*The moment optimization (11.3.32) gives* $\|\mathcal{A}\|_{*,\mathbb{R}} = 13.4164\cdots$. *It is achieved at the following nonsymmetric nuclear decomposition (only four decimal digits are shown):*

$$\begin{bmatrix} 0.4542 \\ -0.4547 \end{bmatrix} \otimes \begin{bmatrix} -0.6384 \\ -0.0746 \end{bmatrix} \otimes \begin{bmatrix} 0.6291 \\ 0.1314 \end{bmatrix} + \begin{bmatrix} 0.1197 \\ 0.5725 \end{bmatrix} \otimes \begin{bmatrix} -0.2946 \\ 0.5053 \end{bmatrix} \otimes \begin{bmatrix} 0.4136 \\ -0.4136 \end{bmatrix}$$
$$+ \begin{bmatrix} 1.4912 \\ 1.8131 \end{bmatrix} \otimes \begin{bmatrix} 1.4675 \\ 1.8344 \end{bmatrix} \otimes \begin{bmatrix} 1.4609 \\ 1.8386 \end{bmatrix}.$$

When $\mathbb{F} = \mathbb{C}$, we can similarly formulate a moment optimization problem like (11.3.32) for computing $\|\mathcal{A}\|_{*,\mathbb{C}}$. Each $x^{(j)}$ is now a complex vector variable. For nonsymmetric tensors, it is usually much more expensive to compute nuclear norms. This is because there are $m$ vector variables $x^{(1)}, \ldots, x^{(m)}$ and the corresponding Moment-SOS relaxations are much more expensive to solve. Computation for the complex case is much harder, because the total number of real variables is doubled.

### 11.3.4 ▪ Exercises

**Exercise 11.3.1.** *([101]) For the tensor*

$$\mathcal{T} = \frac{1}{\sqrt{3}}(e_1 \otimes e_1 \otimes e_2 + e_1 \otimes e_2 \otimes e_1 + e_2 \otimes e_1 \otimes e_1),$$

*apply Lemma 11.3.1 to show that*

$$\|\mathcal{T}\|_{*,\mathbb{C}} = \frac{3}{2} < \|\mathcal{T}\|_{*,\mathbb{R}} = \sqrt{3}.$$

**Exercise 11.3.2.** *For the tensor* $\mathcal{A} \in \mathrm{S}^6(\mathbb{C}^3)$ *given as*

$$\mathcal{A}_{i_1 i_2 i_3 i_4 i_5 i_6} = (1 + \sqrt{-1})^{i_1 + \cdots + i_6 - 6} + (1 - \sqrt{-1})^{i_1 + \cdots + i_6 - 6},$$

*show that* $\|\mathcal{A}\|_{*,\mathbb{C}} = 686$ *and a nuclear decomposition is*

$$\mathcal{A} = \begin{bmatrix} 1 \\ 1 - \sqrt{-1} \\ -2\sqrt{-1} \end{bmatrix}^{\otimes 6} + \begin{bmatrix} 1 \\ 1 + \sqrt{-1} \\ 2\sqrt{-1} \end{bmatrix}^{\otimes 6}.$$

**Exercise 11.3.3.** *([236]) (i) For the tensor* $\mathcal{A} \in \mathrm{S}^4(\mathbb{R}^4)$ *such that* $\mathcal{A}(x) = x_1 x_2 x_3 x_4$, *show that* $\|\mathcal{A}\|_{*,\mathbb{R}} = \|\mathcal{A}\|_{*,\mathbb{C}} = \frac{2}{3}$ *and real and complex symmetric nuclear decompositions are simultaneously given by*

$$\begin{aligned} \mathcal{A}(x) = \ \frac{1}{192} \Big( & (-x_1 - x_2 + x_3 + x_4)^4 + (-x_1 + x_2 - x_3 + x_4)^4 \\ & + (-x_1 + x_2 + x_3 - x_4)^4 + (x_1 + x_2 + x_3 + x_4)^4 \\ & - (-x_1 + x_2 + x_3 + x_4)^4 - (x_1 - x_2 + x_3 + x_4)^4 \\ & - (x_1 + x_2 - x_3 + x_4)^4 - (x_1 + x_2 + x_3 - x_4)^4 \Big). \end{aligned}$$

*(ii) For the tensor $\mathcal{A} \in S^4(\mathbb{R}^2)$ such that $\mathcal{A}(x) = x_1^2 x_2^2$, show that $\|\mathcal{A}\|_{*,\mathbb{R}} = 1$ and a real nuclear decomposition is given by*

$$\mathcal{A}(x) = \frac{1}{12}(-x_1 + x_2)^4 + \frac{1}{12}(x_1 + x_2)^4 - \frac{1}{6}x_1^4 - \frac{1}{6}x_2^4.$$

*Also show $\|\mathcal{A}\|_{*,\mathbb{C}} = 2/3$ and a complex nuclear decomposition is given by*

$$\mathcal{A}(x) = \frac{1}{24}\Big((x_1 - x_2)^4 + (x_1 + x_2)^4 - (x_1 + \sqrt{-1}x_2)^4 - (x_1 - \sqrt{-1}x_2)^4\Big).$$

**Exercise 11.3.4.** *Let $\mathcal{A}$ be the orthogonally decomposable tensor*

$$\mathcal{A} = \lambda_1(q_1)^{\otimes m} + \cdots + \lambda_r(q_r)^{\otimes m},$$

*where $q_1, \ldots, q_r \in \mathbb{C}^n$ are orthonormal vectors, and $\lambda_1, \ldots, \lambda_r$ are complex scalars. Show that $\|\mathcal{A}\|_{*,\mathbb{C}} = |\lambda_1| + \cdots + |\lambda_r|$.*

**Exercise 11.3.5.** *Show that the nuclear norm unit ball $\{\mathcal{X} : \|\mathcal{X}\|_{*,\mathbb{F}} \le 1\}$ is the convex hull of the compact set $U = \{\mathcal{X} : \mathrm{rank}_{\mathbb{F}}(\mathcal{X}) = 1, \|\mathcal{X}\| = 1\}$. Use this fact to show the equalities (11.3.2) and (11.3.3).*

**Exercise 11.3.6.** *Suppose $\mathcal{A} \in S^m(\mathbb{R}^n)$ is a SOEP tensor of even order $m$. If $\mathcal{A} = \sum_{i=1}^{r} \lambda_i(u_i)^m$ is a nuclear decomposition, with $\lambda_i \in \mathbb{R}$ and $u_i \in \mathbb{R}^n$, show that all $\lambda_i \ge 0$.*

**Exercise 11.3.7.** *Let $\mathcal{A} \in S^{2d}(\mathbb{R}^n)$ be the tensor such that*

$$\mathcal{A}(x) = (x_1^2 + \cdots + x_n^2)^d.$$

*Show that $\mathcal{A}$ is SOEP. Express $\|\mathcal{A}\|_{*,\mathbb{R}}$ in terms of $d, n$ and given an explicit real nuclear decomposition.*

**Exercise 11.3.8.** *Prove Lemma 11.3.3.*

## 11.4 ▪ Tensor Entanglements

Tensor entanglement is an important topic in quantum physics. In $m$-partite systems, a $m$-partite state is represented by a tensor of order $m$ and with unit length. A state is called a *product state*, or a *pure state*, if its representing tensor is rank-1. A state is called *entangled* if it is not a product or pure state. Similarly, rank-1 tensors are called *pure* tensors, and higher rank tensors are called *entangled* tensors.

For a tensor $\mathcal{A} \in \mathbb{F}^{n_1 \times \cdots \times n_m}$ with unit length $\|\mathcal{A}\| = 1$, its *geometric measure of entanglement* (GME) is

$$g(\mathcal{A}) := \min_{\mathcal{X} \in \mathbb{F}^{n_1 \times \cdots \times n_m}} \Big\{\|\mathcal{A} - \mathcal{X}\| : \mathrm{rank}(\mathcal{X}) = 1, \|\mathcal{X}\| = 1\Big\}. \qquad (11.4.1)$$

It measures the distance to the set of unit length pure tensors. If we write $\mathcal{X} = x^{(1)} \otimes \cdots \otimes x^{(m)}$ for unit length vectors $x^{(i)} \in \mathbb{F}^{n_i}$, then

$$\|\mathcal{A} - \mathcal{X}\|^2 = 2(1 - \mathrm{Re}(\langle \mathcal{A}, x^{(1)} \otimes \cdots \otimes x^{(m)} \rangle)).$$

Note that maximizing the real part $\mathrm{Re}(\langle \mathcal{A}, x^{(1)} \otimes \cdots x^{(m)} \rangle)$ is equivalent to maximizing the modulus $|\langle \mathcal{A}, x^{(1)} \otimes \cdots x^{(m)} \rangle|$. So, the GME can be expressed as

$$\mathrm{g}(\mathcal{A})^2 = 2\Big(1 - \max_{x^{(i)} \in \mathbb{F}^{n_i}, \|x^{(i)}\|=1} |\langle \mathcal{A}, x^{(1)} \otimes \cdots \otimes x^{(m)} \rangle|\Big).$$

In terms of the spectral norm, we can get

$$\mathrm{g}(\mathcal{A}) = \sqrt{2(1 - \|\mathcal{A}\|_{\sigma,\mathbb{F}})}. \tag{11.4.2}$$

When $\|\mathcal{A}\| \neq 1$, the GME of $\mathcal{A}$ is defined to be $\mathrm{g}(\mathcal{A}/\|\mathcal{A}\|)$. To compute $\mathrm{g}(\mathcal{A})$, we need to know the spectral norm $\|\mathcal{A}\|_{\sigma,\mathbb{F}}$, which depends on the field $\mathbb{F}$. Computing geometric measures of entanglements is typically a hard question. We refer to [81, 131, 207, 278] for related work.

**Example 11.4.1.** *The following examples are from [278].*

(i) *For the tensor $\frac{1}{\sqrt{2}}(e_1^{\otimes 3} + e_2^{\otimes 3})$, the spectral norm is $\frac{1}{\sqrt{2}}$ for both $\mathbb{F} = \mathbb{R}$ and $\mathbb{C}$, so the GME is $\sqrt{2 - \sqrt{2}}$. The closest unit length pure tensor is $e_1^{\otimes 3}$ or $e_2^{\otimes 3}$.*

(ii) *For the tensor $\frac{1}{\sqrt{3}}\Big(e_1^{\otimes 3} + e_2^{\otimes 3} + e_3^{\otimes 3}\Big)$, the spectral norm is $\frac{1}{\sqrt{3}}$ for both $\mathbb{F} = \mathbb{R}$ and $\mathbb{C}$, so the GME is $\sqrt{2(1 - \frac{1}{\sqrt{3}})}$. The closest unit length pure tensor is $e_i^{\otimes 3}$ for each $i = 1, 2, 3$.*

(iii) *For the symmetrization tensor $\sqrt{6} \cdot \mathrm{sym}(e_1 \otimes e_2 \otimes e_3)$, the spectral norm is $\frac{\sqrt{2}}{3}$ for both $\mathbb{F} = \mathbb{R}$ and $\mathbb{C}$, so the GME is $\sqrt{2(1 - \frac{\sqrt{2}}{3})}$. The closest unit length pure tensor is $\frac{1}{3\sqrt{3}}(1, 1, 1)^{\otimes 3}$.*

The maximum tensor entanglement problem (MTEP) concerns the maximum value of $\mathrm{g}(\mathcal{A})$ for all unit length tensors $\mathcal{A}$. A unit length tensor achieving the maximum GME is called the *maximally entangled tensor*. Define the constant

$$\sigma(\mathbb{F}; n_1, \ldots, n_m) := \min_{\substack{\mathcal{A} \in \mathbb{F}^{n_1 \times \cdots \times n_m}, \\ \|\mathcal{A}\|=1}} \|\mathcal{A}\|_{\sigma,\mathbb{F}}. \tag{11.4.3}$$

Then the maximum GME can be expressed as

$$\sqrt{2(1 - \sigma(\mathbb{F}; n_1, \ldots, n_m))}.$$

Interestingly, $\sigma(\mathbb{F}^{n_1 \times \cdots \times n_m})$ is closely related to the following constant

$$\nu(\mathbb{F}^{n_1 \times \cdots \times n_m}) := \max_{\substack{\mathcal{A} \in \mathbb{F}^{n_1 \times \cdots \times n_m}, \\ \|\mathcal{A}\|=1}} \|\mathcal{A}\|_{*,\mathbb{F}}. \tag{11.4.4}$$

Since $\|\mathcal{A}\|_{\sigma,\mathbb{F}} \leq \|\mathcal{A}\|_{*,\mathbb{F}}$ for all $\mathcal{A}$ (see exercise), we have

$$\sigma(\mathbb{F}; n_1, \ldots, n_m) \leq 1 \leq \nu(\mathbb{F}; n_1, \ldots, n_m).$$

The constants $\sigma(\mathbb{F}; n_1, \ldots, n_m)$ and $\nu(\mathbb{F}; n_1, \ldots, n_m)$ are reciprocal to each other and they are achievable in the same tensor.

**Theorem 11.4.2.** *([81]) For $\mathbb{F} = \mathbb{R}$ or $\mathbb{C}$, it always holds that*

$$\sigma(\mathbb{F}^{n_1 \times \cdots \times n_m}) \cdot \nu(\mathbb{F}^{n_1 \times \cdots \times n_m}) = 1. \tag{11.4.5}$$

*Moreover, for a unit length tensor $\mathcal{A} \in \mathbb{F}^{n_1 \times \cdots \times n_m}$, $\|\mathcal{A}\|_{\sigma,\mathbb{F}} = \sigma(\mathbb{F}^{n_1 \times \cdots \times n_m})$ if and only if $\|\mathcal{A}\|_{*,\mathbb{F}} = \nu(\mathbb{F}^{n_1 \times \cdots \times n_m})$.*

The constants $\sigma(\mathbb{F}^{n_1 \times \cdots \times n_m})$ and $\nu(\mathbb{F}^{n_1 \times \cdots \times n_m})$ are generally difficult to compute. For a few cases, however, they can be given by explicit formulae.

**Proposition 11.4.3.** *([81]) For the real binary tensor space of order $m$,*

$$\sigma(\mathbb{R}^{2 \times \cdots \times 2}) = 2^{\frac{1-m}{2}}, \quad \nu(\mathbb{R}^{2 \times \cdots \times 2}) = 2^{\frac{m-1}{2}}. \tag{11.4.6}$$

*The optimal norms are achieved for the following real symmetric tensor:*

$$\frac{1}{\sqrt{2}} \left( \lambda \left( \frac{e_1}{\sqrt{2}} + \frac{\sqrt{-1}e_2}{\sqrt{2}} \right)^{\otimes m} + \bar{\lambda} \left( \frac{e_1}{\sqrt{2}} - \frac{\sqrt{-1}e_2}{\sqrt{2}} \right)^{\otimes m} \right),$$

*where $\lambda$ is any unitary complex scalar.*

**Example 11.4.4.** *The above conclusion does not hold for the complex case.*

(i) *Consider the complex tensor in $\mathbb{C}^{2 \times 2 \times 2}$*

$$\mathcal{W} := \frac{1}{\sqrt{3}} \left( e_2 \otimes e_1 \otimes e_1 + e_1 \otimes e_2 \otimes e_1 + e_1 \otimes e_1 \otimes e_2 \right).$$

*One can show that $\|\mathcal{W}\|_{\sigma,\mathbb{C}} = \frac{2}{3}$, $\|\mathcal{W}\|_{*,\mathbb{C}} = \frac{3}{2}$ (see [81, 101]). Therefore, $\mathrm{g}(\mathcal{W}) = \sqrt{\frac{2}{3}}$. This tensor $\mathcal{W}$ is the most entangled one in $\mathbb{C}^{2 \times 2 \times 2}$, as shown in [49]. In fact, the maximum entangled tensors in $\mathbb{C}^{2 \times 2 \times 2}$ are of the form $(U_1, U_2, U_3) \times \mathcal{W}$, where $U_1, U_2, U_3$ are 2-by-2 unitary matrices. Indeed, we have*

$$\sigma(\mathbb{C}^{2 \times 2 \times 2}) = \frac{2}{3}, \quad \nu(\mathbb{C}^{2 \times 2 \times 2}) = \frac{3}{2}.$$

(ii) *Consider the tensor in $\mathbb{C}^{2 \times 2 \times 2 \times 2}$*

$$\begin{aligned}
\mathcal{M} := \frac{1}{\sqrt{6}} \Big( &e_1 \otimes e_1 \otimes e_2 \otimes e_2 + e_2 \otimes e_2 \otimes e_1 \otimes e_1 \\
&+ \exp\{\tfrac{2\pi}{3}\sqrt{-1}\} \left( e_2 \otimes e_1 \otimes e_2 \otimes e_1 + e_1 \otimes e_2 \otimes e_1 \otimes e_2 \right) \\
&+ \exp\{\tfrac{4\pi}{3}\sqrt{-1}\} \left( e_2 \otimes e_1 \otimes e_1 \otimes e_2 + e_1 \otimes e_2 \otimes e_2 \otimes e_1 \right) \Big).
\end{aligned}$$

*One can show that $\|\mathcal{M}\|_{\sigma,\mathbb{C}} = \frac{\sqrt{2}}{3}$, as in [81]. This tensor is maximumly entangled in $\mathbb{C}^{2 \times 2 \times 2 \times 2}$. Indeed, we have*

$$\sigma(\mathbb{C}^{2 \times 2 \times 2 \times 2}) = \frac{\sqrt{2}}{3}, \quad \nu(\mathbb{C}^{2 \times 2 \times 2 \times 2}) = \frac{3}{\sqrt{2}}.$$

For general cases, the constants $\sigma(\mathbb{F}^{n_1 \times \cdots \times n_m})$ and $\nu(\mathbb{F}^{n_1 \times \cdots \times n_m})$ are mostly unknown. They can be bounded as follows.

**Proposition 11.4.5.** *([81, 278]) If $n_1 \geq n_2 \geq \cdots \geq n_m$, then*

$$\begin{aligned}
\sigma(\mathbb{F}^{n_1 \times \cdots \times n_m}) &\geq (n_2 n_3 \cdots n_m)^{-\frac{1}{2}}, \\
\nu(\mathbb{F}^{n_1 \times \cdots \times n_m}) &\leq (n_2 n_3 \cdots n_m)^{\frac{1}{2}}.
\end{aligned}$$

### 11.4.1 ▪ Exercises

**Exercise 11.4.1.** *Show that* $\|\mathcal{A}\|_{\sigma,\mathbb{F}} \leq \|\mathcal{A}\|_{*,\mathbb{F}}$ *for every tensor* $\mathcal{A}$.

**Exercise 11.4.2.** *([81]) Show that* $\sigma(\mathbb{F}^{n_1 \times \cdots \times n_m})$ *and* $\nu(\mathbb{F}^{n_1 \times \cdots \times n_m})$ *satisfy the following inequalities:*

$$\sigma(\mathbb{F}^{n_1 \times \cdots \times n_m \times n_{m+1}}) \geq \frac{1}{n_{m+1}} \cdot \sigma(\mathbb{F}^{n_1 \times \cdots \times n_m}),$$
$$\nu(\mathbb{F}^{n_1 \times \cdots \times n_m \times n_{m+1}}) \leq n_{m+1} \cdot \nu(\mathbb{F}^{n_1 \times \cdots \times n_m}).$$

**Exercise 11.4.3.** *([278]) For the following symmetric tensor*

$$\tfrac{1}{3}\big(e_1 \otimes e_1 \otimes e_1 \otimes e_1 + e_1 \otimes e_2 \otimes e_2 \otimes e_3 + e_1 \otimes e_3 \otimes e_3 \otimes e_2$$
$$+ e_2 \otimes e_1 \otimes e_2 \otimes e_2 + e_2 \otimes e_2 \otimes e_3 \otimes e_1 + e_2 \otimes e_3 \otimes e_1 \otimes e_3$$
$$+ e_3 \otimes e_1 \otimes e_3 \otimes e_3 + e_3 \otimes e_2 \otimes e_1 \otimes e_2 + e_3 \otimes e_3 \otimes e_2 \otimes e_1\big),$$

*determine its GME and nearest unit length pure tensor.*

**Exercise 11.4.4.** *([81]) For each unit length* $\mathcal{A} \in \mathtt{S}^m(\mathbb{F}^n)$, *show that*

$$\mathrm{g}(\mathcal{A}) = \sqrt{2\Big(1 - \max_{\|x\|=1} |\langle \mathcal{A}, x^{\otimes m}\rangle|\Big)}. \tag{11.4.7}$$

*For the symmetric tensor space, define the constants*

$$\sigma(\mathtt{S}^m(\mathbb{F}^n)) := \min_{\mathcal{A} \in \mathtt{S}^m(\mathbb{F}^n)} \{\|\mathcal{A}\|_{\sigma,\mathbb{F}} : \|\mathcal{A}\| = 1\},$$
$$\nu(\mathtt{S}^m(\mathbb{F}^n)) := \max_{\mathcal{B} \in \mathtt{S}^m(\mathbb{F}^n)} \{\|\mathcal{B}\|_{*,\mathbb{F}} : \|\mathcal{B}\| = 1\}. \tag{11.4.8}$$

(i) *Show that* $\sigma(\mathtt{S}^m(\mathbb{F}^n)) \cdot \nu(\mathtt{S}^m(\mathbb{F}^n)) = 1$ *and*

$$\sigma(\mathtt{S}^m(\mathbb{F}^n)) \geq n^{-\frac{(m-1)}{2}}, \quad \nu(\mathtt{S}^m(\mathbb{F}^n)) \leq n^{\frac{(m-1)}{2}}.$$

(ii) *For unit length* $\mathcal{A} \in \mathtt{S}^m(\mathbb{F}^n)$, *show that*

$$\|\mathcal{A}\|_{\sigma,\mathbb{F}} = \sigma(\mathtt{S}^m(\mathbb{F}^n)) \quad \Leftrightarrow \quad \|\mathcal{A}\|_{*,\mathbb{F}} = \nu(\mathtt{S}^m(\mathbb{F}^n)).$$

**Exercise 11.4.5.** *Prove the values of GMEs for the tensors given in Examples 11.4.1 and 11.4.4.*

## 11.5 ▪ Tensor Positivity

A tensor can be viewed as a multilinear functional. For $\mathcal{A} = (\mathcal{A}_{i_1 i_2 \ldots i_m}) \in \mathbb{R}^{n_1 \times \cdots \times n_m}$ and $u_i \in \mathbb{R}^{n_i}$, the inner product

$$\langle \mathcal{A}, u_1 \otimes \cdots \otimes u_m \rangle = \sum_{i_1, i_2, \cdots, i_m = 1}^{n} \mathcal{A}_{i_1 i_2 \ldots i_m} \cdot (u_1)_{i_1} \cdot (u_2)_{i_2} \cdots (u_m)_{i_m}$$

gives a multilinear functional on $\mathbb{R}^{n_1} \times \cdots \times \mathbb{R}^{n_m}$. Based on this, various notions of positivity can be defined for tensors.

## 11.5.1 ▪ Nonnegative and psd tensors

A matrix is said to be nonnegative if all its entries are nonnegative. The same notion is defined for tensors. A tensor $\mathcal{A} \in \mathbb{R}^{n_1 \times \cdots \times n_m}$ is said to be *nonnegative* if $\langle \mathcal{A}, u_1 \otimes \cdots \otimes u_m \rangle \geq 0$ for all $u_i \in \mathbb{R}_+^{n_i}$. This is equivalent to every entry $\mathcal{A}_{i_1 i_2 \ldots i_m}$ being nonnegative. The cone of nonnegative tensors in $\mathbb{R}^{n_1 \times \cdots \times n_m}$ is denoted as

$$\mathscr{N}^{n_1 \times \cdots \times n_m} := \left\{ \mathcal{A} \in \mathbb{R}^{n_1 \times \cdots \times n_m} : \text{all } \mathcal{A}_{i_1 i_2 \ldots i_m} \geq 0 \right\}. \tag{11.5.1}$$

If $\mathcal{A}$ is nonnegative, we denote that $\mathcal{A} \geq 0$.

A matrix $A \in \mathbb{R}^{n \times n}$ is positive semidefinite (psd) if the quadratic form $x^T A x \geq 0$ for all $x := (x_1, \ldots, x_n) \in \mathbb{R}^n$. This can be similarly defined for tensors. Recall that $\mathrm{T}^m(\mathbb{R}^n)$ denotes the space of hypercubical tensors in $\mathbb{R}^{n \times \cdots \times n}$. For $\mathcal{A} \in \mathrm{T}^m(\mathbb{R}^n)$, recall the notation

$$\mathcal{A}(x) := \langle \mathcal{A}, x^{\otimes m} \rangle. \tag{11.5.2}$$

It is a form of degree $m$ in $x$. When $\mathcal{A}$ is symmetric, $\mathcal{A}$ is uniquely determined by $\mathcal{A}(x)$.

**Definition 11.5.1.** *A tensor $\mathcal{A} \in \mathrm{T}^m(\mathbb{R}^n)$ is said to be* positive semidefinite *(psd) if the form $\mathcal{A}(x) \geq 0$ for all $x \in \mathbb{R}^n$. Further, if $\mathcal{A}(x) > 0$ for all $0 \neq x \in \mathbb{R}^n$, then $\mathcal{A}$ is said to be positive definite (pd). If $\mathcal{A}$ is psd (resp., pd), we denote that $\mathcal{A} \succeq 0$ (resp., $\mathcal{A} \succ 0$).*

Clearly, a nonzero tensor is psd only if the order $m$ is even. When a tensor is nonsymmetric, whether it is psd or not is only determined by its symmetric portion. For $\mathcal{A} \in \mathrm{T}^m(\mathbb{R}^n)$, its symmetrization is the tensor $\mathrm{sym}(\mathcal{A}) \in \mathrm{S}^m(\mathbb{R}^n)$, defined as in (11.1.14). One can show that $\mathcal{A}$ is psd if and only if the symmetric tensor $\mathrm{sym}(\mathcal{A})$ is psd. This is left as an exercise. Therefore, we usually consider symmetric psd tensors of even orders. The set of all psd tensors in $\mathrm{S}^m(\mathbb{R}^n)$ is denoted as

$$\mathscr{P}^{m,n} := \{ \mathcal{A} \in \mathrm{S}^m(\mathbb{R}^n) : \mathcal{A} \succeq 0 \}. \tag{11.5.3}$$

Clearly, $\mathscr{P}^{m,n}$ is a closed convex cone. The following are some examples of psd tensors.

- The Motzkin tensor $\mathcal{M} \in \mathrm{S}^6(\mathbb{R}^3)$ is such that

$$\mathcal{M}(x) = x_1^4 x_2^2 + x_1^2 x_2^4 + x_3^6 - 3 x_1^2 x_2^2 x_3^2.$$

- The Robinson tensor $\mathcal{R} \in \mathrm{S}^6(\mathbb{R}^3)$ is such that

$$\mathcal{R}(x) = x_1^6 + x_2^6 + x_3^6 + 3 x_1^2 x_2^2 x_3^2 - x_1^4(x_2^2 + x_3^2) - x_2^4(x_3^2 + x_1^2) - x_3^4(x_1^2 + x_2^2).$$

- The agi tensor $\mathcal{A} \in \mathrm{S}^{2d}(\mathbb{R}^d)$ is such that

$$\mathcal{A}(x) = x_1^{2d} + \cdots + x_d^{2d} - d x_1^2 \cdots x_d^2.$$

A tensor $\mathcal{A}$ is psd if and only if the polynomial $\mathcal{A}(x)$, given as in (11.1.12), is a psd form. The complexity for checking psd tensors is NP-hard for $m > 2$. People are interested in sufficient conditions for psd tensors. The following is the one based on SOS polynomials.

**Definition 11.5.2.** *A tensor $\mathcal{A} \in \mathrm{T}^m(\mathbb{R}^n)$ is said to be* SOS *if $\mathcal{A}(x) \in \Sigma[x]$, or equivalently, $\mathcal{A}(x) = p_1^2 + \cdots + p_k^2$ for polynomials $p_1, \ldots, p_k \in \mathbb{R}[x]$. If $\mathcal{A}$ is SOS, we denote that $\mathcal{A} \succeq_\sigma 0$.*

The cone of SOS tensors in $\mathrm{T}^m(\mathbb{R}^n)$ is denoted as

$$\Sigma^{m,n} = \{ \mathcal{A} \in \mathrm{S}^m(\mathbb{R}^n) : \mathcal{A} \succeq_\sigma 0 \}. \tag{11.5.4}$$

Clearly, $\Sigma^{m,n} \subseteq \mathscr{P}^{m,n}$. The cones of SOS and PSD tensors are not same, except a few cases of $(m,n)$. This is characterized by Hilbert's theorem, i.e., Theorem 2.4.5.

**Theorem 11.5.3.** *For $n > 1$, $\Sigma^{m,n} = \mathscr{P}^{m,n}$ if and only if*

$$n = 2, \quad or \quad m = 2, \quad or \quad (n,m) = (3,4).$$

A symmetric tensor $\mathcal{A} \in \mathsf{S}^m(\mathbb{R}^n)$ is uniquely determined by the homogeneous truncated multi-sequence (htms) $y = (y_\alpha)_{\alpha \in \overline{\mathbb{N}}_m^n}$ such that

$$y_\alpha = \mathcal{A}_{i_1 \ldots i_m} \quad (\text{for each } x^\alpha = x_{i_1} \cdots x_{i_m}).$$

This gives the linear map $\mathrm{tm} : \mathsf{S}^m(\mathbb{R}^n) \to \mathbb{R}^{\overline{\mathbb{N}}_m^n}$ such that

$$\mathrm{tm}(\mathcal{A}) = y. \tag{11.5.5}$$

In terms of htms, we define two tensor cones (see (2.4.23) and (8.4.9) for the notation $\mathscr{R}_{n,m}^{\mathrm{hom}}$)

$$\mathscr{R}^{m,n} = \left\{ \mathcal{A} \in \mathsf{S}^m(\mathbb{R}^n) : \mathrm{tm}(\mathcal{A}) \in \mathscr{R}_{n,m}^{\mathrm{hom}} \right\}. \tag{11.5.6}$$

We remark that $\mathcal{A} \in \mathscr{R}^{m,n}$ if and only if $\mathcal{A}(x)$ is a SOP form of degree $m$ (see §8.4.3). When the order $m = 2d$ is even, we define the tensor cone

$$\mathscr{S}^{m,n} = \left\{ \mathcal{A} \in \mathsf{S}^m(\mathbb{R}^n) : M_d[\mathrm{tm}(\mathcal{A})] \succeq 0 \right\}. \tag{11.5.7}$$

The following gives the dual relationship between them. The superscript $^\star$ denotes the dual cone.

**Theorem 11.5.4.** *For even $m = 2d$, all the cones $\mathscr{P}^{m,n}$, $\Sigma^{m,n}$, $\mathscr{R}^{m,n}$, $\mathscr{S}^{m,n}$ are proper cones (i.e., they are convex, closed, pointed, and solid). Moreover, they satisfy the following dual relationships:*

$$\left( \mathscr{P}^{m,n} \right)^\star = \mathscr{R}^{m,n}, \quad \left( \mathscr{S}^{m,n} \right)^\star = \Sigma^{m,n}. \tag{11.5.8}$$

***Proof.*** The psd form cone $\mathscr{P}_{n,m}^{\mathrm{hom}}$ is dual to the moment cone $\mathscr{R}_{n,m}^{\mathrm{hom}}$ (see Theorem 8.4.6), and the SOS cone $\Sigma[x]_{n,m}^{\mathrm{hom}}$ is dual to the moment cone $\mathscr{S}_{n,m}^{\mathrm{hom}}$ (see Proposition 2.4.4). They are all proper cones. The conclusions follow directly from the relationship $y = \mathrm{tm}(\mathcal{A})$ as in (11.5.5). $\square$

To check whether $\mathcal{A}$ is psd or not, we need to determine the sign of the minimum value $\nu$ of

$$\begin{cases} \min & \mathcal{A}(x) \\ s.t. & x^T x = 1. \end{cases} \tag{11.5.9}$$

The minimum value $\nu$ can be computed by tight relaxation methods in §6.3 and §6.5. We refer to §8.2.1 for checking memberships for the cone $\mathscr{P}^{m,n}$. An important class of tensors in $\mathscr{R}^{m,n}$ are separable matrices/tensors, which have broad applications in quantum physics. We refer to the work [84, 233, 242].

## 11.5.2 ▪ Copositive and CP tensors

A matrix $A$ is copositive if the quadratic form $x^T A x$ is copositive, i.e., $x^T A x \geq 0$ for all $x \in \mathbb{R}^n_+$. The same can be defined for tensors. This gives the notion of *copositive tensors* (see [276]). Recall that $\mathcal{A}(x)$ is the form determined by $\mathcal{A}$ as in (11.5.2).

**Definition 11.5.5.** *A tensor* $\mathcal{A} \in \mathrm{T}^m(\mathbb{R}^n)$ *is said to be* copositive *if* $\mathcal{A}(x) \geq 0$ *for all* $x \in \mathbb{R}^n_+$. *Furthermore, if* $\mathcal{A}(x) > 0$ *for all* $0 \neq x \in \mathbb{R}^n_+$, *then* $\mathcal{A}$ *is said to be strictly copositive. Denote by* $\mathcal{COP}_n^{\otimes m}$ *the cone of all copositive symmetric tensors in the space* $\mathrm{S}^m(\mathbb{R}^n)$.

Detecting tensor copositivity is typically hard. Its computational complexity is NP-hard (for $m > 1$), since testing matrix copositivity (i.e., $m = 2$) is already NP-hard. A tensor $\mathcal{A}$ is copositive if and only if the associated form $\mathcal{A}(x)$ is copositive. We refer to §9.3 for checking copositive tensors. Related work can be found in [47, 239]. For instance, the following tensor is copositive:

$$\tfrac{1}{3}\left(e_1 \otimes e_1 \otimes e_2 + e_1 \otimes e_2 \otimes e_1 + e_2 \otimes e_1 \otimes e_1\right) + \tfrac{1}{3}\left(e_1 \otimes e_2 \otimes e_2 + e_2 \otimes e_1 \otimes e_2 \right.$$
$$\left. + e_2 \otimes e_2 \otimes e_1\right) + e_3^{\otimes 3} - \tfrac{1}{2}\left(e_1 \otimes e_2 \otimes e_3 + e_1 \otimes e_3 \otimes e_2 + e_2 \otimes e_1 \otimes e_3 \right.$$
$$\left. + e_2 \otimes e_3 \otimes e_1 + e_3 \otimes e_1 \otimes e_2 + e_3 \otimes e_2 \otimes e_1\right).$$

This is because its associated form is $x_1^2 x_2 + x_1 x_2^2 + x_3^3 - 3 x_1 x_2 x_3$, which is nonnegative for all $x \in \mathbb{R}^3_+$.

**Example 11.5.6.** *(i) Consider the tensors* $\mathcal{A}_1, \mathcal{A}_2, \mathcal{A}_3 \in \mathrm{S}^3(\mathbb{R}^3)$ *such that*

| | | |
|---|---|---|
| $\mathcal{A}_1(x)$ | $=$ | $x_1^2 x_2 + x_1 x_2^2 + x_3^3 - 3 x_1 x_2 x_3,$ |
| $\mathcal{A}_2(x)$ | $=$ | $x_1^3 + x_2^3 + x_3^3 - x_1^2 x_2 - x_1 x_2^2 - x_1^2 x_3$ |
| | | $-x_1 x_3^2 - x_2^2 x_3 - x_2 x_3^2 + 3 x_1 x_2 x_3,$ |
| $\mathcal{A}_3(x)$ | $=$ | $x_1^2 x_2 + x_2^2 x_3 + x_3^2 x_1 - 3 x_1 x_2 x_3.$ |

*Note that* $\mathcal{A}_1(x_1^2, x_2^2, x_3^2)$, $\mathcal{A}_2(x_1^2, x_2^2, x_3^2)$, *and* $\mathcal{A}_3(x_1^2, x_2^2, x_3^2)$ *are respectively the Motzkin, Robinson, and Choi-Lam forms, which are nonnegative but not SOS (see §2.4.2). These three tensors are all copositive.*
*(ii) ([239]) Consider the tensor* $\mathcal{B} \in \mathrm{S}^4(\mathbb{R}^4)$ *such that*

$$\mathcal{B}(x) = (x_1 + x_2 + x_3 + x_4)^4 - 16(x_1 x_2 + x_2 x_3 + x_3 x_4)^2.$$

*It is copositive, since* $\mathcal{B}(x)$ *has the factorization*

$$\left((x_1 - x_2 + x_3 - x_4)^2 + 4 x_1 x_4\right) \cdot \left((x_1 + x_2 + x_3 + x_4)^2 + 4(x_1 x_2 + x_2 x_3 + x_3 x_4)\right).$$

Like completely positive matrices, we can similarly define completely positive tensors.

**Definition 11.5.7.** *A symmetric tensor* $\mathcal{C} \in \mathrm{S}^m(\mathbb{R}^n)$ *is said to be* completely positive *(CP) if there exist* $u_1, \ldots, u_r \in \mathbb{R}^n_+$ *such that*

$$\mathcal{C} = (u_1)^{\otimes m} + \cdots + (u_r)^{\otimes m}.$$

*Denote by* $\mathcal{CP}_n^{\otimes m}$ *the cone of CP tensors in* $\mathrm{S}^m(\mathbb{R}^n)$.

The CP tensor cone $\mathcal{CP}_n^{\otimes m}$ can be equivalently given as

$$\mathcal{CP}_n^{\otimes m} = \mathrm{cone}\left(\{u^{\otimes m} : u \in \Delta_n\}\right), \tag{11.5.10}$$

where $\Delta_n$ is the standard simplex in $\mathbb{R}^n$. Each $\mathcal{C} \in \mathrm{S}^m(\mathbb{R}^n)$ is uniquely determined by $\mathrm{tm}\,(\mathcal{C}) \in \mathbb{R}^{\overline{\mathbb{N}}^n_m}$, which is defined as in (11.5.5). The linear mapping

$$\mathrm{tm}\left(\sum_i \lambda_i (u_i)^{\otimes m}\right) = \sum_i \lambda_i [u_i]^{\mathrm{hom}}_m \tag{11.5.11}$$

gives an isomorphism between $\mathrm{S}^m(\mathbb{R}^n)$ and $\mathbb{R}^{\overline{\mathbb{N}}^n_m}$. It holds that

$$\mathrm{tm}\left(\mathcal{CP}_n^{\otimes d}\right) = \mathcal{CP}_{n,d}, \qquad \mathrm{tm}^{-1}(\mathcal{CP}_{n,d}) = \mathcal{CP}_n^{\otimes d}, \tag{11.5.12}$$

where $\mathcal{CP}_{n,d}$ is the CP moment cone introduced in §9.2.1. Theorem 9.2.4 directly implies the following.

**Theorem 11.5.8.** *Both* $\mathcal{COP}_n^{\otimes m}$ *and* $\mathcal{CP}_n^{\otimes m}$ *are proper cones and they are dual to each other:*

$$\left(\mathcal{COP}_n^{\otimes m}\right)^* = \mathcal{CP}_n^{\otimes m}, \qquad \left(\mathcal{CP}_n^{\otimes m}\right)^* = \mathcal{COP}_n^{\otimes m}.$$

In view of (11.5.12) and Theorem 9.2.4, a tensor $\mathcal{C} \in \mathrm{S}^m(\mathbb{R}^n)$ is CP if and only if there is a Borel measure $\mu$ supported in $\Delta_n$ such that

$$\mathcal{C} = \int x^{\otimes m} \mathrm{d}\mu.$$

We refer to §9.3.3 for checking memberships in the cone $\mathcal{CP}_n^{\otimes m}$.

We conclude this section with an application of copositive tensors in hypergraph theory (see [47, 239]). A hypergraph $G = (V, E)$ has a vertex set $V = \{1, \ldots, n\}$ and an edge set $E$, such that each edge in $E$ is an unordered tuple $(i_1, \ldots, i_\ell)$, with $i_1, \ldots, i_\ell \in V$. It is $m$-uniform if each edge is an unordered $m$-tuple $(i_1, \ldots, i_m)$ for distinct $i_1, \ldots, i_m$. A coclique of a $m$-uniform hypergraph $G$ is a subset $K \subseteq V$ such that any subset of $K$ with cardinality $m$ does not give an edge of $G$. The maximum cardinality of cocliques of $G$ is called the *coclique number* of $G$, which we denote as $\omega(G)$. Computing $\omega(G)$ is typically hard. An upper bound for $\omega(G)$ can be obtained by using tensor copositivity, as shown in [47]. The adjacency tensor of a $m$-uniform hypergraph $G$ is the symmetric tensor $\mathcal{C} \in \mathrm{S}^m(\mathbb{R}^n)$ such that

$$\mathcal{C}_{i_1 \ldots i_m} = \begin{cases} \frac{1}{(m-1)!} & (i_1, \ldots, i_m) \in E, \\ 0 & \text{otherwise.} \end{cases}$$

Let $\mathcal{I}$ be the identity tensor (i.e., $\mathcal{I}_{i_1 \ldots i_m} = 1$ for $i_1 = \cdots = i_m$ and $\mathcal{I}_{i_1 \ldots i_m} = 0$ otherwise) and let $\mathcal{E}$ be the tensor of all ones. It is shown in [47] that

$$\omega(G)^{m-1} \le \rho$$

for all $\rho$ such that $\rho(\mathcal{I} + \mathcal{C}) - \mathcal{E}$ is copositive. To get such smallest such $\rho$, one can find the largest $\gamma$ such that $(\mathcal{I} + \mathcal{C}) - \gamma \mathcal{E}$ is copositive. Such largest $\gamma$ equals the minimum value $v^*$ of the optimization

$$\begin{cases} \min & \mathcal{A}(x) \\ s.t. & e^T x = 1,\ x \ge 0, \end{cases}$$

for the tensor $\mathcal{A} := \mathcal{I} + \mathcal{C}$. If $v_k$ is a lower bound for $v^*$, e.g., $v_k$ is given by the $k$th order moment relaxation in §5.2 or §6.3, then

$$\omega(G) \le \left(\frac{1}{v^*}\right)^{\frac{1}{m-1}} \le \left(\frac{1}{v_k}\right)^{\frac{1}{m-1}}.$$

Since $\omega(G)$ is an integer, the above implies that

$$\omega(G) \leq \left\lfloor (\frac{1}{v_k})^{\frac{1}{m-1}} \right\rfloor. \tag{11.5.13}$$

As $k$ increases, a hierarchy of upper bounds for $\omega(G)$ can be obtained. We refer to the work [239] for more details.

### 11.5.3 ▪ Exercises

**Exercise 11.5.1.** *Show that a tensor $\mathcal{A} \in \mathrm{T}^m(\mathbb{R}^n)$ is psd if and only if its symmetrization tensor* $\mathrm{sym}(\mathcal{A})$ *is psd.*

**Exercise 11.5.2.** *For $m = 4$ and $n = 3$, show that $\mathscr{R}^{m,n} = \mathscr{S}^{m,n}$. When $m$ is even, show that $\mathscr{S}^{m,n} \subseteq \Sigma^{m,n}$ (see (11.5.4), (11.5.6), (11.5.7) for the notation).*

**Exercise 11.5.3.** *Compute a basis of symmetric rank-1 tensors for the subspace spanned by CP tensors in $\mathcal{CP}_3^{\otimes 3}$ that are orthogonal to the tensor $\mathcal{A}$ for the following cases:*

(i) $\mathcal{A}(x) := x_1^3 + x_2^3 + x_3^3 - x_1^2 x_2 - x_1 x_2^2 - x_1^2 x_3 - x_1 x_3^2 - x_2^2 x_3 - x_2 x_3^2 + 3x_1 x_2 x_3.$

(ii) $\mathcal{A}(x) := x_1^2 x_2 + x_2^2 x_3 + x_3^2 x_1 - 3x_1 x_2 x_3.$

(iii) $\mathcal{A}(x) := x_1^2 x_2 + x_2^2 x_3 + x_3^3 - 3x_1 x_2 x_3.$

**Exercise 11.5.4.** *For an even order $m$, show that*

$$\mathcal{N}^{n \times \cdots \times n} + \mathscr{P}^{m,n} \subseteq \mathcal{COP}_n^{\otimes m},$$
$$\mathcal{N}^{n \times \cdots \times n} \cap \mathscr{R}^{m,n} \supseteq \mathcal{CP}_n^{\otimes m}.$$

*Give explicit tensors lying in the following complements:*

$$\mathcal{COP}_n^{\otimes m} \backslash (\mathcal{N}^{n \times \cdots \times n} + \mathscr{P}^{m,n}), \quad (\mathcal{N}^{n \times \cdots \times n} \cap \mathscr{R}^{m,n}) \backslash \mathcal{CP}_n^{\otimes m}.$$

**Exercise 11.5.5.** *For even order $m$, use Theorem 9.3.4 to construct a hierarchy of semidefinite representable convex cones $\mathcal{F}_k \subseteq \mathrm{S}^m(\mathbb{R}^n)$ such that*

$$\mathcal{F}_0 \supseteq \cdots \supseteq \mathcal{F}_k \supseteq \cdots, \quad \mathcal{CP}_n^{\otimes m} = \bigcap_{k \geq 0} \mathcal{F}_k,$$
$$\mathrm{int}\,(\mathcal{COP}_n^{\otimes m}) \subseteq \bigcup_{k \geq 0} (\mathcal{F}_k)^\star \subseteq \mathcal{COP}_n^{\otimes m}.$$

**Exercise 11.5.6.** *([93]) Binary CP tensors can be characterized by linear matrix inequalities. Let $T : \mathrm{S}^m(\mathbb{R}^2) \to \mathrm{S}^m(\mathbb{R}^2)$ be the linear map such that*

$$T(\begin{bmatrix} s \\ t \end{bmatrix}^{\otimes m}) = \begin{bmatrix} s+t \\ t \end{bmatrix}^{\otimes m}. \tag{11.5.14}$$

*For $\mathcal{A} \in \mathrm{S}^m(\mathbb{R}^2)$, let*

$$y := (y_0, y_1, \ldots, y_m) = \mathrm{tm}\,(T(\mathcal{A})).$$

*If $m = 2d$ is even, show that $\mathcal{A}$ is CP if and only if $y$ satisfies the following two LMIs:*

$$\begin{bmatrix} y_0 & y_1 & \cdots & y_d \\ y_1 & y_2 & \cdots & y_{d+1} \\ \vdots & \vdots & \ddots & \vdots \\ y_d & y_{d+1} & \cdots & y_{2d} \end{bmatrix} \succeq 0, \tag{11.5.15}$$

$$\begin{bmatrix} y_1 - y_2 & y_2 - y_3 & \cdots & y_d - y_{d+1} \\ y_2 - y_3 & y_3 - y_4 & \cdots & y_{d+1} - y_{d+2} \\ \vdots & \vdots & \ddots & \vdots \\ y_d - y_{d+1} & y_{d+1} - y_{d+2} & \cdots & y_{2d-1} - y_{2d} \end{bmatrix} \succeq 0. \tag{11.5.16}$$

*If $m = 2d + 1$ is odd, show that $\mathcal{A}$ is CP if and only if $y$ satisfies the following two LMIs:*

$$\begin{bmatrix} y_1 & y_2 & \cdots & y_{d+1} \\ y_2 & y_3 & \cdots & y_{d+2} \\ \vdots & \vdots & \ddots & \vdots \\ y_{d+1} & y_{d+2} & \cdots & y_{2d+1} \end{bmatrix} \succeq 0, \tag{11.5.17}$$

$$\begin{bmatrix} y_0 - y_1 & y_1 - y_2 & \cdots & y_d - y_{d+1} \\ y_1 - y_2 & y_2 - y_3 & \cdots & y_{d+1} - y_{d+2} \\ \vdots & \vdots & \ddots & \vdots \\ y_d - y_{d+1} & y_{d+1} - y_{d+2} & \cdots & y_{2d} - y_{2d+1} \end{bmatrix} \succeq 0. \tag{11.5.18}$$

# 11.6 ▪ Tensor Eigenvalues and Singular Values

Let $\mathbb{F} = \mathbb{R}$ or $\mathbb{C}$ and let $x := (x_1, \ldots, x_n)$. Recall that $\mathrm{T}^m(\mathbb{F}^n)$ denotes the space of hypercubical $m$th order tensors over the field $\mathbb{F}$ and $\mathrm{S}^m(\mathbb{F}^n)$ is the subspace of symmetric tensors in $\mathrm{T}^m(\mathbb{F}^n)$. For a power $k \leq m$ and $\mathcal{A} \in \mathrm{T}^m(\mathbb{F}^n)$, denote by $\mathcal{A}x^k$ the tensor in $\mathrm{T}^{m-k}(\mathbb{F}^n)$ such that

$$(\mathcal{A}x^k)_{i_1,\ldots,i_{m-k}} := \sum_{1 \leq j_1,\ldots,j_k \leq n} \mathcal{A}_{i_1\ldots i_{m-k}j_1\ldots j_k} x_{j_1} \cdots x_{j_k}. \tag{11.6.1}$$

If $\mathcal{A} = \sum_{k=1}^{r} u^{k,1} \otimes \cdots \otimes u^{k,m}$, then the above is the same as

$$\mathcal{A}x^k = \sum_{i=1}^{r} u^{i,1} \otimes \cdots \otimes u^{i,m-k} \cdot (u^{i,m-k+1})^T x \cdots (u^{i,m})^T x. \tag{11.6.2}$$

In particular, $\mathcal{A}x^m$ is the form $\mathcal{A}(x)$ as in (11.1.12) and $\mathcal{A}x^{m-1}$ is a vector of dimension $n$. For convenience of notation, we also denote

$$x^{[k]} := \left[ (x_1)^k \quad \cdots \quad (x_n)^k \right]^T.$$

The following are some classical notions of *tensor eigenvalues* (see [177, 273, 277]).

**Definition 11.6.1.** *For $\mathcal{A} \in \mathrm{T}^m(\mathbb{C}^n)$, a pair $(\lambda, u) \in \mathbb{C} \times \mathbb{C}^n$ is called an H-eigenpair of $\mathcal{A}$ if*

$$\mathcal{A}u^{m-1} = \lambda u^{[m-1]}, \quad u \neq 0. \tag{11.6.3}$$

*The $\lambda$ is called an H-eigenvalue of $\mathcal{A}$ and the $u$ is called an H-eigenvector associated to $\lambda$.*

**Definition 11.6.2.** *For $\mathcal{A} \in \mathrm{T}^m(\mathbb{C}^n)$, a pair $(\lambda, u) \in \mathbb{C} \times \mathbb{C}^n$ is called a $\mathbb{Z}$-eigenpair of $\mathcal{A}$ if*

$$\mathcal{A}u^{m-1} = \lambda u, \quad u^T u = 1. \tag{11.6.4}$$

*The $\lambda$ is called an $\mathbb{Z}$-eigenvalue of $\mathcal{A}$ and the $u$ is called an $\mathbb{Z}$-eigenvector associated to $\lambda$.*

In some literature, when $\mathcal{A}$ is a real tensor, the $\lambda$ is called an eigenvalue if $u$ satisfying (11.6.3) is not real, and the $\lambda$ is called an E-eigenvalue if the $u$ in (11.6.4) is not real. For convenience of notions, they are also called H-eigenvalues and Z-eigenvalues respectively. As introduced in [62], various tensor eigenvalues can be given in a unified framework.

**Definition 11.6.3.** *For two tensors $\mathcal{A} \in \mathrm{T}^m(\mathbb{C}^n)$ and $\mathcal{B} \in \mathrm{T}^{m'}(\mathbb{C}^n)$ (the orders $m, m'$ are not necessarily equal), a number $\lambda \in \mathbb{C}$ is a $\mathcal{B}$-eigenvalue of $\mathcal{A}$ if there exists $u \in \mathbb{C}^n$ such that*

$$\mathcal{A}u^{m-1} = \lambda \, \mathcal{B}u^{m'-1}, \quad \mathcal{B}u^{m'} = 1. \tag{11.6.5}$$

*The $u$ is called a $\mathcal{B}$-eigenvector associated to $\lambda$, and $(\lambda, u)$ is called a $\mathcal{B}$-eigenpair.*

When $\mathcal{B}$ is clear in the context, $\mathcal{B}$-eigenvalues (resp., $\mathcal{B}$-eigenvectors, $\mathcal{B}$-eigenpairs) are just simply called eigenvalues (resp., eigenvectors, eigenpairs). The following are some special cases.

- When $m' = m$ and $\mathcal{B}$ is the identity tensor (i.e., $\mathcal{B}x^m = x_1^m + \cdots + x_n^m$), the $\mathcal{B}$-eigenvalues are just the H-eigenvalues.

- When $m' = 2$ and $\mathcal{B}$ is such that $\mathcal{B}x^2 = x_1^2 + \cdots + x_n^2$, the $\mathcal{B}$-eigenvalues are just Z-eigenvalues.

- Let $D \in \mathcal{S}_{++}^n$ be a real symmetric positive definite matrix. When $m' = 2$ and $\mathcal{B}$ is such that $\mathcal{B}x^2 = x^T Dx$, the $\mathcal{B}$-eigenvalues are called the D-eigenvalues [274]. Equivalently, a D-eigenpair $(\lambda, u)$ is given by

$$\mathcal{A}u^{m-1} = \lambda \, Du, \quad u^T Du = 1.$$

When $\lambda$ is real, the eigenvector $u$ is not necessarily real. For convenience of writing, an eigenpair $(\lambda, u)$ is called *real* if both $\lambda$ and $u$ are real, and $\lambda$ is called a *real eigenvalue* if there exists a real eigenvector. By the largest (resp., smallest) eigenvalue, we mean the largest (resp., smallest) real eigenvalue. The complexity of computing extreme (the biggest or smallest) eigenvalues for $m \geq 3$ is NP-hard (see [128]). Tensor eigenvalues have applications in signal processing and diffusion imaging (see [208, 274, 275]). In addition to extreme eigenvalues, middle eigenvalues are also useful in applications (see [50, 62, 175, 275]).

## 11.6.1 ▪ Symmetric tensor eigenvalues

For two symmetric tensors $\mathcal{A} \in \mathrm{S}^m(\mathbb{R}^n)$ and $\mathcal{B} \in \mathrm{S}^{m'}(\mathbb{R}^n)$,

$$\nabla(\mathcal{A}x^m) = m \, \mathcal{A}x^{m-1}, \quad \nabla(\mathcal{B}x^{m'}) = m' \, \mathcal{B}x^{m'-1}, \tag{11.6.6}$$

where $\nabla$ denotes the gradient in $x$. Due to the identities

$$x^T \nabla(\mathcal{A}x^m) = m \, \mathcal{A}x^m, \quad x^T \nabla(\mathcal{B}x^{m'}) = m' \, \mathcal{B}x^{m'},$$

the eigen-equation (11.6.5) is equivalent to

$$\frac{1}{m} \nabla(\mathcal{A}u^m) = \frac{1}{m'} \lambda \nabla(\mathcal{B}u^{m'}), \quad \mathcal{B}u^{m'} = 1.$$

So, $(\lambda, u)$ is a $\mathcal{B}$-eigenpair if and only if $u$ is a critical point for the optimization

$$\begin{cases} \max & \mathcal{A}x^m \\ s.t. & \mathcal{B}x^{m'} = 1 \end{cases} \tag{11.6.7}$$

and $\lambda = \mathcal{A}u^m$. Therefore, $(\lambda, u)$ is a $\mathcal{B}$-eigenpair if and only if $u$ is a critical point of (11.6.7) with the critical value $\lambda$. The optimization (11.6.7) has only finitely many critical values, including both complex and real ones (see Theorem 6.1.1). That is, every symmetric tensor $\mathcal{A}$ has only finitely many complex eigenvalues. So we can order the real eigenvalues monotonically as

$$\lambda_1 > \lambda_2 > \cdots > \lambda_N.$$

The KKT system for (11.6.7) is

$$\nabla \mathcal{A}x^m = \mu \nabla \mathcal{B}x^{m'}, \quad \mathcal{B}x^{m'} = 1.$$

If we premultiply $x^T$ in the left of the above, then

$$m\mathcal{A}x^m = x^T \nabla \mathcal{A}x^m = \mu x^T \nabla \mathcal{B}x^{m'} = \mu \cdot m' \mathcal{B}x^{m'} = m'\mu.$$

This implies that $\mu = \frac{m}{m'} \mathcal{A}x^m$. In view of (11.6.6), the KKT system can be equivalently written as

$$\mathcal{A}x^{m-1} - (\mathcal{A}x^m) \cdot \mathcal{B}x^{m'-1} = 0, \quad \mathcal{B}x^{m'} = 1.$$

Denote the polynomials

$$h_i := \begin{cases} (\mathcal{A}x^{m-1})_i - (\mathcal{A}x^m) \cdot (\mathcal{B}x^{m'-1})_i, & i = 1, \ldots, n, \\ \mathcal{B}x^{m'} - 1, & i = n+1. \end{cases}$$

They give the polynomial tuple

$$h := (h_1, \ldots, h_n, h_{n+1}). \tag{11.6.8}$$

Then, (11.6.7) is equivalent to the optimization

$$\begin{cases} \max & \mathcal{A}x^m \\ s.t. & h(x) = 0. \end{cases} \tag{11.6.9}$$

When $\mathcal{B}x^{m'} = 1$ is a smooth hypersurface (i.e., $\mathcal{B}u^{m'-1} \neq 0$ for every $u$ satisfying $\mathcal{B}u^{m'} = 1$), a point $u$ is feasible for (11.6.9) if and only if $u$ is a critical point of (11.6.7), i.e., $u$ is a $\mathcal{B}$-eigenvector. Therefore, the objective value of (11.6.9) at each feasible point is a $\mathcal{B}$-eigenvalue of $\mathcal{A}$. The objective values are $\lambda_1, \ldots, \lambda_N$. In the following, we show how to compute them sequentially, from $\lambda_1$ to $\lambda_N$.

## The largest eigenvalue

The largest eigenvalue $\lambda_1$ is the maximum value of (11.6.9). Let

$$k_0 := \left\lceil \frac{m+m'-1}{2} \right\rceil. \tag{11.6.10}$$

We can apply the Moment-SOS hierarchy to solve (11.6.9). For $k = k_0, k_0 + 1, \ldots$, solve the moment relaxation

$$\begin{cases} \max & \langle \mathcal{A}x^m, y \rangle \\ s.t. & \mathscr{V}_{h_j}^{(2k)}[y] = 0 \ (j = 1, \ldots, n+1), \\ & y_0 = 1, M_k[y] \succeq 0, y \in \mathbb{R}^{\mathbb{N}_{2k}^n}. \end{cases} \tag{11.6.11}$$

Let $\rho_k^{(1)}$ denote the optimal value of (11.6.11). Then, it holds that the monotonicity relation:

$$\rho_{k_0}^{(1)} \geq \rho_{k_0+1}^{(1)} \geq \cdots \geq \lambda_1.$$

The convergence for the hierarchy of relaxations (11.6.11) is summarized as follows. Recall that a real hypersurface $f(x) = 0$ is said to be smooth if for every real point $u$ satisfying $f(u) = 0$, it holds that $\nabla f(u) \neq 0$.

**Theorem 11.6.4.** *([62]) Suppose the real hypersurface $\mathcal{B}x^{m'} = 1$ is smooth. Let $\lambda_1$ be the largest real $\mathcal{B}$-eigenvalue of $\mathcal{A}$. Then, the following hold:*

(i) *It holds that $\rho_k^{(1)} = \lambda_1$ for all $k$ large enough.*

(ii) *Suppose there are only finitely many real $\mathcal{B}$-eigenvectors associated to $\lambda_1$. If $k$ is large enough, then for every optimizer $y^*$ of (11.6.11), there exists an integer $t \in [k_0, k]$ such that*

$$\text{rank } M_{t-k_0}[y^*] = \text{rank } M_t[y^*]. \tag{11.6.12}$$

*Moreover, for each $k$ such that (11.6.12) holds, we have that $\rho_k^{(1)} = \lambda_1$, the truncation $y^*|_{2t}$ admits a finitely atomic measure $\mu$, and each point in $\text{supp}(\mu)$ is a real $\mathcal{B}$-eigenvector associated to $\lambda_1$.*

The conclusions in Theorem 11.6.4 can be implied by Theorem 6.3.5, since (11.6.9) does not have inequality constraints. If the flat truncation (11.6.12) is met, there exist $\ell := \text{rank } M_t[y^*]$ distinct feasible points $u_1, \ldots, u_\ell$ of (11.6.9) and each $u_i$ is an eigenvector with $\lambda_1 = f(u_i)$. The method in §2.7.2 can be used to get $u_i$. The flat truncation (11.6.12) is sufficient and necessary for detecting the convergence $\rho_k^{(1)} = \lambda_1$. When (11.6.12) holds, if rank $M_k[y^*]$ is maximum among the set of all optimizers of (11.6.11), then we can get all maximizers of (11.6.9). This is shown in Proposition 5.3.6. For such cases, we can get all $\mathcal{B}$-eigenvectors associated to $\lambda_1$. However, if there are infinitely many $\mathcal{B}$-eigenvectors $u$ normalized as $\mathcal{B}u^{m'} = 1$, then (11.6.12) is typically not satisfied. For such special cases, we refer to the method in §6.6.2.

## Middle eigenvalues

Suppose the $r$th largest real eigenvalue $\lambda_r$ of $\mathcal{A}$ is known for some $1 \leq r \leq N$. We discuss how to compute the $(r+1)$th largest real eigenvalue $\lambda_{r+1}$ if it exists, or how to detect its nonexistence.

First, assume $\lambda_{r+1}$ exists. Let $\delta$ be a scalar such that

$$0 < \delta < \lambda_r - \lambda_{r+1}. \tag{11.6.13}$$

Consider the constrained optimization

$$\begin{cases} \max & \mathcal{A}x^m \\ \text{s.t.} & h_j(x) = 0 \ (j = 1, \ldots, n+1), \\ & \mathcal{A}x^m \leq \lambda_r - \delta. \end{cases} \tag{11.6.14}$$

When (11.6.13) holds, the optimal value of (11.6.14) is $\lambda_{r+1}$. Consider the hierarchy of moment relaxations (for $k := k_0, k_0 + 1, \ldots$)

$$\begin{cases} \max & \langle \mathcal{A}x^m, y \rangle \\ \text{s.t.} & \mathscr{V}_{h_j}^{(2k)}[y] = 0 \ (j = 1, \ldots, n+1), \\ & L_p^{(k)}[y] \succeq 0, \ M_k[y] \succeq 0, \\ & y_0 = 1, \ y \in \mathbb{R}^{\mathbb{N}_{2k}^n}. \end{cases} \tag{11.6.15}$$

In the above, $p := \lambda_r - \delta - \mathcal{A}x^m$. Let $\rho_k^{(r+1)}$ denote the optimal value of (11.6.15) when the relaxation order is $k$. Similarly, we have the monotonicity relation:

$$\rho_{k_0}^{(r+1)} \geq \rho_{k_0+1}^{(r+1)} \geq \cdots \geq \lambda_{r+1}.$$

The convergence for the hierarchy of relaxations (11.6.15) is as follows.

**Theorem 11.6.5.** *([62]) Suppose the real hypersurface $\mathcal{B}x^{m'} = 1$ is smooth. Let $\lambda_r$ (resp., $\lambda_{r+1}$) be the rth (resp., $(k+1)$th) largest real $\mathcal{B}$-eigenvalue of $\mathcal{A}$. For every $\delta$ satisfying (11.6.13), we have the following:*

  (i) *For all $k$ big enough, we have $\rho_k^{(r+1)} = \lambda_{r+1}$.*

  (ii) *Suppose there are only finitely many real $\mathcal{B}$-eigenvectors $u$ associated to $\lambda_{r+1}$. If $k$ is large enough, then for every optimizer $y^*$ of (11.6.15), there exists an integer $t \in [k_0, k]$ such that (11.6.12) holds. For each $k$ such that (11.6.12) holds, we have $\rho_k^{(r+1)} = \lambda_{r+1}$, the truncation $y^*|_{2t}$ admits a finitely atomic measure $\mu$, and each point in $\operatorname{supp}(\mu)$ is a real $\mathcal{B}$-eigenvector associated to $\lambda_{r+1}$.*

Theorem 11.6.5 can also be implied by Theorem 6.3.5, since (11.6.14) has a single inequality constraint. The equality $\rho_k^{(r+1)} = \lambda_{r+1}$ can be detected by checking the flat truncation (11.6.12). If it holds, we can get rank $M_t[y^*]$ distinct $\mathcal{B}$-eigenvectors associated to $\lambda_{r+1}$. When there are infinitely many $\mathcal{B}$-eigenvectors for $\lambda_{r+1}$, we refer to the method in §6.6.2 for getting a $\mathcal{B}$-eigenvector.

Second, we discuss how to detect whether $\lambda_{r+1}$ exists or not. When it exists, we show how to choose $\delta$ satisfying (11.6.13). This can be done by considering the following minimization problem:

$$\begin{cases} \min & \mathcal{A}x^m \\ \text{s.t.} & h_j(x) = 0 \, (j = 1, \ldots, n+1), \\ & \mathcal{A}x^m \geq \lambda_r - \delta. \end{cases} \qquad (11.6.16)$$

Let $\chi_r$ denote the optimal value of (11.6.16). The following theorem is useful.

**Theorem 11.6.6.** *([62]) Suppose the real hypersurface $\mathcal{B}x^{m'} = 1$ is smooth. Let $\lambda_r$ be the rth largest real $\mathcal{B}$-eigenvalue and let $\lambda_{min}$ be the smallest real $\mathcal{B}$-eigenvalue for $\mathcal{A}$. For every $\delta > 0$, we have the following:*

  (i) *The relaxation (11.6.15) is infeasible for some $k$ if and only if*

$$\lambda_r - \delta < \lambda_{min}.$$

  (ii) *If $\chi_r = \lambda_r$ and $\lambda_{r+1}$ exists, then $\lambda_{r+1} < \lambda_r - \delta$, so (11.6.13) holds.*

  (iii) *If $\chi_r = \lambda_r$ and (11.6.15) is infeasible for some $k$, then $\lambda_r = \lambda_{min}$ and $\lambda_{r+1}$ does not exist.*

The problem (11.6.16) is polynomial optimization. The Moment-SOS hierarchy is applicable for solving it. This hierarchy also has finite convergence (see [231]). For $\delta > 0$ sufficiently small, we must have $\chi_r = \lambda_r$, no matter whether $\lambda_{k+1}$ exists or not. This is because $\chi_r$ is the smallest $\mathcal{B}$-eigenvalue greater than or equal to $\lambda_r - \delta$.

Third, the existence of $\lambda_{r+1}$ and the relation (11.6.13) can be determined as follows. Choose an initial value for $\delta$ (say, 0.05) and then solve (11.6.16). If $\chi_r < \lambda_r$, decrease the value $\delta$ (say, let $\delta := \delta/5$) and solve (11.6.16) again. Repeat this process until we get $\chi_r = \lambda_r$. This process must terminate when $\delta > 0$ is sufficiently small. After $\chi_k = \lambda_r$ is met, there are only two possibilities:

(i) If $\lambda_{r+1}$ does not exist, then $\lambda_r = \lambda_{min}$. By item (i) of Theorem 11.6.6, the relaxation (11.6.15) must be infeasible for some $k$, which then confirms the nonexistence of $\lambda_{r+1}$, according to item (iii) of Theorem 11.6.6.

(ii) If $\lambda_{r+1}$ exists, then $\lambda_{r+1} < \lambda_r - \delta$, by item (ii) of Theorem 11.6.6, so (11.6.13) holds. Then, by item (i) of Theorem 11.6.5, $\rho_k^{(r+1)} = \lambda_{r+1}$ when $k$ is sufficiently large.

In summary, if $\lambda_{r+1}$ does not exist, we can get a certificate for that; if it exists, we can get $\lambda_{r+1}$ by solving the moment relaxation (11.6.15).

After $\lambda_{r+1}$ is obtained (if it exists), we can repeat the above procedure to determine if $\lambda_{r+2}$ exists or not. Repeating this process, one can eventually get all real eigenvalues. We refer to [62] for more details.

## 11.6.2 ▪ Nonsymmetric tensor eigenvalues

There are major differences between symmetric and nonsymmetric tensor eigenvalues. For symmetric tensors, the eigenvalues are critical values of some polynomial optimization problems. However, for nonsymmetric tensors, this is typically not the case. There are special features for nonsymmetric tensor eigenvalues. A nonsymmetric tensor may have none or infinitely many Z-eigenvalues.

**Example 11.6.7.** *([238]) (i) Consider the tensor $\mathcal{A} \in \mathrm{T}^4(\mathbb{R}^2)$ such that*

$$\mathcal{A}_{1112} = \mathcal{A}_{1222} = 1, \mathcal{A}_{2111} = \mathcal{A}_{2122} = -1$$

*and $\mathcal{A}_{ijkl} = 0$ otherwise. The Z-eigenpair equations are*

$$(x_1^2 + x_2^2)x_2 = \lambda x_1, \quad -(x_1^2 + x_2^2)x_1 = \lambda x_2, \quad x_1^2 + x_2^2 = 1.$$

*The above does not have a solution, so $\mathcal{A}$ has no Z-eigenvalues. The H-eigenvalue system is*

$$(x_1^2 + x_2^2)x_2 = \lambda x_1^3, \quad -(x_1^2 + x_2^2)x_1 = \lambda x_2^3, \quad (x_1, x_2) \neq (0,0).$$

*It does not have a real solution.*
*(ii) Consider the tensor $\mathcal{A} \in \mathrm{T}^4(\mathbb{R}^2)$ such that $\mathcal{A}_{1111} = \mathcal{A}_{2112} = 1$ and $\mathcal{A}_{ijkl} = 0$ for all other $ijkl$. The Z-eigenpair equations are*

$$x_1^3 = \lambda x_1, \quad x_1^2 x_2 = \lambda x_2, \quad x_1^2 + x_2^2 = 1.$$

*Each $\lambda \in [0,1]$ is a real Z-eigenvalue, with the eigenvectors $(\pm\sqrt{\lambda}, \pm\sqrt{1-\lambda})$.*

For generic tensors, there are finitely many complex Z-eigenvalues. If $\mathcal{A} \in \mathrm{T}^m(\mathbb{C}^n)$ has generic entries, it is shown in [44] that each Z-eigenvalue has multiplicity one and the number of complex Z-eigenpairs is

$$\frac{(m-1)^n - 1}{m-2}.$$

When $\mathcal{A}$ is a generic symmetric tensor in $\mathrm{S}^m(\mathbb{C}^n)$, this conclusion is also true (see [44]). Unlike Z-eigenvalues, every tensor has only finitely many complex H-eigenvalues.

**Proposition 11.6.8.** *([238]) Every tensor $\mathcal{A} \in \mathrm{T}^m(\mathbb{C}^n)$ has $n(m-1)^{n-1}$ complex $\mathrm{H}$-eigenvalues, counting their multiplicities.*

The $\mathrm{Z}$- and $\mathrm{H}$-eigenvalues are special cases of $\mathcal{B}$-eigenvalues. Let $h$ be the polynomial tuple as in (11.6.8). Note that $u$ is a $\mathrm{Z}$-eigenvector of $\mathcal{A}$ if and only if $h(u) = 0$. If $u$ is a $\mathcal{B}$-eigenvector, the associated eigenvalue is $\mathcal{A}u^m$, since

$$u^T \mathcal{A} u^{m-1} = \mathcal{A}u^m, \quad u^T \mathcal{B} u^{m'-1} = \mathcal{B}u^{m'} = 1.$$

Therefore, we can still use the optimization (11.6.9) to compute $\mathcal{B}$-eigenvalues. Note that $(\lambda, u)$ is a $\mathcal{B}$-eigenpair if and only if $u$ is a feasible point of (11.6.9) and $\lambda$ is the corresponding objective value. Moment-SOS relaxations in §11.6.1 can be similarly applied to compute $\mathcal{B}$-eigenpairs. When $\mathcal{A}$ has only finitely many $\mathcal{B}$-eigenpairs, there are similar convergence properties as in the previous subsection. We refer to [238] for more details.

### 11.6.3 ▪ Tensor singular values

For a real matrix $A$, its singular values are critical values of the bilinear form $x^T A y$ over the unit spheres $\|x\|_2 = \|y\|_2 = 1$. When $A$ is complex, its singular values are critical values of the modulus $|x^T A y|$ over the spheres. Critical values can be similarly defined for tensors. This gives the notion of singular values for tensors, introduced by Lim [177].

A tensor $\mathcal{A} \in \mathbb{C}^{n_1 \times \cdots \times n_m}$ is uniquely determined by

$$\mathcal{A}(x^{(1)}, \ldots, x^{(m)}) := \sum_{i_1 \in [n_1], \ldots, i_m \in [n_m]} \mathcal{A}_{i_1 \ldots i_m} \left(x^{(1)}\right)_{i_1} \cdots \left(x^{(m)}\right)_{i_m}, \tag{11.6.17}$$

which is a multilinear form in complex variables $x^{(1)} \in \mathbb{C}^{n_1}, \ldots, x^{(m)} \in \mathbb{C}^{n_m}$. For convenience of notation, we write $x^{(i)}_\times$ to mean that $x^{(i)}$ is missing, i.e.,

$$\left(x^{(1)}, \ldots, x^{(i)}_\times, \ldots, x^{(m)}\right) := (x^{(1)}, \ldots, x^{(i-1)}, x^{(i+1)}, \ldots, x^{(m)}). \tag{11.6.18}$$

For each $i = 1, \ldots, m$, define the product

$$\mathcal{A} \times_{(i)} \left(x^{(1)}, \ldots, x^{(i)}_\times, \ldots, x^{(m)}\right) := \nabla_{x^{(i)}} \left(\mathcal{A}(x^{(1)}, \ldots, x^{(m)})\right), \tag{11.6.19}$$

where $\nabla_{x^{(i)}}$ denotes the gradient with respect to $x^{(i)}$. The entries of $x^{(i)}$ are labeled as

$$x^{(i)} = \left((x^{(i)})_1, \ldots, (x^{(i)})_{n_i}\right).$$

To introduce tensor singular values, consider the maximization

$$\begin{cases} \max & \mathcal{A}(x^{(1)}, \ldots, x^{(m)}) \\ s.t. & \left((x^{(i)})_1\right)^2 + \cdots + \left((x^{(i)})_{n_i}\right)^2 = 1, \\ & i = 1, \ldots, m. \end{cases} \tag{11.6.20}$$

The KKT system for (11.6.20) gives the equations

$$\begin{cases} \mathcal{A} \times_{(1)} \left(x^{(1)}_\times, x^{(2)}, \ldots, x^{(m)}\right) = \sigma_1 x^{(1)}, \\ \\ \mathcal{A} \times_{(2)} \left(x^{(1)}, x^{(2)}_\times, \ldots, x^{(m)}\right) = \sigma_2 x^{(2)}, \\ \quad \vdots \\ \mathcal{A} \times_{(m)} \left(x^{(1)}, x^{(2)}, \ldots, x^{(m)}_\times\right) = \sigma_m x^{(m)} \end{cases}$$

for scalars $\sigma_1, \ldots, \sigma_m$. Because of the constraints in (11.6.20), the above further implies

$$\sigma_1 = \cdots = \sigma_m = \mathcal{A}(x^{(1)}, \ldots, x^{(m)}).$$

As in [177], this motivates the tensor singular value equations

$$\begin{cases} \mathcal{A} \times_{(1)} (x^{(1)}, x^{(2)}, \ldots, x^{(m)}) &= \sigma x^{(1)}, \\[2mm] \mathcal{A} \times_{(2)} (x^{(1)}, x^{(2)}, \ldots, x^{(m)}) &= \sigma x^{(2)}, \\[1mm] &\vdots \\[1mm] \mathcal{A} \times_{(m)} (x^{(1)}, x^{(2)}, \ldots, x^{(m)}) &= \sigma x^{(m)}. \end{cases} \qquad (11.6.21)$$

**Definition 11.6.9.** *([177]) For a tensor $\mathcal{A} \in \mathbb{C}^{n_1 \times \cdots \times n_m}$, a scalar $\sigma$ is called a singular value of $\mathcal{A}$ if there exist unit length vectors $x^{(i)} \in \mathbb{C}^{n_i}$, $i = 1, \ldots, m$, such that all equations in (11.6.21) hold. The vector $x^{(i)}$ is called a mod-i singular vector, the tuple $(x^{(1)}, \ldots, x^{(m)})$ is called a singular vector tuple, and $(\sigma, x^{(1)}, \ldots, x^{(m)})$ is called a singular tuple.*

We remark that if $(\sigma, u^{(1)}, \ldots, u^{(m)})$ is a singular tuple, then

$$\left( \frac{\alpha_1 \cdots \alpha_m}{\alpha_i^2} \sigma, \; \alpha_1 u^{(1)}, \ldots, \alpha_m u^{(m)} \right)$$

is also a singular tuple for all unitary scalars $\alpha_i$ such that

$$|\alpha_1| = \cdots = |\alpha_m| = 1, \quad \alpha_1^2 = \cdots = \alpha_m^2.$$

Such two singular tuples are considered to be *equivalent*. For $m \geq 2$, the singular values always appear in $\pm$ pairs. For $m > 3$, there are infinitely many complex singular values. In computation, we often scale singular vectors so that the singular values are real and compute singular tuples that are not equivalent to each other. The number of classes of equivalent singular tuples is counted in [99]. When $\mathcal{A}$ is real, there is at least one real singular tuple. However, not all singular tuples are necessarily real.

**Example 11.6.10.** *Let $\mathcal{A} \in \mathrm{T}^3(\mathbb{C}^2)$ be the tensor such that*

$$\mathcal{A}_{112} = 1, \quad \mathcal{A}_{221} = -1$$

*and all other entries are zero. The singular value equations are*

$$\begin{aligned} (x^{(2)})_1 (x^{(3)})_2 &= \sigma \cdot (x^{(1)})_1, \\ -(x^{(2)})_2 (x^{(3)})_1 &= \sigma \cdot (x^{(1)})_2, \\ (x^{(1)})_1 (x^{(3)})_2 &= \sigma \cdot (x^{(2)})_1, \\ -(x^{(1)})_2 (x^{(3)})_1 &= \sigma \cdot (x^{(2)})_2, \\ -(x^{(1)})_2 (x^{(2)})_2 &= \sigma \cdot (x^{(3)})_1, \\ (x^{(1)})_1 (x^{(2)})_1 &= \sigma \cdot (x^{(3)})_2. \end{aligned}$$

*The equivalent classes of singular tuples $(\sigma, u^{(1)}, u^{(2)}, u^{(3)})$ are*

| $\sigma$ | $u^{(1)}$ | $u^{(2)}$ | $u^{(3)}$ |
|---|---|---|---|
| $1$ | $(1,0)$ | $(1,0)$ | $(0,1)$ |
| $-1$ | $(0,1)$ | $(0,1)$ | $(1,0)$ |
| $\frac{1}{\sqrt{2}}$ | $(\frac{1}{\sqrt{2}}, \frac{1}{\sqrt{2}})$ | $(\frac{1}{\sqrt{2}}, \frac{-1}{\sqrt{2}})$ | $(\frac{1}{\sqrt{2}}, \frac{1}{\sqrt{2}})$ |
| $\frac{1}{\sqrt{2}}$ | $(\frac{1}{\sqrt{2}}, -\frac{-1}{\sqrt{2}})$ | $(\frac{1}{\sqrt{2}}, \frac{1}{\sqrt{2}})$ | $(\frac{1}{\sqrt{2}}, \frac{1}{\sqrt{2}})$ |
| $\frac{1}{\sqrt{2}}$ | $(\frac{1}{\sqrt{2}}, \frac{-1}{\sqrt{2}})$ | $(\frac{1}{\sqrt{2}}, \frac{-1}{\sqrt{2}})$ | $(\frac{-1}{\sqrt{2}}, \frac{1}{\sqrt{2}})$ |
| $\frac{1}{\sqrt{2}}$ | $(\frac{1}{\sqrt{2}}, \frac{1}{\sqrt{2}})$ | $(\frac{1}{\sqrt{2}}, \frac{1}{\sqrt{2}})$ | $(\frac{-1}{\sqrt{2}}, \frac{1}{\sqrt{2}})$ |

## Unitary eigenvalues

When $\mathcal{A}$ is not real, the optimization (11.6.20) may not be well defined. However, we can consider the complex optimization problem

$$\begin{cases} \max & |\langle \mathcal{A}, x^{(1)} \otimes \cdots \otimes x^{(m)} \rangle| \\ s.t. & \|x^{(i)}\|^2 = 1, \ i = 1, \ldots, m. \end{cases} \tag{11.6.22}$$

It is worthy to note that the objective of (11.6.22) may not be the modulus of $\mathcal{A}(x^{(1)}, \ldots, x^{(m)})$. We remark that (11.6.22) is equivalent to optimizing the real part of its objective:

$$\begin{cases} \max & \langle \mathcal{A}, x^{(1)} \otimes \cdots \otimes x^{(m)} \rangle + \overline{\langle \mathcal{A}, x^{(1)} \otimes \cdots \otimes x^{(m)} \rangle} \\ s.t. & \|x^{(i)}\|^2 = 1, \ i = 1, \ldots, m. \end{cases} \tag{11.6.23}$$

(The $\bar{z}$ denotes the complex conjugate of $z$.) The KKT system for (11.6.23) gives the equations

$$\begin{cases} \mathcal{A} \times_{(i)} \overline{(x^{(1)}, \ldots, \underset{\times}{x^{(i)}}, \ldots, x^{(m)})} & = & \mu_i \cdot x^{(i)}, \\ \\ \overline{\mathcal{A}} \times_{(i)} (x^{(1)}, \ldots, \underset{\times}{x^{(i)}}, \ldots, x^{(m)}) & = & \mu_i \cdot \overline{x^{(i)}} \end{cases} \tag{11.6.24}$$

for scalars $\mu_i$, $i = 1, \ldots, m$. Since each $x^{(i)}$ has unit length, the above implies

$$\mu_1 = \cdots = \mu_m = \langle \mathcal{A}, x^{(1)} \otimes \cdots \otimes x^{(m)} \rangle.$$

Therefore, the KKT system (11.6.24) is equivalent to (for $i = 1, \ldots, m$)

$$\begin{cases} \mathcal{A} \times_{(i)} \overline{(x^{(1)}, \ldots, \underset{\times}{x^{(i)}}, \ldots, x^{(m)})} & = & \mu \cdot x^{(i)}, \\ \\ \overline{\mathcal{A}} \times_{(i)} (x^{(1)}, \ldots, \underset{\times}{x^{(i)}}, \ldots, x^{(m)}) & = & \mu \cdot \overline{x^{(i)}}. \end{cases} \tag{11.6.25}$$

This gives the notion of unitary eigenvalues (U-eigenvalues), which is introduced in [207].

**Definition 11.6.11.** *([207]) For a tensor $\mathcal{A} \in \mathbb{C}^{n_1 \times \cdots \times n_m}$, a scalar $\mu$ is called a unitary eigenvalue (or U-eigenvalue) of $\mathcal{A}$ if there exist unit length vectors $x^{(i)} \in \mathbb{C}^{n_i}$ such that (11.6.25) holds for each $i = 1, \ldots, m$. The vector $x^{(i)}$ is called a mod-$i$ U-eigenvector, the tuple $(x^{(1)}, \ldots, x^{(m)})$ is called a U-eigenvector tuple, and $(\mu, x^{(1)}, \ldots, x^{(m)})$ is called a U-eigentuple.*

We remark that all U-eigenvalues are real for all tensors. If $(\mu, u^{(1)}, \ldots, u^{(m)})$ is a U-eigentuple, one can check that

$$\left( \frac{\mu}{\alpha_1 \cdots \alpha_m}, \alpha_1 u^{(1)}, \ldots, \alpha_m u^{(m)} \right)$$

is also a U-eigentuple for all unitary scalars $\alpha_i$ such that

$$|\alpha_1| = \cdots = |\alpha_m| = 1, \quad \alpha_1 \alpha_2 \cdots \alpha_m \in \mathbb{R}.$$

Such two U-eigentuples are said to be *equivalent*. Note that U-eigenvalues appear in $\pm$ pairs. In computation, we focus on U-eigentuples that are not equivalent to each other.

**Example 11.6.12.** *Let $\mathcal{A} \in \mathrm{T}^3(\mathbb{C}^2)$ be the strictly upper triangular tensor such that $\mathcal{A}_{112} = \mathcal{A}_{122} = 1$ and all other entries are zero. The U-eigenvalue equations in (11.6.25) are ($\mu$ is real)*

$$(x^{(2)})_1(x^{(3)})_2 + (x^{(2)})_2(x^{(3)})_2 = \mu \cdot \overline{(x^{(1)})_1},$$
$$0 = \mu \cdot \overline{(x^{(1)})_2},$$
$$(x^{(1)})_1(x^{(3)})_2 = \mu \cdot \overline{(x^{(2)})_1},$$
$$(x^{(1)})_1(x^{(3)})_2 = \mu \cdot \overline{(x^{(2)})_2},$$
$$0 = \mu \cdot \overline{(x^{(3)})_1},$$
$$(x^{(1)})_1(x^{(2)})_1 + (x^{(1)})_1(x^{(2)})_2 = \mu \cdot \overline{(x^{(3)})_2}.$$

*For the case $\mu = 0$, the U-eigenvalue equations become*

$$(x^{(3)})_2 \cdot ((x^{(2)})_1 + (x^{(2)})_2) = 0,$$
$$(x^{(1)})_1(x^{(3)})_2 = 0,$$
$$(x^{(1)})_1 \cdot ((x^{(2)})_1 + (x^{(2)})_2) = 0.$$

*For unit length vectors $\|x^{(i)}\|$, $i = 1, 2, 3$, the $(x^{(1)}, x^{(2)}, x^{(3)})$ is a U-eigenvector tuple if they satisfy any two of the following three equations:*

$$(x^{(1)})_1 = 0, \quad (x^{(2)})_1 + (x^{(2)})_2 = 0, \quad (x^{(3)})_2 = 0.$$

*There are infinitely many U-eigenvectors for $\mu = 0$.*

*For the case $\mu \neq 0$, the U-eigenvalue equations become*

$$(x^{(1)})_2 = (x^{(3)})_1 = 0,$$
$$(x^{(1)})_1(x^{(3)})_2 = \mu \cdot \overline{(x^{(2)})_1} = \mu \cdot \overline{(x^{(2)})_2},$$
$$(x^{(2)})_1(x^{(3)})_2 + (x^{(2)})_2(x^{(3)})_2 = \mu \cdot \overline{(x^{(1)})_1},$$
$$(x^{(1)})_1(x^{(2)})_1 + (x^{(1)})_1(x^{(2)})_2 = \mu \cdot \overline{(x^{(3)})_2}.$$

*Note that $(x^{(2)})_1 = (x^{(2)})_2$. Since $(x^{(1)})_2 = (x^{(3)})_1 = 0$, we can see that*

$$|(x^{(1)})_1| = |(x^{(3)})_2| = 1, \quad |(x^{(2)})_1| = |(x^{(2)})_2| = \frac{1}{\sqrt{2}}.$$

*The above equations imply that*

$$\mu = 2(x^{(1)})_1(x^{(2)})_1(x^{(3)})_2, \quad |\mu| = \sqrt{2}.$$

*There is only one equivalent class of U-eigentuples $(\mu, u^{(1)}, u^{(2)}, u^{(3)})$ for $\mu \neq 0$, which is*

| $\mu$ | $u^{(1)}$ | $u^{(2)}$ | $u^{(3)}$ |
|---|---|---|---|
| $\sqrt{2}$ | $(1,0)$ | $(\frac{1}{\sqrt{2}}, \frac{1}{\sqrt{2}})$ | $(0,1)$ |

*For this tensor, the equivalent classes of singular tuples and U-eigentuples are the same.*

## US-eigenvalues

When $\mathcal{A}$ is symmetric, a similar version of (11.6.22) is

$$\begin{cases} \max & |\langle \mathcal{A}, x^{\otimes m} \rangle| \\ s.t. & \|x\|^2 = 1, \ x \in \mathbb{C}^n. \end{cases} \tag{11.6.26}$$

The above optimization is equivalent to

$$\begin{cases} \max & \langle \mathcal{A}, x^{\otimes m} \rangle + \overline{\langle \mathcal{A}, x^{\otimes m} \rangle} \\ s.t. & \|x\|_2 = 1, \ x \in \mathbb{C}^n. \end{cases} \tag{11.6.27}$$

Similarly, the KKT system for (11.6.27) gives the equations

$$\begin{cases} \mathcal{A} \times_{(1)} \overline{(x, x, \dots, x)} & = & \nu \cdot x, \\[2mm] \overline{\mathcal{A}} \times_{(1)} (x, x, \dots, x) & = & \nu \cdot \overline{x}. \end{cases} \tag{11.6.28}$$

In the left hand sides of the above, $x$ is repeated for $m - 1$ times. This gives unitary symmetric eigenvalues (US-eigenvalues), which are introduced in [207].

**Definition 11.6.13.** *([207]) For a symmetric tensor $\mathcal{A} \in \mathsf{S}^m(\mathbb{C}^n)$, a scalar $\nu$ is called a unitary symmetric eigenvalue (or US-eigenvalue) of $\mathcal{A}$ if there exists a unit length vector $x \in \mathbb{C}^n$ such that (11.6.28) holds. The vector $x$ is called a US-eigenvector, and $(\nu, x)$ is called a US-eigenpar.*

Similar to earlier cases, if $(\nu, u)$ is a US-eigentuple, then $(\frac{\nu}{\alpha^m}, \alpha u)$ is also a US-eigentuple for every unitary scalar $\alpha$ such that

$$|\alpha| = 1, \quad \alpha^m \in \mathbb{R}.$$

Such US-eigentuples are said to be *equivalent*. The US-eigenvalues appear in $\pm$ pairs. In computation, we often consider different classes of equivalent US-eigentuples.

**Example 11.6.14.** *Let $\mathcal{A} \in \mathsf{S}^3(\mathbb{C}^3)$ be the symmetric tensor such that*

$$\mathcal{A}_{111} = \mathcal{A}_{222} = \mathcal{A}_{333} = 1,$$
$$\mathcal{A}_{123} = \mathcal{A}_{132} = \mathcal{A}_{213} = \mathcal{A}_{231} = \mathcal{A}_{312} = \mathcal{A}_{321} = \tfrac{1}{6}$$

*and all other entries are zeros. The US-eigenvalue equations are ($\nu$ is real)*

$$x_1^2 + x_2 x_3/3 = \nu \overline{x_1},$$
$$x_2^2 + x_1 x_3/3 = \nu \overline{x_2},$$
$$x_3^2 + x_1 x_2/3 = \nu \overline{x_3},$$

*plus the unit length condition $\|x\| = 1$. The equivalent classes of US-eigenpairs $(\nu, u)$ for this tensor are listed as follows:*

| $\nu$ | $u$ |
|---|---|
| $\frac{4\sqrt{3}}{9}$ | $(\frac{\sqrt{3}}{3}, \frac{\sqrt{3}}{3}, \frac{\sqrt{3}}{3})$ |
| $1$ | $(1, 0, 0)$ |
| $1$ | $(0, 1, 0)$ |
| $1$ | $(0, 0, 1)$ |
| $\frac{7}{9}$ | $(\frac{2}{3}, \frac{2}{3}, \frac{1}{3})$ |
| $\frac{7}{9}$ | $(\frac{2}{3}, \frac{1}{3}, \frac{2}{3})$ |
| $\frac{7}{9}$ | $(\frac{1}{3}, \frac{2}{3}, \frac{2}{3})$ |

## 11.6.4 ▪ Exercises

**Exercise 11.6.1.** *([44, Example 5.8]) For the tensor $\mathcal{A} \in \mathrm{S}^6(\mathbb{R}^3)$ such that*

$$\mathcal{A}x^6 := x_1^4 x_2^2 + x_1^2 x_2^4 + x_3^6 - 3x_1^2 x_2^2 x_3^2,$$

*determine all its real* Z- *and* H-*eigenvalues.*

**Exercise 11.6.2.** *For the tensor $\mathcal{A} \in \mathrm{T}^3(\mathbb{R}^3)$ whose nonzero entries are*

$$\mathcal{A}_{112} = 1, \quad \mathcal{A}_{121} = 1, \quad \mathcal{A}_{113} = 1, \quad \mathcal{A}_{131} = 1, \quad \mathcal{A}_{211} = 1,$$
$$\mathcal{A}_{222} = 1, \quad \mathcal{A}_{223} = 1, \quad \mathcal{A}_{232} = 1, \quad \mathcal{A}_{333} = 1,$$

*determine all its real* Z- *and* H-*eigenpairs.*

**Exercise 11.6.3.** *Let $\mathcal{A} \in \mathrm{T}^3(\mathbb{R}^2)$ be the upper triangular tensor such that*

$$\mathcal{A}_{111} = \mathcal{A}_{112} = \mathcal{A}_{122} = \mathcal{A}_{222} = 1$$

*and all other entries are zero. Determine all its singular tuples and* U-*eigentuples.*

**Exercise 11.6.4.** *Let $\mathcal{A} \in \mathrm{S}^3(\mathbb{R}^3)$ be the symmetric tensor such that*

$$\mathcal{A}_{ijk} = \begin{cases} 0, & i+j+k \text{ is odd}, \\ 1, & i+j+k \text{ is even}. \end{cases}$$

*Determine all its* US-*eigentuples.*

**Exercise 11.6.5.** *For a polynomial tuple $p = (p_1, \dots, p_s)$, let*

$$\Omega = \{x \in \mathbb{R}^n : p(x) \geq 0\}.$$

*For two symmetric tensors $\mathcal{A}, \mathcal{B}$, show that the largest real $\mathcal{B}$-eigenvalue of $\mathcal{A}$, whose associated eigenvectors belong to $\Omega$, is equal to the optimal value of*

$$\begin{cases} \max & \mathcal{A}x^m \\ s.t. & h(x) = 0, \\ & p(x) \geq 0, \end{cases} \tag{11.6.29}$$

*where $h$ is the polynomial tuple as in (11.6.8).*

**Exercise 11.6.6.** *([62, 231]) Prove Theorem 11.6.6. Show that the Moment-SOS hierarchy for solving (11.6.16) has finite convergence.*

**Exercise 11.6.7.** *Prove Proposition 11.6.8. (Hint: use resultants.)*

**Exercise 11.6.8.** *Show that the optimization problems (11.6.22) and (11.6.23) are equivalent to each other. Show that the KKT system for (11.6.23) implies (11.6.24). Similarly, show that the optimization (11.6.26) is equivalent to (11.6.27) and the KKT system for (11.6.27) implies (11.6.28).*

# Chapter 12

# Special Topics

*There are various special topics about polynomial optimization. This chapter introduces some recent work on the topics of sparse optimization, rational optimization, saddle points, Nash equilibrium, bilevel optimization, distributed robust optimization, and multi-objective optimization.*

## 12.1 ▪ Sparse Polynomial Optimization

A polynomial $p$ in $x \in \mathbb{R}^n$ of degree $2d$ is SOS if and only if there exists a psd matrix $W \succeq 0$ such that
$$p = [x]_d^T W [x]_d.$$
The monomial vector $[x]_d$ has the length $\binom{n+d}{d}$, so $W$ has the dimension $\binom{n+d}{d}$-by-$\binom{n+d}{d}$. The total number of coefficients is $\binom{n+2d}{2d}$. The length grows quickly as $n, d$ increase. For instance, for $n = d = 10$, $\binom{n+d}{d} = 184756$. On the other hand, if $d$ is fixed, then $\binom{n+d}{d}$ is a polynomial of degree $d$ in $n$. In applications, the polynomial $p$ is often sparse, i.e., many coefficients are zeros. We discuss how to exploit sparsity in polynomial optimization.

The *support* of a polynomial $p$, for which we denote $\operatorname{supp}(p)$, is the set of monomial powers whose coefficients are nonzero. The convex hull of $\operatorname{supp}(p)$ is called the *Newton polytope* of $p$, for which we denote $\operatorname{New}(p)$. For two polynomials $p, q$, their Newton polytopes satisfy the homomorphism property:
$$\operatorname{New}(pq) = \operatorname{New}(p) + \operatorname{New}(q). \tag{12.1.1}$$

When $p$ is sparse, i.e., its support cardinality $|\operatorname{supp}(p)|$ is far less than $\binom{n+2d}{2d}$, it is expected to get a sparse SOS representation. The following is a basic result about sparsity.

**Theorem 12.1.1.** *([281, Theorem 1]) If $p = p_1^2 + \cdots + p_m^2$, then the support of each $p_i$ satisfies*

$$\operatorname{supp}(p_i) \subseteq \mathcal{N} := \operatorname{conv}\left(\frac{1}{2}\operatorname{supp}(p)\right). \tag{12.1.2}$$

***Proof.*** Let $P := \bigcup_{i=1}^m \operatorname{supp}(p_i)$ and let $a_1, \ldots, a_\ell$ be extreme points of the convex hull $\operatorname{conv}(P)$. To prove the conclusion, it is enough to show each $a_k \in \mathcal{N}$. Observe that for each $a_k$, the coefficient of the monomial $x^{a_k}$ must be nonzero for at least one $p_i$. In the decomposition $p = p_1^2 + \cdots + p_m^2$, the coefficient of the monomial $x^{2a_k}$ must be nonzero, because any power relation $2a_k = \alpha + \beta$ implies $a_k = \alpha = \beta$. So, we have each $2a_k \in \mathrm{supp}(p)$ and hence $a_k \in \mathcal{N}$. Therefore, $\mathrm{supp}(p_i) \subseteq \mathcal{N}$ for each $i = 1, \dots, m$.  □

When $p$ is a sparse polynomial, the Newton polytope $\mathcal{N}$ may still be big. For instance, for $p = x_1^{2d} + \cdots + x_n^{2d}$, the polytope $\mathcal{N}$ consists of all powers whose degrees equal $d$. It is possible to get a more sparse SOS representation than the one given in (12.1.2). A useful trick is the *correlative sparsity pattern*. A polynomial $p$ is associated to the graph $G = ([n], E)$, where $[n] = \{1, \dots, n\}$ and $(i, j) \in E$ if and only if $x_i x_j$ appears in some monomial of $p$. Let $\{C_1, C_2, \dots, C_K\}$ be the set of all maximal cliques of the graph $G$. Then, one can consider a further sparse SOS representation in the form

$$p = \sum_i \sum_j s_{ij}^2, \quad \mathrm{supp}(s_{ij}) \subseteq C_i.$$

It is typically hard to get all maximal cliques of $G$. As proposed in [316], one can instead try to replace $\{C_1, C_2, \dots, C_K\}$ by the set of all maximal cliques of the *chordal extension* of $G$. A chordal graph is one in which all cycles of four or more vertices have a chord, which is an edge that is not a part of the cycle but connects two vertices of the cycle. We refer to [28] for chordal graphs. There are efficient methods for finding maximal cliques of chordal graphs. Chordal extension is essentially a *sparse symbolic Cholesky factorization*, which can be done as follows. Let $R = (R_{ij})_{1 \le i,j \le n}$ be the *correlative sparsity pattern* (csp) matrix of the polynomial $p$, i.e., $R$ is a random symmetric matrix such that $R_{ij} = 0$ for all $x_i x_j$ ($i \ne j$) not appearing in any monomial of $p$. For instance, a csp matrix for $x_1^4 + x_2^4 + x_3^4 + x_4^4 + (x_1 + x_2 + x_3)x_4$ has the following pattern:

$$R := \begin{bmatrix} * & 0 & 0 & * \\ 0 & * & 0 & * \\ 0 & 0 & * & * \\ * & * & * & * \end{bmatrix},$$

where each $*$ stands for a nonzero entry. Assign random values to these nonzero entries such that $R$ is positive definite. We can apply sparse symbolic Cholesky factorization to get a sparse Cholesky factor $L$ (i.e., $R = LL^T$) and then associate $L$ with a graph $G'$. In MATLAB, one can use the function symamd to get a permutation matrix $P$. Then $P^T R P$ generally has a sparser Cholesky factorization than $R$ does. The function chol can be used to find the Cholesky factor $L$ of $P^T R P$, i.e., $P^T R P = LL^T$. Then $G'$ is usually a chordal extension of $G$. Find maximal cliques $C_1, \dots, C_K$ of the graph $G'$. Then we get a sparse SOS representation of $p$. However, we remark that such a sparse SOS relaxation does not always exist (see Example 12.1.3). We refer to [144, 151, 148, 159, 213, 259, 316, 318, 319, 333] for recent work on sparse polynomial optimization. The software can be found in [186, 317].

We introduce the sparse SOS relaxation method in [213]. Consider the unconstrained optimization problem

$$\min_{x \in \mathbb{R}^n} \quad f(x) := \sum_{i=1}^m f_i(x_{\Delta_i}), \tag{12.1.3}$$

where each $\Delta_i \subseteq [n]$ and $f_i(x_{\Delta_i})$ is a polynomial in the subvector

$$x_{\Delta_i} := (x_j)_{j \in \Delta_i}.$$

The objective $f(x)$ is given with a sparsity pattern. This kind of sparse polynomial optimization frequently appears in applications (see [213, 216]). A sparse SOS relaxation for solving (12.1.3) is to find the maximum $\gamma$ such that

$$f(x) - \gamma = \sum_{i=1}^{m} s_i(x_{\Delta_i}),$$

where each $s_i(x_{\Delta_i})$ is an SOS polynomial in $x_{\Delta_i}$, instead of $x$. Exploiting the sparsity can reduce the computational cost significantly. Let $D$ be the maximum cardinality of $\Delta_i$, i.e.,

$$D := \max_i |\Delta_i|.$$

In applications, we often have $D \ll n$. Let

$$d := \max_{1 \leq i \leq m} \left\{ \left\lceil \frac{1}{2} \deg(f_i(x_{\Delta_i})) \right\rceil \right\}.$$

The above sparse SOS relaxation for solving (12.1.3) is

$$\begin{cases} \max & \gamma \\ s.t. & f(x) - \gamma = \sum_{i=1}^{m} [x_{\Delta_i}]_{d_i}^T W_i [x_{\Delta_i}]_{d_i}, \\ & W_i \in \mathcal{S}_+^{\binom{|\Delta_i|+d_i}{d_i}}, \ i = 1, \ldots, m. \end{cases} \qquad (12.1.4)$$

Each $W_i$ is labeled by a pair $(\eta, \tau)$ such that

$$\eta, \tau \in \mathbb{N}_{d_i}^n, \quad \mathrm{supp}(\eta), \mathrm{supp}(\tau) \subseteq \Delta_i.$$

(For a vector $a = (a_1, \ldots, a_n)$, its support is $\mathrm{supp}(a) := \{i : a_i \neq 0\}$.) For convenience, denote the support sets

$$\boxed{\begin{array}{rcl} \mathcal{F}_i & := & \{\alpha \in \mathbb{N}_{2d}^n : \mathrm{supp}(\alpha) \subseteq \Delta_i\}, \\ & & i = 1, \ldots, m, \\ \mathcal{F} & := & \mathcal{F}_1 \cup \cdots \cup \mathcal{F}_m. \end{array}} \qquad (12.1.5)$$

Let $\mathbb{R}^{\mathcal{F}}$ denote the space of all sparse tms $y := (y_\alpha)_{\alpha \in \mathcal{F}}$ with real entries. Write the objective as

$$f = \sum_{\alpha \in \mathcal{F}} f_\alpha x^\alpha.$$

A comparison of coefficients in (12.1.4) gives the linear equations

$$\begin{cases} f_0 - \gamma & = & \sum_{i=1}^{m} W_i(0,0), \\ f_\alpha & = & \sum_{i=1}^{m} \sum_{\eta+\tau=\alpha} W_i(\eta, \tau) \quad \text{for } \alpha \neq 0. \end{cases} \qquad (12.1.6)$$

For a sparse tms $y \in \mathbb{R}^{\mathcal{F}}$, the sparse moment matrix $M_{d_i}^{\Delta_i}[y]$ is such that

$$\langle q^2, y \rangle = \mathrm{vec}(q)^T \left( M_{d_i}^{\Delta_i}[y] \right) \mathrm{vec}(q)$$

for all $q \in \mathbb{R}[x_{\Delta_i}]_{d_i}$. The dual optimization of (12.1.4) is

$$\begin{cases} \min & \sum_{\alpha \in \mathcal{F}} f_\alpha y_\alpha \\ s.t. & M_{d_i}^{\Delta_i}[y] \succeq 0, \ i = 1, \ldots, m, \\ & y_0 = 1, \ y \in \mathbb{R}^{\mathcal{F}}. \end{cases} \qquad (12.1.7)$$

The relaxation (12.1.7) has $m$ linear matrix inequalities, which are of length at most $\binom{D+d}{d} = O(D^d)$. The sparse tms $y$ has the dimension

$$O\left(m\binom{D+2d}{2d}\right) = O(mD^{2d}).$$

Let $f_{spa}$, $f_{smo}$ denote the optimal values of (12.1.4), (12.1.7) respectively. Let $f_{min}$ be the minimum value of (12.1.3) and $f_{sos}$ be the lower bound given by the classical SOS relaxation (4.1.2). Then, it holds that

$$f_{spa} = f_{smo} \leq f_{sos} \leq f_{min}. \tag{12.1.8}$$

There exist examples such that $f_{spa} < f_{sos}$ (see Example 12.1.3). However, under some assumptions, we have $f_{spa} = f_{sos}$. A frequently used one is the *running intersection property* (RIP; see [159]): the sets $\Delta_1, \Delta_2, \ldots, \Delta_m$ are such that for each $i = 1, \ldots, m-1$,

$$\Delta_{i+1} \cap \left(\bigcup_{j=1}^{i} \Delta_j\right) \subseteq \Delta_k \quad \text{for some} \quad k \leq i. \tag{12.1.9}$$

The following gives a useful condition for $f_{spa} = f_{sos}$.

**Theorem 12.1.2.** *([213]) Suppose (12.1.7) has an optimizer $y^*$ such that each $M_{d_i}^{\Delta_i}[y^*]$ has a representing measure. If the RIP (12.1.9) holds, then $f_{spa} = f_{sos}$.*

The RIP (12.1.9) alone is not sufficient for $f_{spa} = f_{sos}$, shown by the following example.

**Example 12.1.3.** *Let $f(x) = f_1(x_1, x_2) + f_2(x_2, x_3)$, where*

$$f_1 = x_1^4 + (x_1 x_2 - 1)^2, \quad f_2 = x_2^2 x_3^2 + (x_3^2 - 1)^2.$$

*Solving the dense and sparse SOS relaxations, we get that*

$$f_{spa} \approx 5.0 \cdot 10^{-5} < f_{sos} \approx 0.8499.$$

*One can show that the minimum $f_{min}$ is achievable. By evaluation on critical values, one can show that $f_{min} \approx 0.8650$. For this polynomial, both the dense and the sparse SOS relaxations are not tight: $f_{spa} < f_{sos} < f_{min}$.*

When the RIP (12.1.9) fails, it may occur that $f_{spa} < f_{sos}$, even if each $f_i(x_{\Delta_i})$ is quadratic. The following is such an example.

**Example 12.1.4.** *Let $f(x) = f_1(x_1, x_2) + f_2(x_2, x_3) + f_3(x_1, x_3)$, where*

$$f_1 = \frac{1}{2}(x_1^2 + x_2^2) + 2x_1 x_2, \quad f_2 = \frac{1}{2}(x_2^2 + x_3^2) + 2x_2 x_3,$$

$$f_3 = \frac{1}{2}(x_1^2 + x_3^2) + 2x_1 x_3.$$

*Clearly, $\Delta_1 = \{1,2\}$, $\Delta_2 = \{2,3\}$, $\Delta_3 = \{1,3\}$. The RIP (12.1.9) fails and*

$$f_{spa} = -\infty < f_{sos} = f_{min} = 0.$$

In the following, we discuss how to extract a minimizer $x^* = (x_1^*, \ldots, x_n^*)$ from the sparse moment relaxation (12.1.7). Suppose $y^* = (y_\alpha^*)_{\alpha \in \mathcal{F}}$ is an optimizer of (12.1.7). A candidate minimizer to the optimization (12.1.3) is

$$u := (y_{e_1}^*, \ldots, y_{e_n}^*).$$

This choice is good if the relaxation (12.1.7) is tight and (12.1.3) has a unique minimizer. However, if (12.1.3) has more than one minimizer, then it is subtle to get a minimizer. Suppose all the moment matrices $M_d^{\Delta_i}[y^*]$ are flat, that is,

$$\text{rank } M_{\ell_i}^{\Delta_i}[y^*] = \text{rank } M_d^{\Delta_i}[y^*] \qquad (12.1.10)$$

for some degree $\ell_i < d$. By Theorem 2.7.7, we can extract a finite set, say, $\mathcal{X}_{\Delta_i}$, of values for the subvector $(x^*)_{\Delta_i}$. The following theorem tells how to get a minimizer $x^*$ for (12.1.3) from the sets $\mathcal{X}_{\Delta_i}$.

**Theorem 12.1.5.** *([213]) Suppose $y^*$ is a minimizer for the sparse moment relaxation (12.1.7) such that all $M_d^{\Delta_i}[y^*]$ are flat, i.e., (12.1.10) holds for all $i$. Then, each point $x^* = (x_1^*, \ldots, x_n^*)$, such that $(x^*)_{\Delta_i} \in \mathcal{X}_{\Delta_i}$ for all $i$, is a minimizer of (12.1.3).*

For instance, consider the unconstrained optimization

$$\min_{x \in \mathbb{R}^3} \quad \underbrace{(x_1^2 - 1)^2 + (x_1 - x_2)^4}_{f_1(x_{\Delta_1})} + \underbrace{(x_2 - x_3)^4}_{f_2(x_{\Delta_2})}.$$

Clearly, $\Delta_1 = \{1, 2\}$, $\Delta_2 = \{2, 3\}$. Solving the dual problem (12.1.7), we get the minimizer $y^*$, which is such that

$$M_2^{\Delta_1}[y^*] = M_2^{\Delta_2}[y^*] = \begin{bmatrix} 1 & 0 & 0 & 1 & 1 & 1 \\ 0 & 1 & 1 & 0 & 0 & 0 \\ 0 & 1 & 1 & 0 & 0 & 0 \\ 1 & 0 & 0 & 1 & 1 & 1 \\ 1 & 0 & 0 & 1 & 1 & 1 \\ 1 & 0 & 0 & 1 & 1 & 1 \end{bmatrix}.$$

Both $\mathcal{M}_1^{\Delta_1}(y^*)$ and $\mathcal{M}_1^{\Delta_2}(y^*)$ are rank two and flat. By Theorem 2.7.7, we can get that

$$\mathcal{X}_{\Delta_1} = \left\{ \begin{bmatrix} -1 \\ -1 \end{bmatrix}, \begin{bmatrix} 1 \\ 1 \end{bmatrix} \right\}, \quad \mathcal{X}_{\Delta_2} = \left\{ \begin{bmatrix} -1 \\ -1 \end{bmatrix}, \begin{bmatrix} 1 \\ 1 \end{bmatrix} \right\}.$$

Since the $x_2$-entry of $x_{\Delta_1}$ and $x_{\Delta_2}$ must be the same, we can get two minimizers $\pm(1, 1, 1)$.

## 12.1.1 ▪ Exercises

**Exercise 12.1.1.** *Prove the Newton polytope homomorphism (12.1.1).*

**Exercise 12.1.2.** *Prove the bound relations in (12.1.8).*

**Exercise 12.1.3.** *Show that the dual optimization of (12.1.4) is (12.1.7).*

**Exercise 12.1.4.** *Show that the polynomial $f$ in Example 12.1.3 achieves the minimum value.*

**Exercise 12.1.5.** *In the optimization (12.1.3), if each $f_i$ is quadratic and the RIP (12.1.9) holds, show that $f_{sos} = f_{spa}$.*

## 12.2 ▪ Optimization with Rational Functions

Consider the optimization problem

$$
\begin{cases}
\min & R(x) := \frac{f(x)}{g(x)} \\
s.t. & x \in K
\end{cases}
\tag{12.2.1}
$$

for given polynomials $f, g \in \mathbb{R}[x]_d$ and for a given semialgebraic set $K \subseteq \mathbb{R}^n$. The objective $R(x)$ is a rational function. The denominator of $R(x)$ may achieve zero on $K$. At a point $u \in K$ with $g(u) = 0$, the objective value $R(u)$ is then defined as

$$
R(u) := \liminf_{\substack{x \to u, \\ x \in K, \, g(x) > 0}} \frac{f(x)}{g(x)}.
\tag{12.2.2}
$$

Let $\vartheta_{min}$ denote the optimal value of (12.2.1). A point $x^* \in K$ is said to be a minimizer for (12.2.1) if $g(x^*) > 0$ and $f(x^*)/g(x^*) = \vartheta_{min}$. For the case $g(x^*) = 0$, the $x^*$ is said to be an *asymptotic minimizer* for (12.2.1) if there exists a sequence $\{v_l\}_{l=1}^{\infty} \subseteq K$ such that each $g(v_l) > 0$ and

$$
\frac{f(v_l)}{g(v_l)} \to \vartheta_{min} \quad \text{as} \quad v_l \to x^*.
$$

We look for the minimum value $\vartheta_{min}$, as well as a minimizer or an asymptotic one. This contains a broad class of optimization problems (see [40, 139, 169, 212]).

Without loss of generality, we assume that $g \geq 0$ on $K$, because otherwise one can minimize $R$ over the following two subsets

$$
K \cap \{g(x) \geq 0\}, \quad K \cap \{g(x) \leq 0\}
$$

separately, or one can replace $R(x)$ by $\frac{f(x)g(x)}{g(x)^2}$. When $g$ has a zero on $K$, we usually assume

$$
K = \mathrm{cl}\left(K \cap \{g(x) > 0\}\right).
\tag{12.2.3}
$$

Under the assumption (12.2.3), $R(x) \geq \gamma$ on $K$ if and only if $f(x) - \gamma g(x) \geq 0$ on $K$, which is equivalent to the membership $f - \gamma g \in \mathscr{P}_d(K)$.

In this section, we assume $K$ is the semialgebraic set

$$
K = \left\{ x \in \mathbb{R}^n \;\middle|\; \begin{array}{l} c_i(x) = 0 \, (i \in \mathcal{E}), \\ c_j(x) \geq 0 \, (j \in \mathcal{I}) \end{array} \right\}
\tag{12.2.4}
$$

for given polynomials $c_i, c_j \in \mathbb{R}[x]$. Then (12.2.1) is equivalent to the linear conic optimization

$$
\begin{cases}
\max & \gamma \\
s.t. & f - \gamma g \in \mathscr{P}_d(K).
\end{cases}
\tag{12.2.5}
$$

Its dual optimization problem is

$$
\begin{cases}
\min & \langle f, y \rangle \\
s.t. & \langle g, y \rangle = 1, \, y \in \mathscr{R}_d(K).
\end{cases}
\tag{12.2.6}
$$

Denote the constraining polynomial tuples $c_{eq}, c_{in}$ as in (5.2.1). For a degree $k$, the $k$th order SOS relaxation for (12.2.5) is

$$
\begin{cases}
\max & \gamma \\
s.t. & f - \gamma g \in \mathrm{Ideal}[c_{eq}]_{2k} + \mathrm{QM}[c_{in}]_{2k}.
\end{cases}
\tag{12.2.7}
$$

Similarly, its dual optimization is the $k$th order moment relaxation

$$\left\{ \begin{array}{ll} \min & \langle f, y \rangle \\ s.t. & \langle g, y \rangle = 1, \\ & \mathscr{V}_{c_i}^{(2k)}[y] = 0 \, (i \in \mathcal{E}), \\ & L_{c_j}^{(2k)}[y] \succeq 0 \, (j \in \mathcal{I}), \\ & M_k[y] \succeq 0, \\ & y \in \mathbb{R}^{\mathbb{N}_{2k}^n}. \end{array} \right. \tag{12.2.8}$$

Note that (12.2.5) is a special case of the linear conic optimization (8.3.2). The convergence properties in §8.3 are applicable for the Moment-SOS hierarchy of (12.2.7)-(12.2.8).

The optimizers of (12.2.1) can be obtained as follows. Let

$$d_0 := \max_{i \in \mathcal{E} \cup \mathcal{I}} \left\{ 1, \lceil \tfrac{1}{2} \deg(c_i) \rceil \right\}.$$

Suppose $y^*$ is a minimizer of (12.2.8) for a relaxation order $k$ and it satisfies the flat truncation

$$\operatorname{rank} M_{t-d_0}[y^*] = \operatorname{rank} M_t[y^*] \tag{12.2.9}$$

for some integer $t \in [d_0, k]$. Then, there exist $r := M_t[y^*]$ distinct points $u_1, \ldots, u_r \in K$ such that

$$y^*|_{2t} = \lambda_1[u_1]_{2t} + \cdots + \lambda_r[u_r]_{2t} \tag{12.2.10}$$

for positive scalars $\lambda_1, \ldots, \lambda_r > 0$. They satisfy the constraint

$$1 = \langle g, y^* \rangle = \lambda_1 g(u_1) + \cdots + \lambda_r g(u_r).$$

Note that each $g(u_i) \geq 0$ on $K$. There must exist at least one $i$ such that $g(u_i) > 0$. Denote the label sets

$$\boxed{\begin{array}{ll} \mathcal{I}_+ & := \{i \in [r] : g(u_i) > 0\}, \\ \mathcal{I}_0 & := \{i \in [r] : g(u_i) = 0\}. \end{array}} \tag{12.2.11}$$

The constraint $\langle g, y^* \rangle = 1$ implies that

$$\sum_{i \in \mathcal{I}_+} \lambda_i g(u_i) = 1.$$

So, $\mathcal{I}_+ \neq \emptyset$. Consider the parametrization

$$y(\xi) = \sum_{i \in \mathcal{I}_+} \xi_i[u_i]_{2k} + \sum_{i \in \mathcal{I}_0} \xi_i[v_i]_{2k}$$

in the variable $\xi := (\xi_i)_{i \in \mathcal{I}_+ \cup \mathcal{I}_0}$. Observe that $y(\xi)$ is feasible for (12.2.8) for every $\xi \in D$, where

$$D := \left\{ \xi \in \mathbb{R}^{\mathcal{I}_+ \cup \mathcal{I}_0} : \sum_{i \in \mathcal{I}_+} \xi_i g(u_i) = 1, \text{ each } \xi_i \geq 0 \right\}.$$

Observe that

$$\langle f, y(\xi) \rangle = \sum_{i \in \mathcal{I}_+} \xi_i g(u_i) \frac{f(u_i)}{g(u_i)} + \sum_{i \in \mathcal{I}_0} \xi_i f(u_i).$$

Let $\vartheta_1$ denote the optimal value of (12.2.6), then

$$\frac{f(u_i)}{g(u_i)} \geq \vartheta_1 \quad \text{for each } i \in \mathcal{I}_+,$$

$$\vartheta_1 \geq \langle f, y^* \rangle = \sum_{i \in \mathcal{I}_+} \xi_i g(u_i) \frac{f(u_i)}{g(u_i)} + \sum_{i \in \mathcal{I}_0} \xi_i f(u_i).$$

Since $y^*$ is a minimizer of (12.2.8), one can show that (the proof is left as an exercise)

$$\boxed{\begin{aligned} f(u_i)/g(u_i) &= \vartheta_1 \quad \text{for each } i \in \mathcal{I}_+, \\ f(u_i) &= 0 \quad \text{for each } i \in \mathcal{I}_0. \end{aligned}} \tag{12.2.12}$$

Since $\vartheta_1 \leq \vartheta_{min}$ and $\frac{f(u_i)}{g(u_i)} \geq \vartheta_{min}$ for each $i \in \mathcal{I}_+$, we can get that

$$\frac{f(u_i)}{g(u_i)} = \vartheta_{min} \quad \text{for each } i \in \mathcal{I}_+. \tag{12.2.13}$$

This means that for each $i \in \mathcal{I}_+$, the point $u_i$ is a minimizer of (12.2.1). For the case that $g(u_i) = 0$, i.e., $i \in \mathcal{I}_0$, the point $u_i$ may or may not be a minimizer for (12.2.1). We have the following theorem.

**Theorem 12.2.1.** *Assume $g(x) \geq 0$ on $K$ and (12.2.3) holds. Suppose $y^*$ is a minimizer of (12.2.8) such that (12.2.10) holds. Then, for each $i \in \mathcal{I}_+$, the point $u_i$ is a minimizer of (12.2.1); for each $i \in \mathcal{I}_0$, if*

$$\liminf_{\substack{x \to u_i, \\ x \in K, g(x) > 0}} \frac{f(x)}{g(x)} = \vartheta_{min}, \tag{12.2.14}$$

*then $u_i$ is an asymptotic minimizer.*

***Proof.*** Under the assumption (12.2.3), we know $R(x) \geq \gamma$ on $K$ if and only if $f - \gamma g \in \mathscr{P}_d(K)$. So the maximum value of (12.2.5) is $\vartheta_{min}$. In the above, we have already seen that $u_i$ is a minimizer for each $i \in \mathcal{I}_+$. For each $i \in \mathcal{I}_0$, if the condition (12.2.14) holds, then there is a sequence $\{u_i^{(l)}\}_{l=1}^{\infty} \subseteq K$ such that $u_i^{(l)} \to u_i$, $g(u_i^{(l)}) > 0$, and

$$\vartheta_{min} = \lim_{l \to \infty} \frac{f(u_i^{(l)})}{g(u_i^{(l)})}.$$

This means that $u_i$ is an asymptotic minimizer.    □

When (12.2.14) fails, the $u_i$ may not be an asymptotic minimizer. For instance, consider $f = x$, $g = x^2$, and $K = [0, 1]$. We have $\vartheta_{min} = 1$ and $f - \vartheta_{min} g = x - x^2$. A minimizer for the moment relaxation is $y^* = [u_1]_2 + [u_2]_2$ with the points $u_1 = 0$, $u_2 = 1$. The condition (12.2.14) fails at $u_1$. The minimizer for the rational optimization is $u_2$, while $u_1$ is not an asymptotic minimizer. The following is an example of using the Moment-SOS hierarchy to solve rational optimization.

**Example 12.2.2.** *Consider the rational optimization*

$$\begin{cases} \min\limits_{x \in \mathbb{R}^2} & \frac{x_1^3 + x_2^3 + 3x_1 x_2 + 1}{x_1^2(x_2 + 1) + x_2^2(1 + x_1) + x_1 + x_2} \\ s.t. & 2x_1 - x_1^2 \geq 0,\ 2x_2 - x_2^2 \geq 0, \\ & 4 - x_1 x_2 \geq 0,\ x_1^2 + x_2^2 - \frac{1}{2} \geq 0. \end{cases}$$

*The moment relaxation (12.2.8) can be implemented in* GloptiPoly *as follows:*

```
mpol('x', 2);
f = x(1)^3+x(2)^3+3*x(1)*x(2)+1;
g = x(1)^2*(x(2)+1)+x(2)^2*(1+x(1))+x(1)+x(2);
K = [2*x(1)-x(1)^2>=0, 2*x(2)-x(2)^2>=0,
     4-x(1)*x(2) >=0,  x(1)^2+x(2)^2-1/2>=0];
k = 2; % give a value for the relaxation order
MO = msdp( min(mom(f)), [mom(g)==1],K, k);
[status, rmom_k] = msol(MO);
```

*For the order $k = 2$, both (12.2.7) and (12.2.8) return the lower bound 1, which equals the minimum value $\vartheta_{min} = 1$. The syntax* double(x) *returns three minimizers* $(0,1)$, $(1,0)$, $(1,1)$.

Rational function optimization has broad applications. For instance, finding nearest common divisors of two univariate polynomials can be formulated as a rational optimization problem (see [140, 212]). Let $p(z), q(z) \in \mathbb{C}[z]_m$ be two univariate polynomials such that

$$
\begin{aligned}
p(z) &:= z^m + p_{m-1}z^{m-1} + p_{m-2}z^{m-2} + \cdots + p_1 z + p_0, \\
q(z) &:= z^m + q_{m-1}z^{m-1} + q_{m-2}z^{m-2} + \cdots + q_1 z + q_0.
\end{aligned}
$$

If $p(z), q(z)$ have common divisors, their greatest common divisor (GCD) can be computed by the Euclidean algorithm. If they do not, we look for the smallest perturbations in their coefficients such that the perturbed ones have a common divisor.

We measure the perturbations on $p(z), q(z)$ by the 2-norm of their coefficients, i.e.,

$$
\|p\| := \sqrt{\sum_{k=0}^{m-1} |p_k|^2}, \quad \|q\| := \sqrt{\sum_{k=0}^{m-1} |q_k|^2}.
$$

Suppose the perturbed polynomials have the form

$$
\begin{aligned}
\hat{p}(z) &= z^m + \hat{p}_{m-1}z^{m-1} + \hat{p}_{m-2}z^{m-2} + \cdots + \hat{p}_1 z + \hat{p}_0, \\
\hat{q}(z) &= z^m + \hat{q}_{m-1}z^{m-1} + \hat{q}_{m-2}z^{m-2} + \cdots + \hat{q}_1 z + \hat{q}_0,
\end{aligned}
$$

with a common zero $a$, i.e., $\hat{p}(a) = \hat{q}(a) = 0$. The coefficient perturbations are measured as

$$
\mathcal{N}(a, \hat{p}, \hat{q}) := \sum_{i=0}^{m-1} |p_i - \hat{p}_i|^2 + \sum_{j=0}^{m-1} |q_j - \hat{q}_j|^2.
$$

We look for $(a, \hat{p}, \hat{q})$ such that $\mathcal{N}(a, \hat{p}, \hat{q})$ is minimum, subject to $\hat{p}(a) = \hat{q}(a) = 0$. For fixed $a$, the perturbation $\mathcal{N}(a, \hat{p}, \hat{q})$ is a convex quadratic function in $(\hat{p}, \hat{q})$, subject to linear equality constraints. It was shown in [140] that

$$
\min_{(\hat{p},\hat{q}):\hat{p}(a)=\hat{q}(a)=0} \mathcal{N}(a, \hat{p}, \hat{q}) = \frac{|p(a)|^2 + |q(a)|^2}{\sum_{i=0}^{m-1} |a^2|^i}. \tag{12.2.15}
$$

Therefore we get the rational optimization problem

$$
\min_{a\in\mathbb{C}} \frac{|p(a)|^2 + |q(a)|^2}{\sum_{i=0}^{m-1} |a^2|^i}. \tag{12.2.16}
$$

If we let $a = x_1 + \sqrt{-1}x_2$, with $x_1, x_2 \in \mathbb{R}^1$, then (12.2.16) is a rational optimization problem in $x = (x_1, x_2)$. We refer to [212] for more details about optimization for nearest GCDs.

### 12.2.1 ▪ Exercises

**Exercise 12.2.1.** *([212]) Solve the following optimization problem:*

$$\begin{cases} \min\limits_{x \in \mathbb{R}^3} & \frac{x_1^4 + x_2^4 + x_3^4 + x_1^2 + x_2^2 + x_3^2 + 2x_1 x_2 x_3 (x_1 + x_2 + x_3)}{x_1^3 + x_2^3 + x_3^3 + 2x_1 x_2 x_3} \\ s.t. & x_1^4 + x_2^4 + x_3^4 = 1 + x_1^2 x_2^2 + x_2^2 x_3^2 + x_3^2 x_1^2, \\ & x_3 \geq x_2 \geq x_1 \geq 0. \end{cases}$$

**Exercise 12.2.2.** *([212]) Solve the following optimization problem:*

$$\begin{cases} \min\limits_{x \in \mathbb{R}^4} & \frac{x_1^2 + x_2^2 + x_3^2 + x_4^2 + 2(x_2 + x_3 + x_1 x_3 + x_1 x_4 + x_2 x_4) + 1}{x_1 + x_4 + x_1 x_2 + x_2 x_3 + x_3 x_4} \\ s.t. & x_1^2 + x_2^2 - 2x_3 x_4 = 0, \\ & 4 - x_1^2 - x_2^2 - x_3^2 - x_4^2 \geq 0, \\ & (x_1, x_2, x_3, x_4) \geq 0. \end{cases}$$

**Exercise 12.2.3.** *Find the smallest perturbation such that the following two polynomials have a common complex divisor:*

$$p(z) = z^3 - 6z^2 + 11z - 6, \quad q(z) = z^3 - 6.24z^2 + 10.75z - 6.50.$$

**Exercise 12.2.4.** *Find the smallest perturbation such that the following two polynomials have a common real divisor:*

$$p(z) = z^3 + z^2 - 2, \quad q(z) = z^3 + 1.5z^2 + 1.5z - 1.25.$$

**Exercise 12.2.5.** *Prove the equalities in (12.2.12), (12.2.13), and (12.2.15).*

## 12.3 ▪ The Hierarchy of Local Minimums

A polynomial function may have none, one or several, or infinitely many local minimizers. However, the set of all local minimum values is always finite. This is because there are only finitely many critical values for every polynomial (see Theorem 6.1.1). If they exist, we can order the local minimum values of a polynomial $f$ monotonically as

$$\nu_1 < \nu_2 < \cdots < \nu_N.$$

The value $\nu_r$ is the $r$th smallest local minimum value of $f$. The ordered tuple $(\nu_1, \ldots, \nu_N)$ is called the *hierarchy of local minimums*.

   If $u$ is a local minimizer of $f$, then it must satisfy the first and second order necessary conditions:

$$\nabla f(u) = 0, \quad \nabla^2 f(u) \succeq 0. \qquad (12.3.1)$$

If $\nabla f(u) = 0$ and $\nabla^2 f(u) \succ 0$ (positive definite), then $u$ is a strict local minimizer. If $\nabla f(u) = 0$ but $\nabla^2 f(u) \succeq 0$ is singular, we cannot conclude the local optimality. For such cases, higher order derivatives are required to make a judgement. Indeed, it is NP-hard to check whether a critical point is a local minimizer or not (cf. [201]). To see this, consider the special case that $f$ is a form of degree 4. The origin 0 is always a critical point of such $f$, while 0 is a local minimizer if and only if $f$ is nonnegative everywhere. So, checking the local optimality at 0 is equivalent to checking whether $f$ is nonnegative everywhere. It is well known that the latter question is NP-hard.

To compute the hierarchy of local minimums, we introduce the notion of $H$-minimizers, which are also referenced as *second order critical* (or *stationary*) points in the literature.

**Definition 12.3.1.** *([231]) A point $u$ satisfying (12.3.1) is said to be an H-minimizer, and the value $f(u)$ is called an H-minimum.*

For polynomials whose coefficients are generic, the $H$-minimizers are the same as local minimizers. This is shown in §5.5. For each degree $d$, there exists a Zariski closed subset $Z$ of $\mathbb{R}[x]_d$ such that for all $f \in \mathbb{R}[x]_d \backslash Z$, a point $u$ is a local minimizer of $f$ if and only if it is an $H$-minimizer. The set $\mathbb{R}[x]_d \backslash Z$ is open dense in $\mathbb{R}[x]_d$. For special polynomials, there might be no local minimizers, even if they are bounded below. For instance, the polynomial $x_1^2 + (x_1 x_2 - 1)^2$ is bounded below, but it has no local minimizers.

Since each $H$-minimum is a critical value, the set of $H$-minimums is also finite. We order them monotonically as

$$f_1 < f_2 < \cdots < f_{N_1}.$$

The value $f_r$ is the $r$th smallest $H$-minimum of $f$. The number $N_1$ is at least equal to the number of local minimum values. We can compute $f_1$ through $f_{N_1}$ sequentially, from the smallest to the biggest. This is done in the work [231].

The smallest $H$-minimum $f_1$ equals the optimal value of the constrained optimization

$$\begin{cases} \min & f(x) \\ s.t. & \nabla f(x) = 0, \\ & \nabla^2 f(x) \succeq 0. \end{cases} \tag{12.3.2}$$

There is a polynomial matrix inequality constraint. We can apply the hierarchy of Moment-SOS relaxations introduced in §10.4. For each $k = 1, 2, \ldots$, the $k$th order moment relaxation is

$$\begin{cases} \min & \langle f, y \rangle \\ s.t. & \mathscr{V}_{f_{x_i}}^{(2k)}[y] = 0 \, (i = 1, \ldots, n), \\ & L_{\nabla^2 f}^{(k)}[y] \succeq 0, \\ & M_k[y] \succeq 0, \\ & y_0 = 1, \, y \in \mathbb{R}^{\mathbb{N}_{2k}^n}. \end{cases} \tag{12.3.3}$$

In the above, each $f_{x_i}$ denotes the partial derivative of $f$ with respect to $x_i$. The dual optimization of (12.3.3) is the SOS relaxation

$$\begin{cases} \max & \eta \\ s.t. & f - \eta \in \text{Ideal}[\nabla f]_{2k} + \text{QM}[\nabla^2 f]_{2k}. \end{cases} \tag{12.3.4}$$

We refer to (10.2.4) for the notation $\text{QM}[\nabla^2 f]_{2k}$ of truncated quadratic module for the Hessian matrix polynomial. Let $\vartheta_k^{(1)}, \eta_k^{(1)}$ denote the optimal values of (12.3.3), (12.3.4) respectively for the relaxation order $k$. The convergence properties of the hierarchy of (12.3.3)-(12.3.4) are summarized as follows.

**Theorem 12.3.2.** *([231]) Let $\mathcal{H}(f)$ denote the set of H-minimizers of $f$ and let $f_1$ be the smallest H-minimum, if it exists. Let*

$$Q_1 := \text{Ideal}[\nabla f] + \text{QM}[\nabla^2 f].$$

(i) *If (12.3.3) is infeasible for some $k$, then $\mathcal{H}(f) = \emptyset$ and hence $f$ has no local minimizers.*

(ii) *If $Q_1$ is archimedean and $\mathcal{H}(f) = \emptyset$, then (12.3.4) is unbounded above and (12.3.3) is infeasible for all $k$ big enough.*

(iii) *If $Q_1$ is archimedean and $\mathcal{H}(f) \neq \emptyset$, then $\vartheta_k^{(1)} = \eta_k^{(1)} = f_1$ for all $k$ big enough.*

(iv) *If $V_{\mathbb{R}}(\nabla f)$ is finite and $\mathcal{H}(f) \neq \emptyset$, then every optimizer $y^*$ of (12.3.3) has a flat truncation, when $k$ is big enough.*

The above conclusions can be shown by applying the matrix Positivstellensatz, i.e., Theorem 10.1.5. When the polynomial $f$ has generic coefficients, the set $V_{\mathbb{R}}(\nabla f)$ is finite (see §5.5.2). That is, the assumption that $V_{\mathbb{R}}(\nabla f)$ is finite is satisfied generically. We refer to §4.2 and §10.4 for how to extract minimizers.

Suppose the $r$th smallest $H$-minimum $f_r$ is obtained for some $r \geq 1$. We discuss how to check whether the next bigger $H$-minimum $f_{r+1}$ exists or not. For a scalar $\delta > 0$, consider the optimization

$$\begin{cases} \min & f(x) \\ s.t. & \nabla f(x) = 0, \\ & \nabla^2 f(x) \succeq 0, \\ & f(x) \geq f_r + \delta. \end{cases} \tag{12.3.5}$$

Let $h^+(f_r + \delta)$ denote the optimal value of (12.3.5). Note that $h^+(f_r + \delta)$ is the smallest $H$-minimum greater than or equal to $f_r + \delta$. So, if $f_{r+1}$ exists,

$$0 < \delta \leq f_{r+1} - f_r \quad \Rightarrow \quad f_{r+1} = h^+(f_r + \delta).$$

Similarly, we solve (12.3.5) by the moment relaxations:

$$\begin{cases} \min & \langle f, y \rangle \\ s.t. & \mathscr{V}_{f_{x_i}}^{(2k)}[y] = 0 \, (i = 1, \ldots, n), \\ & L_{\nabla^2 f}^{(k)}(y) \succeq 0, \\ & L_{f - f_r - \delta}^{(k)}(y) \succeq 0, \\ & M_k[y] \succeq 0, \\ & y_0 = 1, \, y \in \mathbb{R}^{\mathbb{N}_{2k}^n}. \end{cases} \tag{12.3.6}$$

The dual optimization is the $k$th order SOS relaxation

$$\begin{cases} \max & \gamma \\ s.t. & f - \gamma \in \text{Ideal}[\nabla f]_{2k} + \text{QM}[f - f_r - \delta, \nabla^2 f]_{2k}. \end{cases} \tag{12.3.7}$$

Let $\vartheta_k^{(r+1)}$, $\eta_k^{(r+1)}$ denote the optimal values of (12.3.6), (12.3.7) respectively, for the relaxation order $k$. The following are the convergence properties.

**Theorem 12.3.3.** *([231]) Suppose $f_r$ is the $r$th smallest $H$-minimum of $f$ and $f_{\max}$ is the biggest one of them. Let*

$$Q_2 := \text{Ideal}[\nabla f] + \text{QM}[f - f_r - \delta, \nabla^2 f].$$

(i) *If (12.3.6) is infeasible for some $k$, then (12.3.5) is infeasible and*

$$f_r + \delta > f_{\max}.$$

(ii) *If $Q_2$ is archimedean and $f_r + \delta > f_{\max}$, then (12.3.7) is unbounded above and (12.3.6) is infeasible for all $k$ big enough.*

(iii) *If $Q_2$ is archimedean and $f_r + \delta \leq f_{\max}$, then for all $k$ big enough*

$$\vartheta_k^{(r+1)} = \eta_k^{(r+1)} = h^+(f_r + \delta).$$

(iv) *If the set $V_{\mathbb{R}}(\nabla f) \cap \{f(x) \geq f_r + \delta\}$ is finite and $f_r + \delta \leq f_{\max}$, then for all $k$ big enough, every optimizer $y^*$ of (12.3.6) has a flat truncation.*

The remarks after Theorem 12.3.2 also apply here. Now we discuss how to detect existence of $f_{r+1}$. To this end, we solve the maximization problem

$$\begin{cases} \max & f(x) \\ s.t. & \nabla f(x) = 0, \\ & \nabla^2 f(x) \succeq 0, \\ & f_r + \delta - f(x) \geq 0. \end{cases} \tag{12.3.8}$$

Let $h^-(f_r + \delta)$ denote the optimal value of (12.3.8). The following is the criterion for selecting $\delta$ such that $0 < \delta < f_{r+1} - f_r$ when $f_{r+1}$ exists.

**Lemma 12.3.4.** *([231]) For $\delta > 0$, $h^-(f_r + \delta) = f_r$ if and only if $\delta < f_{r+1} - f_r$.*

The optimal value $h^-(f_r + \delta)$ can be similarly computed by the Moment-SOS hierarchy. The convergence properties are similar to those in Theorems 12.3.2 and 12.3.3. The following is the algorithm for determining $f_{r+1}$.

**Algorithm 12.3.5.** *Determine $f_{r+1}$ as follows:*

**Step 0** *Choose a small positive value for $\delta$ (e.g., 0.01).*

**Step 1** *Compute the optimal value $h^-(f_r + \delta)$ of (12.3.8).*

**Step 2** *Solve the hierarchy of semidefinite relaxations (12.3.6).*

(i) *If (12.3.6) is infeasible for some $k$ and $h^-(f_r + \delta) = f_r$, then $f_r = f_{\max}$ and stop.*

(ii) *If (12.3.6) is infeasible for some $k$ but $h^-(f_r + \delta) > f_r$, then decrease the value of $\delta$ (e.g., let $\delta := \delta/2$) and go to Step 1.*

(iii) *If (12.3.6) is feasible for all $k$, we have $h^+(f_r + \delta)$ for all $k$ big enough. If $h^-(f_r + \delta) = f_r$, then $f_{r+1} = h^+(f_r + \delta)$ and stop; otherwise, decrease the value of $\delta$ (e.g., let $\delta := \delta/2$) and go to Step 1.*

Once $f_{r+1}$ is obtained, one can use the same algorithm to determine the existence and value of $f_{r+2}$. This process can be repeated, until we get all $H$-minimums. After $H$-minimums are obtained, we can determine the hierarchy of local minimums. Note that a point $u$ is a local minimizer if and only if $f(u)$ equals the minimum value of the constrained optimization

$$\begin{cases} \min & f(x) \\ s.t. & \|x - u\|^2 \leq \rho \end{cases} \tag{12.3.9}$$

for some radius $\rho > 0$ small enough. The optimization (12.3.9) can be solved by tight relaxation methods in §6.3 or §6.5. We refer to [231] for more details. The following is an example for the hierarchy of local minimums.

**Example 12.3.6.** *([231]) Consider the polynomial $f$ that is*

$$x_1^6 + x_2^6 + x_3^6 + x_4^6 - 5(x_1^3 x_2^2 + x_2^2 x_3^3 + x_3 x_4^4) + 6(x_1^2 x_2^2 + x_3^3 x_4 + x_1 x_2 x_3 x_4)$$
$$-7(x_1 x_2 x_3 + x_2 x_3 x_4) + (x_1 + x_2 + x_3 + x_4 - 1)^2 - 1.$$

*There are a total of 5 $H$-minimums $f_1, \ldots, f_5$, which are as follows:*

| $r$ | $f_r$ | $H$-minimizers | local optimality |
|---|---|---|---|
| 1 | $-1813.2169$ | $(3.0149, 3.3618, 3.7667, -3.7482)$ | minimizer |
| 2 | $-1515.4286$ | $(-1.1245, -3.0510, 3.6415, -3.6848)$ | minimizer |
| 3 | $-140.8532$ | $(-0.6017, 2.2670, 2.4317, 2.7935)$ | minimizer |
| 4 | $-62.7880$ | $(2.2031, -2.3876, 2.4169, 2.7577)$ | minimizer |
| 5 | $-4.3786$ | $(0.8653, -0.3392, -1.2499, 0.7930)$ | minimizer |

*They are all local minimums, because $\nabla^2 f \succ 0$ at the $H$-minimizers. The local minimum values are $f_1, f_2, \ldots, f_5$.*

### 12.3.1 ▪ Exercises

**Exercise 12.3.1.** *Determine the hierarchy of $H$-minimums and the hierarchy of local minimums for the following polynomials:*

(i) $x_1^2 + (x_1 x_2 - 1)^2$.

(ii) $2x_2^4(x_1 + x_2)^4 + x_2^2(x_1 + x_2)^2 + 2x_2(x_1 + x_2) + x_2^2$.

(iii) $1 + x_1^4 x_2^2 + x_1^2 x_2^4 - 3x_1^2 x_2^2$.

(iv) $21x_2^2 - 92x_1 x_3^2 - 70x_2^2 x_3 - 95x_1^4 - 47x_1 x_3^3 + 51x_2^2 x_3^2 + 47x_1^5 + 5x_1 x_2^4 + 33x_3^5$.

**Exercise 12.3.2.** *Prove Theorems 12.3.2, 12.3.3 and Lemma 12.3.4.*

## 12.4 ▪ Saddle Point Problems

Let $X \subseteq \mathbb{R}^n, Y \subseteq \mathbb{R}^m$ be two sets and let $F(x, y)$ be a continuous function in $(x, y) \in X \times Y$. A pair $(x^*, y^*) \in X \times Y$ is said to be a *saddle point* for $F$ over $X \times Y$ if

$$F(x^*, y) \leq F(x^*, y^*) \leq F(x, y^*) \tag{12.4.1}$$

for all $(x, y) \in X \times Y$. If $(x^*, y^*)$ is a saddle point, one can further show that

$$\min_{x \in X} \max_{y \in Y} F(x, y) = F(x^*, y^*) = \max_{y \in Y} \min_{x \in X} F(x, y). \tag{12.4.2}$$

If it exists, a saddle point can solve the above min-max and max-min optimization simultaneously. The saddle point problem (SPP) concerns existence and computation of saddle points.

When $F$ is convex-concave in $(x, y)$ and $X, Y$ are nonempty compact convex sets, there exists a saddle point (see [24, §2.6]). For convex-concave type SPPs, there exist computational methods based on gradients, subgradients, or variational inequalities (see [116, 117, 203]). When $F$ is not convex-concave or one of $X, Y$ is nonconvex, a saddle point may not exist.

We discuss how to solve saddle point problems when $F, X, Y$ are given by polynomials. If there exists a saddle point, we aim at finding one; if there is none, we look for a certificate for the nonexistence. The results in this section are mostly from the work [245].

Assume $X, Y$ are semialgebraic sets given as

$$X = \left\{ x \in \mathbb{R}^n \ \middle| \ \begin{array}{l} g_i(x) = 0 \ (i \in \mathcal{E}_{11}), \\ g_i(x) \geq 0 \ (i \in \mathcal{E}_{12}) \end{array} \right\}, \tag{12.4.3}$$

$$Y = \left\{ y \in \mathbb{R}^m \ \middle| \ \begin{array}{l} h_j(y) = 0 \ (j \in \mathcal{E}_{21}), \\ h_j(y) \geq 0 \ (j \in \mathcal{E}_{22}) \end{array} \right\}, \tag{12.4.4}$$

where each $g_i$ is a polynomial in $x := (x_1, \ldots, x_n)$ and each $h_j$ is a polynomial in $y := (y_1, \ldots, y_m)$. The $\mathcal{E}_{11}, \mathcal{E}_{12}, \mathcal{E}_{21}, \mathcal{E}_{22}$ are labeling sets of finite cardinalities (possibly empty). For notational convenience, we denote

$$\begin{array}{l|l|l} g_{eq} := (g_i)_{i \in \mathcal{E}_{11}}, & h_{eq} := (h_j)_{j \in \mathcal{E}_{21}}, & \mathcal{E}_1 := \mathcal{E}_{11} \cup \mathcal{E}_{12}, \\ g_{in} := (g_i)_{i \in \mathcal{E}_{12}}, & h_{in} := (h_j)_{j \in \mathcal{E}_{22}}, & \mathcal{E}_2 := \mathcal{E}_{21} \cup \mathcal{E}_{22}. \end{array} \tag{12.4.5}$$

If $(x^*, y^*)$ is a saddle point, then $x^*$ is a minimizer of

$$\left\{ \begin{array}{ll} \min & F(x, y^*) \\ s.t. & g_i(x) = 0 \ (i \in \mathcal{E}_{11}), \\ & g_i(x) \geq 0 \ (i \in \mathcal{E}_{12}) \end{array} \right. \tag{12.4.6}$$

and $y^*$ is a maximizer of

$$\left\{ \begin{array}{ll} \max & F(x^*, y) \\ s.t. & h_j(y) = 0 \ (j \in \mathcal{E}_{21}), \\ & h_j(y) \geq 0 \ (j \in \mathcal{E}_{22}). \end{array} \right. \tag{12.4.7}$$

If the LICQC (see §5.1) holds for (12.4.6)-(12.4.7), there exist Lagrange multipliers $\lambda_i^*, \mu_j^*$ such that

$$\begin{array}{l} \nabla_x F(x^*, y^*) = \sum_{i \in \mathcal{E}_1} \lambda_i^* \nabla_x g_i(x^*), \\ 0 \leq \lambda_i^* \perp g_i(x^*) \geq 0 \ (i \in \mathcal{E}_{12}), \end{array} \tag{12.4.8}$$

$$\begin{array}{l} \nabla_y F(x^*, y^*) = \sum_{j \in \mathcal{E}_2} \mu_j^* \nabla_y h_j(y^*), \\ 0 \geq \mu_j^* \perp h_j(y^*) \geq 0 \ (j \in \mathcal{E}_{22}). \end{array} \tag{12.4.9}$$

In the above, $a \perp b$ means that the product $a \cdot b = 0$, and $\nabla_x$ (resp., $\nabla_y$) denotes the gradient of a function with respect to $x$ (resp., $y$). For convenience, suppose the labeling sets and constraining polynomial tuples are such that

$$\begin{array}{l} \mathcal{E}_1 = \{1, \ldots, \ell_1\}, \quad \mathcal{E}_2 = \{1, \ldots, \ell_2\}, \\ g = (g_1, \ldots, g_{\ell_1}), \quad h = (h_1, \ldots, h_{\ell_2}). \end{array}$$

Then $x^*, y^*, \lambda_i^*, \mu_j^*$ satisfy the equations

$$\underbrace{\begin{bmatrix} \nabla_x g_1(x) & \nabla_x g_2(x) & \cdots & \nabla_x g_{\ell_1}(x) \\ g_1(x) & 0 & \cdots & 0 \\ 0 & g_2(x) & \cdots & 0 \\ \vdots & \vdots & \ddots & \vdots \\ 0 & 0 & \cdots & g_{\ell_1}(x) \end{bmatrix}}_{G(x)} \underbrace{\begin{bmatrix} \lambda_1 \\ \vdots \\ \lambda_{\ell_1} \end{bmatrix}}_{\lambda} = \underbrace{\begin{bmatrix} \nabla_x F(x, y) \\ 0 \\ \vdots \\ 0 \end{bmatrix}}_{F_1(x,y)}, \tag{12.4.10}$$

$$
\underbrace{\begin{bmatrix} \nabla_y h_1(y) & \nabla_y h_2(y) & \cdots & \nabla_y h_{\ell_2}(y) \\ h_1(y) & 0 & \cdots & 0 \\ 0 & h_2(y) & \cdots & 0 \\ \vdots & \vdots & \ddots & \vdots \\ 0 & 0 & \cdots & h_{\ell_2}(y) \end{bmatrix}}_{H(y)} \underbrace{\begin{bmatrix} \mu_1 \\ \vdots \\ \mu_{\ell_2} \end{bmatrix}}_{\mu} = \underbrace{\begin{bmatrix} \nabla_y F(x, y) \\ 0 \\ \vdots \\ 0 \end{bmatrix}}_{F_2(x,y)}.
\tag{12.4.11}
$$

As shown in §6.2, when $g, h$ are nonsingular polynomial tuples, there exist matrix polynomials $G_1(x), H_1(y)$ such that

$$
G_1(x)G(x) = I_{\ell_1}, \quad H_1(y)H(y) = I_{\ell_2}.
\tag{12.4.12}
$$

Then, one can express Lagrange multiplier vectors as

$$
\boxed{\begin{aligned} \lambda^* &= G_1(x^*)F_1(x^*, y^*), \\ \mu^* &= H_1(y^*)F_2(x^*, y^*). \end{aligned}}
$$

For convenience, we denote the Lagrange multiplier vector polynomials

$$
\boxed{\begin{aligned} \lambda(x, y) &:= G_1(x)F_1(x, y), \\ \mu(x, y) &:= H_1(y)F_2(x, y). \end{aligned}}
\tag{12.4.13}
$$

Write entries of the Lagrange multiplier vectors $\lambda(x, y), \mu(x, y)$ as

$$
\boxed{\begin{aligned} \lambda(x, y) &= \big(\lambda_1(x, y), \ldots, \lambda_{\ell_1}(x, y)\big), \\ \mu(x, y) &= \big(\mu_1(x, y), \ldots, \mu_{\ell_2}(x, y)\big). \end{aligned}}
$$

At each saddle point $(x^*, y^*)$, the Lagrange multiplier vectors $\lambda^*, \mu^*$ in (12.4.8)-(12.4.9) can be expressed as

$$
\lambda^* = \lambda(x^*, y^*), \quad \mu^* = \mu(x^*, y^*).
$$

Therefore, every saddle point $(x^*, y^*)$ is a solution to the polynomial system

$$
\boxed{\begin{aligned} & g_i(x) = 0 \, (i \in \mathcal{E}_{11}), \quad h_j(y) = 0 \, (j \in \mathcal{E}_{21}), \\ & \nabla_x F(x, y) = \sum_{i \in \mathcal{E}_1} \lambda_i(x, y) \nabla_x g_i(x), \\ & \nabla_y F(x, y) = \sum_{j \in \mathcal{E}_2} \mu_j(x, y) \nabla_j h_j(x), \\ & 0 \le \lambda_i(x, y) \perp g_i(x) \ge 0 \, (i \in \mathcal{E}_{12}), \\ & 0 \ge \mu_j(x, y) \perp h_j(y) \ge 0 \, (j \in \mathcal{E}_{22}). \end{aligned}}
\tag{12.4.14}
$$

However, not every solution to (12.4.14) is a saddle point, since the KKT conditions are typically not sufficient for ensuring optimality. To get a saddle point, we consider the new optimization problem

$$
\left\{ \begin{aligned} \min_{x \in X, y \in Y} \quad & F(x, y) \\ s.t. \quad & \nabla_x F(x, y) - \sum_{i \in \mathcal{E}_1} \lambda_i(x, y) \nabla_x g_i(x) = 0, \\ & \nabla_y F(x, y) - \sum_{j \in \mathcal{E}_2} \mu_j(x, y) \nabla_y h_j(y) = 0, \\ & 0 \le \lambda_i(x, y) \perp g_i(x) \ge 0 \, (i \in \mathcal{E}_{12}), \\ & 0 \ge \mu_j(x, y) \perp h_j(y) \ge 0 \, (j \in \mathcal{E}_{22}). \end{aligned} \right.
\tag{12.4.15}
$$

Suppose $(x^*, y^*)$ is a minimizer of (12.4.15). If $x^*$ is a minimizer of $F(x, y^*)$ over $X$ and $y^*$ is a maximizer of $F(x^*, y)$ over $Y$, then $(x^*, y^*)$ is a saddle point. Otherwise, $(x^*, y^*)$ is not a saddle point, i.e., there exists $u \in X$ and/or there exists $v \in Y$ such that

$$F(u, y^*) - F(x^*, y^*) < 0, \quad \text{and/or} \quad F(x^*, v) - F(x^*, y^*) > 0.$$

The points $u, v$ can be used to give new constraints

$$F(u, y) - F(x, y) \geq 0, \quad F(x, y) - F(x, v) \geq 0. \tag{12.4.16}$$

Every saddle point $(x, y)$ must satisfy (12.4.16), so it can be safely added to (12.4.15) without losing any saddle points. For generic polynomials $F, g, h$, the problem (12.4.15) has only finitely many feasible points (see [245]). Therefore, by repeatedly adding new inequalities like (12.4.16), we can eventually get a saddle point or certify nonexistence of saddle points. This results in the following algorithm.

**Algorithm 12.4.1.** *Let $K_1 = K_2 = S_a := \emptyset$.*

**Step 1** *If the optimization (12.4.15) is infeasible, then there is no saddle point and stop; otherwise, solve (12.4.15) for a set $K^{(0)}$ of minimizers. Let $k := 0$.*

**Step 2** *For each $(x^*, y^*) \in K^{(k)}$, do the following:*

(a) *Solve the minimization problem*

$$\begin{cases} \min\limits_{x \in X} & F(x, y^*) \\ s.t. & \nabla_x F(x, y^*) - \sum\limits_{i \in \mathcal{E}_1} \lambda_i(x, y) \nabla_x g_i(x^*) = 0, \\ & 0 \leq \lambda_i(x, y^*) \perp g_i(x) \geq 0 \, (i \in \mathcal{E}_{12}) \end{cases} \tag{12.4.17}$$

*and get a set of minimizers $S_1(y^*)$. Let $\vartheta_1(y^*)$ denote the minimum value of (12.4.17). If $F(x^*, y^*) > \vartheta_1(y^*)$, update $K_1 := K_1 \cup S_1(y^*)$.*

(b) *Solve the maximization problem*

$$\begin{cases} \max\limits_{y \in Y} & F(x^*, y) \\ s.t. & \nabla_y F(x^*, y) - \sum\limits_{j \in \mathcal{E}_2} \mu_j(x^*, y) \nabla_y h_j(y) = 0, \\ & 0 \geq \mu_j(x^*, y) \perp h_j(y) \geq 0 (j \in \mathcal{E}_{22}) \end{cases} \tag{12.4.18}$$

*and get a set of maximizers $S_2(x^*)$. Let $\vartheta_2(x^*)$ denote the maximum value of (12.4.18). If $F(x^*, y^*) < \vartheta_2(x^*)$, update $K_2 := K_2 \cup S_2(x^*)$.*

(c) *If $\vartheta_1(y^*) = F(x^*, y^*) = \vartheta_2(x^*)$, update $S_a := S_a \cup \{(x^*, y^*)\}$.*

**Step 3** *If $S_a \neq \emptyset$, then each one in $S_a$ is a saddle point and stop; otherwise go to Step 4.*

**Step 4** *Solve the minimization problem*

$$\begin{cases} \min\limits_{x \in X, y \in Y} & F(x, y) \\ s.t. & \nabla_x F(x, y) - \sum\limits_{i \in \mathcal{E}_1} \lambda_i(x, y) \nabla_x g_i(x) = 0, \\ & \nabla_y F(x, y) - \sum\limits_{j \in \mathcal{E}_2} \mu_j(x, y) \nabla_y h_j(y) = 0, \\ & 0 \leq \lambda_i(x, y) \perp g_i(x) \geq 0 \, (i \in \mathcal{E}_{12}), \\ & 0 \geq \mu_j(x, y) \perp h_j(y) \geq 0 \, (j \in \mathcal{E}_{22}), \\ & F(u, y) - F(x, y) \geq 0 \, (u \in K_1), \\ & F(x, v) - F(x, y) \leq 0 \, (v \in K_2). \end{cases} \tag{12.4.19}$$

*If the optimization (12.4.19) is infeasible, then there is no saddle point and stop; otherwise, compute a set $K^{(k+1)}$ of optimizers for (12.4.19). Let $k := k + 1$ and go to Step 2.*

For generic polynomials, the feasible set $\mathcal{K}_0$ of (12.4.15), as well as each $K^{(k)}$ in Algorithm 12.4.1, is finite. The convergence properties of Algorithm 12.4.1 are summarized as follows.

**Theorem 12.4.2.** *([245]) Assume the Lagrange multipliers can be expressed as in (12.4.13). Let $\mathcal{K}_0$ be the feasible set of (12.4.15), let $S$ be the set of saddle points for (12.4.1), and let $S_a$ be the set produced by Algorithm 12.4.1. If the complement set of $S$ in $\mathcal{K}_0$ (i.e., the set $\mathcal{K}_0 \setminus S$) is finite, then Algorithm 12.4.1 must terminate within finitely many loops. Moreover, if $S_a \neq \emptyset$, then each $(x^*, y^*) \in S_a$ is a saddle point; if $S_a = \emptyset$ after Algorithm 12.4.1 terminates, then there is no saddle point.*

The polynomial optimization problems in Algorithm 12.4.1 can be solved by Moment-SOS relaxations introduced in Chapter 5. They also have finite convergence under some genericity assumptions (see [245]).

**Example 12.4.3.** *([245]) Consider the following sets and the function:*

$$\begin{aligned} X &= \{x \in \mathbb{R}^3 : x_1 \geq 0, x_1 x_2 \geq 1, x_2 x_3 \geq 1\}, \\ Y &= \{y \in \mathbb{R}^3 : y_1 \geq 0, y_1 y_2 \geq 1, y_2 y_3 \geq 1\}, \\ F(x,y) &= x_1^3 y_1 + x_2^3 y_2 + x_3^3 y_3 - 3x_1 x_2 x_3 - y_1^2 - 2y_2^2 - 3y_3^2. \end{aligned}$$

*The Lagrange multipliers can be expressed as*

$$\lambda_1 = (1 - x_1 x_2)F_{x_1}, \quad \lambda_2 = x_1 F_{x_1}, \quad \lambda_3 = -x_1 F_{x_1} + x_2 F_{x_2}.$$

*Similar expressions hold for $\mu_j(x,y)$. After 9 loops by Algorithm 12.4.1, we get the saddle point:*

$$x^* = (1.2599, 1.2181, 1.3032), \quad y^* = (1.0000, 1.1067, 0.9036).$$

## 12.4.1 ▪ Exercises

**Exercise 12.4.1.** *Prove the equalities in (12.4.2).*

**Exercise 12.4.2.** *Determine saddle points for the following sets and functions:*

*(i) $X = Y = \Delta_4$ (the standard simplex in $\mathbb{R}^4$) and*

$$F(x,y) = \sum_{i,j=1}^4 x_i^2 y_j^2 - \sum_{i \neq j}(x_i x_j + y_i y_j).$$

*(ii) $X = Y = [0,1]^3$ and*

$$F(x,y) = \sum_{i=1}^3 (x_i + y_i) + \sum_{i<j}(x_i^2 y_j^2 - y_i^2 x_j^2).$$

*(iii)* $X = Y = [-1, 1]^3$ *and*

$$F(x, y) = \sum_{i=1}^{3}(x_i + y_i) - \prod_{i=1}^{3}(x_i - y_i).$$

*(iv)* $X = Y = \mathbb{S}^2$ *(the unit sphere in $\mathbb{R}^3$) and*

$$F(x, y) = x_1^3 + x_2^3 + x_3^3 + y_1^3 + y_2^3 + y_3^3 + 2(x_1 x_2 y_1 y_2 + x_1 x_3 y_1 y_3 + x_2 x_3 y_2 y_3).$$

## 12.5 ▪ Nash Equilibrium Problems

The Nash equilibrium problem (NEP) is to find strategies for a set of players such that each player's strategy is optimal, when the strategies for other players are fixed. Suppose there are $N$ players and the $i$th player's strategy vector is $x_i \in \mathbb{R}^{n_i}$. For convenience, we write that

$$x_i := (x_{i,1}, \ldots, x_{i,n_i}), \quad x := (x_1, \ldots, x_N).$$

The total dimension of all strategies is $n := n_1 + \cdots + n_N$. When the $i$th player's strategy $x_i$ is considered, the tuple of remaining players' strategy vectors is denoted as

$$x_{-i} := (x_1, \ldots, x_{i-1}, x_{i+1}, \ldots, x_N).$$

When $x_{-i}$ is given, the $i$th player aims at choosing his strategy $x_i$ to be an optimizer for his own decision optimization problem

$$\mathrm{F}_i(x_{-i}): \begin{cases} \min_{x_i \in \mathbb{R}^{n_i}} & f_i(x_i, x_{-i}) \\ s.t. & g_{ij}(x_i) = 0 \, (j \in \mathcal{E}_i), \\ & g_{ij}(x_i) \geq 0 \, (j \in \mathcal{I}_i). \end{cases} \tag{12.5.1}$$

The $i$th player's objective $f_i$ is a function in $(x_1, \ldots, x_N)$ while each constraining function $g_{ij}$ is in $x_i$. The $\mathcal{E}_i$ and $\mathcal{I}_i$ are disjoint labeling sets of finite cardinalities (possibly empty). Note that each $g_{ij}$ does not depend on $x_{-i}$. For convenience of describing the $i$th player's optimization, we also write that

$$x = (x_i, x_{-i}).$$

The feasible strategy set of the $i$th player's optimization $\mathrm{F}_i(x_{-i})$ is

$$X_i := \left\{ x_i \in \mathbb{R}^{n_i} \, \middle| \, \begin{array}{l} g_{ij}(x_i) = 0 \, (j \in \mathcal{E}_i), \\ g_{ij}(x_i) \geq 0 \, (j \in \mathcal{I}_i) \end{array} \right\}. \tag{12.5.2}$$

The tuple $x = (x_1, \ldots, x_N)$ is said to be a feasible point if each $x_i \in X_i$.

**Definition 12.5.1.** *A strategy tuple $x^* = (x_1^*, \ldots, x_N^*)$ is called a Nash Equilibrium (NE) if each $x_i^*$ is a minimizer of the $i$th player's optimization $\mathrm{F}_i(x_{-i}^*)$, where*

$$x_{-i}^* := (x_1^*, \ldots, x_{i-1}^*, x_{i+1}^*, \ldots, x_N^*).$$

The computational task of the NEP is

$$\boxed{\begin{array}{c} \text{Find a strategy tuple } x^* = (x_1^*, \ldots, x_N^*) \\ \text{such that each } x_i^* \text{ is an optimizer of } \mathrm{F}_i(x_{-i}^*). \end{array}} \tag{12.5.3}$$

We discuss how to solve NEPs when the functions $f_i$ and $g_{ij}$ are polynomials. The following are some examples of NEPs.

**Example 12.5.2.** *([249]) (i) Consider the 2-player NEP with*

$$F_1(x_{-1}) : \begin{cases} \min\limits_{x_1 \in \mathbb{R}^2} & x_{1,1}(x_{1,1} + x_{2,1} + 4x_{2,2}) + 2x_{1,2}^2 \\ s.t. & 1 - (x_{1,1})^2 - (x_{1,2})^2 \geq 0, \end{cases}$$

$$F_2(x_{-2}) : \begin{cases} \min\limits_{x_2 \in \mathbb{R}^2} & x_{2,1}(x_{1,1} + 2x_{1,2} + x_{2,1}) \\ & \quad + x_{2,2}(2x_{1,1} + x_{1,2} + x_{2,2}) \\ s.t. & 1 - (x_{2,1})^2 - (x_{2,2})^2 \geq 0. \end{cases}$$

*Each player's objective is strictly convex in its own strategy variable, since their Hessian's are positive definite. The KKT systems for $F_1(x_{-1})$ and $F_2(x_{-2})$ are respectively*

$$\begin{cases} 2x_{1,1} + x_{2,1} + 4x_{2,2} = -2\lambda_1 x_{1,1}, \\ 4x_{1,2} = -2\lambda_1 x_{1,2}, \\ 1 - (x_{1,1})^2 - (x_{1,2})^2 \geq 0, \\ \lambda_1(1 - (x_{1,1})^2 - (x_{1,2})^2) = 0, \ \lambda_1 \geq 0, \end{cases}$$

$$\begin{cases} x_{1,1} + 2x_{1,2} + 2x_{2,1} = -2\lambda_2 x_{2,1}, \\ 2x_{1,1} + x_{1,2} + 2x_{2,2} = -2\lambda_2 x_{2,2}, \\ 1 - (x_{2,1})^2 - (x_{2,2})^2 \geq 0, \\ \lambda_2(1 - (x_{2,1})^2 - (x_{2,2})^2) = 0, \ \lambda_2 \geq 0. \end{cases}$$

*By solving them directly, one can show that this NEP has only 3 NEs. These NEs and the corresponding Lagrange multipliers are*

| | | | |
|---|---|---|---|
| *1st NE:* | $x_1^* = (0,0),$ | $x_2^* = (0,0),$ | $\lambda_1^* = \lambda_2^* = 0;$ |
| *2nd NE:* | $x_1^* = (1,0),$ | $x_2^* = \frac{1}{\sqrt{5}}(-1,-2),$ | $\lambda_1^* = \frac{9\sqrt{5}}{10} - 1, \ \lambda_2^* = \frac{\sqrt{5}}{2} - 1;$ |
| *3rd NE:* | $x_1^* = (-1,0),$ | $x_2^* = \frac{1}{\sqrt{5}}(1,2),$ | $\lambda_1^* = \frac{9\sqrt{5}}{10} - 1, \ \lambda_2^* = \frac{\sqrt{5}}{2} - 1.$ |

*(ii) Consider the 2-player NEP with*

$$F_1(x_{-1}) : \begin{cases} \min\limits_{x_1 \in \mathbb{R}^2} & x_{1,1}^2 + 2x_{2,1}x_{1,1} + x_{1,2}^2 + 2x_{2,2}x_{1,2} \\ s.t. & x_1 \in [-1,1]^2, \end{cases}$$

$$F_2(x_{-2}) : \begin{cases} \min\limits_{x_2 \in \mathbb{R}^2} & x_{2,1}^2 + 2x_{1,1}x_{2,1} + x_{2,2}^2 + 2x_{1,2}x_{2,2} \\ s.t. & x_2 \in [-1,1]^2. \end{cases}$$

*For all $s,t \in [-1,1]$ and $x_1^* = (s,t)$, $x_2^* = (-s,-t)$, one can verify that $(x_1^*, x_2^*)$ is a NE. Note that each player's objective is strictly convex in its own strategy. There are infinitely many NEs for this NEP.*

When each strategy set $X_i$ is finite, the NEP is called a *finite game*. For finite games, Nash equilibria typically do not exist. So people are usually interested in mixed strategies, which are probability distributions on the finite sets of strategies. Mixed strategy solutions are usually targeted for finite games [202]. When

$$f_1 + \cdots + f_N = 0,$$

the NEP is called a *zero sum game*. A 2-player zero sum game is equivalent to a saddle point problem (see §12.4). We refer to [14, 66, 200, 256, 287] for classical work on NEPs. It is generally a hard task to solve NEPs [65, 291]. Recent work for solving NEPs can be found in [1, 67, 153, 157]. This section introduces the method in [249] for solving NEPs given by polynomial functions.

## 12.5.1 ▪ A hierarchy of relaxations

For the $i$th player's optimization $F_i(x_{-i})$, suppose the labels are such that

$$\begin{aligned} \mathcal{E}_i \cup \mathcal{I}_i &:= \{1, \ldots, m_i\}, \\ g_i(x_i) &:= (g_{i1}(x_i), \ldots, g_{im_i}(x_i)). \end{aligned}$$

Under the LICQC, if $x = (x_1, \ldots, x_N)$ is a NE, then there exist Lagrange multipliers $\lambda_{ij}$ such that for each $i = 1, \ldots, N$,

$$\begin{cases} \sum_{j=1}^{m_i} \lambda_{ij} \nabla_{x_i} g_{ij}(x_i) = \nabla_{x_i} f_i(x), \\ 0 \le \lambda_{ij} \perp g_{ij}(x_i) \ge 0 \, (j \in \mathcal{I}_i). \end{cases} \tag{12.5.4}$$

They are KKT conditions for the optimization (12.5.1). Such $x$ is called a KKT point for the NEP. This gives the equations ($i = 1, \ldots, N$)

$$\underbrace{\begin{bmatrix} \nabla_{x_i} g_{i1} & \nabla_{x_i} g_{i2} & \cdots & \nabla_{x_i} g_{im_i} \\ g_{i1}(x) & 0 & \cdots & 0 \\ 0 & g_{i2}(x) & \cdots & 0 \\ \vdots & \vdots & \ddots & \vdots \\ 0 & 0 & \cdots & g_{im_i}(x) \end{bmatrix}}_{G_i(x_i)} \underbrace{\begin{bmatrix} \lambda_{i1} \\ \lambda_{i2} \\ \vdots \\ \lambda_{im_i} \end{bmatrix}}_{\lambda_i} = \underbrace{\begin{bmatrix} \nabla_{x_i} f_i \\ 0 \\ \vdots \\ 0 \end{bmatrix}}_{\hat{f}_i(x)}. \tag{12.5.5}$$

If there exists a matrix polynomial $H_i(x_i)$ in $x_i$ such that

$$H_i(x_i) G_i(x_i) = I_{m_i}, \tag{12.5.6}$$

then we can express the Lagrange multiplier vector $\lambda_i$ as

$$\lambda_i = H_i(x_i) G_i(x_i) \lambda_i = H_i(x_i) \hat{f}_i(x).$$

The matrix polynomial $H_i(x_i)$ satisfying (12.5.6) exists if and only if the constraining tuple $g_i$ is nonsingular, i.e., $G_i(x_i)$ has full column rank for all $x_i \in \mathbb{C}^{n_i}$ (see Proposition 6.2.3). This is a generic assumption (see Theorem 6.2.4). When $g_i$ is nonsingular, the LICQC holds at every minimizer of (12.5.1), so there must exist $\lambda_{ij}$ satisfying (12.5.4) and $\lambda_{ij}$ can be expressed as

$$\lambda_{ij} = \lambda_{ij}(x) := \big( H_i(x_i) \hat{f}_i(x) \big)_j. \tag{12.5.7}$$

We assume that each $g_i$ is nonsingular, so $\lambda_{ij}(x)$ can be expressed as polynomial functions as in (12.5.7). Consequently, each Nash equilibrium $x$ must satisfy the following polynomial system ($i = 1, \ldots, N$):

$$\begin{cases} \sum_{j=1}^{m_i} \lambda_{ij}(x) \nabla_{x_i} g_{ij}(x_i) &= \nabla_{x_i} f_i(x), \\ g_{ij}(x_i) &= 0 \, (j \in \mathcal{E}_i), \\ \lambda_{ij}(x) g_{ij}(x_i) &= 0 \, (j \in \mathcal{I}_i), \\ g_{ij}(x_i) &\ge 0 \, (j \in \mathcal{I}_i), \\ \lambda_{ij}(x) &\ge 0 \, (j \in \mathcal{I}_i). \end{cases} \tag{12.5.8}$$

The above are necessary conditions for NEs, but they are typically not sufficient.

The Cartesian product of all players' strategy sets is

$$X := X_1 \times \cdots \times X_N.$$

A strategy tuple $x = (x_1, \ldots, x_N) \in X$ if and only if each $x_i \in X_i$. Recall the notation

$$[x]_1 := \begin{bmatrix} 1 \\ x \end{bmatrix}.$$

For a generic positive definite matrix $\Theta \in \mathcal{S}^{n+1}$, we consider the optimization problem

$$\begin{cases} \min_{x \in X} & [x]_1^T \cdot \Theta \cdot [x]_1 \\ s.t. & \nabla_{x_i} f_i(x) = \sum_{j=1}^{m_i} \lambda_{ij}(x) \nabla_{x_i} g_{ij}(x_i), \\ & \quad i = 1, \ldots, N, \\ & \lambda_{ij}(x) g_{ij}(x_i) = 0, \ \lambda_{ij}(x) \geq 0, \\ & \quad i = 1, \ldots, N, \ j \in \mathcal{I}_i. \end{cases} \tag{12.5.9}$$

When each $g_i$ is nonsingular, each NE $x$ must be a feasible point of (12.5.9). For every feasible point $x$ of (12.5.9), the $x_i$ is a KKT point for the optimization $\mathrm{F}_i(x_{-i})$. It is important to observe that if (12.5.9) is infeasible, then there are no NEs. If (12.5.9) is feasible, then it must have a minimizer, because the objective is a positive definite quadratic function. Moreover, for generically chosen $\Theta$, the minimizer of (12.5.9) is unique.

Assume $u := (u_1, \ldots, u_N)$ is an optimizer of (12.5.9). Recall the notation

$$u_{-i} := (u_1, \ldots, u_{i-1}, u_{i+1}, \ldots, u_N).$$

If each $u_i$ is a minimizer of $\mathrm{F}_i(u_{-i})$, then $u$ is a NE. To check this for each player, consider the optimization problems $(i = 1, \ldots, N)$:

$$\begin{cases} \omega_i := \min & f_i(x_i, u_{-i}) - f_i(u_i, u_{-i}) \\ s.t. & x_i \in X_i. \end{cases} \tag{12.5.10}$$

If all the optimal values $\omega_i \geq 0$, then $u$ is a NE. If one of them is negative, then $u$ is not a NE. Let $U_i$ be a set of some optimizers of (12.5.10). If $\omega_i < 0$ for some $i$, then the tuple $u$ violates the following inequalities:

$$f_i(x_i, x_{-i}) \leq f_i(v, x_{-i}) \quad (v \in U_i). \tag{12.5.11}$$

It is worthy to note that every NE must satisfy (12.5.11).

When $u$ is not a NE, we look for a new candidate NE by enforcing the inequalities in (12.5.11). This results in the refined optimization problem

$$\begin{cases} \min_{x \in X} & [x]_1^T \cdot \Theta \cdot [x]_1 \\ s.t. & \nabla_{x_i} f_i(x) = \sum_{j=1}^{m_i} \lambda_{ij}(x) \nabla_{x_i} g_{ij}(x_i), \\ & \lambda_{ij}(x) g_{ij}(x_i) = 0, \ \lambda_{ij}(x) \geq 0, \\ & f_i(v, x_{-i}) - f_i(x_i, x_{-i}) \geq 0 \ (v \in \mathcal{K}_i), \\ & \quad i = 1, \ldots, N, \ j \in \mathcal{I}_i. \end{cases} \tag{12.5.12}$$

In the above, each $\mathcal{K}_i$ is a union of optimizer sets $U_i$ in the above. We solve (12.5.12) for an optimizer. If it is verified to be a NE, then we get one. If it is not, we can add more inequalities like (12.5.11). Repeating this, we get the following algorithm.

**Algorithm 12.5.3.** *For the NEP (12.5.3), do the following:*

**Step 0** *Initialize $\mathcal{K}_i := \emptyset$ for all $i$ and let $l := 0$. Choose a generic positive definite matrix $\Theta$ of length $n + 1$.*

**Step 1** *Solve the optimization problem (12.5.12). If it is infeasible, there is no NE and stop; otherwise, solve it for an optimizer $u$.*

**Step 2** *For each $i = 1, \ldots, N$, solve the optimization (12.5.10). If all $\omega_i \geq 0$, $u$ is a NE and stop. If one of $\omega_i$ is negative, go to Step 3.*

**Step 3** *For each $i$ with $\omega_i < 0$, obtain a set $U_i$ of some (may not all) optimizers of (12.5.10) and then update $\mathcal{K}_i := \mathcal{K}_i \cup U_i$. Let $l := l + 1$, then go to Step 1.*

The optimization problems in Algorithm 12.5.3 can be solved by Moment-SOS relaxation methods in Chapter 5. When the polynomials $f_i, g_{ij}$ have generic coefficients, the NEP (12.5.3) has finitely many KKT points, so the feasible set of (12.5.9) is finite (see [249]). The following is the convergence theorem.

**Theorem 12.5.4.** *([249]) Assume each constraining polynomial tuple $g_i$ is nonsingular and let $\lambda_{ij}(x)$ be Lagrange multiplier polynomials as in (12.5.7). Let $\mathcal{G}$ be the feasible set of (12.5.9) and $\mathcal{G}^*$ be the set of all NEs. If the complement $\mathcal{G} \backslash \mathcal{G}^*$ is a finite set, i.e., the cardinality $L := |\mathcal{G} \backslash \mathcal{G}^*| < \infty$, then Algorithm 12.5.3 must terminate within $L$ loops.*

The NEP (12.5.3) is called convex if each player's optimization problem $\mathrm{F}_i(x_{-i})$ is convex, i.e., each objective $f_i(x_i, x_{-i})$ is a convex function in $x_i$ for given $x_{-i}$, the equality constraining function $g_{ij}(x_i)$ ($j \in \mathcal{E}_i$) is linear in $x_i$, and the inequality constraining function $g_{ij}(x_i)$ ($j \in \mathcal{I}_i$) is concave in $x_i$. For convex NEPs, the NEs are equivalent to KKT points, under certain constraint qualification conditions (e.g., Slater's condition or the LICQC). We have the following conclusion.

**Corollary 12.5.5.** *Assume each constraining polynomial tuple $g_i$ is nonsingular and let $\lambda_{ij}(x)$ be Lagrange multiplier polynomials as in (12.5.7). Suppose each $g_{ij}(x_i)$ ($j \in \mathcal{E}_i$) is linear, each $g_{ij}(x_i)$ ($j \in \mathcal{I}_i$) is concave, and each $f_i(x_i, x_{-i})$ is convex in $x_i$ for given $x_{-i}$. Then Algorithm 12.5.3 must terminate at the initial loop $l = 0$, returning a NE or reporting that there is no NE.*

For convex NEPs, it is possible that there are finitely or infinitely many NEs, even if each player's objective is strictly convex in its own strategy. This is shown in Example 12.5.2.

## 12.5.2 ▪ More NEs

Algorithm 12.5.3 aims at finding one NE. In some applications, people may need more NEs, or even all. For the case that there is a unique NE, people may also need verification for the uniqueness. We discuss how to compute more NEs or how to verify the uniqueness.

Assume that $x^*$ is a NE produced by Algorithm 12.5.3. Note that $x^*$ is a minimizer of (12.5.12). Since $\Theta$ is a generic positive definite matrix, if $x^*$ is an isolated NE (e.g., this is the case if there are only finitely many NEs), there exists a scalar $\delta > 0$ such that

$$[x]_1^T \Theta [x]_1 \geq [x^*]_1^T \Theta [x^*]_1 + \delta$$

for every NE $x$ other than $x^*$. For such $\delta$, one can try to find a new NE by solving the optimization

$$
\left\{
\begin{array}{rl}
\min\limits_{x \in X} & [x]_1^T \Theta [x]_1 \\[2mm]
s.t. & \nabla_{x_i} f_i(x) - \sum\limits_{j=1}^{m_i} \lambda_{ij}(x) \nabla_{x_i} g_{ij}(x_i) = 0, \\[2mm]
& \lambda_{ij}(x) g_{ij}(x_i) = 0,\ \lambda_{ij}(x) \geq 0, \\[1mm]
& f_i(v, x_{-i}) - f_i(x_i, x_{-i}) \geq 0\ (\forall\, v \in \mathcal{K}_i), \\[1mm]
& \quad i = 1, \ldots, N,\ j \in \mathcal{I}_i, \\[1mm]
& [x]_1^T \Theta [x]_1 \geq [x^*]_1^T \Theta [x^*]_1 + \delta.
\end{array}
\right.
\tag{12.5.13}
$$

After an optimizer of (12.5.13) is obtained, we can check whether it is a NE or not by solving (12.5.10). If it is, we get a new NE other than $x^*$. If it is not, we can update $\mathcal{K}_i$ to a larger set. Repeating this, we are able to get more NEs.

We should choose a small enough $\delta > 0$ for (12.5.13) such that there is no other Nash equilibrium $u$ satisfying

$$
[u]_1^T \Theta [u]_1 \leq [x^*]_1^T \Theta [x^*]_1 + \delta.
$$

To check this, we need to solve the maximization problem:

$$
\left\{
\begin{array}{rl}
\max\limits_{x \in X} & [x]_1^T \Theta [x]_1 \\[2mm]
s.t. & \nabla_{x_i} f_i(x) - \sum\limits_{j=1}^{m_i} \lambda_{ij}(x) \nabla_{x_i} g_{ij}(x_i) = 0, \\[2mm]
& \lambda_{ij}(x) g_{ij}(x_i) = 0,\ \lambda_{ij}(x) \geq 0, \\[1mm]
& f_i(v, x_{-i}) - f_i(x_i, x_{-i}) \geq 0\ (\forall\, v \in \mathcal{K}_i), \\[1mm]
& \quad i = 1, \ldots, N,\ j \in \mathcal{I}_i, \\[1mm]
& [x]_1^T \Theta [x]_1 \leq [x^*]_1^T \Theta [x^*]_1 + \delta.
\end{array}
\right.
\tag{12.5.14}
$$

It is important to observe that if $x^*$ is also a maximizer of (12.5.14), then the feasible set of (12.5.13) contains all NEs except $x^*$, under the genericity assumption on $\Theta$.

**Proposition 12.5.6.** *([249]) Assume $\Theta$ is a generic positive definite matrix and $x^*$ is a minimizer of (12.5.12).*

(i) *If $x^*$ is also a maximizer of (12.5.14), then there is no other Nash equilibrium $u$ satisfying $[u]_1^T \Theta [u]_1 \leq [x^*]_1^T \Theta [x^*]_1 + \delta$.*

(ii) *If $x^*$ is an isolated KKT point, then there exists $\delta > 0$ such that $x^*$ is also a maximizer of (12.5.14).*

Proposition 12.5.6 shows the existence of $\delta > 0$ such that (12.5.12) and (12.5.14) have the same optimal value. However, it does not give a concrete value for such $\delta$. In computational practice, we can first give a priori value for $\delta$. If it does not work, then we can decrease $\delta$ to a smaller value (e.g., let $\delta := \delta/5$). By repeating this, the optimization (12.5.14) will eventually have $x^*$ as a maximizer. The following is the algorithm for determining whether or not there exists a NE other than $x^*$.

**Algorithm 12.5.7.** *Give an initial value for $\delta$ (say, 0.01).*

**Step 1** *Solve the maximization (12.5.14). If its optimal value $\eta$ equals $\upsilon := [x^*]_1^T \Theta [x^*]_1$, then go to Step 2. If $\eta$ is bigger than $\upsilon$, then let $\delta := \min(\delta/5, \eta - \upsilon)$ and repeat this step.*

**Step 2** *Solve the minimization (12.5.13). If it is infeasible, then there are no additional NEs; if it is feasible, solve it for a minimizer $u$.*

**Step 3** *For each $i = 1, \ldots, N$, solve the optimization (12.5.10). If all optimal values $\omega_i \geq 0$, then $u$ is a NE and stop; if one of $\omega_i$ is negative, go to Step 4.*

**Step 4** *For each $i = 1, \ldots, N$ with $\omega_i < 0$, update $\mathcal{K}_i := \mathcal{K}_i \cup U_i$ and go to Step 2. Here, $\mathcal{K}_i$ and $U_i$ are from Algorithm 12.5.3.*


When $x^*$ is not an isolated KKT point, there may not exist a satisfactory $\delta > 0$ in Step 1. For such a case, more investigation is required to verify uniqueness of the NE or to find other NEs. However, for NEPs given by generic polynomials, there are only finitely many KKT points (see [249]). The following is the convergence property of Algorithm 12.5.7.

**Theorem 12.5.8.** *([249]) Under the same assumptions as in Theorem 12.5.4, if $\Theta$ is a generic positive definite matrix and $x^*$ is an isolated KKT point, then Algorithm 12.5.7 must terminate within finitely many loops, returning a NE other than $x^*$ or reporting nonexistence of other NEs.*

Once a new NE is obtained, we can repeatedly apply Algorithm 12.5.7, to compute more NEs, if they exist. For this purpose, we need to use the same generic matrix $\Theta$. In particular, if there are finitely many NEs, we can eventually get all of them.

**Example 12.5.9.** *([249]) Consider the following 2-player NEP:*

$$
\mathrm{F}_1(x_{-1}) : \left\{ \begin{array}{ll} \min\limits_{x_1 \in \mathbb{R}^3} & \sum\limits_{j=1}^{3} x_{1,j}(x_{1,j} - j \cdot x_{2,j}) \\ s.t. & 1 - x_{1,1}x_{1,2} \geq 0,\ 1 - x_{1,2}x_{1,3} \geq 0,\ x_{1,1} \geq 0, \end{array} \right.
$$

$$
\mathrm{F}_2(x_{-2}) : \left\{ \begin{array}{ll} \min\limits_{x_2 \in \mathbb{R}^3} & \prod\limits_{j=1}^{3} x_{2,j} + \sum\limits_{\substack{1 \leq i < j \leq 3 \\ 1 \leq k \leq 3}} x_{1,i}x_{1,j}x_{2,k} + \sum\limits_{\substack{1 \leq i \leq 3 \\ 1 \leq j < k \leq 3}} x_{1,i}x_{2,j}x_{2,k} \\ s.t. & 1 - (x_{2,1})^2 - (x_{2,2})^2 = 0. \end{array} \right.
$$

*The Lagrange multipliers for $\mathrm{F}_1(x_2)$ can be expressed as*

$$
\lambda_{1,1} = (1 - x_{1,1}x_{1,2})\frac{\partial f_1}{\partial x_{1,1}}, \quad \lambda_{1,2} = -x_{1,1}\frac{\partial f_1}{\partial x_{1,2}}, \quad \lambda_{1,3} = x_{1,1}\frac{\partial f_1}{\partial x_{1,1}} - x_{1,2}\frac{\partial f_1}{\partial x_{1,2}}.
$$

*The optimization $\mathrm{F}_2(x_1)$ has sphere constraints, so its Lagrange multipliers can be expressed as in §6.3. Applying Algorithm 12.5.7 repeatedly, we get four NEs:*

| $x_1^*$ | $x_2^*$ |
|---|---|
| $(0.3198, 0.6396, -0.6396)$, | $(0.6396, 0.6396, -0.4264)$; |
| $(0.0000, 0.3895, 0.5842)$, | $(-0.8346, 0.3895, 0.3895)$; |
| $(0.2934, -0.5578, 0.8803)$, | $(0.5869, -0.5578, 0.5869)$; |
| $(0.0000, -0.5774, -0.8660)$, | $(-0.5774, -0.5774, -0.5774)$. |

## 12.5.3 ▪ Generalized NEPs

Generalized Nash equilibrium problems (GNEPs) are extensions of NEPs such that each player's strategy set also depends on other players' strategies. For a GNEP, the $i$th player's optimization problem $F_i(x_{-i})$ is in the form

$$\begin{cases} \min_{x_i \in \mathbb{R}^{n_i}} & f_i(x_i, x_{-i}) \\ s.t. & g_{ij}(x_i, x_{-i}) = 0 \, (j \in \mathcal{E}_i), \\ & g_{ij}(x_i, x_{-i}) \geq 0 \, (j \in \mathcal{I}_i), \end{cases} \qquad (12.5.15)$$

where the constraining function $g_{ij}$ depends on $x_{-i}$. A strategy tuple $x^* = (x_1^*, \ldots, x_N^*)$ is called a generalized Nash equilibrium (GNE) if each $x_i^*$ is a minimizer of (12.5.15) when $x_{-i}$ is fixed to be

$$(x_1^*, \ldots, x_{i-1}^*, x_{i+1}^*, \ldots, x_N^*).$$

The task of a GNEP is to compute one or several GNEs, or to detect nonexistence of GNEs.

The following is an example of 2-player GNEP with the optimization:

| $F_1(x_{-1})$ | | $F_2(x_{-2})$ | |
|---|---|---|---|
| $\min_{x_1 \in \mathbb{R}^1}$ | $x_1$ | $\min_{x_2 \in \mathbb{R}^1}$ | $(x_2)^2 - (x_1 - 1)x_2$ |
| $s.t.$ | $x_2(x_1 - x_2 - 1) \geq 0,$ | $s.t.$ | $(x_1)^2 + (x_2)^2 \leq 3,$ |
| | $x_1 \geq 0,$ | | $x_2 \geq 0.$ |

For the first player, when $x_2 > 0$, its feasible set is $x_1 \geq x_2 + 1$, and its best strategy is $x_1 = x_2 + 1$. When $x_2 = 0$, the first player's best strategy is $x_1 = 0$. For any fixed $x_1$ with $x_1^2 \leq 3$, the second player's problem is feasible and its best strategy is $\max(\frac{x_1-1}{2}, 0)$. One can check that $(0,0)$ is a GNE for this GNEP.

GNEPs have broad applications [14, 194, 287]. We refer to [90, 91, 92] for recent surveys. When GNEPs are given by polynomial or rational functions, computational methods for solving GNEPs are given in [243, 244, 250].

## 12.5.4 ▪ Exercises

**Exercise 12.5.1.** *([249]) Find one or more NEs for the following 2-player NEP:*

$$F_1(x_{-1}) : \begin{cases} \min_{x_1 \in \mathbb{R}^2} & -x_{2,1}(x_{1,1})^2 - x_{2,2}x_{1,1} + (x_{2,2} - \frac{1}{2})x_{1,2} \\ s.t. & 1 - (x_{1,1})^2 - (x_{1,2})^2 \geq 0, \end{cases}$$

$$F_2(x_{-2}) : \begin{cases} \min_{x_2 \in \mathbb{R}^2} & x_{1,2}x_{2,1} + (x_{2,2} - \frac{1}{2})^2 \\ s.t. & 1 - x_{2,1} - x_{2,2} \geq 0, \, x_{2,1} \geq 0, \, x_{2,2} \geq 0. \end{cases}$$

**Exercise 12.5.2.** *([249]) Find one or more NEs for the following 3-player NEP:*

$$F_1(x_{-1}) : \begin{cases} \min_{x_1 \in \mathbb{R}^2} & (2x_{1,1} - x_{1,2} + 3)x_{1,1}x_{2,1} \\ & \quad + [(2x_{1,2})^2 + (x_{3,2})^2]x_{1,2} \\ s.t. & 1 - x_1^T x_1 \geq 0, \end{cases}$$

$$F_2(x_{-2}) : \begin{cases} \min_{x_2 \in \mathbb{R}^2} & [(x_{2,1})^2 - x_{1,2}]x_{2,1} \\ & \quad + [(x_{2,2})^2 + 2x_{3,2} + x_{1,2}x_{3,1}]x_{2,2} \\ s.t. & x_2^T x_2 - 1 = 0, \, x_{2,1} \geq 0, \, x_{2,2} \geq 0, \end{cases}$$

$$F_3(x_{-3}) : \begin{cases} \min_{x_3 \in \mathbb{R}^2} & (x_{1,1}x_{1,2} - 1)x_{3,1} - [3(x_{3,2})^2 + 1]x_{3,2} \\ & \quad + 2[x_{3,1} + x_{3,2}]x_{3,1}x_{3,2} \\ s.t. & 1 - (x_{3,1})^2 \geq 0, \, 1 - (x_{3,2})^2 \geq 0. \end{cases}$$

## 12.6 ▪ Bilevel Polynomial Optimization

A bilevel optimization problem is in the form

$$
\left\{
\begin{array}{rl}
\min\limits_{x \in \mathbb{R}^n, y \in \mathbb{R}^p} & F(x,y) \\
\text{s.t.} & h_i(x,y) = 0 \, (i \in \mathcal{E}_1), \\
& h_j(x,y) \geq 0 \, (j \in \mathcal{I}_1), \\
& y \in S(x),
\end{array}
\right.
\tag{12.6.1}
$$

where $S(x)$ is the set of global optimizer(s) of

$$
\left\{
\begin{array}{rl}
\min\limits_{z \in \mathbb{R}^p} & f(x,z) \\
\text{s.t.} & g_i(x,z) = 0 \, (i \in \mathcal{E}_2), \\
& g_j(x,z) \geq 0 \, (j \in \mathcal{I}_2).
\end{array}
\right.
\tag{12.6.2}
$$

The problem (12.6.1) is called the *upper* level optimization, while (12.6.2) is the *lower* level one. The function $F(x,y)$ is called the upper level objective and $h_i(x,y), h_j(x,y)$ are the upper level constraining functions, while $f(x,z)$ is called the lower level objective and $g_i(x,z), g_j(x,z)$ are the lower level constraining functions. The sets $\mathcal{E}_1, \mathcal{I}_1, \mathcal{E}_2, \mathcal{I}_2$ are finite label sets (possibly empty). The $x$ is called the upper level decision variable, while $y$ is called the lower level decision variable. This section introduces the method in [247] for solving bilevel optimization.

The feasible set of the lower level optimization (12.6.2) is

$$
Z(x) := \left\{ z \in \mathbb{R}^p \; \middle| \; \begin{array}{l} g_i(x,z) = 0 \, (i \in \mathcal{E}_2), \\ g_j(x,z) \geq 0 \, (j \in \mathcal{I}_2) \end{array} \right\}.
\tag{12.6.3}
$$

The bilevel optimization (12.6.1) is said to be *simple* if $Z(x) \equiv Z$ is independent of $x$, and it is called *general* if $Z(x)$ depends on $x$. When all functions are polynomials, we call (12.6.1) a bilevel polynomial optimization problem (BPOP). For traditional work on bilevel optimization, we refer to the monographs [11, 76, 77, 304] and the survey [56]. Bilevel optimization is mathematically challenging, due to the constraint $y \in S(x)$. Even for the case that all functions are linear, the computational complexity for solving bilevel optimization is already NP-hard (see [18]). The preliminary approach of relaxing the constraint $y \in S(x)$ to the first order optimality conditions may not even produce a feasible point for (12.6.1). The following is such an example.

**Example 12.6.1.** *([247]) Consider the BPOP*

$$
\left\{
\begin{array}{rl}
\min\limits_{x \in \mathbb{R}^1, y \in \mathbb{R}^1} & xy - y + \tfrac{1}{2}y^2 \\
\text{s.t.} & 1 - x^2 \geq 0, 1 - y^2 \geq 0, \\
& y \in S(x),
\end{array}
\right.
$$

*where $S(x)$ is the optimizer set of*

$$
\left\{
\begin{array}{rl}
\min\limits_{z \in \mathbb{R}^1} & -xz^2 + \tfrac{1}{2}xz^4 \\
\text{s.t.} & 1 - z^2 \geq 0.
\end{array}
\right.
$$

*If $y \in S(x)$ is relaxed to the KKT conditions, we get*

$$
\left\{
\begin{array}{rl}
\min\limits_{x,y,\lambda \in \mathbb{R}^1} & xy - y + \tfrac{1}{2}y^2 \\
\text{s.t.} & 1 - x^2 \geq 0, \, 1 - y^2 \geq 0, \\
& -xy + xy^3 + \lambda y = 0, \\
& \lambda \geq 0, \, \lambda(1 - y^2) = 0.
\end{array}
\right.
$$

*The above optimization has the optimal value $F^* := -1.5$ and the minimizer $(x^*, y^*) :=$*
*$(-1, 1)$. However, $(x^*, y^*)$ is not feasible for the BPOP, since*

$$y^* \notin S(x^*) = \{0\}.$$

For a point $y \in Z(x)$, the constraint $y \in S(x)$ is equivalent to

$$f(x, y) \leq v(x) := \inf_{z \in Z(x)} f(x, z). \tag{12.6.4}$$

The above $v(x)$ is called the *value function* for the lower level optimization. For convenience of
notation, denote the set

$$\mathcal{U} := \left\{ (x, y) \,\middle|\, \begin{array}{ll} h_i(x, y) &= 0 \ (i \in \mathcal{E}_1), \\ g_i(x, y) &= 0 \ (i \in \mathcal{E}_2), \\ h_j(x, y) &\geq 0 \ (j \in \mathcal{I}_1), \\ g_j(x, y) &\geq 0 \ (j \in \mathcal{I}_2) \end{array} \right\}. \tag{12.6.5}$$

In view of (12.6.4), the bilevel optimization (12.6.1) is equivalent to

$$\left\{ \begin{array}{ll} \min & F(x, y) \\ s.t. & f(x, y) \leq v(x), \\ & (x, y) \in \mathcal{U}. \end{array} \right. \tag{12.6.6}$$

This is a generalized semi-infinite program (GSIP), since $Z(x)$ depends on $x$ and the inequality
constraint in (12.6.6) is requiring

$$f(x, y) \leq f(x, z) \quad \text{for all } z \in Z(x). \tag{12.6.7}$$

To solve (12.6.6), one could construct a sequence of polynomial optimization relaxations
$(P_k)$ that have the same objective $F(x, y)$ but different feasible sets $\mathcal{U}_k$, satisfying the nesting
containment

$$\mathcal{F} \subseteq \cdots \subsetneqq \mathcal{U}_k \subsetneqq \cdots \subsetneqq \mathcal{U}_1 \subsetneqq \mathcal{U}_0 \subsetneqq \mathcal{U}. \tag{12.6.8}$$

In the above, $\mathcal{F}$ is the feasible set of (12.6.1). Let $(x^{(k)}, y^{(k)})$ be an optimizer of the relaxation

$$(P_k): \quad \left\{ \begin{array}{ll} \min & F(x, y) \\ s.t. & (x, y) \in \mathcal{U}_k. \end{array} \right. \tag{12.6.9}$$

If $y^{(k)} \in S(x^{(k)})$, then $(x^{(k)}, y^{(k)})$ is also an optimizer of (12.6.6). If otherwise $y^{(k)} \notin S(x^{(k)})$,
we can add new constraints to get a tighter relaxation $(P_{k+1})$. For the sequence $\{(x^{(k)}, y^{(k)})\}_{k=1}^{\infty}$
produced this way, we expect that its limit or accumulation point is an optimizer of (12.6.6).
This is a kind of *exchange* technique for solving SIPs (see [125]). If we only relax the infinite
constraints as in (12.6.7) to finitely many ones, the convergence rate is extremely slow. This was
shown by numerical experiments in [234].

A more efficient method for solving bilevel optimization was proposed in [247]. It combines
KKT conditions and the exchange technique. Under some constraint qualifications (see §5.1),
each $y \in S(x)$ satisfies the KKT conditions for (12.6.2). If each Lagrange multiplier can be
expressed as a polynomial (or rational) function $\lambda_j(x, y)$ in $(x, y)$, then we can choose the initial
relaxation to be

$$(P_0): \quad \left\{ \begin{array}{ll} \min_{(x,y) \in \mathcal{U}} & F(x, y) \\ s.t. & \nabla_z f(x, y) - \sum_{j \in \mathcal{E}_2 \cup \mathcal{I}_2} \lambda_j(x, y) \nabla_z g_j(x, y) = 0, \\ & \lambda_j(x, y) \geq 0, \ \lambda_j(x, y) g_j(x, y) = 0 \ (j \in \mathcal{I}_2). \end{array} \right.$$

(The above $\nabla_z$ denotes the gradient with respect to the lower level variable $z$.) Suppose $(x^{(k)}, y^{(k)})$ is an optimizer of the $k$th relaxation $(P_k)$. If $y^{(k)} \in S(x^{(k)})$, then $(x^{(k)}, y^{(k)})$ must be an optimizer for (12.6.6). Otherwise, there must exist a point $z^{(k)} \in Z(x^{(k)})$ such that

$$f(x^{(k)}, z^{(k)}) - f(x^{(k)}, y^{(k)}) < 0.$$

To get a better candidate than $(x^{(k)}, y^{(k)})$, we look for a polynomial vector function $q^{(k)}(x, y)$ such that

$$\boxed{\begin{array}{l} q^{(k)}(x^{(k)}, y^{(k)}) = z^{(k)}, \\ q^{(k)}(x, y) \in Z(x) \quad \text{for all } (x, y) \in \mathcal{U}. \end{array}} \qquad (12.6.10)$$

Then we can see that

$$y \in S(x) \quad \Rightarrow \quad f(x, q^{(k)}(x, y)) - f(x, y) \geq 0. \qquad (12.6.11)$$

Therefore, the feasible set of $(P_k)$ can be tightened to

$$\mathcal{U}_{k+1} := \mathcal{U}_k \cap \left\{ f(x, q^{(k)}(x, y)) - f(x, y) \geq 0 \right\}. \qquad (12.6.12)$$

The $(k+1)$th polynomial optimization relaxation can be selected as

$$(P_{k+1}) : \quad \left\{ \begin{array}{ll} \min\limits_{(x,y) \in \mathcal{U}_k} & F(x, y) \\ s.t. & f(x, q^{(k)}(x, y)) - f(x, y) \geq 0. \end{array} \right. \qquad (12.6.13)$$

Repeating this, we can either get an optimal solution of (12.6.1) at some $k$th loop or obtain an infinite sequence $\{(x^{(k)}, y^{(k)})\}_{k=1}^{\infty}$. An accumulation point of this sequence is expected to be a global minimizer of (12.6.1).

## 12.6.1 ▪ A hierarchy of relaxations

We give a hierarchy of polynomial optimization relaxations to solve (12.6.1). For convenience, suppose the constraints are labeled such that

$$\mathcal{E}_2 \cup \mathcal{I}_2 := \{1, \ldots, m_2\}, \quad g := (g_1(x, z), \ldots, g_{m_2}(x, z)).$$

The KKT system for the lower level optimization (12.6.2) implies

$$\underbrace{\begin{bmatrix} \nabla_z g_1(x, y) & \cdots & \nabla_z g_{m_2}(x, y) \\ g_1(x, y) & \cdots & 0 \\ 0 & \cdots & 0 \\ \vdots & \ddots & \vdots \\ 0 & \cdots & g_{m_2}(x, y) \end{bmatrix}}_{G(x,y)} \underbrace{\begin{bmatrix} \lambda_1 \\ \vdots \\ \lambda_{m_2} \end{bmatrix}}_{\lambda} = \underbrace{\begin{bmatrix} \nabla_z f(x, y) \\ 0 \\ \vdots \\ 0 \end{bmatrix}}_{\hat{f}(x,y)}, \qquad (12.6.14)$$

together with the sign conditions $\lambda_j \geq 0$ for $j \in \mathcal{I}_2$. Since it also depends on $x$, the above matrix $G(x, y)$ is typically not full column ranked for all complex pairs $(x, y)$. Hence, there typically does not exist a matrix polynomial $L(x, y)$ such that $L(x, y)G(x, y) = I_{m_2}$. However, rational expressions always exist for Lagrange multipliers. So we make the following assumption.

**Assumption 12.6.2.** *([247]) Suppose the condition (12.6.14) holds for every minimizer of (12.6.1), there exist polynomials $d_1(x, y), \ldots, d_{m_2}(x, y) \geq 0$ on $\mathcal{U}$, and there exist nonzero polynomials $\phi_1(x, y), \ldots, \phi_{m_2}(x, y)$ such that*

$$\lambda_j d_j(x, y) = \phi_j(x, y), \quad j = 1, \ldots, m_2, \qquad (12.6.15)$$

*for all $x, y, \lambda_j$ satisfying (12.6.14).*

The equation (12.6.15) gives the rational expressions

$$\lambda_j = \frac{\phi_j(x,y)}{d_j(x,y)}, \quad j = 1, \ldots, m_2.$$

These expressions can be obtained as follows. Suppose $W(x, y)$ is a matrix polynomial such that

$$W(x,y)G(x,y) = \begin{bmatrix} d_1(x,y) & 0 & 0 \\ \vdots & \ddots & \vdots \\ 0 & 0 & d_{m_2}(x,y) \end{bmatrix}. \tag{12.6.16}$$

Then we can see that

$$W(x,y)G(x,y)\lambda = W(x,y)\hat{f}(x,y), \tag{12.6.17}$$

$$\lambda_j = \frac{\left(W(x,y)\hat{f}(x,y)\right)_j}{d_j(x,y)} \tag{12.6.18}$$

for $j = 1, \ldots, m_2$. (The subscript $j$ denotes the $j$th entry.) Let $D(x, y)$ be the least common multiple (lcm) of the denominators

$$D(x,y) := \mathrm{lcm}\big(d_1(x,y), \ldots, d_{m_2}(x,y)\big)$$

and let $D_j(x, y)$ be the quotient

$$D_j(x,y) := \frac{D(x,y)}{d_j(x,y)}.$$

Under Assumption 12.6.2, every optimizer of (12.6.1) is contained in the set

$$\mathcal{K} := \left\{ (x,y) \; \middle| \; \begin{array}{l} D \cdot \nabla_z f(x,y) = \sum_{j \in [m_2]} D_j \cdot \phi_j \cdot \nabla_z g_j(x,y), \\ \phi_j(x,y) \geq 0, \; \phi_j(x,y)g_j(x,y) = 0 \; (j \in \mathcal{I}_2) \end{array} \right\}. \tag{12.6.19}$$

When each $d_j(x, y) > 0$ on $\mathcal{U}$, a pair $(x, y) \in \mathcal{U}$ is a KKT point for (12.6.2) if and only if $(x, y)$ belongs to $\mathcal{K}$. If some $d_j(\hat{x}, \hat{y}) = 0$ for some $(\hat{x}, \hat{y}) \in \mathcal{U}$, every KKT point of (12.6.2) is also contained in $\mathcal{K}$.

**Assumption 12.6.3.** *([247]) For every pair $(\hat{x}, \hat{y}) \in \mathcal{U} \cap \mathcal{K}$ and for every $\hat{z} \in S(\hat{x})$, there exists a polynomial vector $q(x, y) := (q_1(x, y), \ldots, q_p(x, y))$ such that*

$$\boxed{\begin{array}{c} q(\hat{x}, \hat{y}) = \hat{z}, \\ q(x,y) \in Z(x) \text{ for all } (x,y) \in \mathcal{U}. \end{array}} \tag{12.6.20}$$

*The function $q(x, y)$ is called a polynomial extension of $\hat{z}$ at $(\hat{x}, \hat{y})$.*

Assumption 12.6.3 is readily satisfiable for many cases. For instance, if the lower level optimization (12.6.2) has linear constraints (for given vectors $a_i, b_i \in \mathbb{R}^n$ and scalars $c_i, d_i \in \mathbb{R}$) such that (note $z = (z_1, \ldots, z_p)$)

$$a_i^T x + c_i \leq z_i \leq b_i^T x + d_i, \quad i = 1, \ldots, p,$$

then $q$ can be selected as

$$q_i(x,y) := \mu_i(a_i^T x + c_i) + (1 - \mu_i)(b_i^T x + d_i) \tag{12.6.21}$$

for the scalars

$$\mu_i := \frac{b_i^T \hat{x} + d_i - \hat{z}_i}{(b_i - a_i)^T \hat{x} + (d_i - c_i)} \in [0,1].$$

We will discuss Assumptions 12.6.2 and 12.6.3 with more details in §12.6.2 and §12.6.3.

Based on the above, we get the following algorithm for solving the bilevel optimization (12.6.1).

**Algorithm 12.6.4.** *([247]) For given $F, h_i, f, g_j$ as in (12.6.1), do the following:*

**Step 0** *Find Lagrange multiplier expressions as in (12.6.15). Let $\mathcal{U}_0 := \mathcal{U} \cap \mathcal{K}$ and $k := 0$, where $\mathcal{K}$ is the set in (12.6.19).*

**Step 1** *Solve the polynomial optimization*

$$(P_k): \begin{cases} F_k^* := \min & F(x,y) \\ s.t. & (x,y) \in \mathcal{U}_k. \end{cases} \tag{12.6.22}$$

*If $(P_k)$ is infeasible, then either (12.6.1) has no optimizers, or none of its optimizers is a point of the set $\mathcal{K}$. If it is feasible and has a minimizer, solve $(P_k)$ for a minimizer $(x^{(k)}, y^{(k)})$.*

**Step 2** *Solve the lower level optimization*

$$(Q_k): \begin{cases} v_k^* := \min_{z \in \mathbb{R}^p} & f(x^{(k)}, z) - f(x^{(k)}, y^{(k)}) \\ s.t. & z \in Z(x^{(k)}), \ (x^{(k)}, z) \in \mathcal{K}, \end{cases} \tag{12.6.23}$$

*for an optimizer $z^{(k)}$. If the optimal value $v_k^* = 0$, then $(x^{(k)}, y^{(k)})$ is an optimizer for (12.6.1) and stop. Otherwise, go to Step 3.*

**Step 3** *Construct a polynomial vector function $q^{(k)}(x,y)$ such that*

$$\boxed{\begin{array}{c} q^{(k)}(x^{(k)}, y^{(k)}) = z^{(k)}, \\ q^{(k)}(x,y) \in Z(x) \quad \text{for all} \quad (x,y) \in \mathcal{U}. \end{array}}$$

*Update the set $\mathcal{U}_{k+1}$ as*

$$\mathcal{U}_{k+1} := \mathcal{U}_k \cap \left\{ f(x, q^{(k)}(x,y)) - f(x,y) \geq 0 \right\}.$$

*Let $k := k + 1$ and go to Step 1.*

The polynomial optimization problems $(P_k), (Q_k)$ in Algorithm 12.6.4 can be solved by Moment-SOS relaxations. We refer to the methods in §5.2 and §5.3 for how to solve them.

**Example 12.6.5.** *([247]) Consider the general BPOP:*

$$\begin{cases} \min_{x,y \in \mathbb{R}^4} & x_1^2 y_4^2 - x_2 y_3^2 + x_3 y_1 - x_4 y_2 \\ s.t. & (4 - x_1^2 - x_2^2, \ -x_1 - x_2^2, \ y_1 - x_1, \ e^T x) \geq 0, \\ & (x_3 + x_4 - 3, \ 1 + x_3 - x_4, \ 3 - x_3, \ x_4) \geq 0, \\ & y \in S(x), \end{cases}$$

*where $S(x)$ is the set of optimizer(s) of*

$$\begin{cases} \min_{z \in \mathbb{R}^4} & (x_1 - z_1)^2 + (x_2 - z_2)^2 + z_3 - z_4 \\ \text{s.t.} & 4x_3^2 - x_1^2 - x_2^2 + 2x_1 z_1 + 2x_2 z_2 - z^T z \geq 0, \\ & (z_3, x_3 - z_3, z_4, x_4 - z_4) \geq 0. \end{cases}$$

*The matrix polynomial $W(x, y)$ satisfying (12.6.16) is*

$$\begin{bmatrix} -1 & 0 & 0 & 0 & & & & \\ -(x_3 - y_3)y_3 & 0 & (x_3 - y_3)(y_1 - x_1) & 0 & & & & \\ y_3^2 & 0 & -y_3(y_1 - x_1) & 0 & & & & \\ -(x_4 - y_4)y_4 & 0 & 0 & (x_4 - y_4)(y_1 - x_1) & & & & \\ y_4^2 & 0 & 0 & -y_4(y_1 - x_1) & & & & \\ & & & & 0 & 0 & 0 & 0 & 0 \\ & & & & 0 & 0 & y_1 - x_1 & 0 & 0 \\ & & & & 0 & y_1 - x_1 & 0 & 0 & 0 \\ & & & & 0 & 0 & 0 & 0 & y_1 - x_1 \\ & & & & 0 & 0 & 0 & y_1 - x_1 & 0 \end{bmatrix}$$

*for the denominator vector*

$$d(x, y) = (y_1 - x_1) \cdot \left( 2, x_3 - y_3, y_3, x_4 - y_4, y_4 \right).$$

*Note that $d(x, y) \geq 0$ for all feasible $(x, y)$. The lower level feasible set $Z(x)$ is a mixture of separable and ball type constraints:*

$$Z(x) = \left\{ z \in \mathbb{R}^4 \,\middle|\, \begin{array}{l} (z_1 - x_1)^2 + (z_2 - x_2)^2 + z_3^2 + z_4^2 \leq 4x_3^2, \\ 0 \leq z_3 \leq x_3, \, 0 \leq z_4 \leq x_4 \end{array} \right\}.$$

*The polynomial vector function $q := (q_1, q_2, q_3, q_4)$ in Assumption 12.6.3 can be selected as*

$$q_1 = x_1 + \mu_1 x_3, \, q_2 = x_2 + \mu_2 x_3, \, q_3 = \mu_3 x_3, \, q_4 = \mu_4 x_4, \tag{12.6.24}$$

*where (for the given triple $(\hat{x}, \hat{y}, \hat{z})$)*

$$\mu_1 = \frac{\hat{z}_1 - \hat{x}_1}{\hat{x}_3}, \, \mu_2 = \frac{\hat{z}_2 - \hat{x}_2}{\hat{x}_3}, \, \mu_3 = \frac{\hat{z}_3}{\hat{x}_3}, \, \mu_4 = \frac{\hat{z}_4}{\hat{x}_4}.$$

*Since $1 \leq \hat{x}_3 \leq 3$ and $0 \leq \hat{x}_4 \leq 1 + \hat{x}_3$, we select $\mu_4 = 0$ for the special case that $\hat{x}_4 = 0$. This bilevel optimization was solved by Algorithm 12.6.4 in the loop $k = 1$. The computational results are shown as below.*

The convergence of Algorithm 12.6.4 is summarized as follows. First, if the optimization (12.6.2) is convex for each given $x$, then Algorithm 12.6.4 will find a global optimizer of (12.6.1) in the initial loop $k = 0$.

**Proposition 12.6.6.** *([247]) Suppose that Assumptions 12.6.2 and 12.6.3 hold and all $d_j(x, y) > 0$ on $\mathcal{U}$. For every given $x$, assume that $f(x, z)$ is convex with respect to $z$, $g_i(x, z)$ is linear in $z$ for $i \in \mathcal{E}_2$, and $g_j(x, z)$ is concave in $z$ for $j \in \mathcal{I}_2$. Assume the KKT conditions hold for (12.6.2) for all feasible $x$. Then, the bilevel optimization (12.6.1) is equivalent to $(P_0)$ and Algorithm 12.6.4 terminates at the initial loop $k = 0$.*

| | |
|---|---|
| $(P_0)$ | $F_0^* = -41.7143,$ |
| | $x^{(0)} = (-1.5616, 1.2496, 3.0000, 4.0000),$ |
| | $y^{(0)} = (-1.5616, 6.4458, 3.0000, 0.0008),$ |
| $(Q_0)$ | $v_0^* = -33.9991,$ |
| | $z^{(0)} = (-1.5615, 1.2496, 0.0000, 4.0000),$ |
| | $q^{(0)} = (x_1, x_2, 0, x_4)$ as in (12.6.24). |
| $(P_1)$ | $F_1^* = -6.0000,$ |
| | $x^{(1)} = (-2.0000, 0.0001, 3.0000, 0.0001),$ |
| | $y^{(1)} = (-2.0000, 0.0001, -0.0000, 0.0001),$ |
| $(Q_1)$ | $v_1^* = -2.7612 \cdot 10^{-9} \to$ stop. |
| Time | 3.42 seconds, |
| Output | $F^* = F_1^*,\ x^* = x^{(1)},\ y^* = y^{(1)}.$ |

Second, if Algorithm 12.6.4 terminates at some loop, then it produces an optimizer for (12.6.1).

**Proposition 12.6.7.** *([247]) Under Assumptions 12.6.2 and 12.6.3, if Algorithm 12.6.4 terminates at some kth loop, then $(x^{(k)}, y^{(k)})$ is an optimizer of (12.6.1).*

Last, we have the asymptotic convergence for Algorithm 12.6.4.

**Theorem 12.6.8.** *([247]) For Algorithm 12.6.4, we assume the following:*

*(a) All optimization problems $(P_k)$ and $(Q_k)$ have global minimizers.*

*(b) Algorithm 12.6.4 does not terminate at any loop, so it produces the infinite sequence $\{(x^{(k)}, y^{(k)}, z^{(k)})\}_{k=0}^{\infty}$.*

*(c) Suppose $(x^*, y^*, z^*)$ is an accumulation point of $\{(x^{(k)}, y^{(k)}, z^{(k)})\}_{k=0}^{\infty}$ and the value function $v(x)$ is continuous at $x^*$.*

*(d) The polynomial functions $q^{(k)}(x, y)$ converge to $q^{(k)}(x^*, y^*)$ uniformly for $k \in \mathbb{N}$ as $(x, y) \to (x^*, y^*)$.*

*Then, $(x^*, y^*)$ is a global minimizer for the bilevel optimization (12.6.1).*

To ensure that the sequence $\{(x^{(k)}, y^{(k)}, z^{(k)})\}$ has an accumulation point, a sufficient condition is that the set $\mathcal{U}$ is bounded or the upper level objective $F(x, y)$ satisfies the growth condition that the sublevel set

$$\left\{ (x, y) \in \mathcal{U} \cap \mathcal{K} : F(x, y) \leq \vartheta \right\}$$

is bounded for all value $\vartheta$. The condition (d) can be checked on polynomial functions $q^{(k)}(x, y)$. For instance, it must hold if the degrees and coefficients of the polynomials $q^{(k)}(x, y)$ are uniformly bounded. We also need to assume that the value function $v(x)$ is continuous at $x^*$. This is the case under the so-called *restricted inf-compactness* (RIC) condition (see, e.g., [106, Definition 3.13]) and either $Z(x)$ is independent of $x$ or the MFCQ holds at some $\bar{z} \in Z(x^*)$. The RIC holds at $x^*$ for the value function $v(x)$ if the value $v(x^*)$ is finite and there exist a compact set $\Omega$ and a positive number $\epsilon_0$, such that for all $\|x - x^*\| < \epsilon_0$ with $v(x) < v(x^*) + \epsilon_0$, there exists $z \in S(x) \cap \Omega$. For instance, $v(x)$ satisfies the RIC at $x^*$ (see [54, §6.5.1]) under any one of the following conditions.

- The set $Z(x)$ is uniformly compact around $x^*$ (i.e., there is a neighborhood $N(x^*)$ of $x^*$ such that the closure of $\bigcup_{x \in N(x^*)} Z(x)$ is compact).

- The lower level objective $f(x, z)$ satisfies the growth condition, i.e., there exists a positive constant $\delta > 0$ such that the set

$$\left\{ z \in \mathbb{R}^p \left| \begin{array}{rcl} g_i(x^*, z) & = & \alpha_i (i \in \mathcal{E}), \\ g_j(x^*, z) & = & \alpha_j (j \in \mathcal{I}), \\ \sum_{i \in \mathcal{E} \cup \mathcal{I}} \alpha_i^2 & \leq & \delta, \\ f(x^*, z) & \leq & \vartheta \end{array} \right. \right\}$$

is bounded for every real scalar $\vartheta$.

- The objective $f(x, z)$ is weakly coercive in $z$ with respect to $Z(x)$ for all $x$ sufficiently close to $x^*$, i.e., there is a neighborhood $N(x^*)$ of $x^*$ such that

$$\lim_{z \in Z(x), \|z\| \to \infty} f(x, z) = \infty \qquad \forall x \in N(x^*).$$

### 12.6.2 ▪ Lagrange multiplier expressions

Recall the matrix polynomial $G(x, y)$ in (12.6.14). If $G(x, y) = G(y)$ is independent of $x$ (this is the case if (12.6.1) is a simple BPOP), then there generally exists a matrix polynomial $W(y)$ such that $W(y)G(y) = I_{m_2}$. For instance, this is the case if $g(x, z) = g(z)$ is a nonsingular polynomial tuple. However, if $G(x, y)$ depends on $y$, there typically does not exist a matrix polynomial $W(x, y)$ such that $W(x, y)G(x, y) = I_{m_2}$. This is because $G(x, y)$ in (12.6.14) usually does not have full column rank for all complex pairs $(x, y)$. We generally do not expect polynomial expressions for Lagrange multipliers when (12.6.1) is a general BPOP.

There always exists a matrix polynomial $W(x, y)$ satisfying (12.6.16) for denominators $d_j(x, y) \geq 0$ on $\mathcal{U}$. Note that the product

$$H(x, y) := G(x, y)^T G(x, y)$$

is a psd matrix polynomial. If the determinant $\det H(x, y)$ is not identically zero (this is the general case), then the adjoint matrix $\mathrm{adj}\big(H(x, y)\big)$ satisfies

$$\mathrm{adj}\big(H(x, y)\big) \cdot H(x, y) = \det H(x, y) \cdot I_{m_2}.$$

The equation (12.6.16) is then satisfied for the choice

$$\boxed{\begin{array}{rcl} W(x, y) & = & \mathrm{adj}\big(H(x, y)\big) \cdot G(x, y)^T, \\ d_j(x, y) & = & \det H(x, y), \ j = 1, \ldots, m_2. \end{array}}$$

The above choices for $W(x, y), d_j(x, y)$ typically have high degrees, so they may not be computationally efficient. There often exist low degree choices for them. The following is such an example.

**Example 12.6.9.** *([247]) Consider the lower level optimization problem*

$$\left\{ \begin{array}{ll} \min_{z \in \mathbb{R}^2} & x_1 z_1 + x_2 z_2 \\ s.t. & 2z_1 - z_2 \geq 0, \ x_1 - z_1 \geq 0, \ z_2 \geq 0, x_2 - z_2 \geq 0. \end{array} \right.$$

*The matrix $G(x, y)$ and $\hat{f}(x, y)$ in (12.6.14) are*

$$G(x, y) = \begin{bmatrix} 2 & -1 & 0 & 0 \\ -1 & 0 & 1 & -1 \\ 2y_1 - y_2 & 0 & 0 & 0 \\ 0 & x_1 - y_1 & 0 & 0 \\ 0 & 0 & y_2 & 0 \\ 0 & 0 & 0 & x_2 - y_2 \end{bmatrix}, \quad \hat{f}(x, y) = \begin{bmatrix} x_1 \\ x_2 \\ 0 \\ 0 \\ 0 \\ 0 \end{bmatrix}.$$

*The equation (12.6.16) holds for the denominators*

$$d_1(x, y) = 2x_1 - y_2, \quad d_2(x, y) = 2x_1 - y_2,$$
$$d_3(x, y) = x_2(2x_1 - y_2), \quad d_4(x, y) = x_2(2x_1 - y_2)$$

*and for the matrix $W(x, y)$, that is,*

$$\begin{bmatrix} x_1 - y_1 & 0 & 1 & 1 & 0 & 0 \\ y_2 - 2y_1 & 0 & 2 & 2 & 0 & 0 \\ (x_2 - y_2)(x_1 - y_1) & (x_2 - y_2)(2x_1 - y_2) & x_2 - y_2 & x_2 - y_2 & 2x_1 - y_2 & 2x_1 - y_2 \\ y_2(y_1 - x_1) & y_2(y_2 - 2x_1) & -y_2 & -y_2 & 2x_1 - y_2 & 2x_1 - y_2 \end{bmatrix}.$$

*Note that each $d_j(x, y) \geq 0$ for all $(x, y) \in \mathcal{U}$.*

In computational practice, we often choose $W(x, y), d_j(x, y)$ in (12.6.16) to have low degrees and each $d_j(x, y) > 0$ on $\mathcal{U}$ (or $d_j(x, y) \geq 0$ on $\mathcal{U}$). Although we prefer explicit expressions for $W(x, y)$ and $d(x, y)$, it may be inconvenient to do that for some BPOPs. The following is a numerical method for computing such $W(x, y)$ and $d(x, y)$. Select a general point $(\hat{x}, \hat{y}) \in \mathcal{U}$. For a priori degree $\ell$, we consider the following convex optimization problem in the coefficients of $W(x, y), d_1(x, y), \ldots, d_{m_2}(x, y)$:

$$\begin{cases} \max\limits_{W, d_j, \gamma_j} & \gamma_1 + \cdots + \gamma_{m_2} \\ \text{s.t.} & W(x, y)G(x, y) = \text{diag}[d_1(x, y), \ldots, d_{m_2}(x, y)], \\ & d_j(\hat{x}, \hat{y}) = 1, \ \gamma_j \geq 0, \ j = 1, \ldots, m_2, \\ & W(x, y) \in \left(\mathbb{R}[x, y]_{2\ell - \deg(G)}\right)^{m_2 \times (p + m_2)}, \\ & d_j(x, y) - \gamma_j \in \text{Ideal}[\Phi]_{2\ell} + \text{QM}[\Psi]_{2\ell} \ (j \in [m_2]). \end{cases} \quad (12.6.25)$$

In the above, the polynomial tuples $\Phi, \Psi$ are

$$\begin{array}{rl} \Phi & := \{h_i\}_{i \in \mathcal{E}_1} \cup \{g_i\}_{i \in \mathcal{E}_2}, \\ \Psi & := \{h_j\}_{j \in \mathcal{I}_1} \cup \{g_j\}_{j \in \mathcal{I}_2}. \end{array} \quad (12.6.26)$$

The first equality constraint in (12.6.25) is the same as (12.6.16), which gives a set of linear equations about coefficients of $W(x, y)$ and $d_j(x, y)$. The last constraint implies that for each $j$

$$d_j(x, y) \geq \gamma_j \geq 0 \quad \text{on} \quad \mathcal{U}.$$

The equality $d_j(\hat{x}, \hat{y}) = 1$ ensures that $d_j(x, y)$ is not identically zero. In computational practice, we often begin with a low degree $\ell$. If (12.6.25) is infeasible, then we increase the value of $\ell$, until it becomes feasible. We remark that (12.6.25) must have a solution when the degree $\ell$ is large enough. This is shown before Example 12.6.9.

### 12.6.3 ▪ Polynomial extensions

We show how to choose the polynomial extension $q(x,y)$ for Assumption 12.6.3. If (12.6.2) has linear equality constraints, we can get rid of them by eliminating variables. If (12.6.2) has nonlinear equality constraints, generally there is no polynomial $q(x,y)$ satisfying Assumption 12.6.3, unless the corresponding algebraic variety is rational. So, we consider the case that (12.6.2) has no equality constraints, i.e., the equality labeling set $\mathcal{E}_2 = \emptyset$. Moreover, we assume the polynomials $g_j(x,z)$ are linear in $z$ for each $j \in \mathcal{E}_2$. Recall the polynomial tuples $\Phi$, $\Psi$ given as in (12.6.26). For a priori degree $\ell$ and for given $(\hat{x}, \hat{y}, \hat{z})$, we consider the following polynomial system about the polynomial tuple $q$:

$$\begin{cases} q(\hat{x}, \hat{y}) = \hat{z}, \\ g_j(x, q) \in \text{Ideal}[\Phi]_{2\ell} + \text{QM}[\Psi]_{2\ell} & \text{for } j \in \mathcal{I}_2, \\ q = (q_1, \ldots, q_p) \in \big(\mathbb{R}[x,y]\big)^p. \end{cases} \qquad (12.6.27)$$

The second constraint in (12.6.27) implies that

$$g_j(x, q(x,y)) \geq 0 \quad \text{on} \quad \mathcal{U}$$

for every $j \in \mathcal{I}_2$. Clearly, every $q$ as in (12.6.27) must satisfy Assumption 12.6.3.

**Example 12.6.10.** *([247]) Consider Example 12.6.9 with*

$$\hat{x} = (1,0), \quad \hat{y} = (1,0), \quad \hat{z} = (0,0),$$

$$h(x,y) = (3x_1 - x_2, x_2, x_2 - x_1 + 1), \quad g(x,y) = (2y_1 - y_2, x_1 - y_1, y_2, x_2 - y_2).$$

*For $\ell = 2$, a satisfactory $q := (q_1, q_2)$ for (12.6.27) is*

$$q_1(x,y) = \frac{x_2}{3}, \quad q_2(x,y) = \frac{2x_2}{3},$$

*because $g(x,q) = \frac{1}{3}(0, 3x_1 - x_2, 2x_2, x_2)$ and*

$$3x_1 - x_2, x_2 \in \text{Ideal}[\Phi]_4 + \text{QM}[\Psi]_4.$$

In computational practice, we often prefer explicit expressions for $q(x,y)$. If the feasible set $Z(x)$ of (12.6.2) is independent of $x$, then we can just simply choose $q(x,y) := \hat{z}$ for Assumption 12.6.3. The $q$ is a constant polynomial function. Therefore, Assumption 12.6.3 is satisfied for all simple BPOPs. In the following, we give explicit expressions for $q$, for the box and simplex type constraints.

**Example 12.6.11.** *([247]) (i) Suppose the feasible set $Z(x)$ of (12.6.2) is*

$$l(x) \leq Az \leq u(x),$$

*where $A := \begin{bmatrix} a_1 & \cdots & a_{m_2} \end{bmatrix}^T$ is a full row rank matrix and $m_2 \leq p$. Select vectors $a_{m_2+1}, \ldots, a_p$ such that the matrix*

$$B := \begin{bmatrix} a_1 & \cdots & a_{m_2} & a_{m_2+1} & \cdots & a_p \end{bmatrix}^T$$

*is invertible. Then the linear coordinate transformation $z = B^{-1}w$ makes the constraints become the box constraints*

$$l_j(x) \leq w_j \leq u(x)_j, \ j = 1, \ldots, m_2.$$

*Hence we can choose $q = B^{-1}q'$, where $q' := (q_1', \ldots, q_p')$ as*

$$q_j'(x, y) := \begin{cases} \mu_j l_j(x) + (1 - \mu_j)u_j(x), & j = 1, \ldots, m_2, \\ (By)_j, & j = m_2 + 1, \ldots, p, \end{cases}$$

*where each scalar*

$$\mu_j := \frac{u_j(\hat{x}) - (B\hat{z})_j}{u_j(\hat{x}) - l_j(\hat{x})} \in [0, 1].$$

*For the special case that $u_j(\hat{x}) - l_j(\hat{x}) = 0$, we just set $\mu_j = 0$. One can verify that $q(x, y) \in Z(x)$ for all $(x, y) \in \mathcal{U}$.*

*(ii) Suppose that the feasible set $Z(x)$ of (12.6.2) is*

$$a^T z \leq u(x), \; z_j \geq l_j(x), \; j = 1, \ldots, p,$$

*where $a := (a_1, \ldots, a_p) \geq 0$, $u(x)$, and all $l_j(x)$ are polynomials in $x$. We can choose $q = (q_1, \ldots, q_p)$ such that*

$$q_j(x, y) := c_j \cdot \big( u(x) - a^T l(x) \big) + l_j(x),$$

*where each scalar*

$$c_j := \frac{\hat{z}_j - l_j(\hat{x})}{u(\hat{x}) - a^T l(\hat{x})} \geq 0.$$

*In particular, we set $c_j = 0$ for the case $u(\hat{x}) - a^T l(\hat{x}) = 0$. Note that*

$$q_j(\hat{x}, \hat{y}) = l_j(\hat{x}) + c_j \cdot (u(\hat{x}) - a^T l(\hat{x})) = \hat{z}_j.$$

*For all $(x, y) \in \mathcal{U}$, it is clear that $q(x, y) \geq l(x)$. Moreover,*

$$a^T q(x, y) = a^T l(x)(1 - \sum_{j=1}^{p} a_j c_j) + (\sum_{j=1}^{p} a_j c_j)u(x) \leq u(x),$$

*since $a^T l(x) \leq u(x)$ and $a_1 c_1 + \cdots + a_p c_p \leq 1$. Therefore, we have $q(x, y) \in Z(x)$ for all $(x, y) \in \mathcal{U}$.*

## 12.6.4 ▪ Exercises

**Exercise 12.6.1.** *([247]) Solve the general BPOP:*

$$\begin{cases} \min_{x, y \in \mathbb{R}^2} & x_1 y_1^3 + x_2 y_2^3 - x_1^2 x_2^2 \\ s.t. & (x_1 x_2 - 1, \; x_1, \; x_2, \; 4 - x_1^2 - x_2^2 - y_1^2 - y_2^2) \geq 0, \\ & y \in S(x), \end{cases}$$

*where $S(x)$ is the optimizer set of*

$$\begin{cases} \min_{z \in \mathbb{R}^2} & z_1^2 + z_2^2 - 2x_2 z_1 - x_1 x_2 z_2 \\ s.t. & (z_1, \; z_2 - x_2 z_1, \; 2x_1 - x_2 z_1 - z_2) \geq 0. \end{cases}$$

**Exercise 12.6.2.** *([247]) Solve the general BPOP:*

$$\begin{cases} \min_{x, y \in \mathbb{R}^4} & x_1^2 y_3^2 - 2x_3 x_4 + 1.2 x_1 x_3 - x_4^2(y_3 + 2y_4) \\ s.t. & (\mathbf{1}^T x, 8 - \mathbf{1}^T x, 4x_1 x_2 - y_1^2 - y_2^2) \geq 0, \\ & (x_1 - y_1, x_2 - y_2, 4 - x_1 - x_2, 4 - x_3^2 - x_4^2) \geq 0, \\ & y \in S(x), \end{cases}$$

*where $S(x)$ is the set of optimizer(s) of*

$$
\begin{cases}
\min_{z \in \mathbb{R}^4} & x_1 z_1^2 + x_2 z_2^2 + x_3 z_3 - x_4 z_4 \\
\text{s.t.} & (z_1 - z_2 - x_2, x_1 - z_1 + z_2, z_1 + z_2 + x_1 + x_2) \geq 0, \\
& (4x_1 - 2x_2 - z_1 - z_2, z_3, z_4, 3 - z_3 - z_4) \geq 0.
\end{cases}
$$

## 12.7 ▪ Distributionally Robust Optimization

The distributionally robust optimization (DRO) is frequently used for decision making problems under uncertainty. A typical DRO problem is

$$
\begin{cases}
\min & f(x) \\
\text{s.t.} & \inf_{\mu \in \mathcal{M}} \mathbb{E}_\mu[h(x, \xi)] \geq 0, \\
& x \in X,
\end{cases}
\tag{12.7.1}
$$

where $x$ is the decision variable, $X \subseteq \mathbb{R}^n$ is the constraining set, and $\xi \in \mathbb{R}^p$ is a random variable obeying the distribution of a measure $\mu \in \mathcal{M}$. The notation $\mathbb{E}_\mu$ stands for the expectation value with respect to the distribution measure of $\xi$. The $\mathcal{M}$ is a set of measures whose distributions have uncertainties. In applications, people often consider Borel measures whose supports are contained in a given set $S \subseteq \mathbb{R}^p$. We call $\mathcal{M}$ the *ambiguity set*, which describes the uncertainty of the measure $\mu$.

In applications, the ambiguity set $\mathcal{M}$ is often given by moments, i.e., the measure set $\mathcal{M}$ is specified by the first, second, or higher order moments [48, 73, 111, 168]. For instance, the Markov ambiguity set is given by expectations [320]. Higher order moments are also often used [48, 107]. When the ambiguity set is given by moments, the problem (12.7.1) is called the distributionally robust optimization of moments.

We refer to [72, 241, 264, 322, 336] for related work on DRO. The distributionally robust optimization has broad applications, e.g., portfolio management [73, 335], network design [195, 326], inventory problems [26, 320], and machine learning [85, 108]. For more detailed introductions to DRO, we refer to the survey [279]. The results in this section are mostly from the work [246].

### 12.7.1 ▪ Reformulation for DRO

We reformulate the DRO (12.7.1) equivalently as a conic optimization problem with moment constraints. Assume that the objective $f(x)$ is a polynomial in $x$ and $h(x, \xi)$ is a polynomial in $\xi$ whose coefficients are linear in $x$, i.e., $h(x, \xi)$ can be written as

$$
h(x, \xi) := (Ax + b)^T [\xi]_d
\tag{12.7.2}
$$

for a matrix $A \in \mathbb{R}^{\binom{p+d}{d} \times n}$ and a vector $b \in \mathbb{R}^{\binom{p+d}{d}}$. The integer $d$ is the degree of $h(x, \xi)$ in $\xi$. Suppose measures in $\mathcal{M}$ are supported in the set

$$
S := \{\xi \in \mathbb{R}^p : g_1(\xi) \geq 0, \ldots, g_{m_1}(\xi) \geq 0\}
\tag{12.7.3}
$$

for a given tuple $g := (g_1, \ldots, g_{m_1})$ of polynomials in $\xi$. Assume the ambiguity set $\mathcal{M}$ is given as

$$
\mathcal{M} := \{\mu : \ \text{supp}(\mu) \subseteq S, \ \mathbb{E}_\mu[\xi]_d \in Y \ \},
\tag{12.7.4}
$$

where $Y$ is the constraining set for moments of $\mu$. The set $Y$ is often closed and convex, but not necessarily. Its conic hull is $\operatorname{cone}(Y)$ and the closure is denoted as $\operatorname{cl}(\operatorname{cone}(Y))$. In computation, it is often a Cartesian product of linear, second order, or semidefinite cones. Suppose the constraining set $X$ for $x$ is

$$X := \{x \in \mathbb{R}^n : c_1(x) \geq 0, \ldots, c_{m_2}(x) \geq 0\} \tag{12.7.5}$$

for a given tuple $c = (c_1, \ldots, c_{m_2})$ of polynomials in $x$.

The DRO (12.7.1) can be equivalently reformulated as polynomial optimization with moment cones. Observe that

$$\inf_{\mu \in \mathcal{M}} \mathbb{E}_\mu h(x, \xi) \geq 0 \quad \Leftrightarrow \quad (Ax + b)^T y \geq 0 \,\, \forall y \in \mathscr{R}_d(S) \cap \operatorname{cone}(Y).$$

Consider the intersection

$$K := \mathscr{R}_d(S) \cap \operatorname{cone}(Y). \tag{12.7.6}$$

Then, it holds that

$$\inf_{\mu \in \mathcal{M}} \mathbb{E}_\mu[h(x, \xi)] \geq 0 \quad \Leftrightarrow \quad Ax + b \in K^\star, \tag{12.7.7}$$

where $K^\star$ is the dual cone of $K$. The dual cone of $Y$ is

$$Y^\star = \{\phi \in \mathbb{R}[\xi]_d : \langle \phi, z \rangle \geq 0 \,\forall z \in Y\}. \tag{12.7.8}$$

Note $Y^*$ is a convex cone of the polynomial space $\mathbb{R}[\xi]_d$. Observe that

$$\mathscr{R}_d(S)^\star = \mathscr{P}_d(S), \quad \mathscr{P}_d(S)^\star = \operatorname{cl}(\mathscr{R}_d(S)),$$
$$\left(\mathscr{P}_d(S) + Y^\star\right)^\star = \operatorname{cl}(\mathscr{R}_d(S)) \cap \operatorname{cl}(\operatorname{cone}(Y)).$$

When both $\mathscr{R}_d(S)$ and $\operatorname{cone}(Y)$ are closed, we have

$$\operatorname{cl}(\mathscr{R}_d(S) \cap \operatorname{cone}(Y)) = \operatorname{cl}(\mathscr{R}_d(S)) \cap \operatorname{cl}(\operatorname{cone}(Y)). \tag{12.7.9}$$

If one of them is not closed, the above may or may not hold. Note that $(C^\star)^\star = C$ if $C$ is a closed convex cone. When (12.7.9) holds and the sum $\mathscr{P}_d(S) + Y^\star$ is closed, we can express the dual cone $K^\star$ as

$$K^\star = \mathscr{P}_d(S) + Y^\star. \tag{12.7.10}$$

Since $h(x, \xi) = (Ax + b)^T[\xi]_d$, the membership $Ax + b \in K^\star$ means that as a polynomial in $\xi$, $h(x, \xi)$ belongs to the cone $K^\star$. We have the following theorem.

**Theorem 12.7.1.** *([246]) Assume the set $X$ is given as in (12.7.5). If the equality (12.7.10) holds, then (12.7.1) is equivalent to the following optimization problem:*

$$\begin{cases} \min_{x \in \mathbb{R}^n} & f(x) \\ s.t. & c_1(x) \geq 0, \ldots, c_{m_2}(x) \geq 0, \\ & h(x, \xi) \in \mathscr{P}_d(S) + Y^\star. \end{cases} \tag{12.7.11}$$

The membership constraint in (12.7.11) means that $h(x, \xi)$, as a polynomial in $\xi$, is the sum of a polynomial in $\mathscr{P}_d(S)$ and a polynomial in $Y^\star$. When $f, c_1, \ldots, c_{m_2}$ are all linear, (12.7.10) is a linear conic optimization problem. When they are polynomials, we can apply Moment-SOS relaxations to solve (12.7.11). Denote the degree

$$d_1 := \max\left\{\left\lceil \frac{1}{2}\deg(f)\right\rceil, \left\lceil \frac{1}{2}\deg(c)\right\rceil\right\}.$$

Define the projection map $\pi : \mathbb{R}^{N_{2d_1}^n} \to \mathbb{R}^n$ such that

$$\pi(w) := (w_{e_1}, \ldots, w_{e_n}), \quad w \in \mathbb{R}^{N_{2d_1}^n}. \tag{12.7.12}$$

A moment relaxation for the optimization (12.7.11) is

$$\begin{cases} \min_{(x,w)} & \langle f, w \rangle \\ s.t. & w_0 = 1, M_{d_1}[w] \succeq 0, \\ & L_{c_i}^{(d_1)}[w] \succeq 0 \, (i \in [m_2]), \\ & h(x, \xi) \in \mathscr{P}_d(S) + Y^\star, \\ & x = \pi(w), \, w \in \mathbb{R}^{N_{2d_1}^n}. \end{cases} \tag{12.7.13}$$

Under the SOS-convexity assumption (see §7.1.1), relaxation (12.7.13) is equivalent to (12.7.11). We have the following theorem.

**Theorem 12.7.2.** *([246]) Suppose the ambiguity set $\mathcal{M}$ is given as in (12.7.4) and the set $X$ is given as in (12.7.5). Assume the dual cone $K^\star$ can be expressed as in (12.7.10) and the polynomials $f, -c_1, \ldots, -c_{m_2}$ are SOS-convex. Then, the optimization problems (12.7.13) and (12.7.11) are equivalent in the following sense: they have the same optimal value, and $w^*$ is a minimizer of (12.7.13) if and only if $x^* := \pi(w^*)$ is a minimizer of (12.7.11).*

The dual optimization of (12.7.13) can be shown to be

$$\begin{cases} \max_{(\gamma,y)} & \gamma - \langle b, y \rangle \\ s.t. & f(x) - y^T A x - \gamma \in \mathrm{QM}[c]_{2d_1}, \\ & \gamma \in \mathbb{R}, \, y \in \mathrm{cl}\,(K). \end{cases} \tag{12.7.14}$$

The first membership in (12.7.14) means that $f(x) - y^T A x - \gamma$, as a polynomial in $x$, belongs to the truncated quadratic module $\mathrm{QM}[c]_{2d_1}$.

### 12.7.2 ▪ Moment-SOS relaxations

We have shown that (12.7.11) is equivalent to the linear conic optimization (12.7.13) under certain assumptions. It is still hard to solve (12.7.13) directly, due to the membership constraint $h(x, \xi) \in \mathscr{P}_d(S) + Y^\star$. For its dual problem (12.7.14), it is similarly difficult to deal with the conic membership constraint $y \in \mathrm{cl}\,(K)$. However, both (12.7.13) and (12.7.14) can be solved by Moment-SOS relaxations.

Recall that $S$ is the set as in (12.7.3). For each integer $k \geq \frac{d}{2}$, it holds that the nesting containment

$$(\mathrm{QM}[g]_{2k} \cap \mathbb{R}[\xi]_d) \subseteq (\mathrm{QM}[g]_{2k+2} \cap \mathbb{R}[\xi]_d) \subseteq \cdots \subseteq \mathscr{P}_d(S).$$

So we consider the following restriction of (12.7.13):

$$\begin{cases} \min_{(x,w)} & \langle f, w \rangle \\ s.t. & w_0 = 1, M_{d_1}[w] \succeq 0, \\ & L_{c_i}^{(d_1)}[w] \succeq 0 \, (i \in [m_2]), \\ & h(x, \xi) \in \mathrm{QM}[g]_{2k} + Y^\star, \\ & x = \pi(w), w \in \mathbb{R}^{N_{2d_1}^n}. \end{cases} \tag{12.7.15}$$

Since $(\mathrm{QM}[g]_{2k})^\star = \mathscr{S}[g]_{2k}$, the dual optimization of (12.7.15) is

$$
\left\{
\begin{aligned}
& \max_{(\gamma,y,z)} && \gamma - \langle b, y \rangle \\
& \text{s.t.} && f(x) - y^T A x - \gamma \in \mathrm{QM}[c]_{2d_1}, \\
& && y = z|_d, \\
& && \gamma \in \mathbb{R},\ z \in \mathscr{S}[g]_{2k}, \\
& && y \in \mathrm{cl}\,(\mathrm{cone}\,(Y)).
\end{aligned}
\right.
\tag{12.7.16}
$$

We remark that $\mathrm{QM}[g]$ is a quadratic module in the polynomial ring $\mathbb{R}[\xi]$, while $\mathrm{QM}[c]$ is a quadratic module in the polynomial ring $\mathbb{R}[x]$. The optimization (12.7.16) is a relaxation of (12.7.14), since it has a larger feasible set. There exist both quadratic module and moment constraints in (12.7.16). The primal-dual pair (12.7.15)-(12.7.16) can be solved as a semidefinite program. The following is a basic property.

**Theorem 12.7.3.** *([246]) Assume (12.7.9) holds. Suppose $(\gamma^*, y^*, z^*)$ is an optimizer of (12.7.16) for the relaxation order $k$. Then $(\gamma^*, y^*)$ is a maximizer of (12.7.14) if and only if $y^* \in \mathrm{cl}\,(\mathscr{R}_d(S))$.*

If $\mathscr{R}_d(S)$ is closed, then we only need to check $y^* \in \mathscr{R}_d(S)$ in the above. When $S$ is compact, the moment cone $\mathscr{R}_d(S)$ is closed. As in §8.2, the membership $y^* \in \mathscr{R}_d(S)$ can be checked by Algorithm 8.2.4. Once $(\gamma^*, y^*)$ is confirmed to be a maximizer of (12.7.14), we can obtain a minimizer for (12.7.1). This is given in the following theorem.

**Theorem 12.7.4.** *([246]) Assume (12.7.9) holds. For a relaxation order $k$, suppose $(x^*, w^*)$ is a minimizer of (12.7.15) and $(\gamma^*, y^*, z^*)$ is a maximizer of (12.7.16) such that $y^* \in \mathrm{cl}\,(\mathscr{R}_d(S))$. Assume there is no duality gap between (12.7.15) and (12.7.16), i.e., they have the same optimal value. If the point $x^*$ belongs to the set $X$ and $f(x^*) = \langle f, w^* \rangle$, then $x^*$ is a minimizer of (12.7.11). Moreover, if in addition the dual cone $K^*$ can be expressed as in (12.7.10), then $x^*$ is also a minimizer of (12.7.1).*

In Theorem 12.7.4, we can have $x^* \in X$ and $f(x^*) = \langle f, w^* \rangle$ under the SOS-convexity assumption.

**Theorem 12.7.5.** *([246]) Assume (12.7.9) holds. For a relaxation order $k$, suppose $(x^*, w^*)$ is a minimizer of (12.7.15) and $(\gamma^*, y^*, z^*)$ is a maximizer of (12.7.16) such that $y^* \in \mathrm{cl}\,(\mathscr{R}_d(S))$. Assume there is no duality gap between (12.7.15) and (12.7.16). If $f, -c_1, \ldots, -c_{m_2}$ are SOS-convex polynomials, then $x^* := \pi(w^*)$ is a minimizer of (12.7.11). Moreover, if in addition $K^*$ can be expressed as in (12.7.10), then $x^*$ is also a minimizer of (12.7.1).*

The following is the Moment-SOS algorithm for solving the optimization (12.7.13) and its dual (12.7.14), as well as the DRO (12.7.1).

**Algorithm 12.7.6.** *([246]) For given $f, h, \mathcal{M}, S, Y$, do the following:*

**Step 0** *Get a computational representation for $\mathrm{cl}\,(\mathrm{cone}\,(Y))$ and the dual cone $Y^\star$. Let $k := \lceil \frac{d}{2} \rceil$.*

**Step 1** *Solve (12.7.15) for a minimizer $(x^*, w^*)$ and solve (12.7.16) for a maximizer $(\gamma^*, y^*, z^*)$.*

**Step 2** *Apply Algorithm 8.2.4 to check whether $y^* \in \mathscr{R}_d(S)$ or not. If $y^* \in \mathscr{R}_d(S)$, obtain a finitely atomic representing measure for $y^*$, and stop. If $y^* \notin \mathscr{R}_d(S)$, let $k := k + 1$ and go to Step 1.*

The convergence of Algorithm 12.7.6 is shown in [246]. We conclude this section with a DRO example.

**Example 12.7.7.** *([246]) Consider the DRO*

$$
\begin{cases}
\min_{x \in \mathbb{R}^3} & f(x) = (x_1 - x_3 + x_1 x_3)^2 + (2x_2 + 2x_1 x_2 - x_3^2)^2 \\
s.t. & \inf_{\mu \in \mathcal{M}} \mathbb{E}_\mu[h(x, \xi)] \geq 0, \\
& c_1(x) = 1 - x_1^2 - x_2^2 - x_3^2 \geq 0, \\
& c_2(x) = 3x_3 - x_1^2 - 2x_2^4 \geq 0,
\end{cases}
\tag{12.7.17}
$$

*where (the random variable $\xi = (\xi_1, \xi_2)$ is bivariate)*

$$
\begin{aligned}
h(x, \xi) &= (1 - x_3)\xi_1^2\xi_2^2 + (x_1 - x_2 + x_3 - 1)\xi_1\xi_2^2 \\
&\quad + (x_1 + x_2 + x_3 + 1)\xi_2^2 + (x_1 - x_3)\xi_1^2 - \xi_2,
\end{aligned}
$$

$$
S = \{\xi \in \mathbb{R}^2 : 1 - \xi^T\xi \geq 0\}, \quad g := 1 - \xi^T\xi,
$$

$$
Y = \left\{ y \in \mathbb{R}^{\mathbb{N}_4^2} \middle|
\begin{array}{l}
y_{00} = 1, \ 0.1 \leq y_\alpha \leq 1 \ (0 < |\alpha| \leq 4) \\
\begin{bmatrix}
y_{20} & y_{11} & y_{30} & y_{12} \\
y_{11} & y_{02} & y_{21} & y_{03} \\
y_{30} & y_{21} & y_{40} & y_{22} \\
y_{12} & y_{03} & y_{22} & y_{04}
\end{bmatrix} \preceq 2I_4
\end{array}
\right\}.
$$

*The polynomials $f$ and all $-c_i$ are SOS-convex. Algorithm 12.7.6 terminates in the initial loop with $k = 2$. The optimal value $F^*$ and optimizer $x^*$ of (12.7.11) are respectively*

$$
F^* \approx 0.0160, \quad x^* \approx (0.4060, 0.0800, 0.4706).
$$

*The optimizer for (12.7.16) is*

$$
\begin{aligned}
y^* \approx (&0.3180, 0.2750, 0.1411, 0.2436, 0.1137, 0.0744, 0.2199, 0.0950, \\
&0.0552, 0.0460, 0.2011, 0.0819, 0.0426, 0.0318, 0.0318).
\end{aligned}
$$

*The measure $\mu$ for $y^* = \int [\xi]_4 d\mu$ is $0.2527\delta_{u_1} + 0.7473\delta_{u_2}$, with*

$$
u_1 \approx (0.6325, 0.7745), \quad u_2 \approx (0.9434, 0.3317).
$$

## 12.7.3 ▪ Exercises

**Exercise 12.7.1.** *Assume each $\mu \in \mathcal{M}$ is a probability measure (i.e., $\mathbb{E}_\mu[1] = 1$). Show that the distributionally robust min-max optimization*

$$
\min_{x \in X} \max_{\mu \in \mathcal{M}} \mathbb{E}_\mu[F(x, \xi)]
\tag{12.7.18}
$$

*is equivalent to the following DRO:*

$$
\begin{cases}
\min & x_0 \\
s.t. & \inf_{\mu \in \mathcal{M}} \mathbb{E}_\mu[x_0 - F(x, \xi)] \geq 0, \\
& x \in X, x_0 \in \mathbb{R}.
\end{cases}
\tag{12.7.19}
$$

**Exercise 12.7.2.** *Show that the dual optimization of (12.7.13) is (12.7.14).*

## 12.8 ▪ Multi-objective Optimization

The multi-objective optimization problem (MOP) is to optimize several objectives simultaneously over a common feasible set. A typical MOP is

$$
\begin{cases}
\min & f(x) := (f_1(x), \ldots, f_m(x)) \\
s.t. & c_i(x) = 0 \, (i \in \mathcal{E}), \\
& c_j(x) \geq 0 \, (j \in \mathcal{I}),
\end{cases}
\tag{12.8.1}
$$

where $f_i, c_i, c_j$ are functions in $x := (x_1, \ldots, x_n) \in \mathbb{R}^n$. Let $K$ denote the feasible set of (12.8.1). Generally, there does not exist a common minimizer for all $f_i$. Therefore, people are interested in points such that some or all of the objectives cannot be further optimized. This leads to the following classical concepts (see [188, 196, 133]).

**Definition 12.8.1.** *A point $x^* \in K$ is said to be a Pareto point (PP) if there is no $x \in K$ such that $f_i(x) \leq f_i(x^*)$ for all $i = 1, \ldots, m$ and $f_j(x) < f_j(x^*)$ for at least one $j$. The point $x^*$ is said to be a weakly Pareto point (WPP) if there is no $x \in K$ such that $f_i(x) < f_i(x^*)$ for all $i = 1, \ldots, m$.*

In the literature, Pareto points (resp., weakly Pareto points) are also referenced as Pareto optimizers (resp., weakly Pareto optimizers) or just Pareto solutions (resp., weakly Pareto solutions). A vector $v := (v_1, \ldots, v_m)$ is called a *Pareto value* (resp., weakly Pareto value) for (12.8.1) if there exists a Pareto point (resp., weakly Pareto point) $x^*$ such that $v = f(x^*)$. *Pareto front* is the set of objective values at Pareto points. Every Pareto point is a weakly Pareto point, while the converse is typically not the case. Detecting existence or nonexistence of PPs and WPPs is a major problem for MOPs. Scalarization is a classical method for finding PPs or WPPs. It transforms a MOP into a single objective optimization problem.

**Definition 12.8.2.** *The linear scalarization problem (LSP) for (12.8.1), with a nonzero weight $w := (w_1, \ldots, w_m) \geq 0$, is the optimization*

$$
\min_{x \in K} \quad w_1 f_1(x) + \cdots + w_m f_m(x).
\tag{12.8.2}
$$

Every minimizer of (12.8.2) is a weakly Pareto point for nonzero $w \geq 0$ and it is a Pareto point for $w > 0$. Varying weights in (12.8.2) may give different (weakly) Pareto points. A nonzero weight $w$ is said to be *proper* if the LSP (12.8.2) is bounded below. Otherwise, the $w$ is called *improper*. Not every Pareto point is a minimizer of (12.8.2) for some weight $w$. For instance, Example 12.8.8 has infinitely many Pareto points, but only two of them can be obtained by LSPs. Under some additional assumptions (e.g., convexity), LSPs may give all Pareto points (see [89]).

Another frequently used scalarization is the Chebyshev scalarization. It requires one to use the minimum value of each objective.

**Definition 12.8.3.** *The Chebyshev scalarization problem (CSP) for (12.8.1), with a nonzero weight $w = (w_1, \ldots, w_m) \geq 0$, is*

$$
\min_{x \in K} \max_{1 \leq i \leq m} \quad w_i \big( f_i(x) - f_i^* \big),
\tag{12.8.3}
$$

*where the minimum value $f_i^* := \min_{x \in K} f_i(x) > -\infty$.*

Every minimizer of the CSP (12.8.3) is a weakly Pareto point. Interestingly, every weakly Pareto point is the minimizer of a CSP for some weight (see [196]). However, the minimizer of a CSP may not be a Pareto point. There also exist other scalarizations, such as the lexicographic method [55]. More scalarizations can be found in references [51, 188, 196, 289]. Interesting topics for MOPs are the following:

- What is a convex description for the set of (weakly) Pareto values?

- For LSPs, how do we find proper weights? How can we detect if an LSP is unbounded below? How can we detect nonexistence of proper weights?

- How can we solve CSPs efficiently for (weakly) Pareto points? If some minimum value $f_i^*$ is $-\infty$, how can we get (weakly) Pareto points?

- For a given point, how can we detect if it is a (weakly) Pareto point? How do we detect existence or nonexistence of (weakly) Pareto points?

We discuss these topics when the MOP is given by polynomials. Recent work on polynomial MOPs can be found in [27, 135, 136, 137, 138, 141, 174]. The results in this section are mostly from [251].

## 12.8.1 ▪ Convex geometry of Pareto values

Pareto values (PVs) and weakly Pareto values (WPVs) are related to the epigraph set

$$\mathcal{U} := \{(u_1, \ldots, u_m) \mid u_i \geq f_i(x), \ i = 1, \ldots, m, \ x \in K\}. \tag{12.8.4}$$

The image of $K$ under the objective map $f := (f_1, \ldots, f_m)$ is the set

$$f(K) := \{(f_1(x), \ldots, f_m(x)) : x \in K\}.$$

Note that $\mathcal{U} = f(K) + \mathbb{R}_+^m$ and $\operatorname{conv}(\mathcal{U}) = \operatorname{conv}(f(K)) + \mathbb{R}_+^m$. If $K$ is convex and each objective $f_i$ is convex, the set $\mathcal{U}$ is also convex. When $\mathcal{U}$ is convex, every Pareto point is a minimizer of some LSP (see [89]).

For a nonzero $w \in \mathbb{R}^m$ and $b \in \mathbb{R}$, the set $H = \{u \in \mathbb{R}^m : w^T u = b\}$ is a supporting hyperplane for $\mathcal{U}$ (see §1.2) if

$$b = \inf_{u \in \mathcal{U}} w^T u.$$

The $w$ is the normal of $H$. In particular, if there exists $v \in \mathcal{U}$ such that $w^T u \geq w^T v$ for all $u \in \mathcal{U}$, then $H$ is called a supporting hyperplane through $v$. Since $\mathcal{U}$ contains $f(u) + \mathbb{R}_+^n$, the normal $w$ must be nonnegative for $H$ to be a supporting hyperplane.

To characterize Pareto values, we define a kind of *minimal points*. Let $\pi$ be a permutation of $(1, \ldots, m)$. For a set $T \subseteq \mathbb{R}^m$, construct the following chain of nesting subsets:

$$T = T_0 \supseteq T_1 \supseteq \cdots \supseteq T_m$$

such that for each $k = 1, \ldots, m$, $T_k$ is the subset of vectors in $T_{k-1}$ whose $\pi(k)$th entry is the smallest. If $T_m \neq \emptyset$, then each $v \in T_m$ is called a $\pi$-*minimal* point of $T$. For $u, v \in T_m$, all the entries of $u, v$ must be the same, so $u = v$ and hence $T_m$ consists of a single point, if it is nonempty. In particular, if $T$ is compact, then $T_m \neq \emptyset$ and it consists of a single point. PVs and WPVs can be characterized as follows.

**Proposition 12.8.4.** *([251]) For each $v \in f(K)$, we have the following:*

(i) *The vector $v$ is a WPV if and only if $v$ lies on the boundary of $\mathcal{U}$. Moreover, if $v$ is an extreme point of conv $(\mathcal{U})$, then $v$ is a PV.*

(ii) *Assume $\mathcal{U}$ is convex. If $v$ is a WPV, then there exists a supporting hyperplane for $\mathcal{U}$ through $v$ whose normal is nonnegative, i.e., there exists $0 \neq w \geq 0$ such that $w^T u \geq w^T v$ for all $u \in \mathcal{U}$.*

(iii) *Suppose $H = \{u : w^T u = w^T v\}$ is a supporting hyperplane for $\mathcal{U}$ through $v$, with a normal vector $0 \neq w \geq 0$. If $w > 0$, then $v$ is a PV. When $w$ has a zero entry, if $u \in f(K)$ is a $\pi$-minimal point of $H \cap \mathcal{U}$ for a permutation $\pi$ of $(1, \ldots, m)$, then $u$ is a PV. If $u \in f(K)$ is an extreme point of $H \cap \mathcal{U}$, then $u$ is also a PV.*

The following are useful facts about PVs and WPVs.

- Not every WPV lies on the boundary of conv $(\mathcal{U})$. For instance, consider

$$\begin{cases} \min & (x_1, x_2) \\ s.t. & x_1 \geq 0, x_2 \geq 0, x_1^2 + x_2^2 = 1. \end{cases}$$

For each $t \in (0, 1)$, the point $(t, \sqrt{1 - t^2})$ is a WPP (also a PP), but it does not lie on the boundary of conv $(\mathcal{U})$.

- If $\mathcal{U}$ is not convex, there may not exist a supporting hyperplane through a WPV. For instance, in the above MOP, for every $t \in (0, 1)$, there is no supporting hyperplane for conv $(\mathcal{U})$ through $(t, \sqrt{1 - t^2})$.

- In the item (iii) of Proposition 12.8.4, if $w$ has a zero entry, then $v$ may not be a Pareto value. For instance, consider the MOP with $f_1 = x_1$, $f_2 = x_2^2$, and $K = \mathbb{R}^2$. For $w = (0, 1)$ and $v = (0, 0)$, the equation $w^T u = 0$ gives a supporting hyperplane through $(0, 0)$, but $(0, 0)$ is not a Pareto value.

- If $v$ is a PV, it may not be an extreme point of $\mathcal{U}$ or $H \cap \mathcal{U}$. For instance, consider the MOP

$$\begin{cases} \min & (x_1, x_2) \\ s.t. & x_1 \geq 0, x_2 \geq 0, x_1 + x_2 = 1. \end{cases}$$

The set $\mathcal{U} = \{x_1 \geq 0, x_2 \geq 0, x_1 + x_2 \geq 1\}$. Clearly, for every $t \in (0, 1)$, the vector $(t, 1 - t)$ is a PV, but it is not an extreme point of $\mathcal{U}$. The hyperplane $H = \{x_1 + x_2 = 1\}$ supports $\mathcal{U}$ at $(t, 1 - t)$. However, $(t, 1 - t)$ is not an extreme point of the intersection $H \cap \mathcal{U}$ for every $t \in (0, 1)$.

When $K$ is bounded, there always exist supporting hyperplanes for $\mathcal{U}$. When $K$ is unbounded, they may not exist. For given $v = (v_1, \ldots, v_m) \in f(K)$, how do we determine whether there is a supporting hyperplane through it? For this purpose, we consider the linear optimization in $w_0 \in \mathbb{R}$ and $w = (w_1, \ldots, w_m) \in \mathbb{R}^m$:

$$\begin{cases} \max & w_0 \\ s.t. & 1 - e^T w = 0, \ w_i \geq w_0, \ i = 1, \ldots, m, \\ & \sum_{i=1}^{m} w_i(f_i(x) - v_i) \geq 0 \quad \text{on } K. \end{cases} \tag{12.8.5}$$

Let $\omega^*$ be the optimal value of (12.8.5). Clearly, there is a supporting hyperplane through $v$ if and only if $\omega^* \geq 0$. Let $d$ be the maximum degree of all $f_i$. The third constraint in (12.8.5) is equivalent to the membership

$$\sum_{i=1}^{m} w_i(f_i(x) - v_i) \in \mathscr{P}_d(K).$$

The dual cone of $\mathscr{P}_d(K)$ is the closure $\mathrm{cl}(\mathscr{R}_d(K))$, where $\mathscr{R}_d(K)$ is the moment cone (see (8.1.2)). The dual optimization of (12.8.5) is

$$\begin{cases} \min & t \\ \text{s.t.} & t - \langle f_i - v_i, y \rangle \geq 0,\ i = 1, \ldots, m, \\ & 1 = mt - \sum_{i=1}^{m} \langle f_i - v_i, y \rangle, \\ & y \in \mathrm{cl}(\mathscr{R}_d(K)). \end{cases} \tag{12.8.6}$$

If (12.8.6) has a feasible point with $t < 0$, then there are no nonnegative supporting hyperplanes through $v$. Since each $v_i$ is a scalar, one can see that

$$\langle f_i - v_i, y \rangle = \langle f_i, y \rangle - v_i \langle 1, y \rangle = \langle f_i, y \rangle - v_i y_0.$$

When $t < 0$ is feasible for (12.8.6), there also exists a feasible $y \in \mathscr{R}_d(K)$ with $y_0 > 0$. One can scale such $(t, y)$ so that $y_0 = 1$. Hence, the existence of $t < 0$ in (12.8.6) is equivalent to

$$\begin{cases} \tau = mt' - \sum_{i=1}^{m} (\langle f_i, y \rangle - v_i), \\ t' \geq \langle f_i, y \rangle - v_i,\ i = 1, \ldots, m, \\ \tau > 0 > t', \\ y_0 = 1,\ y \in \mathscr{R}_d(K). \end{cases}$$

The above is then equivalent to

$$\boxed{\begin{array}{l} v_i > \langle f_i, y \rangle,\ i = 1, \ldots, m, \\ y_0 = 1,\ y \in \mathscr{R}_d(K). \end{array}}$$

We consider the set

$$\mathcal{V} := \left\{ v \ \middle|\ \begin{array}{l} v = (v_1, \ldots, v_m) \\ v_i > \langle f_i, y \rangle,\ i = 1, \ldots, m, \\ y_0 = 1,\ y \in \mathscr{R}_d(K) \end{array} \right\}. \tag{12.8.7}$$

**Theorem 12.8.5.** *([251]) Assume $K$ has nonempty interior. Then, the interior of the convex hull conv $(\mathcal{U})$ is the set $\mathcal{V}$ as in (12.8.7). Moreover, when $\mathcal{U}$ is convex, a vector $v \in f(K)$ is a weakly Pareto value if and only if $v$ belongs to the boundary of the closure $\mathrm{cl}(\mathcal{V})$.*

When the polynomials are SOS-convex, we can get a semidefinite representation for $\mathcal{V}$.

**Theorem 12.8.6.** *([251]) Assume there are no equality constraints and $K$ has nonempty interior. If all $f_i$ and $-c_j$ ($j \in \mathcal{I}$) are SOS-convex polynomials, then the interior of $\mathcal{U}$ is equal to the set*

$$\mathcal{V}_1 := \left\{ (v_1, \ldots, v_m) \ \middle|\ \begin{array}{l} \langle c_j, y \rangle \geq 0\ (j \in \mathcal{I}), \\ v_i > \langle f_i, y \rangle,\ i = 1, \ldots, m, \\ M_{d_0}[y] \succeq 0, \\ y_0 = 1,\ y \in \mathbb{R}^{\mathbb{N}_{2d_0}^n} \end{array} \right\}, \tag{12.8.8}$$

*where*

$$d_0 := \max_{j \in \mathcal{I}} \left\{ \left\lceil \frac{d}{2} \right\rceil, \left\lceil \frac{1}{2} \deg(c_j) \right\rceil \right\}.$$

*Moreover, a vector $v \in f(K)$ is a weakly Pareto value if and only if it lies on the boundary of* cl $(\mathcal{V}_1)$.

## 12.8.2 ▪ The linear scalarization

For a weight $w := (w_1, \ldots, w_m)$, denote the weighted sum of objectives

$$f_w(x) := w_1 f_1(x) + \cdots + w_m f_m(x). \tag{12.8.9}$$

We consider the LSP

$$\begin{cases} \min & f_w(x) \\ s.t. & x \in K. \end{cases} \tag{12.8.10}$$

Recall that $w \neq 0$ is a proper weight if (12.8.10) is bounded below. The Moment-SOS hierarchy can be applied to solve (12.8.10). When $K$ is unbounded, the tight relaxation method in §6.3 can be applied to solve (12.8.10), when the constraints are nonsingular.

**Example 12.8.7.** *Consider the objectives*

$$f_1 = \left( \sum_{i=1}^{5} x_i^4 \right) + x_1^2 x_2 + x_1 x_2^2 - 3x_1 x_2 x_3 + x_3 x_4 x_5 + x_3^3,$$

$$f_2 = \left( \sum_{i=1}^{5} x_i^2 \right) - x_1 x_2^2 - x_2 x_3^2 + x_3 x_4^2 + x_4 x_5^2$$

*and the constraint $x_1^2 + \cdots + x_5^2 \geq 1$. The feasible set is unbounded. A list of some weights $w$ and the corresponding Pareto points are given as follows:*

| $w$ | Pareto point |
|---|---|
| $(0.5, 0.5)$ | $(-0.3371, 0.4659, -0.7504, -0.2807, -0.1655)$ |
| $(0.25, 0.75)$ | $(-0.0986, 0.3316, -0.6802, -0.5493, -0.3405)$ |
| $(0.75, 0.25)$ | $(-0.7711, 0.9015, -1.1818, -0.5752, -0.5114)$ |

## Existence of proper weights

It is possible that $f_w(x)$ is unbounded below on $K$ for some weight $w$. For instance, $f_w(x)$ is unbounded below for $w = (0, 1)$ in Example 12.8.7. We refer to §8.5 for how to detect unboundedness. Not every Pareto point is the minimizer of a LSP, shown as follows.

**Example 12.8.8.** *Consider the MOP with*

$$f_1 = -x_1^3 - x_2^3 + (x_3 - x_4)^2, \quad f_2 = x_1^2 - x_2^2 + (x_3 + x_4)^2$$

*and the constraints $0 \leq x_1, x_2 \leq 1$. For $w_1 \geq w_2$, the minimizer is $(1, 1, 0, 0)$. For $w_1 < w_2$, the minimizer is $(0, 1, 0, 0)$. So the LSP can only give two Pareto points, by exploring all possibilities of weights. However, each $(t, 1, 0, 0)$, with $0 \leq t \leq 1$, is a Pareto point.*

When $K$ is bounded, the LSP (12.8.10) is bounded below for all weights. When $K$ is unbounded, (12.8.10) may be unbounded below for some $w$. To find a (weakly) Pareto point, we look for a nonzero weight $w \geq 0$ such that (12.8.10) is bounded below, i.e., $w$ is a proper weight. Denote the set of proper weights:

$$\mathcal{W} := \left\{ w \in \mathbb{R}_+^m : f_w(x) \text{ is bounded below on } K, \ w \neq 0 \right\}. \tag{12.8.11}$$

Clearly, $\mathcal{W}$ is a convex cone. Let $d$ denote the maximum degree of $f_1, \ldots, f_m$. Then, a nonzero weight $w$ is proper if and only if there exists a scalar $\gamma \in \mathbb{R}$ such that $f_w(x) - \gamma \in \mathscr{P}_d(K)$. So,

$$\mathcal{W} = \left\{ w \in \mathbb{R}_+^m : f_w(x) \in \mathscr{P}_d(K) + \mathbb{R}, \ w \neq 0 \right\}. \tag{12.8.12}$$

The cone $\mathscr{P}_d(K)$ contains the sum $\mathrm{Ideal}[c_{eq}] + \mathrm{QM}[c_{in}]$, so

$$\left\{ 0 \neq w \in \mathbb{R}_+^m : f_w(x) \in \mathrm{Ideal}[c_{eq}] + \mathrm{QM}[c_{in}] + \mathbb{R} \right\} \subseteq \mathcal{W}. \tag{12.8.13}$$

In the above $c_{eq} = (c_i)_{i \in \mathcal{E}}$ and $c_{in} = (c_j)_{j \in \mathcal{I}}$.

When $\mathrm{Ideal}[c_{eq}] + \mathrm{QM}[c_{in}]$ is archimedean ($K$ is bounded for this case), the containment in (12.8.13) is an equality. This is because if $f_w(x)$ is bounded below on $K$, then $f_w(x) - \gamma \in \mathrm{Ideal}[c_{eq}] + \mathrm{QM}[c_{in}]$ for $\gamma$ small enough. When $K$ is unbounded, the sum $\mathrm{Ideal}[c_{eq}] + \mathrm{QM}[c_{in}]$ cannot be archimedean, and the containment in (12.8.13) is typically not an equality.

Among all proper weights $w \geq 0$ normalized as $e^T w = 1$, the smallest possibility of the minimum value of (12.8.10) is the minimum of $f_1^*, \ldots, f_m^*$, where $f_i^*$ is the minimum value of $f_i(x)$ on $K$. Some of $f_i^*$ may be $-\infty$. For the choice $w = e_i$, the minimum value of (12.8.10) is $f_i^*$. Beyond them, people are also interested in $w$ such that the minimum value of (12.8.10) is maximum. For the minimum value of $f_w(x)$ on $K$ to be maximum, we consider the optimization

$$\begin{cases} \max & \gamma \\ s.t. & 1 - e^T w = 0, \ w_1 \geq 0, \ldots, w_m \geq 0, \\ & \sum_{i=1}^m w_i f_i - \gamma \in \mathscr{P}_d(K). \end{cases} \tag{12.8.14}$$

The dual cone of $\mathscr{P}_d(K)$ is $\mathrm{cl}\,(\mathscr{R}_d(K))$. (When $K$ is compact, the moment cone $\mathscr{R}_d(K)$ is closed.) The dual optimization of (12.8.14) is

$$\begin{cases} \min & \mu \\ s.t. & \mu - \langle f_i, y \rangle \geq 0, \ i = 1, \ldots, m, \\ & y_0 = 1, \ y \in \mathrm{cl}\,(\mathscr{R}_d(K)). \end{cases} \tag{12.8.15}$$

The Moment-SOS hierarchy in §8.3 can be similarly applied to solve (12.8.14)-(12.8.15). When $\mathrm{Ideal}[c_{eq}] + \mathrm{QM}[c_{in}]$ is archimedean, the convergence of the hierarchy can be similarly shown as in §8.3.2 (also see [161, 232, 251]).

**Example 12.8.9.** *Consider the objectives*

$$f_1 = \left( x_1{}^2 + x_2 + x_3 \right)^2 + \left( x_2{}^2 + x_3 + x_4 \right)^2 - 3\,x_1 x_2 x_3 x_4,$$

$$f_2 = \Big( \sum_{i=1}^4 x_i^4 \Big) - (x_1 - x_2)(x_2 - x_3)(x_3 - x_4)(x_4 - x_1),$$

$$f_3 = 3 \Big( \sum_{i=1}^4 x_i^3 \Big) + x_1^2 \left( x_2{}^2 - x_3{}^2 \right) + x_2^2 \left( x_3{}^2 - x_4{}^2 \right) + x_3^2 \left( x_4{}^2 - x_1{}^2 \right)$$

*and the constraints*

$$x_1 x_2 \geq 1,\ x_2 x_3 \geq 1,\ x_3 x_4 \geq 1,\ x_1 \geq 0.$$

*Each $f_i$ is unbounded below on the feasible set. The optimization (12.8.14) can be solved by the Moment-SOS hierarchy. The optimal weight $w^*$ and Pareto point $x^*$ are respectively*

$$w^* \approx (0.5769, 0.2229, 0.2003),\ x^* \approx (1.0105, 0.9897, 1.0105, 0.9897).$$

*The maximum of the minimum value of $f_w(x)$ on $K$ is $\gamma^* \approx 11.9435$.*

When $K$ is unbounded, the LSP (12.8.10) may be unbounded below for some $w$. For this case, we are often interested in the *most average* weight $w$ such that $f_w(x)$ is bounded below on $K$. Denote the degree-$d$ homogenization

$$\widetilde{f_w}(\tilde{x}) := x_0^d f_w\Big(\frac{x}{x_0}\Big).$$

When $K$ is closed at $\infty$ (see Definition 6.6.1), $f_w(x) - \gamma \geq 0$ on $K$ if and only if $\widetilde{f_w}(\tilde{x}) - \gamma x_0^d \geq 0$ on $\widetilde{K}$, where $\widetilde{K}$ is the homogenization of $K$ (note $\tilde{x} = (x_0, x)$)

$$\widetilde{K} := \left\{ \tilde{x} \left| \begin{array}{ccc} \tilde{c}_i(\tilde{x}) & = & 0\,(i \in \mathcal{E}), \\ \tilde{c}_j(\tilde{x}) & \geq & 0\,(j \in \mathcal{I}), \\ \|\tilde{x}\|^2 & = & 1,\ x_0 \geq 0 \end{array} \right. \right\}. \tag{12.8.16}$$

So, we consider the optimization

$$\begin{cases} \max & \mu \\ \text{s.t.} & e^T w = 1, \\ & w_1 \geq \mu, \ldots, w_m \geq \mu, \\ & \widetilde{f_w} - \gamma x_0^d \in \mathscr{P}_d(\widetilde{K}). \end{cases} \tag{12.8.17}$$

The Moment-SOS hierarchy can be applied to solve (12.8.17).

**Example 12.8.10.** *Consider the objectives*

$$f_1 = -\Big(\sum_{i=1}^{4} x_i^3\Big) + 2x_2^4 - x_1 x_2 x_3 x_4,$$

$$f_2 = \Big(\sum_{i=1}^{4} x_i\Big)^3 - x_1^4 - x_2^4 - x_4^4,$$

$$f_3 = x_1^4 - x_2^4 + x_3^4 + x_4^4 - x_1 x_2 x_3 - x_1 x_2$$

*and the constraints $x_1 \geq 0,\ x_2 \geq 0,\ x_3 \geq 0,\ x_4 \geq 0$. All objectives $f_1, f_2, f_3$ are unbounded below on $K$. By solving the Moment-SOS hierarchy, we get the most average weight $w^* = (0.3529, 0.2964, 0.3507)$. The minimum value of the corresponding $f_{w^*}$ is $-0.0067$. Solving the LSP (12.8.10), we get two Pareto points: $(0, 0, 0, 0.7812)$ and $(0.7812, 0, 0, 0)$.*

## Nonexistence of proper weights

We discuss how to detect nonexistence of proper weights. The nonexistence occurs only if the feasible set $K$ is unbounded. Recall that $d$ is the maximum degree of $f_i$ and $\widetilde{f_w}(\tilde{x}) := x_0^d f_w(\frac{x}{x_0})$.

When $K$ is closed at $\infty$, the optimization (12.8.14) is equivalent to

$$\begin{cases} \max & \gamma \\ \text{s.t.} & w_1 + \cdots + w_m = 1, \\ & w_1 \geq 0, \ldots, w_m \geq 0, \\ & \widetilde{f_w} - \gamma x_0^d \in \mathscr{P}_d(\widetilde{K}). \end{cases} \qquad (12.8.18)$$

The dual optimization of (12.8.18) is

$$\begin{cases} \min & \mu \\ \text{s.t.} & \mu - \langle x_0^d f_i(\frac{x}{x_0}), \tilde{y} \rangle \geq 0, \ i = 1, \ldots, m, \\ & \langle x_0^d, \tilde{y} \rangle = 1, \ \tilde{y} \in \mathscr{R}_d(\widetilde{K}). \end{cases} \qquad (12.8.19)$$

When (12.8.19) is unbounded below, the problem (12.8.18) must be infeasible, and hence there is no proper weight. This is the case if (12.8.19) has a decreasing ray $\Delta\tilde{y}$:

$$\boxed{\begin{array}{l} -1 \geq \langle x_0^d f_i(\frac{x}{x_0}), \Delta\tilde{y} \rangle, \ i = 1, \ldots, m, \\ \langle x_0^d, \Delta\tilde{y} \rangle = 0, \ \Delta\tilde{y} \in \mathscr{R}_d(\widetilde{K}). \end{array}} \qquad (12.8.20)$$

Let $f_i^{(d)}$ denote the homogeneous part of degree $d$ for $f_i$, i.e.,

$$f_i^{(d)} = x_0^d f_i\left(\frac{x}{x_0}\right)\big|_{x_0=0}.$$

Since $\widetilde{K} \subseteq \{x_0 \geq 0\}$, the equality $\langle x_0^d, \Delta\tilde{y} \rangle = 0$ implies that every representing measure for $\Delta\tilde{y}$ must be supported in the hyperplane $x_0 = 0$. Therefore, (12.8.20) can be reduced to

$$\boxed{\begin{array}{l} -1 \geq \langle f_i^{(d)}, \Delta y \rangle, \ i = 1, \ldots, m, \\ \Delta y \in \mathscr{R}_d(K^\circ), \end{array}} \qquad (12.8.21)$$

where $K^\circ$ is the set

$$K^\circ := \left\{ x \in \mathbb{R}^n \ \middle| \ \begin{array}{rcl} c_i^{\text{hom}}(x) & = & 0 \ (i \in \mathcal{E}), \\ c_j^{\text{hom}}(x) & \geq & 0 \ (j \in \mathcal{I}), \\ \|x\|^2 & = & 1 \end{array} \right\}. \qquad (12.8.22)$$

In the above, each $c_i^{\text{hom}}(x) = \tilde{c}_i(0, x)$. We remark that if $\deg(f_i) < d$, then $f_i^{(d)} = 0$ and hence $\langle f_i^{(d)}, \Delta y \rangle = 0$, which implies that (12.8.20) is infeasible. Therefore, the decreasing ray $\Delta\tilde{y}$ as in (12.8.20) exists only if all $f_i$ have the same degree. The following is about nonexistence of proper weights.

**Theorem 12.8.11.** *([251]) Assume (12.8.21) has a feasible point*

$$\Delta y = \lambda_1 [z_1]_d + \cdots + \lambda_r [z_r]_d,$$

*with $\lambda_1, \ldots, \lambda_r > 0$ and $z_1, \ldots, z_r \in K^\circ$. If each $(0, z_i)$ lies on the closure $cl\big(\widetilde{K} \cap \{x_0 > 0\}\big)$, then the LSP (12.8.10) is unbounded below for all nonzero $w \geq 0$ and hence $\mathcal{W} = \emptyset$.*

The moment system (12.8.21) can be similarly solved by the Moment-SOS hierarchy introduced in §8.3.1.

**Example 12.8.12.** *Consider the objectives*

$$f_1 = -\left(\sum_{i=1}^{5} x_i^3\right) - x_2^4 + x_4^4 - x_1 x_2 x_3 - x_3 x_4 x_5,$$

$$f_2 = \left(\sum_{i=1}^{5} x_i\right)^3 - \sum_{i=1}^{4} x_i^4 + x_1 x_2 x_3 x_4 + x_2 x_3 x_4 x_5,$$

$$f_3 = x_1^4 - x_2^4 + x_3^4 + x_4^4 - x_1 x_2 x_3 - x_3 x_4 x_5,$$

$$f_4 = -(x_1 x_2)^2 + (x_2 x_3)^2 + (x_3 x_4)^2 + (x_4 x_5)^2$$

*and the constraints* $x_1^2 \geq 1, \ldots, x_5^2 \geq 1$. *By solving the Moment-SOS hierarchy, we get that* $\Delta y = \lambda[u]_4$ *is feasible for (12.8.21) with*

$$u \approx (-0.7014, -0.7049, 0.0533, -0.0428, 0.0803), \quad \lambda \approx 4.1146.$$

*The point* $(0, u)$ *lies on the closure of* $\widetilde{K} \cap \{x_0 > 0\}$. *Therefore, the LSP (12.8.10) is unbounded below for all nonzero weights* $w \geq 0$ *by Theorem 12.8.11.*

We remark that when no proper weights exist, the system (12.8.21) is still possibly infeasible. This can be shown by Exercise 12.8.3.

### 12.8.3 ▪ The Chebyshev scalarization

The Chebyshev scalarization problem is

$$\min_{x \in K} \max_{1 \leq i \leq m} w_i (f_i(x) - f_i^*) \tag{12.8.23}$$

for a nonzero weight $w := (w_1, \ldots, w_m) \geq 0$. In the above, each $f_i^*$ is the minimum value of $f_i$ on $K$ and it is assumed that $f_i^* > -\infty$.

Each minimizer of (12.8.23) is a weakly Pareto point. Conversely, every weakly Pareto point is a minimizer of the CSP (12.8.23) for some weight, provided each $f_i^* > -\infty$. This is because if $x^*$ is a weakly Pareto point, then there exist weights $w_i \geq 0$ such that all $w_i(f_i(x^*) - f_i^*)$ are equal to each other, since $f_i(x^*) - f_i^* \geq 0$ for each $i$. Then $x^*$ is the minimizer for the corresponding CSP.

With the new variable $x_{n+1}$, the CSP (12.8.23) is equivalent to the polynomial optimization

$$\begin{cases} \min & x_{n+1} \\ s.t. & x_{n+1} \geq w_i(f_i(x) - f_i^*), \ i = 1, \ldots, m, \\ & x \in K. \end{cases} \tag{12.8.24}$$

**Example 12.8.13.** *Consider the MOP with objectives*

$$f_1 = \left(\sum_{i=1}^{4} x_i^2\right) - (x_1 x_2 + x_3 x_4)(x_1 x_3 + x_2 x_4),$$

$$f_2 = \left(\sum_{i=1}^{4} x_i^4\right) + x_1 x_2 x_3 + x_2 x_3 x_4 + x_1 x_2 x_3 x_4,$$

$$f_3 = \left(\sum_{i=1}^{4} x_i^6\right) + (x_1^2 - x_2^2 + 1)(x_2^2 - x_3^2 + 1)(x_3^2 - x_4^2 + 1)$$

*and the constraints* $x_1x_2 \leq 1, x_2x_3 \leq 1, x_3x_4 \leq 1, x_1x_4 \leq 1$. *The minimum values* $f_1^*, f_2^*, f_3^*$
*are* $0.0000, -0.0710, 0.6029$ *respectively. A list of some weights and corresponding weakly
Pareto points (they are also Pareto points) are as follows:*

| $w$ | Pareto point |
|---|---|
| $(1,1,1)$ | $(0.000, 0.000, 0.000, 0.4503)$ |
| $(1,2,2)$ | $(-0.0024, -0.0979, -0.0635, -0.5248)$ |
| $(1,2,3)$ | $(-0.0029, -0.1228, -0.0700, -0.5648)$ |

### 12.8.4 ▪ Detection of PPs and WPPs

To detect if a given point $x^* \in K$ is a Pareto point or not, we consider the optimization

$$\begin{cases} \min & f_e(x) := f_1(x) + \cdots + f_m(x), \\ & f_i(x^*) - f_i(x) \geq 0, \ i = 1, \ldots, m, \\ & x \in K. \end{cases} \quad (12.8.25)$$

Let $z^*$ be a minimizer of (12.8.25), if it exists. Then, $x^*$ is a Pareto point if and only if the minimum value of (12.8.25) is equal to $f_e(x^*)$. Moreover, if $x^*$ is not a Pareto point, the minimizer $z^*$ must be a Pareto point, since all the weights are positive. A Pareto point may be obtained by solving (12.8.25).

**Example 12.8.14.** *Consider the MOP with objectives*

$$\begin{aligned} f_1 &= x_1^2(x_1 - 2)^2 + (x_1 - x_2)^2 + (x_2 - x_3)^2 + (x_3 - x_4)^2, \\ f_2 &= -x_1^2 - x_2^2 - x_3^2 - x_4^2 + x_1x_2 + x_2x_3 + x_3x_4 \end{aligned}$$

*and the constraints* $x_1 \geq 0, x_2 \geq 0, x_3 \geq 0, x_4 \geq 0$. *We first solve the CSP (12.8.23) with* $w = (1,1)$ *and get the weakly Pareto point* $x^* = (0,0,0,0)$. *It is not a Pareto point. By solving (12.8.25), we get the Pareto point* $(2,2,2,2)$.

We can similarly detect if a given $x^* \in K$ is a weakly Pareto point or not. Consider the optimization problem

$$\begin{cases} \min & \max_{1 \leq i \leq m} \left( f_i(x) - f_i(x^*) \right) \\ s.t. & f_i(x^*) - f_i(x) \geq 0, \ i = 1, \ldots, m, \\ & x \in K. \end{cases} \quad (12.8.26)$$

Suppose $z^*$ is a minimizer of (12.8.26). Then, $x^*$ is a weakly Pareto point if and only if the optimal value of (12.8.26) is equal to 0. Moreover, if $x^*$ is not a weakly Pareto point, then $z^*$ is a weakly Pareto point. Similarly, the optimization (12.8.26) is equivalent to

$$\begin{cases} \min & x_{n+1} \\ s.t. & x_{n+1} - f_i(x) + f_i(x^*) \geq 0, \ i = 1, \ldots, m, \\ & f_i(x^*) - f_i(x) \geq 0, \ i = 1, \ldots, m, \\ & x \in K. \end{cases} \quad (12.8.27)$$

The optimal value of (12.8.27) is always less than or equal to 0.

When $K$ is unbounded, existence of PPs and WPPs can be investigated by considering the following min-max optimization problem:

$$\min_{x \in K} \max_{1 \leq i \leq m} f_i(x). \qquad (12.8.28)$$

**Theorem 12.8.15.** *([251]) The min-max optimization (12.8.28) has the following properties:*

(i) *If (12.8.28) is unbounded below, then there is no weakly Pareto point, and hence there is no Pareto point. If (12.8.28) is bounded below, then every minimizer of (12.8.28) is a weakly Pareto point.*

(ii) *Let $S$ be the set of minimizers of (12.8.28). For each $x^* \in S$, if $f(x^*)$ is a $\pi$-minimal point of the image $f(S)$ for a permutation $\pi$ of $(1, \ldots, m)$, then $x^*$ is a Pareto point. In particular, if $S$ is compact, then there exists a Pareto point.*

Each optimizer $x^*$ of (12.8.28) is a weakly Pareto point. One can solve (12.8.25) to check whether $x^*$ is a Pareto point or not. If it is not, each minimizer of (12.8.25) is a Pareto point. Similarly, (12.8.28) is equivalent to the optimization

$$\begin{cases} \min & x_{n+1} \\ s.t. & x_{n+1} \geq f_i(x),\ i = 1, \ldots, m, \\ & x \in K. \end{cases} \qquad (12.8.29)$$

Once an optimizer $x^*$ for (12.8.29) is obtained, we can solve (12.8.25) to detect if it is a Pareto point or not.

**Example 12.8.16.** *Consider the MOP with objectives*

$$f_1 = x_1^3 + x_2^3 - x_3^3 + x_3^2 x_4^2, \quad f_2 = x_2^3 + x_3^3 - x_4^3 + x_4^2 x_1^2,$$
$$f_3 = x_3^3 + x_4^3 - x_1^3 + x_1^2 x_2^2, \quad f_4 = x_4^3 + x_1^3 - x_2^3 + x_2^2 x_3^2$$

*and with the exterior ball constraint $x_1^3 + x_2^3 + x_3^3 + x_4^3 \geq 1$. All $f_1, f_2, f_3, f_4$ are unbounded below on $K$. The CSP (12.8.23) does not exist since each $f_i^* = -\infty$. However, solving (12.8.29) gives the Pareto point $(0.6300, 0.6300, 0.6300, 0.6300)$.*

## Nonexistence of WPPs

We discuss how to detect nonexistence of weakly Pareto points, which is possible only if $K$ is unbounded. Recall that $d_i := \deg(f_i)$. Observe that the optimization (12.8.28) is unbounded below if and only if the following optimization problem is unbounded below:

$$\begin{cases} \min & x_{n+1} \\ s.t. & -(-x_{n+1})^{d_i} - f_i(x) \geq 0\ (i \in [m]), \\ & x \in K. \end{cases} \qquad (12.8.30)$$

Let $K_1$ be the feasible set of (12.8.30) and let (note $\tilde{x} := (x_0, x)$)

$$\widetilde{K}_1 := \left\{ (\tilde{x}, x_{n+1}) \middle| \begin{array}{l} -(-x_{n+1})^{d_i} - \widetilde{f}_i(\tilde{x}) \geq 0\ (i \in [m]), \\ \widetilde{c}_i(\tilde{x}) = 0\ (i \in \mathcal{E}),\ \widetilde{c}_j(\tilde{x}) \geq 0\ (j \in \mathcal{I}), \\ \|\tilde{x}\|^2 + |x_{n+1}|^2 = 1,\ x_0 \geq 0 \end{array} \right\}. \qquad (12.8.31)$$

When $K_1$ is closed at $\infty$, $x_{n+1} \geq \gamma$ on $K_1$ if and only if $x_{n+1} - \gamma x_0 \geq 0$ on $\widetilde{K}_1$. Consider the optimization problem

$$\begin{cases} \max & \gamma \\ s.t. & x_{n+1} - \gamma x_0 \in \mathscr{P}_1(\widetilde{K}_1). \end{cases} \tag{12.8.32}$$

The optimization problem (12.8.30) is unbounded below if and only if (12.8.32) is infeasible, when $K_1$ is closed at $\infty$. The dual optimization of (12.8.32) is

$$\begin{cases} \min & \langle x_{n+1}, \check{y} \rangle \\ s.t. & \langle x_0, \check{y} \rangle = 1, \ \check{y} \in \mathscr{R}_1(\widetilde{K}_1). \end{cases} \tag{12.8.33}$$

Note that (12.8.33) is feasible when $K$ is nonempty. So, it is unbounded below if there exists a decreasing ray $\Delta \check{y}$:

$$\boxed{\begin{array}{c} \langle x_{n+1}, \Delta \check{y} \rangle = -1, \ \langle x_0, \Delta \check{y} \rangle = 0, \\ \Delta \check{y} \in \mathscr{R}_1(\widetilde{K}_1). \end{array}} \tag{12.8.34}$$

Since $x_0 \geq 0$ on $\widetilde{K}_1$, the equality $\langle x_0, \Delta \check{y} \rangle = 0$ implies that every representing measure for $\Delta \check{y}$ is supported in $x_0 = 0$. Therefore, (12.8.34) is equivalent to

$$\boxed{\langle x_{n+1}, \Delta \check{y} \rangle = -1, \quad \Delta \check{y} \in \mathscr{R}_1(K_1^\circ),} \tag{12.8.35}$$

where $K_1^\circ$ is the section $x_0 = 0$ of $\widetilde{K}_1$:

$$K_1^\circ := \left\{ (x, x_{n+1}) \ \middle| \ \begin{array}{l} -(-x_{n+1})^{d_i} - f_i^{hom}(x) \geq 0 \ (i \in [m]), \\ c_i^{hom}(x) = 0 \ (i \in \mathcal{E}), \\ c_j^{hom}(x) \geq 0 \ (j \in \mathcal{I}), \\ \|x\|^2 + x_{n+1}^2 = 1 \end{array} \right\}. \tag{12.8.36}$$

The following theorem is about nonexistence of WPPs.

**Theorem 12.8.17.** *([251]) Suppose $\Delta \check{y} = \lambda v$, with $\lambda > 0$ and $v \in K_1^\circ$, is a feasible point for (12.8.35). If the point $(0, v) \in cl(\widetilde{K}_1 \cap \{x_0 > 0\})$, then (12.8.30) and (12.8.28) must be unbounded below, and hence there are no weakly Pareto points.*

The tms $\Delta \check{y} = \lambda v$ satisfying (12.8.35) can be obtained by solving the Moment-SOS hierarchy as in §8.3.

**Example 12.8.18.** *Consider the MOP with the objectives*

$$\begin{aligned} f_1 &= (x_1 x_2 + x_3 x_4)(x_1 x_4 + x_2 x_3) + x_1^2 + x_2^2 + x_3^2 + x_4^2, \\ f_2 &= x_1^3 x_2^2 + x_2^3 x_3^2 + x_3^3 x_4^2 + x_4^3 x_1^2, \\ f_3 &= x_1^4 - x_2^4 + x_3^4 - x_4^4 + x_1 x_2 x_4 + x_1 x_3 x_4, \\ f_4 &= (x_1 - x_2)(x_3 - x_4)^2 + (x_1 - x_3)(x_2 - x_4)^2 \\ &\quad + (x_1 - x_4)(x_2 - x_3)^2 + x_1 x_2 + x_2 x_3 + x_3 x_4 \end{aligned}$$

*and the constraints $x_1 x_2 x_3 \geq 1$, $x_2 x_3 x_4 \geq 1$. By solving the Moment-SOS hierarchy, we get the feasible point $\Delta \check{y} = 3.3597 [v]_1$ for (12.8.35), with*

$$v = (v_1, v_2, v_3, v_4, v_5) = (-0.2761, 0.8737, 0.0000, -0.2680, -0.2976).$$

*The set $K_1$ is not closed at infinity but $(0, v)$ belongs to $cl\big(\widetilde{K}_1 \cap \{x_0 > 0\}\big)$. This is because* $\lim\limits_{\ell \to \infty} \frac{s_\ell}{\|s_\ell\|} = (0, v)$, *where*

$$s_\ell := \left(-\sqrt[3]{\frac{v_2 v_4}{\ell}}, v_1, v_2, \frac{-1}{\ell}, v_4, v_5\right).$$

*Note that $\frac{s_\ell}{\|s_\ell\|} \in \widetilde{K}_1 \cap \{x_0 > 0\}$ when $\ell$ is big enough. By Theorem 12.8.17, there is no weakly Pareto point for this MOP.*

## Nonexistence of PPs

When there are no weakly Pareto points, there must exist no Pareto points. However, a Pareto point may not exist even if weakly Pareto points exist. We discuss how to detect nonexistence of Pareto points for this case.

We consider the optimization (12.8.25) with $x^* \in K$. A Pareto point exists if and only if (12.8.25) is bounded below and has a minimizer for some $x^* \in K$. The "if" implication is clear. When $x^*$ itself is a Pareto point, then $x^*$ must be a minimizer for (12.8.25). This explains the "only if" implication. Let $K(x^*)$ be the feasible set of (12.8.25), parameterized by $x^*$, and let $\widetilde{K}(x^*)$ be the homogenization of $K(x^*)$ similarly as in (12.8.16). Suppose $K(x^*)$ is closed at $\infty$. Then (12.8.25) is bounded below if and only if

$$\widetilde{f}_e(\tilde{x}) - \gamma x_0^d \in \mathscr{P}_d(\widetilde{K}(x^*))$$

for some $\gamma$. We consider the linear conic optimization

$$\begin{cases} \max & \gamma \\ \text{s.t.} & \widetilde{f}_e(\tilde{x}) - \gamma x_0^d \in \mathscr{P}_d(\widetilde{K}(x^*)). \end{cases} \tag{12.8.37}$$

Pareto points do not exist if (12.8.25) is unbounded below for all $x^* \in K$. This is equivalent to the infeasibility of (12.8.37) for all $x^* \in K$. The dual optimization of (12.8.37) is

$$\begin{cases} \min & \langle \widetilde{f}_e, \tilde{y} \rangle \\ \text{s.t.} & \langle x_0^d, \tilde{y} \rangle = 1, \ \tilde{y} \in \mathscr{R}_d(\widetilde{K}(x^*)). \end{cases} \tag{12.8.38}$$

By the weak duality, (12.8.37) is infeasible if (12.8.38) is unbounded below. The problem (12.8.38) is feasible for all $x^* \in K$. Therefore, (12.8.38) is unbounded below if there exists a decreasing ray $\Delta\tilde{y}$:

$$\boxed{\begin{array}{l} \langle \widetilde{f}_e, \Delta\tilde{y} \rangle = -1, \quad \langle x_0^d, \Delta\tilde{y} \rangle = 0, \\ \Delta\tilde{y} \in \mathscr{R}_d(\widetilde{K}(x^*)). \end{array}} \tag{12.8.39}$$

Since $x_0 \geq 0$ on $\widetilde{K}(x^*)$, $\langle x_0^d, \Delta\tilde{y} \rangle = 0$ if and only if every representing measure for $\Delta\tilde{y}$ is supported in the hyperplane $x_0 = 0$. Hence, the existence of $\Delta\tilde{y}$ satisfying (12.8.39) is equivalent to the existence of $\Delta\check{y}$ satisfying

$$\boxed{\langle f_e^{hom}, \Delta\check{y} \rangle = -1, \quad \Delta\check{y} \in \mathscr{R}_d(\widetilde{K}_0^*),} \tag{12.8.40}$$

where $f_e^{hom}(x) := \widetilde{f}_e(0, x)$ and $K_0^*$ is the section $x_0 = 0$ of $K(x^*)$:

$$K_0^* := \left\{ x \in \mathbb{R}^n \ \middle| \ \begin{array}{l} c_j^{hom}(x) = 0 \, (j \in \mathcal{E}), \\ c_j^{hom}(x) \geq 0 \, (j \in \mathcal{I}), \\ -f_i^{hom}(x) \geq 0 \, (i \in [m]), \\ x^T x = 1. \end{array} \right\}. \tag{12.8.41}$$

It is important to observe that $K_0^*$ and (12.8.40) do not depend on $x^*$. If such $\Delta \breve{y}$ exists, then (12.8.25) is unbounded below for all $x^* \in K$, and hence there are no Pareto points. This implies the following theorem.

**Theorem 12.8.19.** *([251]) Suppose $K(x^*)$ is closed at infinity for all $x^* \in K$. If there is $\Delta \breve{y}$ such that (12.8.40) holds, then (12.8.25) is unbounded below for all $x^* \in K$ and hence Pareto points do not exist.*

When Pareto points do not exist, weakly Pareto points may still exist. This is shown in the following and Exercise 12.8.4.

**Example 12.8.20.** *Consider the objectives*

$$\begin{aligned} f_1 &= x_1^4 + x_3^4 + (x_1 x_2)^2 + (x_2 x_3)^2 + (x_3 x_4)^2 + x_1 x_2 x_3 x_4, \\ f_2 &= x_1^4 + x_2^4 + x_3^4 + x_4^4 - 2x_2^4 - x_1^3 x_2 - x_3^3 x_4 \end{aligned}$$

*and the constraint $x_1 x_2 x_3 x_4 \geq 0$. Since $f_1(0,t,0,0) = 0$ is the minimum value, the point $(0,t,0,0)$ is a weakly Pareto point for all $t \in \mathbb{R}$. Since all the polynomials are homogeneous, $K(x^*)$ is closed at infinity for all $x^* \in K$. By solving the Moment-SOS hierarchy, we get $\Delta \breve{y} = 1.0023[u]_4$ satisfying (12.8.40) for $u = (0.0000, -0.9994, 0.0000, 0.0339)$. Hence, there is no Pareto point for this MOP.*

## 12.8.5 ▪ Exercises

**Exercise 12.8.1.** *For the MOP with objectives*

$$f_1 = x_1^3 - x_1^2 x_2 - x_2, \quad f_2 = x_2^3 - x_1 x_2^2 - x_1$$

*and the constraint $x_1 x_2 \leq 1$, show that the LSP (12.8.10) is unbounded below for all nonzero weights $w \geq 0$.*

**Exercise 12.8.2.** *Consider the MOP with objectives*

$$\begin{aligned} f_1 &= (x_1 - x_2)^4 + (x_2 - x_3)^4, \\ f_2 &= \left(\sum_{i=1}^{3} x_i^4\right) + x_1^2 x_2^2 + x_1^2 x_3^2 + x_2^2 x_3^2 \end{aligned}$$

*and with the ball constraint $\|x\| \leq 1$. Show that $f_1, f_2$ are both SOS-convex. Give a semidefinite representation for the set $\mathcal{V}_1$ as in (12.8.8).*

**Exercise 12.8.3.** *For the MOP with $K = \mathbb{R}^1$, $f_1 = x_1^3 + x_1$, $f_2 = -x_1^3$, show that no proper weights exist but the moment system (12.8.21) is infeasible.*

**Exercise 12.8.4.** *For the following MOP*

$$\begin{cases} \min & (x_1, x_2) \\ s.t. & x_1 \geq 0 \end{cases}$$

*show that there are no Pareto points but there exist weakly Pareto points. Also find a tms $\Delta \breve{y}$ satisfying (12.8.40).*

# Bibliography

[1] A.A. AHMADI AND J. ZHANG, *Semidefinite programming and Nash equilibria in bimatrix games*, INFORMS J. Comput., 33(2), 607–628, 2021. (Cited on p. 407)

[2] A.A. AHMADI AND J. ZHANG, *On the complexity of testing attainment of the optimal value in nonlinear optimization*, Math. Program., 184, 221–241, 2020. (Cited on pp. 120, 138, 211)

[3] A.A. AHMADI AND P. PARRILO, *A convex polynomial that is not sos-convex*, Math. Program., 135(1), 275–292, 2012. (Cited on p. 218)

[4] A.A. AHMADI AND P.A. PARRILO, *A complete characterization of the gap between convexity and SOS-convexity*, SIAM J. Optim., 23(2), 811–833, 2013. (Cited on pp. 218, 221)

[5] A.A. AHMADI, G. BLEKHERMAN, AND P. PARRILO, *Convex ternary quartics are SOS-convex*, in preparation. (Cited on p. 218)

[6] W. AI AND S. ZHANG, *Strong duality for the CDT subproblem: A necessary and sufficient condition*, SIAM J. Optim., 19(4), 1735–1756, 2009. (Cited on p. 137)

[7] F. ALIZADEH, J. HAEBERLY, AND M. OVERTON, *Complementarity and nondegeneracy in semidefinite programming*, Math. Program., 77(1), 111–128, 1997. (Cited on p. 21)

[8] N.I. AKHIEZER, *The Classical Moment Problem*, Hafner, New York, 1965. (Cited on p. 99)

[9] J. ALEXANDER AND A. HIRSCHOWITZ, *Polynomial interpolation in several variables*, J. Algebraic Geom., 4, 201–22, 1995. (Cited on p. 344)

[10] S. BANACH, *Über homogene polynome in $(L^2)$*, Studia Math., 7, 36–44, 1938. (Cited on p. 348)

[11] J. BARD, *Practical Bilevel Optimization: Algorithms and Applications*, Kluwer Academic, Dordrecht, 1998. (Cited on p. 413)

[12] A. BARVINOK, *A Course in Convexity*, Graduate Studies in Mathematics, Vol. 54, AMS, Providence, 2002. (Cited on pp. 8, 9, 11, 12)

[13] H. BASS, E. CONNELL, AND D. WRIGHT, *The Jacobian conjecture: reduction of degree and formal expansion of the inverse*, Bull. Amer. Math. Soc. (N.S.), 7(2), 287–330, 1982. (Cited on p. 185)

[14] T. BAŞAR AND G.J. OLSDER, *Dynamic Noncooperative Game Theory*, SIAM, Philadelphia, 1998. (Cited on pp. 407, 412)

[15] C. BAYER AND J. TEICHMANN, *The proof of Tchakaloff's Theorem*, Proc. Amer. Math. Soc., 134, 3035–3040, 2006. (Cited on pp. 76, 247)

[16] S. BASU, R. POLLACK, AND M.-F. ROY, *Algorithms in Real Algebraic Geometry*, Springer, Berlin, 2006. (Cited on pp. 36, 43, 45, 46, 47, 70)

[17] E.G. BELOUSOV AND D. KLATTE, *A Frank–Wolfe type theorem for convex polynomial programs*, Comput. Optim. Appl., 22(1), 37–48, 2002. (Cited on pp. 120, 138, 223)

[18] A. BEN-TAL AND C.E. BLAIR, *Computational difficulties of bilevel linear programming*, Oper. Res., 38, 556–560, 1990. (Cited on p. 413)

[19] A. BEN-TAL AND A. NEMIROVSKI, *Lectures on Modern Convex Optimization: Analysis, Algorithms, and Engineering Applications*, MPS-SIAM Series on Optimization, SIAM, Philadelphia, 2001. (Cited on pp. 12, 14, 17, 20, 24, 25, 228, 236)

[20] A. BERMAN AND U. ROTHBLUM, *A note on the computation of the CP-rank*, Linear Algebra Appl., 419, 1–7, 2006. (Not cited)

[21] A. BERMAN AND N. SHAKED-MONDERER, *Completely Positive Matrices*, World Scientific, Singapore, 2003. (Cited on p. 283)

[22] D. BERTSEKAS, *Nonlinear Programming*, second edition, Athena Scientific, Belmont, 1995. (Cited on pp. 139, 182, 183, 224)

[23] D.P. BERTSEKAS, *Convex Optimization Theory*, Athena Scientific, Belmont, 2009. (Cited on pp. 8, 9, 11)

[24] D. BERTSEKAS, A. NEDIĆ, AND A. OZDAGLAR, *Convex Analysis and Optimization*, Athena Scientific, Belmont, 2003. (Cited on pp. 8, 9, 400)

[25] D. BERTSIMAS, D.B. BROWN, AND C. CARAMANIS, *Theory and applications of robust optimization*, SIAM Review, 53(3), 464–501, 2011. (Not cited)

[26] D. BERTSIMAS, M. SIM, AND M. ZHANG, *Adaptive distributionally robust optimization*, Management Sci., 65(2), 604–618, 2019. (Cited on p. 424)

[27] V. BLANCO, J. PUERTO, AND S. BEN ALI, *A semidefinite programming approach for solving multiobjective linear programming*, J. Glob. Optim., 58(3), 465–480, 2014. (Cited on p. 430)

[28] J. BLAIR AND B. PEYTON, *An introduction to chordal graphs and clique trees*, in Graph Theory and Sparse Matrix Computations, J. George, J. Gilbert, and J. Liu, eds., pp. 1–30, Springer, Berlin, 1993. (Cited on p. 388)

[29] G. BLEKHERMAN, *There are significantly more nonnegative polynomials than sums of squares*, Israel J. Math., 153, 355–380, 2006. (Cited on pp. 57, 58)

[30] G. BLEKHERMAN, *Convex Forms that are not Sums of Squares*, Preprint, 2009. arXiv:0910.0656 [math.AG] (Cited on p. 220)

[31] G. BLEKHERMAN, P.A. PARRILO, AND R.R. THOMAS (eds.), *Semidefinite Optimization and Convex Algebraic Geometry*, MOS-SIAM Series on Optimization, SIAM, Philadelphia, 2012. (Cited on pp. 32, 45, 59, 60, 74)

[32] G. BLEKHERMAN AND Z. TEITLER, *On maximum, typical and generic ranks*, Math. Ann., 362(3-4), 1021–1031, 2015. (Cited on p. 341)

[33] G. BLEKHERMAN, D. PLAUMANN, R. SINN, AND C. VINZANT, *Low-rank sum-of-squares representations on varieties of minimal degree*, Int. Math. Res. Not., 2019(1), 33–54, 2019. (Cited on p. 302)

[34] J. BOCHNAK, M. COSTE, AND M-F. ROY, *Real Algebraic Geometry*, Springer, Berlin, 1998. (Cited on pp. 36, 43, 45, 46, 47, 70, 71, 72, 73, 159)

[35] I. BOMZE, *Copositive optimization - recent developments and applications*, Eur. J. Oper. Res., 216, 509–520, 2012. (Cited on p. 292)

[36] S. BOYD, L. EL GHAOUI, E. FERON, AND V. BALAKRISHNAN, *Linear Matrix Inequalities in System and Control Theory*, SIAM, Philadelphia, 1994. (Cited on pp. 20, 228)

[37] S. BOYD AND L. VANDENBERGHE, *Convex Optimization*, Cambridge University Press, Cambridge, 2004. (Cited on pp. 8, 9, 228, 230, 242)

[38] W. BRUNS AND U. VETTER, *Determinantal Rings*, Lecture Notes in Mathematics, Vol. 1327, Springer, Berlin, 1988. (Cited on pp. 204, 205)

[39] W. BRUNS AND R. SCHWÄNZL, *The number of equations defining a determinantal variety*, Bull. London Math. Soc., 22(5), 439–445, 1990. (Cited on p. 204)

[40] F. BUGARIN, D. HENRION, AND J. LASSERRE, *Minimizing the sum of many rational functions*, Math. Program. Comput., 8, 83–111, 2016. (Cited on p. 392)

[41] S. BURER, *SDPLR: a C package for solving large-scale semidefinite programming problems*. (Cited on p. 25)

[42] S. BURER, *On the copositive representation of binary and continuous nonconvex quadratic programs*, Math. Program., 120(2), 479–495, 2009. (Cited on p. 292)

[43] A.P. CALDERÓN, *A note on biquadratic forms*, Linear Algebra Appl., 7, 175–177, 1973. (Cited on p. 303)

[44] D. CARTWRIGHT AND B. STURMFELS, *The number of eigenvalues of a tensor*, Linear Algebra Appl., 438, 942–952, 2013. (Cited on pp. 380, 386)

[45] V. CHANDRASEKARAN, B. RECHT, P. PARRILO, AND A. WILLSKY, *The convex geometry of linear inverse problems*, Found. Comput. Math., 12(6), 805–849, 2012. (Cited on p. 356)

[46] B. CHEN, S. HE, Z. LI, AND S. ZHANG, *Maximum block improvement and polynomial optimization*, SIAM J. Optim., 22(1), 87–107, 2012. (Cited on p. 348)

[47] H. CHEN, Z. HUANG, AND L. QI, *Copositive tensor detection and its applications in physics and hypergraphs*, Comput. Optim. Appl. 69, 133–158, 2018. (Cited on pp. 372, 373)

[48] Z. CHEN, M. SIM, AND H. XU, *Distributionally robust optimization with infinitely constrained ambiguity sets*, Oper. Res., 67(5), 1328–1344, 2019. (Cited on p. 424)

[49] L. CHEN, A. XU, AND H. ZHU, *Computation of the geometric measure of entanglement for pure multiqubit states*, Phys. Rev. A, 82, 032301, 2010. (Cited on p. 368)

[50] Y. CHEN, Y. DAI, D. HAN, AND W. SUN, *Positive semidefinite generalized diffusion tensor imaging via quadratic semidefinite programming*, SIAM J. Imaging Sci., 6(3), 1531–1552, 2013. (Cited on p. 376)

[51] J.-H. CHO, Y. WANG, R. CHEN, K. CHAN, AND A. SWAMI, *A survey on modeling and optimizing multi-objective systems*, Commun. Surveys Tutorials, 19(3), 1867–1901, 2017. (Cited on p. 430)

[52] M. CHOI, *Positive semidefinite biquadratic forms*, Linear Algebra Appl., 12, 95–100, 1975. (Cited on pp. 304, 306)

[53] M.D. CHOI, T.-Y. LAM, AND B. REZNICK, *Real zeros of positive semidefinite forms. I*, Math. Z., 17 (1), 1–26, 1980. (Cited on pp. 302, 304, 307)

[54] F.H. CLARKE, *Optimization and Nonsmooth Analysis*, Wiley Interscience, New York, 1983; reprinted as Vol. 5 of Classics in Applied Mathematics, SIAM, Philadelphia, 1990. (Cited on p. 419)

[55] E.R. CLAYTON, W.E. WEBER, AND B.W. TAYLOR III, *A goal programming approach to the optimization of multi response simulation models*, IIE Trans., 14(4), 282–287, 1982. (Cited on p. 430)

[56] B. COLSON, P. MARCOTTE, AND G. SAVARD, *An overview of bilevel optimization*, Ann. Oper. Res., 153, 235–256, 2007. (Cited on p. 413)

[57] P. COMON, G. GOLUB, L.-H. LIM, AND B. MOURRAIN, *Symmetric tensors and symmetric tensor rank*, SIAM J. Matrix Anal. Appl., 30(3), 1254–1279, 2008. (Cited on pp. 340, 342, 343, 344, 345, 346)

[58] J.B. CONWAY, *A Course in Functional Analysis*, second edition, Springer, Berlin, 1990. (Cited on p. 262)

[59] D. COX, J. LITTLE, AND D. O'SHEA, *Ideals, Varieties, and Algorithms: An Introduction to Computational Algebraic Geometry and Commutative Algebra*, Springer, Berlin, 2007. (Cited on pp. 35, 36, 37, 38, 39, 43)

[60] D.A. COX, J. LITTLE, AND D. O'SHEA, *Using Algebraic Geometry*, Springer, New York, 1998. (Cited on pp. 40, 47, 49)

[61] R.M. CORLESS, P.M. GIANNI, AND B.M. TRAGER, *A reordered Schur factorization method for zero-dimensional polynomial systems with multiple roots*, Proceedings of the ACM International Symposium on Symbolic and Algebraic Computation, pp. 133–140, Maui, Hawaii, 1997. (Cited on pp. 41, 42, 82)

[62] C.-F. CUI, Y.-H. DAI, AND J. NIE, *All real eigenvalues of symmetric tensors*, SIAM. J. Matrix Anal. Appl., 35(4), 1582–1601, 2014. (Cited on pp. 349, 350, 376, 378, 379, 380, 386)

[63] R. CURTO AND L. FIALKOW, *Truncated K-moment problems in several variables*, J. Oper. Theory, 54, 189–226, 2005. (Cited on pp. 79, 82, 123, 124)

[64] R. CURTO AND L. FIALKOW, *Recursiveness, positivity, and truncated moment problems*, Houst. J. Math., 17, 603–635, 1991. (Cited on pp. 79, 93, 98, 99, 101)

[65] C. DASKALAKIS, P.W. GOLDBERG, AND C.H. PAPADIMITRIOU, *The complexity of computing a Nash equilibrium*, SIAM J. Comput., 39(1), 195–259, 2009. (Cited on p. 407)

[66] R.S. DATTA, *Universality of Nash equilibria*, Math. Oper. Res., 28(3), 424–432, 2003. (Cited on p. 407)

[67] R.S. DATTA, *Finding all Nash equilibria of a finite game using polynomial algebra*, Econ. Theory, 42(1), 55–96, 2010. (Cited on p. 407)

[68] E. DE KLERK AND D.V. PASECHNIK, *Approximation of the stability number of a graph via copositive programming*, SIAM J. Optim., 12(4), 875–892, 2002. (Cited on pp. 131, 138, 140, 285, 294)

[69] L. DE LATHAUWER, B. DE MOOR AND J. VANDEWALLE, *On the best rank-1 and rank-$(R_1, R_2, ..., R_N)$ approximation of higher-order tensors*, SIAM. J. Matrix Anal. Appl., 21(4), 1324–1342, 2000. (Cited on pp. 347, 348)

[70] E. DE KLERK, *Aspects of Semidefinite Programming: Interior Point Algorithms and Selected Applications*, Applied Optimization, Vol. 65, Kluwer Academic, Dordrecht, 2002. (Cited on pp. 20, 25)

[71] E. DE KLERK AND M. LAURENT, *On the Lasserre hierarchy of semidefinite programming relaxations of convex polynomial optimization problems*, SIAM J. Optim., 21(3), 824–832, 2011. (Cited on p. 224)

[72] E. DE KLERK, D. KUHN, AND K. POSTEK, *Distributionally robust optimization with polynomial densities: theory, models and algorithms*, Math. Program., 181, 265–296, 2020. (Cited on p. 424)

[73] E. DELAGE AND Y. YE, *Distributionally robust optimization under moment uncertainty with application to data-driven problems*, Oper. Res., 58(3), 595–612, 2010. (Cited on p. 424)

[74]  J. DEMMEL, *Applied Numerical Linear Algebra*, SIAM, Philadelphia, 1997. (Cited on p. 18)

[75]  J. DEMMEL, J. NIE, AND V. POWERS, *Representations of positive polynomials on non-compact semialgebraic sets via KKT ideals*, J. Pure Appl. Algebra, 209(1), 189–200, 2007. (Cited on p. 180)

[76]  S. DEMPE, V. KALASHNIKOV, G. PÉREZ-VALDÉS, AND N. KALASHNYKOVA, *Bilevel Programming Problems*, Energy Systems, Springer, Berlin, 2015. (Cited on p. 413)

[77]  S. DEMPE, *Bilevel Optimization: Theory, Algorithms and Applications*, Fakultät für Mathematik und Informatik, TU Bergakademie Freiberg, 2018. (Cited on p. 413)

[78]  E. DE KLERK, M. LAURENT AND P. PARRILO, *On the equivalence of algebraic approaches to the minimization of forms on the simplex*, in Positive Polynomials in Control (D. Henrion and A. Garulli, eds.), Lecture Notes on Control and Information Sciences, Vol. 312, pp. 121–132, Springer, Berlin, 2005. (Cited on pp. 146, 285)

[79]  E. DE KLERK, M. LAURENT, AND Z. SUN, *Convergence analysis for Lasserre's measure-based hierarchy of upper bounds for polynomial optimization*, Math. Program., 162(1), 363–392, 2017. (Not cited)

[80]  V. DE SILVA AND L.-H. LIM, *Tensor rank and the ill-posedness of the best low-rank approximation problem*, SIAM. J. Matrix Anal. Appl., 30(3), 1084–1127, 2008. (Cited on p. 341)

[81]  H. DERKSEN, S. FRIEDLAND, L.-H. LIM, AND L. WANG, *Theoretical and computational aspects of entanglement*, Preprint, 2017. arXiv:1705.07160 [quant-ph] (Cited on pp. 367, 368, 369)

[82]  P. DICKINSON, *An improved characterisation of the interior of the completely positive cone*, Electron. J. Linear Algebra., 20, 723–729, 2010. (Cited on p. 283)

[83]  P. DICKINSON, *Geometry of the copositive and completely positive cones*, J. Math. Anal. Appl., 380(1), 377–395, 2011. (Cited on pp. 278, 283, 284, 285)

[84]  M. DRESSLER, J. NIE, AND Z. YANG, *Separability of Hermitian tensors and PSD decompositions*, Linear Multilinear Algebra, 2021. doi.org/10.1080/03081087.2021.1965078 (Cited on p. 371)

[85]  J. DUCHI AND H. NAMKOONG, *Learning models with uniform performance via distributionally robust optimization*, Ann. Stat., 49(3), 1378–1406, 2021. (Cited on p. 424)

[86]  R. DUDLEY, *Real Analysis and Probability*, Wadsworth, Brooks, and Cole, 1989. (Cited on p. 31)

[87]  M. DÜR AND G. STILL, *Interior points of the completely positive cone*, Electron. J. Linear Algebra, 17, 48–53, 2008. (Cited on p. 283)

[88]  M. DÜR, *Copositive Programming - a survey*, in Recent Advances in Optimization and its Applications in Engineering, M. Diehl, F. Glineur, E. Jarlebring, and W. Michiels (eds.), pp. 3–20, Springer, Berlin, 2010. (Cited on pp. 278, 283, 292)

[89]  M. EMMERICH AND A. DEUTZ, *A tutorial on multiobjective optimization: fundamentals and evolutionary methods*, Natural Comput., 17(3), 585–609, 2018. (Cited on pp. 429, 430)

[90]  F. FACCHINEI AND C. KANZOW, *Generalized Nash equilibrium problems*, 4OR, 5, 173–210, 2007. (Cited on p. 412)

[91]  F. FACCHINEI AND C. KANZOW, *Generalized Nash equilibrium problems*, Ann. Oper. Res., 175(1), 177–211, 2010. (Cited on p. 412)

[92]  A. FISCHE, H. MARKUS, AND K. SCHÖNEFELD, *Generalized Nash equilibrium problems-recent advances and challenges*, Pesquisa Operacional, 34(3), 521–558, 2014. (Cited on p. 412)

[93]  J. FAN, J. NIE, AND A. ZHOU, *Completely positive binary tensors*, Math. Oper. Res., 44(3), 1087–1100, 2019. (Cited on p. 374)

[94]  L. FAYBUSOVICH, *Global optimization of homogeneous polynomials on the simplex and on the sphere*, Frontiers in Global Optimization (C. Floudas and P. Pardalos, editors), Kluwer Academic, Dordrecht, 2003. (Cited on p. 133)

[95]  L. FIALKOW AND J. NIE, *Positivity of Riesz functionals and solutions of quadratic and quartic moment problems*, J. Funct. Anal., 258(1), 328–356, 2010. (Cited on pp. 79, 144, 145)

[96]  L. FIALKOW AND J. NIE, *The truncated moment problem via homogenization and flat extensions*, J. Funct. Anal., 263(6), 1682–1700, 2012. (Cited on pp. 268, 271)

[97]  M. FRANK AND P. WOLFE, *An algorithm for quadratic programming*, Naval Res. Logistics (NRL), 3(1-2), 95–110, 1956. (Cited on p. 138)

[98]  S. FRIEDLAND, *Best rank one approximation of real symmetric tensors can be chosen symmetric*, Front. Math. China, 8(1), 19–40, 2013. (Cited on p. 348)

[99]  S. FRIEDLAND AND G. OTTAVIANI, *The number of singular vector tuples and uniqueness of best rank-one approximation of tensors*, Found. Comput. Math., 14(6), 1209–1242, 2014. (Cited on pp. 348, 382)

[100]  S. FRIEDLAND AND L.-H. LIM, *The computational complexity of duality*, SIAM J. Optim., 26(4), 2378–2393, 2016. (Cited on p. 356)

[101]  S. FRIEDLAND AND L.-H. LIM, *Nuclear norm of higher-order tensors*, Math. Comput., 87(311), 1255–1281, 2018. (Cited on pp. 347, 354, 355, 356, 357, 360, 362, 365, 368)

[102]  I. GEL'FAND, M. KAPRANOV, AND A. ZELEVINSKY, *Discriminants, Resultants, and Multidimensional Determinants*, Mathematics: Theory & Applications, Birkhäuser, Basel, 1994. (Cited on pp. 47, 49, 50)

[103]  M. GOEMANS AND D. WILLIAMSON, *Improved approximation algorithms for maximum cut and satisfiability problems using semidefinite programming*, J. ACM, 42(6), 1115–1145, 1995. (Cited on p. 137)

[104]  D. GONDARD AND P. RIBENBOIM, *Le 17e problème de Hilbert pour les matrices*, Bull. Sci. Math., 98, 49–56, 1974. (Cited on p. 304)

[105]  G.-M. GREUEL AND G. PFISTER, *A Singular Introduction to Commutative Algebra*, Springer, Berlin, 2002. (Cited on pp. 35, 39)

[106]  L. GUO, G-H. LIN, J.J. YE, AND J. ZHANG, *Sensitivity analysis of the value function for parametric mathematical programs with equilibrium constraints*, SIAM J. Optim., 24(3), 1206–1237, 2014. (Cited on p. 419)

[107]  B. GUO, J. NIE, AND Z. YANG, *Learning diagonal Gaussian mixture models and incomplete tensor decompositions*, Vietnam J. Math., 50(2), 421–446, 2022. (Cited on p. 424)

[108]  M. GÜRBÜZBALABAN, A. RUSZCZYŃSKI, AND L. ZHU, *A stochastic subgradient method for distributionally robust non-convex learning*, J. Optim. Theory Appl. 194, 1014–1041, 2022. (Cited on p. 424)

[109]  N. GVOZDENOVIĆ AND M. LAURENT, *The operator $\Psi$ for the chromatic number of a graph*, SIAM J. Optim., 19(2), 572–591, 2008. (Cited on p. 294)

[110]  M. HALL AND M. NEWMAN, *Copositive and completely positive quadratic forms*, Proc. Cambridge Philos. Soc., 59, 329–33, 1963. (Cited on p. 278)

[111] G. HANASUSANTO, V. ROITCH, D. KUHN AND W. WIESEMANN, *A distributionally robust perspective on uncertainty quantification and chance constrained. programming*, Math. Program. 151(1), 35–62, 2015. (Cited on p. 424)

[112] D. HANDELMAN, *Representing polynomials by positive linear functions on compact convex polyhedra*, Pac. J. Math., 132(1), 35–62, 1988. (Cited on p. 71)

[113] J. HARRIS, *Algebraic Geometry: A First Course*, Graduate Textbooks in Mathematics, Springer, Berlin, 1992. (Cited on p. 35)

[114] R. HARTSHORNE, *Algebraic Geometry*, Graduate Texts in Mathematics No. 52, Springer, New York, 1977. (Cited on p. 51)

[115] S. HE, Z. LUO, J. NIE, AND S. ZHANG, *Semidefinite relaxation bounds for indefinite homogeneous quadratic optimization*, SIAM J. Optim., 19(2), 503–523, 2008. (Cited on pp. 144, 146)

[116] B. HE AND X. YUAN, *Convergence analysis of primal-dual algorithms for a saddle-point problem: From contraction perspective*, SIAM J. Imaging Sci., 5(1), 119–149, 2012. (Cited on p. 401)

[117] Y. HE AND R.D.C. MONTEIRO, *Accelerating block-decomposition first-order methods for solving composite saddle-point and two-player Nash equilibrium problems*, SIAM J. Optim., 25(4), 2182–2211, 2015 (Cited on p. 401)

[118] J.W. HELTON AND V. VINNIKOV, *Linear matrix inequality representation of sets*, Commun. Pure Appl. Math., 60(5), 654–674, 2007. (Cited on pp. 228, 229)

[119] J.W. HELTON AND J. NIE, *Sufficient and necessary conditions for semidefinite representability of convex hulls and sets*, SIAM J. Optim., 20(2), 759–791, 2009. (Cited on pp. 230, 231)

[120] J.W. HELTON AND J. NIE, *Semidefinite representation of convex sets*, Math. Program., Ser. A, 122(1), 21–64, 2010. (Cited on pp. 218, 219, 233, 234, 235)

[121] J.W. HELTON AND J. NIE, *A semidefinite approach for truncated K-moment problem*, Found. Comput. Math., 12(6), 851–881, 2012. (Cited on p. 77)

[122] D. HENRION AND J. LASSERRE, *Detecting global optimality and extracting solutions in GloptiPoly*, in Positive Polynomials in Control, pp. 293–310, Lecture Notes in Control and Information Science Vol. 312, Springer, Berlin, 2005. (Cited on pp. 79, 82, 124)

[123] D. HENRION AND J. LASSERRE, *Convergent relaxations of polynomial matrix inequalities and static output feedback*, IEEE Trans. Auto. Control, 51, 192–202, 2006. (Cited on p. 323)

[124] D. HENRION, J. B. LASSERRE, AND J. LOEFBERG, *GloptiPoly 3: moments, optimization and semidefinite programming*, Optim. Methods Softw., 24(4-5), 761–779, 2009. (Cited on p. 32)

[125] R. HETTICH AND K.O. KORTANEK, *Semi-infinite programming: Theory, methods, and applications*, SIAM Review, 35, 380–429, 1993. (Cited on p. 414)

[126] R. HILDEBRAND, *The extreme rays of the $5 \times 5$ copositive cone*, Linear Algebra Appl., 437(7), 1538–1547, 2012. (Cited on p. 278)

[127] C. HILLAR AND J. NIE, *An elementary and constructive proof of Hilbert's 17th Problem for matrices*, Proc. AMS, 136, 73–76, 2008. (Cited on p. 304)

[128] C. HILLAR AND L.-H. LIM, *Most tensor problems are NP-hard*, J. ACM, 60(6), 45, 2013. (Cited on pp. 348, 376)

[129] A. HOFFMAN AND F. PEREIRA, *On copositive matrices with $-1, 0, 1$ entries*, J. Comb. Theory, Ser. B, 14(3), 302–309, 1973. (Cited on p. 278)

[130] R. HORN AND C. JOHNSON, *Matrix Analysis*, Cambridge University Press, Cambridge, 1985. (Cited on p. 18)

[131] B. HUA, G. NI, AND M. ZHANG, *Computing geometric measure of entanglement for symmetric pure states via the Jacobian SDP relaxation technique*, J. Oper. Res. Soc. China, 5(1), 111–121, 2017. (Cited on p. 367)

[132] L. HUANG, J. NIE, AND Y.-X. YUAN, *Homogenization for polynomial optimization with unbounded sets*, Math. Program. (2022). doi.org/10.1007/s10107-022-01878-5 (Cited on pp. 120, 163)

[133] J. JAHN, *Vector Optimization: Theory, Applications, and Extensions*, Springer, Berlin, 2011. (Cited on p. 429)

[134] V. A. JAKUBOVIČ, *Factorization of symmetric matrix polynomials*, Dokl. Akad. Nauk SSSR, 194, 532–535, 1970 (Russian). (Cited on p. 302)

[135] L. JIAO, J. LEE, Y. OGATA, AND T. TANAKA, *Multi-objective optimization problems with SOS-convex polynomials over an LMI constraint*, Taiwanese J. Math., 24(4), 1021–1043, 2020. (Cited on p. 430)

[136] L. JIAO, J. LEE, AND Y. ZHOU, *A hybrid approach for finding efficient solutions in vector optimization with SOS-convex polynomials*, Oper. Res. Lett., 48(2), 188–194, 2020. (Cited on p. 430)

[137] L. JIAO AND J. LEE, *Finding efficient solutions in robust multiple objective optimization with SOS-convex polynomial data*, Ann. Oper. Res., 296, 803–820, 2021. (Cited on p. 430)

[138] L. JIAO, J. LEE, AND N. SISARAT, *Multi-objective convex polynomial optimization and semidefinite programming relaxations*, J. Glob. Optim., 80(1), 117–138, 2021. (Cited on p. 430)

[139] D. JIBETEAN AND E. DE KLERK, *Global optimization of rational functions: A semidefinite programming approach*, Math. Program., 106(1), 93–109, 2006. (Cited on p. 392)

[140] N. KARMARKAR AND Y. LAKSHMAN, *On approximate GCDs of univariate polynomials*, J. Symb. Comput., 26, 653–666, 1998. (Cited on p. 395)

[141] D.S. KIM, T. PHAM, AND N.V. TUYEN, *On the existence of Pareto solutions for polynomial vector optimization problems*, Math. Program., 177, 321–341, 2019. (Cited on p. 430)

[142] D. KIMSEY AND H. WOERDEMAN, *The truncated matrix-valued K-moment problem on $\mathbb{R}^d$, $\mathbb{C}^d$ and $\mathbb{T}^d$*, Trans. Amer. Math. Soc., 365, 5393–5430, 2013. (Cited on p. 314)

[143] D. KIMSEY, *An operator-valued generalization of Tchakaloff's theorem*, J. Funct. Anal., 266(3), 1170–1184, 2014. (Cited on pp. 314, 315)

[144] S. KIM, M. KOJIMA, AND H. WAKI, *Generalized Lagrangian duals and sums of squares relaxations of sparse polynomial optimization problems*, SIAM J. Optim., 15(3), 697–719, 2005. (Cited on p. 388)

[145] D. KLATTE, *On a Frank–Wolfe type theorem in cubic optimization*, Optimization, 68, 539–547, 2019. (Cited on p. 138)

[146] I. KLEP AND M. SCHWEIGHOFER, *Pure states, positive matrix polynomials and sums of hermitian squares*, Indiana Univ. Math. J., 59(3), 857–874, 2010. (Cited on pp. 302, 305, 306)

[147] I. KLEP AND J. NIE, *A Matrix Positivstellensatz with lifting polynomials*, SIAM J. Optim., 30(1), 240–261, 2020. (Cited on pp. 332, 333, 335, 336, 337)

[148] I. KLEP, V. MAGRON AND J. POVH, *Sparse noncommutative polynomial optimization*, Math. Program. 193(2), 789–829, 2022. (Cited on p. 388)

[149] E. KOFIDIS AND P.A. REGALIA, *On the best rank-1 approximation of higher-order supersymmetric tensors*, SIAM. J. Matrix Anal. Appl., 23(3), 863–884, 2002. (Cited on p. 348)

[150] N. KOGAN AND A. BERMAN, *Characterization of completely positive graphs*, Discrete Math., 114, 297–304, 1993. (Cited on p. 283)

[151] M. KOJIMA, S. KIM, AND H. WAKI, *Sparsity in Sums of Squares of Polynomials*, Math. Program., 103(1), 45–62, 2005. (Cited on p. 388)

[152] T.G. KOLDA AND B.W. BADER, *Tensor decompositions and applications*, SIAM Review, 51(3), 455–500, 2009. (Cited on p. 340)

[153] J.B. KRAWCZYK AND U. STANISLAV, *Relaxation algorithms to find Nash Equilibria with economic applications*, Environ. Model. Assess., 5(1), 63–73, 2000. (Cited on p. 407)

[154] M.G. KREIN AND A.A. NUDELMAN, *The Markov Moment Problem and Extremal Problems*, Translations of Mathematical Monographs Vol. 50, AMS, Providence, 1977. (Cited on pp. 71, 98, 99)

[155] J.L. KRIVINE, *Anneaux préordonnés*, J. Anal. Math., 12, 307–326, 1964. (Cited on pp. 71, 73)

[156] J.M. LANDSBERG, *Tensors: Geometry and Applications*, Graduate Studies in Mathematics Vol. 128, AMS, Providence, 2012. (Cited on pp. 340, 341, 342, 345)

[157] R. LARAKI AND J.B. LASSERRE, *Semidefinite programming for min-max problems and games*, Math. Program., 131, 305–332, 2012. (Cited on p. 407)

[158] J.B. LASSERRE, *Global optimization with polynomials and the problem of moments*, SIAM J. Optim., 11(3), 796–817, 2001. (Cited on pp. 53, 119, 120, 139, 140, 143, 145, 167)

[159] J.B. LASSERRE, *Convergent SDP-relaxations in polynomial optimization with sparsity*, SIAM J. Optim., 17(3), 822–843, 2006. (Cited on pp. 388, 390)

[160] J.B. LASSERRE, M. LAURENT, AND P. ROSTALSKI, *Semidefinite characterization and computation of zero-dimensional real radical ideals*, Found. Comput. Math., 8, 607–647, 2008. (Cited on pp. 126, 174)

[161] J.B. LASSERRE, *A semidefinite programming approach to the generalized problem of moments*, Math. Program., 112, 65–92, 2008. (Cited on pp. 258, 434)

[162] J.B. LASSERRE, *Moments, Positive Polynomials and Their Applications*, Imperial College Press, London, 2009. (Cited on pp. 64, 70, 262)

[163] J.B. LASSERRE, *Convex sets with semidefinite representation*, Math. Program., 120, 457–477, 2009. (Cited on p. 234)

[164] J.B. LASSERRE, *Convexity in semialgebraic geometry and polynomial optimization*, SIAM J. Optim., 19(4), 1995–2014, 2009. (Cited on pp. 220, 221, 224)

[165] J.B. LASSERRE, *On representations of the feasible set in convex optimization*, Optim. Lett., 4(1), 1–5, 2010. (Cited on pp. 222, 223)

[166] J.B. LASSERRE, *Introduction to Polynomial and Semi-Algebraic Optimization*, Cambridge University Press, Cambridge, 2015. (Cited on pp. 32, 64, 70)

[167] J.B. LASSERRE, *The Moment-SOS Hierarchy*, in Proceedings of the International Congress of Mathematicians (ICM 2018), Vol. 3, B. Sirakov, P. Ney de Souza, and M. Viana (eds.), pp. 3761–3784, World Scientific, Singapore, 2019. (Cited on pp. 32, 120)

[168] J.B. LASSERRE AND T. WEISSER, *Distributionally robust polynomial chance-constraints under mixture ambiguity sets*, Math. Program., 185, 409–453, 2021. (Cited on p. 424)

[169] J.B. LASSERRE, V. MAGRON, S. MARX, AND O. ZAHM, *Minimizing rational functions: A hierarchy of approximations via pushforward measures*, SIAM J. Optim., 31(3), 2285–2306, 2021. (Cited on p. 392)

[170] M. LAURENT, *Revisiting two theorems of Curto and Fialkow on moment matrices*, Proc. AMS, 133(10), 2965–2976, 2005. (Cited on pp. 79, 82, 84, 124)

[171] M. LAURENT, *Sums of squares, moment matrices and optimization over polynomials*, in Emerging Applications of Algebraic Geometry of IMA Volumes, Mathematics and its Applications Vol. 149, pp. 157–270, Springer, Berlin, 2009. (Cited on pp. 32, 40, 42, 64, 68, 70, 76, 78, 79, 120, 125, 126, 152, 174, 247, 248)

[172] M. LAURENT, *Optimization over polynomials: Selected topics*, in Proceedings of the International Congress of Mathematicians, S.Y. Jang, Y.R. Kim, D.-W. Lee, and I. Yie (eds.), pp. 843–869, 2014. (Cited on pp. 32, 70)

[173] D. LAY, *Linear Algebra and Its Applications*, fourth edition, Pearson Education, London, 2012. (Cited on p. 3)

[174] J. LEE AND L. JIAO, *Solving fractional multicriteria optimization problems with sum of squares convex polynomial data*, J. Optim. Theory Appl., 176, 428–455, 2018. (Cited on p. 430)

[175] G. LI, L. QI, AND G. YU, *The Z-eigenvalues of a symmetric tensor and its application to spectral hypergraph theory*, Numer. Linear Algebra Appl., 20, 1001–1029, 2013. (Cited on p. 376)

[176] C. LING, J. NIE, L. QI, AND Y. YE, *Biquadratic optimization over unit spheres and semidefinite programming relaxations*, SIAM J. Optim., 20(3), 1286–1310, 2009. (Cited on p. 327)

[177] L.-H. LIM, *Singular values and eigenvalues of tensors: a variational approach*, in Proceedings of the IEEE International Workshop on Computational Advances in Multi-Sensor Adaptive Processing (CAMSAP '05), pp. 129–132, 2005. (Cited on pp. 375, 381, 382)

[178] L.-H. LIM AND P. COMON, *Multiarray signal processing: tensor decomposition meets compressed sensing*, C. R. Acad. Sci. Ser. B Mech., 338(6), 311–320, 2010. (Cited on p. 356)

[179] L.-H. LIM, *Tensors and hypermatrices*, in Handbook of Linear Algebra, second edition, L. Hogben (ed.), CRC Press, Boca Raton, 2013. (Cited on pp. 340, 341, 346)

[180] L.-H. LIM AND P. COMON, *Blind multilinear identification*, IEEE Trans. Inf. Theory, 60(2), 1260–1280, 2014. (Cited on p. 356)

[181] J. LÖFBERG, *YALMIP: a toolbox for modeling and optimization in Matlab*, in Proceedings of the IEEE CACSD Symposium, Taiwan, 2004. (Cited on pp. 324, 326)

[182] E. LOOIJENGA, *Isolated Singular Points on Complete Intersections*, London Mathematical Society Lecture Note Series Vol. 77, Cambridge University Press, Cambridge, 1984. (Cited on p. 50)

[183] L. LOVÁSZ, *On the Shannon capacity of a graph*, IEEE Trans. Inf. theory, 25, 1–7, 1979. (Cited on p. 26)

[184] F. LUKÁCS, *Verscharfung der ersten Mittelwersatzes der Integralrechnung für rationale Polynome*, Math. Z., 2, 229–305, 1918. (Cited on pp. 88, 90)

[185] Z.-Q. LUO, N. D. SIDIROPOULOS, P. TSENG, AND S. ZHANG, *Approximation bounds for quadratic optimization with homogeneous quadratic constraints*, SIAM J. Optim., 18(1), 1–28, 2007. (Cited on pp. 144, 146)

[186] V. MAGRON AND J. WANG, *TSSOS: a Julia library to exploit sparsity for large-scale polynomial optimization*, Preprint, 2021. arXiv.org/abs/2103.00915 (Cited on p. 388)

[187] A. MARKOV, *Lecture notes on functions with the least deviation from zero*, 1906. Reprinted in *Markov A.A. Selected Papers*, pp. 244–291, N. Achiezer (ed.), GosTechIzdat, Moscow, 1948 (in Russian). (Cited on pp. 88, 90)

[188] R. MARLER AND J. ARORA, *Survey of multi-objective optimization methods for engineering*, Struct. Multidisc. Optim., 26(6), 369–395, 2004. (Cited on pp. 429, 430)

[189] M. MARSHALL, *Optimization of polynomial functions.*, Canad. Math. Bull., 46(4), 575–587, 2003. (Cited on p. 68)

[190] M. MARSHALL, *Representation of non-negative polynomials with finitely many zeros*, Ann. Faculte Sci. Toulouse, 15, 599–609, 2006. (Cited on pp. 156, 158, 159)

[191] M. MARSHALL, *Positive Polynomials and Sums of Squares*, Mathematical Surveys and Monographs Vol. 146, AMS, Providence, 2008. (Cited on pp. 32, 61, 64, 74, 75, 114, 159)

[192] M. MARSHALL, *Representation of non-negative polynomials, degree bounds and applications to optimization*, Canad. J. Math., 61, 205–221, 2009. (Cited on p. 159)

[193] J.E. MAXFIELD AND H. MINC, *On the Matrix Equation $X'X = A$*, Proc. Edinburgh Math. Soc., 13(2), 125–129, 1962. (Cited on pp. 277, 281, 282, 283, 285)

[194] L. MCKENZIE, *On the existence of a general equilibrium for a competitive market*, Econometrica, 27, 54–71, 1959. (Cited on p. 412)

[195] M. MEVISSEN, E. RAGNOLI, AND J. YU, *Data-driven distributionally robust polynomial optimization*, Adv. Neural Inf. Process. Syst., 26, 37–45, 2013. (Cited on p. 424)

[196] K. MIETTINEN, *Nonlinear Multiobjective Optimization*, Vol. 12, Springer, Berlin, 2012. (Cited on pp. 429, 430)

[197] MOSEK APS, *The MOSEK optimization toolbox for MATLAB manual*, Version 9.0., 2019. http://docs.mosek.com/9.0/toolbox/index.html (Cited on p. 25)

[198] T. MOTZKIN AND E. STRAUS, *Maxima for graphs and a new proof of a theorem of Túran*, Canad. J. Math., 17, 533–540, 1965. (Cited on pp. 138, 140, 294)

[199] N. MUNRO (ED.), *The Use of Symbolic Methods in Control System Analysis and Design*, Control Engineering Series Vol. 56, IEE Books, London, 1999. (Cited on p. 122)

[200] R.B. MYERSON, *Game Theory*, Harvard University Press, Cambridge, 2013. (Cited on p. 407)

[201] K. MURTY AND S. KABADI, *Some NP-complete problems in quadratic and nonlinear programming*, Math. Program., 39(2), 117–129, 1987. (Cited on pp. 278, 396)

[202] J. NASH, *Non-cooperative games*, Ann. Math., 1951, 286–295. (Cited on p. 406)

[203] A. NEDIĆ AND A. OZDAGLAR, *Subgradient methods for saddle-point problems*, J. Optim. Theory Appl., 142(1), 205–228, 2009. (Cited on p. 401)

[204] Y. NESTEROV, *Squared Functional Systems and Optimization Problems*, in High Performance Optimization, H. Frank et al. (eds.), pp. 405–440, Kluwer Academic, Dordrecht, 2000. (Cited on p. 119)

[205] Y. NESTEROV, *Random walk in a simplex and quadratic optimization over convex polytopes*, CORE Discussion Paper, CORE, Catholic University of Louvain, Louvain-la-Neuve, Belgium, 2003. (Cited on p. 131)

[206] T. NETZER, D. PLAUMANN, AND M. SCHWEIGHOFER, *Exposed faces of semidefinitely representable sets*, SIAM J. Optim., 20(4), 1944–1955, 2010. (Cited on p. 234)

[207] G. NI, L. QI, AND M. BAI, *Geometric measure of entanglement and U-eigenvalues of tensors*, SIAM J. Matrix Anal. Appl., 35(1), 73–87, 2014. (Cited on pp. 367, 383, 385)

[208] Q. NI, L. QI, AND F. WANG, *An eigenvalue method for testing positive definiteness of a multivariate form*, IEEE Trans. Automatic Control, 53, 1096–1107, 2008. (Cited on p. 376)

[209] J. NIE AND J. DEMMEL, *Minimum ellipsoid bounds for solutions of polynomial systems via sum of squares*, J. Glob. Optim., 33, 511–525, 2005. (Cited on p. 228)

[210] J. NIE, J. DEMMEL, AND B. STURMFELS, *Minimizing polynomials via sum of squares over the gradient ideal*, Math. Program., Ser. A, 106(3), 587–606, 2006. (Cited on pp. 120, 202)

[211] J. NIE AND M. SCHWEIGHOFER, *On the complexity of Putinar's Positivstellensatz*, J. Complex., 23, 135–150, 2007. (Cited on pp. 114, 261)

[212] J. NIE, J. DEMMEL, AND M. GU, *Global minimization of rational functions and the nearest GCDs*, J. Glob. Optim., 40(4), 697–718, 2008. (Cited on pp. 392, 395, 396)

[213] J. NIE AND J. DEMMEL, *Sparse SOS relaxations for minimizing functions that are summations of small polynomials*, SIAM J. Optim., 19(4), 1534–1558, 2008. (Cited on pp. 388, 390, 391)

[214] J. NIE, P. PARRILO, AND B. STURMFELS, *Semidefinite representation of k-ellipse*, in IMA Volume 146: Algorithms in Algebraic Geometry, A. Dickenstein, F.-O. Schreyer, and A. Sommese (eds.), pp. 117–132, Springer, New York, 2008. (Cited on pp. 228, 236)

[215] J. NIE AND B. STURMFELS, *Matrix cubes parametrized by eigenvalues*, SIAM. J. Matrix Anal. Appl., 31(2), 755–766, 2009. (Cited on p. 228)

[216] J. NIE, *Sum of squares method for sensor network localization*, Comput. Optim. Appl., 43(2), 151–179, 2009. (Cited on pp. 119, 388)

[217] J. NIE AND K. RANESTAD, *Algebraic degree of polynomial optimization*, SIAM J. Optim., 20(1), 485–502, 2009. (Cited on pp. 128, 149, 164, 181)

[218] J. NIE, *Polynomial matrix inequality and semidefinite representation*, Math. Oper. Res., 36(3), 398–415, 2011. (Cited on pp. 327, 329, 330)

[219] J. NIE, *Discriminants and nonnegative polynomials*, J. Symb. Comput., 47(2), 167–191, 2012. (Cited on pp. 50, 211, 250, 251, 252)

[220] J. NIE, *Sum of squares methods for minimizing polynomial forms over spheres and hypersurfaces*, Front. Math. China, 7, 321–346, 2012. (Cited on pp. 131, 133, 135, 353)

[221] J. NIE, *First order conditions for semidefinite representations of convex sets defined by rational or singular polynomials*, Math. Program., 131(1), 1–36, 2012. (Cited on pp. 238, 239, 240, 242, 243)

[222] J. NIE, *Convex hulls of quadratically parameterized sets with quadratic constraints*, in Mathematical Methods in Systems, Optimization, and Control, pp. 247–258, Operator Theory Advances and Applications Vol. 222, Birkhäuser/Springer Basel, 2012. (Cited on p. 234)

[223] J. NIE, *Chapter 6: Semidefinite Representability*, in Semidefinite Optimization and Convex Algebraic Geometry, pp. 251–291, SIAM, Philadelphia, 2013. (Cited on pp. 228, 231)

[224] J. NIE, *An exact Jacobian SDP relaxation for polynomial optimization*, Math. Program., 137, 225–255, 2013. (Cited on pp. 43, 158, 202, 203, 207, 212, 253)

[225] J. NIE, *Certifying convergence of Lasserre's hierarchy via flat truncation*, Math. Program., 142(1-2), 485–510, 2013. (Cited on pp. 120, 123, 124, 125, 126, 147, 148, 149, 153, 154, 158)

[226] J. NIE, *Polynomial optimization with real varieties*, SIAM J. Optim., 23(3), 1634–1646, 2013. (Cited on pp. 72, 74, 168, 169, 171, 173, 175, 176)

[227] J. NIE, *An approximation bound analysis for Lasserre's relaxation in multivariate polynomial optimization*, J. Oper. Res. Soc. China, 1(3), 313–332, 2013. (Cited on p. 143)

[228] J. NIE AND L. WANG, *Semidefinite relaxations for best rank-1 tensor approximations*, SIAM. J. Matrix Anal. Appl., 35(3), 1155–1179, 2014. (Cited on pp. 119, 131, 138)

[229] J. NIE, *The A-truncated K-moment problem*, Found. Comput. Math., 14(6), 1243–1276, 2014. (Cited on pp. 77, 255, 256, 267, 268, 269)

[230] J. NIE, *Optimality conditions and finite convergence of Lasserre's hierarchy*, Math. Program., 146(1-2), 97–121, 2014. (Cited on pp. 144, 155, 158, 159, 164, 165, 166, 168)

[231] J. NIE, *The hierarchy of local minimums in polynomial optimization*, Math. Program., 151(2), 555–583, 2015. (Cited on pp. 379, 386, 397, 398, 399, 400)

[232] J. NIE, *Linear optimization with cones of moments and nonnegative polynomials*, Math. Program., 153, 247–274, 2015. (Cited on pp. 261, 262, 263, 272, 434)

[233] J. NIE AND X. ZHANG, *Positive maps and separable matrices*, SIAM J. Optim., 26(2), 1236–1256, 2016. (Cited on p. 371)

[234] J. NIE, L. WANG, AND J.J. YE, *Bilevel polynomial programs and semidefinite relaxation methods*, SIAM J. Optim., 27(3), 1728–1757, 2017. (Cited on p. 414)

[235] J. NIE, *Generating polynomials and symmetric tensor decompositions*, Found. Comput. Math., 17(2), 423–465, 2017. (Cited on p. 343)

[236] J. NIE, *Symmetric tensor nuclear norms*, SIAM J. Appl. Algebra Geometry, 1(1), 599–625, 2017. (Cited on pp. 357, 358, 359, 360, 363, 364, 365)

[237] J. NIE, *Low rank symmetric tensor approximations*, SIAM. J. Matrix Anal. Appl., 38(4), 1517–1540, 2017. (Cited on pp. 343, 348)

[238] J. NIE AND X. ZHANG, *Real eigenvalues of nonsymmetric tensors*, Comput. Optim. Appl., 70(1), 1–32, 2018. (Cited on pp. 380, 381)

[239] J. NIE, Z. YANG, AND X. ZHANG, *A complete semidefinite algorithm for detecting copositive matrices and tensors*, SIAM J. Optim., 28(4), 2902–2921, 2018. (Cited on pp. 289, 372, 373, 374)

[240] J. NIE, *Tight relaxations for polynomial optimization and Lagrange multiplier expressions*, Math. Program., 178 (1-2), 1–37, 2019. (Cited on pp. 182, 184, 185, 189, 190, 192, 193, 194, 196, 197, 202)

[241] J. NIE, L. YANG, AND S. ZHONG, *Stochastic polynomial optimization*, Optim. Methods Softw., 35(2), 329–347, 2020. (Cited on p. 424)

[242] J. NIE AND Z. YANG, *Hermitian tensor decompositions*, SIAM. J. Matrix Anal. Appl., 41(3), 1115–1144, 2020. (Cited on p. 371)

[243] J. NIE, X. TANG, AND L. XU, *The Gauss-Seidel Method for Generalized Nash Equilibrium Problems of Polynomials*, Comput. Optim. Appl., 78(2), 529–557, 2021. (Cited on p. 412)

[244] J. NIE AND X. TANG, *Convex generalized Nash equilibrium problems and polynomial optimization*, Math. Program., 198, 1485–1518, 2023. (Cited on p. 412)

[245] J. NIE, Z. YANG, AND G. ZHOU, *The saddle point problem of polynomials*, Found. Comput. Math., 22(4), 1133–1169, 2022. (Cited on pp. 401, 403, 404)

[246] J. NIE, L. YANG, S. ZHONG AND G. ZHOU, *Distributionally robust optimization with moment ambiguity sets*, J. Sci. Comput., 94, 12, 2023. (Cited on pp. 424, 425, 426, 427, 428)

[247] J. NIE, L. WANG, J. J. YE, AND S. ZHONG, *A Lagrange multiplier expression method for bilevel polynomial optimization*, SIAM J. Optim., 31(3), 2368–2395, 2021. (Cited on pp. 413, 414, 415, 416, 417, 418, 419, 420, 422, 423)

[248] J. NIE, X. TANG, Z. YANG AND S. ZHONG, *Dehomogenization for completely positive tensors*, Numer. Algebra, Control Optim., 13(2), 340–363, 2023. (Cited on p. 288)

[249] J. NIE AND X. TANG, *Nash Equilibrium Problems of Polynomials*, Preprint, 2020. arXiv:2006.09490 [math.OC] (Cited on pp. 406, 407, 409, 410, 411, 412)

[250] J. NIE, X. TANG, AND Z. ZHONG, *Rational generalized Nash equilibrium problems*, SIAM J. Optim., to appear. (Cited on p. 412)

[251] J. NIE AND Z. YANG, *The Multi-Objective Polynomial Optimization*, Preprint, 2021. arXiv:2108.04336 (Cited on pp. 430, 432, 434, 436, 439, 440, 442)

[252] J. NIE, L. WANG, AND Z. ZHENG, *Low rank tensor decompositions and approximations*, J. Oper. Res. Soc. China (2023). doi.org/10.1007/s40305-023-00455-7 (Cited on p. 348)

[253] J. NOCEDAL AND S. WRIGHT, *Numerical Optimization*, Springer, Berlin, 2006. (Cited on p. 139)

[254] W.T. OBUCHOWSKA, *On generalizations of the Frank–Wolfe theorem to convex and quasi-convex programmes*, Comput. Optim. Appl., 33, 349–364, 2006. (Cited on p. 138)

[255] L. OEDING AND G. OTTAVIANI, *Eigenvectors of tensors and algorithms for Waring decomposition*, J. Symb. Comput., 54, 9–35, 2013. (Cited on p. 345)

[256] M.J. OSBORNE AND A. RUBINSTEIN, *A Course in Game Theory*, MIT Press, Cambridge, 1994. (Cited on p. 407)

[257] P. PARRILO AND B. STURMFELS, *Minimizing polynomial functions*, in Algorithmic and Quantitative Real Algebraic Geometry (Piscataway, NJ, 2001), pp. 83–99, DIMACS Series in Discrete Mathematics and Theoretical Computer Science Vol. 60, AMS, Providence, 2003. (Cited on pp. 53, 120)

[258] P. PARRILO, *Semidefinite Programming relaxations for semialgebraic problems*, Math. Program., Ser. B, 96(2), 293–320, 2003. (Cited on pp. 53, 120, 121, 285)

[259] P. PARRILO, *Exploiting structure in sum of squares programs*, in Proceedings for the 42nd IEEE Conference on Decision and Control, Maui, Hawaii, 2003. (Cited on p. 388)

[260] G. PATAKI, *On the rank of extreme matrices in semidefinite programs and the multiplicity of optimal eigenvalues*, Math. Oper. Res., 23(2), 339–358, 1998. (Cited on p. 26)

[261] I. PÓLIK AND T. TERLAKY, *A survey of the S-lemma*, SIAM Review, 49(3), 371–418, 2007. (Cited on p. 144)

[262] G. PÓLYA, *Über positive Darstellung von Polynomen Vierteljschr*, Naturforsch. Ges. Zuürich 73 (1928), 141–145, in Collected Papers Vol. 2, R.P. Boas (ed.), MIT Press, Cambridge, 1974, pp. 309–313. (Cited on p. 279)

[263] G. PÓLYA AND G. SZEGÖ, *Problems and Theorems in Analysis II*, Springer, New York, 1976 (Cited on pp. 88, 90)

[264] K. POSTEK, A. BEN-TAL, D. DEN HERTOG AND B. MELENBERG, *Robust optimization with ambiguous stochastic constraints under mean and dispersion information*, Oper. Res., 66(3), 814–33, 2018. (Cited on p. 424)

[265] J. POVH AND F. RENDL, *Copositive and semidefinite relaxations of the quadratic assignment problem*, Discrete Optim., 6(3), 231–241, 2009. (Cited on p. 295)

[266] J. POVH AND F. RENDL, *A copositive programming approach to graph partitioning*, SIAM J. Optim., 18(1), 223–241, 2007. (Cited on p. 296)

[267] V. POWERS AND B. REZNICK, *Polynomials that are positive on an interval*, Trans. AMS, 352(10), 4677–4692, 2000. (Cited on pp. 88, 90, 113, 114, 115, 116)

[268] V. POWERS AND B. REZNICK, *A new bound for Pólya's Theorem with applications to polynomials positive on polyhedra*, J. Pure Appl. Algebra, 164(1-2), 221–229, 2001. (Cited on pp. 279, 280, 281, 285)

[269] A. PRESTEL AND C. DELZELL, *Positive Polynomials: From Hilbert's 17th Problem to Real Algebra*, Springer Monographs in Mathematics, Springer, Berlin, 2001. (Cited on pp. 70, 75)

[270] T. JACOBI AND A. PRESTEL, *Distinguished representations of strictly positive polynomials*, J. Reine Angew. Math., 532, 223–235, 2001. (Cited on p. 74)

[271] C. PROCESI AND M. SCHACHER, *A Non-Commutative Real Nullstellensatz and Hilbert's 17th Problem*, Ann. Math., 104 (1976), pp. 395–406. (Cited on p. 304)

[272] M. PUTINAR, *Positive polynomials on compact semi-algebraic sets*, Indiana Univ. Math. J., 42, 203–206, 1993. (Cited on pp. 70, 262)

[273] L. QI, *Eigenvalues of a real supersymmetric tensor*, J. Symb. Comput., 40, 1302–1324, 2005. (Cited on p. 375)

[274] L. QI, Y. WANG, AND E.-X. WU, *D-eigenvalues of diffusion kurtosis tensors*, J. Comput. Appl. Math., 221, 150–157, 2008. (Cited on p. 376)

[275] L. QI, G. YU, AND E.X. WU, *Higher order positive semidefinite diffusion tensor imaging*, SIAM J. Imaging Sci., 3(3), pp. 416–433, 2010. (Cited on p. 376)

[276] L. QI, *Symmetric nonnegative tensors and copositive tensors*, Linear Algebra Appl., 439(1), 228–238, 2013. (Cited on p. 372)

[277] L. QI AND Z. LUO, *Tensor Analysis: Spectral Theory and Special Tensors*, SIAM, Philadelphia, 2017. (Cited on p. 375)

[278] L. QI, G. ZHANG, AND G. NI, *How entangled can a multi-party system possibly be?*, Phys. Lett. A, 382(22), 1465–1471, 2018. (Cited on pp. 367, 368, 369)

[279] H. RAHIMIAN AND S. MEHROTRA, *Distributionally Robust Optimization: A Review*. Preprint, 2019. arXiv:1908.05659 (Cited on p. 424)

[280] K. RANESTAD AND F.-O. SCHREYER, *On the rank of a symmetric form*, J. Algebra, 346(1), 340–342, 2011. (Cited on p. 343)

[281] B. REZNICK, *Extremal psd forms with few terms*, Duke Math. J., 45, 363–374, 1978. (Cited on p. 387)

[282] B. REZNICK, *Sums of even powers of real linear forms*, Mem. Amer. Math. Soc., 96, 463, 1992. (Cited on p. 272)

[283] B. REZNICK, *Some concrete aspects of Hilbert's $17^{th}$ problem*, Contemp. Math., 253, 251–272, 2000. (Cited on pp. 57, 60, 199, 285)

[284] B. REZNICK, *On the absence of uniform denominators in Hilbert's seventeenth problem*, Proc. Amer. Math. Soc., 133, 2829–2834, 2005. (Cited on pp. 60, 129, 131)

[285] R.T. ROCKAFELLAR, *Convex Analysis*, Princeton Mathematical Series, No. 28, Princeton University Press, Princeton, 1970. (Cited on pp. 8, 9)

[286] C.A. ROGERS, *Probability measures on compact sets*, Proc. London Math. Soc., s3-52(2), 328–348, 1986. (Cited on p. 250)

[287] J.B. ROSEN, *Existence and uniqueness of equilibrium points for concave n-person games*, Econometrica, 1965, 520–534. (Cited on pp. 407, 412)

[288] M. ROSENBLUM AND J. ROVNYAK, *The factorization problem for nonnegative operator valued functions*, Bull. Amer. Math. Soc., 77, 287–318, 1971. (Cited on p. 302)

[289] P. RUÍZ-CANALES AND A. RUFIÁN-LIZANA, *A characterization of weakly efficient points*. Math. Program., 68(1-3), 205–212, 1995. (Cited on p. 430)

[290] J. SAUNDERSON, *A convex form that is not a sum of squares*, Math. Oper. Res., 2022. doi.org/10.1287/moor.2022.1273 (Cited on p. 220)

[291] R. SAVANI AND B. VON STENGEL, *Hard-to-solve bimatrix games*, Econometrica, 74(2), 397–429, 2006. (Cited on p. 407)

[292] C. SCHEIDERER, *Sums of squares of functions on real algebraic varieties*, Trans. Amer. Math. Soc., 352, 1039–1069, 1999. (Cited on pp. 144, 158)

[293] C. SCHEIDERER, *Non-existence of degree bounds for weighted sums of squares representations*, J. Complex., 21, 823–844, 2005. (Cited on p. 158)

[294] C. SCHEIDERER, *Distinguished representations of non-negative polynomials*, J. Algebra, 289, 558–573, 2005. (Cited on p. 158)

[295] C. SCHEIDERER, *Positivity and sums of squares: A guide to recent results*, in Emerging Applications of Algebraic Geometry, M. Putinar and S. Sullivant (eds.), IMA Volumes in Mathematics and it Applications 149, Springer, Berlin, 2009, pp. 271–324. (Cited on pp. 32, 70, 159)

[296] C. SCHEIDERER, *Spectrahedral shadows*, SIAM J. Appl. Algebra Geometry, 2, 26–44, 2018. (Cited on p. 231)

[297] C. SCHERER AND C. HOL, *Matrix sum-of-squares relaxations for robust semi-definite programs*, Math. Program., 107, 189–211, 2006. (Cited on pp. 305, 307, 309)

[298] K. SCHMÜDGEN, *The K-moment problem for compact semialgebraic sets*, Math. Ann., 289, 203–206, 1991. (Cited on p. 70)

[299] R. SCHNEIDER, *Convex Bodies: The Brunn-Minkowski Theory*, Encyclopedia of Mathematics and its Applications Vol. 44. Cambridge University Press, Cambridge, 1993. (Cited on p. 10)

[300] M. SCHWEIGHOFER, *On the complexity of Schmüdgen's Positivstellensatz*, J. Complex., 20(4), 529–543, 2004. (Cited on p. 114)

[301] M. SCHWEIGHOFER, *Optimization of polynomials on compact semialgebraic sets*, SIAM J. Optim., 15(3), 805–825, 2005. (Not cited)

[302] M. SCHWEIGHOFER, *Global optimization of polynomials using gradient tentacles and sums of squares*, SIAM J. Optim., 17(3), 920–942, 2006. (Cited on p. 120)

[303] I. SHAFAREVICH, *Basic Algebraic Geometry 1: Varieties in Projective Space*, second edition, Springer-Verlag, 1994. (Cited on p. 35)

[304] K. SHIMIZU, Y. ISHIZUKA, AND J.F. BARD, *Nondifferentiable and Two-level Mathematical Programming*, Kluwer Academic, Dordrecht, 1997. (Cited on p. 413)

[305] A.J. SOMMESE AND C.W. WAMPLER, *The Numerical Solution of Systems of Polynomials*, World Scientific, Singapore, 2005. (Cited on p. 181)

[306] G. STENGLE, *A Nullstellensatz and a Positivstellensatz in semialgebraic geometry*, Math. Ann., 207, 87–97, 1974. (Cited on p. 73)

[307] G. STENGLE, *Complexity estimates for the Schmüdgen Positivstellensatz*, J. Complexity, 12(2), 167–174, 1996. (Cited on pp. 113, 177)

[308] J.F. STURM, *Using SeDuMi 1.02: A MATLAB toolbox for optimization over symmetric cones*, Optim. Methods Softw., 11&12, 625–653, 1999. doi.org/10.1080/10556789908805766 (Cited on pp. 25, 27, 148)

[309] B. STURMFELS, *Solving Systems of Polynomial Equations*, CBMS Regional Conference Series in Mathematics Vol. 97, AMS, Providence, 2002. (Cited on pp. 39, 40, 47, 48, 49, 52, 73)

[310] B. STURMFELS AND C. UHLER, *Multivariate Gaussians, semidefinite matrix completion, and convex algebraic geometry*, Ann. Inst. Stat. Math., 62(4), 603–638, 2010. (Cited on p. 231)

[311] G. TANG AND P. SHAH, *Guaranteed tensor decomposition: a moment approach*, in Proceedings of the 32nd International Conference on Machine Learning (ICML-15), pp. 1491–1500, 2015. J. Mach. Learn. Res.: W&CP Vol. 37. (Cited on pp. 356, 357, 358)

[312] V. TCHAKALOFF, *Formules de cubatures mécanique à coefficients non négatifs*, Bull. Sci. Math., 81(2), 123–134, 1957. (Cited on pp. 78, 248)

[313] M. TODD, *Semidefinite optimization*, Acta Numerica, 10, 515–560, 2001. (Cited on pp. 24, 25, 26)

[314] K.C. TOH , M. TODD, AND R. TUTUNCU, *SDPT3: a MATLAB software for semidefinite-quadratic-linear programming*, Optim. Methods Softw., 11(1-4), 545–581. (Cited on p. 25)

[315] H.H. VUI AND P.T. SÓN, *Global optimization of polynomials using the truncated tangency variety and sums of squares*, SIAM J. Optim., 19(2), 941–951, 2008. (Cited on p. 120)

[316] H. WAKI, S. KIM, M. KOJIMA, AND M. MURAMATSU, *Sums of squares and semidefinite program relaxations for polynomial optimization problems with structured sparsity*, SIAM J. Optim., 17(1), 218–242, 2006. (Cited on p. 388)

[317] H. WAKI, S. KIM, M. KOJIMA, M. MURAMATSU, AND H. SUGIMOTO, *SparsePOP: A sparse semidefinite programming relaxation of polynomial optimization problems*, ACM Trans. Math. Softw. (TOMS), 35(2), 15, 2008. (Cited on p. 388)

[318] J. WANG, V. MAGRON, AND J.-B. LASSERRE, *TSSOS: A Moment-SOS hierarchy that exploits term sparsity*, SIAM J. Optim., 31(1), 30–58, 2021. (Cited on p. 388)

[319] J. WANG, V. MAGRON, AND J.-B. LASSERRE, *Chordal-TSSOS: A moment-SOS hierarchy that exploits term sparsity with chordal extension*, SIAM J. Optim., 31(1), 114–141, 2021. (Cited on p. 388)

[320] W. WIESEMANN, D. KUHN, AND M. SIM, *Distributionally robust convex optimization*, Oper. Res., 62(6), 1358–1376, 2014. (Cited on p. 424)

[321] H. WOLKOWICZ, R. SAIGAL, AND L. VANDENBERGHE, *Handbook of Semidefinite Programming*, Kluwer Academic, Dordrecht, 2000. (Cited on pp. 19, 20, 21, 22, 24, 25, 26)

[322] H. XU, Y. LIU, AND H. SUN, *Distributionally robust optimization with matrix moment constraints: Lagrange duality and cutting plane methods*, Math. Program., 169(2), 489–529, 2018. (Cited on p. 424)

[323] V.A. YAKABOVICH, *The S-procedure in non-linear control theory*, Vestnik Leningrad Univ. Math., 4, 73–93, 1977. In Russian, 1971. (Cited on p. 144)

[324] M. YAMASHITA, K. FUJISAWA, K. NAKATA, M. NAKATA, M. FUKUDA, K. KOBAYASHI, AND K. GOTO, *A high-performance software package for semidefinite programs: SDPA 7*, Research Report B-460, Dept. of Mathematical and Computing Science, Tokyo Institute of Technology, Tokyo, Japan, September, 2010. (Cited on p. 25)

[325] L. YANG, D. SUN, AND K. TOH, *SDPNAL+: a majorized semismooth Newton-CG augmented Lagrangian method for semidefinite programming with nonnegative constraints*, Math. Program. Comput., 7, 331–366, 2015. (Cited on p. 25)

[326] Y. YANG AND W. WU, *A distributionally robust optimization model for real-time power dispatch in distribution networks*, IEEE Trans. Smart Grid, 10(4), 3743–3752, 2019. (Cited on p. 424)

[327] J. YU, *Do most polynomials generate a prime ideal?*, J. Algebra, 459, 468–474, 2016. (Cited on p. 163)

[328] M. YUAN AND C.-H. ZHANG, *On tensor completion via nuclear norm minimization*, Found. Comput. Math., 16(4), 1031–1068, 2016. (Cited on p. 356)

[329] Y. YUAN, *Recent advances in trust region algorithms*, Math. Program., 151(1), 249–281, 2015. (Cited on pp. 137, 139)

[330] Q. ZHANG, *Completely positive cones: Are they facially exposed?*, Linear Algebra Appl., 558, 195–204, 2018. (Cited on p. 284)

[331] T. ZHANG AND G. H. GOLUB, *Rank-one approximation to high order tensors*, SIAM. J. Matrix Anal. Appl., 23(2), 534–550, 2001. (Cited on p. 348)

[332] X. ZHANG, C. LING, AND L. QI, *The best rank-1 approximation of a symmetric tensor and related spherical optimization problems*, SIAM. J. Matrix Anal. Appl., 33(3), 806–821, 2012. (Cited on p. 348)

[333] Y. ZHENG AND F. GIOVANNI, *Sum-of-squares chordal decomposition of polynomial matrix inequalities*, Math. Program. (2021). doi.org/10.1007/s10107-021-01728-w (Cited on p. 388)

[334] A. ZHOU AND J. FAN, *The CP-matrix completion problem*, SIAM. J. Matrix Anal. Appl., 35(1), 127–142, 2014. (Cited on pp. 291, 297)

[335] S. ZHU AND M. FUKUSHIMA, *Worst-case conditional value-at-risk with application to robust portfolio management*, Oper. Res., 57(5), 1155–1168, 2009. (Cited on p. 424)

[336] S. ZYMLER, D. KUHN, AND B. RUSTEM, *Distributionally robust joint chance constraints with second-order moment information*, Math. Program., 137(1-2), 167–198, 2013. (Cited on p. 424)

# Glossary of Notation

| | |
|---|---|
| $\mathbb{N}$ | The set of nonnegative integers |
| $\mathbb{Z}$ | The set of integers |
| $\mathbb{Q}$ | The set of rational numbers |
| $\mathbb{R}$ | The field of real numbers |
| $\mathbb{R}_+$ | The set of nonnegative real numbers |
| $\mathbb{R}_{++}$ | The set of positive real numbers |
| $\mathbb{C}$ | The field of complex numbers |
| $\mathbb{R}^n$ | The $n$-dimensional real Euclidean space |
| $\mathbb{C}^n$ | The $n$-dimensional complex Euclidean space |
| $\mathbb{P}^n$ | The $n$-dimensional projective space |
| $\mathbb{N}_d^n$ | The set of powers in $n$ variables and degrees up to $d$ |
| $\overline{\mathbb{N}}_d^n$ | The set of powers in $n$ variables and degrees equal to $d$ |
| $[k]$ | The set $\{1, 2, \ldots, k\}$ |
| $\lceil t \rceil$ | The ceiling of a real number $t$ |
| $\lfloor t \rfloor$ | The floor of a real number $t$ |
| inf | The infimum of a function over a set |
| sup | The supremum of a function over a set |
| $\emptyset$ | The empty set |
| $\mathbb{E}$ | The expectation of a function in a random variable |
| $e_i$ | The $i$th canonical basis vector in the Euclidean space |
| $e$ | The vector of all ones (the length depends on the operation on it) |
| $I_n$ | The $n$-by-$n$ identity matrix |
| $I$ | The identity matrix (the length depends on product) |
| $\mathbf{1}_n$ | The $n$-dimensional vector of all ones |
| $\mathbf{1}_{n \times n}$ | The $n$-by-$n$ matrix of all ones |
| $\Delta_n$ | The standard simplex in $\mathbb{R}^n$ |
| $\mathbb{S}^{n-1}$ | The unit sphere in $\mathbb{R}^n$ |
| $B(c, r)$ | The 2-norm ball of center $c$ and radius $r$ |
| $X \times Y$ | The Cartesian product of $X$ and $Y$ |
| $\mathbb{R}^{m \times n}$ | The space of $m$-by-$n$ real matrices |
| $\mathbb{C}^{m \times n}$ | The space of $m$-by-$n$ complex matrices |
| $\mathcal{S}^n$ | The space of $n$-by-$n$ real symmetric matrices |
| $\mathcal{S}_+^n$ | The cone of $n$-by-$n$ real symmetric psd matrices |
| $\mathcal{S}_{++}^n$ | The set of positive definite matrices in $\mathcal{S}_+^n$ |
| $\mathcal{Q}_n$ | The Lorentz (second order) cone of $\mathbb{R}^{n+1}$ |
| $\mathcal{N}^{n \times n}$ | The cone of $n$-by-$n$ symmetric nonnegative matrices |

| | |
|---|---|
| $\mathcal{N}^n$ | The cone of $n$-by-$n$ symmetric nonnegative matrices |
| $a \in A$ | $a$ belongs to the set $A$ |
| $a \notin A$ | $a$ does not belong to the set $A$ |
| $\emptyset$ | The empty set |
| $A \subseteq B$ | $A$ is a subset of $B$ |
| $A \subsetneq B$ | $A$ is a proper subset of $B$ |
| $a \perp b$ | $a$ is perpendicular to $b$ |
| $(a)_{1:m}$ | The subvector of $a$ whose labels are $1, \ldots, m$ |
| $\bar{a}$ | The complex conjugate of an array $a$ |
| $A \succeq 0$ | The matrix $A$ is positive semidefinite |
| $A \succ 0$ | The matrix $A$ is positive definite |
| $A^{-1}$ | The inverse of an invertible matrix $A$ |
| $\det(A)$ | The determinant of a matrix $A$ |
| $\mathrm{diag} A$ | The diagonal vector of a matrix $A$ |
| $\mathrm{diag} a$ | The diagonal matrix whose diagonal vector is $a$ |
| $\mathrm{diag} A_1, \ldots, A_m$ | The block diagonal matrix whose diagonal blocks are $A_1, \ldots, A_m$ |
| $\mathrm{trace}(A)$ | The trace of a matrix $A$ |
| $\mathrm{Range}\,(A)$ | The range space of a matrix $A$ |
| $\lambda(A)$ | The set of eigenvalues of a matrix $A$ |
| $\mathrm{Range}\,(\mathcal{A})$ | The range of a map $\mathcal{A}$ |
| $\ker \mathcal{A}$ | The kernel of $\mathcal{A}$ |
| $\mathrm{Null}\,(\mathcal{A})$ | The null space of $\mathcal{A}$ |
| $\mathcal{A}^{-1}(b)$ | The preimage of $b$ under the map $\mathcal{A}$ |
| $\mathcal{A}^*$ | The adjoint map of $\mathcal{A}$ |
| $A \bullet B$ | The inner product of two matrices $A, B$ |
| $\|A\|_F$ | The Frobenius norm of a matrix $A$ |
| $\|A\|$ | The spectral 2-norm of a matrix $A$ |
| $\|A\|_*$ | The nuclear norm of a matrix $\mathcal{A}$ |
| $X^T$ | The transpose of $X$ |
| $X^*$ (or $X^H$ | The Hermitian (or conjugate) transpose of $X$ |
| $x$ | $(x_1, \ldots, x_n)$ |
| $[x]_d$ | The vector of all monomials in $x$ with degrees up to $d$ |
| $[x]_d^{\mathrm{hom}}$ | The vector of all monomials in $x$ with degrees equal to $d$ |
| $[x]_B$ | The vector of polynomials from the set $B$ |
| $[K]_d$ | The set of all vectors $[x]_d$ with $x \in K$ |
| $\mathbb{R}[x]$ | The ring of real coefficient polynomials in $x$ |
| $\mathbb{R}[x]_d$ | The set of polynomials in $\mathbb{R}[x]$ of degrees at most $d$ |
| $\mathbb{R}[x]^{\mathrm{hom}}$ | The set of homogeneous polynomials in $\mathbb{R}[x]$ |
| $\mathbb{R}[x]_d^{\mathrm{hom}}$ | The set of homogeneous polynomials in $\mathbb{R}[x]$ of degrees at most $d$ |
| $\mathbb{C}[x]$ | The ring of complex coefficient polynomials in $x$ |
| $\mathbb{C}[x]_d$ | The set of polynomials in $\mathbb{C}[x]$ of degrees at most $d$ |
| $\mathbb{Q}[x]$ | The ring of complex coefficient polynomials in $x$ |
| $\Sigma[x]$ | The cone of SOS polynomials in $x$ |
| $\Sigma[x]_d$ | The set of SOS polynomials in $x$ and of degrees at most $d$ |
| $\Sigma[x]_{n,d}$ | The set of SOS polynomials in $n$ variables and degrees up to $d$ |
| $\Sigma[x]^{\mathrm{hom}}$ | The set of homogeneous SOS polynomials in $x$ |
| $\Sigma[x]_d^{\mathrm{hom}}$ | The set of homogeneous SOS polynomials in $\Sigma[x]_d$ |
| $\Sigma_{n,d}^{\mathrm{hom}}$ | The set of homogeneous SOS polynomials in $n$ variables and degree $d$ |
| $\Sigma^d[x]$ | The set of degree-$d$ SOP polynomials in $x$ |
| $\Sigma^d[x]^{\mathrm{hom}}$ | The set of degree-$d$ SOP forms in $x$ |

| | |
|---|---|
| $\mathbf{LM}(f)$ | Leading monomial of $f$ |
| $\mathbf{LT}(f)$ | Leading term of $f$ |
| $\mathrm{vec}(f)$ | The coefficient vector of $f$ |
| $[f]_J$ | The equivalent class of $f$ with respect to an ideal $J$ |
| $\sqrt{I}$ | The radical ideal of an ideal $I$ |
| $V_{\mathbb{C}}(I)$ | The complex variety of an ideal $I$ |
| $V_{\mathbb{R}}(I)$ | The real variety of an ideal $I$ |
| $\mathbb{R}[x]/I$ | The quotient space of $\mathbb{R}[x]$ over an ideal $I$ |
| $\mathbb{C}[x]/I$ | The quotient space of $\mathbb{C}[x]$ over an ideal $I$ |
| $M_{x_i}$ | The companion matrix with respect to multiplication of $x_i$ |
| $I(V)$ | The vanishing ideal of $V$ |
| $R[V]$ | The coordinate ring of $V$ |
| $T_u(V)$ | The tangent space of $V$ at $u$ |
| $\dim(S)$ | The dimension of a set $S$ |
| $\mathrm{int}\,(S)$ | The interior of $S$ |
| $\mathrm{cl}\,(S)$ | The closure of $S$ |
| $\partial S$ | The boundary of the set $S$ |
| $|S|$ | The cardinality of $S$ |
| $F^{\triangle}$ | The complementary face of $F$ |
| $\mathrm{lin}\,(G)$ | The linear span of $G$ |
| $K^{\circ}$ | The polar of a set $K$ |
| $K^{\star}$ | The dual cone of a set $K$ |
| $\mathrm{conv}\,(S)$ | The convex hull of a set $S$ |
| $\mathrm{cone}\,(S)$ | The conic hull of a set $S$ |
| $\mathrm{aff}\,(S)$ | The affine hull of a set $S$ |
| $S^{\star}$ | The dual cone of a set $S$ |
| $\mathrm{dist}(S,T)$ | The distance between two sets $S$ and $T$ |
| $\mathrm{face}\,(T,C)$ | The minimum face of a convex set $C$ containing $T$ |
| $V^{*}$ | The dual space of a vector space $V$ |
| $\mathrm{supp}(\mu)$ | The support of a measure $\mu$ |
| $meas(y,K)$ | The set of representing measures for $y$ supported in $K$ |
| $\mathrm{supp}(p)$ | The support of a polynomial $p$ |
| $\mathrm{supp}(p)$ | The support of a polynomial $p$ |
| $\mathrm{New}(p)$ | The Newton polytope of a polynomial $p$ |
| $\Delta(p)$ | The discriminant of a polynomial tuple $p$ |
| $\mathrm{Res}(p)$ | The resultant of a polynomial tuple $p$ |
| $\mathscr{P}_d(K)$ | The cone of polynomials that are nonnegative on $K$ |
| $\mathscr{P}_d(K)$ | The cone of degree-$d$ polynomials that are nonnegative on $K$ |
| $\mathscr{R}_d(K)$ | The cone of degree-$d$ tms whose representing measure is supported in $K$ |
| $\mathscr{R}_{n,d}$ | The cone of vectors $[x]_d$ with $x \in \mathbb{R}^n$ |
| $\mathscr{R}_{n,d}^{\mathrm{hom}}$ | The cone of vectors $[x]_d^{\mathrm{hom}}$ with $x \in \mathbb{R}^n$ |
| $\mathrm{Ideal}[h]$ | The ideal generated by $h$ |
| $\mathrm{Ideal}[h]_d$ | The degree-$d$ truncation of $\mathrm{Ideal}[h]$ |
| $\mathscr{Z}[h]_d$ | The subspace of tms vanishing on $\mathrm{Ideal}[h]_d$ |
| $\mathrm{QM}[g]$ | The quadratic module generated by $g$ |
| $\mathrm{QM}[g]_d$ | The degree-$d$ truncation of $\mathrm{QM}[g]$ |
| $\mathrm{Pre}[g]$ | The preordering generated by $g$ |
| $\mathrm{Pre}[g]_d$ | The degree-$d$ truncation of $\mathrm{Pre}[g]$ |
| $\mathscr{S}[g]_{2k}$ | The cone of degree-$2k$ tms determined by $g$ |
| $\mathscr{Z}[g]_d$ | The subspace of degree-$d$ tms vanishing on $\mathrm{Ideal}[h]_d$ |

| | |
|---|---|
| $M_k[y]$ | The $k$th order moment matrix determined by $y$ |
| $L_p^{(k)}[y]$ | The $k$th localizing matrix of $y$ with respect to $p$ |
| $\mathcal{V}_p^{(d)}[y]$ | The degree-$d$ localizing vector of $y$ with respect to $p$ |
| $\mathscr{L}_y$ | The Riesz functional determined by $y$ |
| $\otimes$ | The Kronecker prouduct or tensor product |
| $\mathcal{COP}_n$ | The cone of $n$-by-$n$ copositive symmetric matrices |
| $\mathcal{COP}_{n,d}$ | The cone of degree-$d$ copositive forms in $n$ variables |
| $\mathcal{CP}_n$ | The cone of $n$-by-$n$ completely positive matrices |
| $\mathcal{CP}_{n,d}$ | The cone of degree-$d$ completely positive moments |
| $\mathcal{COP}_n^{\otimes m}$ | The cone of copositive tensors in $\mathrm{T}^m(\mathbb{R}^n)$ |
| $\mathcal{CP}_n^{\otimes m}$ | The cone of completely positive tensors in $\mathrm{S}^m(\mathbb{R}^n)$ |
| $\mathcal{NPP}_n$ | The cone of nonnegative plus psd matrices in $\mathcal{S}^n$ |
| $\mathcal{DNN}_n$ | The cone of doubly nnonnegative matrices in $\mathcal{S}^n$ |
| $\mathrm{S}^m(\mathbb{F}^n)$ | The space of order-$m$ symmetric tensors over $\mathbb{F}^n$ |
| $\mathrm{T}^m(\mathbb{F}^n)$ | The space of order-$m$ cubical tensors over $\mathbb{F}^n$ |
| $\mathscr{P}^{m,n}$ | The cone of psd tensors in $\mathrm{S}^m(\mathbb{F}^n)$ |
| $\Sigma^{m,n}$ | The cone of SOS tensors in $\mathrm{T}^m(\mathbb{R}^n)$ |
| $\|\mathcal{A}\|_{\sigma,\mathbb{F}}$ | The spectral norm of a tensor $\mathcal{A}$ over the field $\mathbb{F}$ |
| $\|\mathcal{A}\|_{*,\mathbb{F}}$ | The nuclear norm of a tensor $\mathcal{A}$ over the field $\mathbb{F}$ |
| $\mathrm{g}(\mathcal{A})$ | The geometric measure of entanglement (GME) of $\mathcal{A}$ |

# Index